FADS AND FANCIES
OF ELEMENTARY PARTICLE PHYSICS

SELECTED WORKS OF

KAMESHWAR C. WALI

FADS AND FANCIES
OF ELEMENTARY PARTICLE PHYSICS

SELECTED WORKS OF
KAMESHWAR C. WALI
Syracuse University, USA

World Scientific
NEW JERSEY • LONDON • SINGAPORE • BEIJING • SHANGHAI • HONG KONG • TAIPEI • CHENNAI • TOKYO

Published by

World Scientific Publishing Co. Pte. Ltd.
5 Toh Tuck Link, Singapore 596224
USA office: 27 Warren Street, Suite 401-402, Hackensack, NJ 07601
UK office: 57 Shelton Street, Covent Garden, London WC2H 9HE

British Library Cataloguing-in-Publication Data
A catalogue record for this book is available from the British Library.

FADS AND FANCIES OF ELEMENTARY PARTICLE PHYSICS
Selected Works of Kameshwar C Wali

ISBN 978-981-123-690-7 (hardcover)

For any available supplementary material, please visit
https://www.worldscientific.com/worldscibooks/10.1142/12276#t=suppl

Printed in Singapore

For my wife, Kashi

Preface

Kameshwar C. Wali

I have included in this volume an edited collections of my scientific papers published in the *Physical Review* and other physics journals over sixty years of working in the field of high energy theoretical physics. I selected the papers that I feel provide the greatest general interest both from a historical and research perspective to physicists and students pursuing advanced degrees in physics and related fields. In most cases, I have presented these papers in general colloquia at many institutes. My hope is that the scope of these papers, published between 1957 and 2017, will offer insight into some of the predominant challenges that occupied theoretical physicists during the latter half of the 20th century and into the 21st.

I arrived in the United States as a young doctoral student in 1955 having completed a Master's degree in physics and mathematics from Banaras Hindu University in India. There, I was inspired by Professor of Mathematics Jayant Narlikar who introduced me to the theory of relativity and new advances in physics that were far from the regular curriculum. With his encouragement and others at BHU, I sought to travel to America. My advisor at the University of Wisconsin, Madison was Professor Robert G. Sachs, a well-known nuclear physics theorist.

Without waiting for the conventional tests for research, I immediately began working with Dr. Sachs, and another student, J. Enoch, on a research paper titled *Thoery of Photo Production of Pions with Nucleaons* (Enouch *et al.*, 1957), followed by *Electromagnetic Form Factors* (Ernst *et al.*, 1960). Thus began my research career.

My early work involved efforts to unravel the structure of the growing list of elementary particles and their strong interactions. When the SU(3) symmetry of the particles was proposed, I was quick to explore its implications for strong interactions and found a long list of "hadrons", the strongly interacting particles. This quickly led to the discovery of new hadrons.

Several years later, I was involved with a proposal by Bunji Sakita, and simultaneously by F. Gursey and L. A. Radicati, to incorporate spin into the symmetry through SU(6) called SU(6,6)(7). At the same time, Abdus Salam and co-workers discovered the same symmetry group. The discovery of this extension made news at the time, promising to be a fundamental breakthrough. Thus, it provided a rich set of predictions for the spectra of strong interactions of the particles that were known at the time. I applied this to hadron scattering interactions. It turned out that there were some fundamental problems with the structure of such symmetries. However, the introduction of supersymmetry circumvented those limitations.

Not long after this work, the notion of duality — local and semi-local — was introduced as a means to incorporate both the growing list of hadronic states and resonances, the scattering states and the regi poles that described high-energy interactions. I, along with several collaborators, considered some important implications of these ideas.

By this time, it was being established that the fundamental particles were quarks and gluons and that their weak and electromagnetic interactions were described by non-Abelian gauge theories with broken symmetries.

As many physicists have experienced, theoretical fancies are not always borne out by experimental research — an important collaborative face of elementary particle physics. This book includes my work of over sixty years of following important and exciting fads and fancies that were current in theoretical physics. Some of these ideas were not supported by experimentation, but others bore fruit.

Contents

Introduction

Milne and Eddington claimed that to the mind well trained in mathematics and epistemology, the laws of nature manifest themselves without appeal to experiment. In refuting this philosophy, Max Born has written a short but delightful essay in which he has probed the role of theory and experiment in the development of contemporary classical and quantum physics. As is to be expected, Born emphasizes the strong under-pinning of experimental facts preceding theoretical abstractions leading to new observable consequences which, in turn, are put to experimental test. As illustrations of the constant interplay between theory and experiment, Born points out the empirical edifices on which Maxwell built his theory edifices provided by the work of Faraday, Ampere and others. The striking empirical equality between the gravitational and inertial masses, known from Newton's time but ignored for two centuries, was at the heart of Einstein's equivalence principle and led to the subsequent development of his general theory of relativity. And if it were not for the rich and accurate spectroscopic data confronting the old quantum theory of Bohr (and leading to direct contradictions with experiments in some instances), the new quantum mechanics of Heisenberg, Schrodinger, Dirac and Born would perhaps not have been born. We could have lived in peace with the *ad hoc* rules and decrees of Bohr as we live in peace today with the new quantum mechanics, notwithstanding the objections of some on philosophical and epistemological grounds. Born thus argues that physical laws are derived "by a long inductive process, in which flashes of imagination have alternated with painful observation and interpretation of facts," and not by following the philosophy of Milne and Eddington. "I believe there is no philosophical high road in science, with epistemological sign posts," says Born. "No, we are in a jungle and find our way by trial and error, building our road behind as we proceed. We do not find signposts at crossroads, but our own scouts erect them. To help the rest." He concludes his essay by saying,

> "My advice to those who wish to learn the art of scientific prophecy is not to rely on abstract reason, but to decipher the secret language of Nature's documents the facts of experience."

In pure mathematics, one might think the situation could be different. But, at a conference in 1931 on topology and abstract algebra, speaking about abstracting and axiomatizing in mathematics, Herman Weyl had this to say:

> "... all the general notions do not fall into our laps by themselves. But definite concrete problems were first conquered in their undivided complexity, single-handed by brute force, so to speak. Only afterwards, the axiomaticians came along and stated: Instead of breaking in the door with all your might and bruising your hands, you should have constructed such and such a key of skill, and by it, you would have been able to open the door quite smoothly. However, they can construct the key only because they are able, after the breaking in was successful, to study the lock from within and without. Before you can generalize, formalize, and axiomatize, there must be a mathematical substance."

Born and Weyl are saying the same thing: First find what works — what is in accordance with Nature's book of documents.

At the present time, however, the situation in high energy physics is not what they envisioned. Nature's book of documents is fast becoming inaccessible to human ingenuity. The interplay between theory and experiment, which Born describes so accurately in his essay, has led us to a seemingly impenetrable barrier that prevents the best of our theoretical ideas by direct experimentation. In our quest for the ultimate constituents of all matter and their interactions, we have arrived at what seems to me to be an impasse.

The hope that our deeper and deeper probes into the structure of matter will ultimately lead to a few basic constituents with properties derivable from space-time symmetries has disappeared. How delightful it could have been if we could say about elementary particles what Chandrasekhar says often about black holes:

> "... both in the underlying mathematics and in the underlying physics, black holes are simple objects ... All the standard equations of mathematical physics can be solved exactly in the Kerr space-time ... They are by their simplicity beautiful..."
>
> "... The study of black holes is, at least in my judgement, the best manifestation of an ancient motto: The simple is the seal of the 'true' and 'beauty' is the splendor of truth."

Instead, as we probe more deeply, we encounter more and more complexity. Quarks and leptons which comprise the elementary particle world of today come in three identical (with respect to $SU(2)_L \times U(1)$) families, each consisting of fifteen helicity states. While no observed phenomenon disagrees with, nor does it demand structure beyond, the Standard Model, there are vexing problems and puzzles: the mass spectrum of quarks and leptons, the pattern of intra-generational mixings, the origin and magnitude of CF-violation, the number of families, and so on. Quantum chromodynamics — the theory of strong interactions — is yet to be established on a firm quantitative basis, especially in its non-perturbative features which are supposed to explain confinement and provide the correct, observed hadronic spectrum.

So far, however, theory and experiment have played their traditional roles. Elegance, uniqueness, naturalness, charm, etc., are accepted theoretical criteria, provided they are buttressed by experiments.

From this stage on, however, the path followed so far — moving up in energy step by step, making models and testing their consequences — will soon, if it has not already, come to an end. The obvious next milestone down the road from the standard theory is the unification of electroweak and strong interactions (Grand Unification). But relatively simple arguments tell us that this grand unification takes place at $\sim 10^{15}$ GeV, an energy way beyond the reach of our present of future accelerators. Consequently, the theory cannot be tested directly. One has to look for low energy manifestations of the extrapolated, mutilated theory.

The extrapolation of the high-energy symmetric theory from the GUT scale to the low energy region is beset with ambiguities. There are "definite concrete problems to be conquered," but the overriding concern of the theorists is to make predictions amenable to experimental tests. In the process, additional assumptions make their covert entry. Nucleon decay is an example. If theory based on minimal SU(5) GUT had not claimed the proton lifetime to be in the 10^{30}–10^{32} year range, the immense experimental effort would not have begun. Now that there is no confirmation of the original optimistic predictions, one can find any number of holes in the theoretical arguments and invent patchworks to predict the lifetime to be just beyond the limit set by current experiments. The result: a credibility gap!

If theory cannot make precise predictions in the experimentally accessible region, what are its "signposts" which one can believe? As is well known, the Higgs mechanism is an indispensable part of a renormalizable gauge theory whose symmetries are broken spontaneously. The surviving Higgs meson of the Standard Model is yet to be discovered. In grand unified theories, one demands a complex Higgs structure to obtain the desired symmetry-breaking patterns. While many definite concrete problems of spontaneous symmetry-breaking remain to be conquered even in the simplest GUT, the hierarchy and fine-tuning problems capture our attention. Supersymmetry makes its entrance. In its global form, it is neither a gauge theory nor the generalization of any known physical theory. Every elementary fermion (boson) is accompanied by a boson (fermion): multitudes of new particles (squarks, sleptons, gluinos, photinos, winos, zinos, Higgsinos) whose properties supersymmetry alone cannot predict. Local gauge supersymmetry goes much further: it embraces gravity as well. Things look good for supergravity in a few low orders of perturbation (but not in higher orders). In any case, although the exact theory governs phenomena around and above Planckian energy, the problems and ambiguities in extrapolating it to low energy remain. No new insights into the mysteries of the known particles emerge. Higgs structure is arbitary. A new hidden matter sector must be postulated.

Is superstring theory the answer to all the theoretical worries? To some wanderers in the jungle of theoretical ideas, it is. They see this as a sign post which symbolizes elegance, harmony, and a confluence of new and diverse mathematical branches. They envision a finite quantum theory of gravity necessarily accompanying other interactions and hence a truly unified theory with "everything determined: no free parameters." "It is a miracle; it is the theory of the world!" Such euphoric statements naturally create sharp divisions — the believers and the skeptics. The skeptics ask, "What world?" The observed and the observable world? Or a world that has nothing to do with it? Superstring theory has a long way to go from one-dimensional strings in ten-dimensional space-time to the four-dimensional elementary particles. No one would disagree that it has yet to demonstrate that is has in it the necessary gumption to implement its own descent from the majestic heights of Planck scale down to the earth, to explain the observed, and to predict new but observable, consequences which, after all, are the hallmarks of great theories of the past.

Astrophysics, cosmology, the early universe — all are the natural arenas for such far-reaching ideas, we are told. At an earlier time, the laws of physics were established first in the laboratory and then were extrapolated both in space and time to account for the extreme conditions of matter in the astrophysical world. We then had good reasons to be happy if we reached agreement within a few orders of magnitude. The reverse — to rely on astrophysical numbers as more than mere guidelines (or for the companionship of astrophysicists in our midst) — is fraught with uncertainties. After all, there is only one universe we know, and that is the universe we live in. We are supposed to establish its initial conditions by establishing the physical laws.

It appears that the traditional step-by-step confrontation between theory and experiment in high energy physics cannot be maintained indefinitely. There is naturally a disparity between the capacities of theory and experiment, but it is held in check by the requirement that theory explains the observed world. If this check cannot be maintained, what other criteria should one adopt? What restraints should theorists put upon themselves?

Should we disregard the observed world and postulate a world of one's own in which elegance, simplicity, and beauty reign supreme? That would be the ultimate anthropic principle.

Electromagnetic Form Factors of the Nucleon*

F. J. Ernst
University of Wisconsin, Madison, Wisconsin and Columbia University, New York, New York

R. G. Sachs†
University of Wisconsin, Madison, Wisconsin and University of Paris, Paris, France

AND

K. C. Wali
University of Wisconsin, Madison, Wisconsin

(Received March 18, 1960)

The physical interpretation of the electromagnetic form factors is discussed with special reference to the gauge invariance of particular theories. A distinction is made between the condition that the one nucleon matrix element satisfy the equation of continuity ("weak gauge invariance") and the stronger condition imposed by the generalized Ward identity ("strong gauge invariance"). The former is shown to be a consequence of covariance under the improper Lorentz transformations, and hence it has no new content concerning the functional behavior of the form factors. The latter implies restrictions on the current operator which may have an important effect on the results of calculations of form factors.

In connection with the physical interpretation, it is noted that the moments of the charge and current distribution are determined by $F_{ch}=F_1-(q^2/2M)F_2$ and $F_{mag}=(1/2M)F_1+F_2$. Specifically the second moment of the charge distribution, $-6F_{ch}'(0)$, is found, in the case of the neutron, to be directly measured by the neutron-electron interaction *without* the intervening subtraction of the Foldy term.

These matters are investigated in detail by means of a specific model of the nucleon which is a covariant generalization of the fixed source static model having the property that it gives results identical with the static model in the limit $M\rightarrow\infty$. It is found that strong gauge invariance requires the addition of line currents which make significant contributions to the form factors in general and, in particular, to the proton charge radius even in the static approximation. This suggests that as a consequence of strong gauge invariance, important contributions to the charge radius must arise in any theory from intermediate states of large mass. The model also provides a means of consistently calculating recoil corrections to the static model. They are found to be large.

1. INTRODUCTION

THE electromagnetic interactions of the nucleon provide, in principle, a direct source of information concerning the structure of the nucleon. Although a prodigious amount of experimental information concerning the electromagnetic interactions is available,[1] a satisfactory interpretation of all of the data has not been possible.[2] The failure of any theoretical treatment of the electromagnetic form factors

* Work supported in part by the University of Wisconsni Research Committee with funds granted by the Wisconsin Alumni Research Foundation, in part by the U. S. Atomic Energy Commission, and in part by the U. S. Air Force, through the Research Contracts Division and through the European Office, Air Research and Development Command.

† Guggenheim Fellow, 1959–60.

[1] For a summary of experimental data see R. Hofstadter, F. Bumiller, and M. R. Yearian, Revs. Modern Phys. **30**, 482 (1958).

[2] A summary of the theoretical interpretations predating the use of dispersion relations is provided by D. R. Yennie, M. M. Lévy, and D. G. Ravenhall, Revs. Modern Phys. **29**, 144 (1957).

to give quantitative agreement with the data has been interpreted as an indication of a serious difficulty with the theory. However, there are several general questions concerning the nature of the approximations used and their interpretation which seem to warrant careful investigation before drawing any such conclusion. The purpose of this work is to study in some detail two such questions, one being the problem of gauge invariance and the other the physical interpretation of the form factors.

The requirement of gauge invariance is usually limited to the condition that the one-nucleon matrix element of the current should satisfy the equation of continuity. However, it is found that this condition is automatically satisfied for any matrix element which is covariant under the improper Lorentz transformations. Hence, this condition, which will be referred to as "weak gauge invariance," does not, in fact, guarantee the conservation of charge. We suggest here that the minimum requirement is the generalized Ward identity, which is a statement of the equation of continuity for the photon vertex off the nucleon mass shell. This will be called the condition of "strong gauge invariance." It will be shown, by making use of a specific model, that strong gauge invariance requires additions to the current operators which make important contributions to the form factors.

The model used here is a covariant generalization of the fixed source static model, designed in such a way as to give the same results as the static model in the limit of large nucleon mass. Strong gauge invariance is established by introducing "line currents" which can be used to estimate the effect of charge conservation on the form factors. A particularly surprising result is that there is a large line-current contribution to the isoscalar second moment of the charge distribution even in the static limit. This is of particular interest since the large value of the experimental isoscalar second moment has been a major source of difficulty in earlier work.[2]

The model also provides a consistent procedure for calculating recoil corrections to the static model. The fact that recoil effects are included means that another question of some interest can be investigated. In 1951, Foldy[3] showed that the neutron-electron interaction could be described as the sum of two terms, one of them being given exactly in terms of the known anomalous magnetic moment of the nucleon. He interpreted the other term as the second moment of the pion charge cloud distribution. The contribution of the magnetic moment, called the "Foldy term" is inversely proportional to the nucleon mass. Therefore it was argued, in particular by Salzman,[4] that it is not taken into account by the static model, which corresponds to the limit $M\rightarrow\infty$. Hence, the neutron-electron interaction as calculated by the static model has usually been compared to the difference between the observed interaction and the Foldy term. The disagreement measured in that way is enormous.

The interpretation of Foldy's result is re-examined below, by making use of the fact that the limit $M\rightarrow\infty$ can be handled systematically. We find that in a consistent treatment of the static model, the result of the static model is to be compared directly with the observed neutron-electron interaction, without making the subtraction described above. Therefore the discrepancy between the static model and observation is not nearly so great as had been thought, nor is the experimental result as mysterious as it seemed.

As an illustration of the possibility of calculating recoil corrections, we shall obtain the terms of order M^{-1} in the magnetic moments. These will turn out to be large, of the same order as the static contributions, so that any success of the static model is probably fortuitous. Of particular interest is the isoscalar magnetic moment, which is strictly a recoil effect. Its evaluation to order M^{-1} by means of the present model gives an expression in close agreement with that obtained from the static model on the basis of the mirror theorem.[5] Perhaps this agreement is not so surprising since the early result is just a consequence of the conservation of angular momentum. As is well known, the result disagrees seriously with the observed value of the isoscalar magnetic moment.

2. ELECTROMAGNETIC FORM FACTORS

The interactions under discussion involve the emission and absorption of a single photon and may be described in terms of the matrix element $\langle P',s'|j_\mu(0)|P,s\rangle$ of the current density operator between nucleon states of 4-momentum P and P' and spin s and s'. The operator is evaluated at the space-time point $x_\mu=0$. It is well known[4] that this matrix element may be written in the form

$$\langle P',s'|j_\mu(0)|P,s\rangle=\frac{ie}{(2\pi)^3}\frac{M}{(P_0P_0')^{\frac{1}{2}}}\bar{u}_{s'}(\mathbf{P}')\times[F_1(q^2)\gamma_\mu-F_2(q^2)\sigma_{\mu\nu}q_\nu]u_s(\mathbf{P}),\quad(1)$$

where $q_\mu=P_\mu'-P_\mu$, $\sigma_{\mu\nu}=(1/2i)[\gamma_\mu,\gamma_\nu]$, and $u_s(\mathbf{P})$ is the positive energy Dirac spinor of a free nucleon of mass M, momentum $\mathbf{P}$ and spin s.

The matrix element is restricted to the form Eq. (1) by the requirement of covariance under the improper Lorentz group. In particular, it is shown in the Appendix that the additional term of the form $F_3(q^2)q_\mu$ is excluded by the requirements imposed on the current density under the time-reversal transformation. The condition imposed on the matrix element by the equation of

[3] L. L. Foldy, Phys. Rev. **83**, 688 (1951); **87**, 688 (1952); **87**, 693 (1952).
[4] G. Salzman, Phys. Rev. **99**, 973 (1955).
[5] R. G. Sachs, Phys. Rev. **87**, 1100 (1952).

continuity, namely,

$$q_\mu\langle P',s'|j_\mu(0)|P,s\rangle=0 \tag{2}$$

is satisfied by Eq. (1). Thus the condition of weak gauge invariance, Eq. (2), is a direct consequence of the covariance[6] and does not, in fact, contain any information concerning the consequences of charge conservation. Therein lies its weakness. It is satisfied for any choice of the functions $F_1(q^2)$ and $F_2(q^2)$. On the other hand, it is clear that the form of the functions, as calculated in any given model, is affected by the condition of charge conservation. This results from the restriction on the form of the current operator imposed by the equation of continuity,

$$\partial j_\mu(x)/\partial x_\mu=0. \tag{3}$$

Not only does Eq. (3) imply the condition Eq. (2) on the one-nucleon matrix element, but it also implies an extension of the condition off the nucleon mass shell, namely, the generalized Ward identity.[7] This condition of strong gauge invariance would appear to be a minimum requirement on any calculation of electromagnetic properties of the nucleon.

It is of interest to discuss the physical meaning of the form factors $F_1(q^2)$ and $F_2(q^2)$ which describe, in some sense, the distribution of charge and magnetization in the nucleon. The various moments of these distributions are often used to give an insight into the form of the distribution; examples are the magnetic moment, second moment of the charge distribution, and so on. The simplest moment is the zero-order moment of F_1, namely,

$$F_1(0)=\tfrac{1}{2}(1+\tau_3) \tag{4}$$

is just the charge of the nucleon in units of e.

To obtain the magnetic moment, for example, we wish to calculate the expectation value of the operator $\frac{1}{2}\int d^3r\, \mathbf{r}\times\mathbf{j}(x)$ in a state in which the nucleon is at rest. This expectation value is most easily determined for a wave packet $\int d^3P\, a(\mathbf{P})|P,s\rangle$. Since

$$\langle P'|j_\mu(x)|P\rangle=e^{-i(P'-P)\cdot x}\langle P'|j_\mu(0)|P\rangle, \tag{5}$$

we find for the magnetic moment

$$\langle\mathfrak{M}\rangle=\frac{1}{2}\int d^3P'\, a^*(P')\int d^3P\, a(P)\int d^3r\, e^{-i(P'-P)\cdot x}\mathbf{r} \times\langle P',s'|\mathbf{j}(0)|P,s\rangle$$
$$=-(2\pi)^3 i\frac{1}{2}\int d^3P'\, a^*(P')\int d^3P\, a(P)\nabla_P\delta(\mathbf{P}-\mathbf{P}') \times\langle P',s'|\mathbf{j}(0)|P,s\rangle e^{i(P_0'-P_0)t}.$$

We now substitute for the matrix element from Eq. (1), integrate by parts, and finally take as the wave packet

$$|a(\mathbf{P})|^2=\delta(\mathbf{P}),$$

with the result

$$\langle\mathfrak{M}\rangle=e[F_1(0)/2M+F_2(0)]\bar{u}_{s'}(0)\boldsymbol{\sigma}u_s(0). \tag{6}$$

Since $eF_1(0)/2M$ is, according to Eq. (4), the Dirac moment, we see that $eF_2(0)$ is the anomalous moment of the nucleon.

The second moment of the charge distribution may be calculated in a similar manner; the expectation value of the quantity $\int d^3r\, r^2 j_4(x)$ is required. Proceeding in just the manner described above, we find for the second moment

$$\langle\mathfrak{R}^2\rangle=-6\left[F_1'(0)-\frac{F_2(0)}{2M}\right]+F_1(0)\left[-\frac{3}{4M^2}+\int d^3P\frac{P_0}{M}\nabla_P a(\mathbf{P})\cdot\nabla_P a(\mathbf{P})\right], \tag{7}$$

where the prime denotes differentiation with respect to q^2. The ambiguous term proportional to $F_1(0)$, which vanishes for the neutron, is just the uninteresting effect of distributing the charge of the proton in a wave packet. It will be dropped henceforth.[8]

The results Eq. (6) and Eq. (7) suggest that we introduce new form factors, the charge form factor

$$F_{ch}(q^2)=F_1(q^2)-(q^2/2M)F_2(q^2), \tag{8}$$

and the magnetic form factor

$$F_{mag}(q^2)=(1/2M)F_1(q^2)+F_2(q^2). \tag{9}$$

Note that this usage differs from the usual, but the difference has an important physical significance. F_{ch} measures the charge distribution, not F_1 as is so often assumed. Also, F_{mag} rather than F_2 measures the distribution of magnetization. This point was surmised by Yennie, Lévy, and Ravenhall,[2] and a clear demonstration that it is a reasonable interpretation has been given by Walecka,[9] who shows that F_{ch} and F_{mag} measure the interaction with static electric and magnetic fields, respectively.

We now have from Eqs. (7) and (8)

$$\langle\mathfrak{R}^2\rangle=-6F_{ch}'(0), \tag{10}$$

after dropping the uninteresting term, and from Eqs. (6) and (9),

$$\mathfrak{M}=eF_{mag}(0)\boldsymbol{\sigma}. \tag{11}$$

We are now in a position to comment on the Foldy term. Foldy[3] showed that the neutron-electron interaction is proportional to $F_1'(0)-F_2(0)/2M$, which we now see is just $F_{ch}'(0)$. Thus a measurement of the

[6] Note that this result would not obtain if the initial and final states described particles of different mass.

[7] Y. Takahashi, Nuovo cimento 6, 370 (1957).

[8] Yennie, Lévy, and Ravenhall, reference 2, include part of this term in their expression for the second moment but its contribution is too small to have a significant effect on their discussion.

[9] J. P. Walecka, Nuovo cimento 11, 821 (1959).

neutron-electron interaction is simply a measurement of $F_{ch}'(0)$, which is also seen to be directly proportional to the second moment of the total charge distribution. A calculation of the second moment should then be directly comparable to the measurement, without subtracting the Foldy term. In particular, it will be shown in Sec. 7 that the limit as $M\to\infty$ of $-6F_{ch}'(0)$ is identical with the second moment of the pion charge distribution as calculated in the static model. Hence, a consistent interpretation of the static model would associate the entire neutron-electron interaction with the second moment of the pion charge distribution.

3. DESCRIPTION OF THE MODEL

The model to be considered is a covariant generalization of the fixed-source static model in its most primitive form. No effort will be made here to determine rescattering corrections. Thus our starting point is the lowest-order covariant perturbation theory. The modification is simply to insert at the pion-nucleon vertices a general vertex function, as indicated in Fig. 1, and to calculate the form factors in terms of the vertex function. The vertex function will be found to play the same role as the source function in the static model.

There are several terms in the general vertex function corresponding to the several possible vertex form factors. For the sake of simplicity, only the pseudovector term is included; it is the one having the closest correspondence to the static model. This has the form $\gamma_5(\boldsymbol{P}_2-\boldsymbol{P}_1)$ multiplied by a form factor, where $\boldsymbol{P}\equiv\gamma_\mu P_\mu$. The vertex form factor is, in general, an invariant function of P_1 and P_2 (see Fig. 1), thus any three combinations of the three invariant variables P_1^2, P_2^2, $(P_1\cdot P_2)$ may appear in it. On the other hand, in the static model the source function depends on only one variable. In order to study the approach to the static limit, the vertex function is also taken to be a function of a single invariant variable $\xi(P_1,P_2)$ the variable being chosen to be such a combination of P_1^2, P_2^2 and $P_1\cdot P_2$ as to make the correspondence with the static model as close as possible. This choice will be specified below. The vertex part, Fig. 1 is therefore taken to be of the form

$$i(8\pi)^{\frac{1}{2}}fv(\xi)\gamma_5(\boldsymbol{P}_2-\boldsymbol{P}_1), \tag{12a}$$

for a charged pion and

$$i(4\pi)^{\frac{1}{2}}fv(\xi)\gamma_5(\boldsymbol{P}_2-\boldsymbol{P}_1) \tag{12b}$$

for a neutral pion. We shall normalize v in such a way

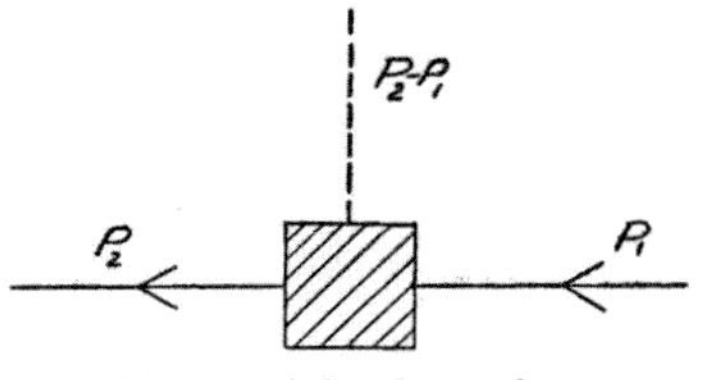

FIG. 1. Diagram of the pion-nucleon vertex.

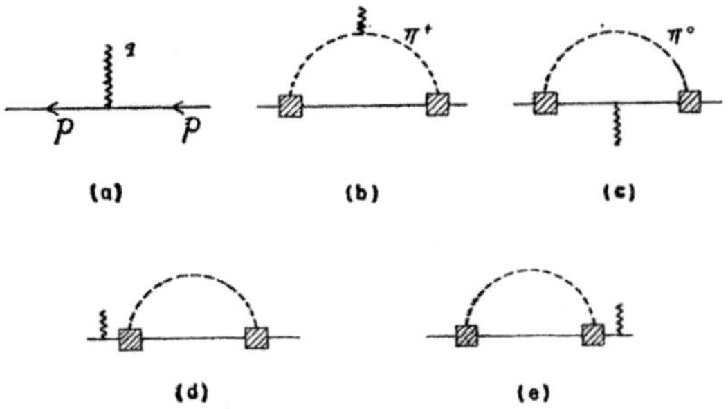

FIG. 2. Diagrams representing the electromagnetic interaction of the proton.

that f corresponds to the usual renormalized, unrationalized coupling constant, $f^2\approx 0.08$.

The calculation of the electromagnetic form factor consists of substituting the vertex part Eq. (12a) or Eq. (12b) in the matrix element obtained from the diagrams Fig. 2 for the proton and Fig. 3 for the neutron. For a given function $v(\xi)$, these diagrams could be evaluated directly without further assumptions or approximations. However, in order to avoid introducing a specific function and to study the expansion in powers of M^{-1}, we proceed in the following manner. Consider for example the term corresponding to Fig. 2(c): The contribution to $\langle P',s'|j_\mu(0)|P,s\rangle$ is

$$\frac{4\pi ef^2}{(2\pi)^7}\frac{M}{(P_0P_0')^{\frac{1}{2}}}\bar{u}_{s'}(\mathbf{P}')\int d^4k\,\gamma_5\boldsymbol{k}v(\xi_1)S_F(P'-k)$$
$$\times\gamma_\mu S_F(P-k)\gamma_5\boldsymbol{k}v(\xi_2)\Delta_F(k)u_s(\mathbf{P}),$$

where

$$\xi_1=\xi(P',P'-k),\quad \xi_2=\xi(P-k,P),$$
$$S_F(p)=(i\boldsymbol{p}-M)/(p^2+M^2-i\epsilon),$$

and

$$\Delta_F(k)=(k^2+\mu^2-i\epsilon)^{-1}.$$

Our purpose is to reduce this to an integral in the 3-dimensional $\mathbf{k}$ space, similar to the type of integral occurring in the static model. Therefore we assume that $v(\xi_1)$ and $v(\xi_2)$ are analytic in the upper half of the k_0 plane, vanishing on the infinite semicircle, and integrate k_0 over a contour closed in the upper half plane.[10] Contributions from the three poles on the negative k_0 axis are thereby obtained.

When the variable $\xi(P_1,P_2)$ is chosen to be

$$\xi(P_1,P_2)=\left[\frac{4M^2+\mu^2+(P_2+P_1)^2}{4M\mu}\right]^2+\frac{\mu^2+(P_2-P_1)^2}{\mu^2}, \tag{13}$$

very simple results are obtained for the contributions of these three poles and the poles appearing from the other diagrams. In fact, this choice leads directly to the static model in the no-recoil limit.

[10] It is not implied that this simple behavior is to be expected for the vertex function arising from any realistic, relativistic theory. These assumptions are introduced only as a prescription for reproducing the results of the static model.

An expansion in powers of $|\mathbf{k}|/M$ and $|\mathbf{q}|/M$ is now possible if it is assumed that at each pole

$$v(\xi)\approx 0, \tag{14}$$

except for

$$|k|^2/M^2\ll 1.$$

This rather unrealistic assumption is in the spirit of the static model and will be the basis for our discussion. Evaluation of the diagrams of Figs. 2 and 3 to any order in M^{-1} is now a straightforward but tedious matter. The results to lowest order are presented in Sec. 7 and some of the first order corrections in Sec. 8.

4. CONSEQUENCES OF GAUGE INVARIANCE

It is immediately recognized that the diagrams in Figs. 2 and 3 do not present a gauge invariant result. As long as the black boxes represent anything other than a local pseudoscalar interaction, diagrams of the type shown in Fig. 4 are expected to contribute appreciably to the electromagnetic interactions. For example, if the black box represents vertex parts contributed by higher order diagrams, each of the internal lines carrying charge must, in turn, be permitted to produce the photon.

Nevertheless, the gauge invariance condition Eq. (2) *is* satisfied by the matrix elements obtained from Figs. 2 and 3. Indeed, it is easily shown that *each of the terms* indicated by Figs. 2(a), 2(b), 2(c) and 3 satisfies this condition, as does the sum of diagrams Figs. 2(d) and 2(e), as long as the function $v(\xi)$ is symmetric under interchange of P_1 and P_2. Note that for a local pseudovector vertex, not even the usual catastrophic terms arising from the momentum dependence are required to satisfy the condition Eq. (2). The reason for this is simply that Eq. (2) follows from covariance under a Lorentz transformation, as remarked in Sec. 2, and each of these terms satisfies separately the covariance condition. These remarks serve to illustrate the weakness of Eq. (2).

To satisfy the requirements of strong gauge invariance, the behavior of the photon vertex off the nucleon mass shell must be considered. Then it is found that the diagrams indicated in Fig. 4 are required. We shall give a prescription for writing down terms of the type indicated in Fig. 4 for an arbitrary vertex function $v(\xi)$. However, there is no unique prescription for determining these terms, there are always many ways to satisfy the condition of gauge invariance. That given

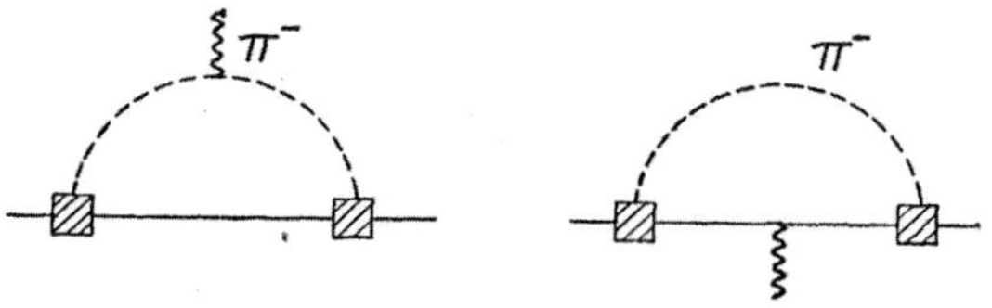

FIG. 3. Diagrams representing the electromagnetic interaction of the neutron.

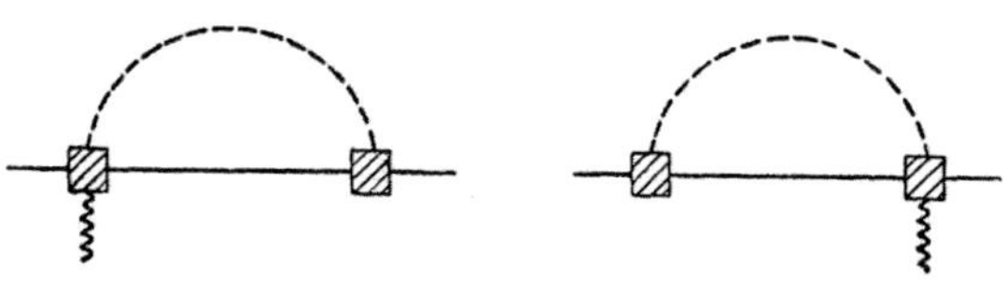

FIG. 4. Vertex current contributions to the nucleon form factors.

here seems to be the simplest, it makes use of line currents which are a natural generalization[11] of the usual substitution of $P_\mu-(e/c)A_\mu$ for P_μ.

The currents to be introduced may be described in configuration space as currents flowing along a line connecting two points in space-time which are coupled by the nonlocal pion-nucleon vertex function. In this way it is seen that they serve to transfer the electric charge across the nonlocal vertex. We shall not go into this representation in configuration space here, but merely remark that the form of the line is arbitrary. It will be taken to be a straight line for the sake of simplicity. The corresponding matrix element is given below in momentum representation and we shall simply demonstrate that the term used does indeed satisfy the condition of strong gauge invariance.

If we write

$$\langle P',s'|j_\mu(0)|P,s\rangle=\frac{4\pi ief^2}{(2\pi)^7}\frac{M}{(P_0P_0')^{\frac{1}{2}}}\bar{u}_{s'}(\mathbf{P}')J_\mu u_s(\mathbf{P}), \tag{15}$$

the contributions to J_μ from the diagrams in Fig. 4 are as follows: For the proton with intermediate π^0,

$$\begin{aligned}J_\mu{}^L(\pi^0)=\int d^4k\Big\{\gamma_5\Big[\int_0^1 d\alpha\Big(\frac{\partial}{\partial P_\mu'}+\frac{\partial}{\partial P_\mu}\Big)\\ \times g(P'-\alpha q,\, P-k+(1-\alpha)q)\Big]\\ \times S_F(P-k)\gamma_5 g(P-k,\, P)\\ +\gamma_5 g(P',\, P'-k)S_F(P'-k)\gamma_5\\ \times\Big[\int_0^1 d\alpha\Big(\frac{\partial}{\partial P_\mu'}+\frac{\partial}{\partial P_\mu}\Big)\\ \times g(P'-k-\alpha q,\, P+(1-\alpha)q)\Big]\Big\}\Delta_F(k);\end{aligned} \tag{16a}$$

for the proton with intermediate π^+,

$$\begin{aligned}J_\mu{}^L(\pi^+)=2i\int d^4k\Big\{\gamma_5\Big[\int_0^1 d\alpha\frac{\partial}{\partial P_\mu'}g(P'-\alpha q,\, P-k)\Big]\\ \times S_F(P-k)\gamma_5 g(P-k,\, P)\\ +\gamma_5 g(P',\, P'-k)S_F(P'-k)\gamma_5\\ \times\Big[\int_0^1 d\alpha\frac{\partial}{\partial P_\mu}g(P'-k,\, P+(1-\alpha)q)\Big]\Big\}\Delta_F(k);\end{aligned} \tag{16b}$$

[11] R. G. Sachs, Phys. Rev. **74**, 433 (1948); **75**, 1605 (1949).

and, finally, for the neutron with intermediate π^-,

$$J_\mu{}^L(\pi^-)=2\int d^4k\Big\{\gamma_5\Big[\int_0^1 d\alpha\frac{\partial}{\partial P_\mu}$$
$$\times g(P',P-k+(1-\alpha)q)\Big]S_F(P-k)$$
$$\times\gamma_5 g(P-k,P)+\gamma_5 g(P',P'-k)S_F(P'-k)\gamma_5$$
$$\times\Big[\int_0^1 d\alpha\frac{\partial}{\partial P_\mu{}'}g(P'-k-\alpha q,P)\Big]\Big\}\Delta_F(k). \quad (16c)$$

The definition of the function g is

$$g(p_2,p_1)=(p_2-p_1)v(\xi), \quad (17)$$

so that the indicated differentiations of the factor (p_2-p_1) lead to the usual "catastrophic" terms associated with the pseudovector coupling, while the derivatives of $v(\xi)$ provide the pure line currents.

The test of strong gauge invariance may be taken to be

$$q_\mu J_\mu=0, \quad (18)$$

where J_μ includes all contributions other than that of Fig. 2(a). The point is that Eq. (18) applies off the mass shell for the nucleons. It will be found to be equivalent to the generalized Ward identity. Before applying this test, it is helpful to take account of mass renormalization, which also has an effect on the electromagnetic interactions.

5. MASS RENORMALIZATION

Since our procedure is analogous to the lowest order perturbation diagram in meson theory, the mass renormalization consists merely of the subtraction of a term having the form suggested by the diagram Fig. 5. We may make the usual Dyson expansion of this diagram

$$\Sigma(p)=A+(ip+M)B+(ip+M)^2\Sigma_f(p), \quad (19)$$

where A and B are constants. Insertion of $\Sigma(p)$ between external lines leaves only the constant A to be subtracted. However, we wish to perform the renormalization without reference to the external lines, that is, in J_μ. Hence, we introduce the renormalization diagram Fig. 6, where the $\times$ denotes the momentum-dependent mass correction

$$\delta M=-[A+(ip+M)B]. \quad (20)$$

The introduction of the electromagnetic interaction

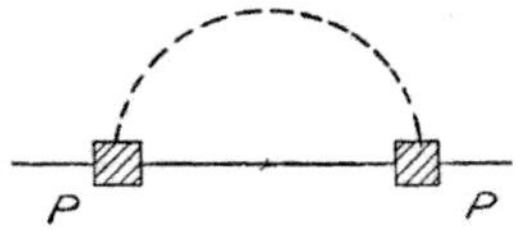

FIG. 5. Self-energy contribution to the nucleon mass.

now must be made in such a way as to maintain the gauge invariance of this momentum-dependent mass correction. Since δM is in effect a nonlocal operator, this can again be accomplished by means of line currents. Hence, in the one-photon matrix element for the proton, we must introduce terms corresponding to the diagrams of Fig. 7, where the contribution to J_μ of Fig. 7(c) is just

$$J_\mu{}^{\delta M}=-\frac{\partial}{\partial P_\mu{}'}\int_0^1 d\alpha\,\Sigma(P'-\alpha q), \quad (21)$$

and

$$\Sigma(p)=3\int d^4k\,\gamma_5 g(P,P-k)S_F(P-k)$$
$$\times\gamma_5 g(P-k,P)\Delta_F(k) \quad (22)$$

may be expanded in the form of Eq. (19). The term in Σ_f does not contribute to the matrix element, Eq. (15), but it is required in Eq. (21) for strong gauge invariance. There is no contribution of the type $J_\mu{}^{\delta M}$ for the neutron.

It is immediately evident that diagrams Fig. 7(a) and Fig. 7(b) cancel those of Figs. 2(d) and 2(e), respectively. Hence, the sum of all proton diagrams except Fig. 2(a) is given by inserting into Eq. (15), the

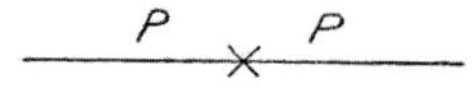

FIG. 6. Mass correction.

expression

$$J_\mu(\text{proton})=J_\mu{}^N(\pi^0)+J_\mu{}^\pi(\pi^+)+J_\mu{}^L(\pi^0)$$
$$+J_\mu{}^L(\pi^+)+J_\mu{}^{\delta M}, \quad (23a)$$

where the last three terms are given by Eqs. (16) and (21) while the nucleonic contribution [Fig. 2(c)] is

$$J_\mu{}^N(\pi^0)=i\int d^4k\,\gamma_5 g(P',P'-k)S_F(P'-k)$$
$$\times\gamma_\mu S_F(P-k)\gamma_5 g(P-k,P)\Delta_F(k), \quad (24a)$$

and the pionic contribution [Fig. 2(b)] is

$$J_\mu{}^\pi(\pi^+)=-2\int d^4p\,\gamma_5 g(P',p)S_F(p)\gamma_5 g(p,P)$$
$$\times\Delta_F(P'-p)(P_\mu{}'+P_\mu-2p_\mu)\Delta_F(P-p). \quad (24b)$$

For the neutron, the sum of the diagrams in Figs. 3 and 4 is

$$J_\mu(\text{neutron})=J_\mu{}^N(\pi^-)+J_\mu{}^\pi(\pi^-)+J_\mu{}^L(\pi^-), \quad (23b)$$

where $J_\mu{}^L(\pi^-)$ is given by Eq. (16c) and

$$J_\mu{}^N(\pi^-)=2J_\mu{}^N(\pi^+), \quad (24c)$$

$$J_\mu{}^\pi(\pi^-)=-J_\mu{}^\pi(\pi^+). \quad (24d)$$

The condition of strong gauge invariance, Eq. (18)

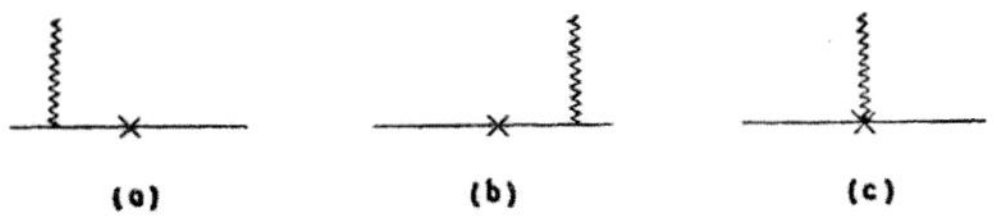

FIG. 7. Contributions of the mass renormalization to the electromagnetic interaction of the proton.

can now be easily verified by noting that

$$(i\not{p}+M)S_F(p)=-1,$$

whence

$$S_F(P'-k)i\mathbf{q}S_F(P-k)=S_F(P'-k)-S_F(P-k).$$

Similarly

$$q_\mu\Delta_F(P'-p)(P_\mu'+P_\mu-2p_\mu)\Delta_F(P-p)=\Delta_F(P-p)-\Delta_F(P'-p).$$

The following relationships are also used:

$$\int_0^1 d\alpha\, q_\mu\left(\frac{\partial}{\partial P_\mu'}+\frac{\partial}{\partial P_\mu}\right)\Phi(P'-\alpha q,\, P+(1-\alpha)q)=-\int_0^1 d\alpha\,\frac{d\Phi}{d\alpha}=\Phi(P',P')-\Phi(P,P),$$

$$\int_0^1 d\alpha\, q_\mu\frac{\partial}{\partial P_\mu'}\Phi(P'-\alpha q,\, P)=\Phi(P',P)-\Phi(P,P),$$

and

$$\int_0^1 d\alpha\, q_\mu\frac{\partial}{\partial P_\mu}\Phi(P',\, P+(1-\alpha)q)=\Phi(P',P')-\Phi(P',P),$$

where $\Phi(x,y)$ is an arbitrary function of the two 4-dimensional variables x, y.

Since

$$q_\mu J_\mu{}^{\delta M}=-q_\mu\frac{\partial}{\partial P_\mu'}\int_0^1 d\alpha\,\Sigma(P'-\alpha q)=\Sigma(P)-\Sigma(P'),$$

we see that strong gauge invariance for the proton is equivalent to the generalized Ward identity

$$(P_\mu'-P_\mu)\Lambda_\mu(P',P)=\Sigma(P')-\Sigma(P), \qquad (25)$$

where $\Lambda_\mu(P',P)$ is the photon vertex function

$$\Lambda_\mu(P',P)=J_\mu{}^N(\pi^0)+J_\mu{}^\pi(\pi^+)+J_\mu{}^L(\pi^0)+J_\mu{}^L(\pi^+). \qquad (26)$$

According to Eqs. (21) and (19)

$$\bar{u}_{s'}(\mathbf{P}')J_\mu{}^{\delta M}u_s(\mathbf{P})=-iB\bar{u}_{s'}(\mathbf{P}')\gamma_\mu u_s(\mathbf{P}). \qquad (27)$$

Comparison of Eq. (15) with Eq. (1) therefore yields

$$\bar{u}_{s'}(\mathbf{P}')[F_1(q^2)\gamma_\mu-F_2(q^2)\sigma_{\mu\nu}q_\nu]u_s(\mathbf{P})=4\pi f^2\bar{u}_{s'}(\mathbf{P}')\{\gamma_\mu[1-(2\pi)^{-4}iB]+(2\pi)^{-4}\Lambda_\mu(P',P)\}u_s(\mathbf{P}). \qquad (28)$$

But the Ward identity

$$\Lambda_\mu(P,P)=\partial\Sigma/\partial P_\mu$$

combined with Eq. (19) shows that

$$\bar{u}_{s'}(\mathbf{P}')\Lambda_\mu(P,P)u_s(\mathbf{P})=iB\bar{u}_{s'}(\mathbf{P}')\gamma_\mu u_s(\mathbf{P}),$$

whence it follows from Eq. (28) that $F_1(0)=1$, in accordance with the required normalization for the proton, Eq. (4). To circumvent the calculation of B, we may first obtain an unrenormalized ${}^uF_1(q^2)$ defined by

$$\bar{u}_{s'}(\mathbf{P}')[{}^uF_1(q^2)\gamma_\mu-\sigma_{\mu\nu}q_\nu F_2(q^2)]u_s(\mathbf{P})=4\pi^2(2\pi)^{-4}f^2\bar{u}_{s'}(\mathbf{P}')\Lambda_\mu(P',P)u_s(\mathbf{P}), \qquad (29)$$

and then determine $F_1(q^2)$ from

$$F_1(q^2)={}^uF_1(q^2)-{}^uF_1(0)+1. \qquad (30)$$

For the neutron, the condition of strong gauge invariance is satisfied by Eq. (23b) without the introduction of a mass normalization term. Hence, Eq. (25) is replaced by

$$(P_\mu'-P_\mu)\Lambda_\mu(P',P)=0, \qquad (31)$$

and Eq. (4) follows directly by noting that Eq. (31) is equivalent to setting $B=0$, although B does not vanish, of course, for the mass renormalization of the neutron.

6. CHOICE OF REFERENCE FRAME

It is convenient to perform the computations in the Breit frame defined by the condition

$$\mathbf{P}'=-\mathbf{P}. \qquad (32)$$

In this case the left-hand side of Eq. (29) can be simplified by making use of the Dirac equation to give

$$i\,{}^uF_{\text{ch}}(q^2)=4\pi(2\pi)^{-4}f^2\bar{u}(-\mathbf{P})\Lambda_4(P',P)u(\mathbf{P}), \qquad (33a)$$

and

$$i(\boldsymbol{\sigma}\times\mathbf{q})\;{}^uF_{\text{mag}}(q^2)=4\pi(2\pi)^{-4}f^2\bar{u}(-\mathbf{P})\boldsymbol{\Lambda}(P',P)u(\mathbf{P}), \qquad (33b)$$

where a 2×2 spin matrix notation has been substituted for the labeling of the spin states. Here

$${}^uF_{\text{ch}}(q^2)={}^uF_1(q^2)-(q^2/2M)F_2(q^2),$$

and

$${}^uF_{\text{mag}}(q^2)=(1/2M)\,{}^uF_1(q^2)+F_2(q^2)$$

are the unrenormalized charge and magnetic form factors defined in accordance with Eqs. (8) and (9). The renormalized form factors are then found from Eq. (30) to be

$$F_{\text{ch}}(q^2)={}^uF_{\text{ch}}(q^2)-{}^uF_{\text{ch}}(0)+1, \qquad (34a)$$

and

$$F_{\text{mag}}(q^2)={}^uF_{\text{mag}}(q^2)+(1/2M)[1-{}^uF_{\text{ch}}(0)]. \qquad (34b)$$

7. THE STATIC LIMIT

The calculation of $F_{\text{ch}}(q^2)$ and $F_{\text{mag}}(q^2)$ on the basis of Eqs. (33), (26), (24), and (16) may now be carried out by the method outlined in Sec. 3, which produces an expansion in powers of M^{-1}. In this section, attention is limited to the lowest order, or static terms. The results are as follows (we set $\mu=1$ as well as $\hbar=c=1$):

For the neutron,

$$[F_{ch}^{(0)}(q^2)]_{neutron}=-\frac{f^2}{\pi^2}\int d^3k\Big[\Big(\mathbf{k}^2-\frac{\mathbf{q}^2}{4}\Big)$$

$$\times\frac{v_+v_-}{\omega_+\omega_-(\omega_++\omega_-)}-\frac{1}{2}\frac{\mathbf{k}^2v_0^2}{\omega^3}\Big], \quad (35a)$$

where

$$\omega_\pm=[1+|\mathbf{k}\pm\tfrac{1}{2}\mathbf{q}|^2]^{\frac{1}{2}},$$
$$\omega=[1+\mathbf{k}^2]^{\frac{1}{2}}. \quad (36)$$

$$v_\pm=v(\omega_\pm^2);\quad v_0=v(\omega^2). \quad (37)$$

Note that $q^2=\mathbf{q}^2$ in the Breit frame. It is found immediately that for $q^2=0$

$$[F_{ch}^0(0)]_{neutron}=0,$$

in accordance with Eq. (4) [note that Eq. (8) gives $F_{ch}(0)=F_1(0)$, in general]. Also

$$[F_{mag}^{(0)}(q^2)]_{neutron}=-\frac{f^2}{2\pi^2}\int d^3k\Big\{\Big[\mathbf{k}^2-\frac{(\mathbf{q}\cdot\mathbf{k})^2}{q^2}\Big]$$

$$\times\frac{v_+v_-}{\omega_+^2\omega_-^2}\Big\}. \quad (38a)$$

For the proton,

$$[^uF_{ch}^{(0)}(q^2)]_{proton}=\frac{f^2}{\pi^2}\int d^3k\Big((\mathbf{k}^2-q^2/4)$$

$$\times\frac{v_+v_-}{\omega_+\omega_-(\omega_++\omega_-)}+\frac{1}{4}\frac{\mathbf{k}^2}{\omega^3}v_0^2-4\int_0^1 d\alpha\frac{v_0}{\omega}\{\tfrac{1}{2}\mathbf{k}^2v_0'$$

$$+[\mathbf{k}^2+\alpha(\mathbf{q}\cdot\mathbf{k})]v_\alpha'\}\Big), \quad (35b)$$

where

$$v_\alpha=v(1+[\mathbf{k}+\alpha\mathbf{q}]^2),$$
$$v_\alpha'=v'(1+[\mathbf{k}+\alpha\mathbf{q}]^2), \quad (39)$$

and

$$v'(\xi)=dv/d\xi.$$

The quantity required by Eqs. (34) for renormalization is

$$[^uF_{ch}^{(0)}(0)]_{proton}=-\frac{3}{4}\frac{f^2}{\pi^2}\int\frac{d^3k}{\omega}\Big[\frac{\mathbf{k}^2}{\omega^2}v_0^2-8\mathbf{k}^2v_0v_0'\Big]. \quad (40)$$

The second term can be integrated by parts to yield

$$[^uF_{ch}^{(0)}(0)]_{proton}=-\frac{3}{4}\frac{f^2}{\pi^2}\int d^3k\frac{v_0^2}{\omega}\Big[6-\frac{\mathbf{k}^2}{\omega^2}\Big]. \quad (41)$$

It can be seen from Eq. (34b) that the renormalization of F_{mag} is of order M^{-1}, hence it is not to be included in the static limit. Then, for the proton,

$$[F_{mag}^{(0)}(q^2)]_{proton}=-[F_{mag}^{(0)}(q^2)]_{neutron}, \quad (38b)$$

where the expression on the right-hand side may be obtained from Eq. (38a).

In Eqs. (35) and (38), the terms proportional to v_+v_- are the contributions of the pion current J_μ^π. They are identical with the Born terms obtained by Walecka[9] in his calculation of the form factors in the static model. The terms proportional to v_0^2 [in Eqs. (35) only] arise from the nucleon current J_μ^N and they are independent of $\mathbf{q}$. Hence, their role is just to maintain charge conservation. The line-current contributions are easily recognized by their α dependence. The only such term in the static approximation occurs for $F_{ch}^{(0)}$ of the proton.[12]

From $F_{ch}^{(0)}$ we may now calculate the second moment of the charge distribution $(\mathcal{R}^2)^{(0)}=-6(dF_{ch}^{(0)}/dq^2)_{q^2=0}$ in the static limit. Since the form factor is the same, the result for the neutron is identical with that obtained in the static model by Salzman,[4] for example:

$$(\langle\mathcal{R}^2\rangle^{(0)})_{neutron}=-\frac{f^2}{\pi}\int_0^\infty dk\frac{\mathbf{k}^4}{\omega^7}\Big[2\omega^4\Big(\frac{dv_0}{dk}\Big)^2$$

$$+5(3\omega^2-2k^2)v_0^2\Big]. \quad (42a)$$

This is the anticipated result which shows that the second moment of the pion charge distribution calculated in the static model is to be compared to the complete neutron-electron interaction without subtracting the Foldy turn. The static model therefore gives much closer agreement with experiment than had previously been thought. For the usual choice of a cutoff close to the nucleon mass, the calculated value is about twice the experimental value.[13] However, it is to be noted that, with this high a cutoff, the neglect of recoil terms is clearly not justified for our procedure.

The second moment of the proton charge distribution is made up of two terms, one is the pionic contribution which is the negative of the neutron second moment and the other is the line-current contribution:

$$(\langle\mathcal{R}^2\rangle^{(0)})_{proton}=-(\langle\mathcal{R}^2\rangle^{(0)})_{neutron}+4\frac{f^2}{\pi}\int_0^\infty dk\frac{\mathbf{k}^2}{\omega^7}$$

$$\times\Big[5v_0^2+\omega^4(1+\tfrac{2}{3}\omega^2)\Big(\frac{dv_0}{dk}\Big)^2\Big]. \quad (42b)$$

It is clear that the second term, due entirely to the line currents, is somewhat larger than the first.

It has been remarked by Yennie, Lévy, and Ravenhall[2] that application of the mirror theorem[5] to $\langle\mathcal{R}^2\rangle$ indicates that the isoscalar quantity $\langle\mathcal{R}^2\rangle_{proton}$

[12] To demonstrate that the line-current contributions to $F_{mag}^{(0)}$ vanish, an integration by parts in $\cos\theta=(\mathbf{k}\cdot\mathbf{q})/|\mathbf{k}||\mathbf{q}|$ has been performed. The fact that these terms vanish is probably related to the fact that the line currents flow radially in the static model and therefore do not give rise to any magnetic multipole moments. See R. H. Capps and W. G. Holladay, Phys. Rev. **99**, 931 (1955).

[13] S. Treiman and R. G. Sachs, Phys. Rev. **103**, 435 (1956).

$+\langle \mathcal{R}^2\rangle_{\text{neutron}}$ depends only on contributions from the nucleon (core) charge distribution. Hence, the line-current contribution is presumably nucleonic in origin. For a usual value of the cutoff, its order of magnitude agrees with the experimental value[2]

$$[\langle \mathcal{R}^2\rangle_{\text{proton}}+\langle \mathcal{R}^2\rangle_{\text{neutron}}]^{\frac{1}{2}}\approx 0.75\times 10^{-13}\ \text{cm},$$

which is, of course, a surprisingly large number. Evidently, in the process of introducing a cutoff, the requirements of charge conservation are such as to cause a large shift in the electric charge, even if the cutoff is limited to a small spatial region. In this connection, it should be remarked that the line-current contribution is obtained in the form given in Eq. (42b) only after several integrations by parts. As can be seen from Eq. (35b), the contribution originally involves at least one derivative of the source function, hence for a constant v, the line-current term would vanish.

From Eq. (38b) it is evident that only pionic contributions to the magnetic form factor occur in the static limit. The neutron and proton magnetic moments are oppositely equal in this approximation and they agree exactly with the values obtained in the usual static model. Hence, the magnetic moments are identical with those obtained from the static model. As Miyazawa has shown,[14] these may be brought into reasonable agreement with the isotopic vector part of the observed nucleon moment by including rescattering corrections.

8. RECOIL CORRECTIONS TO THE MAGNETIC MOMENT

Although it is actually a recoil effect, it is customary in applications of the static model to include a nucleon (core) spin contribution to the magnetic moment. This has been a constant source of failure for the static model since it leads to much too large a value of the isoscalar moment.[14] However, it has often been suggested that a consistent treatment of recoil effects would overcome this difficulty. We have now made a consistent calculation on the basis of the procedure outlined above with the following results:

For the neutron, the contribution of the pion current is

$$[\Delta_\pi F_{\text{mag}}^{(1)}(0)]_{\text{neutron}}=\frac{1}{6M}\frac{f^2}{\pi^2}\int d^3k\frac{\mathbf{k}^2}{\omega}\times\left[\frac{3}{2}\frac{\mathbf{k}^2}{\omega^4}v_0^2-2v_0v_0'\right],\quad (43\text{a})$$

that of the nucleon (core) current is

$$[\Delta_N F_{\text{mag}}^{(1)}(0)]_{\text{neutron}}=-\frac{1}{12M}\frac{f^2}{\pi^2}\int d^3k\frac{\mathbf{k}^2}{\omega^3}v_0^2,\quad (44\text{a})$$

and that of the line current is

$$[\Delta_L F_{\text{mag}}^{(1)}(0)]_{\text{neutron}}=-\frac{1}{6M}\frac{f^2}{\pi^2}\int\frac{d^3k}{\omega}\times\left[\left(3+\frac{\mathbf{k}^2}{\omega^2}\right)v_0^2+2\mathbf{k}^2v_0v_0'\right].\quad (45\text{a})$$

The total $1/M$ effect on the neutron moment is therefore

$$[F_{\text{mag}}^{(1)}(0)]_{\text{neutron}}=-\frac{1}{2M}\frac{f^2}{\pi^2}\int\frac{d^3k}{\omega}\times\left[\left(1+\frac{1}{2}\frac{\mathbf{k}^2}{\omega^4}\right)v_0^2+\tfrac{4}{3}\mathbf{k}^2v_0v_0'\right].\quad (46\text{a})$$

For the proton, the pion contribution and the nucleon contributions to the unrenormalized form factors are related to those of the neutron as follows:

$$[\Delta_\pi\, {}^uF_{\text{mag}}(0)]_{\text{proton}}=-[\Delta_\pi\, {}^uF_{\text{mag}}^{(1)}(0)]_{\text{neutron}},\quad (43\text{b})$$

and

$$[\Delta_N\, {}^uF_{\text{mag}}^{(1)}(0)]_{\text{proton}}=\tfrac{1}{2}[\Delta_N {}^uF_{\text{mag}}^{(1)}(0)]_{\text{neutron}}.\quad (44\text{b})$$

The line-current contribution is

$$[\Delta_L\, {}^uF_{\text{mag}}^{(1)}(0)]_{\text{proton}}=\frac{1}{6M}\frac{f^2}{\pi^2}\int\frac{d^3k}{\omega}\times\left[\left(3+\frac{\mathbf{k}^2}{\omega^2}\right)v_0^2+8\mathbf{k}^2v_0v_0'\right].\quad (45\text{b})$$

Also for the proton, there are the additional terms produced by renormalization in accordance with Eq. (34b):

$$F_{\text{mag}}^{(1)}(0)={}^uF_{\text{mag}}^{(1)}(0)+\frac{1}{2M}[1-{}^uF_{\text{ch}}^{(0)}(0)],\quad (47)$$

where ${}^uF_{\text{ch}}^{(0)}(0)$ is given by Eq. (40) or Eq. (41). In this way, we find for the total recoil correction to the magnetic moment of the proton:

$$[F_{\text{mag}}^{(1)}(0)]_{\text{proton}}=\frac{1}{2M}+\frac{1}{2M}\frac{f^2}{\pi^2}\int\frac{d^3k}{\omega}\times\left[\left(\frac{1}{2}\frac{\mathbf{k}^2}{\omega^4}+1-\frac{1}{2}\frac{\mathbf{k}^2}{\omega^2}\right)v_0^2+\frac{14}{3}\mathbf{k}^2v_0v_0'\right].\quad (46\text{b})$$

The order of magnitude of the recoil correction to the neutron or proton moment is a sensitive (quadratic) function of the cutoff in v_0. For the usual choice of a cutoff close to the nucleon mass, the correction is very large, of the order of several nuclear magnetons. This is not very surprising since the expansion in powers of $1/M$ is based on the assumption that the cutoff is much smaller than M. Hence we come to the same

[14] H. Miyazawa, Phys. Rev. **101**, 1564 (1956).

conclusion as Walecka[9] and others, that the static model is a very poor approximation in view of the probable importance of the recoil corrections. The apparent agreement with experiment (of the isovector magnetic moment, for example) is evidently fortuitous.

Despite this conclusion, it is of some interest to consider the sum of neutron and proton moments. From Eqs. (38b), (46a), and (46b), we find to order $1/M$,

$$[F_{\text{mag}}(0)]_{\text{proton}}+[F_{\text{mag}}(0)]_{\text{neutron}} = (1/2M)[1-\tfrac{4}{3}\,{}^uF_{\text{ch}}(0)], \quad (48)$$

where ${}^uF_{\text{ch}}(0)$ is given by Eq. (40). This is evidently the same result obtained from the mirror theorem in the static model,[5] since ${}^uF_{\text{ch}}(0)$ may be interpreted as the probability for finding the nucleon accompanied by a pion field. The success of the static model in this respect is not very surprising since the derivation of the result relied mainly on the conservation of angular momentum.

Because of its general nature, Eq. (48) poses a serious problem for any theory of the electromagnetic form factors. From the observed magnetic moments it is found by means of Eq. (48) that

$$[{}^uF_{\text{ch}}(0)]\approx 0.09. \quad (49)$$

If one calculates ${}^uF_{\text{ch}}{}^{(0)}$ from Eq. (41) using the accepted value of $f^2\approx 0.08$ and a cutoff corresponding to the nucleon mass, a much larger value than that given by Eq. (49) is obtained. Hence, this is a clear point of failure of the first-order recoil corrections to the static model.

9. CONCLUSION

The importance of taking into account the condition of strong gauge invariance in this work suggests that it must be kept in mind in any attempt to calculate electromagnetic form factors. We have found, for example, that the separation of the contribution to the matrix element of the pion current from that of the nucleon current is not really gauge invariant although it satisfies the condition of weak gauge invariance. The latter condition was noted by Federbush, Goldberger, and Treiman[15] in connection with the corresponding separation of the integral equations arising from dispersion relations. Therefore, the gauge invariance of their approximate form of the integral equations is subject to serious doubt.

Furthermore, any other mutilation of the exact equations is subject to doubt on the same grounds. Such approximations provide integral equations involving strong (pion-nucleon, etc.) vertex functions and we have seen how the condition of charge conservation requires that a means must be provided for transferring the charge across such a nonlocal vertex. The importance of this consideration is illustrated by the result that the charge radius of the proton turns out to be very large through the influence of line currents. The line currents are just a manifestation of the condition of charge conservation. They represent the electromagnetic effects of higher order diagrams. In this connection, it is interesting that even for a cutoff near the nucleon mass, the line-current contribution to the charge form factor of the proton is most significant. This indicates that the electromagnetic contributions of intermediate particles of large mass may, through conditions imposed by gauge invariance, be comparable to the pion contribution, in spite of the large denominators appearing with such terms.

[15] P. Federbush, M. L. Goldberger, and S. B. Treiman, Phys. Rev. 112, 642 (1958).

APPENDIX

We wish to demonstrate that the time reversal properties of $j_\mu(x)$ are such that the matrix element $\langle P',s'|j_\mu(0)|P,s\rangle$ cannot contain a term of the form $\bar{u}_{s'}(\mathbf{P}')q_\mu F_3(q^2)u_s(\mathbf{P})$. Under the Wigner time inversion operator K, the transformation of the current density is

$$K\mathbf{j}(0)K^{-1}=-\mathbf{j}(0),\qquad K\rho(0)K^{-1}=\rho(0), \quad (\text{A-1})$$

where $\rho(x)=-ij_4(x)$. Phases of the state vectors are chosen so that

$$K|P,s\rangle=|P_-,-s\rangle, \quad (\text{A-2})$$

where $P_-=(-\mathbf{P},P_0)$ if $P=(\mathbf{P},P_0)$. Since K is an antiunitary operator, and $\mathbf{j}(0)$, $\rho(0)$ are Hermitian operators, we may write

$$\langle K(P',s')|K\mathbf{j}(0)|P,s\rangle=\langle P,s|\mathbf{j}(0)|P',s'\rangle,$$
$$\langle K(P',s')|K\rho(0)|P,s\rangle=\langle P,s|\rho(0)|P',s'\rangle. \quad (\text{A-3})$$

But also

$$\langle K(P',s')|K\mathbf{j}(0)|P,s\rangle=\langle K(P',s')|K\mathbf{j}(0)K^{-1}K|P,s\rangle = -\langle K(P',s')|\mathbf{j}(0)K|P,s\rangle,$$

by Eq. (A-1). Then from Eqs. (A-2) and (A-3) we have

$$\langle P,s|\mathbf{j}(0)|P',s'\rangle=-\langle P_-',-s'|\mathbf{j}(0)P_-,-s\rangle. \quad (\text{A-4})$$

The corresponding result for $\rho(0)$ is

$$\langle P,s|\rho(0)|P',s'\rangle=\langle P_-',-s'|\rho(0)|P_-,-s\rangle. \quad (\text{A-5})$$

It is now easily seen by means of a straightforward substitution that $\bar{u}_{s'}(\mathbf{P}')q_\mu u_s(\mathbf{P})$ does not satisfy the conditions (A-4) and (A-5) while Eq. (1) does do so.

PHYSICAL REVIEW VOLUME 119, NUMBER 4 AUGUST 15, 1960

Pion-Pion Interactions in τ and τ' Decays*

R. F. SAWYER AND K. C. WALI
University of Wisconsin, Madison, Wisconsin
(Received April 11, 1960)

The final-state interactions in τ and τ' decays of K^+ mesons are studied by means of a Mandelstam representation. It is assumed that only the S-wave pion-pion scattering amplitude is large enough to have an appreciable imaginary part. Coupled, linear, singular, integral equations are found for the amplitudes describing $K^+ + \pi \rightarrow \pi + \pi$ in the physical and unphysical regions. From the solutions to these equations the τ and τ' decay matrix elements may be constructed. In an approximation the equations are solved in terms of pion-pion phase shifts. Comparison with experiment is made using Chew and Mandelstam's solution of the corresponding nonlinear equations for $\pi-\pi$ scattering. Consistency with experiment is found for values of the coupling constant implying repulsive $T=0$ and $T=2$ phase shifts. Independently of the approximation the equations show that the implications in τ and τ' decay of the rule, $\Delta T=\frac{1}{2}, \frac{3}{2}$, are identical with those of the $\Delta T=\frac{1}{2}$ rule. Hence the energy spectra in τ' decay cannot be a critical test of the $\Delta T=\frac{1}{2}$ rule. Also, independently of the approximation, it is shown that the decay matrix element cannot be expanded in a series of integral powers of the pion kinetic energies.

1. INTRODUCTION

IN a recent analysis of 900 τ^+ decays of K^+ mesons, $K^+ \rightarrow \pi^+ + \pi^+ + \pi^-$, it was found that the π^- energy spectrum differs significantly from that predicted from the density of states alone.[1] The minimum pion wavelength in this process is about one pion Compton wavelength. If the weak interaction proceeds in some way through the heavy fermion pairs and has a range determined by the intermediate mass, one would expect very little dependence upon the pion momenta in the matrix element, because the pion wavelength is large compared to the radius of interaction. Hence the observed deviation from a constant matrix element is some evidence for final-state interactions which extend the spatial region of interaction for the outgoing pions. This is an attractive possibility in view of the current interest in pion-pion interactions. The system of three low-energy pions should be an ideal analyzer for such effects.

Thomas and Holladay[2] have investigated the effect of an attractive, $T=2$, pion-pion interaction in τ decay by using Watson's final-state interaction formalism.[3] This method is suited to discussion of the final-state scattering of a single pair of particles. It is applicable to the problem of τ^+ decay with a $T=2$, $\pi-\pi$ force because the state of two π^+ is pure $T=2$ and the other pairs, $\pi^+\pi^-$, are predominantly $T=0$ in the S wave. If, however, there are strong $T=0$ pion-pion effects, then a formulation is required which is capable of dealing simultaneously with interactions between various pairs. The Mandelstam representation can serve as the basis for such an analysis, just as an ordinary single variable dispersion relation can be used to discuss a single final-state interaction.

In the present work Mandelstam representations are assumed for the τ and τ' decay amplitudes.[3a] It is assumed that only the S-wave pion-pion scattering amplitudes are large (have an imaginary part). Linear integral equations are obtained for the $\tau+\pi \rightarrow \pi+\pi$ scattering amplitudes in the physical and unphysical regions. These equations involve the pion-pion S-wave phase shifts. From the solutions the Mandelstam representations for τ and τ' decays may be constructed. In general, two new parameters are needed to characterize τ decay. The $\Delta T=\frac{1}{2}, \frac{3}{2}$ rule can be applied to yield a simpler set of equations with only one parameter, which, if solved,

* This work was supported in part by the U. S. Atomic Energy Commission and in part by the Research Committee of the University of Wisconsin with funds provided by the Wisconsin Alumni Research Foundation.

[1] S. McKenna, S. Natali, M. O'Connell, J. Tietge, and N. C. Varshneya, Nuovo cimento 10, 763 (1958).

[2] B. S. Thomas and W. G. Holladay, Phys. Rev. 115, 1329 (1958).

[3] Also, A. N. Mitra has investigated the effects of a $T=2$ pion-pion resonance in the τ decay [Nuclear Phys. 6, 404 (1958)].

[3a] N. N. Khuri and S. B. Treiman have also considered this problem (Phys. Rev. to be published).

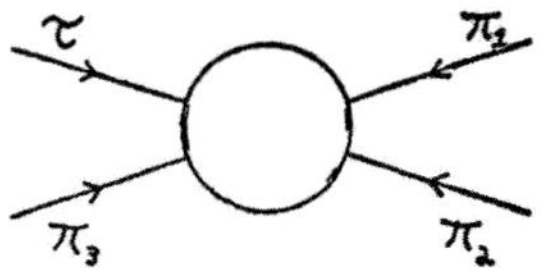

FIG. 1. The τ-decay interaction, $\tau+\pi\rightarrow\pi+\pi$.

should give τ and τ' spectra free of new parameters. The implications of $\Delta T=\frac{1}{2}$ and $\Delta T=\frac{1}{2},\frac{3}{2}$ are identical in our approach. In an approximation these equations have been solved in terms of pion-pion phase shifts and integrals over pion-pion phase shifts. The phases themselves are obtained from nonlinear integral equations which follow from the same approach to the $\pi-\pi$ problem.[4] These equations (for large S waves) have been solved[5] by Chew, Mandelstam, and Noyes, and we have used their solutions in analyzing our results.

2. THE MANDELSTAM REPRESENTATION

We begin by considering the problem of $\tau+\pi\rightarrow\pi+\pi$ scattering. The variables s_1, s_2, and s_3 are defined by drawing all three pion momenta inward (Fig. 1).

$$\begin{aligned} s_1&=-(p_2+p_3)^2,\\ s_2&=-(p_1+p_3)^2,\\ s_3&=-(p_1+p_2)^2. \end{aligned}\tag{1}$$

The three pions are referred to as π_1, π_2, and π_3. The assignment of charge will be deferred until the next section. For the process $\tau+\pi_3\rightarrow\pi_1+\pi_2$ the variables s_1, s_2, and s_3 are expressed in terms of the initial three momentum, q', the final momentum, q, and a scattering angle, θ, in the center-of-mass coordinate system,

$$\begin{aligned} s_3&=4(q^2+\mu^2),\\ s_2&=(M_K^2/2)-(\mu^2/2)-2q^2+2qq'\cos\theta,\\ s_1&=(M_K^2/2)-(\mu^2/2)-2q^2-2qq'\cos\theta, \end{aligned}$$

with

$$2(q^2+\mu^2)^{\frac{1}{2}}=(q'^2+\mu^2)^{\frac{1}{2}}+(q'^2+M_K^2)^{\frac{1}{2}}.\tag{2}$$

Here M_K is the mass of K^+ and μ is the pion mass. We shall take, for the moment, $M_K=3\mu$. This will simplify the contributions to the unphysical region. It will eventually be argued that the τ-decay matrix element does not vary strongly with the K-meson mass. For $M_K=3\mu$ the physical region extends to $q=(3\mu^2)^{\frac{1}{2}}$. Between $q=(3\mu^2)^{\frac{1}{2}}$ and $q=0$, q' is imaginary.

The amplitude $A(s_1s_2s_3)$ for the reaction $\tau+\pi_3\rightarrow\pi_1+\pi_2$ (with s_3 the energy variable) is given by,

$$\langle\pi_1\pi_2\,\text{out}|\pi_3\tau\,\text{in}\rangle =i(2\pi)^5\delta^4(p_\tau+p_3-p_1-p_2)\frac{A(s_1,s_2,s_3)}{(\omega_\tau\omega_1\omega_2\omega_3)^{\frac{1}{2}}}.\tag{3}$$

The continuation of this function $A(s_1,s_2,s_3)$ to all values of s_1, s_2, and s_3, constrained by the condition $s_1+s_2+s_3=3\mu^2+M_K^2\approx12\mu^2$, provides amplitudes for other channels of $\tau+\pi\rightarrow\pi+\pi$ scattering and also for τ decay. Since s_1, s_2, and s_3 stand for energies of two-pion systems in their center-of-mass system, we expect the analytic structure to be the same as for $\pi-\pi$ scattering.[4] In the case of each s the lowest mass intermediate state is that of two pions. We write therefore,

$$\begin{aligned} A(s_1,s_2,s_3) &=\frac{1}{\pi^2}\int_{(2\mu)^2}^{\infty}ds_1'\int_{(2\mu)^2}^{\infty}ds_2'\frac{\rho_{12}(s_1',s_2')}{(s_1'-s_1)(s_2'-s_2)}\\ &+\frac{1}{\pi^2}\int_{(2\mu)^2}^{\infty}ds_2'\int_{(2\mu)^2}^{\infty}ds_3'\frac{\rho_{23}(s_2',s_3')}{(s_2'-s_2)(s_3'-s_3)}\\ &+\frac{1}{\pi^2}\int_{(2\mu)^2}^{\infty}ds_3'\int_{(2\mu)^2}^{\infty}ds_1'\frac{\rho_{13}(s_1',s_3')}{(s_3'-s_3)(s_1'-s_1)}. \end{aligned}\tag{4}$$

This representation may be rearranged to give

$$\begin{aligned} A(s_1,s_2,s_3) &=\frac{1}{\pi^2}\int ds_1'\frac{1}{(s_1'-s_1)}\bigg[\int ds_2'\frac{\rho_{12}(s_1',s_2')}{s_1'+s_2'+s_3-3\mu^2-M_K^2}\\ &\qquad+\int ds_3'\frac{\rho_{13}(s_1',s_3')}{s_1'+s_3'+s_2-3\mu^2-M_K^2}\bigg]\\ &+\frac{1}{\pi^2}\int ds_2'\frac{1}{(s_2'-s_2)}\bigg[\int ds_3'\frac{\rho_{23}(s_2',s_3')}{s_2'+s_3'+s_1-3\mu^2-M_K^2}\\ &\qquad+\int ds_1'\frac{\rho_{12}(s_2',s_1')}{s_2'+s_1'+s_3-3\mu^2-M_K^2}\bigg]\\ &+\frac{1}{\pi^2}\int ds_3'\frac{1}{(s_3'-s_3)}\bigg[\int ds_1'\frac{\rho_{13}(s_1',s_3')}{s_1'+s_3'+s_2-3\mu^2-M_K^2}\\ &\qquad+\int ds_2'\frac{\rho_{23}(s_2',s_3')}{s_3'+s_2'+s_1-3\mu^2-M_K^2}\bigg]. \end{aligned}\tag{5}$$

From the region of nonvanishing ρ determined by Mandelstam[4] we see that the functions $\rho(x,y)$ vanish unless $x+y>36\mu^2$. For small s, therefore, the typical inside integral of (5),

$$I(x,s)=\int dy\frac{\rho(x,y)}{x+y+s-3\mu^2-M_K^2},$$

should be nearly independent of s. Applying this argument to each term in (5) we arrive at the approximate form,[6]

[4] G. F. Chew and S. Mandelstam, University of California Radiation Laboratory Report UCRL-8728 (unpublished).

[5] G. F. Chew, S. Mandelstam, and H. P. Noyes, University of California Radiation Laboratory Report UCRL-9001 (unpublished).

[6] A representation similar to (6) has been used by M. Gourdin and A. Martin (CERN preprint) and has been attributed by them to M. Cini and S. Fubini.

$$A(s_1,s_2,s_3)=\frac{1}{\pi}\int_{4\mu^2}^{\infty}ds'\frac{\sigma_1(s')}{s'-s_1}+\frac{1}{\pi}\int_{4\mu^2}^{\infty}ds'\frac{\sigma_2(s')}{s'-s_2}+\frac{1}{\pi}\int_{4\mu^2}^{\infty}ds'\frac{\sigma_3(s')}{s'-s_3}. \quad (6)$$

This simplification in the case of pion-pion scattering leads to exactly the Chew-Mandelstam nonlinear integral equations for S-wave dominant amplitudes (that is, the equations in which the imaginary part is taken as pure S-wave). Chew and Mandelstam[5] have shown that within a wide range of pion-pion coupling constants consistent solutions with a small P wave exist.

Now defining, for the process $\tau+\pi_3\to\pi_1+\pi_2$

$$A_s(q^2)=\frac{1}{2}\int_{-1}^{+1}A(s_1,s_2,s_3+i\epsilon)d(\cos\theta), \quad (7)$$

we see that (in spite of the fact that s_1 and s_2 become complex in the region $4\mu^2<s_3<16\mu^2$),

$$\mathrm{Im}A_s(q^2)=\sigma_3[4(q^2+\mu^2)] \quad (8)$$

for all $q^2>0$. Similar expressions relate σ_1 and σ_2 to the other two S-wave scattering amplitudes. Therefore, knowledge of the S-wave amplitudes for the three processes, $\tau_1+\pi_1\to\pi_2+\pi_3$, $\tau_1+\pi_2\to\pi_1+\pi_3$, and $\tau_1+\pi_3\to\pi_1+\pi_2$, is sufficient for construction of the single-variable Mandelstam relation (6).

In the physical region, $q^2>3\mu^2$, the S-wave unitarity condition including only 2π intermediate states is,

$$\mathrm{Im}A_s(\tau\pi_3,\pi_1\pi_2)=[q^2/(q^2+1)]^{\frac{1}{2}}\sum A_s(\tau\pi_3,2\pi)t_s^{\dagger}(2\pi,\pi_1\pi_2), \quad (9)$$

where t_s is the S-wave pion-pion scattering amplitude. The sum is to be extended over all intermediate charge states. q is the pion momentum in the intermediate and final two-pion states. For an intermediate and final state of definite total isotopic spin, T, we have

$$t_s=[(q^2+1)/q^2]^{\frac{1}{2}}e^{i\delta_T(q)}\sin\delta_T(q).$$

To show the applicability of this unitarity condition also in the unphysical region, $q^2<3$, we consider the quantity,

$$\langle 0|K(0)|\pi_3\tau\ \mathrm{in}\rangle=\frac{i}{(2\omega_3)^{\frac{1}{2}}}\int e^{-ip_3\cdot x}d^4x\langle 0|T(K(0)J_3(x))|\tau\rangle,$$

where $K(0)$ is some operator such that $\langle 0|K(0)|2\pi\rangle\neq 0$. From standard methods it follows that

$$\mathrm{Im}\langle 0|K(0)|\pi_3\tau\ \mathrm{in}\rangle=[\pi/(2\omega_3)^{\frac{1}{2}}]\sum\delta^4(p_n-p_\tau-p_3)\times\langle 0|K(0)|n\rangle\langle n|J_3(0)|\tau\rangle, \quad (10)$$

where J_3 is the current associated with π_3. For $-(p_\tau+p_3)^2$ less than the four π-meson threshold the only intermediate states are those of two pions. The operator K may be chosen to connect with a two-pion state of any particular angular momentum or isotopic spin.

We note that $\langle 0|K(0)|2\pi\ \mathrm{out}\rangle_{T,J}$ has the phase $\exp[-i\delta_{TJ}(q)]$. If we require $\mathrm{Im}\langle 0|K(0)|\pi_3\tau\ \mathrm{in}\rangle$ to be real, we find from (10) that $\langle 2\pi\ \mathrm{out}|\tau_J J_3(0)|\tau\rangle$ has the phase, $\exp[i\delta_{TJ}(q)]$. This implies the conditions (9).

3. τ^+ AND τ'^+ DECAY

For τ and τ' decays, π_1 and π_2 will be taken as the identical pions and π_3 as the odd pion. Thus for the τ^+-decay representation the process, $\tau+\pi_3\to\pi_1+\pi_2$, is $\tau^++\pi^+\to\pi^++\pi^+$. Then $\sigma_3(s')$ is determined from $\mathrm{Im}A_s(\tau^+\pi^+,\pi^+\pi^+)$. The weight functions σ_1 and σ_2 are identical by the Bose principle and are both equal to the imaginary part of $A_s(\tau^+\pi^-,\pi^+\pi^-)$. In the unitarity condition for $A_s(\tau^+\pi^-,\pi^+\pi^-)$ one encounters the intermediate amplitude $A_s(\tau^+\pi^-,\pi^0\pi^0)$. This must be determined from a separate dispersion relation for τ' decay.

Setting $\sigma_1=\sigma_2=\sigma$, $\sigma_3=\rho$, we write

$$A^\tau(s_1,s_2,s_3)=\lambda+\frac{s_3-s_0}{\pi}\int ds'\frac{\rho(s')}{(s'-s_3)(s'-s_0)}+\frac{s_2-s_0}{\pi}\int ds'\frac{\sigma(s')}{(s'-s_2)(s'-s_0)}+\frac{s_1-s_0}{\pi}\int ds'\frac{\sigma(s')}{(s'-s_1)(s'-s_0)}, \quad (11)$$

and

$$A^{\tau'}(s_1,s_2,s_3)=\lambda'+\frac{s_3-s_0}{\pi}\int ds'\frac{\rho'(s')}{(s'-s_3)(s'-s_0)}+\frac{s_2-s_0}{\pi}\int ds'\frac{\sigma'(s')}{(s'-s_2)(s'-s_0)}+\frac{s_1-s_0}{\pi}\int ds'\frac{\sigma'(s')}{(s'-s_1)(s'-s_0)}. \quad (12)$$

In the case of τ' decay, s_3 stands for the energy in the center-of-mass system of the two π^0. A subtraction has been here inserted into the Mandelstam relation itself, rather than into the partial wave dispersion relations as in reference 4. This is essential since our aim is to discuss the continuation to τ decay by constructing the complete Mandelstam relation. Solution of the S wave, $\tau+\pi\to\pi+\pi$ problem will give only the densities ρ and σ. The point s_0 is chosen as the symmetrical point in pion-pion scattering, $s_1=s_2=s_3=\frac{4}{3}\mu^2$. It is of no importance that this point does not obey the constraint between s_1, s_2, and s_3 in τ decay. For any given τ mass the function A is uniquely defined as an independent function of s_1, s_2, and s_3, if one demands the analytic properties in (6). Equation (11) then results from defining λ as $A(s_0,s_0,s_0)$.

In determination of ρ, σ, ρ', σ', the S-wave amplitudes of four processes will be used,

$$\begin{aligned} A_1(q^2)&=A(\tau^+\pi^+,\pi^+\pi^+)_s, & \mathrm{Im}A_1(q^2)&=\rho[4(q^2+\mu^2)],\\ A_2(q^2)&=A(\tau^+\pi^-,\pi^+\pi^-)_s, & \mathrm{Im}A_2(q^2)&=\sigma[4(q^2+\mu^2)],\\ A_3(q^2)&=A(\tau^+\pi^-,\pi^0\pi^0)_s, & \mathrm{Im}A_3(q^2)&=\rho'[4(q^2+\mu^2)],\\ A_4(q^2)&=A(\tau^+\pi^0,\pi^+\pi^0)_s, & \mathrm{Im}A_4(q^2)&=\sigma'[4(q^2+\mu^2)]. \end{aligned} \tag{13}$$

Here q is always the momentum in the two-pion system. The unitarity conditions for the four amplitudes are given from (9) in terms of pion-pion S-wave amplitudes $t_0(q^2)$ and $t_2(q^2)$. They are

$$\begin{aligned} &[(q^2+\mu^2)/q^2]^{\frac{1}{2}}\rho[4(q^2+\mu^2)]=A_1t_2^\dagger,\\ &[(q^2+\mu^2)/q^2]^{\frac{1}{2}}\sigma[4(q^2+\mu^2)]\\ &\qquad=A_2(\tfrac{1}{3}t_2^\dagger+\tfrac{2}{3}t_0^\dagger)+A_3(-\tfrac{1}{3}t_2^\dagger+\tfrac{1}{3}t_0^\dagger),\\ &[(q^2+\mu^2)/q^2]^{\frac{1}{2}}\rho'[4(q^2+\mu^2)]\\ &\qquad=A_2(\tfrac{1}{3}t_2^\dagger-\tfrac{1}{3}t_0^\dagger)+A_3(\tfrac{2}{3}t_2^\dagger+\tfrac{1}{3}t_0^\dagger),\\ &[(q^2+\mu^2)/q^2]^{\frac{1}{2}}\sigma'[4(q^2+\mu^2)]=A_4t_2^\dagger. \end{aligned} \tag{14}$$

From (12) we derive S-wave dispersion relations for the various A. When the angular integrations are performed an amplitude is expressed in terms of a subtraction constant, an integral which gives a cut from $q^2=0$ to $q^2=\infty$, and an integral which has singularities to the left of the imaginary axis. These contributions can be expressed in terms of functions $R(\rho;\nu)$ and $L(\rho;\nu)$ with $\nu=(q^2/\mu^2)$,

$$\begin{aligned} R(\rho;\nu)&=\frac{\nu-\nu_0}{\pi}\int_0^\infty d\nu'\frac{\rho[4(\nu'+1)\mu^2]}{(\nu'-\nu)(\nu'-\nu_0)},\\ L(\rho;\nu)&=\frac{\nu-\nu_0}{\pi}\frac{1}{\nu}\left(\frac{\nu+1}{\nu-3}\right)^{\frac{1}{2}}\int_0^\infty d\nu'\frac{\rho[4(\nu'+1)\mu^2]}{(\nu'-\nu_0)}\\ &\qquad\times\ln\frac{\nu'+\frac{1}{2}\nu+\frac{1}{2}\nu[(\nu-3)/(\nu+1)]^{\frac{1}{2}}}{\nu'+\frac{1}{2}\nu-\frac{1}{2}\nu[(\nu-3)/(\nu+1)]^{\frac{1}{2}}}. \end{aligned} \tag{15}$$

The dispersion relations become,

$$\begin{aligned} A_1(\nu)&=\lambda+R(\rho;\nu)+2L(\sigma;\nu),\\ A_2(\nu)&=\lambda+R(\sigma;\nu)+L(\rho+\sigma;\nu),\\ A_3(\nu)&=\lambda'+R(\rho';\nu)+2L(\sigma';\nu),\\ A_4(\nu)&=\lambda'+R(\sigma';\nu)+L(\rho'+\sigma';\nu). \end{aligned} \tag{16}$$

Taken with the unitarity conditions, (14), these comprise a set of four coupled linear integral equations for the four amplitudes A.

4. ΔT RULES IN τ AND τ' DECAY

The $\Delta T=\frac{1}{2}$ rule, which relates τ' to τ decay, would reduce this set of four equations to two independent equations. For our amplitudes, A, the $\Delta T=\frac{1}{2}$ rule implies the relations,

$$A_4=A_1/2,\quad A_3=A_2-A_1/2. \tag{17}$$

Essentially equivalent relations have been given by Weinberg.[7]

The four equations, (16), have been written down in order to show that the implications of $\Delta T=\frac{1}{2}$, $\frac{3}{2}$ are the same as those of $\Delta T=\frac{1}{2}$. To prove this we first note that $\Delta T=\frac{1}{2}$, $\frac{3}{2}$ implies that $\lambda'=\lambda/2$. This follows from the definitions of λ as $A^\tau(s_0,s_0,s_0)$ and λ' as $A^{\tau'}(s_0,s_0,s_0)$. When A is divided into a completely symmetric function of the three variables plus a function of mixed symmetry as in reference 7, one sees that the function of mixed symmetry vanishes at the symmetric point. For the completely symmetrical function we need only $\Delta T<\frac{5}{2}$ to prove $A_{\mathrm{sym}}{}^{\tau'}=\frac{1}{2}A_{\mathrm{sym}}{}^\tau$ as in reference 7. Hence, $\lambda'=\lambda/2$.

When $\lambda'=\lambda/2$ is inserted into the integral equations (16) and when the unitarity conditions (14) are invoked it is possible to see that the substitutions,

$$A_4=A_1/2,\quad A_3=A_2-A_1/2,$$

make the last two equations linearly dependent upon the first two. Hence, if the four equations have unique solutions (assuming particular behavior at infinity) and if $\lambda'=\lambda/2$ the solutions must be such that the relations, (17), are satisfied. $\Delta T=\frac{1}{2}$, $\frac{3}{2}$ therefore implies the same conclusions as does $\Delta T=\frac{1}{2}$. Henceforth we shall adopt the relations (17) implied by the $\Delta T=\frac{1}{2}$, $\frac{3}{2}$ rule.

5. THE RELATION TO PION-PION SCATTERING

Under the conditions (17) our equations become quite similar in form to the nonlinear equations describing pion-pion S-wave scattering. This is apparent if we write the two independent equations in terms of the combinations,

$$B_2=A_1,\quad B_0=3A_2-\tfrac{1}{2}A_1.$$

Equations (16) become (using 14),

$$\begin{aligned} B_2(\nu)&=\lambda+R(B_2t_2^\dagger;\nu)+L(\tfrac{1}{3}B_2t_2^\dagger+\tfrac{2}{3}B_0t_0^\dagger;\nu),\\ B_0(\nu)&=\tfrac{5}{2}\lambda+R(B_0t_0^\dagger;\nu)+L(10/3B_2t_2^\dagger+\tfrac{2}{3}B_0t_0^\dagger;\nu). \end{aligned} \tag{18}$$

This structure reflects the fact that under the $\Delta T=\frac{1}{2}$ rule the K^+ must decay into a state with $T=1$ and therefore the K^+ couples to the pions with exactly the same coefficients as would π^+. This enables us to define our "$T=2$" and "$T=0$" combinations of τ and π.

Equations (18) may be converted into nonsingular integral equations by well-known methods.[8] However, these equations are of little value unless one knows pion-pion phase shifts accurately at all energies and is willing to do extensive numerical work. Instead, we look for further approximations.

The nonlinear equations for $\pi-\pi$ scattering[4] have exactly the form of (18), but with B_0 replaced by t_0, B_2 by t_2, and λ replaced by $-2\lambda_\pi$, λ_π being the coupling constant of reference 4. The form of the function R is identical in the two problems but the function L is

[7] S. Weinberg, Phys. Rev. Letters 4, 87 (1960).
[8] R. Omnes, Nuovo cimento 8, 316 (1958).

different. The function $L(2\sigma;\nu)$ is given, e.g., for $T=2$, in the case of τ decay by

$$L(2\sigma;\nu)=\frac{\nu-\nu_0}{\pi}\frac{1}{\nu}\left(\frac{\nu+1}{\nu-3}\right)^{\frac{1}{2}}\int_0^\infty d\nu'\frac{1}{\nu'-\nu_0}$$
$$\times\left[\ln\frac{\nu'+\frac{1}{2}\nu+\frac{1}{2}\nu[(\nu-3)/(\nu+1)]^{\frac{1}{2}}}{\nu'+\frac{1}{2}\nu-\frac{1}{2}\nu[(\nu-3)/(\nu+1)]^{\frac{1}{2}}}\right] \tag{19a}$$
$$\times[\tfrac{1}{3}B_2(\nu')t_2^\dagger(\nu')+\tfrac{2}{3}B_0(\nu')t_0^\dagger(\nu')]\left(\frac{\nu'}{\nu'+1}\right)^{\frac{1}{2}}.$$

and in the case of $\pi-\pi$ scattering by

$$\bar{L}(2\sigma;\nu)=\frac{\nu-\nu_0}{\pi}\frac{1}{\nu}\int_0^\infty\frac{d\nu'}{\nu'-\nu_0}\left(\frac{\nu'}{\nu'+1}\right)^{\frac{1}{2}}$$
$$\times[\tfrac{1}{3}t_2(\nu')t_2^\dagger(\nu')+\tfrac{2}{3}t_0(\nu')t_0^\dagger(\nu')]\ln\frac{\nu'+1+\nu}{\nu'+1}. \tag{19b}$$

If $\bar{L}(\rho;\nu)$ were the same function of ρ and ν as $L(\rho;\nu)$ then the solutions of Eqs. (18) could be expressed in terms of pion-pion solutions as

$$B_2=-\tfrac{1}{2}(\lambda/\lambda_\pi)t_2,\quad B_0=-\tfrac{1}{2}(\lambda/\lambda_\pi)t_0. \tag{20}$$

We shall take these solutions as a basis for further analysis. With given $\pi-\pi$ phase shifts one may determine whether $\bar{L}(\rho;\nu)$ is indeed a good approximation to $-2\lambda^{-1}\lambda_\pi L(\rho;\nu)$ when the relations (20) are assumed. We have investigated this question using integrable phase shifts and the results are discussed in the Appendix.

These are the solutions we would have obtained had we begun with a K meson of mass μ. That is, in our approximation when the τ-decay amplitude is expressed in terms of the pion variables s_1, s_2, and s_3 in the standard way, the weight functions are independent of the τ mass.

There is a diagrammatic interpretation that independently suggests this result. Consider all diagrams of the general form of Fig. 2, where one pair of pions interacts locally. The contribution of this diagram is a function only of the variable $s_3=-(p_1+p_2)^2$. The sum of such diagrams is of the form of Eq. (6). We note that each such diagram depends only upon the center-of-mass energy of some one pair of pions, and not upon the τ mass.

Substituting (20) and the unitarity conditions (14)

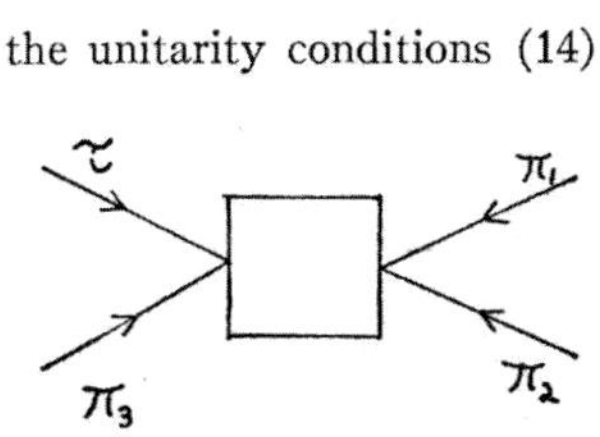

FIG. 2. One type of τ-decay diagram.

into the subtracted representation (11) we find, in terms of $s_i=4(\nu_i^2+1)\mu^2$,

$$A^\tau(\nu_1,\nu_2,\nu_3)$$
$$=-\frac{\lambda}{2\lambda_\pi}\left[-2\lambda_\pi+\frac{\nu_1-\nu_0}{\pi}\right.$$
$$\times\int_0^\infty d\nu'\frac{\frac{1}{6}t_2(\nu')t_2^\dagger(\nu')+\frac{1}{3}t_0(\nu')t_0^\dagger(\nu')}{(\nu'-\nu_1)(\nu'-\nu_0)}\left(\frac{\nu'}{\nu'+1}\right)^{\frac{1}{2}}$$
$$+\frac{\nu_2-\nu_0}{\pi}\int_0^\infty d\nu'\frac{\frac{1}{6}t_2(\nu')t_2^\dagger(\nu')+\frac{1}{3}t_0(\nu')t_0^\dagger(\nu')}{(\nu'-\nu_2)(\nu'-\nu_0)}\left(\frac{\nu'}{\nu'+1}\right)^{\frac{1}{2}}$$
$$\left.+\frac{\nu_3-\nu_0}{\pi}\int_0^\infty d\nu'\frac{t_2(\nu')t_2^\dagger(\nu')}{(\nu'-\nu_3)(\nu'-\nu_0)}\left(\frac{\nu'}{\nu'+1}\right)^{\frac{1}{2}}\right]. \tag{21}$$

For physical τ decay the ν_1, ν_2, ν_3 are the three relative pion squared momenta in the final state,

$$\nu_1=|\mathbf{p}_2-\mathbf{p}_3|^2/4,\quad \nu_2=|\mathbf{p}_1-\mathbf{p}_3|^2/4,\quad \nu_3=|\mathbf{p}_1-\mathbf{p}_2|^2/4.$$

The pion-pion scattering equations may be used once more to express the integrals in (21) in terms of $\pi-\pi$ phase shifts at the momenta ν_i, subtraction constants, and left-hand cut integrals. These left-hand cut contributions are real. We obtain, finally,

$$-[(2\lambda_\pi)/\lambda]A^\tau(\nu_1,\nu_2,\nu_3)$$
$$=4\lambda_\pi+t_2(\nu_3)+\tfrac{1}{6}[t_2(\nu_1)+t_2(\nu_2)]$$
$$+\tfrac{1}{3}[t_0(\nu_1)+t_0(\nu_2)]+(7/6)[I_2(\nu_1)+I_2(\nu_2)]$$
$$+\tfrac{1}{3}[I_0(\nu_1)+I_0(\nu_2)]+\tfrac{1}{3}[I_2(\nu_3)+2I_0(\nu_3)], \tag{22}$$

where

$$I_{2,0}(\nu)=\frac{1}{\pi}\int_0^\infty d\nu'\left(\frac{\nu'+1}{\nu'}\right)^{\frac{1}{2}}\sin^2\delta_{2,0}(\nu')$$
$$\times\frac{1}{\nu'-\nu_0}\left[1+\frac{\nu_0-\nu'}{\nu}\ln\left(1+\frac{\nu}{\nu'+1}\right)\right].$$

A similar expression for τ' decay follows from (17).

6. NUMERICAL RESULTS

Chew and Mandelstam[5] have numerically solved the S-wave dominant pion-pion equations. We have used their solutions for $\lambda_\pi=-0.1$, $+0.1$, $+0.3$ in order to analyze our Eq. (22). The integrals, I (which are nonsingular) have been expanded to first order in ν and have been evaluated using Gauss' three-point technique, the points being read directly off the curves of reference 5. To evaluate the $t(\nu)$ terms a linear fit to the curves of reference 5 was made for the quantities, $[\nu/(\nu+1)]^{\frac{1}{2}}\cot\delta(\nu)$. Then the $t(\nu)=[(\nu+1)/\nu]^{\frac{1}{2}}\exp[i\delta(\nu)]\sin\delta(\nu)$ were expanded in powers of $(\nu)^{\frac{1}{2}}$, with terms to order ν retained. The resulting π^- and π^+ energy spectra in τ decay were calculated and the results were compared with the spectra predicted on the basis of a constant matrix element.

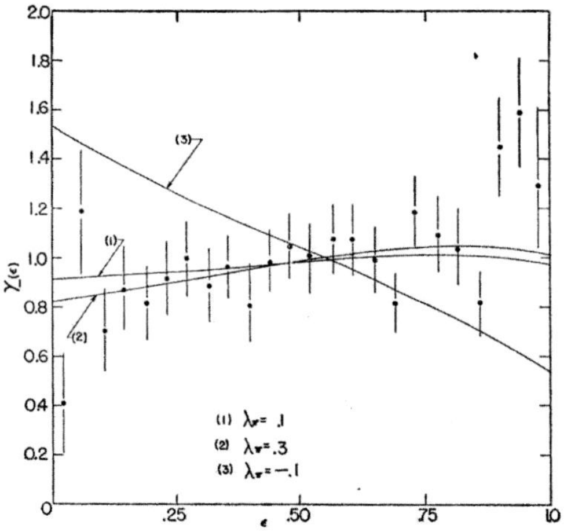

FIG. 3. The π^- distribution in τ^+ decay as a function of energy ϵ (in units of the maximum π^- energy). The experimental points are from reference 1. The curves are theoretical curves for indicated values of λ_π.

Defining the ratios of the π^+, π^- spectra predicted from (22) to the density of states spectra as $r_{+,-}(\omega)$, we plot the results for $r_-(\omega)$ for $\lambda_\pi=-0.1$, $+0.1$, $+0.3$ in Fig. 3. The experimental points are taken from reference 1.

The two repulsive values of λ_π, 0.1 and 0.3, are seen to give a $r_-(\omega)$ which increases with energy through most of the range of ω, in qualitative agreement with experiment. $\lambda_\pi=0.3$ is a better fit to the data. The scattering lengths for $\lambda_\pi=0.3$ are about $a_2=-0.48\mu^{-1}$, $a_0=-0.8\mu^{-1}$.

The spectrum ratio, $r_+(\omega)$, is plotted in Fig. 4 for $\lambda_\pi=0.1$ and $\lambda_\pi=0.3$. It decreases with increasing energy, also in agreement with experiment.

The attractive value $\lambda_\pi=-0.1$ gives π^- and π^+ spectra sloping violently the wrong way.

The energy spectrum for the π^+ in τ' decay has been calculated for $\lambda_\pi=0.1$ and $\lambda_\pi=0.3$ and shows a decrease (Fig. 5) with increasing π^+ energy, compared to the density of states prediction. This is qualitatively in accord with Weinberg's prediction[7]; however, the curvature of the plots for $r(\omega)$ are not compatible with Weinberg's analysis. Unfortunately, there are not enough experimental points near the high-energy end of the π^- spectrum to check the markedly curved aspect of our $r(\omega)$ in that region.

Weinberg[7] has analyzed the effect of the $\Delta T=\frac{1}{2}$ rule in the τ and τ' decay energy spectra on the basis of a development of the matrix elements in a power series in the invariants. He takes the invariants essentially to be the individual pion kinetic energies. They can as well be taken as our three ν's, being the squares of the pion relative momenta. Weinberg takes the first two terms in an expansion of the matrix element in integral powers of $\mathbf{p}_1^2$, $\mathbf{p}_2^2$, $\mathbf{p}_3^2$ and shows that they lead to a linear form for $r_{+,-}(\omega)$ (up to relativistic corrections). Our deviation from this linear form is due to the presence of terms in $|A|^2$ which cannot be so expanded. These terms arise from the fact that $(\nu)^{-\frac{1}{2}}\exp[i\delta(\nu)]\sin\delta(\nu)$ is an even function of $\nu^{\frac{1}{2}}$ only insofar as it is real. There are imaginary terms of the form $i\nu^{\frac{1}{2}}$ in an expansion. Since we have several phases entering (22), when we form $|A|^2$ and expand for small ν's we find terms of the form, e.g.,

$$(\nu_1\nu_2)^{\frac{1}{2}}=(|\mathbf{p}_2-\mathbf{p}_3|\,|\mathbf{p}_1-\mathbf{p}_3|)^{\frac{1}{2}}.$$

This kind of term is not expansible in powers of $\mathbf{p}_1^2$, $\mathbf{p}_2^2$, $\mathbf{p}_3^2$ and this effect is seen to be a somewhat important one in our analysis.

It is probably true in general that matrix elements leading to three-particle states with strong interactions cannot be expanded in integral powers of the invariants, $\mathbf{p}_\alpha\cdot\mathbf{p}_\beta$. In this problem we may see this property without making our approximation of neglecting the dependence upon the mass by looking back to the Mandelstam representation (12). For s_1, s_2, s_3 in the physical region (as in τ decay) each integral may be expressed as a sum of subtraction terms, an amplitude $A_s(\nu_i)$ for $\tau+\pi\to\pi+\pi$ at the momentum, $(\nu_i)^{\frac{1}{2}}$, and a real contribution from left-hand cuts. Since the amplitudes $A_s(\nu_i)$ have the pion-pion phases we obtain, as in (22) a sum of terms with various pion-pion phases plus some real terms. The threshold behavior of the phases alone invalidates Weinberg's expansion.

7. CONCLUSION

We have seen how Chew and Mandelstam's S-wave-dominant equations may serve as a basis for predicting the pion energy spectra in τ and τ' decay, under the assumption that final-state interactions are responsible for the shapes of these spectra. Linear integral equations have been derived for functions from which the τ and τ' decay matrix elements may be constructed.

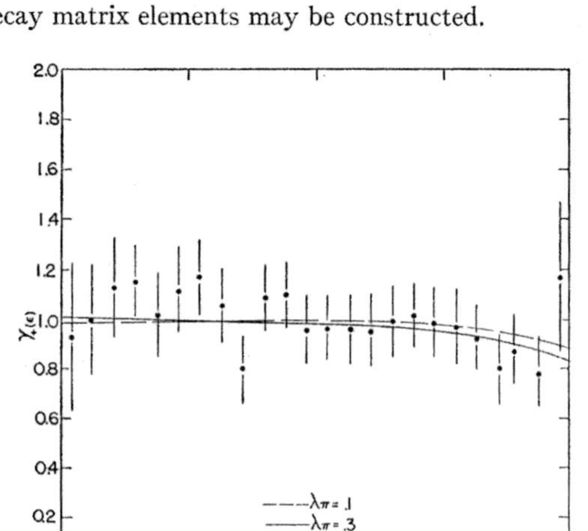

FIG. 4. The π^+ energy distribution in τ^+ decay. The experimental points are from reference 1. The curves are theoretical ones for indicated values of λ_π.

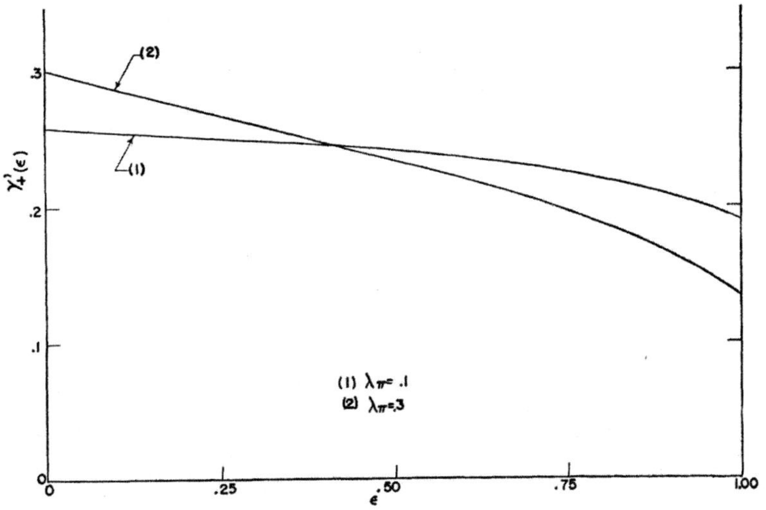

FIG. 5. The π^+ energy distribution in τ'^+ decay.

Under the rather drastic assumptions we have made in approximating the solution of the integral equations, it would seem that a repulsive S-wave force is required to explain the observed features of the τ^+-decay spectrum. Since we are using the Chew-Mandelstam approach to τ decay we must take seriously the similar equations for $\pi-\pi$ scattering. For this reason we may not take arbitrary phase shifts, but must take the ones from reference 5. Thus we cannot duplicate Thomas and Holladay's result[2] for an attractive $T=2$ force, because this is seen to imply an attractive $T=0$ force of greater strength.

From the original equations, without approximation, we have derived some results. We have shown that the implications of $\Delta T=\frac{1}{2}$, $\frac{3}{2}$ are identical to those of $\Delta T=\frac{1}{2}$. This is in spite of the fact that the matrix element deviates markedly from a constant. (A constant matrix element picks out the $T=1$ or $T=3$ final states and makes $\Delta T=\frac{3}{2}$ unobservable.)

The qualitative explanation for this result is that one may imagine a matrix element in the absence of final state interactions which is a constant. $\Delta T=\frac{1}{2}$, $\frac{3}{2}$ then implies a pure $T=1$ final state. The $\pi-\pi$ coupling is now turned on and may induce some momentum dependence in the matrix element; but it preserves the $T=1$ nature of the final state. Unfortunately this means that the spectrum in τ' decay, like the branching ratio, cannot be a critical test for $\Delta T=\frac{1}{2}$.

The other result of our equations which is independent of approximation is that the matrix element may not be expanded in integral powers of the invariants as is assumed in reference 7.

ACKNOWLEDGMENT

We wish to thank Professor W. G. Holladay for interesting conversations.

APPENDIX

The integrals in (19a) and (19b) have been performed, using (20), for a particular $t(\nu)$ corresponding to a scattering length approximation to the phase shift [omitting the relativistic factor $(\nu+1)^{\frac{1}{2}}$]

$$t(\nu)t^\dagger(\nu)=a^2/(a^2\nu+1).$$

The results for $0<\nu<3$ are

$$-2(\lambda_\pi/\lambda)L(\nu)=\frac{2}{\nu}\left(\frac{\nu+1}{\nu-3}\right)^{\frac{1}{2}}\frac{a^2}{1-\nu_0a^2}\times\left(\frac{1}{a}\ln\frac{1+ar}{a+ar^*}-(\nu_0)^{\frac{1}{2}}\ln\frac{\nu_0^{\frac{1}{2}}+r}{\nu_0^{\frac{1}{2}}+r^*}\right),$$

with

$$r=\left(\frac{\nu}{2}\right)^{\frac{1}{2}}\left[1+\left(\frac{\nu-3}{\nu+1}\right)^{\frac{1}{2}}\right]^{\frac{1}{2}},$$

and

$$\bar{L}(\nu)=\frac{2}{\nu}\frac{a^2}{1-\nu_0a^2}\left(\frac{1}{a}\ln\frac{1+a(1+\nu)^{\frac{1}{2}}}{1+a}-(\nu_0)^{\frac{1}{2}}\ln\frac{\nu_0^{\frac{1}{2}}+(1+\nu)^{\frac{1}{2}}}{\nu_0^{\frac{1}{2}}+1}\right).$$

We have examined these results for values of a between 0.2 and 1 in units of the pion Compton wavelength. In the region $1<\nu<3$ there is in every case a 20% to 30% difference between $-2\lambda_\pi\lambda^{-1}L(\nu)$ and $\bar{L}(\nu)$. In the region $0<\nu<1$ the discrepancy is much larger and depends on the value of a. We conclude that the approximation leading to (20), while crude, does not neglect altogether the contributions from $L(\nu)$.

Some Consequences of the Decay Modes and Production of the New Heavy Bosons (*).

G. Feldman, T. Fulton and K. C. Wali
The Johns Hopkins University - Baltimore, Md.

(ricevuto il 27 Dicembre 1961)

Summary. — An investigation is made of the consequences following from the production and decay of the recently discovered heavy bosons. It is found that various correlation experiments can determine their spins and parities and their polarization states arising from the production process. Further, it is shown that a study of leptonic decay modes can throw significant light on the electromagnetic structure of the leptons.

1. – Introduction.

A number of resonances [1-5] have been discovered recently in π-meson systems. It is of interest to investigate what information can be obtained from the production and decays of bosons of this type. In this paper we analyse a number of different decay modes. We find that the observation of various energy and angular distributions can lead to a determination of some properties not only of these bosons, but also of the μ-meson.

(*) Supported in part by the National Science Foundation.

(1) J. A. Anderson, V. X. Bang, P. G. Burke, D. D. Carmony and N. Schmitz: *Phys. Rev. Lett.*, **6**, 365 (1961).

(2) D. Stonehill, C. Baltay, H. Courant, W. Fickinger, E. C. Fowler, H. Kraybill, J. Sandweiss, J. Sanford and H. Taft: *Phys. Rev. Lett.*, **6**, 624 (1961).

(3) A. R. Erwin, R. March, W. D. Walker and E. West: *Phys. Rev. Lett.*, **6**, 628 (1961).

(4) B. Maglić, L. Alvarez, A. Rosenfeld and M. L. Stevenson: *Phys. Rev. Lett.*, **7**, 178 (1961).

(5) A. Pevsner, R. Kraemer, M. Nussbaum, C. Richardson, P. Schlein, R. Strand, T. Toohig, M. Block, A. Engler, R. Gessaroli and C. Meltzer: *Phys. Rev. Lett.*, **7**, 421 (1961).

In Section **2** we list the various distributions in decay modes (correlated to the production of the boson) which follow directly from invariance principles. In Section **3** we consider Dalitz plots for the 3π decay mode and the effect of some final state interaction. In Section **4** we analyse leptonic decay modes. Section **5** is devoted to an estimate of some branching ratios.

2. – Correlations between production and decay directions.

Let us consider the production and decay of a vector boson U (*i.e.* spin 1, parity + or —). Results for spin 0 bosons will be discussed as the need arises. The most general density matrix describing the state of a vector meson [6-8]

$$\varrho_{in} = \tfrac{1}{3}[1 + P_i S_i + Q_{ij}(S_i S_j + S_j S_i)]\,, \tag{2.1}$$

where P_i are the components of a real vector (the polarization vector), Q_{ji} are elements of a real symmetric traceless tensor (the polarization tensor) and the S_i are the spin-one matrices. The pseudovector $\boldsymbol{P}$ is determined by the production process for the boson and has the form

$$\boldsymbol{P} = f\,\boldsymbol{k}\times\boldsymbol{k}'\,, \tag{2.2}$$

where $\boldsymbol{k}$ and $\boldsymbol{k}'$ are two independent vectors associated with the production process and f is a scalar function of the invariants. If one observes only the direction of the boson, U, in the production process, Q_{ij} has the simple form

$$\bar{Q}_{ij} = g(\hat{\boldsymbol{k}}_i\hat{\boldsymbol{k}}_j - \tfrac{1}{3}\delta_{ij})\,, \tag{2.3}$$

where $\hat{\boldsymbol{k}}$ is a unit vector in the direction of U, and g is a scalar function of the boson energy.

Let $|U\rangle$ be the state vector for the boson and $U_i \equiv \langle i|U\rangle$ its components in a Cartesian representation. The transition matrix element for the decay of U into any available final state is

$$\langle\alpha, \boldsymbol{a}|M|U\rangle = \boldsymbol{U}\cdot\boldsymbol{A}^{\alpha}\,, \tag{2.4}$$

where $\boldsymbol{A}^{\alpha}$ is a vector function of the vectors ($\boldsymbol{a}$) available in the final state, and α runs over the other degrees of freedom (*e.g.* photon polarization, lepton

(6) R. H. Dalitz: *Proc. Phys. Soc.* (*London*), A **65**, 175 (1952).
(7) L. Wolfenstein and J. Ashkin: *Phys. Rev.*, **85**, 947 (1952).
(8) W. Lakin: *Phys. Rev.*, **98**, 139 (1955).

spins, etc.). The final distribution function ϱ_f is given by ([9])

$$\varrho_f = M\varrho_{in}M^+ . \tag{2.5}$$

Using eqs. (2.4) and (2.1), we obtain in the α representation

$$\langle\alpha|\varrho_f|\beta\rangle = \tfrac{1}{3}[\boldsymbol{A}^\alpha\cdot\boldsymbol{A}^{+\beta} - i\boldsymbol{P}\cdot\boldsymbol{A}^\alpha\times\boldsymbol{A}^{+\beta} - 2\boldsymbol{A}^\alpha\cdot Q\cdot\boldsymbol{A}^{+\beta}] , \tag{2.6}$$

where $\boldsymbol{A}^\alpha\cdot Q\cdot\boldsymbol{A}^{+\beta} = Q_{ij}A_i^\alpha A_j^{+\beta}$, and we have used the fact that

$$\langle l|S_m|n\rangle = i\varepsilon_{lmn} . \tag{2.7}$$

We now consider the application of eq. (2.6) to various decay processes ([10]).

a) $U \to 3\pi$.

Let $\hat{\boldsymbol{p}}_1$, $\hat{\boldsymbol{p}}_2$ be unit vectors in the direction of two of the π-mesons in the U rest frame. There are no spin degrees of freedom in the final state. In this case $\boldsymbol{A}^\alpha$ is given by $\boldsymbol{A}_\pm$, where

$$\boldsymbol{A}_- = F\hat{\boldsymbol{p}}_1\times\hat{\boldsymbol{p}}_2 = F\boldsymbol{n} , \tag{2.8}$$

if U is a polar vector, and

$$\boldsymbol{A}_+ = F_1\hat{\boldsymbol{p}}_1 + F_2\hat{\boldsymbol{p}}_2 , \tag{2.9}$$

for U a pseudovector. The functions F, F_1 and F_2 are invariant functions of the meson energies ω_1 and ω_2.

Use of (2.6) leads to

$$\varrho_- = |F|^2\,\tfrac{1}{3}\,\boldsymbol{n}^2[1 - 2\hat{\boldsymbol{n}}\cdot Q\cdot\hat{\boldsymbol{n}}] , \tag{2.10}$$

and

$$\varrho_+ = \frac{|G|^2}{3}\left\{1 + 2\,\mathrm{Im}\frac{F_1F_2^+}{|G|^2}\boldsymbol{P}\cdot\boldsymbol{n} - \frac{2}{|G|^2}[|F_1|^2\,\hat{\boldsymbol{p}}_1\cdot Q\cdot\hat{\boldsymbol{p}}_1 + |F_2|^2\,\hat{\boldsymbol{p}}_2\cdot Q\cdot\hat{\boldsymbol{p}}_2 + 2\,\mathrm{Re}\,F_1F_2^+\,\hat{\boldsymbol{p}}_1\cdot Q\cdot\hat{\boldsymbol{p}}_2]\right\} , \tag{2.11}$$

where

$$|G|^2 = |F_1\hat{\boldsymbol{p}}_1 + F_2\hat{\boldsymbol{p}}_2|^2 . \tag{2.12}$$

([9]) In terms of (2.5) the final state density matrix is given by $\varrho_f/\mathrm{Tr}\,\varrho_f$.

([10]) Most of these decay processes have been listed by NAMBU (Y. NAMBU: *Proceedings of the Seventh Annual Rochester Conference* (New York), pp. 1-26).

Integrating over one of the π-meson momenta leads to the expressions

$$\bar{\varrho}_- = \frac{\overline{|\boldsymbol{F}\boldsymbol{n}|^2}}{3}\,[1 + \hat{\boldsymbol{p}}_1\cdot Q\cdot\hat{\boldsymbol{p}}_1]\,, \tag{2.13}$$

and

$$\bar{\varrho}_+ = \frac{\overline{|G|^2}}{3}\left[1 + \left(1 - \frac{3}{2}\,\frac{\overline{|\boldsymbol{F}_2\boldsymbol{n}|^2}}{\overline{|G|^2}}\right)\hat{\boldsymbol{p}}_1\cdot Q\cdot\hat{\boldsymbol{p}}_1\right]. \tag{2.14}$$

If the boson were a scalar, the distribution ϱ would be of the above forms with $\boldsymbol{P}$ and Q taken equal to zero.

b) $\mathrm{U}\to 2\pi$.

In this case, U must be a polar vector and $\boldsymbol{A}^\alpha$ is given by $\boldsymbol{A}_{2\pi}$, where

$$\boldsymbol{A}_{2\pi} = D\hat{\boldsymbol{p}}\,, \tag{2.15}$$

and $\hat{\boldsymbol{p}}$ is a unit vector in the direction of one of the π's in the U rest system. The distribution is

$$\varrho_- = \frac{|D|^2}{3}\,[1 - 2\hat{\boldsymbol{p}}\cdot Q\cdot\hat{\boldsymbol{p}}]\,. \tag{2.16}$$

c) $\mathrm{U}\to\pi^0 + \gamma$.

Let $\hat{\boldsymbol{q}}$ be a unit vector in the photon direction in the U rest frame. In addition, there are photon polarization degrees of freedom. $\boldsymbol{A}^\alpha$ is given by $\boldsymbol{A}^\alpha_\pm$ where

$$\boldsymbol{A}^\alpha_- = H(\boldsymbol{\epsilon}^{(\alpha)}\times\hat{\boldsymbol{q}})\,, \tag{2.17}$$

and

$$\boldsymbol{A}^\alpha_+ = H\boldsymbol{\epsilon}^{(\alpha)}\,. \tag{2.18}$$

The index α runs over 1 and 2. The unit vectors $\boldsymbol{\epsilon}^{(\alpha)}$ describe the two transverse polarizations of the photon. Substitution into (2.6) leads to

$$\varrho^+_{11} = \varrho^-_{22} = \frac{|H|^2}{3}\,[1 - 2\boldsymbol{\epsilon}^{(1)}\cdot Q\cdot\boldsymbol{\epsilon}^{(1)}]\,; \tag{2.19}$$

$$\varrho^+_{22} = \varrho^-_{11} = \frac{|H|^2}{3}\,[1 - 2\boldsymbol{\epsilon}^{(2)}\cdot Q\cdot\boldsymbol{\epsilon}^{(2)}]\,; \tag{2.20}$$

$$\varrho^+_{12} = \varrho^{*+}_{21} = -\varrho^{*-}_{12} = -\varrho^-_{21} = -\frac{|H|^2}{3}\,[2\boldsymbol{\epsilon}^{(1)}\cdot Q\cdot\boldsymbol{\epsilon}^{(2)} + i\boldsymbol{P}\cdot\hat{\boldsymbol{q}}]\,. \tag{2.21}$$

We note that in general the photons are in states of elliptical polarization. Summing over polarizations leads to the angular distribution

$$\mathrm{Tr}\,\varrho^+ = \mathrm{Tr}\,\varrho^- = \tfrac{2}{3}|H|^2[1 + \hat{\boldsymbol{q}}\cdot Q\cdot\hat{\boldsymbol{q}}]\,. \tag{2.22}$$

The above expression is identical in form to (2.13).

If the boson were a scalar, it could not, of course, decay by this mode.

d) U→lepton pair.

The leptons in this process are electrons or μ-mesons. If we assume their interactions are purely electrodynamic, to lowest order in the charge, these modes can occur only for a 1^- boson.

In this case $\boldsymbol{A}^{(\alpha)}$ is given by

$$A^{tt'} = C\boldsymbol{\Gamma}^{tt'}, \tag{2.23}$$

where

$$\Gamma_\mu^{tt'} = \bar{u}_t(r)\gamma_\mu v_{t'}(s)\,, \tag{2.24}$$

and $u_t(r)$ $(v_{t'}(s))$ are the particle (antiparticle) spinors for four-momenta $r(s)$. Substitution of (2.23) into (2.6) and summing over final spins leads to a distribution similar to that given by (2.22). If we observe the polarization $\mathscr{P}$ of one of the final leptons (say the μ^+), we obtain

$$\mathscr{P}\,\mathrm{Tr}\,\varrho_f = \mathrm{Tr}\,\varrho_f\,\boldsymbol{\sigma} = C'\left[\boldsymbol{P} + \left(\frac{M}{2\mu} - 1\right)\hat{\boldsymbol{r}}\cdot\boldsymbol{P}\,\hat{\boldsymbol{r}}\right], \tag{2.25}$$

where μ is the mass of the lepton.

e) U→π^0 + lepton pair.

In this case $\boldsymbol{A}^\alpha$ is given by the expressions

$$\boldsymbol{A}_-^{tt'} = B\,\boldsymbol{q}\times\boldsymbol{\Gamma}^{tt'}\,; \tag{2.26}$$

$$\boldsymbol{A}_+^{tt'} = B(\boldsymbol{q}\,\Gamma_4^{tt'} - q_4\boldsymbol{\Gamma}^{tt'})\,, \tag{2.27}$$

where q_μ is the total energy-momentum vector of the pair. The various distributions can be obtained like those of the previous modes.

The distributions listed above can be used to determine some properties of the boson. For example, if the parity of U were unknown, it could be determined by various experiments or combinations of experiments. From eq. (2.22) we see that a measurement of the angular correlation between the decay photon in process *c*) and the boson will determine the tensor polarization Q. If Q is nonzero, the eqs. (2.19)–(2.21) indicate that photons emitted at right angles to the direction of U are linearly polarized. Knowing Q, a measurement of this polarization will determine the parity of U. An alternative to the measurement of the linear polarization is to study the correlations between the direction of U and the plane of the lepton pair which is emitted in process e).

A less definitive experiment for determining the parity makes use of process a). A measurement of a nonzero correlation in the angle φ (where $\cos\varphi = \hat{\boldsymbol{k}}\cdot\hat{\boldsymbol{p}}_1$) for those events for which $\hat{\boldsymbol{k}}\cdot\hat{\boldsymbol{n}} = 0$ implies that the parity of U is positive. The converse is not true.

If the vector polarization $\boldsymbol{P}$ is not zero, then a measurement of the polarization of one of the leptons in process e) would provide still another (unambiguous) determination of the parity.

We observe that none of the above experiments could distinguish between a spin 0 boson and a spin 1 particle produced without polarization (*i.e.* $\boldsymbol{P} = Q = 0$).

Alternatively, if the boson is known to have spin 1, a number of the above distributions could be used to determine the polarization states of U.

3. – Distributions in the 3π decay mode.

Two of the presently observed bosons (ω and η (4,5)) decay into three π-mesons. If one assumes that the final state interactions can be neglected, the energy distributions (Dalitz plots (11,12)) can be used to determine the spin and parity of the boson. In the case of the ω, these distributions have been obtained (4) assuming the isotopic spin, I, is zero. For completeness, we list in Table I the distributions for particles of spin 0^-, $1^\pm$ and I-spin 0 or 1.

The variables r, φ of Table I are given in terms of ω_1, ω_2, ω_3, the energies of the π-mesons in the boson rest frame by

$$\omega_1 = \frac{M}{3}\left[1 + \varepsilon r \cos\left(\varphi - \frac{2\pi}{3}\right)\right]; \tag{3.1}$$

$$\omega_2 = \frac{M}{3}\left[1 + \varepsilon r \cos\left(\varphi + \frac{2\pi}{3}\right)\right]; \tag{3.2}$$

$$\omega_3 = \frac{M}{3}\left[1 + \varepsilon r \cos\varphi\right], \tag{3.3}$$

(11) R. H. Dalitz: *Phil. Mag.*, **44**, 1068 (1953).
(12) E. Fabri: *Nuovo Cimento*, **11**, 479 (1954).

TABLE I. – *Dalitz plots for $\pi^+\pi^-\pi^0$ decays.*

$$I = 0$$

$J = 0^-$: $r^6 \sin^2 3\varphi\,,$

$J = 1^-$: $1 - \left[1 + \dfrac{\frac{1}{2}\varepsilon}{(1-\frac{1}{2}\varepsilon)^2}\right] r^2 - \dfrac{\frac{1}{2}\varepsilon}{(1-\frac{1}{2}\varepsilon)^2}\, r^3 \cos 3\varphi\,,$

$J = 1^+$: $r^2[(1-\frac{1}{2}\varepsilon) - r\cos 3\varphi]\,,$

$$I = 1$$

$J = 0^-$: $1\,,$

$J = 1^-$: $r^2 \sin^2\varphi \left\{1 - \left[1 + \dfrac{\frac{1}{2}\varepsilon}{(1-\frac{1}{2}\varepsilon)^2}\right] r^2 - \dfrac{\frac{1}{2}\varepsilon}{(1-\frac{1}{2}\varepsilon)^2}\, r^3 \cos 3\varphi\right\},$

$J = 1^+$: $(1-\frac{1}{2}\varepsilon) + r\cos\varphi + \frac{1}{2}\varepsilon r^2 \cos^2\varphi\,.$

See text for definitions of r, φ and ε.

where M is the boson mass, m the π-meson mass and

$$\varepsilon = 1 - \frac{3m}{M}\,. \tag{3.4}$$

Note that for $I=0$, the Dalitz plots have a six-fold symmetry in φ, corresponding to the symmetry of the decay distributions under the interchange of any two π's. For $I=1$, the symmetry is lower. The decay distributions are the same if we let $\varphi \to -\varphi$, corresponding to symmetry under interchange of π^+ and π^-.

It is conceivable that the existence of the ρ-meson ($I=1$, $J=1^-$) would affect the decay of the ω into three π's. As a model for studying possible effects of final state interactions we have considered the 3π decay of the ω through the emission of a virtual ρ. This effect is illustrated ([13]) for the case of an ω of $I=0$, $J=0^-$, 1^-. The modified distributions are

$$\frac{r^6}{[1-\delta r^2]^2}\,, \qquad I=0,\ J=0^-; \tag{3.5}$$

$$(1-Cr^2)\left|\frac{1-\eta r^2}{1-\delta r^2}\right|^2, \qquad I=0,\ J=1^-, \tag{3.6}$$

([13]) Equations (3.5) and (3.6) follow after integration over φ in the circle inscribing the kinematically defined region.

where, using the known masses ρ of and ω, we obtain

$$\delta = .15\,, \qquad \eta = .06\,, \qquad C = 1.4\,,$$

and the expressions in square brackets contain the corrections due to final state interactions.

The effect of this kind of final state interaction would be much enhanced if there existed a ρ-type particle of somewhat lower mass.

4. – Distributions for π^0 and lepton pair modes.

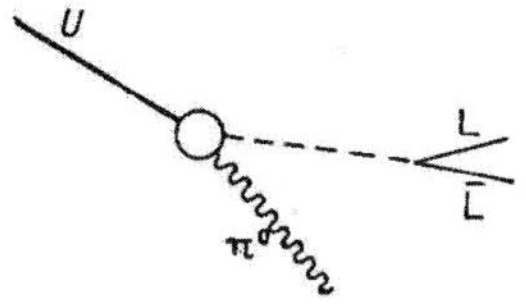

Fig. 1. – Illustration of the decay process $U \to \pi^0 + \bar{L} + L$.

In this section we are concerned with the mode *e*).

The decay distribution for this mode can provide some useful information not only about the boson, but also about the leptons. Figure 1 illustrates the decay into lepton pairs, assuming conventional electromagnetic interaction for the leptons. The matrix elements for the various spin possibilities are listed below:

$$\mathcal{M}_0 = \frac{e'}{\varkappa^2} \frac{G(q^2) q^2 k_\mu}{\sqrt{4M\omega}} \Lambda_\mu\,; \tag{4.1}$$

$$\mathcal{M}_- = \frac{e'}{\varkappa} G(q^2) \frac{\varepsilon_{\mu\nu\lambda\varrho}}{\sqrt{4M\omega}} U_\mu k_\nu q_\lambda \Lambda_\varrho\,; \tag{4.2}$$

$$\mathcal{M}_+ = \frac{e'}{\varkappa} \frac{G(q^2)}{\sqrt{4M\omega}} (\delta_{\mu\lambda}\delta_{\varrho\nu} - \delta_{\mu\varrho}\delta_{\nu\lambda}) U_\mu k_\nu q_\lambda \Lambda_\varrho\,, \tag{4.3}$$

where

$$\Lambda_\varrho = \frac{e\mu}{\sqrt{r_0 s_0}} \frac{F(q^2)}{q^2} (\bar{u}(r)\gamma_\varrho v(s))\,, \tag{4.4}$$

k_μ, r_μ, s_μ are the four-vectors for the boson and leptons, respectively, $q = r + s$, M is the boson mass, ω the π^0 energy and $\varkappa$ a characteristic mass associated with the $(\omega\pi\gamma)$ vertex. The dimensionless parameter e' is a coupling constant of order e, the electronic charge. The dimensionless functions $G(q^2)$, $F(q^2)$ are the form factors associated with the $(\omega\pi\gamma)$ and lepton vertices, respectively, and $G(0) = F(0) = 1$. For point leptons, $F(q^2) = 1$. The vector U_μ is the spin vector of the boson.

The angular distributions for the electron pair (in the limit $\mu = 0$ and

1489

$G(q^2) = F(q^2) = 1$ are given as follows:

$$\frac{dw_0}{d\cos\theta\, dr_0} = K \frac{r_0^2 s_0^2 (1+\cos\theta)}{M - r_0(1-\cos\theta)}\,; \tag{4.5}$$

$$\frac{dw_{\mp}}{d\cos\theta\, dr_0} = K' \frac{1}{M - r_0(1-\cos\theta)} \left[\frac{r_0^2 + s_0^2}{1-\cos\theta} \mp r_0 s_0\right], \tag{4.6}$$

where

$$\frac{s_0}{M} = \left[\frac{1}{2} M\left(1 - \frac{m^2}{M^2}\right) - r_0\right] \Big/ [M - r_0(1-\cos\theta)]\,, \tag{4.7}$$

and θ is the angle between the leptons in the boson rest frame. The striking difference between spin one and spin zero is that the angular distribution of the former is sharply peaked in the forward direction. The distributions could be modified significantly only in the backward direction by $F(q^2)$ and $G(q^2)$, which of course are unknown. Integration over r_0 leads to the angular distributions illustrated in Fig. 2.

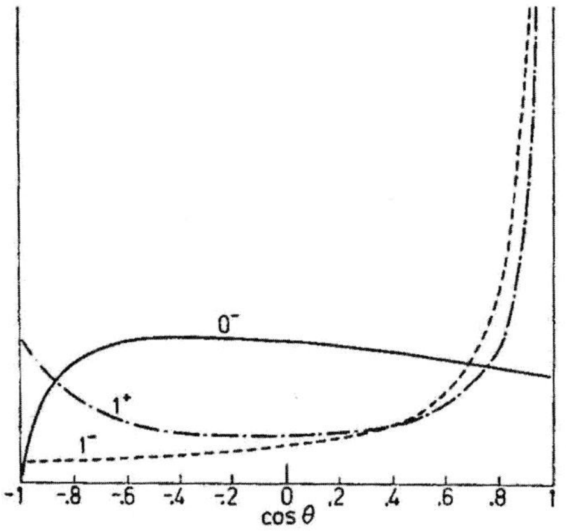

Fig. 2. – Distributions as a function of the angle between the leptons for different spins and parities (normalized to the same area).

Alternatively, consideration of the distribution as a function of q^2 could, in principle, lead to a determination of the μ-meson form factor over the broad region $4\mu^2 < q^2 < (M-m)^2$. We define $R(q^2)$ as the ratio of μ-meson to electron distributions for fixed q^2. Taking the 1^- case as an example, we obtain

$$R(q^2) = \left(1 + \frac{2\mu^2}{q^2}\right)\left(1 - \frac{4\mu^2}{q^2}\right)^{\frac{1}{2}} F(q^2)\,, \tag{4.8}$$

where we have assumed $F_{(e)}(q^2) = 1$ and have neglected the electron mass. Note that this ratio is independent of the unknown form factor $G(q^2)$.

5. – Branching ratios.

We now estimate the branching ratios for some of the decay modes. (We shall restrict ourselves to the case of 1^-.) Assuming that electrons and μ-mesons behave as point charges, we obtain the following decay rates [14]:

$$w_{2L} = \frac{(e'')^2}{4\pi}\frac{e^2}{4\pi}\frac{1}{3}\left(1+\frac{2\mu^2}{M^2}\right)\left(1-\frac{4\mu^2}{M^2}\right)^{\frac{1}{2}} M; \tag{5.1}$$

$$w_{\pi^0,2L} = \frac{e'^2}{4\pi}\left[\frac{e^2}{4\pi}\frac{1}{3\pi}\left(\frac{M}{\varkappa}\right)^2\frac{1}{3M^2}\int\limits_{4\mu^2}^{(M-m)^2}\frac{|\boldsymbol{q}|^3}{q^2}\left(1+\frac{2\mu^2}{q^2}\right)\left(1-\frac{4\mu^2}{q^2}\right)^{\frac{1}{2}}\mathrm{d}q^2\right.; \tag{5.2}$$

$$w_{\pi^0,\gamma} = \frac{e'^2}{4\pi}\left(\frac{M}{\varkappa}\right)^2\frac{1}{24}\left(1-\frac{m^2}{M^2}\right)^3 M; \tag{5.3}$$

$$w_{3\pi} = \frac{g^2}{4\pi}\left(\frac{M}{\varkappa}\right)^6\frac{\varepsilon^4}{2^4\cdot 3^3\cdot\sqrt{3}\pi}\left(1-\frac{3}{2}\varepsilon+\frac{1}{4}\varepsilon^2\right)M; \tag{5.4}$$

$$w_{2\pi} = \frac{g'^2}{4\pi}\frac{1}{12}\left(1-\frac{4m^2}{M^2}\right)^{\frac{3}{2}} M, \tag{5.5}$$

where

$$\boldsymbol{q}^2 = \frac{(M^2-m^2)^2-2(M^2+m^2)q^2+q^4}{4M^2}, \tag{5.6}$$

and ε has been defined previously. In obtaining these formulae we have set $G(q^2)=1$, and assumed a point interaction in the 3π and 2π decay modes. Also we have assumed that the mass $\varkappa$ is a measure of the boson structure.

The most interesting branching ratio [15] follows from (5.1). It is $w_{2\mu}/w_{2e}$. One can predict it exactly, provided only that electrons and μ-mesons behave as point charges,

$$\frac{w_{2\mu}}{w_{2e}} \approx 1, \tag{5.7}$$

for ρ, η and ω.

Assuming further that $G(q^2)=1$, we obtain:

$$\left(\frac{w_{\pi^0,2\mu}}{w_{\pi^0,2e}}\right)_\omega \approx \frac{1}{20}; \tag{5.8}$$

$$\left(\frac{w_{\pi^0,2\mu}}{w_{\pi^0,2e}}\right)_\eta \approx \frac{1}{60}; \tag{5.9}$$

$$\left(\frac{w_{\pi^0,2e}}{w_{\pi^0,\gamma}}\right)_\omega \approx \frac{1}{100}; \tag{5.10}$$

$$\left(\frac{w_{\pi^0,2e}}{w_{\pi^0,\gamma}}\right)_\eta \approx \frac{1}{90}. \tag{5.11}$$

---

[14] Some of these decay rates have been discussed by MATTHEWS (P. T. MATTHEWS: *Proceedings of the Conference on Elementary Particles* (Aix-en-Provence, 1961), to be published).

[15] The consequences of these modes were first considered by NAMBU and SAKURAI (private communication, A. PEVSNER).

The remaining branching ratios require some assumptions about coupling constants and the mass $\varkappa$. We shall assume $e'^2/4\pi \approx e''^2/4\pi \approx e^2/4\pi$ and $g^2/4\pi \approx g'^2/4\pi \approx 1$. We determine $\varkappa$ by fitting the 3π width of ω to be 30 MeV ($\varkappa = 1.3\, m$). It then follows that

$$\left(\frac{w_{3\pi}}{w_{\pi^0\gamma}}\right)_\omega = \frac{8}{1}, \tag{5.12}$$

$$\left(\frac{w_{3\pi}}{w_{\pi^0\gamma}}\right)_\eta = \frac{1}{3}, \tag{5.13}$$

$$\left(\frac{w_{2L}}{w_{\pi^0\gamma}}\right)_\omega = \frac{1}{280}; \tag{5.14}$$

$$\left(\frac{w_{2L}}{w_{\pi^0\gamma}}\right)_\eta = \frac{1}{130}. \tag{5.15}$$

With the above assumptions (independent of $\varkappa$)

$$(w_{2\pi})_\rho = 52 \text{ MeV}. \tag{5.16}$$

6. – Discussion.

The most interesting consequence of the above analyses is that the leptonic decay modes of ω and η could be used to study either a possible breakdown of electrodynamics or the structure of the μ-meson. Indeed (assuming ω and η are 1^-) any departure from that predicted for the ratio $w_{2\mu}/w_{2e}$ would be clear indication of the non point charge behavior of the μ and/or e. A detailed study of the μ and e form factors could be made by observing the ratio of the decay into $\pi^0 + \mu^+ + \mu^-$ and $\pi^0 + e^+ + e^-$ for fixed q^2. We emphasize again that q^2 is timelike and ranges between $4\mu^2$ and $(M-m)^2$. Our estimate of the number of these events relative to the 3π mode in the relevant region of q^2 is roughly 10^{-4} for the ω and about 5 times larger for the η.

By studying various distributions in the decay processes we can obtain some interesting information about the ρ, η, ω particles themselves. In particular by measuring suitable angular distributions (and polarizations) with respect to the direction and plane of production of the bosons, one can confirm their spin and parity assignments. Alternatively, assuming prior knowledge of the spins and parities, these distributions provide information about the polarization and orientation of the produced boson.

$3\pi^0$ TO $\pi^+\pi^-\pi^0$ BRANCHING RATIO OF A 0^{-+} η MESON*

K. C. Wali
The Johns Hopkins University, Baltimore, Maryland
(Received June 18, 1962)

The assignment of zero spin, odd spatial parity, and even G parity (0^{-+}) to the recently discovered η meson[1] has been suggested by several authors.[2] The purpose of this note is to consider, under certain assumptions which will be specified in the following discussion, evidence in favor of this assignment based on a comparison of the Fabri-Dalitz plot for the 3π decay modes of η and K^+, K_2^0, and to point out that the $3\pi^0$ to $\pi^+\pi^-\pi^0$ branching ratio of η is uniquely related to the π^0 energy spectrum in the decay $\eta \to \pi^+ + \pi^- + \pi^0$.

It was shown by Sawyer and Wali[3] that as a consequence of the $\Delta T = \frac{1}{2}$ selection rule and invariance under CP, K^+ and K_2^0 decay into a final state of 3π's in isotopic spin $T = 1$ state. A single scalar function of three variables $f(\omega_1, \omega_2, \omega_3)$, where ω_1, ω_2, ω_3 are the energies of the pions, then determines the decay amplitude into various charged and uncharged modes of the final state 3π's. Further, without any dynamical assumptions regarding the specific form of the function $f(\omega_1, \omega_2, \omega_3)$, one can conclude that the $\pi^+(\pi^0)$ energy spectrum in the τ'-decay mode of the K^+ ($K^+ \to \pi^0 + \pi^0 + \pi^+$) should be the same as the π^0 (π^+ or π^-) spectrum in the decay $K_2^0 \to \pi^+ + \pi^- + \pi^0$. Experimentally the data[4] on $K^+ \to \pi^+ + \pi^+ + \pi^-$ indicate that a power series expansion in ω_1, ω_2, ω_3 for the decay amplitude with the retention of only the linear terms agrees reasonably well with the data. If such a "linear approximation"[5,6] is assumed, then only one parameter is needed to fit the shapes of the various spectra in both K^+ and K_2^0 decays. The relevant point of interest to our discussion is the prediction that the number of decays $K_2^0 \to \pi^+ + \pi^- + \pi^0$ as a function of π^0 energy should decrease with the increase in the π^0 energy. This and similar predictions have been confirmed by experiments.[7] Therefore, it is interesting to note that the π^0 spectrum in the charged mode of decay of η ($\eta \to \pi^+ + \pi^- + \pi^0$) has the same qualitative features as the π^0 in $K_2^0 \to \pi^+ + \pi^- + \pi^0$. Since η is assigned even G parity, it decays into 3π's by violating charge independence. If we consider the nonvanishing lowest order in the electromagnetic interaction, invariance under C requires that the final state of the 3π's be in the isotopic spin-one state.

Given that both K and η decay into $T = 1$ states of three pions, there is a very general dynamical assumption under which the spectra may be related. This is the assumption that in both cases the deviation of the matrix element from a constant is due to final-state pion-pion interactions. In either case one begins, in the absence of final-state interactions, with a constant matrix element and the completely symmetric $T = 1$ state; the spectra induced by rescattering will differ only by the effect of the small K, η mass difference.[8]

Following reference 3, therefore, we write the Lorentz invariant decay amplitude, $\vec{A}(\omega_1, \omega_2, \omega_3)$, in the $T = 1$ state, which satisfies Bose statistics for the pions as

$$\vec{A}(\omega_1, \omega_2, \omega_3) = (\vec{\varphi}_1 \cdot \vec{\varphi}_2)\vec{\varphi}_3 f(\omega_1, \omega_2, \omega_3) + (\vec{\varphi}_2 \cdot \vec{\varphi}_3)\vec{\varphi}_1 f(\omega_2, \omega_3, \omega_1) + (\vec{\varphi}_3 \cdot \vec{\varphi}_1)\vec{\varphi}_2 f(\omega_3, \omega_1, \omega_2), \quad (1)$$

where $f(x, y, z) = f(y, x, z)$. The $\vec{\varphi}$'s are the isotopic spin vectors of the pions. The matrix elements for the various decay modes can then be written down from (1). We assume an expansion of the form [9]

$$\langle \pi^+\pi^-\pi^0 | T | \eta \rangle = f(\omega_+, \omega_-, \omega_0) = 1 + A(\omega_+ + \omega_-) + B\omega_0, \quad (2)$$

which for calculational simplicity can be cast in an equivalent form,

$$f(\omega_+, \omega_-, \omega_0) = X[1 + \beta(3\omega_0 - M)], \quad (3)$$

where A, B, X, and β are related constants and M is the mass of η.[10]

From (1) and (3),

$$\langle \pi^0\pi^0\pi^0 | T | \eta \rangle = (3/\sqrt{6})X. \quad (4)$$

The branching ratio which results is

$$R\left[\frac{\eta \to 3\pi^0}{\eta \to \pi^+ + \pi^- + \pi^0}\right] = 1.5\frac{P}{1 + (\beta M\epsilon/2)^2}, \quad (5)$$

where $\epsilon = (M - 3)/M$ and P is a space factor which is equal to unity if the mass difference between the neutral and the charged pions is neglected, but is equal to ≈1.15 if it is taken into account. Thus if $\beta = 0$, we get a maximum value for R, $R_{max} \approx 1.73$. The shape of the π^0 spectrum in the decay $\eta \to \pi^+ + \pi^- + \pi^0$, on the other hand, is determined by β. Normalizing the lower end of the spectrum to one and dividing out the phase-space factor, the π^0 energy spectrum, $S(T_0)$, is given by

$$S(T_0) = \left[1 + \frac{3\beta}{1 + \beta(3 - M)}T_0\right]^2, \quad (6)$$

where T_0 is the kinetic energy of the pion.

In Fig. 1 we plot the π^0 energy spectrum for different values of R, using Eqs. (5) and (6). The present information concerning this spectrum dictates the choice of negative sign for β in Eq. (6). As another measure of the asymmetry in the Fabri-Dalitz plot we have also calculated[11] using the matrix element in (3), percentages for the N_1, N_2, and N_3 numbers for different sectors shown in Fig. 2 for the decay $\eta \to \pi^+ + \pi^- + \pi^0$. These results are contained in Table I. Table I also contains the values of a parameter γ which determines the shape of the π^- spectrum, $S_-(T_-)$, in the τ decay mode of the K^+ ($K^+ \to \pi^+ + \pi^+ + \pi^-$),

$$S_-(T_-) = 1 + \gamma T_- + \tfrac{1}{4}\gamma^2 T_-^2, \quad (7)$$

where T_- is the kinetic energy of the π^-. If an expression analogous to (3) is used for $K_2^0 \to \pi^+ + \pi^- + \pi^0$, with M replaced by the K mass M_K, then from Eq. (12) in reference 3, it follows that

$$\frac{1}{\gamma} = -\frac{M_K - 3}{2} - \frac{1 + \beta'(3 - M_K)}{3\beta'}, \quad (8)$$

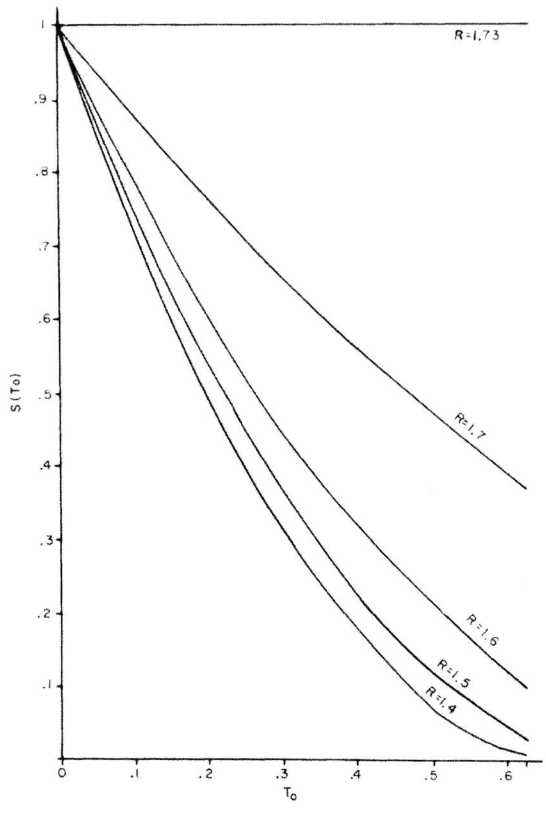

FIG. 1. π^0 energy spectrum as a function of its kinetic energy in the decay $\eta \to \pi^+ + \pi^- + \pi^0$ for different values of R.

where β' is a parameter which plays the same role in $K_2^0 \to \pi^+ + \pi^- + \pi^0$ as β in $\eta \to \pi^+ + \pi^- + \pi^0$.

The presently available information on the η does not warrant a detailed comparison between the theoretical and experimental results. However, we would like to point out that if we consider the total number of events $\eta \to \pi^+ + \pi^- + \pi^0$ from various groups,[12] then the numbers N_1, N_2, N_3 indicate that R lies between 1.6 and 1.7. When $R < 1.6$, the middle sector (N_2) gets "depopulated" relative to the top and the bottom ones. Presently there does not seem to be any convincing evidence for this.[13] Also the observed π^0 spectrum from the combined data is certainly consistent with the behavior shown in Fig. 1. Further, if we use a value for β' the same as β corresponding to R in the range 1.6 to 1.7, Eq. (8) gives a value for γ which agrees rather well with the data on τ decay ($\gamma = 1.9 \pm 0.3$).[4]

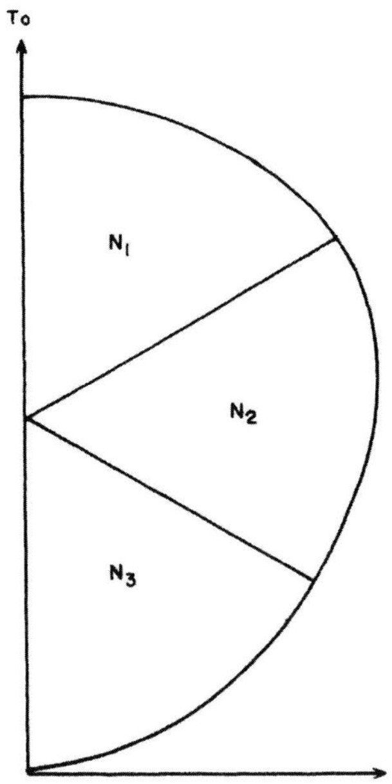

FIG. 2. Sketch of the Fabri-Dalitz plot for $\eta \to \pi^+ + \pi^- + \pi^0$ to define the numbers N_1, N_2, N_3.

To conclude, based on the presently available information on the η, a "linear approximation" like the one used in explaining the Fabri-Dalitz plot for K^+ and $K_2{}^0 \to 3\pi$ seems to be adequate to explain also the data on $\eta \to \pi^+ + \pi^- + \pi^0$. Moreover, if the branching ratio R of $\eta \to 3\pi^0$ to $\eta \to \pi^+ + \pi^- + \pi^0$ lies in the range 1.6 to 1.7, the parameter which determines the shapes of the various spectra in K^+ and $K_2{}^0 \to 3\pi$ and the one which determines the shape of the π^0 spectrum in $\eta \to \pi^+ + \pi^- + \pi^0$ are roughly the same. These considerations, therefore, constitute a strong argument in favor of the assignment of the quantum numbers 0^{-+} to the η. It is also to be noted that the value of R in the range specified above implies that a considerable fraction of the neutral mode of decay of η (the observed ratio for the total neutral to the charged mode lies in the range 3 to 5) consists of $3\pi^0$. The remaining neutral decays have to be explained by the other dominant decay mode $\eta \to 2\gamma$. This, however, raises a difficulty. If the electromagnetic interaction is responsible for $\eta \to 3\pi$, the matrix elements for both the decay modes $\eta \to 3\pi$ and $\eta \to 2\gamma$ have the same power α, where α is the fine-structure constant. The latter mode of decay, however, is favored considerably by phase space. Simple calculations based on invariance arguments,[14] using reasonable values for the "range parameter," predict a large ratio ($\approx 10^2$) for $\eta \to 2\gamma$ to $\eta \to 3\pi$.[15,16]

Table I. The percentages for the numbers N_1, N_2, and N_3 as a function of R. The observed values for these percentages are 24.2, 27.3, and 48.5.[a] The last column contains the parameter γ which determines the shapes of the various spectra in $K^+, K_2{}^0 \to 3\pi$, obtained from Eq. (8) as explained in the text.

R	$(-\beta)$	N_1	N_2	N_3	γ
1.73	0	33.3%	33.3%	33.3%	0
1.7	0.25	27.9	29.1	43.0	0.84
1.6	0.57	23.0	23.0	54.0	1.99
1.5	0.78	20.6	19.4	60	2.91
1.4	0.97	19.4	15.8	64.8	3.84

[a] See reference 12.

It is a great pleasure to thank Professor G. Feldman, Professor T. Fulton, and Professor R. F. Sawyer for many valuable suggestions. The author also had very helpful discussions with Professor A. Pevsner concerning the experimental data.

---

*Work supported by the National Science Foundation.

[1] A. Pevsner et al., Phys. Rev. Letters 7, 421 (1961).

[2] P. L. Bastien et al., Phys. Rev. Letters 8, 114 (1962); M. Gell-Mann, D. Sharp, and W. G. Wagner, Phys. Rev. Letters 8, 261 (1962); L. M. Brown and P. Singer, Phys. Rev. Letters 8, 115 (1962).

[3] R. F. Sawyer and K. C. Wali, Nuovo cimento 17, 938 (1960).

[4] S. McKenna et al., Nuovo cimento 10, 763 (1958).

[5] S. Weinberg, Phys. Rev. Letters 4, 87 (1960).

[6] Such a linear approximation is not strictly valid if the decay amplitude contains terms due to strong final-state π-π interactions. The decay amplitude then necessarily contains an imaginary part involving square roots of the energies. But extensive data is necessary to detect the presence of such terms. See reference 3 and also R. F. Sawyer and K. C. Wali, Phys. Rev. 119, 1429 (1960).

[7] D. Luers et al., Phys. Rev. Letters 7, 255 (1961).

[8] G. Barton and S. P. Rosen, Phys. Rev. Letters 8, 414 (1962). These authors consider a specific model for the decay amplitude.

[9] Such a form for $f(\omega_1, \omega_2, \omega_3)$ also arises if we attempt to write down directly simple interactions responsible for the decay, using invariance arguments. The simplest interaction with no derivatives gives only the constant term which is not consistent with the data. Addition of two derivatives gives the energy-dependent terms.

[10] Masses and energies are expressed in terms of the pion mass, which is taken to be unity.

[11] In calculating the branching ratio and the numbers N_1, N_2, N_3 we have made use of the Fabri variables and

the nonrelativistic approximation of a circle for the allowed kinematical region. E. Fabri, Nuovo cimento 11, 479 (1954). Relativistic corrections are not too important for our arguments.

[12]The combined number of events of $\eta \to \pi^+ + \pi^- + \pi^0$ from Berkeley, Brookhaven, Yale, and Johns Hopkins is 165. The numbers for N_1, N_2, N_3 from this combined analysis are 40,45,80. A. Pevsner (private communication).

[13]The data from the Yale group alone [H. Foelsche et al., in Proceedings of the Eleventh International Conference on High-Energy Physics at CERN (to be published)], indicates such a "depopulation" of the middle sector. Their numbers for N_1, N_2, N_3 are 10, 4, and 12. However, these numbers are too small to be statistically significant.

[14]Taking into account angular momentum barriers, phase space, etc. See G. Feinberg, Phys. Rev. Letters 8, 151 (1962), G. Feldman, T. Fulton, and K. C. Wali, Nuovo cimento 24, 278 (1962).

[15]Brown and Singer propose enhancement of the charged mode of decay of η due to strong final-state interactions. Their mechanism, however, does not explain the Dalitz plot. L. M. Brown and P. Singer, Phys. Rev. Letters 8, 460 (1962).

[16]Gell-Mann, Sharp, and Wagner, reference 2, and also Brown and Singer, reference 15, consider the problem relating to the $\eta \to 2\gamma$ to $\eta \to \pi^+ + \pi^- + \gamma$ ratio. They show assuming "ρ-dominance" model that $\eta \to 2\gamma$ predominates over $\eta \to \pi^+ + \pi^- + \gamma$. Our calculations based on invariance arguments give essentially the same results.

Coupled Channel Approach to $J=\frac{3}{2}^+$ Resonances in the Unitary Symmetry Model*

A. W. MARTIN AND K. C. WALI
Argonne National Laboratory, Argonne, Illinois
(Received 28 December 1962)

An analysis is made of the $p_{3/2}$ pseudoscalar meson-baryon scattering amplitudes in all isotopic spin and strangeness states. In analogy with pion-nucleon scattering, the basic assumption that the single baryon exchange contribution provides the dominant force in these amplitudes is made, but all the coupled two-particle channels in each isotopic spin and strangeness state are included. A relativistic scattering matrix for each state is constructed by a matrix formulation of the N/D method so that it satisfies the requirements of unitarity and symmetry. Being fully relativistic, unlike the static model, the theory does not contain adjustable cutoff parameters. The known masses of the mesons and baryons and the Yukawa-type meson-baryon coupling constants are the only parameters that enter into the calculation. The octet model of Gell-Mann and Ne'eman is used to define the coupling constants in terms of a single parameter f and the known pion-nucleon coupling constant. Then the requirement that $N^*(T=\frac{3}{2};S=0)$, $Y_1^*(T=1;S=-1)$, the recently observed $\Xi^*(T=\frac{1}{2};S=-2)$, and a yet-to-be-discovered $Z^-(T=0,S=-3)$ exist as a tenfold representation of the unitary symmetry model restricts f to a rather narrow range. There are no other resonances in the $p_{3/2}$ state for f in this range, except possibly one in the $T=0$, $S=-1$ system. The positions and widths of these resonances are discussed.

I. INTRODUCTION

RECENTLY, a resonance in the cascade-pion system, in the isotopic spin $T=\frac{1}{2}$ state, has been observed.[1] Although the spin and parity of this resonance have not been measured, the indications are that the spin is greater than $\frac{1}{2}$. If it is assumed that the resonance is in the $p_{3/2}$ state, being the analog of the well-known $p_{3/2}$ resonance in the pion-nucleon system, then one encounters an interesting problem. Experimentally the cascade particle appears to have the same spin,[2] as well as the same isotopic spin, as the nucleon. The Chew-Low static model[3] then, which successfully explained the pion-nucleon 3, 3 resonance, should be even more appropriate in the cascade-pion system because of the increased cascade mass. This reasoning, however, leads us to expect a resonance in the $T=\frac{3}{2}$ cascade-pion state and not in the $T=\frac{1}{2}$ state.

On the other hand, the observed $T=\frac{1}{2}$ resonance was predicted by, and neatly fits into, the group of baryon-meson $p_{3/2}$ resonances known as the tenfold representation of the unitary symmetry group $SU(3)$.[4] The main purpose of this paper is to investigate whether predictions based on dynamical considerations, to be specified below, can be reconciled with those based on symmetry arguments.

The basic assumption in our dynamical treatment is, as is generally believed from analyses of pion-nucleon scattering, the importance of the Born approximation term arising from the single baryon exchange diagram. However, in treating the Born approximation term we depart from the static model[5] with its reliance on an arbitrary cutoff parameter. Our calculation is, in general, closer to the fully relativistic treatments of Baker,[6] and Frautschi and Walecka,[7] as will be discussed in more detail in the following sections. But this does not alter the fact that one cannot obtain a resonance in the $T=\frac{1}{2}$ cascade-pion system since the Born approximation term is repulsive in that state.

One possible approach, suggested immediately by the treatments of the Y^* resonances in the strangeness -1 channels,[8] is based on the idea that the coupled two-particle inelastic channels are important. This notion is reinforced when one notes that the $\Sigma\bar{K}$ and $\Lambda\bar{K}$ thresholds are much closer to that of the $\Xi\pi$ system than are their analogs, ΣK and ΛK, in the pion-nucleon problem. In this way we are led to a matrix formulation for our dynamical considerations. When one turns to a coupled-channel calculation, however, one immediately encounters the problem of an overabundance of coupling constants which, especially in this system, are so poorly defined by experiment that they assume the roles of undetermined parameters. Since we prefer to work with as few arbitrary constants as possible, we turn to the group-theoretical considerations for a model of the coupling constants.

The octet model of the strongly interacting par-

* Work performed under the auspices of the U. S. Atomic Energy Commission.

[1] G. M. Pjerrou, D. J. Prowse, P. Schlein, W. E. Slater, D. H. Stork, and H. K. Ticho, Phys. Rev. Letters **9**, 114 (1962); L. Bertanza, V. Brisson, P. L. Connolly, E. L. Hart, I. S. Mittra, G. C. Moneti, R. R. Rau, N. P. Samios, I. O. Skillicorn, S. S. Yamamoto, M. Goldberg, L. Gray, J. Leitner, S. Lichtman, and J. Westgard, *ibid.* **9**, 180 (1962).

[2] L. Bertanza, V. Brisson, P. L. Connolly, E. L. Hart, I. S. Mittra, G. C. Moneti, R. R. Rau, N. P. Samios, I. O. Skillicorn, S. S. Yamamoto, M. Goldberg, L. Gray, J. Leitner, S. Lichtman, and J. Westgard, Phys. Rev. Letters **9**, 229 (1962).

[3] G. F. Chew and F. E. Low, Phys. Rev. **101**, 1570 (1956).

[4] R. E. Behrends, J. Dreitlein, C. Fronsdal, and B. W. Lee, Rev. Mod. Phys. **34**, 1 (1962); S. L. Glashow and J. J. Sakurai, Nuovo Cimento **25**, 337 (1962); R. Cutkosky, J. Kalckar, and P. Tarjanne, Phys. Letters **1**, 93 (1962).

[5] While this work was in progress, work by R. H. Capps [Nuovo Cimento **27**, 1208 (1963)] was brought to our attention. In it he carries out a similar analysis based on the static model.

[6] M. Baker, Ann. Phys. (N. Y.) **4**, 271 (1958).

[7] S. C. Frautschi and J. D. Walecka, Phys. Rev. **120**, 1486 (1960).

[8] D. Amati, A. Stanghellini, and B. Vitale, Nuovo Cimento **13**, 1143 (1959); K. C. Wali, T. Fulton, and G. Feldman, Phys. Rev. Letters **6**, 644 (1961).

ticles proposed by Gell-Mann, and independently by Ne'eman,[9] leads to a two-parameter set of expressions for the baryon-pseudoscalar meson coupling constants. These are conveniently summarized by Sakurai.[10] Furthermore, the requirement that the pion-nucleon coupling constant be $g_{N\pi}^2/4\pi=15$ reduces the dependence to a single arbitrary parameter. We adopt the octet-model set of coupling constants as a working hypothesis.

As mentioned in the opening remarks, the octet model forms the basis for predictions of the interactions of elementary particles. These predictions go far beyond the simple coupling constant considerations outlined above. Among the more striking of these is the suggestion that a family of $p_{3/2}$ resonances can be identified as belonging to the tenfold representation in $SU(3)$. The members of this family[4] are taken to be the familiar pion-nucleon 3, 3 resonance, the strangeness -1 Y_1^*, the recently discovered Ξ^*, and a yet-to-be-discovered resonance or bound state in the strangeness -3 channel with isospin zero, tentatively named the Z^- by Glashow and Sakurai.[11] For the prediction to be fulfilled, all of these resonances must have the angular-momentum quantum numbers of the 3, 3 resonance.

If we take these suggestions seriously, then it is clear that an explanation of the cascade-pion resonance is too narrow in scope. Rather we should attempt to treat all of the baryons and pseudoscalar mesons on an equal footing, apart from their mass differences which will obviously affect dynamical considerations, and see whether an underlying structure suggestive of all four resonances in the representation 10 exists. In this approach it is, of course, necessary to make the usual assumptions about the spins and parities of the particles involved. The baryons N, Λ, Σ, Ξ are taken to be spin-$\frac{1}{2}$ particles with positive relative parities while the mesons π, K, $\bar{K}$, η are assumed to be pseudoscalar.

Therefore, in each state of isotopic spin and strangeness, we consider coupled systems of the allowed two-particle states, ignoring the inelastic channels consisting of three or more particles. Finally, we restrict our attention to the $p_{3/2}$ partial-wave state. We may summarize our dynamical model as follows. The relativistic scattering matrix for each state will be constructed by a matrix formulation of the N/D method so that it satisfies the requirements of unitarity and symmetry and contains the singularities arising from the single baryon exchange graphs in all two-particle channels. From the N/D formulation the condition for a resonance or bound state in a given channel is easily isolated and, in this approximate model, the dynamical structure underlying the unitary symmetry group is understood. Because the approach taken is relativistic and, therefore, requires no cutoff parameter, the calculation contains only the unknown coupling constant parameter which is then restricted to a small range by comparison with experiment.

In Sec. II we review the pertinent kinematical considerations and describe our choice of variable and scattering amplitude. Section III is devoted to the description of the many-channel N/D formalism and the comparison of our model with the calculations of pion-nucleon scattering mentioned above. In Sec. IV we turn to the relationship of our treatment to the predictions of the unitary symmetry group. Section V contains the analysis of the mass-difference corrections and a brief discussion of dynamic factors not included in our model. The octet-model single-parameter expressions for the coupling constants are given in Appendix I, while Appendix II contains the tabulation of the appropriate isotopic spin coefficients in the different channels.

II. KINEMATICS AND CHOICE OF AMPLITUDE

The kinematical considerations for baryon-meson scattering have been developed extensively in the literature.[7,12] The principal modifications required in the analysis to follow stem from the fact that we are concerned with inelastic two-body scattering processes as well as elastic ones. This necessitates a more careful treatment of the scattering amplitude in order to eliminate the kinematical singularities inherent in inelastic processes. The following equations, therefore, are intended mainly as a guide to our final choice of partial-wave amplitudes.

We begin with the general two-body scattering amplitude for a process involving a spin-$\frac{1}{2}$ baryon and a spin-zero meson in the initial and final states. Let the initial baryon and meson have masses M_1 and μ_1 and four-momenta p_1 and q_1, respectively; and let M_2 and μ_2, p_2 and q_2, represent the corresponding quantities for the final baryon and meson (Fig. 1). The Lorentz invariant T matrix for this process is defined by[13]

$$S_{fi}=\delta_{fi}-(2\pi)^4 i\delta^4(p_2+q_2-p_1-q_1)\times\left[\frac{M_1M_2}{4E_1E_2\omega_1\omega_2}\right]^{1/2}\bar{u}_2Tu_1, \quad (1)$$

where $E_{1,2}$ and $\omega_{1,2}$ represent the energies of the appropriate baryons and mesons, respectively, and the

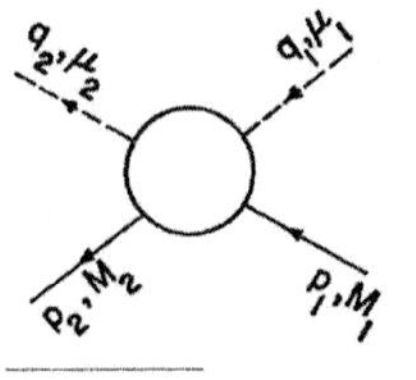

FIG. 1. General meson-baryon scattering diagram.

---

[9] M. Gell-Mann, Phys. Rev. **125**, 1067 (1962); Y. Ne'eman, Nucl. Phys. **26**, 222 (1961).

[10] J. J. Sakurai (to be published).

[11] S. L. Glashow and J. J. Sakurai, Nuovo Cimento **26**, 622 (1962).

[12] G. F. Chew, M. L. Goldberger, F. E. Low, and Y. Nambu, Phys. Rev. **106**, 1337 (1957).

[13] We use natural units $\hbar=c=1$ and the energies are measured in terms of the pion mass μ.

Dirac spinors are normalized so that $\bar{u}u=1$. For the two-body processes we are considering, T has the general structure

$$T=-A(s,t,u)+\frac{q_1+q_2}{2}B(s,t,u), \tag{2}$$

where A and B are scalar functions of the familiar Mandelstam variables

$$s=-(p_1+q_1)^2,$$
$$t=-(q_1-q_2)^2,$$
$$u=-(p_1-q_2)^2,$$
$$s+t+u=M_1^2+M_2^2+\mu_1^2+\mu_2^2.$$

The invariant amplitudes A and B are linear combinations of the amplitudes referring to the total isotopic spin states involved in the scattering.

In the center-of-mass system the differential cross section is given by

$$\frac{d\sigma}{d\Omega}=\frac{q_f}{q_i}\sum_{\text{spins}}\left|\left\langle f\left| f_1+\frac{(\boldsymbol{\sigma}\cdot\mathbf{q}_f)(\boldsymbol{\sigma}\cdot\mathbf{q}_i)}{q_fq_i}f_2\right| i\right\rangle\right|^2, \tag{3}$$

where q_i and q_f are the magnitudes of the three-momenta $\mathbf{q}_i$ and $\mathbf{q}_f$ of the initial and final mesons, respectively. The scalar functions f_1 and f_2 in the two-component representation implied by Eq. (3) are related to the invariant amplitudes by

$$8\pi Wf_1=(E_1+M_1)^{1/2}(E_2+M_2)^{1/2}\times\left[A+\left(W-\frac{M_1+M_2}{2}\right)B\right],$$
$$8\pi Wf_2=(E_1-M_1)^{1/2}(E_2-M_2)^{1/2}\times\left[-A+\left(W+\frac{M_1+M_2}{2}\right)B\right]. \tag{4}$$

In the expressions above, W is the total energy in the c.m. system. For a given baryon-meson state with masses M and μ, the quantities E, ω, and q are related to W by the expressions

$$E=(W^2+M^2-\mu^2)/2W,$$
$$\omega=(W^2-M^2+\mu^2)/2W, \tag{5}$$
$$q=[W^4-2(M^2+\mu^2)W^2+(M^2-\mu^2)^2]^{1/2}/2W.$$

As is well known,[7] the partial-wave scattering amplitude, $f_{l\pm}$, for a state with total angular momentum $J=l\pm\frac{1}{2}$ and parity $(-1)^{l+1}$ is given by the projection

$$f_{l\pm}(W)=\frac{1}{2}\int_{-1}^{1}dx[f_1(W,x)P_l(x)+f_2(W,x)P_{l\pm1}(x)], \tag{6}$$

where $x=\cos\theta$, θ being the scattering angle. From Eqs. (4), (5), and (6)

$$f_{l\pm}=\frac{1}{16\pi W^2}[(W+M_1)^2-\mu_1^2]^{1/2}[(W+M_2)^2-\mu_2^2]^{1/2}\times\left[A_l+\left(W-\frac{M_1+M_2}{2}\right)B_l\right]$$
$$+\frac{1}{16\pi W^2}[(W-M_1)^2-\mu_1^2]^{1/2}[(W-M_2)^2-\mu_2^2]^{1/2}\times\left[-A_{l\pm1}+\left(W+\frac{M_1+M_2}{2}\right)B_{l\pm1}\right], \tag{7}$$

where

$$A_l=\frac{1}{2}\int_{-1}^{1}AP_l(x)dx,$$

$$B_l=\frac{1}{2}\int_{-1}^{1}BP_l(x)dx.$$

For our purposes the coupled-channel scattering matrix for given quantum numbers of strangeness, total isotopic spin, angular momentum, and parity is limited to two-particle states. In order to keep the notation as simple as possible, the partial-wave amplitude for the process $M_i+\mu_i\rightarrow M_j+\mu_j$ in the isotopic spin state T will be denoted by $f_{l\pm}{}^T(j,i)$. Time-reversal invariance requires the scattering matrix to be symmetric, i.e.,

$$f_{l\pm}{}^T(j,i)=f_{l\pm}{}^T(i,j).$$

And the unitarity condition in the physical region demands

$$\operatorname{Im}f_{l\pm}{}^T(j,i)=\sum_n f_{l\pm}{}^{T*}(j,n)q_nf_{l\pm}{}^T(n,i), \tag{8}$$

where the summation extends over the allowed two-particle states in our "elastic" approximation. It should be noted that the phase space factor q_n appearing in the unitarity condition implicitly contains a step function which vanishes below the physical threshold in channel n and is unity above that threshold. In a single-channel case, Eq. (8) leads to the familiar form $q^{-1}\sin\delta_{l\pm}{}^T\exp(i\delta_{l\pm}{}^T)$ for the partial-wave amplitude in the physical region.

Now, in the choice of our amplitudes, we are guided by the considerations discussed by Frautschi and Walecka, namely, we wish to retain as many features of the correct relativistic theory as possible. It is evident from Eq. (7) that the amplitudes $f_{l\pm}{}^T(j,i)$ contain kinematic poles and branch cuts that are in no way related to the dynamic singularities contained in the invariant amplitudes A and B. Furthermore, these kinematic singularities cannot conveniently be removed if we chose to work in the W^2 plane. On the other hand, they are easily eliminated by taking the amplitudes to be functions of the complex variable W.

The appropriate partial-wave amplitudes, which are

devoid of kinematic singularities in the W plane and ensure the correct threshold behavior of the $f_{l\pm}{}^T$, are obtained by defining

$$G_{l+}{}^T(j,i)=\frac{2W}{q_j{}^l[(W+M_j)^2-\mu_j{}^2]^{1/2}}\times f_{l+}{}^T(j,i)\frac{2W}{q_i{}^l[(W+M_i)^2-\mu_i{}^2]^{1/2}},$$
$$G_{l-}{}^T(j,i)=\frac{2W}{q_j{}^{l-1}[(W-M_j)^2-\mu_j{}^2]^{1/2}}\times f_{l-}{}^T(j,i)\frac{2W}{q_i{}^{l-1}[(W-M_i)^2-\mu_i{}^2]^{1/2}}. \quad (9)$$

These definitions are based on the assumption that in the vicinity of the thresholds the projections of the invariant amplitudes behave as

$$A_l(j,i)\propto q_j{}^l q_i{}^l,$$
$$B_l(j,i)\propto q_j{}^l q_i{}^l.$$

The unitarity condition satisfied by the $G_{l\pm}{}^T$, obtained from Eqs. (8) and (9), is given by

$$\mathrm{Im}G_{l\pm}{}^T(j,i)=\sum_n G_{l\pm}{}^{T*}(j,n)\rho_{l\pm}(n)G_{l\pm}{}^T(n,i), \quad (10)$$

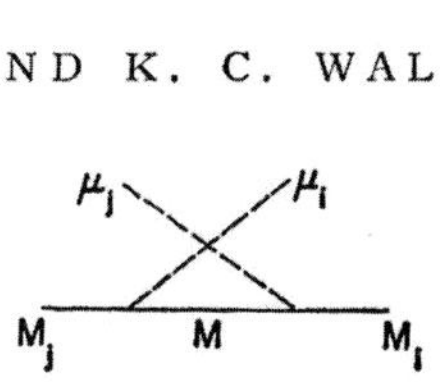

FIG. 2. Single baryon exchange diagram. M_i and μ_i (M_j and μ_j) refer to the initial (final) baryon and meson masses, respectively. M is the mass of the intermediate baryon.

where the phase space factor $\rho_{l\pm}(n)$ is

$$\rho_{l+}(n)=[(W+M_n)^2-\mu_n{}^2]q_n{}^{2l+1}/4W^2,$$
$$\rho_{l-}(n)=[(W-M_n)^2-\mu_n{}^2]q_n{}^{2l-1}/4W^2. \quad (11)$$

An additional property of the partial-wave amplitudes $f_{l\pm}{}^T(j,i)$ and $G_{l\pm}{}^T(j,i)$ in the W plane is the "crossing relation"

$$G_{l+}{}^T(-W)=-G_{l+1,-}{}^T(W),$$
$$\rho_{l+}(-W)=-\rho_{l+1,-}(W), \quad (12)$$

which implies a set of cuts on the negative real axis, whose discontinuities are bounded by unitarity.

Finally, as pointed out in the introduction and as is further discussed in the following section, in our analysis of scattering in the $p_{3/2}$ state we will approximate the dynamical singularities of $G_{1+}{}^T(j,i)$ by those arising from the single baryon exchange graph (Fig. 2). If the isotopic spin factors and coupling constants are suppressed, the general expression for the $p_{3/2}$ contribution of this diagram is[14]

$$G_{B,1+}(j,i)=\frac{(W+M_j+M_i-M)}{4q_j{}^2q_i{}^2}\Bigg[\Bigg\{\frac{3\alpha_{ji}(W)}{[(W+M_j)^2-\mu_j{}^2][(W+M_i)^2-\mu_i{}^2]}-2\frac{(W-M_j-M_i+M)}{(W+M_j+M_i-M)}\Bigg\}$$
$$-\frac{1}{2\beta_{ji}(W)}\Bigg\{\frac{3\alpha_{ji}(W)}{[(W+M_j)^2-\mu_j{}^2][(W+M_i)^2-\mu_i{}^2]}-2\frac{(W-M_j-M_i+M)}{(W+M_j+M_i-M)}$$
$$-\frac{[(W-M_j)^2-\mu_j{}^2][(W-M_i)^2-\mu_i{}^2]}{\alpha_{ji}(W)}\Bigg\}\ln\left(\frac{1+\beta_{ji}(W)}{1-\beta_{ji}(W)}\right)\Bigg], \quad (13)$$

where

$$\alpha_{ji}(W)=W^4-(M_j{}^2+M_i{}^2+\mu_j{}^2+\mu_i{}^2-2M^2)W^2-(M_j{}^2-\mu_j{}^2)(M_i{}^2-\mu_i{}^2),$$
$$\beta_{ji}(W)=4W^2q_jq_i/\alpha_{ji}(W).$$

The singularities of the Born approximation term are contained in the logarithmic factor and can be chosen to consist of two branch cuts on the real axis and a third branch cut running the length of the imaginary axis. The real-axis branch cuts are reflections of each other across the imaginary axis and, in the case of elastic scattering $(M_i,\mu_i)=(M_j,\mu_j)$, the branch points are given by the simple expressions

$$W_1=(M_j{}^2-\mu_j{}^2)/M,$$
$$W_2=(2M_j{}^2+2\mu_j{}^2-M^2)^{1/2}. \quad (14)$$

This choice of cuts is the simplest one consistent with the crossing relation and the reality condition on $G_{1+}(j,i)$. Equation (14) forms the basis of the pole approximation for processes such as pion-nucleon scattering, for which the real-axis cuts are roughly one-third of a pion mass in length and can reasonably be replaced by poles. However, in scattering processes involving the heavy mesons, the cuts may extend up to four pion masses in length and the pole approximation becomes increasingly difficult to justify.

It will be convenient later to have a more tractable expression for $G_{B,1+}(j,i)$ than the exact functional de-

[14] Later when we have occasion to refer to the Born approximation term [Eq. (13)] for particular particle states, we will employ the notation $G_B(BC,A,DE)$ where the letters A through E will be replaced by the appropriate particle symbols, with D and E (B and C) representing the initial (final) baryon and meson, respectively, while A indicates the intermediate baryon.

pendence (13). Such an expression is readily obtained for the low-energy region by expanding the logarithm in powers of q_jq_i, the result being

$$G_{B,1+}(j,i)\approx 8W^4(W-M_j-M_i+M)/3\alpha_{ji}^2(W). \quad (15)$$

A further simplification results from noting that $\alpha_{ji}(W)$ can be factored into the form

$$\alpha_{ji}(W)=(W-c)(W+c)(W^2+d^2),$$

where c and d are real quantities. In Eq. (15), the second-order pole at $W=c$ lies on the branch cut and marks the position of the dominant contribution arising from the cut.

III. PION-NUCLEON SCATTERING AND THE MATRIX N/D FORMATION

It is clear that the minimum requirement for any dynamical model to be used in the understanding of resonances in baryon-meson systems is that it reproduce the experimentally well-established features of pion-nucleon scattering in the $p_{3/2}$ state. As a guide, therefore, in our choice of a dynamical coupled-channel formalism, we turn to three of the well-known theoretical treatments of this N-π problem.

One of the earliest successful approaches to pion-nucleon scattering was the Chew-Low static model and its dispersion-theoretic refinements.[12] In this model the single nucleon exchange diagram and unitarity form the basis of the dynamics; but the approximation of taking the nucleon mass to be very large compared with the pion mass has the effect of keeping only one part of the Born approximation singularities given by Eqs. (13) and (14)—namely, the short branch cut nearest the physical region. This short branch cut becomes the pole at the origin in the static model. In addition, the static model treatment requires an arbitrary cutoff because the integrals involved are divergent.

The static model is easily generalized to the relativistic case by replacing the nearby cut by a pole whose residue (as pointed out by Frautschi and Walecka) is just that of the static model, and by ignoring the other cuts in the Born approximation. Unitarity is satisfied by using the N/D method[15] with a once-subtracted (at the pole) dispersion relation for D. This relativistic formulation requires no cutoff parameter and, in fact, contains no arbitrary constants. When this program is carried through, it is found that no resonance is predicted in the 3, 3 state for pion-nucleon scattering; that is, the short cut in the Born approximation is not sufficiently attractive to cause the phase shift to go through 90°. From this viewpoint it is evident that the primary factor in the static model's successful prediction of the 3, 3 resonance is the necessity for an arbitrary cutoff which allows the integrals to be made as large as required.

The results of the static model were placed on a sounder footing by Frautschi and Walecka, who showed that the single nucleon exchange term did indeed produce a resonant structure for the 3, 3 amplitude if the branch cut along the imaginary axis was included in the dynamic singularities along with the short branch cut. Their single channel calculation, fully relativistic and devoid of cutoff parameters, was based on a many-pole approximation N/D technique which required the solution of simultaneous algebraic equations. This point will be important when we consider the coupled-channel formalism. For our purposes the central result of their analysis is the fact that the Born approximation, treated relativistically, does reproduce the general features of pion-nucleon scattering in the $p_{3/2}$ state. It is also pertinent to note their comment to the effect that the single nucleon exchange singularities are too strong (i.e., the 3, 3 resonance occurs too close to the physical threshold). They remark that a reduction of the discontinuity across the imaginary axis cut by roughly a factor of 2 leads to closer agreement with experiment.

The third pion-nucleon treatment to be discussed is Baker's application of the determinantal method.[6] His formalism, based primarily on perturbation expansions, bears a remarkable similarity to the N/D method. The first-order determinantal result for the partial-wave amplitude may be considered to be an N/D representation, where N is the full Born approximation term [Eq. (13)] with appropriate isotopic spin factors and coupling constants, and D is expressed by a once subtracted dispersion integral over the physical cut. The integrand in the latter expression contains N in the standard way. The subtraction point is chosen to be the nucleon mass. This choice means that the full amplitude is normalized so that its discontinuity across the short cut is essentially equal to the Born approximation discontinuity. These points will be discussed in more detail in the application of our model to pion-nucleon scattering. The many-channel formalism we employ is, apart from small modifications arising from crossing, a matrix generalization of Baker's work. Before proceeding to the matrix N/D formulation, it is important to point out that the resonant structure of the 3, 3 amplitude was also obtained by Baker, with the position of the resonance about 1.2 μ higher than the experimentally observed position.

The success of these single-channel treatments makes it plausible that a relativistic coupled-channel formalism based on unitarity and the single baryon exchange term can be useful in understanding baryon-meson resonances in the $p_{3/2}$ state. We do not expect to be able to predict the exact locations of the resonances, since resonance positions are known to be sensitive to many dynamical effects we cannot include.

The most convenient manner in which to ensure the unitarity of the scattering matrix is by means of a matrix generalization of the N/D method. Such a generalization is not unique but depends, rather, on the approximations to be used. An extension of the single-pole approximation to the many-channel case has been

[15] G. F. Chew and S. Mandelstam, Phys. Rev. 119, 467 (1960).

given by Feldman, Matthews, and Salam,[16] while a more general N/D coupled integral equation formalism has been presented by Bjorken.[17] The single-pole approximation allows one to extend the static model to coupled channels without difficulty. However, the static model requires either the nonrelativistic treatment in which an arbitrary cutoff is introduced or the relativistic treatment in which one cannot obtain even the pion-nucleon resonance. In either case, there is difficulty involved in the replacement of very long branch cuts by single poles. For these reasons, the single-pole approximation is inadequate for our purposes.

A matrix generalization of the many-pole approximation used by Frautschi and Walecka would be quite acceptable as a dynamical model, apart from the extensive work involved in determining the correct residues for the poles. The limitation in this case, however, stems from the difficulty in going from the single-channel method to the matrix form as illustrated by Costa and Ferrari,[18] who based their analysis on Bjorken's matrix N/D formalism. An explicit representation for the many-pole approximation to a scattering matrix of arbitrary dimensionality exists,[19] but suffers from the fact that the integrals contained in the denominator matrix D must be subtracted n times, where n is the number of poles. For these reasons we will not employ the generalization of the many-pole approximation.

Fortunately, a straightforward generalization of Baker's single-channel first-order determinantal method is easily carried out. We take the numerator matrix N to be given by the full Born approximation terms [Eq. (13)] with appropriate isotopic spin factors and coupling constants, and construct D so that the two requirements of symmetry and unitarity of the T matrix are satisfied. Since the numerator matrix in this model is symmetric, Bjorken's matrix formalism, while ensuring unitarity, will not in general satisfy the symmetry requirement unless one proceeds to solve the coupled matrix integral equations.[20] For our approximation scheme such an exact treatment is unwarranted.

Instead, we will utilize the "symmetrized" many-channel N/D method[19] defined as follows:

$$T(W)=N(W)D^{-1}(W),$$

$$D(W)=1-\frac{1}{2\pi}\int\frac{\rho(W')N(W')}{W'-W}dW'-\frac{N^{-1}(W)}{2\pi}\times\int\frac{N(W')\rho(W')}{W'-W}dW'\cdot N(W), \tag{16}$$

where $\rho(W)$ is a diagonal matrix of phase space factors and the ranges of the integrals are over the physical region as determined by the step functions implied by the elements of $\rho(W)$.[21] It is easily verified that the above formulation satisfies unitarity in the physical region and that for symmetric $N(W)$ it leads to a symmetric T matrix. In specializing Eq. (16) to the $p_{3/2}$ state we introduce the following modifications. The crossing relation [Eq. (12)] indicates the existence of physical cuts (for the $d_{3/2}$ amplitudes in this case) running along the negative real axis in the W plane. The unitarity condition bounds the discontinuities across these cuts and it is proper, therefore, to construct the scattering matrix so that unitarity is satisfied on both sets of physical branch cuts. We also write once-subtracted dispersion relations for $D(W)$, subtracting at the point $W=S$, in order to define suitably renormalized coupling constants. In each amplitude, S will be chosen in the region of the right-hand dynamic branch cuts, the singularities nearest to the physical region in our model. The subtraction also ensures the convergence of the integrals so that, in contrast to the static model, we require no cutoff parameters. When these modifications are incorporated, the scattering matrix $G_{1+}(W)$, whose elements are defined by Eq. (9), is written

$$G_{1+}(W)=G_{B,1+}(W)D_{1+}^{-1}(W),$$

$$D_{1+}(W)=1-\frac{(W-S)}{2\pi}\int_{W_0}^{\infty}dW'\left[\frac{\rho_{1+}(W')G_{B,1+}(W')}{(W'-S)(W'-W)}+\frac{\rho_{2-}(W')G_{B,1+}(-W')}{(W'+S)(W'+W)}\right]$$
$$-G_{B,1+}^{-1}(W)\frac{(W-S)}{2\pi}\int_{W_0}^{\infty}dW'\left[\frac{G_{B,1+}(W')\rho_{1+}(W')}{(W'-S)(W'-W)}+\frac{G_{B,1+}(-W')\rho_{2-}(W')}{(W'+S)(W'+W)}\right]G_{B,1+}(W), \tag{17}$$

[16] G. Feldman, P. T. Matthews, and A. Salam, Nuovo Cimento **16**, 549 (1960).
[17] J. D. Bjorken, Phys. Rev. Letters **4**, 473 (1960).
[18] G. Costa and F. Ferrari, Nuovo Cimento **22**, 214 (1961).
[19] A. W. Martin, Ph.D. thesis, Stanford University, Stanford, California, 1962 (unpublished).
[20] J. D. Bjorken and M. Nauenberg, Phys. Rev. **121**, 1250 (1961).
[21] See discussion following Eq. (8).

where the elements of $G_{B,1+}(W)$ are given by Eq. (13) with the appropriate isotopic spin factors and coupling constants, to be found in Appendixes I and II. The integrals in Eq. (17) run only over the positive real axis with W_0 representing the various physical thresholds.

The relationship of our approximation for $G_{1+}(W)$ to the exact solution of the coupled integral equation formalism, taking only the single baryon exchange graphs for the dynamical singularities, may be viewed in the following way. Both scattering matrices satisfy unitarity and are symmetric. Both have the dynamical branch cuts contained in $G_{B,1+}(W)$. However, the exact solution would have the precise discontinuities of G_B across the cuts, while our model provides only a reasonable approximation to these discontinuities for the distant cuts. The nearby branch cuts, on the other hand, are treated accurately in our formalism because of our choice of subtraction point. These statements will be made quantitative in the analysis of pion-nucleon scattering to follow.

Having defined our dynamical model for coupled-channel processes, we turn to the resonance (or bound state) condition in the many-channel case. This condition, which has been discussed by several authors,[22] can be conveniently stated in terms of the inverse K matrix defined by

$$G_{l\pm}^{-1}=K_{l\pm}^{-1}-i\rho_{l\pm}, \tag{18}$$

where $K_{l\pm}^{-1}$ is a real symmetric matrix both above and below thresholds. The vanishing of $\det K_{l\pm}^{-1}$ then gives the resonance location. In the matrix formulation of the N/D method this corresponds to

$$\det K^{-1}(W_r)=\det[\mathrm{Re}D(W_r)\cdot N^{-1}(W_r)]=0, \tag{19}$$

where W_r is the position of the resonance. It should be noted that only the vanishing of det ReD gives what one considers the dynamical resonances. In a single-channel case Eq. (19) reduces to the familiar condition $\rho(W_r)\cot\delta(W_r)=0$.

Finally, in the single-channel case one defines the width Γ of a resonance by noting that near the resonance position

$$\tan\delta=(\Gamma/2)/(W_r-W),$$

which leads to

$$\Gamma/2=\lim_{W\to W_r}(W_r-W)\rho(W)K(W). \tag{20}$$

The extension of these considerations to the many-channel case utilizes the definition of a partial width Γ_i for an elastic channel

$$\Gamma_i/2=\lim_{W\to W_r}(W_r-W)\rho_i(W)K_{ii}(W). \tag{21}$$

The sum of the partial widths, $\Gamma=\sum_i\Gamma_i$, is the full width of the resonance, while the relative branching ratios into the different channels can be expressed in terms of the partial widths and phase space factors.

[22] See, for example, R. H. Dalitz, Rev. Mod. Phys. 33, 471 (1961), and also reference 8.

We conclude this section with an application of our model to pion-nucleon scattering, considered as a single-channel process, in order to fully illustrate the nature of our approximation. Using the isotopic spin factors listed in Appendix II, the two isotopic spin amplitudes, $T=\frac{3}{2}$ and $T=\frac{1}{2}$, for N-π scattering in the $p_{3/2}$ partial-wave state are given by the single-channel form of Eq. (17) as

$$G^{3/2}=\frac{2g^2G_B(N\pi,N,N\pi)}{1-2g^2I(N\pi,N,N\pi)-2ig^2\rho(N\pi)G_B(N\pi,N,N\pi)},$$
$$G^{1/2}=\frac{-g^2G_B(N\pi,N,N\pi)}{1+g^2I(N\pi,N,N\pi)+ig^2\rho(N\pi)G_B(N\pi,N,N\pi)}, \tag{22}$$

where for simplicity the dependence on the total energy variable W has been suppressed and the angular momentum subscript has been dropped. $I(N\pi,N,N\pi)$ represents the real part of the integral in the denominator function,

$$I(N\pi,N,N\pi)=\mathrm{Re}\frac{(W-N)}{\pi}\int_{M+\mu}^{\infty}dW'\left[\frac{\rho_{1+}(W')G_B(W')}{(W'-N)(W'-W-i\epsilon)}+\frac{\rho_{2-}(W')G_B(-W')}{(W'+N)(W'+W+i\epsilon)}\right], \tag{23}$$

and is plotted in Fig. 4. The subtraction point is chosen to be the nucleon mass in keeping with our treatment of the nearby branch cuts. As discussed in Appendix I, it is convenient to absorb the usual factor of 4π into the definition of the coupling constant so that g^2, the square of the pion-nucleon coupling constant appearing in Eq. (22) has the value $g^2=15$.

From the resonance condition and Eq. (22), it is clear that a resonance in the low energy region is possible only in the $T=\frac{3}{2}$ state. The position of the 3, 3 resonance in this model is found to be 2.2 μ higher than the observed position. That it is also higher than the location determined by Baker is due to the inclusion of the integral over the left-hand physical cut. This integral has the effect of reducing the value of $I(N\pi,N,N\pi)$ by roughly 10% in the vicinity of the resonance. The source of the discrepancy with the result of Frautschi and Walecka, who obtain the resonance at too low an energy, is found in the treatment of the distant singularities. In the $T=\frac{3}{2}$ amplitude the discontinuity across the imaginary axis cut, in our model, is only 56% of the actual Born approximation discontinuity in the vicinity of the origin, while the actual discontinuity across the short left-hand branch cut is reduced by slightly more than a factor of 2. In brief, the attractive Born approximation term is made less attractive in this model.

TABLE I. Coupled two-particle pseudoscalar meson-baryon states for each strangeness and isotopic spin channel. The last column contains the factorized form for Det Re$D(W)$ in each case. The quantities S, P, Q, R, $\bar{R}$, and U are defined in Eq. (27).

Strangeness S	Isotopic spin T	Coupled two-particle states	Det Re$D(W)$
1	0	NK	$\bar{R}$
	1	NK	U
0	$\frac{1}{2}$	$N\pi, \Lambda K, \Sigma K, N\eta$	$PQ\bar{R}U$
	$\frac{3}{2}$	$N\pi, \Sigma K$	RU
	0	$N\bar{K}, \Sigma\pi, \Xi K, \Lambda\eta$	$SPQU$
-1	1	$N\bar{K}, \Sigma\pi, \Lambda\pi, \Xi K, \Sigma\eta$	$PQR\bar{R}U$
	2	$\Sigma\pi$	U
-2	$\frac{1}{2}$	$\Xi\pi, \Lambda\bar{K}, \Sigma\bar{K}, \Xi\eta$	$PQRU$
	$\frac{3}{2}$	$\Xi\pi, \Sigma\bar{K}$	$\bar{R}U$
-3	0	$\Xi\bar{K}$	R
	1	$\Xi\bar{K}$	U

In the $T=\frac{1}{2}$ amplitude, on the other hand, the effect of the distant singularities is enhanced, the imaginary axis cut by a factor of roughly $1\frac{1}{2}$ and the left-hand cut by a factor of $2\frac{1}{4}$. As pointed out by Frautschi and Walecka, the $T=\frac{1}{2}$ amplitude is not very sensitive to these distant singularities because of the strong repulsion of the nearby branch cut. Hence, it is not surprising that our values for the $T=\frac{1}{2}$ phase shift differ very little from their results (excluding the contribution due to the π-π singularities). In both approaches, then, qualitative agreement with experiment is obtained in each of the isotopic spin states.

In computing the resonance width, however, which from Eq. (20) is given by

$$\frac{\Gamma}{2}=\frac{\rho(W_r)G_B(W_r)}{(dI/dW)_{W=W_r}}, \tag{24}$$

we obtain too large a value. This is due primarily to the phase space factor in Eq. (24), proportional to q^3, to which the width is very sensitive in the low-energy region. If the phase space factor to be evaluated is taken arbitrarily at the observed resonance position, then Eq. (24) predicts the half-width at half-maximum to be about 100 MeV, as compared to the experimental value of roughly 70 MeV.

IV. SYMMETRY CONSIDERATIONS

As stated in the introduction, all of the Yukawa type coupling constants required for the single baryon exchange diagrams can be expressed in terms of a single parameter, denoted by f, once the octet-model expressions are normalized to the experimentally measured pion-nucleon coupling constant. The explicit single-parameter forms derived from the unitary symmetry model are given in Appendix I. Using these coupling constants and the formalism described in the previous sections, we wish to examine the resonance conditions in the various $p_{3/2}$ states corresponding to different isotopic spin and strangeness. However, to illustrate clearly the implications of adopting the octet-model coupling constant scheme in our calculations, we first consider a simplified model in which all baryons have the same mass B and all pseudoscalar mesons have the mass m.

In this extreme model all channels have the same threshold and phase space factor $\rho(W)$, and the matrices $G_B(W)$ and $D(W)$[Eq. (17)] assume the simple forms

$$\begin{aligned} G_B(W)&=h(W)\mathcal{B},\\ D(W)&=1-I(W)\mathcal{B}-i\rho(W)h(W)\mathcal{B}, \end{aligned} \tag{25}$$

where $h(W)$ represents the scalar Born approximation function $G_B(Bm,B,Bm)$ from Eq. (13), and $I(W)$ is the real part of the integral

$$I(W)=\mathrm{Re}\frac{W-B}{\pi}\int_{B+m}^{\infty}dW'\left[\frac{\rho_{1+}(W')h(W')}{(W'-B)(W'-W-i\epsilon)}+\frac{\rho_{2-}(W')h(-W')}{(W'+B)(W'+W+i\epsilon)}\right]. \tag{26}$$

In Eq. (25), $\mathcal{B}$ is a constant symmetric matrix containing the isotopic spin factors and coupling constants. The relevant information needed to construct $\mathcal{B}$ for a given isotopic spin and strangeness state is contained in the appendices.

From Eq. (25) it is clear that $G_B(W)$ and $D(W)$ can be simultaneously diagonalized by the orthogonal transformation which diagonalizes $\mathcal{B}$. This means that, in this model, the full scattering matrix $G(W)$ for each isotopic spin and strangeness state is diagonalized (put into the eigenphase representation) by a transformation determined solely by coupling constants and isotopic spin factors. Now it is well known[4] that the two-particle baryon-pseudoscalar meson states can be expressed as linear combinations of states forming the basis of the representations 1, 8, 8, 10, $\overline{10}$, and 27 of the $SU(3)$ group. The consequence of choosing the octet model coupling constant scheme is that the elements of the diagonal forms of the various scattering matrices correspond precisely to these six representations. In other words, the orthogonal matrix relating the particle states to the unitary symmetry basis states is exactly the matrix which diagonalizes $\mathcal{B}$ in each strangeness and isotopic spin channel.

For our purposes the primary result following from these considerations is that the determinant of the real part of $D(W)$, being invariant under orthogonal transformations, can be expressed in terms of n factors, where n is the dimensionality of $D(W)$. Each such factor has the general structure of the denominator function in a single-channel case and can be identified with one of the six representations of $SU(3)$ listed above. Thus, the determination of the resonance condition amounts to determining the range of values for the coupling constant parameter f such that one or several of these factors can vanish for a given value of $I(W)$. Table I contains the factorized expressions for det Re$D(W)$ in each isotopic spin and strangeness state, where we have introduced the convenient notation:

$$\begin{aligned} S&=1-(4/3)(5-10f-4f^2)g^2I(W),\\ P&=1+(2/3)(5-10f-4f^2)g^2I(W),\\ Q&=1+2(1-2f+4f^2)g^2I(W),\\ R&=1-(8/3)(1+f-2f^2)g^2I(W),\\ \bar{R}&=1-(8/3)(1-5f+4f^2)g^2I(W),\\ U&=1-(4/3)(1-2f+4f^2)g^2I(W). \end{aligned} \tag{27}$$

It is clear from Table I that the factors S, P, Q, R, $\bar{R}$, and U are to be identified in that order with the 1-, 8-, 8-, 10-, $\overline{10}$-, and 27-fold representations of $SU(3)$ and that the vanishing of any one of them corresponds to the realization, in this oversimplified model, of a family of resonances belonging to that representation. In order to find the restrictions on the values of the parameter f such that resonances are possible in some states and not in others it is necessary to evaluate the integral $I(W)$ of Eq. (26). For this purpose we utilize the baryon and pseudoscalar meson mass formula[9,23] to determine the masses B and m:

$$B=\frac{N+\Xi}{2}=\frac{3\Lambda+\Sigma}{4}=8.2\ \mu,$$

$$m^2=K^2=\frac{3\eta^2+\pi^2}{4}=(3.5\ \mu)^2.$$

The real part of the integral, $I(W)=I(Bm,B,Bm)$, is plotted in Fig. 3, from which it follows that the quantity $g^2I(W)$ in Eq. (27) lies in the range 0 to roughly $\frac{2}{3}$ in the low-energy region. The range of values for the parameter f which will give resonances in each of the six representations are then found to be

$$\begin{array}{ll} S & -2.84<f<0.34,\\ P & f<-3.1,\quad 0.59<f,\\ Q & \text{no possibility},\\ R & -0.28<f<0.78,\\ \bar{R} & f<0.15,\quad 1.15<f,\\ U & f<-0.06,\quad 0.56<f. \end{array} \tag{28}$$

In applying this idealized model to the currently available experimental data, we use the apparent absence of $p_{3/2}$ resonances in the NK system (strangeness $+1$) to exclude the vanishing of $\bar{R}$ and U. The lack of a $p_{3/2}$ resonance in the $N\pi$ $T=\frac{1}{2}$ system then further excludes the vanishing of P. These arguments leave us with a relatively narrow range of allowed values for f, namely, $0.15<f<0.56$. It is amusing to note that this

[23] S. Okubo, Progr. Theoret. Phys. (Kyoto) **27**, 949 (1962).

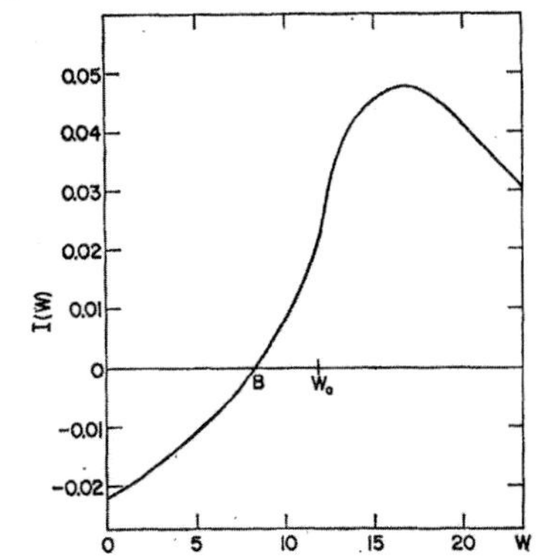

FIG. 3. $I(W)=I(Bm,B,Bm)$ defined in Eq. (26), plotted as a function of the total energy W in units of the pion mass μ. $B=8.2\ \mu$, $m=3.5\ \mu$. The threshold, W_0, is 11.7 μ and the subtraction point is at the baryon mass B.

range of values definitely predicts the vanishing of R, giving resonances in the tenfold representation of the baryon-meson states, while the vanishing of S, corresponding to a $p_{3/2}Y_0{}^*$ resonance, may or may not occur.

From Eq. (27) it is seen that the maximum attraction in the tenfold representation, given by the factor R, occurs at $f=\frac{1}{4}$. This value of f predicts the unitary singlet $Y_0{}^*$ resonance at exactly the same location. These two cases differ, however, in that the resonance position given by R is relatively insensitive to variations of f about the value $\frac{1}{4}$, while the position determined by S is strongly dependent on these variations.

It remains to be verified that the results of this idealized model, particularly with regard to the values of f which allow a resonant structure in the states of the tenfold representation, hold as well when the actual particle masses are incorporated into the calculation. This question, as well as the related questions of resonance locations, widths, and branching ratios, is treated in the following section.

V. NUMERICAL RESULTS AND DISCUSSION

The symmetry model analysis described above has the virtue of explicitly showing the possibility, for certain ranges of f, that resonant structures can occur in different sets of scattering amplitudes. It is clear, however, that the mass differences between the elementary particles cannot reasonably be neglected in the low-energy physical region. Not only are the positions and widths of resonances and the relative branching ratios dependent on the actual particle masses, but also the location of the dynamical singularities[24] as is evident from Eq. (14). In this section we apply the relativistic formalism developed in Secs. II and III to the various

[24] From the considerations discussed in Sec. III, the subtraction points for the different strangeness states were chosen to be: $S=+1$, 5.25 μ; $S=0$, $N=6.80\ \mu$; $S=-1$, $\Sigma=8.57\ \mu$; $S=-2$, $\Xi=9.57\ \mu$; $S=-3$, 11.1 μ.

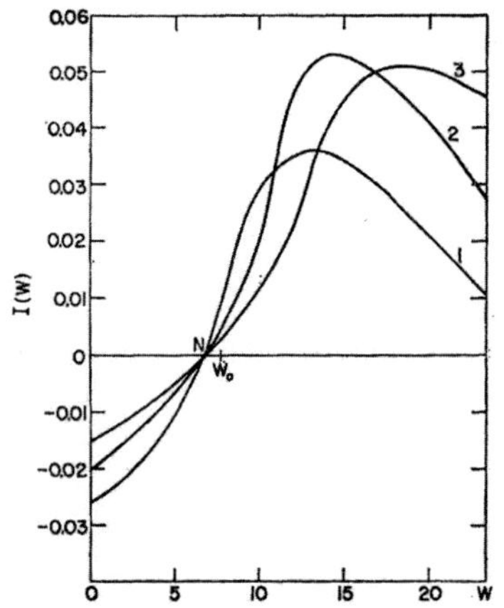

FIG. 4. The elastic integrals necessary in the coupled system with strangeness $S=0$, plotted as functions of the total energy W. Curves 1, 2, and 3 refer to $I(N\pi,N,N\pi)$, $I(N\eta,N,N\eta)$, and $I(\Sigma K,\Xi,\Sigma K)$, respectively. $I(\Lambda K,\Xi,\Lambda K)$ differs very little from curve 3, and has not been plotted. The subtraction point N is the nucleon mass, $6.80\,\mu$, and W_0 is the $N\pi$ threshold, $7.80\,\mu$.

scattering matrices composed of the two-particle states listed in Table I. The only input from the unitary symmetry model, as mentioned in the introduction, is the choice of coupling constants; and the parameter f arising from that choice is the only arbitrary constant appearing in our calculation.

The first point to be made is that the ranges for f that lead to resonant structures [Eq. (28)] in the appropriate amplitudes are not significantly modified by the introduction of the actual particle masses. This is most easily illustrated by the following treatment of the symmetrized N/D method [Eq. (16)]. We rewrite the matrix equation for D in the condensed notation

$$D=1-\tfrac{1}{2}I-\tfrac{1}{2}N^{-1}\tilde{I}N,$$

where I represents the matrix of integrals and $\tilde{I}$ is its transpose. Then defining the antisymmetric matrix R as $R=\tilde{I}N-NI$, we obtain the equivalent form

$$D=1-I-\tfrac{1}{2}N^{-1}R.$$

In the "equal mass" model of Sec. IV the matrix R is identically zero. When the correct particle masses are used in determining R, it is found that this term makes a negligible contribution to the determinant of D.

Thus, the primary modification in determining the resonance conditions is due to the variation of the integrals for different particle states and subtraction points. Several examples of the integrals involved in the calculation are plotted in Figs. 4 and 5. Figure 4, which contains three of the integrals required in the analysis of pion-nucleon scattering in the $p_{3/2}$ partial-wave state, is intended as an illustration of the dependence on the particle state when the subtraction point is held fixed. Figure 5, on the other hand, contains the three integrals necessary in the five single-channel calculations (Table I) and illustrates the dependence upon the choice of subtraction point as well as the particle state. The integrals, which are of the principal-value type above the physical threshold, were evaluated on a computer without approximating either the phase space factor [Eq. (11)] or the Born approximation term [Eq. (13)].

It is worthwhile to point out the following features of the integrals. As would be expected, there is little difference in the low energy region between integrals involving the Λ mass and those with the corresponding dependence on the Σ mass. Second, integrals with a higher physical threshold and the same subtraction point lie below those (in the low energy region) with a lower threshold. This conforms, in part, with expectations regarding the relative importance of inelastic channels with high thresholds. But at the same time it is clear from Fig. 4 that the ΣK channel cannot meaningfully be ignored in the calculation of the 3, 3 pion-nucleon resonance. Third, the integrals for particle states involving the heavier mesons attain a greater maximum value than do those for the states containing pions. Fourth, the integrals for inelastic amplitudes do not differ appreciably from the related "elastic" integrals and, in the low-energy region, their behavior is governed by the particle state associated with the phase space factor. Finally, the integrals tend logarithmically to $-\infty$ as W tends to ∞ along the positive real axis. For consistency, we have verified that no ghosts occur near the physical region.

Incorporating the computed integrals into the symmetrized N/D formalism and evaluating the determinant of the real part of D, we obtain resonant structures in the four scattering matrices corresponding to the tenfold representation. The calculated locations of the resonances and (in parentheses) the experimentally

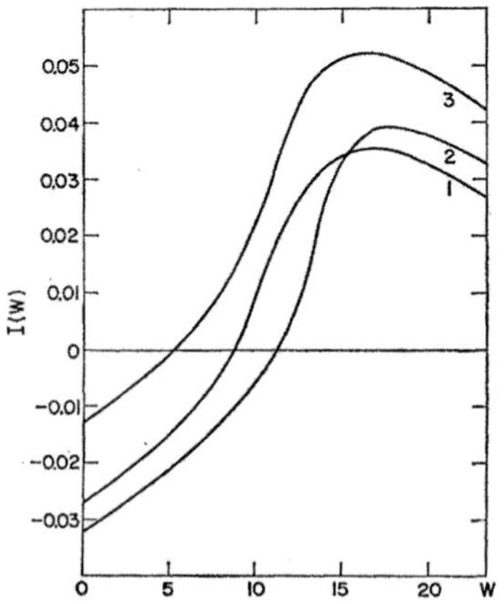

FIG. 5. The integrals necessary in the single channel cases plotted as functions of W. Curves 1, 2, and 3 refer to $I(\Sigma\pi,\Sigma,\Sigma\pi)$, $I(\Xi\bar{K},\Sigma,\Xi\bar{K})$, and $I(NK,\Sigma,NK)$, respectively. The corresponding integrals with the intermediate Λ particle differ very little from the above curves.

observed positions are

$$\begin{array}{lll} N_{3,3}^* & 10.1\,\mu & (9.0\,\mu) \\ Y_1^* & 11.6\,\mu & (10.0\,\mu) \\ \Xi^* & 12.6\,\mu & (11.1\,\mu) \\ Z^- & 13.6\,\mu & (?). \end{array} \quad (29)$$

In calculating these values we have taken $f=\frac{1}{4}$ which corresponds to the lowest possible resonance positions in our model. For f in the range $\frac{1}{6}$ to $\frac{1}{2}$, no resonances are predicted in the other strangeness and isotopic spin systems except the $S=-1$, $T=0(Y_0^*)$ state. As mentioned in the previous section, the position of this unitary singlet resonance is strongly dependent on the value of f, and the resonance vanishes for f greater than roughly $\frac{1}{3}$. With $f=\frac{1}{4}$, the position of the $p_{3/2}Y_0^*$ is predicted to be very close to that of the Y_1^* resonance, namely, $11.6\,\mu$.

Concentrating on the pion-nucleon system for the moment, we note that the inclusion of the ΣK channel lowers the position of the 3, 3 resonance by roughly a pion mass in this model. Furthermore, the width of the resonance is computed to be $\Gamma/2=73$ MeV if we again arbitrarily evaluate the phase space factor at the observed position. This value is to be compared with the 100 MeV half-width obtained in the single-channel model. Both of these effects strongly suggest that the inelastic channels will be of importance in more exact treatments of pion-nucleon scattering. In addition, the effect of the two-particle inelastic channels in the $T=\frac{1}{2}$ system is to reduce the magnitude of the small negative $N\pi$ phase shift and improve the agreement with experiment.[25]

In calculating widths and branching ratios for the remaining $p_{3/2}$ resonances, we consistently evaluate the sensitive phase space terms at the observed resonance locations. The $T=\frac{1}{2}$ cascade-pion resonance width is computed to be $\Gamma/2=5.8$ MeV in this model, in close agreement with experiment.[1] For the Y_1^*, with $f=\frac{1}{4}$, the branching ratios into the $\Lambda\pi$ and $\Sigma\pi$ channels were calculated both for the equal mass model of Sec. IV and the correct mass treatment of this section. In the first case, we obtain $\Lambda\pi$ 91%, $\Sigma\pi$ 9%, while introducing the actual particle masses modifies this to $\Lambda\pi$ 94%, $\Sigma\pi$ 6%, in relatively good agreement with experiment.[26] The width of the Z^- resonance was not evaluated because of the lack of experimental data, while the Y_0^* width was omitted because of its sensitive dependence on the parameter f.

It is, of course, difficult to gauge the effects of dynamical singularities not included in our approximate model. However, qualitative remarks concerning the influence of such diagrams as single ρ or ω vector meson exchange between the baryon and meson lines can be made if ρ and ω are assumed to couple, respectively, to the conserved isotopic spin current and the conserved hypercharge current. Using the simple rules[27] that ρ exchange is repulsive when the isotopic spins are parallel, and attractive otherwise, and that like hypercharges repel in the case of ω exchange while unlike hypercharges attract, we may draw the following rough conclusions. In the pion-nucleon system the ρ exchange is not sufficiently repulsive in the $T=\frac{3}{2}$ state to prevent the 3, 3 resonance from occurring, nor is it attractive enough in the $T=\frac{1}{2}$ state to make the $p_{3/2}$ phase shift positive in the low-energy region.

Although the vector meson exchange terms are not the dominant interactions in the $p_{3/2}$ amplitudes, they are likely to be important in determining resonance positions. In this regard we note that the vector meson terms are attractive in both the Y_1^* and Ξ^* systems, as contrasted with the repulsion in the $N_{3,3}^*$ system noted above. In the Z^- amplitude we may suppose the vector meson effects to be small because of a cancellation between the ρ and ω. On the basis of these extremely sketchy arguments it is interesting to speculate on the possibility of a dispersion theoretic verification of the "equal spacing" rule[28] for the resonances of the tenfold representation. The resonance locations obtained in this calculation [Eq. (29)] show a general equal spacing structure which might easily be improved by including the vector meson contributions.

Finally, we wish to point out two possible experiments relating to the coupling constant parameter f. The first concerns the obvious question of the existence of a Y_0^* resonance in the $p_{3/2}$ state. In particular, if the known Y_0^* at the total energy $10.2\,\mu$ turns out to have these quantum numbers, then the value $f=\frac{1}{4}$ would seem very reasonable. If, however, the predicted resonance is found much higher or does not exist at all, then f must be made larger in this model. As mentioned previously, the prediction of the tenfold family of resonances is relatively insensitive to such adjustments of the coupling constant parameter.

An independent experimental check on f follows from the observation that in this model, if the Σ-Λ mass difference is ignored, the effect of the single baryon exchange term in the $T=0$ NK system is very sensitive to f. For f greater than $\frac{1}{4}$ (and less than 1) the interaction is repulsive in the $p_{3/2}$ state, while values of f less than $\frac{1}{4}$ lead to an attractive interaction. On the basis of rough arguments, we expect the vector meson exchange diagrams effectively to cancel each other in this isotopic spin state. Hence, the sign of the $p_{3/2}$ phase shift in the low-energy region may indicate the appropriate value of f. Similarly, if these oversimplified considerations are valid, we would expect the $p_{1/2}$ phase shift for $T=0$ NK

[25] W. D. Walker, W. D. Shephard, and J. Davis, Phys. Rev. **118**, 1612 (1960).

[26] M. Alston, L. W. Alvarez, P. Eberhard, M. L. Good, W. Graziano, H. K. Ticho, and S. G. Wojcicki, Phys. Rev. Letters **5**, 520 (1960).

[27] J. J. Sakurai, Ann. Phys. (N. Y.) **11**, 1 (1960).

[28] M. Gell-Mann, in *Proceedings of the 1962 International Conference on High-Energy Physics at CERN* (CERN, Geneva, 1962).

scattering to have the opposite sign to that of the $p_{3/2}$ state.

Of course, if the Ξ^* resonance is not in the $p_{3/2}$ state, or if the predicted Z^- is not found, then the basic assumption of this paper, namely, the octet-model expressions for the Yukawa-type coupling constants, must be re-examined. On the other hand, if these predictions are borne out experimentally, then one may feel reasonably confident that the coupling-constant expressions evaluated in the neighborhood of $f=\frac{1}{4}$ will remain meaningful in more refined dynamical treatments. When the usual factor of 4π is absorbed into the coupling constants and the definitions of Appendix I are used, the numerical values obtained with $f=\frac{1}{4}$ are

$$g_{N\pi}^2=h_{\Sigma K}^2=g^2\approx 15,$$
$$g_{\Lambda\pi}^2=g_{\Lambda K}^2=g_{\Lambda\eta}^2=g_{\Sigma\eta}^2=g_{\Xi\eta}^2=\tfrac{3}{4}g^2\approx 11,$$
$$g_{\Sigma\pi}^2=g_{\Xi\pi}^2=g_{\Sigma K}^2=\tfrac{1}{4}g^2\approx 4,$$
$$h_{\Lambda K}^2=g_{N\eta}^2=0.$$

These values appear to be consistent with the experimental data now available.

In conclusion we would like to thank Dr. J. Uretsky and Professor J. J. Sakurai for several helpful discussions.

APPENDIX I

Octet-Model Coupling Constants

In order to establish the notation for coupling constants used in this paper, we write the interaction Lagrangian density for baryons and pseudoscalar mesons in the form

$$\mathcal{L}_{\rm int}=(4\pi)^{1/2}[g_{N\pi}\boldsymbol{\pi}\cdot\bar{N}\boldsymbol{\tau}N+g_{\Lambda\pi}(\boldsymbol{\pi}\cdot\bar{\Lambda}\boldsymbol{\Sigma}+\text{H.c.})-ig_{\Sigma\pi}\boldsymbol{\pi}\cdot\bar{\boldsymbol{\Sigma}}\times\boldsymbol{\Sigma}+g_{\Xi\pi}\boldsymbol{\pi}\cdot\bar{\Xi}\boldsymbol{\tau}\Xi+g_{\Lambda K}(\bar{N}\Lambda K+\text{H.c.})+g_{\Sigma K}(\bar{N}\boldsymbol{\tau}\cdot\boldsymbol{\Sigma}K+\text{H.c.})$$
$$+h_{\Lambda K}(\bar{\Xi}\Lambda K^c+\text{H.c.})+h_{\Sigma K}(\bar{\Xi}\boldsymbol{\tau}\cdot\boldsymbol{\Sigma}K^c+\text{H.c.})+g_{N\eta}\eta\bar{N}N+g_{\Lambda\eta}\eta\bar{\Lambda}\Lambda+g_{\Sigma\eta}\eta\bar{\boldsymbol{\Sigma}}\cdot\boldsymbol{\Sigma}+g_{\Xi\eta}\eta\bar{\Xi}\Xi],$$

where the space-time dependence of the interaction has been suppressed. The particle symbols refer to the field operators in the standard way and H.c. means Hermitian conjugate. The charge structure is contained in the conventional isotopic spin representation, where

$$K^c=-i\tau_2K^*=\begin{pmatrix}-\bar{K}^0\\ K^-\end{pmatrix}$$

is the isotopic spinor of the second kind for the $\bar{K}$ meson. It is convenient to include the factor of $(4\pi)^{1/2}$ in the definition of the coupling constants so that, for example, we have $g_{N\pi}^2\approx 15$ as the experimental value for the pion-nucleon coupling constant.

In the Gell-Mann–Ne'eman octet model for the strongly interacting particles, there are only two independent ways,[10] known as D- and F-type coupling, in which to write Yukawa couplings which are invariant under unitary-symmetry transformations. It follows that the most general form for these couplings is an arbitrary linear combination of the D- and F-type expressions. If we let dg be the coefficient of the D-type form, and fg that for the F-type form, then the resulting expression for the pion-nucleon coupling constant is $g_{N\pi}=(d+f)g$. Without loss of generality we may choose $g=(15)^{1/2}$, in which case experiment restricts the coefficients to $(d+f)=1$.

The combination, therefore, of the known value of $g_{N\pi}^2$ and the octet-model assumption limits the twelve Yukawa-type coupling constants to a one parameter set of expressions. Setting $d=1-f$,[29] we obtain the following representation for the baryon-pseudoscalar meson coupling constants:

$$g_{N\pi}=g,\qquad g_{\Lambda K}=-(1/\sqrt{3})(1+2f)g,\qquad g_{N\eta}=-(1/\sqrt{3})(1-4f)g,$$
$$g_{\Lambda\pi}=(2/\sqrt{3})(1-f)g,\qquad g_{\Sigma K}=(1-2f)g,\qquad g_{\Lambda\eta}=-(2/\sqrt{3})(1-f)g,$$
$$g_{\Sigma\pi}=2fg,\qquad h_{\Lambda K}=-(1/\sqrt{3})(1-4f)g,\qquad g_{\Sigma\eta}=(2/\sqrt{3})(1-f)g,$$
$$g_{\Xi\pi}=-(1-2f)g,\qquad h_{\Sigma K}=-g,\qquad g_{\Xi\eta}=-(1/\sqrt{3})(1+2f)g.$$

APPENDIX II

Isotopic Spin Factors

The isotopic spin structure given in the interaction Lagrangian density of Appendix I allows one to determine easily the appropriate coefficients for the Born approximation terms arising from the single baryon exchange diagram (Fig. 2). We list the isotopic spin factors for each strangeness and total isotopic spin state, using the notation (BC,A,DE) to represent the particles involved in each diagram (cf., footnote 14). The symmetry of the matrices permits us to suppress half of the off-diagonal terms in each case. Finally, for the single channel cases, we use the notation (BC,BC) to represent the total isotopic spin amplitude for the Born approximation contribution.

[29] It should be noted that the parameter f is related to α in Gell-Mann's notation (reference 9) by $f=1-\alpha$.

The results are as follows:

$S=1,\ T=0$

$$(NK,NK)=3(NK,\Sigma,NK)-(NK,\Lambda,NK),$$

$S=1,\ T=1$

$$(NK,NK)=(NK,\Sigma,NK)+(NK,\Lambda,NK),$$

$S=0,\ T=\frac{1}{2}$

$$\begin{array}{cc} & \begin{array}{cccc} N\pi & \Lambda K & \Sigma K & N\eta \end{array} \\ \begin{array}{c} N\pi \\ \Lambda K \\ \Sigma K \\ N\eta \end{array} & \left[\begin{array}{cccc} -(N\pi,N,N\pi) & \sqrt{3}(N\pi,\Sigma,\Lambda K) & 2(N\pi,\Sigma,\Sigma K)+(N\pi,\Lambda,\Sigma K) & \sqrt{3}(N\pi,N,N\eta) \\ & (\Lambda K,\Xi,\Lambda K) & -\sqrt{3}(\Lambda K,\Xi,\Sigma K) & (\Lambda K,\Lambda,N\eta) \\ & & -(\Sigma K,\Xi,\Sigma K) & \sqrt{3}(\Sigma K,\Sigma,N\eta) \\ & & & (N\eta,N,N\eta) \end{array}\right] \end{array},$$

$S=0,\ T=\frac{3}{2}$

$$\begin{array}{cc} & \begin{array}{cc} N\pi & \Sigma K \end{array} \\ \begin{array}{c} N\pi \\ \Sigma K \end{array} & \left[\begin{array}{cc} 2(N\pi,N,N\pi) & (N\pi,\Lambda,\Sigma K)-(N\pi,\Sigma,\Sigma K) \\ & 2(\Sigma K,\Xi,\Sigma K) \end{array}\right] \end{array},$$

$S=-1,\ T=0$

$$\begin{array}{cc} & \begin{array}{cccc} N\bar{K} & \Sigma\pi & \Xi K & \Lambda\eta \end{array} \\ \begin{array}{c} N\bar{K} \\ \Sigma\pi \\ \Xi K \\ \Lambda\eta \end{array} & \left[\begin{array}{cccc} 0 & \sqrt{6}(N\bar{K},N,\Sigma\pi) & (N\bar{K},\Lambda,\Xi K)-3(N\bar{K},\Sigma,\Xi K) & \sqrt{2}(N\bar{K},N,\Lambda\eta) \\ & (\Sigma\pi,\Lambda,\Sigma\pi)-2(\Sigma\pi,\Sigma,\Sigma\pi) & \sqrt{6}(\Sigma\pi,\Xi,\Xi K) & \sqrt{3}(\Sigma\pi,\Sigma,\Lambda\eta) \\ & & 0 & \sqrt{2}(\Xi K,\Xi,\Lambda\eta) \\ & & & (\Lambda\eta,\Lambda,\Lambda\eta) \end{array}\right] \end{array},$$

$S=-1,\ T=1$

$$\begin{array}{cc} & \begin{array}{ccccc} N\bar{K} & \Sigma\pi & \Lambda\pi & \Xi K & \Sigma\eta \end{array} \\ \begin{array}{c} N\bar{K} \\ \Sigma\pi \\ \Lambda\pi \\ \Xi K \\ \Sigma\eta \end{array} & \left[\begin{array}{ccccc} 0 & 2(N\bar{K},N,\Sigma\pi) & \sqrt{2}(N\bar{K},N,\Lambda\pi) & -(N\bar{K},\Lambda,\Xi K)-(NK,\Sigma,\Xi K) & \sqrt{2}(N\bar{K},N,\Sigma\eta) \\ & (\Sigma\pi,\Sigma,\Sigma\pi)-(\Sigma\pi,\Lambda,\Sigma\pi) & -\sqrt{2}(\Sigma\pi,\Sigma,\Lambda\pi) & 2(\Sigma\pi,\Xi,\Xi K) & \sqrt{2}(\Sigma\pi,\Sigma,\Sigma\eta) \\ & & (\Lambda\pi,\Sigma,\Lambda\pi) & \sqrt{2}(\Lambda\pi,\Xi,\Xi K) & (\Lambda\pi,\Lambda,\Sigma\eta) \\ & & & 0 & \sqrt{2}(\Xi K,\Xi,\Sigma\eta) \\ & & & & (\Sigma\eta,\Sigma,\Sigma\eta) \end{array}\right] \end{array},$$

$S=-1,\ T=2$

$$(\Sigma\pi,\Sigma\pi)=(\Sigma\pi,\Lambda,\Sigma\pi)+(\Sigma\pi,\Sigma,\Sigma\pi),$$

$S=-2,\ T=\frac{1}{2}$

$$\begin{array}{cc} & \begin{array}{cccc} \Xi\pi & \Lambda\bar{K} & \Sigma\bar{K} & \Xi\eta \end{array} \\ \begin{array}{c} \Xi\pi \\ \Lambda\bar{K} \\ \Sigma\bar{K} \\ \Xi\eta \end{array} & \left[\begin{array}{cccc} -(\Xi\pi,\Xi,\Xi\pi) & \sqrt{3}(\Xi\pi,\Sigma,\Lambda\bar{K}) & 2(\Xi\pi,\Sigma,\Sigma\bar{K})+(\Xi\pi,\Lambda,\Sigma\bar{K}) & \sqrt{3}(\Xi\pi,\Xi,\Xi\eta) \\ & (\Lambda\bar{K},N,\Lambda\bar{K}) & -\sqrt{3}(\Lambda\bar{K},N,\Sigma\bar{K}) & (\Lambda\bar{K},\Lambda,\Xi\eta) \\ & & -(\Sigma\bar{K},N,\Sigma\bar{K}) & \sqrt{3}(\Sigma\bar{K},\Sigma,\Xi\eta) \\ & & & (\Xi\eta,\Xi,\Xi\eta) \end{array}\right] \end{array},$$

$S=-2,\ T=\frac{3}{2}$

$$\begin{array}{cc} & \begin{array}{cc} \Xi\pi & \Sigma\bar{K} \end{array} \\ \begin{array}{c} \Xi\pi \\ \Sigma\bar{K} \end{array} & \left[\begin{array}{cc} 2(\Xi\pi,\Xi,\Xi\pi) & (\Xi\pi,\Lambda,\Sigma\bar{K})-(\Xi\pi,\Sigma,\Sigma\bar{K}) \\ & 2(\Sigma\bar{K},N,\Sigma\bar{K}) \end{array}\right] \end{array},$$

$S=-3,\ T=0$

$$(\Xi\bar{K},\Xi\bar{K})=3(\Xi\bar{K},\Sigma,\Xi\bar{K})-(\Xi\bar{K},\Lambda,\Xi\bar{K}),$$

$S=-3,\ T=1$

$$(\Xi\bar{K},\Xi\bar{K})=(\Xi\bar{K},\Sigma,\Xi\bar{K})+(\Xi\bar{K},\Lambda,\Xi\bar{K}).$$

Breaking of SU(3) Symmetry in the $\frac{3}{2}^+$ Meson-Baryon Decuplet

K. C. Wali*
Argonne National Laboratory, Argonne, Illinois

AND

Robert L. Warnock
Illinois Institute of Technology, Chicago, Illinois

(Received 14 May 1964)

Breaking of SU(3) invariance is studied in the case of the meson-baryon decuplet with $J^P=\frac{3}{2}^+$. In a simple theory based on single-baryon exchange, the evolution of the broken decuplet from a degenerate origin is traced in detail. Symmetry breaking is introduced by allowing initially degenerate meson and baryon masses to approach continuously their physical values, while obeying at every stage the Gell-Mann–Okubo sum rules. The following conclusions are reached: (i) The Okubo equal spacing rule for the decuplet levels follows from the validity of the Gell-Mann–Okubo rules for the meson and baryon octuplets. (ii) The main effect of the symmetry breaking may be characterized as a mixing of the 10- and 27-dimensional representations. The mixing is small enough so that the resonances can be unambiguously associated with the 10 representation, but at the same time large enough to imply coupling-constant ratios differing appreciably from the values for pure symmetry. (iii) In the approach to pure symmetry through reduction of mass differences, there are no difficulties of the type pointed out by Oakes and Yang. Resonances cross thresholds smoothly, and a degenerate decuplet of bound states is obtained in the limit.

1. INTRODUCTION

MESON-BARYON resonances with $J^P=\frac{3}{2}^+$ have been assigned tentatively[1] to the (3,0) decuplet representation of the group SU(3).[2] The resonant states in question are $N_{3/2}^*$ (1238 MeV, $T=\frac{3}{2}$, $Y=1$), Y_1^* (1385 MeV, $T=1$, $Y=0$), $\Xi_{1/2}^*$ (1530 MeV, $T=\frac{1}{2}$, $Y=-1$). It now appears likely that these states all have the correct $\frac{3}{2}^+$ spin-parity values.[3] To complete the decuplet, a particle Ω_0 (sometimes called Ω_-) with $T=0$, $Y=-2$ was predicted. The recent discovery[4] of such a particle constitutes strong evidence for both the decuplet assignment and the general scheme of the "eightfold way."[2] The discovery is all the more remarkable, since the observed mass (1686±12 MeV) of Ω_0 agrees very well with the prediction of the Gell-Mann–Okubo mass formula.[2,5] For the (3,0) decuplet, the

* Work performed under the auspices of the U. S. Atomic Energy Commission.

1 R. E. Behrends, J. Dreitlein, C. Fronsdal, and B. W. Lee, Rev. Mod. Phys. 34, 1 (1962); S. Glashow and J. J. Sakurai, Nuovo Cimento 26, 622 (1962); M. Gell-Mann, *Proceedings of the 1962 International Conference on High Energy Physics at CERN*, edited by J. Prentki (CERN, Geneva, 1962); R. Cutkosky, J. Kalckar, and P. Tarjanne, Phys. Letters 1, 93 (1962); R. H. Capps, Nuovo Cimento 27, 1208 (1963).

2 M. Gell-Mann, Phys. Rev. 125, 1067 (1962); California Institute of Technology Report CTSL-20, 1961 (unpublished); Y. Ne'eman, Nucl. Phys. 26, 222 (1961).

3 A summary of the experiments and a bibliography is given by R. H. Dalitz, Ann. Rev. Nucl. Sci. 13, 339 (1963).

4 V. E. Barnes, P. L. Connolly, D. J. Grennell, B. B. Culwick, *et al.*, Phys. Rev. Letters 12, 204 (1964). We prefer the notation Ω_0 to Ω_-, since the isotopic spin has been used as a subscript for the other decuplet states. The Ω_0 notation appears also in Ref. 7.

5 S. Okubo, Progr. Theoret. Phys. (Kyoto) 27, 949 (1962).

latter implies that the masses should be a linear function of Y; i.e., that they should be equally spaced.

The agreement of experiment with this equal spacing rule, as well as the agreement of the pseudoscalar meson and baryon masses with the analogous rule for the octuplet representation, deserves close scrutiny from the theoretical side. Gell-Mann and Okubo assume that the perturbation which upsets SU(3) symmetry has the transformation properties of the $T=0$, $Y=0$ member of an SU(3) octuplet, and that departures from symmetry can be calculated as first-order effects of this perturbation. Although this hypothesis results in a good description of mass splittings within the 0^- and $\frac{1}{2}^+$ octuplets and the $\frac{3}{2}^+$ decuplet, one should still ask whether it will be sufficient to describe *all* major departures from strict SU(3) symmetry. For example, may the perturbations of coupling constants be calculated along the same lines as mass perturbations? Putting the matter another way, the sense in which the perturbation is to be regarded as weak needs clarification.

In this paper, we study the symmetry breaking of the decuplet in several of its aspects. Our method is to regard the mass splittings within the decuplet as a direct consequence of the mass splittings within the meson and baryon octuplets. For the latter, we insert the observed masses. This procedure is carried out within a simple theory in which the meson-baryon composite states of the decuplet are produced by single-baryon exchange forces. The theory amounts to a simplification of the model used by Martin and Wali.[6] The simplification is mainly the recognition that the meson and baryon mass differences have their strongest influence in centrifugal barrier effects. With an appropriate choice of coupling parameters, the observed decuplet mass splittings can be reproduced. Thus, without using a perturbation method to calculate the splittings, we conclude that the validity of the mass formulas for the meson and baryon octuplets implies the validity of the formula for the decuplet. This confirms a result of Tarjanne and Cutkosky[7] which is based on a similar model, but on a relatively rough calculation. We also show that from an appropriate point of view the perturbation can be regarded as weak, at least in the problem of mass shifts. Specifically, the off-diagonal perturbations of the initially diagonal D matrix can almost be neglected. On the other hand, when we calculate the perturbations of the coupling constants of decuplet states to their constituents, we find that the perturbation can no longer be regarded as weak. The departures from pure symmetry are pronounced, and show very little resemblance to the formulas of Dullemond, Macfarlane, and Sudarshan.[8] The departures from symmetry are such as to improve the agreement with experiment, particularly with regard to the branching ratio of Y_1^* decays. An alternative way to judge the degree of symmetry breaking is to calculate all transition amplitudes in the basis of those states which, in their degenerate limits, transform according to pure group representations. Evaluation of such amplitudes at the resonance energies shows that group invariance arguments are quite adequate to predict which amplitudes will be large and which small. However, the relative magnitudes of small amplitudes are not well predicted. The amplitudes to which we refer are T matrix elements divided by the initial and final momenta.

In order to trace the development of the broken multiplet from its degenerate origin, we perform a gradual reduction of the meson and baryon mass differences, while maintaining the corresponding Gell-Mann–Okubo relations. As one expects, the Okubo relation for the decuplet is satisfied with more and more precision as the symmetry breaking is reduced. The degenerate level is a bound state with a binding energy of about one pion mass. The coupling constant ratios do not approach their pure symmetry values very rapidly.

We are especially concerned with some criticisms of the decuplet assignment raised by Oakes and Yang.[9] These authors have posed the following questions: (i) If one imagines that the symmetry breaking perturbation is gradually reduced, can the poles representing the decuplet states move in some reasonably simple way to arrive at a common real or complex energy in the degenerate limit? (ii) Is there any theoretical basis for applying the Okubo formula to the decuplet? (iii) Is the symmetry breaking slight enough so that group invariance arguments are meaningful? Our answer to the first two questions is yes, contrary to the contentions of Oakes and Yang. With regard to question (iii), our model shows clearly that group considerations do play a useful role. However, group arguments must be used with caution, particularly in the derivation of coupling constant ratios.

In Sec. 2 we review the structure of the Riemann surface of the many-channel scattering matrix. The ND^{-1} representation is employed.[10] An examination of the motion of poles as symmetry breaking is reduced shows that when a resonance crosses a threshold there are inevitably two poles at complex energies near that threshold. One lies on the sheet reached from the physical sheet by crossing the cut above the threshold, and the other on the sheet reached by crossing the cut below the threshold. One pole causes the resonance peak when the peak is above the threshold, and the other,

---

[6] A. W. Martin and K. C. Wali, Phys. Rev. **130**, 2455 (1963).
[7] P. Tarjanne and R. E. Cutkosky, Phys. Rev. **133**, B1292 (1964); see also S. Frautschi, Phys. Letters **8**, 141 (1964).
[8] C. Dullemond, A. J. Macfarlane, and E. C. G. Sudarshan, Phys. Rev. Letters **10**, 423 (1963); see also E. C. G. Sudarshan, *Proceedings of the Athens Topical Conference on Recently Discovered Resonance Particles* (Ohio University Press, Athens, Ohio, 1963). *Note added in proof*. The formulas of Dullemond *et al.*, have been criticized recently by V. Gupta and V. Singh (to be published).
[9] R. J. Oakes and C. N. Yang, Phys. Rev. Letters **11**, 174 (1963).
[10] R. L. Warnock, Nuovo Cimento **32**, 255 (1964).

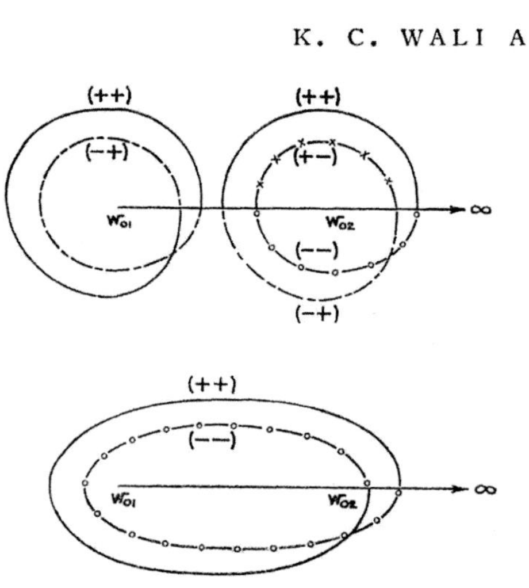

FIG. 1. Diagrams of the course of a point moving over the Riemann surface of a two-channel scattering matrix. The plus and minus signs indicate the four different sheets through which the point may move, as explained in the text.

when it is below. In this conclusion we agree with many other authors who have discussed the problem in different formalisms.[11] In Sec. 3 our model and some methods of calculation are described. Section 4 contains the main results of a numerical evaluation of the theory. We tabulate resonance energies, coupling constants, matrix elements, and so on. In the Appendix a fictitious two-channel problem is analyzed in order to illustrate more concretely the behavior of poles. The model involves a resonance which is first located between the two thresholds, and which crosses the upper threshold as the perturbation is turned off. There is a pair of complex conjugate poles on each of the three unphysical sheets. A calculation shows that the poles have nearly the same position on all three sheets, so it is indeed appropriate to refer to the poles as "shadows" of one another.[12]

2. THE RIEMANN SURFACE AND THE MOTION OF POLES

We are concerned with spin-0 particles scattered from spin-$\frac{1}{2}$ targets, so it is appropriate to work in the complex plane of w, the energy in the center-of-mass frame.[13] We neglect channels with more than two particles. In the w plane, a single partial-wave scattering matrix $T(w)=[T_{ij}^{\gamma}(w)]$, $\gamma=(J,T,Y)$, yields both orbital states $l=J\pm\frac{1}{2}$ corresponding to a given total angular momentum J. T has left and right "physical" branch cuts $(-\infty,-w_0)$ and (w_0,∞), where w_0 is the energy of the lowest state having quantum numbers γ. We denote these cuts collectively by the letter P. By MacDowell's relation,[13]

$$f_{l=J-1/2}(w)=T(w+i0)\,,\qquad w>w_0\,,$$
$$f_{l=J+1/2}(w)=-T(-w-i0)\,,\quad w>w_0\,,\qquad (2.1)$$

where T has been normalized so that in the single-channel case $f_l(w)$ reduces to $f_l=\sin\delta_l\exp(i\delta_l)/q$; here q is the center-of-mass momentum. The structure of the Riemann surface is very clearly brought out by means of the matrix ND^{-1} representation.[14] Suppressing the index γ, we have $T=ND^{-1}$, where $N=[N_{ij}(w)]$ is analytic in a region including the physical cuts P, and[15]

$$D_{ij}(w)=\delta_{ij}-\frac{w-\hat{w}}{\pi}\int_P\frac{\rho_i(w')N_{ij}(w')dw'}{(w'-\hat{w})(w'-w)}=\delta_{ij}+I_{ij}.\qquad (2.2)$$

The factor ρ_i is equal to the i channel momentum for $|w|$ greater than the i-channel threshold, and is zero otherwise.

$$\rho_i(w)=q_i(|w|)\theta(|w|-w_{0i})\,.\qquad (2.3)$$

Because of the analyticity of $N(w)$ around P, we learn all about the sheets which are connected along P by direct inspection of Eq. (2.2). For the problem at hand we are concerned only with the $p_{3/2}$ state, which is associated with the right physical cut (w_0,∞). Each row of the D matrix has just two sheets which are connected by this right cut. We denote these two sheets by the symbols $(+)$ and $(-)$. The representations of the ith row of D on these two sheets are

$$D_{ij}^{(+)}(w)=\delta_{ij}+I_{ij}(w)\,,$$
$$D_{ij}^{(-)}(w)=\delta_{ij}+2iq_i(w)N_{ij}(w)+I_{ij}(w)\,,\qquad (2.4)$$

where $j=1,2,\cdots n$ and I_{ij} is given by Eq. (2.2). Taking into account $q_i(w+i0)=-q_i(w-i0)$, $w>w_{0i}$, it is easy to check that $D_{ij}^{(+)}(w\pm i0)=D_{ij}^{(-)}(w\mp i0)$. Since $D_{ij}^{(+)}$ and $D_{ij}^{(-)}$ are both analytic in the cut plane, it follows that one is the analytic continuation of the other, and that only two sheets of D_{ij} are connected through the right cut. Thus, if there are n channels, T has 2^n sheets connected through the right cut, corresponding to all possible choices for the rows of the D matrix.

In the two-channel case the situation can be visualized easily in the diagram shown in Fig. 1. The notation for sheets follows a suggestion by Yang.[16] Sheet $(+-)$

[11] M. Ross, Phys. Rev. Letters **11**, 450 (1963); R. J. Eden and J. R. Taylor, Phys. Rev. Letters **11**, 516 (1963); M. Nauenberg and J. C. Nearing, *ibid.* **12**, 63 (1964); R. H. Dalitz and G. Rajasekaran, Phys. Letters **5**, 373 (1963); G. Rajasekaran, Nuovo Cimento **31**, 697 (1964); C. R. Hagen, Phys. Rev. Letters **12**, 153 (1964); D. Amati, Phys. Letters **7**, 290 (1963).

[12] R. J. Eden and J. R. Taylor, Phys. Rev. **133**, B1575 (1964).

[13] S. W. MacDowell, Phys. Rev. **116**, 774 (1960); W. R. Frazer and J. R. Fulco, *ibid.* **119**, 1420 (1960); G. Frye and R. L. Warnock, *ibid.* **130**, 478 (1963).

[14] J. D. Bjorken, Phys. Rev. Letters **4**, 473 (1960).

[15] We neglect the possibility that "unphysical" branch points (i.e., those not associated with unitarity in the direct channel) might lie on the physical cuts P. This neglect is not really justified in all the problems treated in this paper; for example, the branch point associated with two-pion exchange in $\bar{K}$-N scattering appears above the π-Λ threshold in all $T=1$, $Y=0$ amplitudes. This problem can be overcome by deforming the paths of integrals in the D matrix so as to dodge the branch points. The model described in Sec. 3 has no such branch points.

[16] C. N. Yang, Proceedings of the Argonne User's Group (to be published).

is the sheet obtained by taking $D_{1j}^{(+)}$ for the first row of D and $D_{2j}^{(-)}$ for the second, and so on. Three other diagrams like that of Fig. 1 may be drawn, since the solid curve can correspond to any one of the four sheets. Now suppose there is a resonance above the upper threshold w_{02}. We make the usual assumption that it is represented by a pole on the sheet reached from the physical sheet by crossing the cut above w_{02}; i.e., there is a pole on sheet $(--)$, near the real axis. If the parameters of the theory are altered so that the resonance moves to a different energy, can it cross the threshold w_{02}? The Y_1^* pole must undergo such a crossing as the decuplet passes to its degenerate limit. If the resonance lies at an energy between w_{01} and w_{02}, it must correspond to a pole on sheet $(-+)$. As Oakes and Yang pointed out, a pole can move continuously from $(--)$ to $(-+)$ only by making a clockwise circuit around w_{02} through the sheet $(+-)$. Such a complicated pole motion seems unlikely, particularly if one expects that the motion of poles can be calculated as a first-order effect of a symmetry-breaking perturbation. Fortunately, the pole need not travel from $(--)$ to $(-+)$, since there is already a different pole on $(-+)$ which takes over to represent the resonance when the latter moves below w_{02}. To show that this new pole is practically inevitable, we note that all poles of T near the physical cut are given by zeros of $\det D$. Let w_R be the position of the original pole on $(--)$: $\det D^{(--)}(w_R)=0$. As $\mathrm{Re}(w_R)$ decreases through w_{02}, $\det D^{(--)}(w_R)$ and $\det D^{(-+)}(w_R)$ become nearly equal, since they differ only by terms involving $q_2(w_R)N_{2j}(w_R)$ as a factor. The explicit formulas are

$$\det D^{(-+)}=(1+2iq_1N_{11}+I_{11})(1+I_{22})-(2iq_1N_{12}+I_{12})I_{21},$$

$$\det D^{(--)}=(1+2iq_1N_{11}+I_{11})(1+2iq_2N_{22}+I_{22})-(2iq_1N_{12}+I_{12})(2iq_2N_{21}+I_{21}).$$

In the case of a narrow resonance crossing the threshold we have $\mathrm{Re}(w_R)=w_{02}$ and $q_2(w_R)\simeq q_2(w_{02})=0$. Therefore, the function $\det D^{(-+)}(w)$ must have a zero at a point very close to w_R. Note that the resultant pole of T on $(-+)$ is quite distinct from the pole on $(--)$, since these two sheets are not directly connected. The same mechanism operates if the resonance crosses a threshold moving upward. Also, the inclusion of additional two-particle channels causes no difficulty. The calculations described in the following confirm this behavior of poles. Not only do we have the poles that are necessary to allow resonances to cross thresholds; we also expect poles on *all* of the unphysical sheets, since the various determinants differ only by small terms. The pole positions on all of the sheets are calculated explicitly in the two-channel model treated in the Appendix.

3. DESCRIPTION OF THE MODEL

As remarked in the Introduction, we assume that the forces responsible for the $J^P=\frac{3}{2}^+$ decuplet are mainly due to baryon exchanges in meson-baryon scattering. For given isotopic spin and hypercharge, we retain only the coupled two-particle channels and construct the matrix ND^{-1} representation for the partial-wave scattering matrix $G(w)$. G is related to the $J=\frac{3}{2}$ partial-wave T matrix defined in the previous section by $G_{ij}=T_{ij}/q_iq_j$. For the matrix N we take the Born approximation for single baryon exchange computed with degenerate masses. Thus, N has the form

$$N(w)=h(w)N_0. \tag{3.1}$$

N_0 is an energy-independent matrix, consisting of Yukawa-type meson-baryon coupling constants given by exact SU(3) symmetry. Thus, N_0 depends on only two parameters—a parameter that determines the $D-F$ mixing ratio[2,6,17] and an over-all coupling constant $g^2/4\pi$. The elements of N_0 can be found from the SU(3) coupling constants and the isotopic spin factors given in Ref. 6 (Appendixes I and II). The energy-dependent factor $h(w)$ is a single function defined by

$$h(w)=\frac{1}{16\pi w}\left\{\frac{1}{E-M}[A_1+(w-M)B_1]+\frac{1}{E+M}[-A_2+(w+M)B_2]\right\},$$

where

$$A_l=-g^2\frac{M-M_i}{q^2}Q_l(x);\quad B_l=\frac{g^2}{q^2}Q_l(x),$$

$$x=1+[2(M^2+m^2)-M_i^2-w^2]/2q^2,$$

$$E\pm M=[(w\pm M)^2-m^2]/2w. \tag{3.2}$$

Q_l denotes the Legendre function of the second kind. The masses of the meson, the external baryon, and the exchanged baryon are m, M, and M_i, respectively. E is the baryon energy, so the squared center-of-mass momentum is $q^2=(E+M)(E-M)$. The rationalized coupling constant is denoted by g.

In computing $h(w)$ we take $m=m_\eta$, $M=M_\Lambda$, and a mass larger than that of Λ for the exchanged baryon: $M_i=10.5m_\pi c^2$. The Λ and η masses represent the Gell-Mann–Okubo degenerate limits for the baryon and meson masses in the absence of the SU(3) symmetry breaking interactions. The rather large mass for the exchanged baryon is a device to move the real axis branch points of the Born approximation to the left of the lowest threshold of the problem (viz., the π-N threshold), while keeping the η and Λ masses for the external lines. By this device we make N analytic on the right, and thereby avoid spurious difficulties which have to do only with the artificial assumption of degenerate masses in N. The N constructed this way actually has a behavior on the right cut roughly similar to that of the Born matrix evaluated with physical masses throughout.

[17] R. E. Cutkosky, Ann. Phys. (N. Y.) 23, 415 (1963).

The D matrix takes the form

$$D(w)=1-\frac{w-\hat{w}}{\pi}\int_P \frac{\rho(w')h(w')dw'}{(w'-\hat{w})(w'-w)}N_0, \quad (3.3)$$

where ρ is a diagonal matrix with elements

$$\rho_{ij}(w)=\delta_{ij}q_i^3(|w|)\theta(|w|-w_{0i}). \quad (3.4)$$

The subtraction point $\hat{w}$ lies in the interval $(-w_0,w_0)$; its value has an influence on the effective strengths of the left singularities, and thereby on the positions of resonances and bound states predicted by the theory. The "centrifugal barrier" factor $\rho(w)$ is computed from the physical masses in the case of full symmetry breaking, or from nondegenerate masses with smaller splittings as the symmetry breaking is decreased. The way in which the symmetry breaking parameter x is introduced is discussed in the next section. The symmetry violation in the present model is due only to $\rho(w)$ and the factor q_iq_j in $T_{ij}=q_iq_jG_{ij}$, which is also computed from degenerate masses. Therefore, our model attributes a higher symmetry to the N matrix than to the D matrix. This may have some general validity, but as far as numerical results are concerned it reproduces those of Ref. 6 in which N was calculated with physical masses. In this sense it is a justifiable simplification especially suited to investigate the questions mentioned earlier.

In the limit of exact SU(3) symmetry, each meson-baryon state specified by isotopic spin T and hypercharge Y can be expressed in terms of the eigenstates that correspond to the irreducible representations of $8\otimes 8$. For a set of coupled two particle channels, there exists a unitary transformation U between the definite particle states and the states that transform by SU(3) representations. The matrix N_0 is diagonalized by U, so that

$$U^\dagger N_0 U=\Lambda, \quad (3.5)$$

where $\Lambda=[\lambda_i\delta_{ij}]$. Further, from

$$G=ND^{-1}=h(w)$$

$$\times\left[N_0^{-1}-\frac{w-\hat{w}}{\pi}\int_P \frac{\rho(w')h(w')}{(w'-\hat{w})(w'-w)}dw'\right]^{-1}, \quad (3.6)$$

it follows that

$$U^\dagger G U=h(w)\left[\Lambda^{-1}-\frac{w-\hat{w}}{\pi}\int_P \frac{U^\dagger\rho(w')Uh(w')}{(w'-\hat{w})(w'-w)}dw'\right]^{-1},$$

$$=h(w)[E(w)]^{-1}. \quad (3.7)$$

The requirement that the scattering matrix be symmetric is satisfied in our approximation scheme, as is evident from Eq. (3.6).

Bound states and resonances correspond to zeros of $\det E$ on the appropriate sheets of that function, as explained in Sec. 2. To investigate the zeros of $\det E$, we introduce ρ_0, which is the unit matrix times $q_0^3(|w|)\theta(|w|-w_0)$, where q_0 and w_0 are the momentum and the threshold, respectively, for the "unperturbed" meson and baryon masses m_η and M_Λ. Then

$$E(w)=\Lambda^{-1}+I_0(w)+\Delta(w), \quad (3.8)$$

where

$$I_0(w)=-\frac{w-\hat{w}}{\pi}\int_P \frac{\rho_0(w')h(w')dw'}{(w'-\hat{w})(w'-w)} \quad (3.9)$$

is the unperturbed unitarity integral and

$$\Delta(w)=-\frac{w-\hat{w}}{\pi}\int_P \frac{(U^\dagger\rho U-\rho_0)(w')h(w')dw'}{(w'-\hat{w})(w'-w)} \quad (3.10)$$

represents the symmetry-breaking perturbation. If the perturbation Δ is zero, the determinant factors:

$$\det(\Lambda^{-1}+I_0)=(\lambda_1^{-1}+I_0)$$
$$\times(\lambda_2^{-1}+I_0)\cdots(\lambda_n^{-1}+I_0). \quad (3.11)$$

Each factor corresponds to one of the representations contained in the direct product $8\otimes 8$. In this case of pure symmetry, the coupling constant and the $D-F$ mixing parameter can be chosen so that the factor corresponding to the $\frac{3}{2}^+$ decuplet vanishes at some bound state or resonance energy w_0^*, and so that the other factors do not vanish at any energy that would be reasonable for a resonance or bound state. Now when Δ is included, the numerical evaluations reported in the next section show that the determinant still factors to a fairly good approximation. We have

$$\det E=\det(\Lambda^{-1}+I_0+\Delta)\simeq(\lambda_1^{-1}+I_0+\Delta_{11})$$
$$\times(\lambda_2^{-1}+I_0+\Delta_{22})\cdots(\lambda_n^{-1}+I_0+\Delta_{nn}). \quad (3.12)$$

The off-diagonal elements of Δ do not contribute appreciably, and in this sense *the perturbation is weak*. The same choice of mixing parameter that was used in the degenerate case still guarantees that only the $\frac{3}{2}^+$ decuplet factor vanishes.

In calculating positions of resonances or bound states, we are concerned with $\det E(w+i0)$, w real, where the limit is taken from the physical sheet. A bound state corresponds to $\det E(w+i0)=0$, and a narrow resonance to a zero of $\det E(w)$ at a complex energy z_R such that $\mathrm{Re}\det E(w_R+i0)=0$, where $\mathrm{Re}z_r\simeq w_R$. Near w_R we may write $\mathrm{Re}\det E(w+i0)=c(w_R-w)$, $c=$constant, and

$$G_{ij}(w)=\frac{\mathfrak{N}_{ij}(w_R)}{c(w_R-w)+i\,\mathrm{Im}\det E(w_R+i0)}, \quad (3.13)$$

$$\mathrm{Im}G_{ij}(w)=\frac{\mathrm{Re}\mathfrak{N}_{ij}(w_R)}{c}\frac{\Gamma/2}{(w_R-w)^2+(\Gamma/2)^2}, \quad (3.14)$$

where $\Gamma/2=-\mathrm{Im}\det E(w_R+i0)/c$. With one minor exception, mentioned in the following section, the resonances of our theory are sufficiently narrow so that Eq. (3.14) is an acceptable approximation.

TABLE I. Positions and widths in units of the pion mass (139.6 MeV) as functions of the parameter x. Quantities in parentheses in the last row are the corresponding experimental numbers.

	$N^*_{3/2}$		Y^*_1		$\Xi^*_{1/2}$		Ω_0	
x	Position	Γ	Position	Γ	Position	Γ	Position	Γ
0.0	11.0	0.0	11.0	0.0	11.0	0.0	11.0	0.0
0.025	10.96	0.0	10.98	0.0	11.0	0.0	11.02	0.0
0.05	10.92	0.0	10.96	0.0	11.0	0.0	11.04	0.0
0.10	10.86	0.0	10.94	0.0	11.02	0.0	11.10	0.0
0.25	10.67	0.0	10.86	0.0	11.05	0.0	11.24	0.0
0.50	10.23	0.0	10.68	0.0	11.13	0.0	11.50	0.0
0.75	9.62	0.24	10.41	0.05	11.10	0.0	11.71	0.0
1.00	9.02	1.06	10.09	0.31	11.09	0.07	12.01	0.0
	(8.87)	(0.72)	(9.92)	(0.36)	(10.96)	(0.05)	(12.07 ±0.09)	(0.0)

The Eq. (3.14) can also be used to define effective coupling constants of decuplet states to meson-baryon states. If one treats the resonant states as though they had infinitely narrow width, the contribution of a resonance to a dispersion integral can be evaluated by treating it as a pole at the mass of the resonance. By comparing the residue of this pole with the corresponding residue in the Born approximation of an appropriate Lagrangian, the conventional coupling constant is related to the resonance parameters.[18] We take the Lagrangian density

$$G_{B^*BP}\tfrac{1}{2}(\partial_\mu\bar{\psi}\phi-\bar{\psi}\partial_\mu\phi)\Psi^\mu+\text{H.c.} \tag{3.15}$$

where ψ, ϕ, and Ψ^μ correspond to baryon, meson, and decuplet fields, respectively. Isopin is suppressed. Ψ^μ is a Rarita-Schwinger field. We find for the coupling of a decuplet state B^* to pseudoscalar meson P and baryon B

$$\frac{G_{B^*BP}{}^2}{4\pi}=\frac{12w_R{}^2}{(w_R+M)^2-m^2}\,\frac{\mathrm{Re}\mathfrak{N}_{ii}}{c}\beta_i, \tag{3.16}$$

where $\mathrm{Re}\mathfrak{N}_{ii}$ is computed from the diagonal matrix element G_{ii} for $B+P\rightarrow B+P$ according to Eq. (3.14). β_i is the appropriate isotopic spin factor. From the explicit Lagrangian given by Martin and Wali,[19] one can find the ratios of the decuplet coupling constants in the limit of exact SU(3). The quantity of particular experimental interest, the Y_1^* branching ratio, is given by

$$\frac{Y_1^*\rightarrow\Sigma+\pi}{Y_1^*\rightarrow\Lambda+\pi}=\frac{2}{3}\,\frac{(\mathrm{Re}\mathfrak{N})_{\Sigma\pi}}{(\mathrm{Re}\mathfrak{N})_{\Lambda\pi}}\,\frac{(q_R)_{\Sigma\pi}{}^3}{(q_R)_{\Lambda\pi}{}^3}. \tag{3.17}$$

In the numerical evaluation of the theory we obtained precise values for the zeros w_R of Re $\det E(w+i0)$, and then observed that nearly the same zeros could be obtained from the factored approximation (3.12). The lowest order corrections to the approximate zeros may be obtained as follows. Let w_A be an approximate zero; i.e., a zero of $E_{11}(w)$, where we have taken the index 1 to correspond to the 10 representation. Then

$$\det E=\begin{vmatrix}\beta(w_A-w) & \delta_{12} & \delta_{13}\cdots\delta_{1n}\\ \delta_{21} & \alpha_2 & \delta_{23}\cdots\delta_{2n}\\ \delta_{31} & \delta_{32} & \alpha_3\cdots\delta_{3n}\\ \vdots & \vdots & \vdots\cdots\vdots\\ \delta_{n1} & \delta_{n2} & \delta_{n3}\cdots\alpha_n\end{vmatrix}. \tag{3.18}$$

One finds that $|\delta/\alpha|<0.1$ over the energy range of interest. To lowest order in δ/α, the correction to the approximate zero w_A is

$$\Delta w_A=-\frac{1}{\beta}\left[\frac{\delta_{12}{}^2}{\alpha_2}+\frac{\delta_{13}{}^2}{\alpha_3}+\cdots+\frac{\delta_{1n}{}^2}{\alpha_n}\right]. \tag{3.19}$$

Another interesting way to analyze $\det E$ is by approximating the eigenvalues of E, since we know the eigenvalues in the case of pure symmetry. We solve the secular equation $\phi(\mu)=\det(E-\mu 1)=0$ by Newton's method. A first approximation is simply $\mu_i=E_{ii}=\alpha_i$, since (3.9) suggests that off-diagonal elements of E are

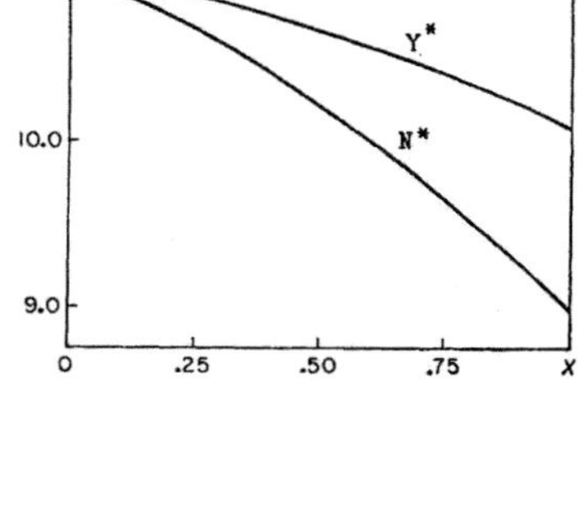

FIG. 2. The positions of the decuplet states plotted as a function of the parameter x.

[18] E. Abers and C. Zemach, Phys. Rev. **131**, 2305 (1963).

[19] A. W. Martin and K. C. Wali, Nuovo Cimento **31**, 1324 (1964).

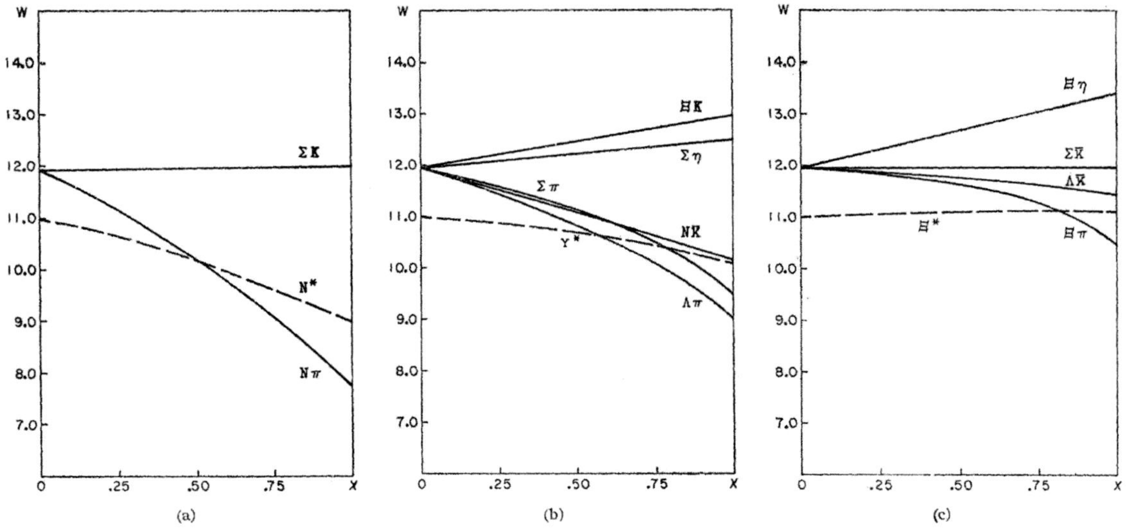

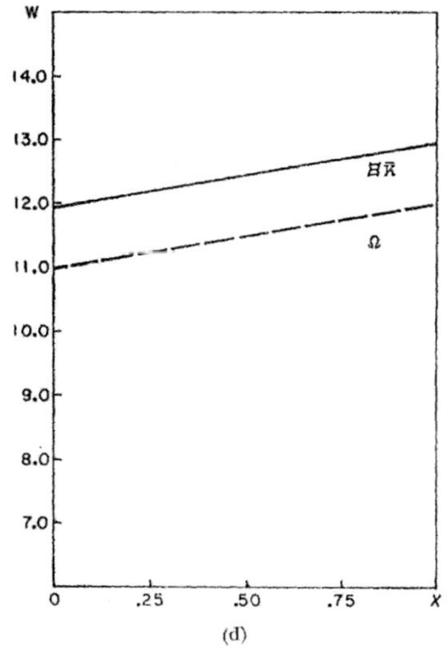

FIG. 3. (a)–(d). The positions of the poles corresponding to the decuplet states relative to the thresholds of their respective coupled channels plotted as a function of the parameter x.

not so very important. Expanding about α_i, we have

$$\phi(\mu)=\phi'(\alpha_i)\left(\mu-\alpha_i+\frac{\phi(\alpha_i)}{\phi'(\alpha_i)}+\cdots\right). \qquad (3.20)$$

The first correction $-\phi/\phi'=-1/(\ln\phi)'$ may be calculated from the formula $\ln \det A=\mathrm{tr}\ln A$. Thus,

$$\mu_i=E_{ii}-\frac{1}{\mathrm{tr}(E_{ii}-E)^{-1}}+\cdots. \qquad (3.21)$$

4. NUMERICAL RESULTS AND CONCLUSIONS

It was stated in the previous section that for fixed masses of the constituent particles the free parameters of our model are the $D-F$ mixing ratio determined by the parameter f, the subtraction point energy $\hat{w}$, and an over-all coupling constant $g^2/4\pi$. We choose the parameters so that the calculated decuplet states correspond closely to the observed ones when the constituent mesons and baryons have their observed masses. Taking account of the results of Ref. 6, we choose the parameter f to be 0.35; this ensures that the decuplet representation is the only one that is resonant in the low-energy region. Then with $\hat{w}=3m_\pi c^2$ and $g^2/4\pi=19$, the decuplet state energies agree fairly well with observation. We wish to emphasize that the approximately uniform level spacing (i.e., the validity of the Okubo formula) is substantially independent of the parameters, provided the resonances appear at all. The average value of the level spacings depends mostly on $\hat{w}$. This is only reasonable, since variations of $\hat{w}$ change the relative strengths of the different left-hand singularities, and hence the range of the attractive forces responsible for the resonance.

We characterize the symmetry violation by means of a parameter x, $0\leq x\leq 1$, introduced directly in the mass formula. For baryons,

$$M=M_0[1+xaY+xb(I(I+1)-Y^2/4)]. \qquad (4.1)$$

When $x=1$, $M(I,Y)$ takes on the observed values; $x=0$ corresponds to exact symmetry. The constants a and b are known, and $M_0=M_\Lambda$. For mesons one has a similar formula for masses squared.

$$m^2=m_0^2[1+xa'Y+xb'(I(I+1)-Y^2/4)], \qquad (4.2)$$

where $m_0^2=m_\eta^2$.

Table I gives the calculated resonance or bound-state positions and widths and the corresponding experimental values. It also gives the positions and widths as functions of x. In this table and in the following energies are given in units of $m_\pi c^2$ (139.6 MeV). The positions

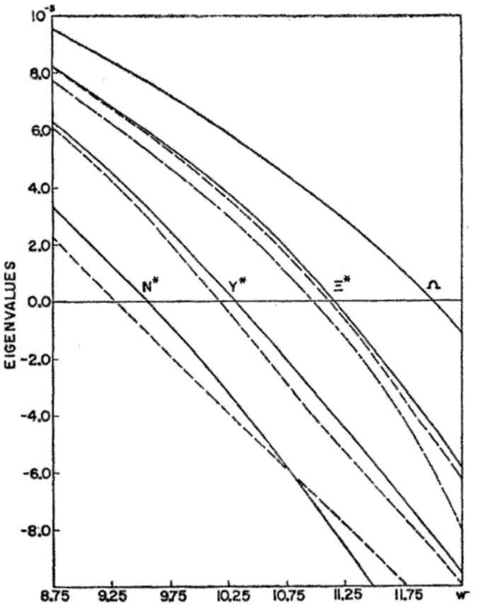

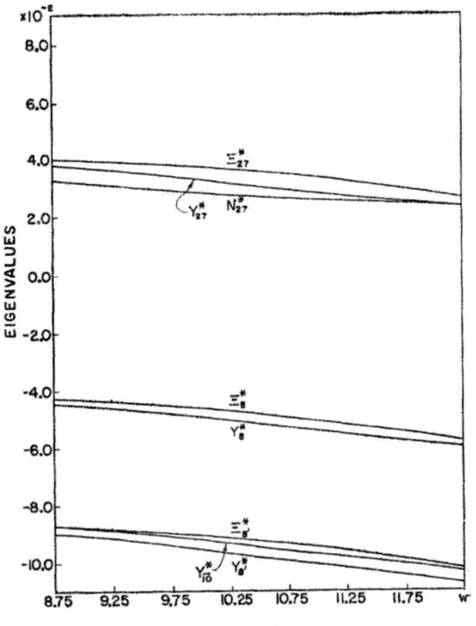

FIG. 4. (a) The eigenvalues corresponding to the 10-fold representation plotted as a function of the energy w. The particle labels used on these curves are merely to indicate the quantum numbers of the coupled channels in which they appear. The dashed curves represent the eigenvalues evaluated including only the perturbations of the diagonal elements [Eq. (3.12)]. The solid curves represent the same eigenvalues evaluated by using the approximation given by Eq. (3.21). The long dash-short dash curve represents the limit of these curves in the unperturbed situation. (b) The "nonresonant" eigenvalues corresponding to the indicated representations plotted as a function of the energy w.

are plotted as functions of x in Fig. 2, which shows that the Okubo equal spacing rule is satisfied to reasonable accuracy. The spacings become more nearly equal as x is decreased. The calculated widths agree well enough with experiment, except in the case of $N_{3/2}^*$. The $N_{3/2}^*$ width being too large seems to be typical of theories based on single baryon exchange.

For the branching ratio $R=(Y_1^*\rightarrow\Sigma+\pi)/(Y_1^*\rightarrow\Lambda+\pi)$ we calculated 0.10 from formula (3.17). This figure is to be compared to $R=0.20$, which is obtained from (3.17) if the factor $(\mathrm{Re}\mathfrak{N})_{\Sigma\pi}/(\mathrm{Re}\mathfrak{N})_{\Lambda\pi}$ is set equal to one.[20] The departure of this factor from one represents effects of symmetry breaking beyond those involved in q^3 phase-space factors. These effects were ignored in earlier estimates of the branching ratio.[21]

Figures 3(a) through 3(d) show the courses of the levels as compared to the courses of the thresholds. The energies plotted are the values of w_R for which $\mathrm{Re}\,\det E(w_R+i0)=0$. The motion of a resonance in crossing a threshold is perfectly smooth and presents no problem of the type discussed by Oakes and Yang.[9]

[20] These two values for R were both calculated at $w=10.085$, the energy of Y_1^* in our theory. Better agreement with the experimental indications (Ref. 3) that $R<0.05$ would be obtained if the parameters of the theory were altered to put Y_1^* at exactly its observed position $w=9.92$. This is because $(q^3)_{\Sigma\pi}/(q^3)_{\Lambda\pi}$ increases rapidly with energy. If we calculate this momentum ratio at $w=9.92$, but keep the same value for the slowly varying factor $(\mathrm{Re}\mathfrak{N})_{\Sigma\pi}/(\mathrm{Re}\mathfrak{N})_{\Lambda\pi}$ that it had at $w=10.085$, we obtain $R=0.07$.

[21] S. L. Glashow and J. J. Sakurai, Nuovo Cimento 25, 337 (1962).

A word about the definition of resonance position is in order. Instead of our definition one could take $\det \mathrm{Re}E(w_R+i0)=0$, as suggested by Dalitz.[22] The position of the peak of the scattering cross section might be taken as a third definition. In the present calculation the three energies differ negligibly for Y_1^* and $\Xi_{1/2}^*$, and all lie within an interval of 15 MeV in the case of $N_{3/2}^*$ at $x=1$. For $N_{3/2}^*$ the difference between the zero of $\mathrm{Re}\,\det E$ and the peak of the cross section means a slight inaccuracy in Eq. (3.14).

The eigenvalues μ_i of the matrix E were computed in the approximation (3.21), over a range of energies.[23] Their real parts are shown in Figs. 4(a) and 4(b), for $x=1$. The diagonal elements $\mathrm{Re}E_{ii}$, which form the zeroth order approximations to the $\mathrm{Re}\mu^i$, are also plotted. As was mentioned in the preceeding section, E factorizes in the limit of exact symmetry. Each factor corresponds to an irreducible representation of SU(3). The quantities $\mathrm{Re}\mu_i$ are proportional to the cotangents of the eigen phase shifts, with or without symmetry breaking. Our choice of the $D-F$ mixing parameter is such that in the pure symmetry case only the phase corresponding to the 10 representation passes through

[22] R. H. Dalitz, Rev. Mod. Phys. 33, 471 (1961).

[23] The approximation (3.21) is not extremely accurate, as can be seen from the poor values for the resonance positions obtained from the approximate eigenvalues. We nevertheless believe that the accuracy is sufficient for our purposes, since $\det E$ agrees well with $\Pi_{i=1}^{n}\mu_i$, except very near the resonances where both quantities are small.

TABLE II. $|U^\dagger GU|^2$ at resonance positions. The normalization is such that $|(U^\dagger GU)_{10,10}|^2=1$.

			$N_{3/2}^*$		
	10	27			
10	1	3.8×10^{-2}			
27		4.4×10^{-3}			
			Y_1^*		
	8	8′	10	$\overline{10}$	27
8	2.8×10^{-4}	7.6×10^{-9}	1.2×10^{-4}	5.1×10^{-8}	1.2×10^{-6}
8′		7.5×10^{-5}	6.9×10^{-5}	2.6×10^{-8}	1.7×10^{-6}
10			1	3.8×10^{-5}	1.7×10^{-2}
$\overline{10}$				8.0×10^{-5}	5.0×10^{-8}
27					6.7×10^{-4}
			$\Xi_{1/2}^*$		
	8	8′	10	27	
8	1.0×10^{-5}	2.8×10^{-10}	1.8×10^{-5}	6.5×10^{-8}	
8′		3.0×10^{-6}	6.2×10^{-5}	3.7×10^{-7}	
10			1	5.2×10^{-3}	
27				4.8×10^{-5}	

$\pi/2$. From Fig. 4 it is clear that this feature persists when the mass differences are introduced. The 10 eigenvalues split apart, as expected, giving the level spacings. The other eigenvalues exhibit smaller splittings and remain large throughout the energy range considered. Therefore, we are justified in associating the resonant states with the 10-dimensional representation, even when when the symmetry is broken. From Fig. 4(a) and Table I it is seen that the zeros of the appropriate diagonal elements $\text{Re}E_{ii}$, agree roughly with the actual level positions. This fact was already indicated in Eq. (3.12). However, it is interesting to note Eq. (3.19), which shows that this agreement will be less if the nonvanishing diagonal elements of E are reduced in magnitude.

In the face of substantial symmetry breaking the meaning of group invariance arguments is not well defined. Perturbations may affect different types of observables in different degrees. Besides finding the perturbations of eigen phase shifts, we have considered some other ways of judging the degree of symmetry violation. The various ways are closely related, but they serve to point up different aspects of the situation.

In Table II we list the squared absolute values of the transition amplitudes in the unitary spin representation; i.e., $|(U^\dagger GU)_{\alpha\beta}|^2$ for $x=1$. As expected, the nonresonant diagonal elements are small. The off-diagonal elements (which are zero in the case of exact symmetry) are of the same order as the nonresonant diagonal terms. The 10–27 transition is by far the largest of the off-diagonal processes, which suggests that the main symmetry breaking effects could be attributed to 10–27 mixing.

To determine the mixtures of various representations in the resonant states, we diagonalize the imaginary part of the matrix of residues of G at the resonance pole. Actually, we diagonalize the matrix $\text{Re}\mathfrak{N}(w_R+i0)$ defined in Eq. (3.13); for narrow resonance width, this is essentially the same as the imaginary part of the residue matrix. The diagonalization is a trivial operation, since $\text{Re}\mathfrak{N}$ is practically degenerate: $\text{Re}\mathfrak{N}_{ij}\simeq v_iv_j$. This fact, which is familiar from the theory of nuclear reactions, may be proved very simply by means of the ND^{-1} representation. We have $\mathfrak{N}=G\,\det E$, so $\det\mathfrak{N}=(\det E)^n\det G$ if there are n channels. But $\det G=\det(N\Lambda^{-1}/\det E$, so $\det\mathfrak{N}$ has a zero of order $n-1$, where $\det E$ vanishes. Thus, $\mathfrak{N}$ has rank 1 at the pole, and consequently every 2×2 subdeterminant of $\mathfrak{N}$ vanishes. By considering 2×2 subdeterminants of the type

$$\begin{vmatrix} \mathfrak{N}_{ii} & \mathfrak{N}_{ij} \\ \mathfrak{N}_{ji} & \mathfrak{N}_{jj} \end{vmatrix} = 0,$$

we find that $\mathfrak{N}_{ij}=v_iv_j$, because of the symmetry $\mathfrak{N}_{ij}=\mathfrak{N}_{ji}$. If w_P is the pole position, and w_R is the corresponding point on the real axis where $\text{Re}\det E(w_R+i0)=0$, we have $\mathfrak{N}(w_P)\simeq\mathfrak{N}(w_R+i0)$. Numerical evaluation of $\mathfrak{N}(w_R+i0)$ shows that $\text{Im}\mathfrak{N}(w_R+i0)\ll\text{Re}\mathfrak{N}(w_R+i0)$. This latter circumstance is essentially the statement that the imaginary parts of the integrals occurring in E are much less than the corresponding real parts. Thus, $\text{Re}\mathfrak{N}_{ij}(w_R+i0)\simeq\mathfrak{N}_{ij}(w_P)=v_iv_j$, and $v_i\simeq v_i^*$. The factorization of $\text{Re}\mathfrak{N}_{ij}$ holds numerically to excellent accuracy. The eigenvector of $\text{Re}\mathfrak{N}$ corresponding to the one nonzero eigenvalue is $\mathbf{v}=[v_i]$. Since $\mathfrak{N}$ was defined as a matrix in the particle representation, the v_i are the "probability amplitudes" for finding various particle states in the resonant state. Of course, since certain channels are closed, v_i^2 does not have direct observational significance. It merely represents the closest mathematical analog to a probability for decay into particle channel i. We state the results in Table III for $x=1$. The channels with thresholds lying close to the resonance energy are generally enhanced with respect to exact SU(3) predictions, while the channels with distant thresholds are depleted. The $\Sigma\pi$ channel does not follow this rule.

Alternatively, we may write $(U^\dagger\,\text{Re}\mathfrak{N}U)_{ij}\simeq u_iu_j$, where the components of $\mathbf{u}=[u_i]$ are the probability amplitudes for finding group representation states in the resonant state. We express the resonant state $|\psi\rangle$

TABLE III. "Probabilities" for broken and unbroken symmetry.

	"Probabilities" v_i^2	v_i^2 by exact SU(3)
$T=\frac{3}{2}$, $Y=1$		
$N\pi$	68.20%	50%
ΣK	31.80	50
$T=1$, $Y=0$		
$\Lambda\pi$	31.84	25
$\Sigma\pi$	15.97	16.67
$N\bar{K}$	21.82	16.67
$\Sigma\eta$	18.00	25
ΞK	12.37	16.67
$T=\frac{1}{2}$, $Y=-1$		
$\Xi\pi$	26.19	25
$\Lambda\bar{K}$	30.33	25
$\Sigma\bar{K}$	22.84	25
$\Xi\eta$	20.64	25

TABLE IV. Ratios of the coupling constants as a function of the parameter x.

$g^2_{B^*BP}$ / $g^2_{N^*N\pi}$ (x)	0.0	0.025	0.05	0.1	0.25	0.5	0.75	1
$N^*N\pi$	6.0	6.0	6.0	6.0	6.0	6.0	6.0	6.0
$N^*\Sigma K$	6.0	5.87	5.75	5.50	4.80	3.76	2.90	2.34
$Y^*\Lambda\pi$	3.0	3.02	2.95	2.90	2.88	3.10	2.35	1.82
$Y^*\Sigma\pi$	1.0	0.992	0.962	0.933	0.871	0.857	0.601	0.431
$Y^*N\bar{K}$	1.0	1.01	0.988	0.981	0.993	1.12	0.905	0.750
$Y^*\Sigma\eta$	3.0	2.97	2.86	2.72	2.47	2.28	1.48	1.01
$Y^*\Xi K$	1.0	0.988	0.950	0.906	0.813	0.738	0.467	0.310
$\Xi^*\Xi\pi$	1.0	0.980	0.957	0.941	0.877	0.880	0.633	0.403
$\Xi^*\Lambda\bar{K}$	3.0	2.95	2.91	2.92	2.85	3.06	2.41	1.68
$\Xi^*\Sigma\bar{K}$	1.0	0.979	0.953	0.934	0.862	0.854	0.613	0.399
$\Xi^*\Xi\eta$	3.0	2.90	2.83	2.76	2.49	2.34	1.63	0.987
$\Omega\Xi\bar{K}$	6.0	5.91	5.66	5.52	5.16	5.25	3.79	2.42

as a normalized linear combination of the group states.

$T=\frac{3}{2}$, $Y=1$

$$|\psi\rangle=0.98|10\rangle+0.18|27\rangle,$$

$T=1$, $Y=0$

$$|\psi\rangle=0.99|10\rangle+0.13|27\rangle+0.008|8'\rangle+0.003|\overline{10}\rangle+0.0007|8\rangle,$$

$T=\frac{1}{2}$, $Y=-1$

$$|\Psi\rangle=0.997|10\rangle+0.07|27\rangle+0.007|8'\rangle+0.001|8\rangle. \quad (4.3)$$

It is important to remember that in these formulas the states on the right represent linear combinations [with coefficients given by exact SU(3) symmetry] of definite particle states with physical masses. As was already apparent in Table II, the principal contamination of the 10 representation comes from the 27. However, the mixing is small enough so that there is no doubt about associating the resonance primarily with the 10 representation. Tarjanne and Cutkosky[7] obtained much larger mixing in a calculation which is difficult to compare with ours. Note that we have diagonalized the residue of $G_{ij}=T_{ij}/q_iq_j$. Had we not factored out q_iq_j, the symmetry might have been less apparent. In any case, our model certainly proves the existence of a set of left singularities which imply both the observed decuplet mass differences and relatively small mixing, provided the mixing is measured in the way specified. In judging the degree of mixing, it is appropriate to consider directly the coefficients of Eq. (4.3), rather than their squares. For example, in the case $T=\frac{3}{2}$, $Y=1$, a comparison of $(0.98)^2$ with $(0.18)^2$ might suggest that the amount of $|27\rangle$ is negligible. However, from $|10\rangle=2^{-1/2}(|\pi N\rangle|-K\Sigma\rangle)$ and $|27\rangle=2^{-1/2}(|\pi N\rangle+|K\Sigma\rangle)$ it follows that $(v_{\pi N}/v_{K\Sigma})^2=[(0.98+0.18)/(0.98-0.18)]^2=2.1$. This ratio, which is unity in the case of no 10–27 mixing, is involved in $g^2_{N^*N\pi}/g^2_{N^*\Sigma K}$.

Finally, we discuss the coupling constants defined by Eq. (3.16). Although the properties of these constants are roughly similar to those of the v_i^2, the two sets of constants are related by factors in which appreciable effects of symmetry breakdown are manifest. For each x we compute $g^2_{B^*BP}$ in units such that $g^2_{N^*N\pi}=6.0$. The results are shown in Table IV. For large values of x the ratios deviate considerably from the values given by exact symmetry. The deviations do not agree with the sum rules of Dullemond *et al.*[8] and Sudarshan,[8] even at the smaller x values.

ACKNOWLEDGMENTS

We wish to acknowledge the generous and expert help of Mary Wrenn, who advised us in programming the IBM-7090 computer at the IIT Research Institute.

We should also like to thank Professor E. C. G. Sudarshan for helpful discussions of the coupling constant sum rules.

APPENDIX: QUANTITATIVE STUDY OF POLES IN A TWO-CHANNEL EXAMPLE

In order to make perfectly explicit the observations of Sec. 2, we calculate the positions of poles on all sheets in a simple, two-channel problem. We wish to illustrate the phenomenon of a resonance crossing a threshold other than the lowest threshold of the problem. The $\pi N-K\Sigma$ problem of the last section is not suitable for this purpose, since the $N_{3/2}^*$ crosses only the lower threshold. The latter crossing occurs merely by the pole on the $(-+)$ sheet passing through the threshold. To obtain a case in which a resonance crosses the upper threshold, we modify the $\pi N-K\Sigma$ problem by reducing the coupling constant so that $N_{3/2}^*$ moves up to an energy just below the $K\Sigma$ threshold. Then, as the symmetry-breaking parameter x is reduced, the resonance moves upward through the $K\Sigma$ threshold.

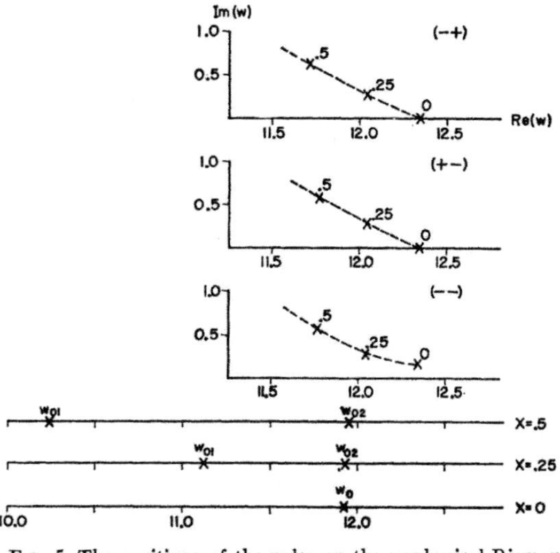

FIG. 5. The positions of the poles on the unphysical Riemann sheets for $0\leq x\leq 0.5$. The corresponding values for the channel thresholds are shown at the bottom.

We take only the values $x=0.5$, 0.25, and 0; for larger x the resonance width becomes ridiculously large. With $g^2/4\pi=12.8$ and $x=0.5$, we have $w_{01}=10.23$, $w_{02}=11.95$, and $w_R=11.72$, in units of the pion mass. As x goes to zero, the thresholds coalesce at $w_0=12.35$.

The trajectories of the poles on the three unphysical sheets are plotted in Fig. 5. The positions of the thresholds for each x are shown (on the same energy scale) below the graphs. As was expected, for each x there is a pair of complex conjugate poles on each of the sheets; the graphs show just the member of each pair that lies in the upper half-plane. For $x=0.5$, the poles have nearly identical positions on the various sheets, and the resonance peak is due to the poles on sheet $(-+)$. For $x=0.25$, the poles again have similar positions on all three sheets, but the resonance has crossed the threshold. The peak is now mainly associated with the poles on sheet $(--)$. In the degenerate limit $x=0$ the "original" resonance poles of $(-+)$ have moved onto the real axis of $(-+)$. Since $(-+)$ is now completely inaccessible from the physical sheet $(++)$, this pole has no physical effect whatever. The same may be said of the poles on $(+-)$, which have also moved onto the real axis. The poles on $(--)$ end up at $w=12.35\pm0.16i$, and therefore represent a resonance with half-width $\Gamma/2=0.16$.

The pole positions as shown in Fig. 5 were calculated in the usual approximation that takes advantage of the fact that the poles are all close to the real axis. If $E^{(\alpha)}$ is the E matrix on sheet (α), then near a zero w_R of $\mathrm{Re}\det E^{(\alpha)}$ we have

$$\det E^{(\alpha)}(w\pm i0)=c(w_R-w)\pm i\,\mathrm{Im}\det E^{(\alpha)}(w_R+i0)$$

and the poles on sheet α are at

$$w\approx w_R\pm i(1/c)\,\mathrm{Im}\det E^{(\alpha)}(w_R+i0).$$

We list the formulas for $\det E^{(\alpha)}(w\pm i0)$ for w on and just below the right-hand physical cut. $\mathcal{E}$ stands for $\mathrm{Re}\det E(w+i0)$, and $\lambda^{-1}=\frac{1}{2}(\lambda_1^{-1}+\lambda_2^{-1})$, where the λ_i are the eigenvalues of N_0 as defined in Eq. (3.5). The integrals I_i are defined by

$$I_i(w)=-\frac{w-\hat{w}}{\pi}\int_P\frac{\rho_i(w')h(w')dw'}{(w'-\hat{w})(w'-w)},$$

and the other notations are as in Sec. 3.

Sheet $(-+)$

$$\begin{aligned}
w<w_{01}:&\quad \mathcal{E}+2|q_1|^3h(I_2+\lambda^{-1})\\
w_{01}<w<w_{02}:&\quad \mathcal{E}\pm iq_1^3h(I_2+\lambda^{-1})\\
w>w_{02}:&\quad \mathcal{E}+(q_1q_2)^3h^2\\
&\qquad\pm ih[q_1^3(I_2+\lambda^{-1})-q_2^3(I_1+\lambda^{-1})];
\end{aligned}$$

Sheet $(+-)$

$$\begin{aligned}
w<w_{01}:&\quad \mathcal{E}+2|q_2|^3h(I_1+\lambda^{-1})\\
w_{01}<w<w_{02}:&\quad \mathcal{E}+|q_2|^3h(I_1+I_2+2\lambda^{-1})\\
&\qquad\mp iq_1^3h(2|q_2|^3h+I_2+\lambda^{-1})\\
w>w_{02}:&\quad \mathcal{E}+(q_1q_2)^3h^2\\
&\qquad\mp ih[q_1^3(I_2+\lambda^{-1})-q_2^3(I_1+\lambda^{-1})];
\end{aligned}$$

Sheet $(--)$

$$\begin{aligned}
w<w_{01}:&\quad \mathcal{E}+4|q_1q_2|^3h^2\\
&\qquad+2h[|q_1|^3(I_1+\lambda^{-1})+|q_2|^3(I_2+\lambda^{-1})]\\
w_{01}<w<w_{02}:&\quad \mathcal{E}+|q_2|^3h(I_1+I_2+2\lambda^{-1})\\
&\qquad\pm iq_1^3h(2|q_2|^3h+I_2+\lambda^{-1})\\
w>w_{02}:&\quad \mathcal{E}-(q_1q_2)^3h^2\\
&\qquad\pm ih[q_1^3(I_2+\lambda^{-1})+q_2^3(I_1+\lambda^{-1})].
\end{aligned}$$

From these formulas one can check that the sheets are connected as shown in Fig. 1; e.g., $\det E^{(-+)}(w\pm i0)=\det E^{(+-)}(w\mp i0)$ for $w>w_{02}$, so $(-+)$ and $(+-)$ are connected along the real axis above the second threshold.

Relativistic Formulation of the $SU(6)$ Symmetry Scheme*

BUNJI SAKITA AND KAMESHWAR C. WALI
Argonne National Laboratory, Argonne, Illinois
(Received 22 March 1965)

A relativistic formulation of the $SU(6)$ symmetry scheme is presented, starting with the basic assumption that the fields corresponding to elementary particles are tensors of $M(12)$ [or $\tilde{U}(12)$ or $SU(12)_{\mathcal{L}}$]. In particular a mixed second-rank tensor and a totally symmetric third-rank tensor are associated with the meson and baryon fields, respectively. It is shown that if these fields are required to satisfy prescribed free-field equations of motion, then one is led to a particle supermultiplet structure which corresponds to the $35\oplus1$ and 56-dimensional representations of $SU(6)$ for the mesons and baryons. It is also shown that the spin-dependent and $SU(3)$-spin-dependent mass splittings can be included in the theory and that solutions in terms of physical particle fields can be obtained. Effective trilinear meson-meson and meson-baryon vertex functions, using these solutions and an interaction Lagrangian which is invariant under $M(12)$, are calculated in the lowest order perturbation. We would like to note especially the following results: (a) From the known pion-nucleon coupling constant, the width of the pion-nucleon (3,3) resonance is calculated to be 94 MeV. (b) The ratio of the magnetic form factors for the neutron and proton is $-\frac{2}{3}$ for all momentum transfers and $\mu_P=(1+2M_P/m_\rho)$ nuclear magnetons. (c) The charge form factor of the neutron is zero for all momentum transfers.

I. INTRODUCTION

THERE has been considerable interest recently in the $SU(6)$ symmetry scheme for elementary particles.[1] It is conceived as an extension of Wigner's nuclear-supermultiplet theory[2] to elementary-particle phenomena. Unlike other higher symmetry schemes,[3] the $SU(6)$ theory proposes to treat the ordinary spin on the same footing as the isotopic spin and hypercharge. Clearly such a formulation is possible only if the space-time variables and the spin variables are completely decoupled. This is possible only in a nonrelativistic theory as in the case of Wigner's supermultiplet theory. Since Lorentz transformations mix the intrinsic spin and the orbital angular momentum in an intricate manner, it is not obvious whether the $SU(6)$ theory can be extended to the relativistic domain. It is therefore not surprising that several attempts[4] have been made towards an understanding of this problem.

The problems connected with a relativistic formulation of the $SU(6)$ theory may be discussed from a purely mathematical point of view of finding an appropriate group of invariance. For this purpose, we recall that the irreducible representations of $SU(6)$ can be decomposed into irreducible representations of $SU(2)\otimes SU(3)$, where the $SU(2)$ can be identified as the ordinary spin group and the $SU(3)$ as the familiar internal symmetry group $SU(3)$. If the theory has to contain orbital angular momentum and spin mixed in a Lorentz-invariant manner, the spin groups $SU(2)$ has to be extended to $SL(2,C)$ which is the covering group of the restricted Lorentz group. A fully relativistic $SU(6)$ theory must include in addition to the homogeneous Lorentz transformations, space and time translations. The required group G therefore must contain $SU(6)$ and the Poincare group as subgroups in such a manner that the intersection of $SL(2,C)\otimes SU(3)$ and $SU(6)$ is $SU(2)\otimes SU(3)$. It has been shown[5] that G must then contain the group $SL(6,C)$. Now depending on how the translations are imbedded in the group, one obtains two types of structures for G: (i) G is given by $P'\otimes Q$, where P' is a group isomorphic to the physical Poincaré group and $Q\supset SL(6,C)$[6]; (ii) G is a semidirect product of T_{36} by $SL(6,C)$ where T_{36} is the group of translations in a 36-dimensional space.[7]

Once a group G is given, its unitary representations on Hilbert space provide a set of symmetry transformations on the physical states which are characterized by the bases of the representations. The basis of an irreducible representation gives a set of physical states which are commonly identified as the particles belonging to a

* Work performed under the auspices of the U. S. Atomic Energy Commission.

[1] F. Gürsey and L. A. Radicati, Phys. Rev. Letters **13**, 173 (1964); A. Pais, *ibid.* **13**, 175 (1964); B. Sakita, Phys. Rev. **136**, B1756 (1964); F. Gürsey, A. Pais, and L. A. Radicati, Phys. Rev. Letters **13**, 299 (1964).

[2] E. Wigner, Phys. Rev. **51**, 106 (1937).

[3] J. Schwinger, Phys. Rev. Letters **12**, 237 (1964); Phys. Rev. **135**, B816 (1964); **136**, B1821 (1964); F. Gürsey, T. D. Lee, and M. Nauenberg, *ibid.* **135**, B468 (1964); P. G. O. Freund and Y. Nambu, Phys. Rev. Letters **12**, 714 (1964); **13**, 221 (1964); M. Gell-Mann, Physics **1**, 63 (1964); Z. Maki, Progr. Theoret. Phys. (Kyoto) **31**, 331 (1964); P. Tarjanne and V. L. Teplitz, Phys. Rev. Letters **11**, 447 (1963); Y. Hara, Phys. Rev. **134**, B701 (1964).

[4] R. P. Feynman, M. Gell-Mann and G. Zweig, Phys. Rev. Letters **13**, 678 (1964); K. Bardaci, J. M. Cornwall, P. G. O. Freund, and B. W. Lee, *ibid.* **13**, 698 (1964) and **14**, 48 (1965); S. Okubo and R. E. Marshak, *ibid.* **13**, 818 (1964) and **14**, 156 (1965); W. Rühl, Phys. Letters **13**, 349 (1964); **14**, 334 (1965); A. Salam, *ibid.* **13**, 354 (1964); T. Fulton and J. Wess, *ibid.* **14**, 57 (1965); P. Roman and J. J. Aghassi, *ibid.* **14**, 68 (1965); M. A. B. Bég and A. Pais, Phys. Rev. Letters **14**, 267 (1965); Y. Né eman, Phys. Letters **14**, 327 (1965); F. Gürsey, *ibid.* **14**, 330 (1965); K. T. Mahanthappa and E. C. G. Sudarshan, Phys. Rev. Letters **14**, 458 (1965); Riazuddin and L. K. Pandit, *ibid.* **14**, 462 (1965).

[5] L. Michel and B. Sakita, Ann. Inst. Henri-Poincaré **11**, 167 (1965).

[6] L. Michel, Second Coral Gables Conference on Symmetry Principles at High Energy, January 1965 (to be published).

[7] L. Michel and B. Sakita (Ref. 5); W. Rühl (Ref. 4), T. Fulton and J. Wess (Ref. 4).

supermultiplet of the system. One can construct a unitary representation of G in case (i) by using Wigner's method for P'. Since G is a direct product of P' by Q, the basis of such a representation must be a tensor product of the basis of P' and the basis of the unitary representation of Q. But $Q\supset SL(6,C)$ which is noncompact. An irreducible unitary representation of Q is of infinite dimensions. This corresponds, therefore, physically to an infinite number of particles belonging to a supermultiplet.

In case (ii), G contains additional translations other than the usual four space-time ones. Clearly the physical interpretation of the extra translations is not easy. Further if one identifies the usual space-time translations with four of the translations in T_{36}, the physical mass is no longer invariant under $SL(6,C)$. It will change continuously under the transformations of $SL(6,C)$. Physically this corresponds to a continuous mass distribution for a given particle state. Since the physical world does not appear to admit either an infinite number of one-particle states for fixed four-momentum or a continuous mass distribution for a given particle state, we are forced to conclude that there is no physically interesting group of invariance which contains the Poincaré group and the $SU(6)$ group in a nontrivial way.[8]

A more physical description of the $SU(6)$ theory is provided by the quark model. A relativistic quark model can be constructed along the lines of the three-field Thirring model. The fundamental field in this model can be described by a 12-component spinor ψ_A. A pair of indices $i\alpha$ can be assigned to A, where i runs from 1 to 4 and can be identified as the Dirac spinor index. α takes the values 1 to 3 and corresponds to the $SU(3)$-spin index. If one decomposes a Dirac spinor field into two two-component spinors (Weyl decomposition), ψ decomposes into two six-component fields ϕ and χ. The fields ϕ and χ then provide vector and conjugate-vector representations of $SL(6,C)$ (Sec. II). As pointed out by several authors,[9] one can construct an interaction Lagrangian which is invariant under $SL(6,C)$. However, it is impossible to construct a free Lagrangian which is also invariant under $SL(6,C)$ without encountering the difficulties mentioned earlier in connection with the group of invariance. Without a free Lagrangian, the standard quantization procedures and the particle interpretation of the fields cannot be carried out.

In spite of this apparent difficulty, the quark model suggests an alternative approach. If we consider a model of noninteracting quarks and construct a free Lagrangian in terms of ψ which is invariant under $P\otimes SU(3)$ ($P\equiv$Poincaré group), we can obtain free field equations of motion (Dirac equation) and commutation relations. The solutions to these equations can be interpreted as particle states which form a basis of an irreducible representation of $SU(6)$ for a fixed momentum $\mathbf{q}$. This suggests the possibility that the basic fields are tensors of $SL(6,C)$ whereas the solutions to appropriate equations of motion for these fields give the desired particle-multiplet structure, even though the equations themselves are not covariant under $SL(6,C)$. The purpose of the present paper is to examine the possibility and consequences of such an approach.[10,11]

If one assumes that the elementary particles are the bound states of one or several quarks and antiquarks, the bound-state wave function (or field in a phenomenological Lagrangian theory) can be described as a product of the fundamental fields ψ's and $\bar{\psi}$'s. In the following discussion, however, it is not necessary to assume explicitly such a quark model. We shall only assume that the fields associated with the elementary particles transform like the products of ψ's and $\bar{\psi}$'s. In particular, the meson field is represented by a second rank mixed tensor $\Phi_A{}^B$ (144 components). A totally symmetric third rank tensor Ψ_{ABC} (364 components) is associated with the baryon field. These tensor representations of $SL(6,C)$ together with their properties under space reflections are discussed in Sec. II. It also contains the interaction Lagrangian which is assumed to be invariant under $SL(6,C)$ and space reflections. The interaction Lagrangian assumed in the present discussion is invariant under a larger group of transformations $M(12)$ [or $\tilde{U}(12)$ or $SU(12)_{\mathcal{L}}$].[12] Section III is devoted to the decomposition of Φ and Ψ into appropriate auxiliary fields and to the discussion of the symmetry properties with respect to the interchange of Dirac and $SU(3)$ spin indices. In Sec. IV, the free field equations of motion for the meson and baryon fields are given. It is shown that the meson field equations admit solutions which correspond to a nonet of 0^- and a nonet of 1^- mesons. The baryon field equations lead to solutions which correspond to a decuplet of $\frac{3}{2}^+$ and an octet of $\frac{1}{2}^+$ baryons. The desired mass splittings are introduced and the solutions for Φ and Ψ are obtained in terms of physical particle fields. These solutions are used to calculate effective vertex parts in the lowest order perturbation calculation in Sec. V. The relations between different coupling constants and some of their consequences are also discussed. Finally, the concluding section is devoted to a summary and the discussion of some of the difficulties of the theory.

[8] S. Coleman, Phys. Rev. **138**, B1262 (1965).

[9] K. Bardakci, J. M. Cornwall, P. G. O. Freund, and B. W. Lee (Ref. 4); S. Okubo and R. E. Marshak (Ref. 4).

[10] B. Sakita and K. C. Wali, Phys. Rev. Letters **14**, 404 (1965). The present paper is an extended version of this letter.

[11] A. Salam, *Proceedings of the Second Coral Gables Conference on Symmetry Principles at High Energy, 1965* (W. H. Freeman and Company, San Francisco, 1965); A. Salam, R. Delbourgo and J. Strathdee, Proc. Roy. Soc. (London) **284**, 146 (1965); A. Salam, R. Delbourgo, J. Strathdee, and M. A. Rashid, Proc. Roy. Soc. (London) **285**, 312 (1965); M. A. B. Bég and A. Pais, Phys. Rev. Letters **14**, 267 (1965); W. Rühl, Phys. Letters **14**, 334 (1965); **15**, 99 (1965); **15**, 101 (1965).

[12] K. Bardakci, J. M. Cornwall, P. G. O. Freund, and B. W. Lee, Phys. Rev. Letters **14**, 48 (1965); R. Delbourgo, A. Salam, and J. Strathdee (to be published).

II. ELEMENTARY FIELDS AND INTERACTIONS

Consider the group 6×6 complex matrices of determinant one, which we shall denote $SL(6,C)$. To every matrix A ($\det A=1$) belonging to this group, there corresponds a linear unimodular transformation $T(A)$ over a six-dimensional complex vector space. A vector ϕ in this space is a fundamental representation of $SL(6,C)$ and undergoes the transformation

$$\phi \to \phi' = A\phi. \tag{2.1}$$

Tensors in this vector space are called tensors of $SL(6,C)$. Let us also consider a conjugate vector space, in which a vector χ is transformed under $SL(6,C)$ as follows:

$$\chi \to \chi' = (A^{-1})^\dagger \chi. \tag{2.2}$$

Tensors in the latter vector space are called conjugate tensors of $SL(6,C)$. The scalar products $\phi^*\chi$ and $\chi^*\phi$ are invariant under the transformations of $SL(6,C)$.

Let these tensors be functions of a space-time point so that they can be regarded as tensor fields. One can then construct a twelve-component tensor field ψ from ϕ and χ:

$$\psi = \begin{pmatrix} \phi \\ \chi \end{pmatrix}. \tag{2.3}$$

ψ is then a 12-component tensor representation of $SL(6,C)$. If we want to identify ψ as the quark field mentioned in the previous section, we must identify the $SL(2,C)$ group of $SL(2,C)\otimes SU(3)$ as the covering group of the restricted Lorentz group[13] and the $SU(3)$ group as the internal symmetry $SU(3)$ group. The decomposition of $\psi_A (A=1, 2, \cdots, 12)$ into the representations of $SL(2,C)\otimes SU(3)$ can be done by assigning a pair of indices $i\alpha$ to A. The index i ($i=1, \cdots, 4$) is the Dirac spinor index and α ($\alpha=1, \cdots, 3$) is the $SU(3)$-spin index. Since to every i there corresponds a four-component spinor, the properties of ψ under spatial reflections can be easily included in the formalism in the usual fashion:

$$\psi \to \psi' = \eta\gamma_4\psi, \tag{2.4}$$

where η is a phase. We also note that if we use the explicit representations

$$\gamma_4 = \begin{pmatrix} 0 & 1 \\ 1 & 0 \end{pmatrix}, \quad \gamma_5 = \begin{pmatrix} 1 & 0 \\ 0 & -1 \end{pmatrix}, \tag{2.5}$$

then $\bar{\psi}\psi$ and $\bar{\psi}\gamma_5\psi$ are invariant under $SL(6,C)$, where

$$\bar{\psi}^A = \bar{\psi}^{i\alpha} = \psi_{i'\alpha}{}^*(\gamma_4)_{i'}{}^i. \tag{2.6}$$

As stated in the previous section, we assume that the fields corresponding to physical particles transform as products of ψ's and $\bar{\psi}$'s under the $SL(6,C)$ transformations. We consider, in particular, a second-rank mixed tensor-field $\Phi_A{}^B$ for the mesons and a totally symmetric third-rank tensor-field Ψ_{ABC} for the baryons. These fields are assumed to be local and their properties under space reflections are given by

$$\Phi \to \Phi' = \gamma_4 \otimes \tilde{\gamma}_4 \Phi, \tag{2.7}$$

$$\Psi \to \Psi' = \gamma_4 \otimes \gamma_4 \otimes \gamma_4 \Psi. \tag{2.8}$$

Written out explicitly, the above equations imply

$$\Phi_{i\alpha}'^{j\beta} = (\gamma_4)_i{}^{i'} \Phi_{i'\alpha}{}^{j'\beta} (\gamma_4)_{j'}{}^j \tag{2.9}$$

$$\Psi_{i\alpha,j\beta,k\gamma}' = (\gamma_4)_i{}^{i'}(\gamma_4)_j{}^{j'}(\gamma_4)_k{}^{k'} \Psi_{i'\alpha,j'\beta,k'\gamma}. \tag{2.10}$$

We construct the interaction Lagrangian L_{int} in terms of these fields by requiring that it be invariant under the $SL(6,C)$ transformations and also under the space reflections. If we restrict ourselves to trilinear meson-meson and meson-baryon interactions, L_{int} can be written in the form[14]

$$L_{\text{int}} = \tfrac{1}{3} m_0 ig \,\text{Tr}(\Phi\Phi\Phi) + iG\bar{\Psi}^{ADC}\Phi_A{}^B\Psi_{BDC}. \tag{2.11}$$

L_{int} in (2.11) is invariant under a larger group of transformations $M(12)$ [or $\tilde{U}(12)$],[12] which is the group of 12×12 complex matrices M satisfying the following condition:

$$M^\dagger[\gamma_4 \otimes \mathbf{1}]M = \gamma_4 \otimes \mathbf{1}, \tag{2.12}$$

where $\mathbf{1}$ is a 3×3 unit matrix so that $[\gamma_4\otimes\mathbf{1}]$ is a 12×12 matrix and $M^\dagger$ is the Hermitian conjugate of M. $M(12)$ is a 144 parameter Lie group which contains $SL(6,C)$ in its subgroups. The quark spinor ψ is a fundamental representation of $M(12)$ and $\bar{\psi}$ is the contragradient representation so that $\bar{\psi}\psi$ is invariant. However $\bar{\psi}\gamma_5\psi$ which is invariant under $SL(6,C)$ is not invariant under $M(12)$.

III. DECOMPOSITION OF THE ELEMENTARY FIELDS

The meson field Φ consists of 144 components which are complex. In order to have antiparticles in the same multiplet as the particles, we must impose the reality condition:

$$\gamma_4 \otimes \tilde{\gamma}_4 \Phi = -\Phi^\dagger. \tag{3.1}$$

For fixed $SU(3)$ indices, Φ is a 4×4 matrix in the Dirac space. Consequently it can be expanded in terms of the sixteen independent Dirac matrices:

$$\Phi = 1\otimes\mathbf{S} + \gamma_\mu\otimes\mathbf{V}_\mu + \tfrac{1}{2}\sigma_{\mu\nu}\otimes\mathbf{T}_{\mu\nu} + \gamma_\mu\gamma_5\otimes\mathbf{A}_\mu + \gamma_5\otimes\mathbf{P}. \tag{3.2}$$

$\mathbf{S}$, $\mathbf{V}_\mu$, $\cdots$, $\mathbf{P}$ are 3×3 matrices in $SU(3)$-spin space.[15] Their properties under space-reflections can be easily

[13] R. F. Streater and A. S. Wightman, *PCT, Spin and Statistics, and All That* (W. A. Benjamin, Inc., New York, 1964).

[14] Invariance under $SL(6,C)$ and space reflections permits also the interactions $\text{Tr}(\Phi\gamma_5\Phi\gamma_5\Phi)$, $\bar{\Psi}(\gamma_5\Phi\gamma_5\otimes 1\otimes 1)\Psi$, $\bar{\Psi}(\gamma_5\Phi\otimes\gamma_5\otimes 1)\Psi$ and $\bar{\Psi}(\Phi\otimes\gamma_5\otimes\gamma_5)\Psi$, which are not $M(12)$ invariant. However, for the present discussion we restrict our phenomenological L_{int} to the form given in (2.11).

[15] We use the bold-face letters for matrices acting in $SU(3)$-spin space.

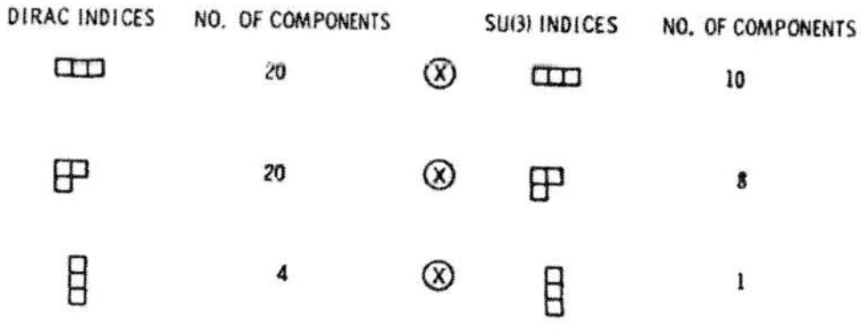

FIG. 1. Young diagrams corresponding to a totally symmetric third-rank tensor.

deduced from those of Φ defined in (2.7). They represent a nonet of scalar, vector, tensor, axial vector, and pseudoscalar fields, respectively.

The baryon field Ψ has 364 components, since it is a totally symmetric third rank tensor, i.e.,

$$\Psi_{i\alpha,j\beta,k\gamma}=\Psi_{j\beta,i\alpha,k\gamma}=\Psi_{i\alpha,k\gamma,j\beta}. \tag{3.3}$$

The condition (3.3) can be satisfied by suitably combining different symmetry properties of the Dirac and $SU(3)$ spin indices. It is convenient to describe the different possible combinations with the help of Young diagrams. The number of possible combinations is three and they are as shown in Fig. 1. The first line in Fig. 1 represents a set of decuplet fields which are completely symmetric with respect to the interchange of the $SU(3)$ indices as well as Dirac spinor indices. We denote these by $D[=D_{ijk,\alpha\beta\gamma}]$. The second line represents a set of octet fields each of which is a third-rank Dirac spinor with mixed symmetry properties. We denote these by $\chi[=\chi_{ijk,\alpha}{}^{\delta},\ \chi_{ijk,\alpha}{}^{\alpha}=0]$. The mixed symmetry implies

$$\chi_{ijk}=-\chi_{ikj}, \tag{3.4}$$

$$\chi_{ijk}+\chi_{jki}+\chi_{kij}=0. \tag{3.5}$$

Finally the last line represents a singlet field which is a completely antisymmetric third-rank Dirac spinor. We denote this by $G[=G_{ijk}]$ where

$$G_{ijk}=-G_{jik}=-G_{ikj}. \tag{3.6}$$

One can, therefore, decompose Ψ uniquely as follows:

$$\Psi_{i\alpha,j\beta,k\gamma}=D_{ijk,\alpha\beta\gamma}+B_{ijk,\alpha\beta\gamma}+\frac{1}{\sqrt{6}}\epsilon_{\alpha\beta\gamma}G_{ijk}, \tag{3.7}$$

where

$$B_{ijk,\alpha\beta\gamma}=\tfrac{1}{3}[\chi_{ijk,\alpha}{}^{\delta}\epsilon_{\delta\beta\gamma}+\chi_{jki,\beta}{}^{\delta}\epsilon_{\delta\gamma\alpha}+\chi_{kij,\gamma}{}^{\delta}\epsilon_{\delta\alpha\beta}]. \tag{3.8}$$

It is also easy to construct the projection operators $\mathbf{P}_D$, $\mathbf{P}_B$, and $\mathbf{P}_G$ by means of which we can obtain the fields D, B, and G from Ψ. To construct these operators, we note that

$$\tfrac{1}{2}\sum_{i=0}^{8}\boldsymbol{\lambda}_i\otimes\boldsymbol{\lambda}_iV=V^T, \tag{3.9}$$

where λ_i $(i=1\cdots 8)$ are the 3×3 matrix representations of the $SU(3)$ generators[16] and $\boldsymbol{\lambda}_0=(\sqrt{\tfrac{2}{3}})1$. V is a second rank tensor in $SU(3)$ space and V^T is its transpose. Then

$$\mathbf{P}_D=\tfrac{1}{6}\sum_{i=0}^{8}[\boldsymbol{\lambda}_i\otimes\boldsymbol{\lambda}_i\otimes 1+1\otimes\boldsymbol{\lambda}_i\otimes\boldsymbol{\lambda}_i+\boldsymbol{\lambda}_i\otimes1\otimes\boldsymbol{\lambda}_i]+\tfrac{1}{6}\boldsymbol{\varepsilon},$$
$$\mathbf{P}_G=\tfrac{1}{6}\boldsymbol{\varepsilon}, \tag{3.10}$$

and

$$\mathbf{P}_B=1\otimes1\otimes1-P_D-P_G, \tag{3.11}$$

where $\boldsymbol{\varepsilon}$ is given by

$$(\boldsymbol{\varepsilon})_{\alpha\beta\gamma}{}^{\alpha'\beta'\gamma'}=\epsilon_{\alpha\beta\gamma}\epsilon^{\alpha'\beta'\gamma'}. \tag{3.12}$$

To make our computations more explicit,

$$(\boldsymbol{\varepsilon})_{\alpha\beta\gamma}{}^{\alpha'\beta'\gamma'}D_{ijk,\alpha'\beta'\gamma'}=\epsilon_{\alpha\beta\gamma}\epsilon^{\alpha'\beta'\gamma'}D_{ijk,\alpha'\beta'\gamma'}=0,$$

since D is symmetric in α, β, γ. Also

$$(\boldsymbol{\varepsilon})_{\alpha\beta\gamma}{}^{\alpha'\beta'\gamma'}B_{ijk,\alpha'\beta'\gamma'}=\epsilon_{\alpha\beta\gamma}[\chi_{ijk,\alpha}{}^{\alpha}]=0$$

and

$$\tfrac{1}{6}(\boldsymbol{\varepsilon})_{\alpha\beta\gamma}{}^{\alpha'\beta'\gamma'}\epsilon_{\alpha'\beta'\gamma'}G_{ijk}=\epsilon_{\alpha\beta\gamma}G_{ijk}.$$

Consequently,

$$(\mathbf{P}_D)_{\alpha\beta\gamma}{}^{\alpha'\beta'\gamma'}D_{ijk,\alpha'\beta'\gamma'}=\tfrac{1}{3}[D_{ijk,\beta\alpha\gamma}+D_{ijk,\alpha\gamma\beta}+D_{ijk,\gamma\beta\alpha}]=D_{ijk,\alpha\beta\gamma},$$

$$(\mathbf{P}_D)_{\alpha\beta\gamma}{}^{\alpha'\beta'\gamma'}B_{ijk,\alpha'\beta'\gamma'}=\tfrac{1}{3}[B_{ijk,\beta\alpha\gamma}+B_{ijk,\alpha\gamma\beta}+B_{ijk,\gamma\beta\alpha}]=\chi_{ijk,\alpha\beta\gamma}+\chi_{jki,\alpha\beta\gamma}+\chi_{kij,\alpha\beta\gamma}=0,$$

$$(\mathbf{P}_D)_{\alpha\beta\gamma}{}^{\alpha'\beta'\gamma'}\epsilon_{\alpha'\beta'\gamma'}G_{ijk}=-\epsilon_{\alpha\beta\gamma}G_{ijk}+\epsilon_{\alpha\beta\gamma}G_{ijk}=0.$$

We also note that the projection operator that gives an octet component $\chi_{ijk,\alpha}{}^{\delta}$ from B is given by

$$(\mathbf{P}_0)_{\alpha}{}^{\alpha'\beta'\gamma'\delta}=\delta_{\alpha}{}^{\alpha'}\epsilon^{\delta\beta'\gamma'}-\tfrac{1}{3}\delta_{\alpha}{}^{\delta}\epsilon^{\alpha'\beta'\gamma'}. \tag{3.13}$$

The symmetry properties of D, χ, and G with respect to the Dirac indices i, j, and k can be exhibited by a further decomposition using the charge conjugation matrix C which has the following properties:

$$C^T=-C,\quad C^{-1}\gamma_\mu C=-\gamma_\mu{}^T. \tag{3.14}$$

From (3.14) it follows that $\gamma_\mu C$ and $\sigma_{\mu\nu}C$ are symmetric matrices whereas $\gamma_\mu\gamma_5C$ and γ_5C are antisymmetric matrices. One can therefore write D_{ijk} in the form

$$D_{ijk}=\tfrac{1}{2}\psi_{\mu,i}(\gamma_\mu C)_{jk}+\tfrac{1}{4}\psi_{\mu\nu,i}(\sigma_{\mu\nu}C)_{jk}, \tag{3.15}$$

which insures the symmetry of D_{ijk} under the interchange of j and k. Since D_{ijk} is totally symmetric,

$$(C^{-1})^{ij}D_{ijk}=(C^{-1}\gamma_\mu\gamma_5)^{ij}D_{ijk}=(C^{-1}\gamma_5)^{ij}D_{ijk}=0. \tag{3.16}$$

From (3.15) and (3.16) we obtain

$$\gamma_\mu\psi_\mu=0;\quad \psi_\mu=\gamma_\rho\psi_{\rho\mu}. \tag{3.17}$$

In (3.15) ψ_μ has 16 components and antisymmetric tensor $\psi_{\mu\nu}$ has 24 components. There are 20 relations among them on account of (3.17). Hence (3.15) and (3.17) correctly represent a totally symmetric third-rank

[16] M. Gell-Mann, Phys. Rev. **125**, 1067 (1962).

Dirac spinor D_{ijk} with 20 independent components. Considerations along similar lines permit the expansion of χ and G as follows:

$$\chi_{ijk}=\tfrac{1}{2}[\xi_i C_{jk}+\eta_{\mu,i}(\gamma_\mu\gamma_5 C)_{jk}+\psi_i(\gamma_5 C)_{jk}],$$

where

$$\xi=\gamma_\mu\gamma_5\eta_\mu+\gamma_5\psi, \tag{3.18}$$

and

$$G_{ijk}=\tfrac{1}{2}[\xi_i' C_{jk}+\eta_{\mu,i}'(\gamma_\mu\gamma_5 C)_{jk}+\psi_i'(\gamma_5 C)_{jk}],$$

where

$$\eta_\mu'=\gamma_\mu\gamma_5\xi', \quad \psi'=-\gamma_5\xi'. \tag{3.19}$$

IV. FREE-FIELD EQUATIONS OF MOTION FOR MESONS AND BARYONS

The spinor fields described so far can lead to no physical consequences unless they are made to represent physical particle states. As stated in the Introduction, we may accomplish this by requiring that Φ and Ψ satisfy prescribed free field equations of motion. The guiding principle in the choice of such equations is provided by one of the most remarkable successes of the $SU(6)$ symmetry, namely the assignment of representations for the low lying baryonic and mesonic states. The octet of $\frac{1}{2}^+$ and the decuplet of $\frac{3}{2}^+$ baryons can be identified as belonging to the 56-dimensional representation. The octet 0^- and the nonet 1^- mesons can be fitted into the 35-dimensional representation whereas the X^0 meson[17] can belong to the singlet representation. We shall choose the wave equations which lead to solutions that correspond to the $SU(6)$ supermultiplet states.

The meson field Φ is required to satisfy

$$\tfrac{1}{2}[\gamma\cdot\partial\otimes 1'\otimes 1\otimes 1'-1\otimes\tilde{\gamma}\cdot\partial\otimes 1\otimes 1']\Phi+m\Phi=0, \tag{4.1}$$

which is the Duffin-Kemmer equation[18] rewritten in a form more convenient for our purpose. To introduce the desired mass splittings, we regard m as a matrix in both Dirac-spin and $SU(3)$-spin space. It can be chosen in a number of different ways. One of the forms which leads to the well-known empirical relations for the masses of the pseudoscalar and vector mesons is as follows:

$$\begin{aligned} m=m_0(1\otimes 1'\otimes 1\otimes 1') &+\tfrac{1}{2}m_1(\gamma_\lambda\otimes\tilde{\gamma}_\lambda+\gamma_\lambda\gamma_5\otimes\tilde{\gamma}_\lambda\tilde{\gamma}_5)\otimes(1\otimes 1') \\ &+\tfrac{1}{2}m'(1\otimes 1'+\gamma_5\otimes\tilde{\gamma}_5)\otimes(\mathbf{\delta}\otimes 1'+1\otimes\mathbf{\delta}') \\ &+m_S\sum_{A=1}^{16}\gamma_A\otimes\tilde{\gamma}_5\tilde{\gamma}_A\tilde{\gamma}_5\otimes\sum_{i=0}^{8}\lambda_i\otimes\lambda_i, \end{aligned} \tag{4.2}$$

where $\mathbf{\delta}$ is given by

$$\mathbf{\delta}=\begin{pmatrix}0&0&0\\0&0&0\\0&0&1\end{pmatrix}.$$

We shall continue to use the direct-product notation with the convention regarding matrix multiplication as defined in (2.7) and (2.9). (Also see footnote 15.) In Eq. (4.2), the first term gives a common mass m_0 to all mesons. The second term is a spin-dependent term and is introduced to split the vector mesons from the pseudoscalar ones. The third term is responsible for Gell-Mann–Okubo (GMO)[19] splittings as well as singlet and octet mixing. The last term splits the singlet pseudoscalar meson from others, since it is proportional to the projection operator of the singlet, pseudoscalar term. If we assume a sufficiently large value for m_S, there would be little mixing between the singlet $0^-(X^0)$ and the octet $0^-(\eta^0)$ states. Since the GMO mass sum rule is satisfied very well for the octet of 0^- mesons, we assume this to be the case.

If we insert the expansion (3.2) for Φ in (4.1), multiply by the Dirac matrices 1, γ_μ, $\sigma_{\mu\nu}$, $\gamma_\mu\gamma_5$, and γ_5, respectively, and take trace, we obtain the following set of equations in $SU(3)$ space:

$$[(m_0+4m_1)1\otimes 1'+m'(\mathbf{\delta}\otimes 1'+1\otimes\mathbf{\delta}')]\mathbf{S}=0, \tag{4.3}$$

$$[m_0 1\otimes 1']\mathbf{V}_\mu+1\otimes 1'\partial_\lambda\mathbf{T}_{\lambda\mu}=0, \tag{4.4}$$

$$[m_0 1\otimes 1'+m'(\mathbf{\delta}\otimes 1'+1\otimes\mathbf{\delta}')]\mathbf{T}_{\mu\lambda}+1\otimes 1'(\partial_\mu\mathbf{V}_\lambda-\partial_\lambda\mathbf{V}_\mu)=0, \tag{4.5}$$

$$[m_0 1\otimes 1']\mathbf{A}_\mu+1\otimes 1'\partial_\mu\mathbf{P}=0, \tag{4.6}$$

$$[(m_0-4m_1)1\otimes 1'+m'(\mathbf{\delta}\otimes 1'+1\otimes\mathbf{\delta}')]\mathbf{P}+1\otimes 1'\partial_\mu\mathbf{A}_\mu=0. \tag{4.7}$$

From these equations it follows that

$$\mathbf{S}=0, \tag{4.8}$$

$$\partial_\mu\mathbf{V}_\mu=0, \tag{4.9}$$

$$[(1\otimes 1')\Box-m_0\{m_0 1\otimes 1'+m'(\mathbf{\delta}\otimes 1'+1\otimes\mathbf{\delta}')\}]\mathbf{V}_\mu=0, \tag{4.10}$$

$$[(1\otimes 1')\Box-m_0\{(m_0-4m_1)1\otimes 1'+m'(\mathbf{\delta}\otimes 1'+1\otimes\mathbf{\delta}')\}]\mathbf{P}=0. \tag{4.11}$$

Equations (4.9), (4.10), and (4.11) represent the free wave equations of motion for a nonet of 1^- mesons and an octet of 0^- mesons. The physical states can be obtained by diagonalizing the mass matrix in (4.10) and

[17] G. R. Kalbfleisch *et al.*, Phys. Rev. Letters **12**, 527 (1964); M. Gundzik *et al.*, *ibid.* **12**, 546 (1964).

[18] R. J. Duffin, Phys. Rev. **54**, 1114 (1938); N. Kemmer, Proc. Roy. Soc. (London) **A173**, 91 (1939). Please note that (for typographical reasons) primes are used with identity and $\mathbf{\delta}$ matrices instead of tildes to indicate the multiplication from the right [Eqs. (2.7)–(2.10)].

[19] M. Gell-Mann, Phys. Rev. **125**, 1067 (1962); S. Okubo, Progr. Theoret. Phys. (Kyoto) **27**, 949 (1962).

(4.11), and are described by

$$\mathbf{V}=\begin{pmatrix} \frac{\rho^0+\omega^0}{\sqrt{2}} & \rho^+ & K^{*+} \\ \rho^- & \frac{\omega^0-\rho^0}{\sqrt{2}} & K^{*0} \\ K^{*-} & \bar{K}^{*0} & \phi^0 \end{pmatrix},$$

$$\mathbf{P}=\begin{pmatrix} \frac{\pi^0}{\sqrt{2}}+\frac{\eta}{\sqrt{6}} & \pi^+ & K^+ \\ \pi^- & -\frac{\pi^0}{\sqrt{2}}+\frac{\eta}{\sqrt{6}} & K^- \\ K^- & \bar{K}^0 & -\frac{2\eta}{\sqrt{6}} \end{pmatrix}.$$

The masses corresponding to these particles are as follows:

$$m_\rho^2=m_\omega^2=m_0^2;\quad m_{K^*}^2=m_0(m_0+m_0');\quad m_\phi^2=m_0(m_0+2m_0'),$$
$$m_\pi^2=m_0(m_0-4m_1);\quad m_K^2=m_0(m_0-4m_1+m_0');\quad m_\eta^2=m_0(m_0-4m_1+\tfrac{4}{3}m_0'). \tag{4.12}$$

It is clear that the squares of the 0^- meson masses satisfy the GMO sum rule for an octet and that

$$m_\phi^2-m_{K^*}^2=m_{K^*}^2-m_\rho^2=m_K^2-m_\pi^2. \tag{4.13}$$

From Eqs. (4.5), (4.6), and (4.12),

$$T_{\mu\lambda,\alpha}{}^\beta=[1/(m_T)_\alpha{}^\beta](\partial_\lambda V_{\mu,\alpha}{}^\beta-\partial_\mu V_{\lambda,\alpha}{}^\beta),$$
$$A_{\mu,\alpha}{}^\beta=(1/m_\rho)\partial_\mu P_\alpha{}^\beta,$$

where $(m_T)_\alpha{}^\beta$ are given by the elements of the matrix

$$m_T=\begin{pmatrix} m_\rho & m_\rho & m_{K^*}^2/m_\rho \\ m_\rho & m_\rho & m_{K^*}^2/m_\rho \\ m_{K^*}^2/m_\rho & m_{K^*}^2/m_\rho & m_\phi^2/m_\rho \end{pmatrix}. \tag{4.14}$$

The solution of the wave equation (4.1) can therefore be explicitly written as

$$\Phi_{i\alpha}{}^{j\beta}=\Big[(\gamma_\mu)_i{}^jV_{\mu,\alpha}{}^\beta-\frac{1}{2(m_T)_\alpha{}^\beta}(\sigma_{\mu\nu})_i{}^jF_{\mu\nu,\alpha}{}^\beta+(\gamma_5)_i{}^j\Big(P_\alpha{}^\beta+\frac{X^0\delta_\alpha{}^\beta}{\sqrt{3}}\Big)-(\gamma_\mu\gamma_5)_i{}^j\Big(\frac{P_\alpha{}^\beta}{m_\rho}+\frac{X^0}{m_S}\frac{\delta_{\alpha\beta}}{\sqrt{3}}\Big)\Big], \tag{4.15}$$

where m_S is the mass of X^0 and $F_{\mu\nu}=\partial_\mu V_\nu-\partial_\nu V_\mu$.

For the baryons, we assume that Ψ satisfies the generalized Bargmann-Wigner equations[20]:

$$[\gamma\cdot\partial+M]\Psi=0, \tag{4.16}$$

where M is again a matrix. The desired mass splittings can be introduced by choosing M to be a matrix in $SU(3)$ space[21] alone. The required form is

$$\mathbf{M}=[M_0\mathbf{1}\otimes\mathbf{1}\otimes\mathbf{1}+(M'-M_1-3M_2)\mathbf{P}_D+(\mathbf{P}_D+\mathbf{P}_B)(M_1\mathbf{G}_1+M_2\mathbf{G}_2)], \tag{4.17}$$

where $\mathbf{P}_D$ and $\mathbf{P}_B$ are the projection operators defined in Eqs. (3.10) and (3.11). The first term gives a common mass to all the baryons. The second term splits the decuplet masses from the octet masses. $\mathbf{G}_1$ and $\mathbf{G}_2$ in the third term are introduced so that they not only give the GMO splittings, but also give the observed relations between the decuplet spacings and the octet spacings. They are given by

$$\mathbf{G}_1=\boldsymbol{\delta}\otimes\mathbf{1}\otimes\mathbf{1}+\mathbf{1}\otimes\boldsymbol{\delta}\otimes\mathbf{1}+\mathbf{1}\otimes\mathbf{1}\otimes\boldsymbol{\delta},$$
$$\mathbf{G}_2=\sum_{i=0}^{8}[\boldsymbol{\lambda}_i\otimes(\boldsymbol{\delta\lambda}_i\otimes\mathbf{1}+\mathbf{1}\otimes\boldsymbol{\delta\lambda}_i)+\text{symmetrizing terms}]. \tag{4.18}$$

To be more explicit,

$$(G_1\Psi)_{\alpha\beta\gamma}=[\Psi_{3\beta\gamma}\delta_\alpha{}^3+\Psi_{\alpha3\gamma}\delta_\beta{}^3+\Psi_{\alpha\beta3}\delta_\gamma{}^3], \tag{4.19}$$

$$(G_2\Psi)_{\alpha\beta\gamma}=[(\Psi_{\beta3\gamma}+\Psi_{\gamma\beta3})\delta_\alpha{}^3+(\Psi_{\alpha\gamma3}+\Psi_{3\alpha\gamma})\delta_\beta{}^3+(\Psi_{3\beta\alpha}+\Psi_{\alpha3\beta})\delta_\gamma{}^3], \tag{4.20}$$

where for convenience we have suppressed the Dirac indices.

As in the meson case, if we insert the expansion (3.7) in (4.16), it follows that

$$\epsilon_{\alpha\beta\gamma}G_{ijk}=0, \tag{4.21}$$

$$[\gamma\cdot\partial+(M_0+M'-M_1-3M_2)+(M_1+2M_2)(\delta_\alpha{}^3+\delta_\beta{}^3+\delta_\gamma{}^3)]_i{}^{i'}D_{i'jk,\alpha\beta\gamma}=0, \tag{4.22}$$

$$[\gamma\cdot\partial+(M_0-M_2)]_i{}^{i'}\chi_{i'jk,\alpha}{}^\beta+(M_1+\tfrac{1}{2}M_2)(\delta_\alpha{}^3-\delta_3{}^\beta)\chi_{ijk,\alpha}{}^\beta+\tfrac{3}{2}M_2[(\delta_\alpha{}^3+\delta_3{}^\beta)\chi_{ijk,\alpha}{}^\beta-\tfrac{2}{3}\delta_\alpha{}^\beta\chi_{ijk,3}{}^3]=0. \tag{4.23}$$

With the identification,

$D_{111}=N^{*++}$, $D_{112}=N^{*+}/\sqrt{3}$, $D_{122}=N^{*0}/\sqrt{3}$,
$D_{222}=N^{*-}$, $D_{113}=Y^{*+}/\sqrt{3}$, $D_{123}=Y^{*0}/\sqrt{6}$,
$D_{223}=Y^{*-}/\sqrt{3}$, $D_{133}=\Xi^{*0}/\sqrt{3}$, $D_{233}=\Xi^{*-}/\sqrt{3}$,
$D_{333}=\Omega^-$,

[20] V. Bargmann and E. P. Wigner, Proc. Natl. Acad. Sci. U. S. 34, 211 (1948).

[21] We split the decuplet mass from the octet by using the $SU(3)$ spin-dependent term. We may, of course, use the ordinary spin-dependent splitting. Because of the over-all symmetry of Ψ, however, it can be proved that both methods are equivalent when the mass matrix is operated on Ψ.

$$\chi=\begin{pmatrix} \frac{\Sigma^0}{\sqrt{2}}+\frac{\Lambda}{\sqrt{6}} & \Sigma^+ & P \\ \Sigma^- & -\frac{\Sigma^0}{\sqrt{2}}+\frac{\Lambda}{\sqrt{6}} & N \\ \Xi^- & \Xi^0 & -\frac{2\Lambda}{\sqrt{6}} \end{pmatrix},$$

the masses of the decuplet and the octet states are given by the mass formulas

$$M_D=(M_0+M_2)+M'-(M_1+\tfrac{1}{2}M_2)Y-M_2[I(I+1)-\tfrac{1}{4}Y^2], \quad (4.24)$$

$$M_O=(M_0+M_2)-(M_1+\tfrac{1}{2}M_2)Y-M_2[I(I+1)-\tfrac{1}{4}Y^2]. \quad (4.25)$$

If we rewrite (4.22) as [no summation is intended for repeated $SU(3)$ indices].

$$[\gamma\cdot\partial+M_{\alpha\beta\gamma}]_i^{i'}D_{i'jk,\alpha\beta\gamma}=0, \quad (4.26)$$

where

$$M_{\alpha\beta\gamma}=(M_0+M'-M_1-3M_2)+(M_1+2M_2)(\delta_\alpha{}^3+\delta_\beta{}^3+\delta_\gamma{}^3),$$

and substitute (3.15) in (4.26), we obtain

$$[\gamma\cdot\partial+M_{\alpha\beta\gamma}]\psi_{\mu,\alpha\beta\gamma}=0; \quad (4.27)$$
$$[\gamma\cdot\partial+M_{\alpha\beta\gamma}]\psi_{\mu\nu,\alpha\beta\gamma}=0.$$

Since $D_{ijk,\alpha\beta\gamma}$ it totally symmetric, Eq. (4.26) can also be written as

$$[\gamma\cdot\partial+M_{\alpha\beta\gamma}]_i^{i'}D_{ji'k,\alpha\beta\gamma}=0.$$

If we now substitute the expansion

$$D_{ji'k,\alpha\beta\gamma}=[\tfrac{1}{2}\psi_{\mu,j}(\gamma_\mu C)_{i'k}+\tfrac{1}{4}\psi_{\mu\nu,j}(\sigma_{\mu\nu}C)_{i'k}]_{\alpha\beta\gamma},$$

$$[(\gamma\cdot\partial+M_{\alpha\beta\gamma})\gamma_\mu C]_{ik}\psi_{\mu,\alpha\beta\gamma}+[(\gamma\cdot\partial+M_{\alpha\beta\gamma})\sigma_{\mu\nu}C]_{ik}\psi_{\mu\nu,\alpha\beta\gamma}=0. \quad (4.28)$$

From (4.28), if we multiply by $C^{-1}\sigma_{\mu\nu}$ and take the trace we obtain

$$\psi_{\mu\nu,\alpha\beta\gamma}=\frac{1}{M_{\alpha\beta\gamma}}(\partial_\nu\psi_\mu-\partial_\mu\psi_\nu). \quad (4.29)$$

The equations

$$[\gamma\cdot\partial+M_{\alpha\beta\gamma}]\psi_{\mu,\alpha\beta\gamma}=0;\quad \gamma_\mu\psi_{\mu,\alpha\beta\gamma}=0 \quad (4.30)$$

are the Rarita-Schwinger equations for a $J=\frac{3}{2}$ particle. Equation (4.29) enables us to express the antisymmetric tensor $\psi_{\mu\nu}$ in terms of the Rarita-Schwinger fields ψ_μ.[22]

Thus $D_{ijk,\alpha\beta\gamma}$ is given by

$$D_{ijk,\alpha\beta\gamma}=\left[\tfrac{1}{2}(\gamma_\mu C)_{jk}\psi_{\mu i,\alpha\beta\gamma}-\frac{1}{4M_{\alpha\beta\gamma}}(\sigma_{\mu\nu}C)_{jk}\Lambda_{\mu\nu i,\alpha\beta\gamma}\right],$$

where

$$\Lambda_{\mu\nu}=(\partial_\mu\psi_\nu-\partial_\nu\psi_\mu). \quad (4.31)$$

Similar considerations lead to

$$\chi_{ijk,\alpha}{}^\beta=\left[\tfrac{1}{2}(\gamma_5 C)_{jk}\psi_{i,\alpha}{}^\beta-\frac{1}{2M_\alpha{}^\beta}(\gamma_\mu\gamma_5 C)_{jk}\partial_\mu\psi_{i,\alpha}{}^\beta\right], \quad (4.32)$$

where ψ satisfies the Dirac equation

$$(\gamma\cdot\partial+M_\alpha{}^\beta)\psi_{i,\alpha}{}^\beta+\tfrac{1}{2}(M_\Sigma-M_\Lambda)\delta_\alpha{}^\beta\psi_{i,3}{}^3=0 \quad (4.33)$$

and $M_\alpha{}^\beta$ are given by the elements of the matrix

$$\begin{pmatrix} M_\Sigma & M_\Sigma & M_N \\ M_\Sigma & M_\Sigma & M_N \\ M_\Xi & M_\Xi & \tfrac{1}{2}(3M_\Lambda-M_\Sigma) \end{pmatrix}.$$

V. EFFECTIVE VERTEX FUNCTIONS

Since we have the explicit solutions of the equations of motion for free fields, we can obtain effective trilinear, quadrilinear, $\cdots$interactions in the lowest order perturbation. For the present discussion we confine ourselves to trilinear meson-meson and meson-baryon interactions. The results without the mass dissymmetries were given in Ref. 10. The solutions for the meson and baryon fields in the previous section provide a natural way of considering the modifications of the effective interactions due to mass differences. The main purpose of this section is to consider these modifications and some of their consequences. The relations between the various coupling constants are summarized in a more explicit manner in the Appendix.

Meson-Meson Interactions

If the solution (4.15) is inserted in L_{int}

$$L_{\text{int}}(\text{mesons})=m_\rho ig\tfrac{1}{4}\,\text{Tr}(\Phi\Phi\Phi),$$

the following effective VPP, VVP, and VVV interactions are obtained:

$$L_{\text{int}}(VPP)=\tfrac{9}{2}ig\,\text{Tr}(\mathbf{V}_\mu\mathbf{P}\overleftrightarrow{\partial}_\mu\mathbf{P}), \quad (5.1)$$

$$L_{\text{int}}(VVP)=3ig\epsilon_{\mu\nu\rho\sigma}\left\{\text{Tr}\left[\partial_\mu\mathbf{V}_\nu\left(\frac{\partial_\rho\mathbf{V}_\sigma}{m}\right)\mathbf{P}\right]+\text{Tr}\left[\left(\frac{\partial_\mu\mathbf{V}_\nu}{m}\right)\partial_\rho\mathbf{V}_\sigma\mathbf{P}\right]+m_\rho\,\text{Tr}\left[\left(\frac{\partial_\mu\mathbf{V}_\nu}{m}\right)\left(\frac{\partial_\rho\mathbf{V}_\sigma}{m}\right)\mathbf{P}\right]\right\}, \quad (5.2)$$

[22] The equivalence between the Bargmann-Wigner wave equation and the Rarita-Schwinger equation for the spin-$\frac{3}{2}$ particle has been shown by C. G. Oliveira and A. Vidal, Notas de Fisica IX, No. B, 226 (1962).

TABLE I. Decuplet-decuplet and decuplet-baryon currents. $H=[(M_i+M_f)^2+q^2]/2M_iM_f$; $I_{\mu\nu}=(H\delta_{\mu\nu}+q_\mu q_\nu/M_iM_f)$.

	$\bar{D}D$	$\bar{D}B$
S	$I_{\mu\nu}\bar{\psi}_\mu\psi_\nu$	0
V	$I_{\mu\nu}\bar{\psi}_\mu\gamma_\rho\psi_\nu$	$\frac{2}{3}H\bar{\psi}_\rho\gamma_5\psi+\frac{2}{3M_i}p_\mu\left(\frac{p_\rho'}{M_f}\bar{\psi}_\mu\gamma_5\psi-i\bar{\psi}_\mu\gamma_\rho\gamma_5\psi\right)$
T	$I_{\mu\nu}\bar{\psi}_\mu\sigma_{\rho\sigma}\psi_\nu$	$\frac{1}{3}H(\bar{\psi}_\sigma\gamma_\rho\gamma_5\psi-\bar{\psi}_\rho\gamma_5\psi)+\frac{1}{3M_iM_f}p_\mu\bar{\psi}_\mu(p_\sigma'\gamma_\rho-p_\rho'\gamma_\sigma)\gamma_5\psi$
A	$I_{\mu\nu}\bar{\psi}_\mu\gamma_5\gamma_\rho\psi_\nu$	$\frac{2}{3}H\bar{\psi}_\rho\psi+\frac{2}{3}\frac{p_\rho'p_\mu}{M_iM_f}\bar{\psi}_\mu\psi$
P	$I_{\mu\nu}\bar{\psi}_\mu\gamma_5\psi_\nu$	$-\frac{2}{3M_i}i\bar{\psi}_\mu p_\mu\psi$

$$L_{\text{int}}(VVV)=3g\Bigg[m_\rho\,\text{Tr}\left(\left[\frac{\mathbf{F}_{\mu\nu}}{m}\right]\mathbf{V}_\mu\mathbf{V}_\nu\right)+\tfrac{1}{2}m_\rho^2\Bigg\{\text{Tr}\left[\left(\frac{\mathbf{F}_{\mu\nu}}{m}\right)\left(\frac{\mathbf{V}_\mu}{m}\right)\mathbf{V}_\nu\right]+\text{Tr}\left[\left(\frac{\mathbf{F}_{\mu\nu}}{m}\right)\mathbf{V}_\mu\left(\frac{\mathbf{V}_\nu}{m}\right)\right]-\text{Tr}\left[\mathbf{F}_{\mu\nu}\left(\frac{\mathbf{V}_\mu}{m}\right)\left(\frac{\mathbf{V}_\nu}{m}\right)\right]\Bigg\}-\tfrac{2}{3}m_\rho\,\text{Tr}\left[\left(\frac{\partial_\mu\mathbf{V}_\nu}{m}\right)\left(\frac{\partial_\nu\mathbf{V}_\rho}{m}\right)\left(\frac{\partial_\rho\mathbf{V}_\mu}{m}\right)\right]\Bigg], \tag{5.3}$$

where $\mathbf{V}/m$ is a 3×3 matrix whose elements are given by

$$\left(\frac{\mathbf{V}}{m}\right)_\alpha^{\ \beta}=\frac{\mathbf{V}_\alpha^{\ \beta}}{(m_T)_\alpha^{\ \beta}}$$

with $(m_T)_\alpha^{\ \beta}$ as defined in (4.14).

If the $SU(3)$ mass splittings are neglected, the vector meson field is coupled to a conserved $SU(3)$ meson current. Among the many relations between the different coupling constants listed in the Appendix, we consider

$$g_{\omega\rho\pi}^2=\frac{4}{m_\rho^2}g_{\rho\pi\pi}^2. \tag{5.4}$$

From the Gell-Mann, Sharp, and Wagner[23] model for the $\omega\to3\pi$ decay, the known $\rho\to2\pi$ decay rate and (5.4), the width Γ of $\omega\to3\pi$ is $\Gamma=5.4$ MeV. This is to be compared with the experimental value of 9.4 MeV. It would be interesting to see whether the difference could be accounted for by a direct $\omega\to3\pi$ contribution that can arise from $\text{Tr}(\Phi\Phi\Phi)$ interaction.

To obtain the effective meson-baryon couplings, it is convenient to define a current

$$J_A^{\ B}=\bar{\Psi}^{BDC}\Psi_{ADC}. \tag{5.5}$$

If we substitute the free field solutions (3.7), (4.31), and (4.32) for Ψ in (5.5), then

$$J_A^{\ B}=J_A^{\ B}(\bar{D}D)+J_A^{\ B}(\bar{D}B)+J_A^{\ B}(\bar{B}D)+J_A^{\ B}(\bar{B}B). \tag{5.6}$$

Each term in Eq. (5.6) can be separated into the $SU(3)$ singlet and octet parts. Further the space-time properties of each of the currents allows the decomposition into the usual scalar, vector, tensor, axial vector, and pseudoscalar parts:

$$J_{i\alpha}^{\ j\beta}(\ )=\tfrac{1}{4}[(1)_i^{\ j}J^S(\ )+(\gamma_\mu)_i^{\ j}J_\mu^{\ V}(\ )+(\sigma_{\mu\nu})_i^{\ j}J_{\nu\mu}^{\ T}(\ )+(\gamma_\mu\gamma_5)_i^{\ j}J_\mu^{\ A}(\ )+(\gamma_5)_i^{\ j}J^P(\ )]_\alpha^{\ \beta}. \tag{5.7}$$

The results[24] are given in Tables I and II. Table I contains the contributions to the current $J_A^{\ B}(\bar{D}D)$ and $J_A^{\ B}(\bar{D}B)$. $J_A^{\ B}(\bar{B}D)$ can be obtained from Table I since $J_A^{\ B}(\bar{B}D)=[J_A^{\ B}(\bar{D}B)]^\dagger$. Table II contains the baryonic current separated into $SU(3)$ parts that transform like the symmetric (D) octet, antisymmetric (F) octet and the singlet (S). The effective $\bar{D}DV$, $\bar{D}DP$, $\bar{D}BV$, $\bar{D}BP$, $\bar{B}BV$, and $\bar{B}BP$ couplings are obtained by combining these currents with the free-field solutions for mesons (4.15). Thus,

$$L_{\text{int}}(\text{baryons-mesons})=L_{\text{int}}(\bar{D}DM)+L_{\text{int}}(\bar{D}BM)+L_{\text{int}}(\bar{B}DM)+L_{\text{int}}(\bar{B}BM),$$

where

$$L_{\text{int}}(\ )=iG\left[J^P(\ )P+\frac{1}{m_\rho}J^A(\ )\partial_\mu P+J_\mu^{\ V}(\ )V_\mu-J_{\mu\nu}^{\ T}(\ )\left(\frac{F_{\mu\nu}}{m}\right)\right]. \tag{5.8}$$

From Tables I and II and (5.8),

$$\frac{1}{iG}L_{\text{int}}(\bar{D}DM)=\left(H\delta_{\mu\nu}+\frac{q_\mu q_\nu}{M_iM_f}\right)\left\{\left(1+\frac{M_i+M_f}{m_\rho}\right)[\bar{\psi}_\mu\gamma_5\psi_\nu P]+[\bar{\psi}_\mu\gamma_\rho\psi_\nu V_\rho]+2i\left[\bar{\psi}_\mu\sigma_{\rho\sigma}q_\sigma\psi_\nu\left(\frac{V_\rho}{m}\right)\right]\right\}, \tag{5.9}$$

[23] M. Gell-Mann, D. Sharp, and W. G. Wagner, Phys. Rev. Letters 8, 261 (1962).

[24] The expressions for different currents on Tables I and II reduce to the corresponding ones in Ref. 10 when the mass splittings are neglected. Please note the following errors in Table I of Ref. 10. In the axial vector (A) current of $\bar{D}D$, it should be $\gamma_5\gamma_\rho$ instead of $\gamma_\rho\gamma_5$. In the $\bar{D}B$ contribution there should be a negative sign before the pseudoscalar (P) current.

TABLE II. Baryon-baryon current.

	F	D	S
S	$\frac{1}{3}H\bar{\psi}\psi$	0	$\frac{1}{3}H\bar{\psi}\psi$
V	$\frac{1}{3}\left(H-\frac{1}{6}\frac{q^2}{M_iM_f}\right)\bar{\psi}\gamma_\rho\psi-\frac{(M_i+M_f)}{18M_iM_f}i\bar{\psi}\sigma_{\rho\sigma}q_\sigma\psi$ $+\frac{(M_f-M_i)}{18M_iM_f}iq_\rho\bar{\psi}\psi$	$\frac{q^2}{6M_iM_f}\bar{\psi}\gamma_\rho\psi+\frac{(M_i+M_f)}{6M_iM_f}i\bar{\psi}\sigma_{\rho\sigma}q_\sigma\psi$ $-\frac{(M_f-M_i)}{6M_iM_f}iq_\rho\bar{\psi}\psi$	$\frac{1}{3}\left(H-\frac{1}{3}\frac{q^2}{M_iM_f}\right)\bar{\psi}\gamma_\rho\psi-i\frac{(M_i+M_f)}{9M_iM_f}\bar{\psi}\sigma_{\rho\sigma}q_\sigma\psi$ $+\frac{M_f-M_i}{9M_iM_f}iq_\rho\bar{\psi}\psi$
T	$\frac{1}{9}H\bar{\psi}\sigma_{\rho\sigma}\psi-\frac{1}{18M_iM_f}(p_\rho'p_\sigma-p_\sigma'p_\rho)\bar{\psi}\psi$	$\frac{1}{6}H\bar{\psi}\sigma_{\rho\sigma}\psi+\frac{1}{6M_iM_f}(p_\rho'p_\sigma-p_\sigma'p_\rho)\bar{\psi}\psi$	$\frac{1}{18}H\bar{\psi}\sigma_{\rho\sigma}\psi-\frac{1}{9M_iM_f}(p_\rho'p_\sigma-p_\sigma'p_\rho)\bar{\psi}\psi$
A	$(2/9)H\bar{\psi}\gamma_5\gamma_\rho\psi$	$\frac{1}{3}H\bar{\psi}\gamma_5\gamma_\rho\psi$	$\frac{1}{9}H\bar{\psi}\gamma_5\gamma_\rho\psi$
P	$(2/9)H\bar{\psi}\gamma_5\psi$	$\frac{1}{3}H\bar{\psi}\gamma_5\psi$	$\frac{1}{9}H\bar{\psi}\gamma_5\psi$

$$\frac{1}{iG}\{L_{\text{int}}(\bar{D}BM)+L_{\text{int}}(\bar{B}DM)\}$$
$$=-\frac{2}{3}\left(1+\frac{M_i+M_f}{m_p}\right)i\left\{\frac{p_\mu}{M_i}[\bar{\psi}_\mu\psi P]+\frac{p_\mu'}{M_f}[\bar{\psi}\psi_\mu P]\right\}+\frac{2}{3}\left(1+\frac{M_i+M_f}{m}\right)$$
$$\times\left\{H([\bar{\psi}_\mu\gamma_5\psi V_\mu]+[\bar{\psi}\gamma_5\psi_\mu V_\mu])-\frac{q_\mu(p_\rho'+p_\rho)}{2M_iM_f}([\bar{\psi}_\mu\gamma_5\psi V_\rho]-[\bar{\psi}\gamma_5\psi_\mu V_\rho])\right.$$
$$\left.+iq_\mu\left(\frac{1}{M_i}[\bar{\psi}_\mu\gamma_\rho\gamma_5\psi V_\rho]+\frac{1}{M_f}[\bar{\psi}\gamma_\rho\gamma_5\psi_\mu V_\rho]\right)\right\}, \quad (5.10)$$

$$\frac{1}{iG}L_{\text{int}}(\bar{B}BM)=\tfrac{1}{3}H\left(1+\frac{M_i+M_f}{m_p}\right)\{[\bar{\psi}\gamma_5\psi P]_D+\tfrac{2}{3}[\bar{\psi}\gamma_5\psi P]_F+\tfrac{1}{3}[\bar{\psi}\gamma_5\psi P]_S\}$$
$$+\frac{\sqrt{3}}{9}H\left(1+\frac{M_i+M_f}{m_S}\right)\text{Tr}[\bar{\psi}\gamma_5\psi]X^0+\sum_{W=F,D,S}\{f_1{}^W[\bar{\psi}\gamma_\rho\psi V_\rho]_W+f_2{}^Wi[\bar{\psi}\sigma_{\rho\sigma}q_\sigma\psi V_\rho]_W\}. \quad (5.11)$$

In order to write the above interactions in as simple a form as possible we have used the following definitions and abbreviations:

$$H=((M_i+M_f)^2+q^2)/2M_iM_f, \quad (5.12)$$

where M_i and M_f refer to the initial and final baryon masses. $q_\mu=p_\mu'-p_\mu$ where p_μ and p_μ' are the four momenta of the initial and final baryons.

$$[\bar{\psi}_\mu O\psi_\nu M]=\bar{\psi}^{\alpha\beta\gamma}O\psi_{\nu,\alpha\beta'\gamma}M_\alpha{}^{\alpha'},$$
$$[\bar{\psi}_\mu O\psi M]=\bar{\psi}_\mu{}^{\alpha\beta\gamma}O\psi_\beta{}^\delta\epsilon_{\delta\gamma\alpha'}M_\alpha{}^{\alpha'},$$
$$[\bar{\psi}O\psi_\mu M]=\epsilon^{\delta\gamma\alpha}\bar{\psi}_\delta{}^\beta O\psi_{\alpha'\beta\gamma}M_\alpha{}^{\alpha'},$$
$$[\bar{\psi}O\psi M]_F=+\text{Tr}(\bar{\psi}OM\psi)-\text{Tr}(\bar{\psi}O\psi M) \quad (5.13)$$
$$[\bar{\psi}O\psi M]_D=\text{Tr}(\bar{\psi}O\psi M)+\text{Tr}(\bar{\psi}OM\psi)-\tfrac{2}{3}\text{Tr}(\bar{\psi}O\psi)\,\text{Tr}M,$$
$$[\bar{\psi}O\psi M]_S=\text{Tr}[\bar{\psi}O\psi]\,\text{Tr}M,$$

where O is any Dirac matrix and M represents either the pseudoscalar mesons P or the vector mesons V. It should be noted that in the present model, for the pseudoscalar mesons,

$$M_\alpha{}^{\alpha'}=P_\alpha{}^{\alpha'}+(1/\sqrt{3})X_0\delta_\alpha{}^{\alpha'}$$

and for the vector mesons,

$$M_\alpha{}^{\alpha'}=V_\alpha{}^{\alpha'},$$

where P and V are defined in the previous section. In the case of vector-meson interactions,

$$(1/m)[\bar{\psi}O\psi V]=[\bar{\psi}O\psi(V/m)],$$

and $f_{1,2}{}^W$ are given in Table III.

If the $SU(3)$ mass splittings are neglected, the Yukawa-type meson-baryon coupling constants have a D/F ratio of $\frac{3}{2}$ which agrees with the corresponding results in the nonrelativistic $SU(6)$ theory.[25] From the

[25] F. Gürsey, A. Pais, L. A. Radicati, Phys. Rev. Letters **13**, 299 (1964).

TABLE III. Expressions for $f_{1,2}^W$ defined in Eq. (5.11).

$i \backslash W$	F	f_i^W D	S
1	$\frac{H}{3}-\frac{q^2}{18M_iM_f}\left(1+\frac{M_i+M_f}{m}\right)$	$\frac{q^2}{6M_iM_f}\left(1+\frac{M_i+M_f}{m}\right)$	$\frac{H}{3}-\frac{q^2}{9M_iM_f}\left(1+\frac{M_i+M_f}{m}\right)$
2	$-\frac{M_i+M_f}{18M_iM_f}+\frac{2}{9}\frac{1}{m}\left(H+\frac{q^2}{4M_iM_f}\right)$	$\frac{M_i+M_f}{6M_iM_f}\left(1+\frac{M_i+M_f}{m}\right)$	$-\frac{1}{9}\frac{M_i+M_f}{M_iM_f}+\frac{1}{9}\frac{1}{m}\left(H+\frac{q^2}{M_iM_f}\right)$

results in the Appendix,

$$g_{N^{*++},P\pi^+}{}^2=\frac{4}{9}\frac{[1+(M_{N^*}+M_N)/m_\rho]^2}{M_N{}^2}G^2, \qquad (5.14)$$

$$g_{PP,\pi^0}{}^2=\frac{50}{81}\left(1-\frac{m_\pi{}^2}{4M^2}\right)^2\left(1+\frac{2M_N}{m_\rho}\right)^2G^2, \qquad (5.15)$$

$$g_{PP,\rho^0}{}^2=\frac{2}{9}\left[1-\frac{1}{3}\frac{m_\rho}{M_N}-\frac{5}{12}\left(\frac{m_\rho}{M_N}\right)^2\right]^2G^2. \qquad (5.16)$$

The requirement that $g_{PP,\pi^0}{}^2=15$ leads to a value of $G^2=2.05$. The width of N^* and $g_{PP,\rho^0}{}^2$ calculated using this value of G^2 are 94 MeV and 0.21. In spite of significant changes in the expressions for these coupling constants due to relativistic effects, the values obtained are in good agreement with the corresponding ones in the nonrelativistic $SU(6)$.[25] Since in the present considerations, the vertex functions are characterized by momentum-dependent form factors, a direct comparison with the experimental values needs further investigation.

It is interesting to note that in the zero-momentum-transfer limit, γ_μ-type coupling of the vector meson to the baryons is pure F-type. Thus the vector-meson field is coupled to a conserved $SU(3)$ current. Further, both the F- and D-type strangeness non-changing currents are conserved even in the presence of $SU(3)$ mass splittings.

Although our treatment has been without reference to quarks, it is helpful to consider the quark model to specify the electromagnetic interaction. In the quark model, the electromagnetic current is described by a bilinear form of the quark fields. As is evident from Table III, if we take only the minimal electromagnetic interaction for the quarks, we shall not obtain the correct value for the magnetic moment of the proton. We assume, therefore, that there is in addition an anamalous magnetic moment interaction. Now the bilinear form in quark fields is a 144-dimensional tensor representation of $M(12)$ and the minimal and anamalous parts transform as a vector and a tensor, respectively. If we assume these transformation properties under $M(12)$, then the electromagnetic interaction has the general form

$$F^V(q^2)J_\rho{}^{(V)}A_\rho-\frac{1}{m}F^T(q^2)J_{\rho\sigma}{}^T\mathfrak{F}_{\rho\sigma}. \qquad (5.17)$$

The contributions of these terms to the charge $[F_{\rm ch}(q^2)]$ and magnetic $[F_{\rm mag}(q^2)]$ form factors[26] are given in Table IV. From these results it follows that[27]

$$\frac{F_{\rm mag}{}^N(q^2)}{F_{\rm mag}{}^P(q^2)}=\frac{\mu_N}{\mu_P}=-\tfrac{2}{3} \qquad (5.18)$$

and

$$F_{\rm ch}{}^N(q^2)=0, \qquad (5.19)$$

for all values of q^2. These consequences are certainly consistent with the presently available experimental information. Another very interesting property of these form factors is that they satisfy the required threshold condition.[28]

In the low momentum transfer region, the electromagnetic structure of the proton and neutron may be dominated by the vector mesons. If we therefore make the further assumption that the electromagnetic field couples to the baryon in exactly the same manner as the $V_1{}^1$ component of the vector meson octet in the low momentum transfer region, we can calculate the absolute values of the total magnetic moments of all the baryons from Table III. We only note here that

$$\mu_P=1+\frac{2M_N}{m_\rho} \quad \text{and} \quad \mu_N=-\tfrac{2}{3}\mu_P \qquad (5.20)$$

in units of nuclear magnetons. Equation (5.20) gives $\mu_\rho\simeq 3.4$ as compared with the experimental value of 2.79.

VI. SUMMARY AND DISCUSSION

The starting point of our formulation was the assumption that fields corresponding to elementary particles are tensors of $M(12)$. The particle multiplet structure itself, however, was derived by the requirement that these fields satisfy prescribed free-field equations of motion which are not covariant with respect to $M(12)$.

---

[26] F. J. Ernst, R. G. Sachs and K. C. Wali, Phys. Rev. **119**, 1105 (1960).

[27] K. J. Barnes, P. Carruthers, and Frank von Hippel, Phys. Rev. Letters 14, 82 (1965).

[28] i.e., $F_{\rm ch}(q^2)=2MF_{\rm mag}(q^2)$ at $q^2=-4M^2$. V. Barger and R. Carhart, Phys. Rev. 136, B281 (1964). This condition is trivially satisfied since $F_{\rm ch}(-4M^2)=F_{\rm mag}(-4M^2)=0$.

TABLE IV. q^2 is the square of the momentum transfer. The form vectors $F^V(q^2)$ and $F^T(q^2)$ that multiply the vector and tensor contributions respectively are omitted from the table.

	Contribution of J^V		Contribution of J^T	
	F	D	F	D
$F_{\mathrm{ch}}(q^2)$	$\frac{(M_i+M_f)^2}{6M_iM_f}\left(1+\frac{q^2}{(M_i+M_f)^2}\right)$	0	$-\frac{(M_i+M_f)}{6M_iM_f}\frac{q^2}{m}\left(1+\frac{q^2}{(M_i+M_f)^2}\right)$	0
$F_{\mathrm{mag}}(q^2)$	$\frac{M_i+M_f}{9M_iM_f}\left(1+\frac{q^2}{(M_i+M_f)^2}\right)$	$\frac{M_i+M_f}{6M_iM_f}\left(1+\frac{q^2}{(M_i+M_f)^2}\right)$	$\frac{(M_i+M_f)^2}{9M_iM_f}\frac{1}{m}\left(1+\frac{q^2}{(M_i+M_f)^2}\right)$	$\frac{(M_i+M_f)^2}{6M_iM_f}\frac{1}{m}\left(1+\frac{q^2}{(M_i+M_f)^2}\right)$

It is remarkable that there exist tensors[29] and equations of motion (Bargmann-Wigner equations) which lead to a supermultiplet structure which corresponds exactly to that of $SU(6)$ symmetry. It was also shown that the observed mass splittings can be easily incorporated in the equations of motion leading to solutions in terms of physical masses.

The interaction Lagrangian was assumed to be invariant under $M(12)$ and effective vertex functions were obtained in the lowest order perturbation calculation using the solutions of the field equations. The relations between various coupling constants were discussed in the text and are also summarized in the Appendix in some cases of interest. In the few cases where an attempt has been made to compare the results with experiments, the results are certainly consistent with available information.

So far we have restricted our attention mainly to vertex parts. The problem of obtaining effective matrix elements for scattering processes remains to be discussed. Let us consider meson-baryon scattering as an example. There is a set of one-particle exchange diagrams such as vector-meson exchange diagrams, which have singularities near the physical region. If we take the $M(12)$ invariant vertex, this set of diagrams does not have an over-all $M(12)$ invariance except in the special case when the four-momentum in the propagator is zero (i.e., forward scattering). Thus, in general, even for the Born terms we do not obtain $M(12)$ invariant results. There are more complicated diagrams, which are in general responsible for the short-range forces. It is reasonable to assume that there are stronger symmetries in the short-range forces. From this basis we would like to propose the following model for the scattering problem. The Born terms for two-particle scattering amplitudes are to be obtained from one-particle exchange graphs which represent long-range forces and from the $M(12)$ invariant direct graphs which represent short-range forces. The latter contributions can be derived from the following effective Lagrangian:

$$L=G_1\bar{\Psi}^{ABC}\bar{\Psi}_{ABC}\Phi_D{}^E\Phi_E{}^D+G_2\bar{\Psi}^{ABC}\Phi_A{}^{A''}\Phi_{A''}{}^{A'}\bar{\Psi}_{A'BC}+G_3\bar{\Psi}^{ABC}\Phi_A{}^{A'}\Phi_B{}^{B'}\bar{\Psi}_{A'B'C}.$$

The complete scattering amplitude can then be obtained from these Born terms by using unitarity relations in a suitable way, e.g., dispersion relations with subtractions.

Perhaps it is also worth pointing out some of the difficulties in the formulation. $M(12)$ contains $SL(6,C)$ which is noncompact. Hence, it is not possible to consider this group as the group of transformations acting on physical states because of the difficulties discussed in the Introduction. Since the free-field equations are not covariant with respect to $M(12)$, the physical states do not form a unitary representations of $M(12)$. On the other hand, our procedure of calculating the effective vertex functions is equivalent to assuming formal invariance under $M(12)$ for these functions. Clearly such an invariance is not maintained when higher order corrections are taken into account, since the internal lines violate $M(12)$ invariance. Hence, it is hard to understand the various relationships between different coupling constants from the point of view of conventional renormalization procedures. Our assumption therefore has to be regarded only as a working hypothesis and its justification has to be sought on some dynamical basis.

ACKNOWLEDGMENT

The authors would like to thank Dr. W. D. McGlinn for many helpful discussions and a critical reading of the manuscript.

APPENDIX

The purpose of this Appendix is to collect together the relationships between diffeerent coupling constants for some cases of interest. In each case an interaction Lagrangian which conserves ordinary isotopic spin and hypercharge is written. The coupling constants are identified with the appropriate vertex functions of the present model with all the particles on the mass shell.

[29] Similarly, other tensors with appropriate symmetry properties have the content of $SU(6)$ multiplets; R. Delbourgo and M. A. Rashid, International Centre for Theoretical Physics, Trieste (to be published).

VPP Interactions

$$L_{\rm int}=i\Big[\frac{1}{i}g_{\rho\pi\pi}\boldsymbol{\varrho}_\mu\cdot\boldsymbol{\pi}\times\overleftrightarrow{\partial}_\mu\boldsymbol{\pi}+g_{\rho KK}\boldsymbol{\varrho}_\mu\cdot K^\dagger\boldsymbol{\tau}\overleftrightarrow{\partial}_\mu K+g_{K^*K\pi}(K_\mu^{*\dagger}\boldsymbol{\tau}K\cdot\overleftrightarrow{\partial}_\mu\boldsymbol{\pi}-\text{H.c.})$$
$$+G_{K^*K\eta}(K_\mu^{*\dagger}K\overleftrightarrow{\partial}_\mu\eta-\text{H.c.})+g_{\omega KK}\omega_\mu K^\dagger\overleftrightarrow{\partial}_\mu K+g_{\phi KK}\phi_\mu K^\dagger\overleftrightarrow{\partial}_\mu K\Big].\qquad\text{(A1)}$$

Comparison with (5.1) gives

$$g_{\rho\pi\pi}=(9/\sqrt{2})g\qquad\text{(A2)}$$

and

$$g_{\rho\pi\pi}=2g_{\rho KK}=2g_{K^*K\pi}=-(2/\sqrt{3})g_{K^*K\eta}=-2g_{\omega KK}=\sqrt{2}g_{\phi KK}.\qquad\text{(A3)}$$

Relations (A3) are consequences of just $SU(3)$ symmetry and ω-ϕ mixing.

VVP Interactions

$$L_{\rm int}=i\epsilon_{\mu\nu\lambda\delta}[g_{\rho\omega\pi}\partial_\mu\boldsymbol{\varrho}_\nu\partial_\lambda\omega_\delta\cdot\boldsymbol{\pi}+g_{K^*K^*\pi}\partial_\mu K_\nu^{*\dagger}\boldsymbol{\tau}\partial_\lambda K_\delta^*\cdot\boldsymbol{\pi}+g_{\rho K^*K}(\partial_\mu\boldsymbol{\varrho}_\nu\cdot\partial_\lambda K_\delta^{*\dagger}\boldsymbol{\tau}K-\text{H.c.})+g_{\omega K^*K}(\partial_\mu\omega_\nu\partial_\lambda K^{*\dagger}K-\text{H.c.})$$
$$+g_{\phi K^*K}(\partial_\mu\phi_\nu\partial_\lambda K_\delta^{*\dagger}K-\text{H.c.})+g_{\rho\rho\eta}\partial_\mu\boldsymbol{\varrho}_\nu\cdot\partial_\lambda\boldsymbol{\varrho}_\delta\eta+g_{\omega\omega\eta}\partial_\mu\omega_\nu\partial_\lambda\omega_\delta\eta+g_{\phi\phi\eta}\partial_\mu\phi_\nu\partial_\lambda\phi_\delta\eta+g_{K^*K\eta}\partial_\mu K^{*\dagger}\partial_\lambda K_\delta^*\eta]\qquad\text{(A4)}$$
$$+g_{\rho\rho X_0}\partial_\mu\boldsymbol{\varrho}_\nu\cdot\partial_\lambda\boldsymbol{\varrho}_\delta X_0+g_{\omega\omega X_0}\partial_\mu\omega_\nu\partial_\lambda\omega_\delta X_0+g_{\phi\phi X_0}\partial_\mu\phi_\nu\partial_\lambda\phi_\delta X_0+g_{K^*K^*X_0}\partial_\mu K^*\partial_\lambda K^*_\delta X_0.$$

Comparison with (5.2) gives

$$g_{\rho\omega\pi}=(9\sqrt{2}/m_\rho)g\qquad\text{(A5)}$$

and

$$g_{K^*K^*\pi}=\frac{1}{6}\left(\frac{m_\rho}{m_{K^*}}\right)^2\left[2+\left(\frac{m_\rho}{m_{K^*}}\right)^2\right]g_{\rho\omega\pi},\qquad g_{\phi\phi\eta}=-\frac{1}{3\sqrt{3}}\left(\frac{m_\rho}{m_\phi}\right)^2\left[2+\left(\frac{m_\rho}{m_\phi}\right)^2\right]g_{\rho\omega\pi},$$

$$g_{\rho K^*K}=\frac{1}{6}\left[1+2\left(\frac{m_\rho}{m_{K^*}}\right)^2\right]g_{\rho\omega\pi},\qquad g_{K^*K^*\eta}=-\frac{1}{6\sqrt{3}}\left(\frac{m_\rho}{m_{K^*}}\right)^2\left[2+\left(\frac{m_\rho}{m_{K^*}}\right)^2\right]g_{\rho\omega\pi},$$

$$g_{\omega K^*K^*}=+\frac{1}{6}\left[1+2\left(\frac{m_\rho}{m_{K^*}}\right)^2\right]g_{\rho\omega\pi},\qquad g_{\rho\rho X_0}=g_{\omega\omega X_0}=\frac{1}{3\sqrt{6}}\left[1+2\left(\frac{m_\rho}{m_S}\right)\right]g_{\rho\omega\pi},\qquad\text{(A6)}$$

$$g_{\phi K^*K}=\frac{1}{3\sqrt{2}}\left[\left(\frac{m_\rho}{m_{K^*}}\right)^2+\left(\frac{m_\rho}{m_\phi}\right)^2+\left(\frac{m_\rho}{m_{K^*}}\right)^2\left(\frac{m_\rho}{m_\phi}\right)^2\right]g_{\rho\omega\pi},\qquad g_{\phi\phi X_0}=\frac{1}{3\sqrt{6}}\left[\left(\frac{m_\rho}{m_\phi}\right)^2+2\left(\frac{m_\rho}{m_S}\right)\left(\frac{m_\rho}{m_\phi}\right)^2\right]g_{\rho\omega\pi},$$

$$g_{\rho\rho\eta}=g_{\omega\omega\eta}=\frac{1}{2\sqrt{3}}g_{\rho\omega\pi},\qquad g_{K^*K^*X_0}=\frac{2}{3\sqrt{6}}\left[\left(\frac{m_\rho}{m_{K^*}}\right)^2+2\left(\frac{m_\rho}{m_S}\right)\left(\frac{m_\rho}{m_{K^*}}\right)^2\right]g_{\rho\omega\pi},$$

VPP and **VVP** coupling constants are related because of (A2) and (A5):

$$g_{\rho\omega\pi}=(2/m_\rho)g_{\rho\pi\pi}.\qquad\text{(A7)}$$

$\bar{D}BP$ Interactions

The space-time structure of the interaction is given by

$$\bar{\psi}_\mu(\partial\psi/\partial x_\mu)\phi,\qquad\text{(A8)}$$

where ϕ represents the pseudoscalar meson field. If the coupling of the baryons and pseudoscalar mesons is decomposed into the conventional isotopic-spin representation,[30] then from (5.10)

$$g_{B^*,BP}=g_{B^*,BP}{}^0\frac{2}{3}\frac{1}{M_B}\left(1+\frac{M_{B^*}+M_B}{m_\rho}\right),\qquad\text{(A9)}$$

[30] For the isotopic-spin decomposition of $\bar{D}BP$ and $B\bar{B}P$ interactions see, for example, A. W. Martin and K. C. Wali, Nuovo Cimento **31**, 1324 (1964); Phys. Rev. **130**, 2455 (1963).

where M_{B^*} and M_B refer to the masses of the members of the decuplet and the baryon octet respectively. $g_{B^*,BP}{}^0$ are determined by $SU(3)$ symmetry and they are given by

$$
\begin{aligned}
G = g_{N^*,N\pi}{}^0 &= -g_{N^*,\Sigma K}{}^0 = -(\sqrt{6})g_{Y^*,\Sigma\pi}{}^0 = -\sqrt{2}g_{Y^*,\Lambda\pi}{}^0 = (\sqrt{6})g_{Y^*,N\bar{K}}{}^0 = (\sqrt{6})g_{Y^*,\Xi K}{}^0 \\
&= \sqrt{2}g_{Y^*,\Sigma\eta}{}^0 = (\sqrt{6})g_{\Xi^*,\Sigma\bar{K}}{}^0 = -(\sqrt{6})g_{\Xi^*,\Xi\pi}{}^0 = \sqrt{2}g_{\Xi^*,\Lambda\bar{K}}{}^0 = -\sqrt{2}g_{\Xi^*,\Xi\eta}{}^0 = g_{\Omega,\Xi\bar{K}}{}^0 .
\end{aligned}
\tag{A10}
$$

$\bar{B}BP$ Interactions

With the usual decomposition of the interactions into isotopic spin representation and (5.11), we have

$$
g_{BB'P} = g_{BB'P}{}^0 \frac{1}{18}\left(\frac{(M_B+M_{B'})^2}{M_B M_{B'}} - \frac{m_P{}^2}{M_B M_{B'}}\right)\left(1+\frac{M_B+M_{B'}}{m_\rho}\right),
\tag{A11}
$$

where $g_{BB'P}{}^0$ are determined by exact $SU(3)$ symmetry and a D/F ratio of $\frac{3}{2}$. Thus

$$
\begin{array}{lll}
g_{NN\pi}{}^0 = 5(G/\sqrt{2}), & g_{\Lambda NK}{}^0 = -3\sqrt{3}(G/\sqrt{2}), & g_{NN\eta} = \sqrt{3}(G/\sqrt{2}), \\
g_{\Lambda\Sigma\pi}{}^0 = 2\sqrt{3}(G/\sqrt{2}), & g_{\Sigma NK}{}^0 = G/\sqrt{2}, & g_{\Lambda\Lambda\eta} = -2\sqrt{3}(G/\sqrt{2}), \\
g_{\Sigma\Sigma\pi}{}^0 = 4(G/\sqrt{2}), & h_{\Lambda\Xi K} = \sqrt{3}(G/\sqrt{2}), & g_{\Sigma\Sigma\eta} = 2\sqrt{3}(G/\sqrt{2}), \\
g_{\Xi\Xi\pi}{}^0 = -G/\sqrt{2}, & h_{\Sigma\Xi K} = -5(G/\sqrt{2}), & g_{\Xi\Xi\eta} = -3\sqrt{3}(G/\sqrt{2}).
\end{array}
$$

The γ_μ- and $\sigma_{\mu\nu}$-type $\mathbf{\bar{B}BV}$ interactions can be easily read from Table III.

Model for Low-Energy Meson-Baryon Scattering

Gary R. Goldstein*
Department of Physics, University of Chicago, Chicago, Illinois
and
Argonne National Laboratory, Argonne, Illinois
AND
Kameshwar C. Wali†
Argonne National Laboratory, Argonne, Illinois
(Received 29 September 1966)

A model for low-energy meson-baryon scattering within the framework of $U(6,6)$ symmetry is studied. The effective $U(6,6)$-symmetric terms are treated as background terms representing distant singularities. These are supplemented by the appropriate one-particle exchange contributions. The model is used to calculate the real parts of the S, P, and D partial-wave amplitudes in the low-energy region of πN and KN elastic scattering. It is found that the results are in reasonably good agreement with the experimental information in $T=\frac{3}{2}$ and $T=\frac{1}{2}$ πN and in $T=1$ KN scattering. However, the model predicts large S-wave scattering in the $T=0$ KN channel.

I. INTRODUCTION

In dispersion theoretic treatments of scattering processes, it has been customary to treat the long-range forces (or, equivalently, singularities near the physical region) by appropriate one-particle exchange graphs. The short-range forces (or, equivalently, distant singularities) are approximated by low-order polynomials in the usual Mandelstam variables s and t. The resulting representation for the scattering amplitude is known as the Cini-Fubini approximation[1] and has proved successful in semiphenomenological analyses of πN and KN elastic scattering in a restricted energy region.[2,3]

The present paper is concerned with the investigation of a specific model for meson-baryon scattering, which leads to Cini-Fubini–type amplitudes in a very natural way. The model is based on a symmetry scheme which is presently known as $U(6,6)$ symmetry.[4–6] The main ideas leading to the model were briefly discussed in I. As regards the development of $U(6,6)$ symmetry, its successes and failures, the interested reader is referred to a number of recent review articles.[7] What is relevant for our present purposes is the consideration that $U(6,6)$ symmetry provides a satisfactory description of three-point functions.[8] This implies in particular that the trilinear meson-baryon coupling constants, where the mesons and baryons belong respectively to the 35-dimensional (octet of pseudoscalar and nonet of vector mesons) and 56-dimensional (octet of $J^P=\frac{1}{2}^+$ and decuplet of $J^P=\frac{3}{2}^+$ baryons) representations of $SU(6)$, are determined in terms of the πN and ρ-$\pi\pi$ coupling constants. Thus, all the one-particle exchange contributions involving these particles are determined in terms of these two coupling constants.

In addition to the one particle exchange contributions, there are "effective" $U(6,6)$ symmetric terms which are polynomials in the variables s and t.[9,10] These terms may be considered as arising from more complicated diagrams. Taken by themselves as representing the full scattering amplitude, they lead to some consequences which are in total disagreement with experimental results.[9,10] Our viewpoint is to regard these terms, supplemented by the one-particle exchange contributions, as input Born terms for dynamical calculations based on dispersion theoretic methods. To see whether such an approach is meaningful, we consider in this paper pseudoscalar meson-baryon scattering in partial wave amplitudes f_l, where $l=0$, 1, 2. As no unitarity corrections are taken into account, the scattering amplitudes are real and the model is applicable only to small phase shifts in the low-energy region.

In Sec. II, we analyze the structure of the $U(6,6)$ symmetric terms. Section III is devoted to the discussion of free-particle propagators and the calculation of one-particle exchange contributions. The general features and the comparison of the numerical results

* Associated Midwest Universities—Argonne National Laboratory Predoctoral Fellow.

† Work done under the auspices of the U. S. Atomic Energy Commission.

1 M. Cini and S. Fubini, Ann. Phys. (N. Y.) 3, 352 (1960).

2 J. Bowcock, W. M. Cottingham, and D. Lurié, Nuovo Cimento **16**, 918 (1960).

3 R. L. Warnock and G. Frye, Phys. Rev. **138**, B947 (1965).

4 A. Salam, R. Delbourgo, and J. Strathdee, Proc. Roy. Soc. (London) **284A**, 146 (1965); A. Salam, R. Delbourgo, M. A. Rashid, and J. Strathdee, *ibid.* **285A**, 312 (1965).

5 M. A. B. Bég and A. Pais, Phys. Rev. Letters **14**, 267 (1965).

6 B. Sakita and K. C. Wali, Phys. Rev. Letters **14**, 404 (1965); Phys. Rev. **139**, B1355 (1965). The second paper will hereafter be referred to as I.

7 R. Delbourgo, M. A. Rashid, A. Salam, and J. Strathdee, in *Proceedings of the Seminar Conference on High-Energy Physics and Elementary Particles, Trieste, 1965* (International Atomic Energy Agency, Vienna, 1965); A. Pais, Rev. Mod. Phys. **38**, 215 (1966); B. Sakita, Advances in High-Energy Physics (to be published).

8 It should be noted that only the collinear subgroup of $U(6,6)$, $SU(6)_W$ is relevant for the three-point functions. H. J. Lipkin and S. Meshkov, Phys. Rev. Letters **14**, 670 (1965).

9 R. Blankenbecler, M. L. Goldberger, K. Johnson, and S. Trieman, Phys. Rev. Letters **14**, 518 (1965).

10 J. M. Cornwall, P. G. O. Freund, and K. T. Mahanthappa, Phys. Rev. Letters **14**, 515 (1965).

of the model in the case of low-energy πN and KN scattering are presented in Sec. IV. The final section contains the summary and the discussion of the results.

II. $U(6,6)$-SYMMETRIC FOUR-POINT FUNCTIONS

In the $U(6,6)$ symmetry scheme, one can construct effective interaction terms that are formally $U(6,6)$ invariant. However, the kinetic terms or the free Lagrangian parts break this symmetry. This was the main reason that led Bég and Pais[11] to suggest that such invariant terms represent effective S-matrix elements. As remarked in the Introduction, we interpret the four-point interaction terms only as approximations to distant singularities in two-body scattering.

For meson-baryon scattering there are four independent $U(6,6)$ symmetric amplitudes and, in momentum space, the effective scattering matrix element is given by

$$\begin{aligned} M_S=&\alpha\bar{\Psi}^{ABC}(p')\Psi_{ABC}(p)\bar{\Phi}_E{}^D(-k')\Phi_D{}^E(k) \\ &+\beta\bar{\Psi}^{ABC}(p')\Psi_{ABD}(p)\bar{\Phi}_E{}^D(-k')\Phi_C{}^E(k) \\ &+\tilde{\beta}\bar{\Psi}^{ABC}(p')\Psi_{ABD}\Phi_E{}^D(k)\bar{\Phi}_C{}^E(-k') \\ &+\gamma\bar{\Psi}^{ABC}(p')\Psi_{ADE}(p)\bar{\Phi}_B{}^D(-k')\Phi_C{}^E(k)\,, \end{aligned} \tag{2.1}$$

where Ψ_{ABC} and $\Phi_A{}^B$ are, respectively, the totally symmetric third-rank and mixed second-rank tensors of $U(6,6)$. As in the case of three-point functions in I, we calculate M_S by using free-field solutions of Bargmann-Wigner equations for Ψ and Duffin-Kemmer equations for Φ. Here α, β, $\tilde{\beta}$, γ are functions of s, t, and u subject to the restrictions due to crossing symmetry:

$$\begin{gathered} \alpha(s,t,u)=\alpha(u,t,s)\,; \quad \beta(s,t,u)=\tilde{\beta}(u,t,s)\,; \\ \gamma(s,t,u)=\gamma(u,t,s)\,. \end{gathered} \tag{2.2}$$

In the present investigation, we make the simplest choice for these arbitrary functions, namely, we regard them as constants. The restrictions in (2.2) then imply that $\beta=\tilde{\beta}$, and hence we have effectively three free parameters.

To obtain M_S, it is convenient to define the "currents"

$$\begin{aligned} j_C{}^D&=\bar{\Phi}_E{}^D(-k')\Phi_C{}^E(k)+\Phi_E{}^D(k)\bar{\Phi}_C{}^E(-k')\,, \\ J_D{}^C&=\bar{\Psi}^{ABC}(p')\Psi_{ABD}(p)\,, \end{aligned} \tag{2.3}$$

and

$$\begin{aligned} J_{ABC}&=\Phi_A{}^D(k)\Psi_{BCD}(p)\,, \\ \bar{J}^{ABC}&=\bar{\Psi}^{ABD}(p')\bar{\Phi}_D{}^C(-k')\,. \end{aligned} \tag{2.4}$$

M_S is then given by

$$M_S=\alpha J_C{}^C j_D{}^D+\beta J_D{}^C j_C{}^D+\gamma\bar{J}^{ABC}J_{ABC}\,. \tag{2.5}$$

As we are interested in pseudoscalar-meson scattering from the octet of baryons, we consider only the pseudoscalar-meson and baryon parts of Φ and Ψ in the currents. From I, we have[12]

$$\Phi_i{}^j{}_\alpha{}^\beta(k):\quad \left[\gamma_5\left(1-\frac{ik}{m_0}\right)\right]_i^j P_\alpha{}^\beta\,,$$

$$\begin{aligned} \Psi_{ijk,\alpha\beta\gamma}:\quad B_{ijk,\alpha\beta\gamma}=\tfrac{1}{3}[&\chi_{ijk,\alpha}{}^\delta\epsilon_{\delta\beta\gamma} \\ &+\chi_{jki,\beta}{}^\delta\epsilon_{\delta\gamma\alpha}+\chi_{kij,\gamma}{}^\delta\epsilon_{\delta\alpha\beta}]\,, \end{aligned}$$

where

$$\chi_{ijk,\alpha}{}^\delta=\tfrac{1}{2}[\gamma_5 C(1+i\not{p}/M_\alpha{}^\delta)]_{jk}u_i(p)B_\alpha{}^\delta\,. \tag{2.6}$$

By using these free-field solutions, we compute the currents (2.3) and (2.4) and hence M_S. The results are summarized in the Appendix (A6) in terms of the conventional invariant amplitudes $A(s,t)$ and $B(s,t)$.[13]

Although M_S has been calculated in this fashion, it is instructive and convenient for the purposes of the next section to decompose the currents $j_C{}^D$ and J_{ABC} in a manner similar to that for $J_C{}^D$ in I [Eqs. (5.6, 5.7)].

$$\begin{aligned} j_C{}^D(\bar{P}P)=j_i{}^j{}_\gamma{}^\delta(\bar{P}P)=\tfrac{1}{4}[&(1)_i{}^j j^S(\bar{P}P) \\ &+(\gamma_\mu)_i{}^j j^V{}_\mu(\bar{P}P)+(\sigma_{\mu\nu})_i{}^j j^T{}_{\nu\mu}(\bar{P}P) \\ &+(\gamma_\mu\gamma_5)_i{}^j j^A{}_\mu(\bar{P}P)+(\gamma_5)_i{}^j j^P(\bar{P}P)]_\gamma{}^\delta\,, \end{aligned}$$

where

$$\begin{aligned} j^S(\bar{P}P)&=4\left(1-\frac{k'\cdot k}{m_0{}^2}\right)(\bar{P}P+P\bar{P})\,, \\ j^V{}_\mu(\bar{P}P)&=\frac{4i}{m_0}(k_\mu+k_\mu')(\bar{P}P-P\bar{P})\,, \\ j^T{}_{\mu\nu}(\bar{P}P)&=\frac{i}{4m_0}(q_\nu j^V{}_\mu(\bar{P}P)-q_\mu j^V{}_\nu(\bar{P}P))\,, \\ j^P(\bar{P}P)&=j^A(\bar{P}P)=0\,, \end{aligned} \tag{2.7}$$

where $q_\mu=k_\mu'-k_\mu$, and P and $\bar{P}$ are $SU(3)$ matrices for incoming and outgoing pseudoscalar mesons.

From $J_{ABC}(BP)$, we can project the parts $J^D{}_{ABC}$ and $J^B{}_{ABC}$ that have the $SU(3)$ transformation property of the decuplet and the octet and expand them as follows:

$$\begin{aligned} J^D{}_{ABC}(BP)&=J^D{}_{ijk,\alpha\beta\gamma}(BP)=\tfrac{1}{2}[(\gamma_\mu C)_{jk}J^D{}_{\mu,i}+\tfrac{1}{2}(\sigma_{\mu\nu}C)_{jk}J^D{}_{\mu\nu,i}]_{\alpha\beta\gamma}\,, \\ J^B{}_{ABC}(BP)&=J^B{}_{ijk,\alpha\beta\gamma}(BP)=\tfrac{1}{3}[J^B{}_{ijk,\alpha}{}^\delta\epsilon_{\delta\beta\gamma}+J^B{}_{jki,\beta}{}^\delta\epsilon_{\delta\gamma\alpha}+J^B{}_{kij,\gamma}{}^\delta\epsilon_{\delta\alpha\beta}]\,, \\ J^B{}_{ijk,\alpha}{}^\delta&=\tfrac{1}{2}[J^\xi{}_i C_{jk}+J^\eta{}_{\mu,i}(\gamma_\mu\gamma_5 C)_{jk}+J^\psi{}_i(\gamma_5 C)_{jk}]_\alpha{}^\delta\,. \end{aligned} \tag{2.8}$$

[11] M. A. B. Bég and A. Pais, Phys. Rev. **137**, B1514 (1965).

[12] Because of the specific way in which mass splittings were introduced in I, $m_0=m_\rho$ and $M_\alpha{}^\delta$ corresponds to the physical mass of the baryon $B_\alpha{}^\delta$.

[13] In the absence of mass splittings these results contain the results of Cornwall *et al.* (Ref. 10) as special cases.

Then,

$$J^D{}_{\lambda,\alpha\beta\gamma}=\frac{-1}{54}\bigg[\bigg(1-2\frac{ik}{m_0}-\frac{p\cdot k}{Mm_0}\bigg)\gamma_\lambda+\frac{2ip_\lambda}{M}\bigg(2-\frac{ik}{M}\bigg)-\frac{2ik_\lambda}{m_0}\bigg]u(p)\sum_{P(\alpha\beta\gamma)}\epsilon_{\delta\sigma\alpha}B_\beta{}^\delta P_\gamma{}^\sigma,$$

$$J^D{}_{\mu\nu,\alpha\beta\gamma}=\frac{-1}{54}\bigg[\bigg(1+\frac{k\cdot p}{Mm_0}\bigg)\sigma_{\mu\nu}-2i\bigg(1+\frac{ik}{m_0}\bigg)\bigg(\frac{p_\mu\gamma_\nu}{M}-\frac{p_\nu\gamma_\mu}{M}\bigg)+\frac{2}{Mm_0}(k_\mu p_\nu-k_\nu p_\mu)\bigg]u(p)\sum_{P(\alpha\beta\gamma)}\epsilon_{\delta\sigma\alpha}B_\beta{}^\delta P_\gamma{}^\sigma,$$

$$\gamma_\mu J^D{}_\mu=0,\quad \gamma_\mu J^D{}_{\mu\nu}=J^D{}_\nu. \tag{2.9}$$

$$J^\psi{}_\alpha{}^\delta=\frac{1}{18}\bigg[\bigg(3-5\frac{ik}{m_0}-2\frac{p\cdot k}{Mm_0}\bigg)\gamma_5u(p)(P_\alpha{}^\sigma B_\sigma{}^\delta-\tfrac{1}{3}\delta_\alpha{}^\delta\,\mathrm{Tr}PB)-\bigg(\frac{ik}{m_0}+\frac{p\cdot k}{Mm_0}\bigg)\gamma_5u(p)(B_\alpha{}^\sigma P_\sigma{}^\delta-\tfrac{1}{3}\delta_\alpha{}^\delta\,\mathrm{Tr}BP)\bigg],$$

$$J^\eta{}_{\mu,\alpha}{}^\delta=\frac{1}{18}\bigg[\bigg\{2\bigg(1+\frac{p\cdot k}{Mm_0}\bigg)\gamma_\lambda-\frac{5ip_\lambda}{M}\bigg(1-\frac{ik}{m_0}\bigg)\bigg\}\gamma_5u(p)(P_\alpha{}^\sigma B_\sigma{}^\delta-\tfrac{1}{3}\delta_\alpha{}^\delta\,\mathrm{Tr}BP)$$
$$+\bigg\{\bigg(1+\frac{p\cdot k}{Mm_0}\bigg)\gamma_\lambda-\frac{ip_\lambda}{M}\bigg(1-\frac{ik}{m_0}\bigg)\bigg\}\gamma_5u(p)(B_\alpha{}^\sigma P_\sigma{}^\delta-\tfrac{1}{3}\delta_\alpha{}^\delta\,\mathrm{Tr}BP)\bigg],$$

and

$$J^\xi=\gamma_5J^\psi+\gamma_\mu\gamma_5J^\eta{}_\mu. \tag{2.10}$$

The decompositions (2.7), (2.9), and (2.10) show more clearly the structure of the $U(6,6)$ symmetric terms (J^Sj^S, $J^V{}_\mu j^V{}_\mu$, $J^\psi{}_BJ^\psi{}_B$, etc.). They also isolate the parts of the current that would couple to the vector-meson, baryon, and decuplet.

III. PROPAGATORS FOR Φ AND Ψ; ONE-PARTICLE EXCHANGES

There are two procedures that one can adopt in computing the one-particle exchange contributions. The first procedure is to regard $U(6,6)$ symmetry for the three-point functions (such as $\mathrm{Tr}\bar{\Psi}\Psi\Phi$, $\mathrm{Tr}\Phi\Phi\Phi$) as the symmetry that relates only the coupling constants and not the vertex functions. It is then a straightforward matter to use the conventional propagators for the physical particles that Ψ and Φ describe and compute the Born terms. Such a procedure [$SU(6)_W$ approach] has been the basis of several bootstrap models.[14] An alternative approach which is more consistent with the original formulation[4,6] of $U(6,6)$ symmetry is to compute the Born terms from the trilinear interaction terms and the propagators for Φ and Ψ derived from a Lagrangian.

Several authors[15] have discussed the Lagrangian formulation within the framework of $U(6,6)$ symmetry. In the present investigation, we follow Salam, Delbourgo, and Strathdee[4] with the appropriate modifications necessary to include the mass splittings. The essential difference between different approaches is in the contact terms (or nonpole terms) that invariably appear in the propagators. The contact terms are unique in the case of the field Φ because of the fact that Φ obeys the Duffin-Kemmer equations which can be derived from a Lagrangian. In the case of the Ψ field, the total symmetry of Ψ_{ABC} requires the presence of auxiliary variables in the Lagrangian. Different procedures lead to different contact terms, although the residues of the poles are the same in all cases. A prescription is therefore necessary to determine the contact terms. We obtain the required prescription by examining the meson-exchange contributions.

If we ignore the mass splittings, the propagator for Φ is given by

$$\langle\Phi_A{}^B,\Phi_C{}^D\rangle_+=\mathfrak{P}_{i\alpha,k\gamma}{}^{j\beta,l\delta}=\frac{1}{2}\bigg[\delta_A{}^D\delta_C{}^B+\frac{(-iq+m)_A{}^D(iq+m)_C{}^B}{m^2+q^2}\bigg]. \tag{3.1}$$

It is easy to verify that

$$\tfrac{1}{2}[(iq+m)_l{}^n\delta_m{}^k+(-iq+m)_m{}^k\delta_l{}^n]\times\mathfrak{P}_{i,k}{}^{j,l}=\frac{1}{m}\delta_i{}^n\delta_m{}^j, \tag{3.2}$$

where Φ satisfies the Duffin-Kemmer equation

$$\tfrac{1}{2}[(iq+m)_l{}^n\delta_m{}^k+(-iq+m)_m{}^k\delta_l{}^n]\Phi_n{}^m=0, \tag{3.3}$$

and where, for convenience, we have suppressed the $SU(3)$ variables. Now to obtain the one-meson exchange contribution to meson-baryon scattering we take the time-ordered product of the trilinear $U(6,6)$-invariant terms $iG\bar{\Psi}^{EFA}\Psi_{EFB}\Phi_A{}^B$ and $(ig/4m_0)\Phi_C{}^D\times\Phi_D{}^G\Phi_G{}^C$, and contract once on the meson fields. The result in terms of (2.3), (2.4), and (3.1) is

$$-\frac{gG}{4m_0}\bar{J}_B{}^A\mathfrak{P}_{AC}{}^{BD}j_D{}^C.$$

[14] R. Gatto and G. Veneziano, Phys. Letters **19**, 512 (1965); R. H. Capps, Phys. Rev. Letters **16**, 1066 (1966).

[15] C. S. Guralnik and T. W. B. Kibble, Phys. Rev. **139**, B712 (1965); S. Kamefuchi and Y. Takahashi, Nuovo Cimento **44**, A1 (1966); A. Salam *et al.* (Ref. 4).

It is clear that the terms in $\mathfrak{P}_{AC}{}^{BD}$ proportional to $\delta_A{}^D\delta_C{}^B$ will produce a $U(6,6)$ invariant contribution to the one-meson exchange (obtained by letting $q_\mu=0$) which is proportional to the β term of Eq. (2.5). The parts of the propagator containing iq's will break the $U(6,6)$ symmetry. Therefore, it is natural to redefine the propagator so that the resulting "one-particle exchange" is a pure symmetry-breaking amplitude. This is accomplished by adding the contact term $-\delta_A{}^D\delta_C{}^D$ to the original propagator (3.1) to obtain the symmetry-breaking propagator

$$\langle\Phi_A{}^B,\bar{\Phi}_C{}^D\rangle_+ =\frac{1}{2}\left[-\delta_A{}^D\delta_C{}^B+\frac{(-iq+m)_A{}^D(iq+m)_C{}^B}{m^2+q^2}\right]. \quad (3.4)$$

Similar considerations lead us to define the propagator corresponding to the Ψ field as

$$\langle\Psi_{ABC},\bar{\Psi}^{DEF}\rangle_+ =\frac{1}{12}\sum_{P(ABC)}\left[\frac{(-iq+M)_A{}^D(-iq+M)_B{}^E(-iq+M)_C{}^F}{M(q^2+M^2)} -\frac{(M-3iq)_A{}^D\delta_B{}^E\delta_C{}^F}{M}\right]. \quad (3.5)$$

Note that both propagators [(3.4) and (3.5)] vanish when the 4-momentum of the exchanged particle goes to zero. From (3.4) and (3.5), the propagators for different components of Φ and Ψ [compare Delbourgo *et al.*, Eqs. (4.29) and (4.30), Ref. 7] can be obtained by multiplying with appropriate combinations of γ matrices and the antisymmetric charge-conjugation matrix C. The inclusion of the mass splittings, based on the work in I, essentially amounts to introducing physical masses in the projected propagators. If we combine the symmetry-breaking propagators (3.4) and (3.5) with the vertex functions, we can write somewhat symbolically the effective meson-exchange and baryon-exchange matrix elements in momentum space as

$$-\tfrac{1}{4}Gg\bar{\Psi}^{ABC}(p')\Psi_{ABD}(p)\langle\Phi_C{}^D\bar{\Phi}_E{}^F\rangle_+ \times[\Phi_F{}^G(k)\bar{\Phi}_G{}^E(-k')+\bar{\Phi}_F{}^G(-k')\Phi_G{}^E(k)],$$

and

$$G^2\bar{\Psi}^{ABA'}(p')\Phi_{A'}{}^C(-k')\langle\Psi_{ABC}\bar{\Psi}^{DEF}\rangle_+\Phi_D{}^{D'}(k)\Psi_{D'EF}(p),$$

or in terms of the currents (2.3) and (2.4),

$$-\tfrac{1}{4}GgJ_D{}^C\langle\Phi_C{}^D\bar{\Phi}_E{}^F\rangle_+j_F{}^E,$$

and

$$G^2\bar{J}^{ABC}\langle\Psi_{ABC}\bar{\Psi}^{DEF}\rangle_+J_{DEF}. \quad (3.6)$$

Finally when we use the decompositions (2.7), (2.9), and (2.10) for the currents, we obtain

for *vector-meson exchange:*

$$-\frac{3Gg}{16}\frac{1}{m_V{}^2+q^2}\left[\frac{3}{2}(q_\lambda q_\mu-\delta_{\mu\lambda}q^2)J^V{}_\mu j^V{}_\lambda -\frac{2i}{m_0}J^T{}_{\mu\nu}(m_V{}^2-\tfrac{1}{2}q^2)q_\mu j_\nu\right], \quad (3.7)$$

for *baryon exchange:*

$$G^2\left[\left(\bar{J}^\psi+\frac{iq_\mu}{M_f}\bar{J}^\eta{}_\mu\right)\frac{M_E-iq}{M_E{}^2+q^2}\left(J^\psi-\frac{iq_\lambda}{M_i}J^\eta{}_\lambda\right) -\bar{J}^\psi\frac{M_E-iq}{M_E}J^\psi+\frac{i}{M_E}(\bar{J}^\psi q_\mu J^\eta{}_\mu-\bar{J}^\eta{}_\mu q_\mu J^\psi)\right]. \quad (3.8)$$

for *decuplet exchange:*

$$G^2\left[\left(\bar{J}^D{}_\mu+\frac{iq_\nu}{M_f}\bar{J}_{\nu\mu}\right) \times\left(\delta_{\mu\lambda}+\frac{q_\mu q_\lambda}{M_E{}^2}\right)\frac{M_E(M_E-iq)}{M_E{}^2+q^2}\left(J_\lambda+\frac{iq_\rho}{M_i}J_{\lambda\rho}\right) -\bar{J}_\mu\frac{M_E-2iq}{M_E}J_\mu+\tfrac{1}{2}\bar{J}_{\mu\nu}\frac{iq}{M_E}J_{\nu\mu}\right]. \quad (3.9)$$

In (3.7), (3.8), and (3.9), a summation over $SU(3)$ variables is implied. The initial, final, and exchange masses M_i, M_f, and M_E, $m_v{}^2$ can be identified easily by the $SU(3)$ variables. The invariant amplitudes A and B obtained from these expressions are summarized in the Appendix [(A7), (A8), and (A9)]. It is to be noted that because of the contact terms in the propagators, the terms that we characterize as "one-particle exchanges" contain, in addition to the conventional exchange terms, some extra nonpole contributions. A characteristic feature of these symmetry-breaking "one-particle exchange" amplitudes is that they vanish at threshold in elastic scattering.

IV. NUMERICAL RESULTS

Before proceeding, certain limitations of the model should be considered. The "one-particle exchanges" that have been included represent only the nearest singularities to the elastic threshold of meson-baryon scattering in the sense that they contain the lowest-lying $SU(6)$ multiplets as intermediate states. Furthermore, these "exchange" terms are real functions of s and t that contain divergent contact parts. The $U(6,6)$-symmetric terms are polynomials in s and t that also diverge for high energies, and if α, β, and γ are chosen to be real constants, the corresponding amplitudes will be real. Hence, without modifications to account for elastic unitarity and higher resonance states, we expect our model to apply only in the low-energy region of the physical cut, where the amplitudes are necessarily small and real. This rules out the $\bar{K}N$ system,

TABLE I. Experimental and theoretical values (using $\alpha=-\beta=-0.0957$) of the S-wave (a_{2T}) and P-wave ($a_{2T,2J}$) scattering lengths for πN and KN elastic scattering.

Scattering length	Experimental values	Ref.	Calculated values
a_3	-0.114 ± 0.003	20	-0.080
	-0.088 ± 0.004	16	
	-0.096	17	
a_1	0.205	20	0.160
	0.171 ± 0.005	16	
	0.157	17	
a_{31}	-0.042 ± 0.004	20	-0.071
	-0.038 ± 0.005	16	
	-0.034	17	
a_{33}	0.215 ± 0.005	16	0.194
a_{11}	-0.101 ± 0.007	16	-0.089
	-0.039	17	
a_{13}	-0.029 ± 0.005	16	-0.047
	-0.030	17	
a_2	-0.205 ± 0.005	18	-0.194
a_0	$0.03\ \pm0.03$	19	0.709
a_{21}			-0.034
a_{23}			-0.002
a_{01}			0.020
a_{03}			-0.026

which is known to be absorptive near threshold, and leaves πN and KN elastic scattering as the most likely candidates for consideration. Since experimental data for these reactions are parameterized as partial-wave amplitudes and scattering lengths, we will compute these from our model. We take the low-energy regions to be $P_{\text{lab}}=0-300$ MeV/c and $P_{\text{lab}}=0-400$ MeV/c for πN and KN, respectively, slightly above the inelastic regions for both (P_{lab} is the laboratory momentum of the incident meson).

The "$U(6,6)$-symmetric" effective matrix elements, as can be seen from (A4) and (A6), contribute to $l=0$, 1, 2 partial waves. In the case of elastic scattering ($M_i=M_f=M$, $m_i=m_f=m$),

$$C(s,t)+D(s,t)+D(u,t)=0 \tag{4.1}$$

and the contribution from the γ term can be written as

$$A^{(\gamma)}(s,t)=(\gamma/36)[(3\mathfrak{D}-\mathfrak{C})D(s,t)+(3\mathcal{E}-\mathfrak{C})D(u,t)],$$
$$B^{\gamma}(s,t)=\tfrac{3}{8}\gamma(\mathcal{E}-\mathfrak{D})b_V(t).$$

It can be verified that at threshold, viz. $t=0$ and $s=(M+m)^2$, $u=(M-m)^2$,

$$A^{(\gamma)}+mB^{(\gamma)}=0. \tag{4.2}$$

It was previously noted that the "one-particle exchanges" also vanish at threshold. Consequently, in elastic scattering at threshold and hence for s-wave scattering lengths, it is only the α and β terms that contribute. It is apparent from (A6) that the α term has the $SU(3)$ and charge-conjugation structure of a singlet scalar meson exchange in the t channel (without a pole). The β term is a combination of two terms, one of which has the structure of an exchange in the t channel of an octet of scalar mesons and the other, an octet of vector mesons. These properties of the α and β terms will be exploited later.

The "one-particle exchanges," in general, contribute to all partial waves. When the nonpole parts are ignored, the "one-particle exchanges" are just the standard vector-meson exchange, baryon exchange, and decuplet exchange. The parameters G and g and hence all the required coupling constants, as discussed in I, are determined from the π-N and ρ-$\pi\pi$ coupling constants:

$$\begin{aligned} g_{PP\pi^0}{}^2&=\frac{50}{81}\left(1-\frac{m_\pi{}^2}{4M_N{}^2}\right)\left(1+\frac{2M_N}{m_\rho}\right)^2G^2,\\ g^2&=\frac{8}{81}g_{\rho\pi\pi}{}^2. \end{aligned} \tag{4.3}$$

From $g_{PP,\pi^0}{}^2/4\pi=15$ and $g_{\rho\pi\pi}{}^2/4\pi=\frac{1}{2}$, we have $G^2/4\pi=2.05$ and $g^2/4\pi=0.0494$. The vector-meson couplings needed in πN and KN scattering are given by the following relations:

$$\begin{aligned} g_{\rho\pi\pi}&=2g_{\rho KK}=(9/2\sqrt{2})g,\\ g_{\rho PP}&=\frac{G}{\sqrt{2}}\frac{2}{3}\left(1-\frac{1}{3}\frac{m_\rho}{M_N}-\frac{5}{12}\frac{m_\rho{}^2}{M_N{}^2}\right),\\ g_{\omega PP}&=\frac{G}{\sqrt{2}}\frac{2}{3}\left(3+\frac{m_\rho}{M_N}-\frac{1}{4}\frac{m_\rho{}^2}{M_N{}^2}\right),\\ g'_{\rho PP}&=\frac{G}{\sqrt{2}}\frac{4}{9}\left(1+\frac{5M_N}{m_\rho}-\frac{3}{4}\frac{m_\rho}{M_N}\right),\\ g'_{\omega PP}&=\frac{G}{\sqrt{2}}\frac{4}{9}\left(-3+\frac{3M_N}{m_\rho}-\frac{9}{4}\frac{m_\rho}{M_N}\right),\\ g_{\phi PP}&=g'_{\phi PP}=0,\quad m_\rho=m_\omega. \end{aligned} \tag{4.4}$$

The Yukawa-type meson-baryon coupling constants and the trilinear decuplet-baryon-meson coupling constant are tabulated in I.

Now we calculate the S-wave scattering lengths for all isospin states of πN and $KN(a_{2T})$. Since a_{2T} only depends on the threshold values of the α and β amplitudes, we have

$$\begin{aligned} a_3&=\frac{-8M_N}{M_N+m_\pi}\left\{\left(1+\frac{m_\pi{}^2}{m_0{}^2}\right)(\alpha+\beta)+\frac{2}{3}\frac{m_\pi}{m_0}\beta\right\},\\ a_1&=\frac{-8M_N}{M_N+m_\pi}\left\{\left(1+\frac{m_\pi{}^2}{m_0{}^2}\right)(\alpha+\beta)-\frac{4}{3}\frac{m_\pi}{m_0}\beta\right\},\\ a_2&=\frac{-8M_N}{M_N+m_K}\left\{\left(1+\frac{m_K{}^2}{m_0{}^2}\right)(\alpha+\tfrac{2}{3}\beta)+\frac{4}{3}\frac{m_K}{m_0}\beta\right\},\\ a_0&=\frac{-8M_N}{M_N+m_K}\left(1+\frac{m_K{}^2}{m_0{}^2}\right)\alpha. \end{aligned} \tag{4.5}$$

(Note that α and β have dimensions of inverse mass.) If we eliminate the parameters and put in the numerical

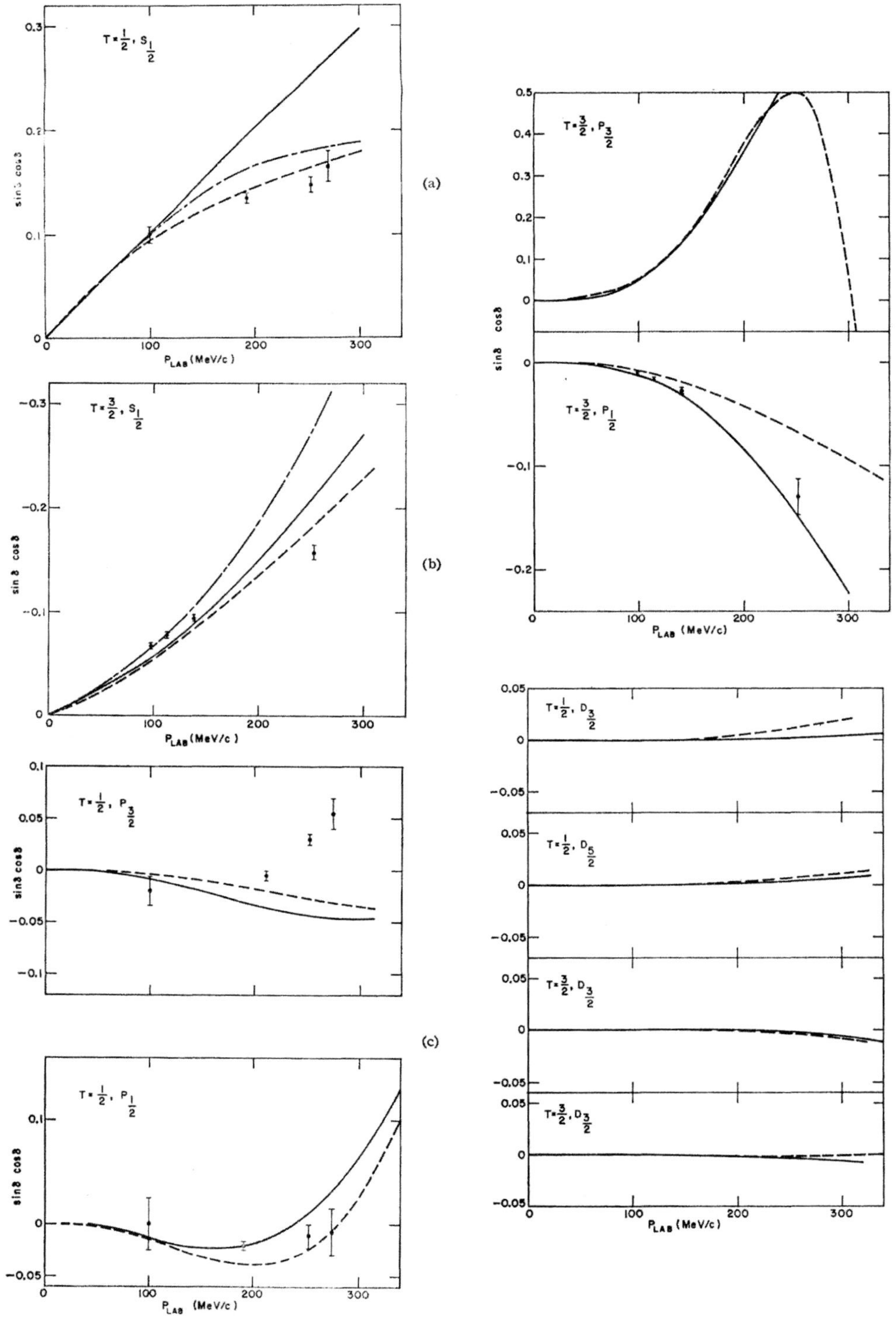

FIG. 1. $\sin\delta_{TJ}{}^{L}\cos\delta_{TJ}{}^{L}$ for πN elastic scattering. The solid curves are our calculated results. The dashed curves are the fits of Roper *et al.* (Ref. 17). The broken dashed curves are from the effective range approximation of Hamilton and Woolcock (Ref. 16). The data points are from the work of Barnes *et al.* (Ref. 20).

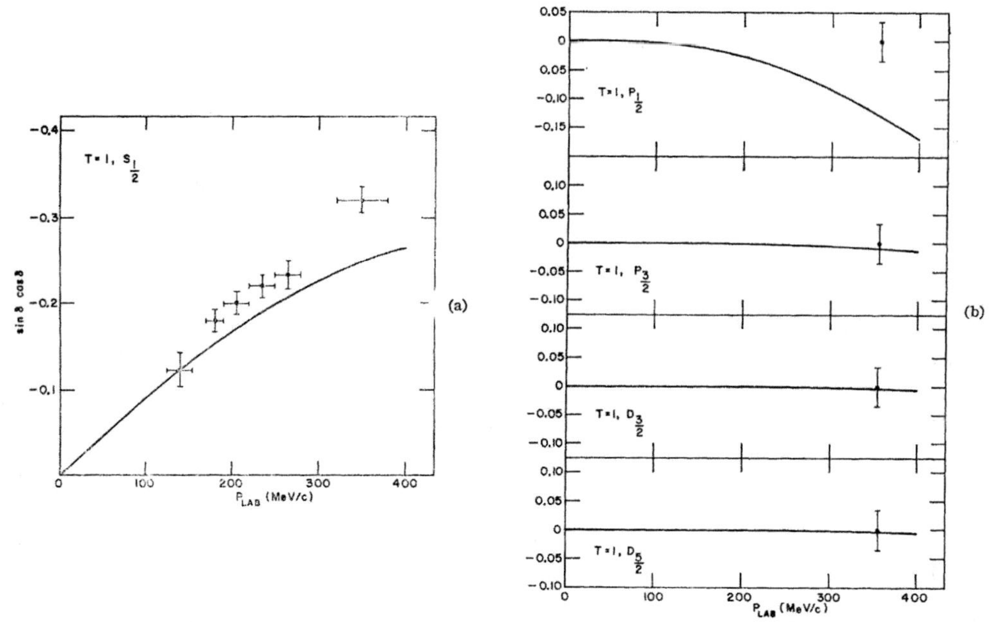

FIG. 2. $\sin\delta_{TJ}{}^L \cos\delta_{TJ}{}^L$ for $T=1$, KN elastic scattering. The solid curves are our calculated results. The data points are from the work of Goldhaber *et al.* (Ref. 18).

values of the masses, we obtain the relations

$$a_2=-0.468a_1+1.497a_3,\qquad a_0=3.303a_1-2.273a_3. \tag{4.6}$$

The scattering lengths determined from experimental data by several authors have been listed in Table I. If we use the Hamilton and Woolcock[16] values for πN S waves, we obtain $a_2=-0.212\pm0.008$ and $a_0=+0.765$ ±0.025. Using the Roper, Wright, and Feld[17] values gives $a_2=-0.217$, $a_0=+0.737$. In both cases, the a_2 values are close to the result of Goldhaber *et al.*[18] The a_0, however, is very large and far from the extrapolation of Stenger *et al.*,[19] whose a_0 is consistent with zero. This serious disagreement for the $T=0$ s wave at zero energy will persist away from the threshold and is the major fault of the model. We will defer further discussion of the $T=0$, KN system to the concluding section, and discuss only $T=\frac{1}{2}, \frac{3}{2}$ for πN and $T=1$ for KN.

Because the scattering lengths in Table I are extrapolated from the lowest-energy experimental data, there is a rather wide variation of values from one set to another. Therefore, it is reasonable to determine our parameters α, β, γ not from the scattering lengths, but by requiring a fit to the S-wave data at the lowest measured energies. One salient feature of the πN S-wave scattering lengths should first be noted: For all three values listed, $a_1\simeq -2a_3$. This is often taken to be evidence for the dominance of vector-meson exchange near threshold. We can simulate this behavior and eliminate one parameter by requiring $\beta=-\alpha$. For πN this has the effect of canceling that part of the β term which has the $SU(3)$ structure of a scalar octet exchange, leaving the part that has the structure of vector exchange. For KN this cancellation does not occur, and in fact for $T=0$ it is only the α term that contributes. To fix the remaining parameters, α and γ, unambiguously we require the $T=\frac{1}{2}$ and $T=1$ S waves to pass through the lowest experimentally determined points. With the definition $L_{2T,2J}=\sin\delta_{TJ}{}^L \cos\delta_{TJ}{}^L=f_{L\pm}{}^T$ (where $L=S, P, D$, etc.), we require

$$S_{11}(P_{\rm lab}=100\ {\rm MeV}/c)=+0.100\,,^{20}$$
$$S_{21}(P_{\rm lab}=140\ {\rm MeV}/c)=-0.123\,.^{18}$$

This fixes the parameters (in units of inverse pion masses),

$$\alpha=-\beta=-0.0957\,,$$
$$\gamma=3.00\,,$$

[16] J. Hamilton and W. S. Woolcock, Rev. Mod. Phys. **35**, 737 (1963).

[17] L. D. Roper, R. M. Wright, and B. T. Feld, Phys. Rev. **138**, B190 (1965).

[18] S. Goldhaber *et al.*, Phys. Rev. Letters **9**, 135 (1962).

[19] V. J. Stenger, W. E. Slater, D. H. Stork, H. K. Ticho, G. Goldhaber, and S. Goldhaber, Phys. Rev. **134**, B1111 (1964).

[20] S. W. Barnes, B. Rose, G. Giacomelli, J. Ring, K. Miyake, and K. Kinsey, Phys. Rev. **117**, 226 (1960); S. W. Barnes, H. Winick, K. Miyake, and K. Kinsey, *ibid.* **117**, 238 (1960).

and allows the calculation of all partial waves that contribute to low-energy πN and KN elastic scattering. The results for the scattering lengths are listed in Table I.

In Figs. 1 and 2, the resulting S, P, and D waves are given as functions of laboratory momentum for $\pi N(T=\frac{1}{2},\frac{3}{2})$ and $KN(T=1)$. For comparison, some of the experimentally determined amplitudes are also given. Note that the πN data of Barnes *et al.*[20] come from fitting differential and total cross sections below $E_\pi=170$ MeV (laboratory kinetic energy), whereas the curves of Roper *et al.*[17] were determined by including polarization data up to $E_\pi=350$ MeV. Because the latter fit includes higher-energy data than Barnes *et al.*, the details of the two fits differ in the region of the N^*, where data are less certain, particularly in the P_{13} and P_{31}. The Hamilton and Woolcock[16] curves for πN S waves are taken from their effective-range fit to data below $E_\pi=45$ MeV. The KN data of Goldhaber *et al.*[18] come from their experiments on K^+p elastic scattering, from $P_L=140-624$ MeV/c.

In general, our results are in reasonably good agreement with the data. All of the πN partial waves have the same sign and qualitative behavior as the data of Roper *et al.* The agreement is particularly good below $P_{\text{lab}}=200$ MeV/c, where N^* no longer dominates the differential and total cross sections. Except for the small negative P_{21}, the KN, $T=1$ partial waves are consistent with the dominance of S-wave scattering reported by Goldhaber *et al.* These results emphasize the importance of the background terms, particularly in the S waves, where the α and β amplitudes dominate, and in the P_{11} partial wave where the large γ term counterbalances the large negative nucleon pole contribution (Table II).

TABLE II. Calculated values of πN partial-wave amplitudes ($\sin\delta_{TJ}^L \cos\delta_{TJ}^L$) with one-particle exchanges and background terms separately tabulated to illustrate their relative importance.

P_{lab} (MeV/c)	Sum of all one-particle exchanges	Sum of all background terms	Total partial-wave amplitude
S_{11}			
60	−0.0010	0.0616	0.0606
100	−0.0044	0.1051	0.1007
140	−0.0114	0.1512	0.1398
200	−0.0227	0.2246	0.2019
240	−0.0369	0.2762	0.2393
300	−0.0618	0.3601	0.2983
S_{31}			
60	0.0022	−0.0340	−0.0318
100	0.0103	−0.0679	−0.0576
140	0.0266	−0.1136	−0.0870
200	0.0544	−0.2049	−0.1505
240	0.0838	−0.2793	−0.1955
300	0.1359	−0.4070	−0.2711
P_{11}			
60	−0.0125	0.0086	−0.0039
100	−0.0508	0.0380	−0.0128
140	−0.1206	0.0991	−0.0215
200	−0.2840	0.2650	−0.0190
240	−0.4310	0.4316	+0.0006
300	−0.7094	0.7718	+0.0624
P_{31}			
60	−0.0022	−0.0013	−0.0035
100	−0.0083	−0.0064	−0.0147
140	−0.0185	−0.0117	−0.0302
200	−0.0386	−0.0477	−0.0863
240	−0.0536	−0.0795	−0.1331
300	−0.0769	−0.1462	−0.2231
P_{13}			
60	−0.0020	−0.0002	−0.0022
100	−0.0077	−0.0009	−0.0086
140	−0.0164	−0.0016	−0.0180
200	−0.0323	−0.0015	−0.0338
240	−0.0427	+0.0007	−0.0420
300	−0.0558	+0.0088	−0.0470
P_{33}			
60	0.0070	0.0030	0.0100
100	0.0305	0.0131	0.0436
140	0.0813	0.0330	0.1143
200	0.2679	0.0850	0.3529
240	0.6190	0.1347	0.7537

V. DISCUSSION AND SUMMARY

The physical motivation for our model is provided by the fact that tensors of $U(6,6)$ subjected to Bargmann-Wigner equations lead to a supermultiplet structure which corresponds exactly to that of $SU(6)$ symmetry. As emphasized by Salam *et al.*,[7] this is entirely equivalent to assuming that the known particles at rest correspond to the representations of a compact $U(6)\otimes U(6)$ group. Given the possibility of such an assumption, one can write down tensor combinations of three-, four-, and higher-point functions which are formally invariant under $U(6,6)$. In the lowest-order perturbation theory, they can be evaluated by substituting the free-field solution for the tensors. Such a procedure has considerable support from experimental observations in the case of three-point functions constructed from Ψ_{ABC} and $\Phi_A{}^B$ which describe the presently known low-lying baryonic and mesonic states. When applied to four-point functions, however, it leads to contradictions and conflict with the basic principle of unitarity. Even in the case of three-point functions, as pointed out in I, there is no theoretical justification for the procedure except as a working hypothesis.

Since, in general, perturbation theory in strong interactions is questionable, we adopt the alternative approach based on dispersion relations. In several calculations[3,16] based on the latter approach, it has been found that it is necessary to introduce background terms which approximate distant singularities in addition to one-particle exchange contributions. It is interesting to investigate whether such background terms possess higher symmetries. Within the framework of $U(6,6)$ symmetry, the contact terms (four-point functions evaluated in the lowest-order perturbation theory) provide a representation of such background terms.[7] If these are supplemented by symmetry-breaking one-particle exchange contributions, one has a starting point to relate a large number of scattering processes. With this point of view, we have investigated only a limited version of the model as applied to pseudoscalar meson-baryon scattering, confining our attention only to the states that form the 35-dimensional and 56-dimensional representation of $SU(6)$ (we could also consider associated production, vector-meson production, and baryonic resonance production in the same context). Further, the model is incomplete for the following reasons. First of all, we have made no attempt to calculate unitarity corrections. Secondly, for simplicity, we have chosen α, β, $\tilde{\beta}$, γ to be real constants. It is more natural to choose functional forms with appropriate cutoff properties in them so that they define a range over which the $U(6,6)$-symmetric terms dominate. This would enable one to use the N/D procedures to obtain unitary scattering amplitudes, thus greatly extending the predictive power of the model. The model then could be tested for higher-energy behavior and polarization predictions.

In spite of these shortcomings, it is encouraging from the results in the previous section that we obtain reasonably good agreement with the presently available experimental information in πN and KN ($T=1$ state only) scattering. There is a serious disagreement in KN ($T=0$ state) in that the model predicts large S-wave scattering in contradiction with the results of Stenger *et al.*[19] Their results, however, are extrapolated from K^+d scattering at comparatively higher energies ($P_{\text{lab}}=350$–812 MeV/c). If further experiments (preferably $K_2{}^0P$) confirm the results of Stenger *et al.*, then our model in its present form is untenable. In the work of Warnock and Frye[3] on the KN system, in which the Stenger results are used, there is an indication of significant contributions from higher resonances. One may then be led to include higher $SU(6)$ multiplets in the scheme considered here. This and other possible modifications discussed above need further investigation.

ACKNOWLEDGMENTS

The authors are indebted to Professor B. Sakita for his continued interest and valuable suggestions during the course of this work. They would also like to thank Dr. R. C. Arnold for many helpful discussions.

APPENDIX

The purpose of this Appendix is to collect together the results for the invariant amplitudes $A(s,t,u)$ and $B(s,t,u)$ from various terms in the model. To this end, we note that A and B are defined by[21]

$$T=-A(s,t,u)+i\frac{q_i+q_f}{2}B(s,t,u)\,, \tag{A1}$$

where T is the T matrix for pseudoscalar meson-baryon scattering, and q_i and q_f are, respectively, the 4-momenta of the initial and final meson. The corresponding 4-momenta of the baryons will be denoted by p_i and p_f. Let m_i, m_f, and M_i, M_f denote the masses of the initial and final meson and baryon masses, respectively. Then, s, t, and u are the Mandelstam variables, $s=-(p_i+q_i)^2$, $t=-(q_i-q_f)^2$, $u=-(p_i-q_f)^2$, and $s+t+u=M_i^2+M_f^2+m_i^2+m_f^2$. When necessary, we denote the mass of the exchanged meson (baryon) by $m_e(M_E)$. The partial-wave projections of A and B are given by

$$A_l(s)=\frac{1}{2}\int_{-1}^{1}AP_l(x)dx\,,\quad B_l(s)=\frac{1}{2}\int_{-1}^{1}BP_l(x)dx\,, \tag{A2}$$

where $x=\cos\theta$, θ being the scattering angle in the c.m. system. The partial-wave scattering amplitude, $f_{l\pm}$, for a state with total angular momentum $J=l\pm\frac{1}{2}$ and parity $(-1)^{l+1}$ are given in terms of A_l and B_l by the following equation:

$$\begin{aligned}f_{l\pm}=&\frac{1}{8\pi W}(E_i+M_i)^{1/2}(E_f+M_f)^{1/2}\\&\times\left[A_l+\left(W-\frac{M_i+M_f}{2}\right)B_l\right]\\&+\frac{1}{8\pi W}(E_i-M_i)^{1/2}(E_f-M_f)^{1/2}\\&\times\left[-A_{l\pm1}+\left(W+\frac{M_i+M_f}{2}\right)B_{l\pm1}\right]\,,\end{aligned} \tag{A3}$$

where W is the c.m. energy and

$$E_{i,f}+M_{i,f}=((W+M_{i,f})^2-m_{i,f}^2)/2W\,.$$

$U(6,6)$-Symmetric Contributions

We would like to emphasize that the term "$U(6,6)$ symmetric" is used for convenience in the sense that the relevant contributions can be derived from an effective Lagrangian which is $U(6,6)$ invariant when expressed in terms of the fields Ψ and Φ. The use of the Bargmann-Wigner equations violates this invariance. In addition, it should be noted that we include $SU(3)$ mass splittings.

To list the invariant amplitudes from different terms

[21] A. W. Martin and K. C. Wali, Nuovo Cimento **31**, 1324 (1964).

in (2.4), we define the following functions of s, t, and u:

$$F(t)=\left(1-\frac{t-M_i^2-M_f^2}{2M_iM_f}\right)\left(1-\frac{t-m_i^2-m_f^2}{2m_\rho^2}\right),$$

$$a_V(s,t)=\frac{2}{3}\frac{(u-s)}{m_\rho^2}\frac{((M_i+M_f)^2-t)}{2M_iM_f},$$

$$\bar{a}_V(s,t)=\frac{1}{9}\frac{(M_i+M_f+2m_\rho)}{m_\rho}\times\left[\frac{(M_i+M_f)(u-s)-(m_i^2-m_f^2)(M_i-M_f)}{M_iM_fm_\rho}\right],$$

$$b_V(t)=\frac{2}{9}\frac{(2m_\rho+M_i+M_f)((M_i+M_f)^2-t)}{M_iM_fm_\rho^2}, \tag{A4}$$

$$C(s,t)=\frac{-2}{M_iM_fm_\rho^2}[m_\rho(M_i-M_f)^2(m_\rho+M_i+M_f)+(M_i+M_f)(m_i^2M_f+m_f^2M_i)-t(M_i+m_\rho)(M_f+m_\rho)-st-(M_f^2+m_f^2-s)(M_i^2+m_i^2-s)],$$

$$D(x,t)=2\left[\left(1-\frac{t-M_i^2-M_f^2}{2M_iM_f}\right)\left(1+\frac{2x-M_i^2-M_f^2}{2m_\rho^2}\right)-2\left(1-\frac{x-M_f^2-m_i^2}{2M_fm_\rho}\right)\left(1-\frac{x-M_i^2-m_f^2}{2M_im_\rho}\right)\right],$$

where $x=u$, s.

To denote the $SU(3)$ dependences, we need the following definitions: $(\bar{B}B)_W$ (where $W=F$, D, or S),

$$(\bar{B}B)_F=B\bar{B}-\bar{B}B;\quad (\bar{B}B)_D=\bar{B}B+B\bar{B}-\tfrac{2}{3}\,\mathrm{Tr}B\bar{B};\quad (\bar{B}B)_S=\mathrm{Tr}B\bar{B}$$

W	F	D	S
a_W	1	0	1
c_W	-1	3	-2

$$\mathcal{C}=\mathrm{Tr}(-\bar{B}BP\bar{P}-\bar{B}B\bar{P}P+\bar{B}\bar{P}BP+\bar{B}PB\bar{P})+\mathrm{Tr}(\bar{B}B)\,\mathrm{Tr}(\bar{P}P),$$

$$\mathfrak{D}=\mathrm{Tr}(-\bar{B}\bar{P}PB)+\mathrm{Tr}(\bar{B}\bar{P})\,\mathrm{Tr}(BP), \tag{A5}$$

$$\mathcal{E}=\mathrm{Tr}(-\bar{B}P\bar{P}B)+\mathrm{Tr}(\bar{B}P)\,\mathrm{Tr}(B\bar{P}).$$

Then,

$$A^{(\alpha)}(s,t)=-4\alpha F(t)\,\mathrm{Tr}(\bar{B}B)\,\mathrm{Tr}(\bar{P}P),$$

$$B^{(\alpha)}(s,t)=0,$$

$$A^{(\beta)}(s,t)=\beta\sum_W(\bar{B}B)_W[-\tfrac{4}{3}a_WF(t)(P\bar{P}+\bar{P}P)-(a_Wa_V(s,t)+c_W\bar{a}_V(s,t))(P\bar{P}-\bar{P}P)],$$

$$B^{(\beta)}(s,t)=-\beta\sum_W(\bar{B}B)_W(3a_W+c_W)b_v(t)(P\bar{P}-P\bar{P}), \tag{A6}$$

$$A^{(\gamma)}(s,t)=\frac{\gamma}{36}[\mathcal{C}C(s,t)+3\mathfrak{D}D(s,t)+3\mathcal{E}D(u,t)],$$

$$B^{(\gamma)}(s,t)=\tfrac{3}{8}\gamma(\mathcal{E}-\mathfrak{D})b_V(t).$$

"One-Particle-Exchange" Contributions

In pseudoscalar-meson–baryon scattering, we have contributions due to vector meson, baryon, and decuplet exchange. We denote the corresponding invariant amplitudes by $A^{(V)}$, $A^{(B)}$, $A^{(D)}$, etc.

Vector-Meson Exchange

$$A^{(V)}(s,t)=-\frac{3Gg}{4}\sum_W\mathrm{Tr}[(\bar{B}B)_W(P\bar{P}-\bar{P}P)]\left[a_W\left(\frac{3m_V^2}{m_V^2-t}a_V-a_V+\frac{[(M_i+M_f)^2-t](M_f-M_i)(m_f^2-m_i^2)}{M_iM_fm_\rho(m_V^2-t)}\right)+c_W\left(\frac{3m_V^2}{m_V^2-t}\bar{a}_V-\bar{a}_V+\frac{(t-4m_V^2)}{9m_\rho M_iM_f}\frac{[(M_i+M_f)(u-s)-(M_i-M_f)(m_i^2-m_f^2)]}{m_V^2-t}\right)\right], \tag{A7}$$

$$B^{(V)}(s,t)=-\frac{3Gg}{4}\sum_W(\bar{B}B)_W(P\bar{P}-\bar{P}P)(3a_W+c_W)\left[\frac{3m_V^2}{m_V^2-t}b_V(t)-b_V(t)+\frac{2}{3}\frac{(t-4m_V^2)}{m_\rho M_iM_f}\frac{[(M_i+M_f)^2-t]}{m_V^2-t}\right].$$

Conventional vector exchange is obtained by taking only the pole terms.

In the case of baryon and decuplet exchanges, it is sufficient to compute the invariant amplitudes $A^{(\ )}$ and $B^{(\ )}$ due to the direct exchanges in the s channel. The corresponding u-channel pole contributions can be obtained by the substitutions $s\leftrightarrow u$, $m_i\leftrightarrow m_f$, $P\leftrightarrow\bar{P}$, and $B(s,t)=-B(u,t)$.

Baryon Exchange

$$A^{(B)}(s,t)=\frac{G^2}{81}\left[M_Ef_i(s)f_f(s)\frac{A_1(s)}{M_E^2-s}-[f_i(s)+f_f(s)-1]A_2(s)-\frac{A_3(s)}{M_E}\right]\times(25\,\mathrm{Tr}\bar{B}\bar{P}PB+5\,\mathrm{Tr}\bar{B}\bar{P}BP+5\,\mathrm{Tr}\bar{B}PB\bar{P}+\mathrm{Tr}\bar{B}BP\bar{P}),$$

$$B^{(B)}(s,t)=\frac{G^2}{81}\left[M_E f_i(s)f_f(s)\frac{B_1^{(s)}}{M_E^2-s}-[f_i(s)+f_f(s)-1]B_2(s)-\frac{B_3(s)}{M_E}\right]$$

$$\times(25\,\mathrm{Tr}\bar{B}\bar{P}PB+5\,\mathrm{Tr}\bar{B}\bar{P}BP+5\,\mathrm{Tr}\,\bar{B}PB\bar{P}+\mathrm{Tr}\bar{B}BP\bar{P}),$$

where

$$f_{i,f}(s)=\frac{2M_{i,f}M_E+M_{i,f}^2-m_{i,f}^2+s}{2M_{i,f}M_E},$$

$$A_1(s)=-\left(1+\frac{M_i+M_E}{m_\rho}\right)\left(1+\frac{M_f+M_E}{m_\rho}\right)\left(M_E-\frac{M_i+M_f}{2}\right)-(4m_\rho+M_i+M_f+2M_E)(s-M_E^2)/2m_\rho^2,$$

$$A_2(s)=-\left(1+\frac{2M_E^2-M_i^2-M_f^2}{2m_\rho^2}+\frac{s-M_E^2}{m_\rho^2}\right),$$

$$B_1(s)=\left(1+\frac{M_i+M_E}{m_\rho}\right)\left(1+\frac{M_f+M_E}{m_\rho}\right)+\frac{s-M_E^2}{m_\rho^2},$$

$$B_2(s)=\left(\frac{2}{m_\rho}+\frac{M_i+M_f}{m_\rho^2}\right),$$

and

$$A_3(s)=A_1(s)-M_EA_2(s),\quad B_3(s)=B_1(s)-M_EB_2(s). \tag{A8}$$

Decuplet Exchange

The results in Eqs. (A7) and (A8) follow from Eqs. (3.7) and (3.8). As explained in Sec. III, these contributions include (i) parts due to the conventional vector or baryon exchanges calculated using the coupling-constant relations implied by $U(6,6)$ or $SU(6)_W$ symmetry [I, Appendix Eqs. (A2), (A3), (A11)]; (ii) parts that arise because of contact terms in the propagators. In the decuplet case, for the sake of conciseness, we present below only the contributions of the type (i), although the complete results from (3.9) have been included in the numerical calculations.

$$A^{(D)}(s,t)=\frac{2}{81}\frac{1}{M_E^2-s}\frac{G^2}{M_iM_f}\left(1+\frac{M_i+M_E}{m_\rho}\right)\left(1+\frac{M_f+M_E}{m_\rho}\right)$$

$$\times\left[\left(M_E-\frac{M_i+M_f}{2}\right)(M_i+C_i)(M_f+C_f)+\frac{3}{2}\left(M_E+\frac{M_i+M_f}{2}\right)(2C_iC_f-M_i^2-M_f^2+t)\right]$$

$$\times\{3(\mathrm{Tr}\bar{B}B\,\mathrm{Tr}\bar{P}P-\mathrm{Tr}\bar{B}P\,\mathrm{Tr}\bar{P}B)+\mathrm{Tr}(3B\bar{B}P\bar{P}-3\bar{B}B\bar{P}P-B\bar{B}\bar{P}P-\bar{B}BP\bar{P}+BP\bar{B}\bar{P}+B\bar{P}\bar{B}P)\},$$

$$B^{(D)}(s,t)=-\frac{2}{81}\frac{1}{M_E^2-s}\frac{G^2}{M_iM_f}\left(1+\frac{M_i+M_E}{m_\rho}\right)\left(1+\frac{M_f+M_E}{m_\rho}\right)$$

$$\times[(M_i+C_i)(M_f+C_f)-\tfrac{3}{2}(2C_iC_f-M_i^2-M_f^2+t)]\{\text{same } SU(3) \text{ dependence as in } A^{(D)}\},$$

where C_i and C_f are constants given by

$$C_i=\frac{M_E^2+M_i^2-m_i^2}{2M_E},\quad C_f=\frac{M_E^2+M_f^2-m_f^2}{2M_E}. \tag{A9}$$

Note that when the partial waves are projected out, all of the $U(6,6)$-symmetric and one-particle exchange terms have the correct threshold dependences, i.e., $f_{l\pm}(q)$ goes as q^{2l} as q (the c.m. momentum) goes to zero.

IMPLICATIONS OF LOCAL DUALITY IN MESON-DECUPLET SCATTERING*

M. J. King and Kameshwar C. Wali
Physics Department, Syracuse University, Syracuse, New York 13210
(Received 6 April 1970)

The consequences of the assumption that the imaginary parts of the direct-channel helicity amplitudes vanish in the forward (backward) scattering if the crossed t channel (u channel) is exotic are examined in the case of meson-decuplet scattering. Confining ourselves to a limited energy region above the threshold, we show that the solutions of the set of equations provided by different external helicities predict the existence of certain patterns of octet and decuplet baryons degenerate in mass and having spin and parity assignments $\frac{5}{2}^{\pm}$, $\frac{3}{2}^{\pm}$, $\frac{1}{2}^{\pm}$.

Recently several authors[1,2] have attempted to predict "exchange degeneracy" patterns for hadronic trajectories on the basis of crossing, SU(3) symmetry, and duality. The scattering amplitude is assumed to consist of two parts: (1) a primarily diffractive background, and (2) a resonant part, the imaginary part of which is expressible at high energies in terms of Regge trajectories in the crossed channels. Hence if a particular scattering amplitude is characterized by internal quantum numbers for which no resonances exist, the Regge trajectories in the crossed channels must exhibit exchange degeneracy so that the corresponding imaginary parts cancel.

In the present note, we investigate the consequences of duality in a local energy region in the case of of the reactions $P+D\to P+D$, where P and D represent, respectively, the octet of pseudoscalar mesons and the decuplet of $J^P=\frac{3}{2}^+$ baryons. Following Mandula, Weyers, and Zweig,[3] we assume that there is a region of s and small t on the one hand, and a region of s and small u on the other, within which the imaginary part can be calculated in two alternate ways, in terms of direct-channel resonances or in terms of Regge trajectories of the crossed channels. Therefore, if we select the s-channel reactions so that their t or u channels are characterized by exotic quantum numbers (viz., 27 in the t channel, 27 and 35 in the u channel), the imaginary parts due to s-channel resonances must add up to zero. We consider a set of direct-channel resonances degenerate in mass but with different spins and parities and examine the constraints on their coupling constants implied by the above requirement in the forward and the backward scattering. We show by including considerations of spin that the particles on leading trajectories must be accompanied by daughters with prescribed ratios of coupling constants between the parent and the daughter particles. We discuss the general pattern of the nontrivial solutions and compare the results with experiments in the case of nonstrange baryons.

The imaginary parts of the six independent s-channel helicity amplitudes free of kinematic singularities[4] can be written as

$$\mathrm{Im}\overline{G}^{R}_{0\mu;0\lambda}(s,\theta_s)=K(s)\sum_{JLL'R'}X^{RR'}(2L'+1)^{1/2}(2L+1)^{1/2}C(L'\tfrac{3}{2}J;0\mu)C(L\tfrac{3}{2}J;0\lambda)\chi_{R'}{}^{J}(L')\chi_{R'}{}^{J}(L)d_{\lambda\mu}{}^{J}(\theta_s), \quad (1)$$

where R characterizes the irreducible SU(3) representation of the crossed t or u channel and $X_{R'}{}^{R}$ are the elements of the corresponding crossing matrix. L,λ (L',μ) denote the orbital angular momentum and helicity of the initial (final) state, and $K(s)$ is a kinematic factor. The χ^J's are related to the partial widths through

$$\chi^J(L)=(\Gamma_{PD}{}^J\Gamma^J)^{1/2}\{[(W^*-W_{\mathrm{c.m.}})/\tfrac{1}{2}\Gamma]^2+1\}^{-1/2}, \quad (2)$$

where Γ^J is the total width and $\Gamma_{PD}{}^J$ the partial width of a resonance of angular momentum J and mass

W^*. $W_{\rm c.m.}$ is the center-of-mass energy. Finally,

$$\hat{d}_{\lambda\mu}{}^J(\theta_s) = (\cos\tfrac{1}{2}\theta_s)^{-|\lambda+\mu|}(\sin\tfrac{1}{2}\theta_s)^{-|\lambda-\mu|}d_{\lambda\mu}{}^J(\theta_s). \tag{3}$$

Our basic equations are

$$\mathrm{Im}\overline{G}^R_{0\mu;0\lambda}(s^*,0) = K(s^*)\sum_{JLL'R'} X^{RR'}_{ts}(2L'+1)^{1/2}(2L+1)^{1/2}C(L'\tfrac{3}{2}J;0\mu)C(L\tfrac{3}{2}J;0\lambda)\chi_{R'}{}^J(L')\chi_{R'}{}^J(L)\hat{d}_{\lambda\mu}{}^J(0) = 0 \tag{4}$$

in the forward direction ($R = 27$), and

$$\mathrm{Im}\overline{G}^R_{0\mu;0\lambda}(s^*,\pi) = K(s^*)\sum_{JLL'R'} X^{RR'}_{us}(-1)^{L'}(2L'+1)^{1/2}(2L+1)^{1/2} \times C(L'\tfrac{3}{2}J;0\mu)C(L\tfrac{3}{2}J;0\lambda)\chi_{R'}{}^J(L')\chi_{R'}{}^J(L)\hat{d}_{\lambda\mu}{}^J(0) = 0 \tag{5}$$

in the backward direction[5] ($R = 27, 35$).

In the s channel we retain contributions due to resonances with nonexotic quantum numbers, i.e., 8 and 10 in the present case. Although our considerations can be generalized to particles of any spin in the intermediate state, for the present purposes we consider a set of particles degenerate in mass ($s^* = W^{*2}$) and $J^P = \frac{5}{2}^\pm, \frac{3}{2}^\pm, \frac{1}{2}^\pm$. This is motivated by the fact that if we specialize our considerations to the reaction $\pi\Delta \to \pi\Delta$ in the energy range from threshold up to 1700 MeV, only these resonances will contribute. In addition, a fair amount of experimental information exists for comparison between theory and experiment. Now define

$$Y^{2J^\pm}_{2\mu,2\lambda} = \sum_{R'} X^{RR'}_{ts}[\chi_{R'}{}^{J^\pm}(L_0)]^2 A^{R',2J^\pm}_{2\mu,2\lambda}, \tag{6}$$

$$Z^{2J^\pm}_{2\mu,2\lambda} = \sum_{R'} X^{RR'}_{us}[\chi_{R'}{}^{J^\pm}(L_0)]^2 A^{R',2J^\pm}_{2\mu,2\lambda}, \tag{7}$$

with $R = 27$ in $X^{RR'}_{ts}$, $R = 27, 35$ in $X^{RR'}_{us}$, and

$$A^{R,2J^\pm}_{2\mu,2\lambda} = \sum_{LL'} \frac{(2L+1)^{1/2}(2L'+1)^{1/2}C(L'\tfrac{3}{2}J;0\mu)C(L\tfrac{3}{2}J;0\lambda)\chi_R{}^{J^\pm}(L)\chi_R{}^{J^\pm}(L')}{(2L_0+1)C(L_0\tfrac{3}{2}J;0\mu)C(L_0\tfrac{3}{2}J;0\lambda)[\chi_R{}^{J^\pm}(L_0)]^2}. \tag{8}$$

Note that for each J, there are two orbital momenta L, L' that contribute. L, L' are odd (even) depending on whether the parity is even (odd). In Eq. (8), L_0 denotes the lesser of the two orbital angular momenta for a given J.[6] Substituting (6) and (7) in (4) and (5), and after some simple algebra, we obtain a set of equations for Y's and Z's. Below we give the set of equations for the Z's and note that the corresponding equations for Y's are obtained by reversing the signs of all the contributions due to odd-parity resonances:

$$3Z^{5+}_{33} + \tfrac{27}{7}Z^{5-}_{33} - Z^{3-}_{33} = 0, \tag{8.1}$$

$$Z^{5+}_{33} + Z^{3+}_{33} + \tfrac{25}{7}Z^{5-}_{33} = 0, \tag{8.2}$$

$$3Z^{5+}_{31} + \tfrac{3}{35}Z^{5-}_{31} - Z^{3-}_{31} = 0, \tag{8.3}$$

$$Z^{5+}_{31} - Z^{3+}_{31} + \tfrac{5}{7}Z^{5-}_{31} = 0, \tag{8.4}$$

$$\tfrac{36}{5}Z^{5+}_{11} - \tfrac{1}{5}Z^{3+}_{11} + 2Z^{1+}_{11} + \tfrac{6}{7}Z^{5-}_{11} - 3Z^{3-}_{11} = 0, \tag{8.5}$$

$$-\tfrac{18}{5}Z^{5+}_{11} + \tfrac{3}{5}Z^{3+}_{11} - \tfrac{12}{7}Z^{5-}_{11} + Z^{3-}_{11} - 2Z^{1-}_{11} = 0. \tag{8.6}$$

From the SU(3) crossing matrix, we can write[7]

$$Z^{2J^\pm}_{2\mu,2\lambda} = [\chi_8{}^{J^\pm}(L_0)]^2 A^{8,J^\pm}_{2\mu,2\lambda} + \tfrac{5}{8}[\chi_{10}{}^{J^\pm}(L_0)]^2 A^{10,J^\pm}_{2\mu,2\lambda}, \tag{9.1}$$

$$Y^{2J^\pm}_{2\mu,2\lambda} = 4[\chi_8{}^{J^\pm}(L_0)]^2 A^{8,2J^\pm}_{2\mu,2\lambda} - 5[\chi_8{}^{J^\pm}(L_0)]^2 A^{10,2J^\pm}_{2\mu,2\lambda}. \tag{9.2}$$

Therefore, $Z^J = 0$ implies that the effective coupling constants of 8 and 10 are separately zero. In other words, no particle of the corresponding J is coupled to the PD system. $Y^J = 0$ does not imply this. We seek nontrivial solutions of the equations for Z's and Y's such that Z's are positive and not all of them are zero.

We assume that the ratios $\chi_R{}^{J^\pm}(L_0+2)/\chi_R{}^{J^\pm}(L_0)$ are small. In the energy region of interest (1.5-1.7 BeV), this is theoretically expected on the basis of centrifugal-barrier effects and appears to be true experimentally. For instance, in the case of $F_{15}(1.69)$ and $D_{15}(1.68)$, the corresponding ratios of χ's are 0.071 and 0.058, respectively.[8] It follows from Eq. (8) and the numerical values of the Clebsch-Gordan coefficients that the $A^{2J^{\pm}}_{2\mu,2\lambda}$ are positive and nonvanishing. Consequently a given resonance, in general, is coupled to all helicity states of the PD system. Without any further assumptions, we can draw the following conclusions:

(i) Resonances belonging to both 8 and 10 representations must contribute. If only 8 or 10 exists,

$$Z^{J^{\pm}}_{2\mu,2\lambda} = CY^{J^{\pm}}_{2\mu,2\lambda}, \tag{10}$$

where C is the same for all Z's. Only a trivial solution exists in this case.

(ii) Likewise, the set of equations are inconsistent unless there are contributions from particles of all three spins $J=\frac{5}{2}$, $\frac{3}{2}$, and $\frac{1}{2}$. Therefore, in general, if the particles on leading trajectories exist, there must be particles on the daughter trajectories.

(iii) Particles of both parities must exist and contribute for a nontrivial solution.

Conclusions (i) and (iii) have been previously derived[1,2] by applying duality considerations to effective Regge residue functions. Conclusion (ii) is new and depends on our specific assumptions concerning the resonance couplings as well as local duality. Present experimental information on some resonances is one motivation for these assumptions. The study of $PB \to PB$ and $PB \to PD$ reactions should provide additional consistency tests.

To see the physical implications more clearly, we make an additional simplifying assumption,[9] viz.

$$A^{8,2J^{\pm}}_{2\mu,2\lambda} = A^{10,2J^{\pm}}_{2\mu,2\lambda} = A^{10,2J^{\pm}}_{2\mu,2\lambda} = A^{2J^{\pm}}_{2\mu,2\lambda}, \tag{11}$$

and give the solutions to Eq. (8) and the corresponding equations for Y's in Table I. These solutions involve two arbitrary parameters p and p', corresponding to the coupling constants or partial widths of $\frac{5}{2}^+$, $\frac{5}{2}^-$ resonances belonging to an octet representation. The constants α, β, γ, δ, and ϵ are known combinations of $A^{J^{\pm}}_{2\mu,2\lambda}$. For further discussion, the only property of these constants we need is that α, β, and γ are positive and of order of magnitude unity. δ and ϵ can be small,[11] but are non-negative.

For experimental comparison, let us consider two energy regions, one at 1700 MeV and the other at 1500 MeV in the $\pi\Delta$ system. $F_{15}(1.69)$ and $D_{15}(1.68)$ are known to couple strongly to the $\pi\Delta$ system.[8] Hence p, $\gamma p'$ must be nonvanishing and, in general, all the other states exist. This corresponds to the four-multiplet $(8^+, 8^-, 10^+, 10^-)$ solution derived from crossing and the absence of exotic states.[2] By requiring special conditions $\alpha p' = p$ $[(9/\alpha)p = p']$, we obtain the three-multiplet solution $8^+, 8^-, 10^-$ $(8^+, 8^-, 10^+)$. However, the experimental evidence based on phase-shift analysis indicates the existence of several particles in this region (indicated in the table) and seems to favor the four-multiplet solution.

In the 1500-MeV region, the known particles are fewer. Of the three particles indicated in the table, it is at least known that $D_{13}(1520)$ has a dominant $\Delta\pi$ decay mode. We note that in addition to the solutions mentioned above, it is also possible to have $\delta = 0$ and/or $\epsilon = 0$, thus reducing the number of particles necessary to satisfy duality.

In any case, a nontrivial solution to the set of equations (in either energy region) exists only if $p \neq 0$ and $p' \neq 0$, and therefore leads to the prediction of the existence and the coupling to $\pi\Delta$ of at least the 8^+ particles with $J^P = \frac{5}{2}^+, \frac{3}{2}^+, \frac{1}{2}^+$, and the 8^- particle with $J^P = \frac{3}{2}^-$. The nonobservance of some of these

Table I. Solutions to Eqs. (4) and (5) in terms of two arbitrary parameters p and p'. α, β, γ, δ, and ϵ are known combinations of $A^{2J^{\pm}}_{2\mu,2\lambda}$ (Eq. 11). Beneath the solutions, we have indicated the (πN) resonances in the two energy regions of interest as revealed through phase-shift analysis (see, e.g., Ref. 10).

$[\chi_8{}^{5/2^+}(1)]^2 = p$ $[F_{15}(1.69)]$	$[\chi_{10}{}^{5/2^+}(1)]^2 = \frac{4}{5}(\alpha p' - p)$	$[\chi_8{}^{5/2^-}(2)]^2 = \delta p'$ $[D_{15}(1.68)]$	$[\chi_{10}{}^{5/2^-}(2)]^2 = \frac{2}{5}\delta[(9/\alpha)p - p']$
$[\chi_8{}^{3/2^+}(1)]^2 = (\beta/\alpha)p$	$[\chi_{10}{}^{3/2^+}(1)]^2 = \frac{2}{5}(\beta/\alpha)(\alpha p' - p)$ $[P_{33}(1.69)]$	$[\chi_8{}^{3/2^-}(0)]^2 = p'$ $[D_{13}(1.73)]$ $[D_{13}(1.52)]$	$[\chi_{10}{}^{3/2^-}(0)]^2 = \frac{2}{5}[(9/\alpha)p - p']$ $[D_{33}(1.69)]$
$[\chi_8{}^{1/2^+}(1)]^2 = (\gamma/\alpha)p$ $[P_{11}(1.75)]$ $[P_{11}(1.47)]$	$[\chi_{10}{}^{1/2^+}(1)]^2 = \frac{2}{5}(\gamma/\alpha)(\alpha p' - p)$	$[\chi_8{}^{1/2^-}(2)]^2 = \epsilon p'$ $[S_{11}(1.71)]$ $[S_{11}(1.55)]$	$[\chi_{10}{}^{1/2^-}(2)]^2 = \frac{2}{5}\epsilon[(9/\alpha)p - p']$ $[S_{31}(1.64)]$

states in the πN system does not rule out their existence. These states may be weakly coupled to the πN system.

Thus the spin considerations[3] in dual models and the examination of the constraints on the s-channel resonances[12] do provide interesting information concerning the particle spectrum and the coupling patterns. We can easily generalize our results to other systems as well as to particles with arbitrary spin. These considerations and a more detailed discussion of the predictions of local duality will be discussed elsewhere.

---

*Work supported by the U. S. Atomic Energy Commission.

[1]R. H. Capps, Phys. Rev. Letters 22, 215 (1969); J. Mandula *et al.*, Phys. Rev. Letters 22, 1147 (1969); V. Barger and C. Michael, Phys. Rev. 186, 1592 (1969).

[2]L. K. Chavda and R. H. Capps, Phys. Rev. (to be published).

[3]J. Mandula, J. Weyers, and G. Zweig, Phys. Rev. Letters 23, 627 (1969).

[4]We have removed the physical boundary singularities by dividing out the appropriate powers of the usual half-angle factors $\cos\frac{1}{2}\theta_s$, $\sin\frac{1}{2}\theta_s$ in order that G will not vanish at $\theta_s = 0$ or π just because of kinematical reasons.

[5]Equation (5) is obtained from Eq. (3) by using the standard symmetry relations for the Clebsch-Gordan coefficients and the d functions.

[6]A particle with arbitrary spin and definite parity has, in general, two couplings to a $(0\frac{3}{2}^+)$ system. The one with the least number of derivatives corresponds to $[\chi_R^{J\pm}(L_0)]^2$. It is assumed that this is the dominant coupling. See, for instance, J. G. Rushbrooke, Phys. Rev. 143, 1345 (1966).

[7]J. J. DeSwart, Nuovo Cimento 31, 420 (1964). For the specific normalizations in Eq. (9), we have followed Chavda and Capps, Ref. 2.

[8]See A. Kernan and H. K. Shepard, Phys. Rev. Letters 23, 1314 (1969).

[9]This can be justified using the "exchange degeneracy" hypothesis.

[10]A. Donnachie, in *Proceedings of the Fourteenth International Conference on High Energy Physics, Vienna, Austria, 1968,* edited by J. Prentki and J. Steinberger (CERN Scientific Information Service, Geneva, Switzerland, 1968), p. 139; see, especially, p. 143.

[11]In α, β, and γ, the leading terms of unity do not cancel. In δ and ϵ the leading terms do cancel, leaving only ratios $\chi(L_0+2)/\chi(L_0)$.

[12]J. L. Rosner, Phys. Rev. Letters 24, 173 (1970).

Properties of Polarization Density Matrix in Regge-Pole Models

Manuel G. Doncel,* Pierre Mery,† Louis Michel, Pierre Minnaert,‡ and Kameshwar C. Wali‖
Institut des Hautes Etudes Scientifiques, 91–Bures sur Yvette, France
(Received 13 April 1972)

We study two-body (or quasi-two-body) parity-conserving reactions with unpolarized beam and target. From the general considerations of angular momentum and parity conservation we derive the structure of the density matrices of the final particles. This structure becomes conspicuous when one uses transversity quantization with a special reordering for the lines and columns. We study the predictions of a single Regge trajectory exchange model. We derive the consequences of reality and factorization of residue functions, and of parity conservation at each vertex. The results are summarized in two tables. For the cases of low-spin particles we give these results in a geometrical form. As an example we plot some experimental data on Q-meson polarization.

I. INTRODUCTION

There is no need to emphasize the importance of the experimental determination of spins and parities of new particles or resonant states in high-energy physics. From the theoretical point of view, the principles involved are the well-known ones, namely, those that originate from the requirement of relativistic invariance. In practice, however, this is not an easy task. One has to make various selections in the data, and also often one has to make assumptions regarding the production mechanism, and so on. Probably the most important problem is the right choice of variables so that one can extract the maximum amount of information with a given set of data. Towards this end, three of the present authors have been engaged in an extensive study[1,2] of the "polarization density matrix and how to present its measurement." The emphasis is on the intrinsic geometrical properties of the density matrix arising from relativistic invariance and the allowed domain for the values of the elements of the density matrix. The experimentally measured values must fall in such a domain to be consistent with the conservation of angular momentum and parity if the particle has a definite spin and parity.

The main purpose of the present paper is to study the implications of the general considerations of Refs. 1 and 2 in a specific model. The model we choose is one in which a single Regge pole with factorizable residues dominates the production mechanism. The results can be generalized to slightly more complicated models which we shall discuss in a later section. The plan of the paper is as follows: In Sec. II, we shall briefly review some general results of Refs. 1 and 2 in the case of a two-particle (or a quasi-two-particle) reaction. Of special importance are the symmetry properties and the rank conditions on the density matrix in helicity and transversity frames. In Sec. III, we specialize to Regge-pole models and discuss in detail two special cases in which the produced particle is (i) a boson with arbitrary spin and parity, the beam particle being a pseudoscalar meson, and (ii) a fermion isobar, the target being a spin-$\frac{1}{2}$ particle. In Sec. IV, we discuss a few specific examples of (i) and (ii). We shall also discuss in this section the recent experimental results[3] on Q production. The final section is devoted to a summary of the main results.

II. MEASUREMENT AND DESCRIPTION OF POLARIZATION

Consider a reaction between four hadrons

$$1+2\to 3+4 \quad (s\text{ channel}),$$

$$4^*+2\to 3+1^* \quad (t\text{ channel}),$$

with p_i and j_i representing the four-momenta and spins of the particles. Particles 3 and/or 4 can be unstable. In the present paper we are interested in the information concerning the polarization state of 3 and/or 4 obtained through the study of angular distributions in parity-conserving decays. It is well known that the polarization state of a particle with positive mass, spin j, and energy-momentum p, is completely specified by a density operator ρ. By an appropriate choice of an orthonormal basis in a $(2j+1)$-dimensional Hilbert space $\mathcal{H}_{2j+1}$,

$$J^2|jm\rangle=j(j+1)|jm\rangle,$$
$$J_3|jm\rangle=m|jm\rangle, \qquad (2.1)$$
$$\langle jm'|jm\rangle=\delta^{m'}{}_m,$$

the density operator can be represented by a

Hermitian, nonnegative, and trace-1 matrix

$$\rho^\dagger = \rho \geq 0, \quad \mathrm{tr}\rho = 1 \tag{2.2}$$

whose matrix elements are

$$\rho^{m'}{}_m = \langle jm' | \rho | jm \rangle \ . \tag{2.3}$$

Conditions (2.2) imply that the eigenvalues of ρ are nonnegative and the diagonal elements satisfy

$$0 \leq \rho^m{}_m \leq 1 \ . \tag{2.4}$$

The joint polarization of particles 3 and 4, with known energy momenta, is described by an $n \times n$ Hermitian, nonnegative, trace-1 matrix $\rho(3,4)$ with $n = (2j_3 + 1)(2j_4 + 1)$.

A. Polarization Domain

The set of all $n \times n$ Hermitian matrices form a $(N+1)$-dimensional real Euclidean space which we shall call $\mathcal{E}_{N+1}$. The set of all trace-1, Hermitian matrices form a hyperplane $\mathcal{E}_N$ in this vector space, where $N = n^2 - 1$. The set of the density matrices which satisfy (2.2) is a subdomain of $\mathcal{E}_N$. This subdomain is called the polarization domain $\mathfrak{D}_j$:

$$\mathfrak{D}_j \subset \mathcal{E}_N \subset \mathcal{E}_{N+1} \ .$$

$\mathfrak{D}_j$ is a *convex domain*, i.e., if

$$\rho_1, \rho_2 \subset \mathfrak{D}_j \ , \tag{2.5}$$

then

$$\alpha\rho_1 + \beta\rho_2 \subset \mathfrak{D}_j \ ,$$

where

$$0 \leq \alpha, \beta, \quad \alpha + \beta = 1 \ .$$

B. Multipole Expansion

1. *One-Particle Polarization*

The density matrix ρ can be expanded in terms of a set of Hermitian matrices $Q_M^{(L)}$ and a set of real parameters $r_M^{(L)}$ as follows:

$$\rho = \rho_0 + \sum_{L=1}^{2j} \rho^{(L)} \ , \tag{2.6}$$

where

$$\rho_0 = \frac{1}{2j+1}\underline{1} \qquad [\underline{1}: (2j+1)\times(2j+1) \text{ identity matrix}]$$

represents the unpolarized state, and

$$\rho^{(L)} = \frac{2j}{2j+1} \sum_{M=-L}^{M=L} Q_M^{(L)} r_M^{(L)} \ . \tag{2.7}$$

The $Q_M^{(L)}$'s and $r_M^{(L)}$'s are related to the more customary expansion of ρ in terms of the multipoles $T_M^{(L)}$'s and the multipole-expansion parameters $t_M^{(L)}$'s by

$$Q_0^{(L)} = \left(\frac{2L+1}{2j}\right)^{1/2} T_0^{(L)} \ ,$$

$$Q_M^{(L)} = (-1)^M \left(\frac{2L+1}{j}\right)^{1/2} \tfrac{1}{2}(T_M^{(L)} + T_M^{(L)\dagger}), \quad M > 0 \ , \tag{2.8}$$

$$Q_{-M}^{(L)} = (-1)^M \left(\frac{2L+1}{j}\right)^{1/2} \frac{1}{2i}(T_M^{(L)} - T_M^{(L)\dagger}), \quad M > 0 \ ,$$

and

$$r_0^{(L)} = \left(\frac{2L+1}{j}\right)^{1/2} t_0^{(L)} \ ,$$

$$r_M^{(L)} = (-1)^M \left(\frac{2L+1}{j}\right)^{1/2} \mathrm{Re}\, t_M^{(L)}, \quad M > 0, \tag{2.9}$$

$$r_{-M}^{(L)} = (-1)^M \left(\frac{2L+1}{j}\right)^{1/2} \mathrm{Im}\, t_M^{(L)}, \quad M > 0 \ .$$

We recall that

$$(T_M^{(L)})^m{}_n = (2j+1)^{1/2} \begin{pmatrix} m & L & j \\ j & M & n \end{pmatrix} \ , \tag{2.10}$$

where $\left(\begin{smallmatrix} m & L & j \\ j & M & n \end{smallmatrix}\right)$ denotes Wigner's $3j$ symbol. The $T_M^{(L)}$'s provide a real non-Hermitian orthonormal basis for the expansion of ρ. The expansion coefficients $t_M^{(L)}$'s in this basis are in general complex, but the Hermiticity of ρ requires

$$t_M^{(L)*} = (-1)^M t_{-M}^{(L)} \ . \tag{2.11}$$

The parameters $r_M^{(L)}$ are more convenient. They are real (proportional to the real and imaginary part of the $t_M^{(L)}$) and they are independent of each other. Furthermore their normalization is such that the degree of polarization of the state is

$$d(\rho) = \left(\sum_{L=1}^{2j} \sum_{M=-L}^{L} (r_M^{(L)})^2\right)^{1/2} \ . \tag{2.12}$$

The parameters $r_M^{(L)}$ are very helpful in plotting the experimental results and to see whether the results are consistent with the requirements of conservation of angular momentum. For details we refer the reader to Refs. 1 and 2.

2. *Two-Particle Polarization*

For joint polarization of a two-particle state, the density operator can be expanded in double multipoles:

$$\rho(3,4) = \sum_{L=0}^{2j_3} \sum_{L'=0}^{2j_4} \rho^{(L,L')} \ , \tag{2.13}$$

with

$$\rho^{(L,L')} = \frac{(2L+1)(2L'+1)}{(2j+1)(2j'+1)} \sum_{M=-L}^{L} \sum_{M'=-L'}^{L'} t_{MM'}^{(L,L')*} \times T(j_3)_M^{(L)} \otimes T(j_4)_{M'}^{(L')} \ . \tag{2.14}$$

The density matrix of each particle when the polarization of the other is not observed is

$$\rho_3 = \mathrm{tr}_4\, \rho(3,4)\ , \qquad \rho_4 = \mathrm{tr}_3\, \rho(3,4)\ . \tag{2.15}$$

The polarization correlation between the two particles 3 and 4 is

$$K(3,4) = \rho(3,4) - \rho_3 \otimes \rho_4\ . \tag{2.16}$$

C. Decay Angular Distribution

1. *Two-Body Decay*

Let $\mathcal{G}_a(\theta,\varphi)$ denote the normalized angular distribution of one of the decay products, in a two-body parity-conserving decay mode a, of a particle of spin j. If the decaying particle is described by the density matrix ρ, then[4]

$$\mathcal{G}_a(\theta,\varphi) = \frac{1}{4\pi} + \sum_{L=1}^{2j}\left(\frac{2L+1}{2j}\right)^{1/2} \lambda_a(L,j) \sum_{M=-L}^{L} t_M^{(L)*}\, Y_M^{(L)}(\theta,\varphi)\ , \tag{2.17}$$

$$\mathcal{G}_a(\theta,\varphi) = \frac{1}{4\pi} + \sum_{L=1}^{2j} \lambda_a(L,j)\left(r_0^{(L)} Y_0^{(L)}(\theta,\varphi) + \sum_{M=1}^{L} (-)^M \sqrt{2}\,[r_M^{(L)}\ \mathrm{Re}Y_M^{(L)}(\theta,\varphi) + r_{-M}^{(L)}\ \mathrm{Im}Y_M^{(L)}(\theta,\varphi)]\right)\ , \tag{2.18}$$

where $Y_M^{(L)}(\theta,\varphi)$ are the usual spherical harmonics, and $t_M^{(L)}$ and $r_M^{(L)}$ are the expansion parameters already defined for the density matrix. The $\lambda_a(L,j)$'s are scalar coefficients (decay coefficients) which depend on L, j and on the particular decay mode a. Their value may depend on the dynamics of the decay mode a, e.g., $1^+ \to 1^- + 0^-$, $j \to \frac{3}{2} + 0$; however, for the most usual two-body hadronic decays,

$$1^- \to 0^- + 0^-,\quad 1^- \to 1^- + 0^-,\quad \tfrac{3}{2}^{\eta} \to \tfrac{1}{2}^+ + 0^- \quad (\eta = \pm), \tag{2.19}$$

they are independent of the dynamics. Indeed the corresponding angular distribution can be written

$$\mathcal{G}_a(\theta,\varphi) = \frac{1}{4\pi}[1 + \lambda_a'(2,j) A^{(2)}(\theta,\varphi)] \tag{2.20}$$

with

$$\begin{aligned} A^{(2)}(\theta,\varphi) &= \left(\frac{4\pi}{2j}\right)^{1/2} \sum_{M=-2}^{M=+2} t_M^{(2)*}\, Y_M^{(2)}(\theta,\varphi) \\ &= \tfrac{1}{2} r_0^{(2)}(3\cos^2\theta - 1) + \tfrac{1}{2}\sqrt{3}\sin^2\theta\,(r_2^{(2)}\cos 2\varphi + r_{-2}^{(2)}\sin 2\varphi) + \tfrac{1}{2}\sqrt{3}\sin 2\theta\,(r_1^{(2)}\cos\varphi + r_{-1}^{(2)}\sin\varphi)\ , \end{aligned} \tag{2.21}$$

and

$$\lambda_a'(2,j) = (20\pi)^{1/2}\lambda_a(2,j)\ .$$

The $\lambda_a'(2,j)$'s are numerical coefficients independent of dynamics. Their value is [cf. Ref. 1 and Ref. 2b with tables giving $\sqrt{4\pi}\,C(L) = \lambda_a'(L,j)/\sqrt{2j}$]:

$$\lambda_a'(2,j) = \begin{cases} -2 & \text{for } 1^- \to 0^- + 0^- \\ 1 & \text{for } 1^- \to 1^- + 0^- \\ -\sqrt{3} & \text{for } \frac{3}{2}^{\eta} \to \frac{1}{2}^+ + 0^-\ . \end{cases} \tag{2.22}$$

Parity conservation in a two-body decay implies that

$$\lambda_a(L,j) = 0 \quad \text{for } L \text{ odd}\ . \tag{2.23}$$

Therefore, by observing a parity-conserving two-body decay, one obtains information concerning only $r_M^{(L)}$'s with L even; i.e., one measures $\rho^{(L)}$, cf. Eqs. (2.6) and (2.7), where L is even. We shall refer to this as the measurement of *even* polarization.

2. *Three-Body Decay*

For three-body decays, the probability distribution is a function of five variables. One can choose for them two of the s, t, u invariants and three angles, θ, φ, ψ, where θ, φ fix the orientation of the normal to the decay plane (in the rest system of the decaying particle) and ψ fixes the rotation of the decay around the normal, e.g., ψ is the azimuth of a chosen one of the final particles. In general both even and odd polarization can be measured; let us be more specific for two usual decays:

(i) $1^- \to 0^- + 0^- + 0^-$. If the dependence of $\mathcal{G}$ on s, t, u is not observed, then we find

$$\mathcal{G}(\theta,\varphi,\psi) = \frac{1}{4\pi}[1 - 2A^{(2)}(\theta,\varphi)]\ . \tag{2.24}$$

This is independent of ψ and of the dynamics. It yields the value of the even polarization (i.e., the coefficients $r^{(2)}$'s).

(ii) $1^+ \to 0^- + 0^- + 0^-$. If only the dependence in

θ, φ is observed, one obtains the even polarization by

$$\mathcal{G}(\theta,\varphi)=\frac{1}{4\pi}[1+A^{(2)}(\theta,\varphi)] \tag{2.25}$$

with $A^{(2)}(\theta,\varphi)$ given in (2.20).

In such a decay it is also possible to measure the odd polarization (i.e., the $r_M^{(1)}$'s) up to a sign, from the complete observation of $\mathcal{G}(\theta,\varphi,\psi,s,t)$.[4]

3. *Correlated Decays*

To observe joint polarization of 3 and 4 one has to measure the joint angular distribution. For instance, for two two-body decays, this is $\mathcal{G}(\theta_3,\varphi_3,\theta_4,\varphi_4)$; it can be expanded in a manner similar to (2.18) with the basis $Y_M^{(L)}(\theta_3,\varphi_3) \times Y_{M'}^{(L')}(\theta_4,\varphi_4)$. If $K(3,4)=0$, i.e., no polarization correlation, then $\mathcal{G}$ factorizes:

$$\mathcal{G}(\theta_3,\varphi_3;\theta_4,\varphi_4)=\mathcal{G}_3(\theta_3,\varphi_3)\mathcal{G}_4(\theta_4,\varphi_4)\ . \tag{2.26}$$

The explicit form of the decay distribution function (and also the explicit form of the ρ matrix) depends on the choice of the reference frame. We deal with this question next.

D. Choice of Reference Frames

To choose an orthonormal basis in $\mathcal{H}_{2j+1}$, one needs to specify a reference frame and an axis of quantization. Following Ref. 2, we associate with each particle a tetrad, i.e., a set of four vectors $n^{(\alpha)}$, $\alpha=0,1,2,3$, which are orthonormal:

$$n^{(\alpha)}\cdot n^{(\beta)}=g^{\alpha\beta}\ . \tag{2.27}$$

(We use the metric $g^{00}=-g^{ii}=1$.) By convention, $n^{(3)}$ is the quantization axis. In a four-particle reaction of the type under consideration, the normal to the reaction plane n (which is defined up to a sign) can be chosen to be an axis common to all particle tetrads. In a given channel, *transversity* quantization is a choice of tetrad ${}^Tn_i^{(\alpha)}$ for each particle i, with ${}^Tn_i^{(3)}=n$ and a conventional choice for ${}^Tn_i^{(2)}$ (then ${}^Tn_i^{(1)}$ is fixed) which is universal but for a sign (see below). In *helicity quantization* ${}^Hn_i^{(2)}=n$; the quantization axis ${}^Hn_i^{(3)}$ is chosen in the following way: In the rest frame of the particle i the space part of ${}^Hn_i^{(3)}$ is collinear to the momentum of the associated particle (e.g., 1 with 3, 2 with 4 in the t channel, 1 with 2, 3 with 4 in the s channel). This defines ${}^Hn_i^{(3)}$ up to a sign. There are different sign conventions in the literature, so every experimental or theoretical paper should be explicit about these conventions. Our conclusions are of course independent of conventions, but our computation will be done with the (channel-independent) convention for every particle:

$${}^Tn^{(3)}=n={}^Hn^{(2)},\quad {}^Tn^{(1)}={}^Hn^{(1)},\quad {}^Tn^{(2)}=-{}^Hn^{(3)}\ . \tag{2.28}$$

That is, the helicity frame is transformed into the transversity frame by a rotation $\tilde{R}$ of $-\frac{1}{2}\pi$ around the first axis: In terms of Euler angles

$$\tilde{R}=(\tfrac{1}{2}\pi,\tfrac{1}{2}\pi,-\tfrac{1}{2}\pi)\ ,$$

and consequently

$$|jm\rangle^T=|jm'\rangle^H D^{(j)}(\tilde{R})^{m'}{}_m\ . \tag{2.29}$$

$D^{(j)}(\tilde{R})$ represents the $(2j+1)$-dimensional, unitary, irreducible representation of the rotation group. In the present case

$$D^{(j)}(\tilde{R})^m{}_{m'}=e^{-i\pi(m-m')/2}\,d^{(j)}(\tfrac{1}{2}\pi)^m{}_{m'}\ . \tag{2.30}$$

From the symmetry properties[5]

$$\begin{aligned} d^{(j)}(\theta)^m{}_{m'}&=d^{(j)}(\theta)^m{}_{m'}{}^*\\ &=(-1)^{m-m'}d^{(j)}(\theta)^{m'}{}_m\\ &=(-1)^{m-m'}d^{(j)}(\theta)^{-m}{}_{-m'}\ ,\\ d^{(j)}(\tfrac{1}{2}\pi)^m{}_{m'}&=(-1)^{j-m}d^{(j)}(\tfrac{1}{2}\pi)^m{}_{-m'}\ , \end{aligned} \tag{2.31}$$

it is easy to obtain for the rotation matrix (2.30)

$$\begin{aligned} D^{(j)}(\tilde{R})^m{}_{m'}&=D^{(j)}(\tilde{R})^{m'}{}_m\\ &=D^{(j)}(\tilde{R})^{-m}{}_{-m'}\\ &=(-)^{m-m'}D^{(j)}(\tilde{R})^m{}_{m'}{}^*\\ &=e^{i\pi j}D^{(j)}(\tilde{R})^m{}_{-m'}{}^*\ . \end{aligned} \tag{2.32}$$

In either transversity or helicity frame, to go from one channel to the other, one has to use crossing matrices which are functions of the variables s, t, and u. Henceforth we shall confine our attention to the t channel. The helicity frame then is the Gottfried-Jackson frame. The density matrices ${}^T\rho$ and ${}^H\rho$ are related by

$${}^T\rho=D^{(j)}(\tilde{R}^{-1})\,{}^H\rho\, D^{(j)}(\tilde{R})\ . \tag{2.33}$$

E. Restrictions on the Density Matrix Due to the Mode of Observation or to the Nature of the Production Reaction

1. *Even Polarization*

As we have seen in Sec. II C, in many cases only the even part (even-L multipoles) of the one-particle density matrix can be measured. What is observed in these cases is the projection ${}^{(E)}\rho$ of ρ on a subspace which is a symmetry plane of $\mathfrak{D}_j$. One can prove[2] that

$${}^{(E)}\rho=\sum_{L\text{ even}}\rho^{(L)}$$

is also a Hermitian, positive, and trace-1 matrix. One also shows[2] that

$$(\rho^{(L)})^T=\rho^{(L)*}$$
$$=(-)^L\Gamma_j\rho^{(L)}\Gamma_j^{-1}, \quad (2.34)$$

where Γ_j is the $(2j+1)\times(2j+1)$ matrix defined, independently of the choice of the quantization axis, by

$$(\Gamma_j)^m{}_n=(-)^{j-m}\delta^{-m}{}_n . \quad (2.35)$$

From (2.34) one easily observes that ${}^{(E)}\rho$ satisfies

$$({}^{(E)}\rho)^T={}^{(E)}\rho^*$$
$$=\Gamma_j({}^{(E)}\rho)\Gamma_j^{-1} . \quad (2.36)$$

Γ_j is the representation of a rotation of $-\pi$ around the $n^{(2)}$ axis,

$$\Gamma_j=D^{(j)}(0,\pi,0) . \quad (2.37)$$

As we shall see, some models are specially interesting by predicting only an even polarization, that is,

$${}^{(E)}\rho=\rho .$$

2. *B Symmetry*

Although in general the polarization domain $\mathfrak{D}_j$ of the density matrix is completely determined by conditions (2.2), in practice, there are further restrictions depending on the nature of the reaction which is responsible for the production of the particle of interest. In this paper, we shall be mainly concerned with a quasi-two-particle reaction which is parity conserving and in which the beam and the target particles are unpolarized. In such a reaction there are only three linearly independent energy momenta. The space-time hyperplane which contains the three observed energy momenta is a symmetry plane for the reaction in which the normal to this hyperplane is the common quantization axis Oz for the tetrads of the four particles. In each tetrad, the B symmetry is represented by the space reflection through the XY plane. Under such a reflection,

$$|jm\rangle^T\to|jm\rangle^{T'}=\eta|jm'\rangle^T D^{(j)}(0,0,\pi)^{m'}{}_m$$
$$=\eta e^{i\pi m}|jm\rangle^T , \quad (2.38)$$

where η is the intrinsic parity of the particle. Consequently,

$$({}^T\rho)^{m'}{}_m=e^{i\pi(m'-m)}({}^T\rho)^{m'}{}_m , \quad (2.39)$$

which implies that

$$({}^T\rho)^{m'}{}_m=0 \ \text{ if } m'-m \text{ is odd.} \quad (2.40)$$

On any one-particle density matrix, with the natural order for the indices (i.e., m',m $=j,j-1,\ldots,-j$) there is a "checkerboard" pattern when we call "white" the elements with $(m'-m)$ odd and "black" those with $(m'-m)$ even. The main diagonal is always black, the second diagonal is black for integer spins and white for half-integer ones. Equation (2.39) means that all the white elements of a B-symmetric density matrix are zero in transversity quantization.

It is clear that by reordering the indices, the nonvanishing (black) elements can be brought together and ${}^T\rho$ can be written in a block form. For this purpose it is convenient to discuss two cases separately:

a. When j is integral. Choose the separation order of indices which separates the black and the white squares: $j,j-2,\ldots,2-j,-j;\ j-1,j-3,\ldots,3-j,1-j$. See for example Fig. 1(a). Then

$${}^T\rho=\begin{bmatrix}A&0\\0&B\end{bmatrix} , \quad (2.41)$$

where A and B are Hermitian matrices of dimensions $(j+1)\times(j+1)$ and $j\times j$, respectively, with

$$\mathrm{tr}A+\mathrm{tr}B=1 . \quad (2.42)$$

Let Δ_k be a $k\times k$ matrix which has 1 on the second diagonal and zero elsewhere,

$$(\Delta_k)^\alpha{}_\beta=\delta^\alpha{}_{-\beta}, \quad \Delta_k{}^2=\underline{1} . \quad (2.43)$$

If ρ contains only even polarization, A and B are symmetric through the second diagonal:

$$A^*=A^T=\Delta_{j+1}A\Delta_{j+1}, \quad B^*=B^T=\Delta_jB\Delta_j . \quad (2.44)$$

The density matrix ${}^H\rho$ is given by (2.33). To study the structure of ${}^H\rho$ we must first study that of the rotation matrix $D^{(j)}(\bar{R})$ in the separation order of indices. Then it can be written in the block form:

$$D^{(j)}(\bar{R})=\begin{bmatrix}X&iW\\iW^T&Y\end{bmatrix}, \quad (2.45)$$

where X and Y are square matrices of the same dimensions as A and B, respectively, and W is a rectangular matrix of dimension $(j+1)\times j$. From (2.32), one can write symmetry properties for X, Y, and W in the following way:

$$X=X^T=X^*=\Delta_{j+1}X\Delta_{j+1}=(-)^j\Delta_{j+1}X=(-)^jX\Delta_{j+1},$$
$$Y=Y^T=Y^*=\Delta_jY\Delta_j=(-)^j\Delta_jY=(-)^jY\Delta_j, \quad (2.46)$$
$$W=W^*=\Delta_{j+1}W\Delta_j=-(-)^j\Delta_{j+1}W=-(-)^jW\Delta_j .$$

From the unitarity of $D^j(\bar{R})$ and the property

$$[D^{(j)}(\bar{R})]^2=(-)^j\Delta , \quad (2.47)$$

where

$$\Delta=\Delta_{j+1}\oplus\Delta_j ,$$

we have

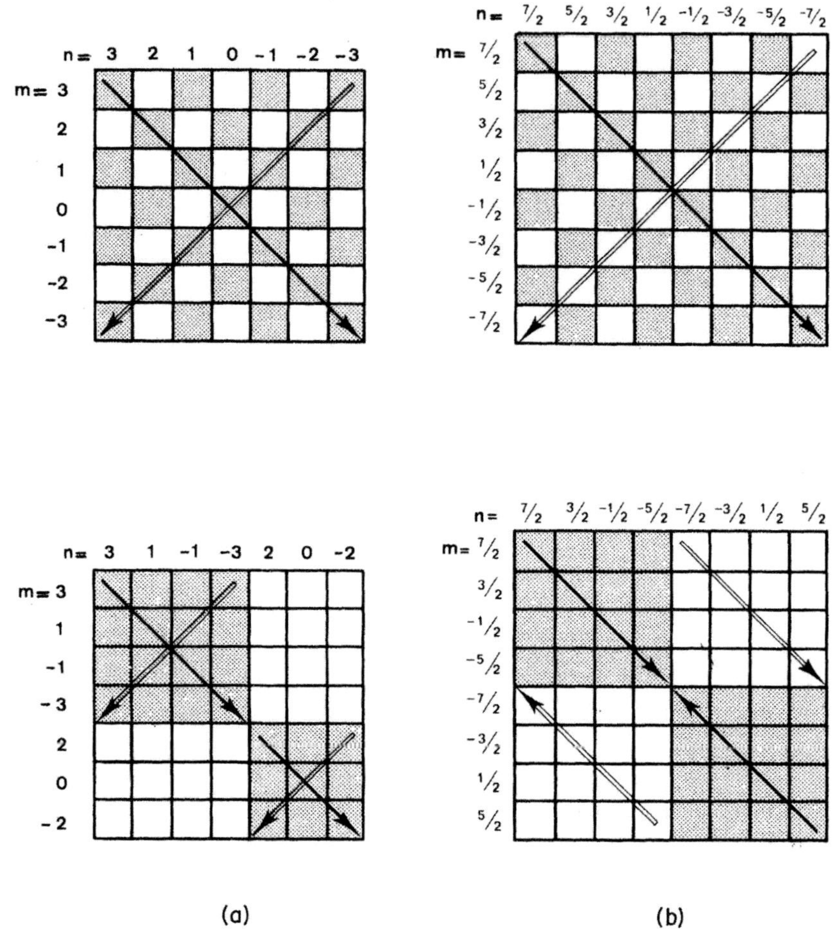

FIG. 1. Structure of Bohr-symmetric density matrices in transversity quantization for the natural order and for the separation order of indices. Part (a) illustrates the case of integer spin ($j=3$) and part (b) the case of half-integer spin ($j=\frac{7}{2}$). The arrows show the transformation of the main and second diagonal in each case.

$$
\begin{aligned}
&X^2+WW^T=\underline{1}\,,\\
&Y^2+W^TW=\underline{1}\,,\\
&X^2-WW^T=(-)^j\Delta_{j+1}\,,\\
&Y^2-W^TW=(-)^j\Delta_j\,,
\end{aligned}
\tag{2.48}
$$

and

$$XW=WY=0\,.$$

In the separation order of indices, the density matrix ${}^H\rho$ can be written

$$
{}^H\rho=\begin{bmatrix} C & D\\ D^\dagger & F\end{bmatrix},
\tag{2.49}
$$

with

$$
\begin{aligned}
&C=C^\dagger=\Delta_{j+1}C\Delta_{j+1}\,,\\
&F=F^\dagger=\Delta_jF\Delta_j\,,\\
&D=-\Delta_{j+1}D\Delta_j\,,\\
&\mathrm{tr}\,C+\mathrm{tr}\,F=1\,.
\end{aligned}
\tag{2.50}
$$

The relation between the submatrices C, D, F and A and B is given by

$$
\begin{aligned}
&C=XAX+WBW^T\,,\\
&F=W^TAW+YBY\,,\\
&D=-i(XAW-WBY)\,,
\end{aligned}
\tag{2.51}
$$

or, conversely,

$$A = XCX + WFW^T + i(XDW^T - WD^\dagger X),$$
$$B = YFY + W^T CW + i(YD^\dagger W - W^T DY). \tag{2.52}$$

Furthermore if ρ contains only even polarization, we get

$$C = C^* = C^T = \Delta_{j+1} C \Delta_{j+1},$$
$$F = F^* = F^T = \Delta_j F \Delta_j, \tag{2.53}$$
$$D = D^* = -\Delta_{j+1} D \Delta_j.$$

The number of parameters necessary to specify ${}^T\rho$ or ${}^H\rho$ is given by k, where

$$k = 2j(j+1) \text{ if } \rho \text{ is } B \text{ symmetric},$$
$$k = j(j+2) \text{ if } \rho \text{ is } B \text{ symmetric and even}. \tag{2.54}$$

b. When j is half-integral. Choose another separation order of indices: $j, j-2, \ldots, -j+1;\ -j, -j+2, \ldots, j-1$. See for example Fig. 1(b). Then a B-symmetric density matrix ${}^T\rho$ can be written in the form

$${}^T\rho = \begin{bmatrix} A & 0 \\ 0 & B \end{bmatrix}, \tag{2.55}$$

where A and B are $(j+\frac{1}{2}) \times (j+\frac{1}{2})$ Hermitian matrices with

$$\mathrm{tr}A + \mathrm{tr}B = 1. \tag{2.56}$$

When ρ contains only even polarization, we get furthermore $B = A^T$ so that

$${}^T\rho = \begin{bmatrix} A & 0 \\ 0 & A^T \end{bmatrix}. \tag{2.57}$$

The unitary transformation $D^{(j)}(\tilde{R})$ in this case can be written in the form

$$D^{(j)}(\tilde{R}) = \frac{1}{\sqrt{2}} \begin{bmatrix} V & e^{i\pi j} V \\ e^{i\pi j} V & V \end{bmatrix}, \tag{2.58}$$

where V is a real, symmetrical, orthogonal transformation:

$$V = V^* = V^T, \quad V^2 = \underline{1}. \tag{2.59}$$

The density matrix ${}^H\rho$ in the helicity frame for a B-symmetric reaction has the form

$${}^H\rho = \begin{bmatrix} C & D \\ -D & C \end{bmatrix}, \tag{2.60}$$

with

$$C = C^\dagger,$$
$$D = -D^\dagger, \tag{2.61}$$
$$\mathrm{tr}C = \tfrac{1}{2}.$$

The submatrices C and D are related to the submatrices A and B by

$$C = \tfrac{1}{2} V(A+B)V,$$
$$D = -\tfrac{1}{2} e^{i\pi j} V(A-B)V, \tag{2.62}$$

or, conversely,

$$A = V(C + e^{i\pi j} D)V,$$
$$B = V(C - e^{i\pi j} D)V. \tag{2.63}$$

If ρ contains only even polarization by putting $B = A^T$ we get the same form (2.60) for the density matrix in the helicity frame but with

$$C = C^* = C^T = \tfrac{1}{2} V(A + A^T)V,$$
$$D = D^* = -D^T = -\tfrac{1}{2} e^{i\pi j} V(A - A^T)V. \tag{2.64}$$

The number of free parameters to specify the density matrix is k, where

$$k = 2j(j+1) - \tfrac{1}{2} \text{ if } \rho \text{ is } B \text{ symmetric},$$
$$k = (j+\tfrac{3}{2})(j-\tfrac{1}{2}) \text{ if } \rho \text{ is } B \text{ symmetric and even}. \tag{2.65}$$

These results are summarized in Table I.

3. *Rank Condition*

The final-state density matrix ρ_f and the cross section σ are related to the initial-state density matrix ρ_i by

$$\sigma \rho_f = T \rho_i T^\dagger, \tag{2.66}$$

where T is the transition matrix. Since the rank of the product of matrices is smaller than or equal to the smallest rank of the matrices of the product, the rank of ρ_f cannot exceed the rank of ρ_i. For example in the reaction

$$\pi + N \to \pi + N^*,$$
$$\mathrm{rank}\ \rho(N^*) \leq 2, \tag{2.67}$$

no matter what N^* spin is. Thus if the spin j of the isobar N^* is $>\frac{1}{2}$ (2.67) imposes constraints on the measured elements of the density matrix. Also we note that the rank of the sum of matrices cannot be larger than the sum of their ranks. Hence in (2.64), for example

$$\mathrm{rank}\ C \leq 2\,\mathrm{rank}\ A, \tag{2.68}$$

but from (2.63)

$$\mathrm{rank}\ (C + e^{i\pi j} D) = \mathrm{rank}\ A. \tag{2.69}$$

As we see, in some reactions angular momentum conservation may impose restriction on the rank of the density matrix (for more examples see Refs. 1 and 2).

Independently of any model, if only the even part ${}^{(E)}\rho$ of the density matrix is measured, one gets, for the rank of this even part,[1]

$$\mathrm{rank}\ {}^{(E)}\rho \leq 2\,\mathrm{rank}\ \rho. \tag{2.70}$$

In models which predict only an even polarization ${}^{(E)}\rho = \rho$ implies of course rank ${}^{(E)}\rho = \mathrm{rank}\ \rho$.

TABLE I. Structure of B-symmetric even-polarization density matrices in the separation order of indices.

	j = integer Separation order of indices: $j, j-2, \ldots, 2-j, -j;\ j-1, j-3, \ldots, 3-j, 1-j$		j = half integer Separation order of indices: $j, j-2, \ldots, -j+1;\ -j, -j+2, \ldots, j-1$	
	B symmetry	B symmetry and even polarization	B symmetry	B symmetry and even polarization
In any transversity frame	$^T\rho = \begin{bmatrix} A & 0 \\ 0 & B \end{bmatrix}$ $A = A^\dagger$ $B = B^\dagger$ $\mathrm{tr}A + \mathrm{tr}B = 1$	$^T\rho = \begin{bmatrix} A & 0 \\ 0 & B \end{bmatrix}$ $A^* = A^T = \Delta_{j+1} A \Delta_{j+1}$ $B^* = B^T = \Delta_j B \Delta_j$ $\mathrm{tr}A + \mathrm{tr}B = 1$	$^T\rho = \begin{bmatrix} A & 0 \\ 0 & B \end{bmatrix}$ $A = A^\dagger$ $B = B^\dagger$ $\mathrm{tr}A + \mathrm{tr}B = 1$	$^T\rho = \begin{bmatrix} A & 0 \\ 0 & A^T \end{bmatrix}$ $A = A^\dagger$ $\mathrm{tr}A = \frac{1}{2}$
In any helicity frame	$^H\rho = \begin{bmatrix} C & D \\ D^\dagger & F \end{bmatrix}$ $C = C^\dagger = \Delta_{j+1} C \Delta_{j+1}$ $F = F^\dagger = \Delta_j F \Delta_j$ $D = -\Delta_{j+1} D \Delta_j$ $\mathrm{tr}C + \mathrm{tr}F = 1$	$^H\rho = \begin{bmatrix} C & D \\ D^T & F \end{bmatrix}$ $C = C^* = C^T = \Delta_{j+1} C \Delta_{j+1}$ $F = F^* = F^T = \Delta_j F \Delta_j$ $D = D^* = -\Delta_{j+1} D \Delta_j$ $\mathrm{tr}C + \mathrm{tr}F = 1$	$^H\rho = \begin{bmatrix} C & D \\ -D & C \end{bmatrix}$ $C = C^\dagger$ $D = -D^\dagger$ $\mathrm{tr}C = \frac{1}{2}$	$^H\rho = \begin{bmatrix} C & D \\ -D & C \end{bmatrix}$ $C = C^* = C^T$ $D = D^* = -D^T$ $\mathrm{tr}C = \frac{1}{2}$
Number k of independent parameters	$k = 2j(j+1)$	$k = j(j+2)$	$k = 2j(j+1) - \frac{1}{2}$	$k = (j+\frac{3}{2})(j-\frac{1}{2})$
Unitary transformation from transversity to helicity, $\tilde{R} = [\frac{1}{2}\pi, \frac{1}{2}\pi, -\frac{1}{2}\pi]$	$D^{(j)}(\tilde{R}) = \begin{bmatrix} X & iW \\ iW^T & Y \end{bmatrix}$ with: $X = X^T = X^* = \Delta_{j+1} X \Delta_{j+1} - (-)^j \Delta_{j+1} X = (-)^j X \Delta_{j+1}$ $Y = Y^T = Y^* = \Delta_j Y \Delta_j = (-)^j \Delta_j Y = (-)^j Y \Delta_j$ $W = W^* = \Delta_{j+1} W \Delta_j = -(-)^j \Delta_{j+1} W = -(-)^j W \Delta_j$ $X^2 + WW^T = \underline{1};\ X^2 - WW^T = (-)^j \Delta_{j+1}$ $Y^2 + W^T W = \underline{1};\ Y^2 - WW^T = (-)^j \Delta_j$ $XW = WY = 0$		$D^{(j)}(\tilde{R}) = \frac{1}{\sqrt{2}} \begin{bmatrix} V & e^{i\pi j} V \\ e^{i\pi j} V & V \end{bmatrix}$ $V = V^* = V^T$ $V^2 = \underline{1}$	
Relations between submatrices in helicity and transversity frames	$A = XCX + WFW^T + i(XDW^T - WD^\dagger X)$ $B = YFY + W^T CW + i(YD^\dagger W - W^T DY)$ $C = XAX + WBW^T$ $F = W^T AW + YBY$ $D = -i(XAW - WBY)$		$A = V(C + e^{i\pi j} D)V$ $B = V(C - e^{i\pi j} D)V$ $C = \frac{1}{2} V(A+B)V$ $D = -\frac{1}{2} e^{i\pi j} V(A-B)V$	
Notations	Δ_k is a $k \times k$ matrix whose elements are $(\Delta_k)^m{}_n = \delta^m{}_{-n}$			

III. POLARIZATION AND REGGE-POLE MODELS

The results discussed in the previous section give the general structure that the density matrices must satisfy from conservation of angular momentum and parity. Any specific model which conserves angular momentum and parity will give density matrices with this structure, but, in addition, if the model has any predictive power on the polarizations, it will give further conditions on the density matrices.

Regge-pole models are such specific models and they are extensively used in the analysis of high-energy scattering. In Sec. III A we shall be mainly concerned with two-body reactions in which a single bosonic Regge trajectory of definite signature is exchanged. At the end of the section we discuss the exchange of several trajectories with the same signature. In Sec. III B we study many-body reactions within the model of a multi-Regge exchange.

A. Two-Body Reactions

Let $G_{\lambda_3\lambda_4;\lambda_1\lambda_2}(s,t)$ and $F_{\lambda_3\lambda_1;\lambda_4\lambda_2}(s,t)$ represent the corresponding s- and t-channel helicity amplitudes for the four-particle reactions mentioned in the previous sections. From the well-known Trueman-Wick crossing relation, we have

$$G_{\lambda_3\lambda_4;\lambda_1\lambda_2} = \sum_{\lambda_4'\lambda_3'\lambda_2'\lambda_1'} d^{j_1}_{\lambda_1'\lambda_1}(\chi_1) d^{j_2}_{\lambda_2'\lambda_2}(\chi_2) \times d^{j_3}_{\lambda_3'\lambda_3}(\chi_3) d^{j_4}_{\lambda_4'\lambda_4}(\chi_4) F_{\lambda_3'\lambda_1';\lambda_4'\lambda_2'}, \quad (3.1)$$

where χ_i is the crossing angle of the ith particle. If the high-s, fixed-t behavior is governed by a single Regge trajectory, we can write

$$F_{\lambda_3\lambda_1;\lambda_4\lambda_2}(s,t) = \beta^{(1)}_{\lambda_3\lambda_1}(t)\beta^{(2)}_{\lambda_4\lambda_2}(t) \times \frac{1+\tau\exp[-i\pi\alpha(t)]}{\sin\pi\alpha(t)} s^{\alpha(t)}, \quad (3.2)$$

where $\beta^{(1)}$ and $\beta^{(2)}$ are the factorized Regge residues, $\alpha(t)$ is the Regge trajectory function, and τ is the signature. The residue functions are real and due to parity conservation at each vertex, they obey the symmetry relations

$$\beta^{(1)}_{-\lambda_3-\lambda_1} = \sigma_E\eta_3\eta_1{*}(-1)^{j_3+j_1}(-1)^{\lambda_3-\lambda_1}\beta^{(1)}_{\lambda_3\lambda_1}, \quad (3.3)$$

$$\beta^{(2)}_{-\lambda_4-\lambda_2} = \sigma_E\eta_4{*}\eta_2(-1)^{j_4+j_2}(-1)^{\lambda_4-\lambda_2}\beta^{(2)}_{\lambda_4\lambda_2}, \quad (3.4)$$

where σ_E is the naturality of the trajectory ($\sigma_E = \tau P$ where P is the parity of the trajectory: $\sigma_E = +1$ or -1 if the trajectory has natural or unnatural parity), and η_i is the intrinsic parity of the ith particle.

Let us consider the polarization state of particle 3 (or 4) when the polarization of 4 (or 3) is not observed. The density matrix in the t-channel helicity frame (or more popularly known as Gottfried-Jackson frame) is given by

$${}^H\rho(s,t)^{m'}{}_m = \frac{\sum_{\lambda_1\lambda_2\lambda_4} F_{m'\lambda_1;\lambda_4\lambda_2}(s,t)F^*_{m\lambda_1;\lambda_4\lambda_2}(s,t)}{\sum_{\lambda_1\lambda_2\lambda_3\lambda_4} F_{\lambda_3\lambda_1;\lambda_4\lambda_2}(s,t)F^*_{\lambda_3\lambda_1;\lambda_4\lambda_2}(s,t)} \quad (3.5)$$

with a similar expression for the density matrix of particle 4. If we use (3.2), we see that the density matrix of particle 3 does not depend on s and has the form

$${}^H\rho(t)^{m'}{}_m = \sum_{\lambda_1} \gamma^{(1)}_{m'\lambda_1}\gamma^{(1)}_{m\lambda_1}, \quad (3.6)$$

where

$$\gamma^{(1)}_{m\lambda_1} = \beta^{(1)}_{m\lambda_1}\Big(\sum_{m\lambda_1}(\beta^{(1)}_{m\lambda_1})^2\Big)^{-1/2}, \quad (3.7)$$

so that

$$\sum_{m\lambda_1}\gamma^{(1)}_{m\lambda_1}\gamma^{(1)}_{m\lambda_1} = 1.$$

Likewise, the density matrix for particle 4 does not depend on s and can be written as

$${}^H\rho(t)^{m'}{}_m = \sum_{\lambda_2}\gamma^{(2)}_{m'\lambda_2}\gamma^{(2)}_{m\lambda_2}, \quad (3.8)$$

with

$$\sum_{m\lambda_2}\gamma^{(2)}_{m\lambda_2}\gamma^{(2)}_{m\lambda_2} = 1.$$

It is clear that the $\gamma^{(i)}$'s obey the symmetry properties of the corresponding $\beta^{(i)}$'s expressed in Eqs. (3.3) and (3.4).

Before entering into any detail concerning the spin of the particles involved into the reaction, one may derive some general consequences from the expressions (3.6) and (3.8) of the density matrices in terms of the *factorized* residue functions $\gamma_{m\lambda}$.

(i) The reality of the residue functions implies the reality of the B-symmetric density matrix in helicity quantization which, as we have seen in the previous section, means that we have only even polarization.

(ii) Because of assumed factorizability of the residues at the two vertices, the ranks of the density matrices of particles 3 and 4 are determined only by the spins of their associated particles in the t channel, i.e., particles 1 and 2, respectively. Thus the density matrices satisfy

$$\text{rank}\,(\rho_3) \leq 2j_1+1, \qquad \text{rank}\,(\rho_4) \leq 2j_2+1. \quad (3.9)$$

(iii) Furthermore, if we consider the joint polarization between the two particles 3 and 4, then factorization implies that there is no polarization correlation between the two particles, i.e.,

$$K(3,4) = 0 \Longleftrightarrow \rho(3,4) = \rho(3)\otimes\rho(4). \quad (3.10)$$

(iv) Finally, consider several reactions which have one common vertex, i.e., they are dominated by the same trajectory and particles 1 and 3 (or 2 and 4) are respectively the same for all reactions. Then the density matrix of the final particle 3 (or 4) is the same for all these reactions, i.e., it is independent of the nature of the particles at the other vertex.

It should be noted here that although the exchange of a single Regge pole dominates a number of processes, it is by no means sufficient to account for all the features of the data. Generally one modifies the simple pole model by including absorptive corrections on contributions due to Regge cuts. There is considerable arbitrariness in introducing such modifications and consequently there is a variety of recipes and models[6] in the literature. Very few general model-independent remarks can be made concerning the density matrices in this context. In general corrections to pole models lead to nonreal residue functions and consequently property (i) disappears. In many models the factorization of the residue functions is also not preserved. But it may be possible to introduce the necessary modifications without destroying the factorization property, in which case the properties (ii), (iii), and (iv) discussed above are still valid.

1. *Production of a Boson*

Let us consider the production of a boson of spin and parity j^η using a pseudoscalar meson as the beam particle:

$$0^- + j_2^{\eta_2} \to j^\eta + j_4^{\eta_4},$$

where the target particle 2 is unpolarized and the polarization state of 4 is not observed.

The density matrix for the observed boson must have rank 1 according to (3.9). It is clear from (3.6) that we can write ${}^H\rho$ in the form (since $\lambda_1 = 0$)

$$^H\rho = \begin{bmatrix} c \\ d \end{bmatrix} [c^T \; d^T] = \begin{bmatrix} cc^T & cd^T \\ dc^T & dd^T \end{bmatrix}, \qquad (3.11)$$

where

$$c = \begin{bmatrix} \gamma_j \\ \gamma_{j-2} \\ \vdots \\ \gamma_{2-j} \\ \gamma_{-j} \end{bmatrix}, \quad d = \begin{bmatrix} \gamma_{j-1} \\ \vdots \\ \gamma_{1-j} \end{bmatrix} \qquad (3.12)$$

are two column vectors with $(j+1)$ and j elements, respectively. The dyadic form (3.11) of ${}^H\rho$ exhibits clearly the rank-1 condition which, expressed in terms of density-matrix elements, reads

$$^H\rho^m{}_n \, {}^H\rho^{m'}{}_{n'} = {}^H\rho^m{}_{n'} \, {}^H\rho^{m'}{}_n . \qquad (3.13)$$

The symmetry property (3.3) in this special case can be written as

$$\gamma_{-\lambda} = -\eta\sigma_E(-1)^{j+\lambda}\gamma_\lambda \qquad (3.14)$$

or

$$\Delta_{j+1}c = -\eta\sigma_E c, \quad \Delta_j d = \eta\sigma_E d. \qquad (3.15)$$

Note that (3.11) has the structure (2.49) with the identification

$$C = cc^T, \quad D = cd^T, \text{ and } F = dd^T.$$

From (3.14) we get

$$\Delta_{j+1}C = C\Delta_{j+1} = -\eta\sigma_E C, \quad \Delta_j F = F\Delta_j = +\eta\sigma_E F, \quad \Delta_{j+1}D = -D\Delta_j = -\eta\sigma_E D. \qquad (3.16)$$

These relations expressed in terms of the matrix elements of ${}^H\rho$ read

$$^H\rho^m{}_n = -\sigma_E\eta(-)^{j+n}\,{}^H\rho^m{}_{-n} = -\sigma_E\eta(-)^{j+m}\,{}^H\rho^{-m}{}_n . \qquad (3.17)$$

The above relations represent the generalization of those found by Ader *et al.*[7] for the diagonal elements of ${}^H\rho$.

The density matrix in the transversity frame takes an even simpler form. As noted in (2.41), ${}^T\rho$ is the direct sum of two matrices A and B. The rank of the total matrix has to be unity. Hence the rank of A or B has to be zero, i.e., A or B has to vanish identically. We show that vanishing of A or B depends on the signature σ_E of the exchanged trajectory and the naturality or unnaturality of the produced boson. Equation (2.52) can be written

$$A = X\Delta_{j+1}{}^2 CX + W\Delta_j{}^2 FW^T + i(X\Delta_{j+1}{}^2 DW^T - WD^T\Delta_{j+1}{}^2 X),$$

and using (2.46) and (3.16) one gets

$$A = -(-)^j\eta\sigma_E A. \qquad (3.18)$$

In the same way, one can write

$$B = Y\Delta_j{}^2 FY + W^T C\Delta_{j+1}{}^2 W + i(YD^T\Delta_{j+1}{}^2 W - W^T\Delta_{j+1}{}^2 DY)$$

and get

$$B = +(-)^j\eta\sigma_E B. \qquad (3.19)$$

Therefore parity and rank conditions imply

$$\sigma_E\eta(-1)^j = -1 \Rightarrow {}^T\rho = \begin{bmatrix} A & 0 \\ 0 & 0 \end{bmatrix}, \quad \text{rank } A = 1 \qquad (3.20)$$

and

$$\sigma_E\eta(-1)^j = +1 \Rightarrow {}^T\rho = \begin{bmatrix} 0 & 0 \\ 0 & B \end{bmatrix}, \quad \text{rank } B = 1. \qquad (3.21)$$

Expressed in terms of density-matrix elements (3.20) and (3.21) read

$$^T\rho^m{}_n = 0 \text{ if } (-)^m \text{ or } (-)^n = \sigma_E\eta, \qquad (3.22)$$

$$^T\rho^m{}_n \, {}^T\rho^{m'}{}_{n'} = {}^T\rho^m{}_{n'} \, {}^T\rho^{m'}{}_n . \qquad (3.23)$$

Equations (3.22) and (3.23) are the translation, in the transversity frame, of Eqs. (3.17) and (3.13).

The usefulness of (3.20) and (3.21) in practice should be apparent. We shall discuss a few specific cases in the next section.

We note that when several trajectories of the same naturality dominate the production mechanism, Eqs. (3.17), (3.18), (3.19), and (3.22) still hold while in general the absence of odd polarization and the rank condition disappear.

2. *Production of a Fermion Isobar*

Consider first the case when the fermion is produced from a spin-$\frac{1}{2}$ target

$$j_1+\tfrac{1}{2}\to j_3+j\,.$$

In this case,

$$^{H}\rho^{m'}{}_{m}=\sum_{\lambda}\gamma^{(2)}_{m'\lambda}\gamma^{(2)}_{m\lambda}, \quad \lambda=\tfrac{1}{2},-\tfrac{1}{2}, \tag{3.24}$$

with the symmetry properties given by Eq. (3.4).

Equation (3.24) exhibits that the rank of $^{H}\rho$ must be 2, independently of the spin and parity of the produced isobar. From (2.57) it follows that, in the transversity frame, the rank of A is unity, i.e., the elements of $^{T}\rho$ satisfy the conditions

$$^{T}\rho^{m}{}_{n}\,{}^{T}\rho^{m'}{}_{n'}={}^{T}\rho^{m}{}_{n'}\,{}^{T}\rho^{m'}{}_{n}\,, \tag{3.25}$$

where

$$m,n,m',n'=j,j-2,\ldots,-j+1\,. \tag{3.26}$$

The equivalent condition in helicity quantization is, from (2.69), that the rank of $(C+e^{i\pi j}D)$ is unity. This condition is equivalent to the vanishing of the determinants of every 2×2 submatrix of $(C+e^{i\pi j}D)$. With the identification

$$^{H}\rho^{m'}{}_{m}=C^{m'}{}_{m}={}^{H}\rho^{-m'}{}_{-m}$$

and (3.27)

$$^{H}\rho^{n'}{}_{-n}=D^{n'}{}_{n}=-\,{}^{H}\rho^{-n'}{}_{n}$$

for m,n,m',n' given by (3.26), the rank condition reads

$$^{H}\rho^{m'}{}_{m}\,{}^{H}\rho^{n'}{}_{n}+{}^{H}\rho^{m'}{}_{-n}\,{}^{H}\rho^{n'}{}_{-m}={}^{H}\rho^{m'}{}_{n}\,{}^{H}\rho^{n'}{}_{m}+{}^{H}\rho^{m'}{}_{-m}\,{}^{H}\rho^{n'}{}_{-n}\,, \tag{3.28}$$

$$^{H}\rho^{m'}{}_{-m}\,{}^{H}\rho^{n'}{}_{n}+{}^{H}\rho^{m'}{}_{m}\,{}^{H}\rho^{n'}{}_{-n}={}^{H}\rho^{m'}{}_{-n}\,{}^{H}\rho^{n'}{}_{m}+{}^{H}\rho^{m'}{}_{n}\,{}^{H}\rho^{n'}{}_{-m} \tag{3.29}$$

for the values of the indices given by (3.26).

The above relations are those found by Ringland and Thews[8] for $m=m'$ and $n=n'$, i.e., for (3.28),

$$(^{H}\rho^{m}{}_{n})^2+(^{H}\rho^{m}{}_{-n})^2={}^{H}\rho^{m}{}_{m}\,{}^{H}\rho^{n}{}_{n}\,, \tag{3.30}$$

while (3.29) becomes an identity.

Note that the above conditions are also valid if the spin-$\frac{1}{2}$ particle associated with the fermion isobar is the beam particle instead of the target.

B. Many-Body Reactions

Up to now we have been interested only in two-body (or quasi-two-body) reactions. We would like to recall some results about multiparticle reactions when the production mechanism is dominated by a multi-Regge exchange and show how our analysis can be extended to such reactions.

Let us first consider the simplest case of a three-particle reaction:

$$1+2\to 3+5+4\,, \tag{3.31}$$

where the production mechanism is dominated by a double Regge exchange described by Fig. 2(a).

We define

$$s_{ij}=(p_i+p_j)^2\,,$$
$$t_{ij}=(p_i-p_j)^2\,,$$

where p_i represents the four-momentum of the ith particle.

There are five independent kinematical variables. In order to preserve the symmetry of Fig. 2(a), they can be chosen as follows:

$$s'=s_{35},\quad s''=s_{45},\quad t'=t_{13},\quad t''=t_{24},\quad \varphi, \tag{3.32}$$

where φ is the angle between the normals to the three planes defined by p_1,p_3,p_5 and p_2,p_4,p_5.

The amplitude for the crossed channel,

$$4^*+2\to 3+5+1^*, \tag{3.33}$$

can be written assuming factorization as[9]

$$F_{\lambda_3\lambda_4\lambda_1,\lambda_5\lambda_2}=\beta^{(1)}_{\lambda_3\lambda_1}(t')\beta^{(2)}_{\lambda_4\lambda_2}(t'')\beta^{(5)}_{\lambda_5}(t',t'',\varphi)\,\frac{1+\tau_1 e^{i\pi\alpha_1(t')}}{\sin\pi\alpha_1(t')}\,\frac{1+\tau_2 e^{i\pi\alpha_2(t'')}}{\sin\pi\alpha_2(t'')}\,s'^{\alpha_1(t')}s''^{\alpha_2(t'')}\,, \tag{3.34}$$

where τ_1 (τ_2) is the signature of the Regge trajectory α_1 (α_2);$\beta^{(1)}(t')$ [$\beta^{(2)}(t'')$] represents the coupling of the Regge trajectory α_1 [α_2] to the particles 1, 3 [2, 4]; and $\beta^{5}(t',t'',\varphi)$ represents the coupling of particle 5 to the Regge trajectories α_1 and α_2.

Parity conservation at each vertex gives, for $\beta^{(1)}$ and $\beta^{(2)}$, symmetry properties analogous to those given in (3.3) and (3.4), while we get for $\beta^{(5)}$

$$\beta^{(5)}_{-\lambda_5}=\sigma_{E1}\sigma_{E2}\eta_5(-)^{j+\lambda}\beta^{(5)}_{\lambda_5}\,, \tag{3.35}$$

where σ_{E1} and σ_{E2} are the naturalities of the α_1 and α_2 trajectories.

As we have already seen, factorization implies that the polarization of a particle is independent of the vertices where it does not appear. So one gets the following relation:

$$_{(a)}\rho_3(t')={}_{(b)}\rho_3(t')\,,$$
$$_{(a)}\rho_4(t'')={}_{(c)}\rho_4(t'')\,, \tag{3.36}$$

where $_{(a)}\rho_3$ ($_{(a)}\rho_4$) is the density matrix of particle

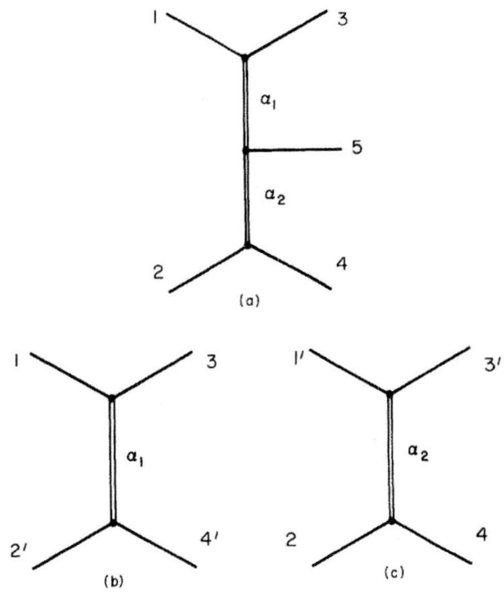

FIG. 2. Trajectory-exchange diagrams for (a) double Regge exchange, (b) and (c) single Regge exchange. The model predicts the same polarization for particles 3 in (a) and (b). It also predicts the same polarization for particles 4 in (a) and (c).

3 (4) for the process described by (3.31), ${}_{(b)}\rho_3$ is the density matrix for particle 3 in the following process:

$$1+2'\to 3+4',$$

with the single trajectory α_1 exchanged in the t channel [see Fig. 2(b)]. In this process we call p_i' the four-momentum of particle i and

$$t'=(p_1'-p_3')^2\,. \tag{3.37}$$

In a similar way ${}_{(c)}\rho_4$ is the density matrix for particle 4 in the following process:

$$1'+2\to 3'+4,$$

when the single trajectory α_2 is exchanged in the t channel [see Fig. 2(c)]. In this process, we call p_i'' the four-momentum of particle i and

$$t''=(p_2''-p_4'')^2\,. \tag{3.38}$$

Relations (3.36) have been obtained by Capella and Ranft.[10] Our analysis of two-body reactions can be extended immediately to three-body reactions when the production mechanism is dominated by a double Regge exchange.

Furthermore if one looks at the polarization of particle 5, it follows from (3.35) that relations (3.17) and (3.22) hold with the following substitutions:

$$-\sigma_B\to\sigma_{B1}\sigma_{B2}\,,$$
$$\eta\to\eta_5\,, \tag{3.39}$$
$$j\to j_5\,.$$

One obtains also the rank condition

$$\operatorname{rank}\,\rho_5(t',t'',\varphi)=1\,. \tag{3.40}$$

This pure state is not easily detected since in order to measure the polarization of particle 5 (if it exists) corresponding to a small bin of t', t'', and φ, a very good statistics is required.

This analysis can be extended in an obvious way to the multiparticle production,

$$1+2\to 3+5+\cdots+n+4\,.$$

All the results of this section are summarized in Table II.

IV. APPLICATIONS TO $j=1, 2, \frac{3}{2}, \frac{5}{2}$

We shall now discuss in greater detail the results of the previous sections in a few specific cases of current experimental interest. Prior to that we would like to make some general remarks, concerning polarization measurements and polarization domains, which are independent of production mechanisms (see Refs. 1 and 2).

A. Polarization Measurement

The quantities which are determined directly from the experimental study of a normalized decay distribution $\mathcal{G}(\theta,\varphi)$ are the coefficients $y_M^{(L)}$ of the spherical harmonics $Y_M^{(L)}(\theta,\varphi)$:

$$\mathcal{G}(\theta,\varphi)=\frac{1}{4\pi}+\sum_{L=1}^{2j}\sum_{M=-L}^{L}y_M^{(L)*}\,Y_M^{(L)}(\theta,\varphi)\,. \tag{4.1}$$

(i) If the decay is parity-conserving, $y_M^{(L)}=0$ for odd L. One must check this first experimentally and if this is not satisfied, it means either parity is not conserved or more likely there are interference effects between the resonance channel and the background; clearly for further analysis of polarization measurements, one must account for the background (see Sec. II B).

(ii) If the reaction is B symmetric, one must check that the components $y_M^{(L)}$ satisfy this symmetry; i.e., in any transversity frame

$$y_M^{(L)}=0\quad\text{for } M \text{ odd}\,, \tag{4.2}$$

and in any helicity frame

$$y_M^{(L)}=(-1)^L y_M^{(L)*}\,. \tag{4.3}$$

Again, if conditions (4.2) or (4.3) are not satisfied, one is not studying the decay of a particle with definite spin and parity; more likely this is due to interferences with the background.

TABLE II. Predictions on polarization from Regge-pole models.

1	For two-body reactions $1+2\to 3+4$ independently of any model, if beam 1 and target 2 are unpolarized.	
	Parity conservation in the reaction $\Rightarrow$	B symmetry for ρ_3, ρ_4; ${}^T\rho^m{}_n = 0$ if $(-)^{m-n} = -1 \Leftrightarrow {}^H\rho^m{}_n = (-)^{m-n}\, {}^H\rho^{-m}{}_{-n}$
2	The single trajectory exchange model adds three relevant hypotheses, whose implications are given separately.	
2.1	Relative reality of the amplitudes $\Rightarrow$	(a) Reality of ${}^H\rho_3$ and ${}^H\rho_4$ (b) Assuming 1, even character of ρ_3, ρ_4: $\rho_3 = {}^{(E)}\rho_3,\ \rho_4 = {}^{(E)}\rho_4 \Leftrightarrow \rho^m{}_n = (-)^{m-n}\rho^{-n}{}_{-m}$
2.2	Factorization of the amplitudes $\Rightarrow$	(a) Rank restrictions: rank $\rho_3 \le 2j_1+1$, rank $\rho_4 \le 2j_2+1$ (b) No correlation in the joint matrix: $\rho(3,4) = \rho_3 \otimes \rho_4$ (c) Vertex dependence of the polarization: for reactions (a) $1+2\to 3+4$, (b) $1+2'\to 3+4'$, and (c) $1'+2\to 3'+4$, with the same trajectory exchange, we have ${}_{(a)}\rho_3(t) = {}_{(b)}\rho_3(t)$, ${}_{(a)}\rho_4(t) = {}_{(c)}\rho_4(t)$
2.3	Parity conservation at each vertex $\Rightarrow$	For $j_1 = 0$, and assuming 1, signature constraints for $\rho(3)$: ${}^T\rho^m{}_n = 0$ if $(-)^m$ or $(-)^n = -\sigma_E\,\eta_1\eta_3 \Leftrightarrow {}^H\rho^m{}_n = \sigma_E\,\eta_1\eta_3(-)^{j_3+n}\,{}^H\rho^m{}_{-n}$
	These hypotheses can be combined, e.g.:	
	2.1 (b) 2.2 (a) for $j_2 = \frac{1}{2}$ $\Rightarrow$	Stronger rank restrictions for ρ_4: ${}^T\rho^{m'}{}_m\,{}^T\rho^{n'}{}_n = {}^T\rho^{m'}{}_n\,{}^T\rho^{n'}{}_m \Leftrightarrow$ $(m, m', n, n' = j, j-2, \ldots, -j+1)$ $\Leftrightarrow \begin{cases} {}^H\rho^{m'}{}_m\,{}^H\rho^{n'}{}_n + {}^H\rho^{m'}{}_{-n}\,{}^H\rho^{n'}{}_{-m} = {}^H\rho^{m'}{}_n\,{}^H\rho^{n'}{}_m + {}^H\rho^{m'}{}_{-m}\,{}^H\rho^{n'}{}_{-n} \\ {}^H\rho^{m'}{}_{-m}\,{}^H\rho^{n'}{}_n + {}^H\rho^{m'}{}_m\,{}^H\rho^{n'}{}_{-n} = {}^H\rho^{m'}{}_{-n}\,{}^H\rho^{n'}{}_m + {}^H\rho^{m'}{}_n\,{}^H\rho^{n'}{}_{-m} \end{cases}$
	The implications of these hypotheses can be used separately in other models.	
3	For several-trajectory exchange with the same naturality σ_E the prediction 2.3 still holds.	
4	For reaction $1+2\to 3+5+4$ with the multi-Regge exchange of Fig. 1(a), all the predictions in 2 hold for particles 1, 2, 3, 4. For particles 5, we have the rank restriction, rank $\rho_5 = 1$ and the signature constraints, as given in 2.3 with the substitutions $\sigma_E\eta_1\eta_3 \to \sigma_{E_1}\sigma_{E_2}\eta_5$, $j_3 \to j_5$.	

(iii) We note that

$$y_M^{(L)} = \left(\frac{2L+1}{2j}\right)^{1/2} \lambda_a(L,j)\, t_M^{(L)}\,, \tag{4.4}$$

and as discussed in Sec. II C 1, $\lambda_a(L,j)$ are "dynamics" independent for the decays listed in (2.19). If they depend on the dynamics of the decay mode one only obtains the $t_M^{(L)}$'s of the density matrix from the $y_M^{(L)}$'s up to an L-dependent factor. A simple example of such a case is

$$j^\eta \to 1^- + 0^-\,, \quad \text{where } \eta = -(-1)^j\,.$$

Indeed, it is easy to see that there are two orbital angular momenta involved ($l = j-1$, $l = j+1$) in the decay and $\lambda_a(L,j)$ will depend on the ratio of the two corresponding amplitudes. Thus one cannot construct the density matrix for A_1 ($j^\eta = 1^+$) by studying the angular distribution of $\pi^\pm$ in $A_1^\pm \to \rho^0 + \pi^\pm$ without further dynamical assumptions (relative strength of s and d waves in the decay). We shall remark later concerning the decay of A_1 into three pseudoscalar mesons.

B. Polarization Domains

The geometrical plots are for polarization what Dalitz plots are for energy momenta. Each polarization state is represented by a point of the polarization domain. We give below for spin 1 how to obtain this point from the polarization parameters, either the $\rho^m{}_n$'s or the $t_M^{(L)}$'s or better the $r_M^{(L)}$'s. Values of the polarization parameters which yield points outside the allowed domain are physically meaningless (in some cases they even require negative angular distribution). Such plots which use the intrinsic properties of the density matrix are also very useful in testing special mechanisms and special properties such as whether the s- or t-channel helicities are conserved.

For example, the even polarization domain of a B-symmetric spin-1 particle[2] is shown in Fig. 3. It is an axially symmetric cone whose meridian section is an equilateral triangle. Each point of this domain represents a polarization state. For a given channel it is possible to define a correspondence independent of the quantization frame and of

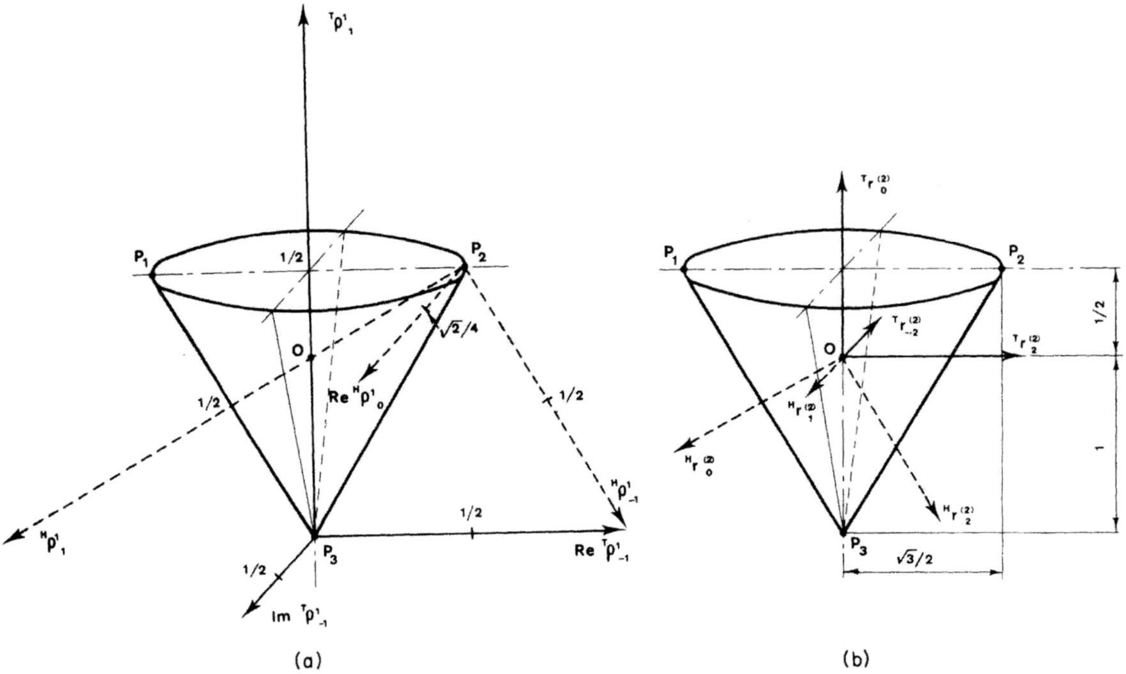

FIG. 3. Domain of B-symmetric even polarization for spin-1 particle. In (a) the orthogonal coordinate system with origin at P_2 (P_3) corresponds to parametrization of the density-matrix element in the transversity (helicity) frame. The length of the coordinate axes represents the unity for the corresponding parameters. In these coordinate systems the equations of the cone are: $|{}^T\rho^1{}_{-1}| \leq {}^T\rho^1{}_1 \leq \frac{1}{2}$; $1 \geq {}^H\rho^1{}_1 + {}^H\rho^1{}_{-1} \geq 0$, ${}^H\rho^0{}_0({}^H\rho^1{}_1 - {}^H\rho^1{}_{-1}) \geq 2(\mathrm{Re}\,{}^H\rho^1{}_0)^2$. In (b) the two orthonormal coordinate systems with origin at O correspond to multipole parameter $r_M^{(2)}$ parametrization in transversity and in helicity frames. In these coordinate systems the equations of the cone are: $-1 \leq {}^Tr_0^{(2)} \leq \frac{1}{2}$, $({}^Tr_2^{(2)})^2 + ({}^Tr_{-2}^{(2)})^2 \leq \frac{1}{3}({}^Tr_0^{(2)} + 1)^2$; $|{}^Hr_2^{(2)}| \leq \sqrt{\frac{1}{3}}\,({}^Hr_0^{(2)} + 1)$, $({}^Hr_1^{(2)})^2 \leq \frac{1}{3}(1 - 2\,{}^Hr_0^{(2)})(1 + {}^Hr_0^{(2)} - \sqrt{3}\,{}^Hr_2^{(2)})$.

the coordinate system of the polarization space; for instance, if the matrix elements of ${}^H\rho$ are used as a set of coordinates, the corresponding axes are drawn on Fig. 3(a). Note that they do not form an orthonormal system of coordinates, the drawn length of each axis represents the value unity.

In Fig. 3(b), the three nonvanishing $r_M^{(2)}$ multipole parameters form an orthonormal system of coordinates in any frame; they are drawn for transversity and helicity quantization. Let us suppose we have chosen the reference frame in the t channel. The axes and coordinates for the s channel are deduced from those in the t channel by a rotation of twice the crossing angle (a function of s and t) around the axis of the cone.

For forward reactions or when the production is helicity-conserving in a specific channel, the density matrix in the helicity frame of this channel must be diagonal, and the representative point must be in the interval of the ${}^H\rho^1{}_1$ axis which is inside the cone.

The point O represents the unpolarized state. The pure states (i.e., totally polarized $\Longleftrightarrow$ rank $\rho = 1$) are at a distance unity from O. For even polarization they are only P_3, the vertex of the cone and the points of the circle bounding the basis of the cone.

The point P_3 represents the state of longitudinal polarization along the normal n, i.e., the state $|1, 0\rangle^T$, and hence ${}^T\rho^0{}_0 = 1$.

The points of the circle represent the transverse polarization state in transversity quantization. For such state the polarization vector e is real (up to a phase) and orthogonal to n, the normal to the reaction plane: $e \cdot n = 0$. So there is a helicity quantization with $n^{(3)} = e$, where such a state is longitudinal and the only nonvanishing matrix element is ${}^H\rho^0{}_0 = 1$.

The even polarization domain for spin $\frac{3}{2}$ is three-dimensional. Its boundary is the two-dimensional sphere drawn in Fig. 4. The matrices which are represented by points lying on this sphere have rank 2. For further details concerning the polarization domain for spin $\frac{3}{2}$ see Ref. 2.

PROPERTIES OF POLARIZATION DENSITY MATRIX...

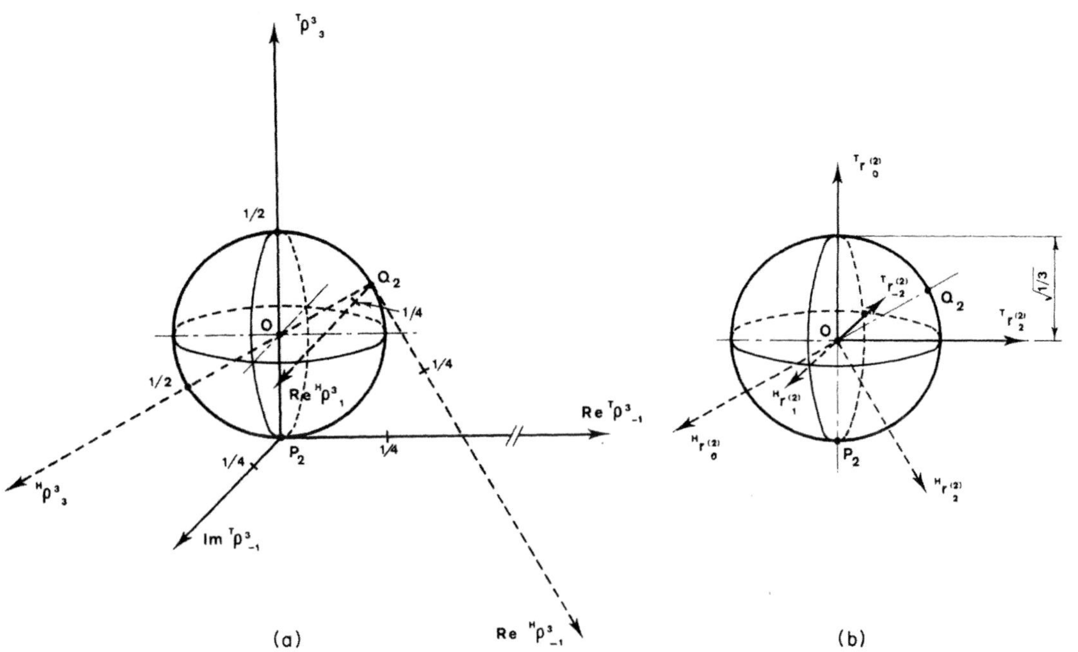

FIG. 4. Domain of B-symmetric even polarization for spin-$\frac{3}{2}$ particle. In (a) the orthogonal coordinate system with origin at P_2 (Q_2) corresponds to density-matrix-element parametrization in the transversity (helicity) frame. The length of the coordinate axes represents the unity for the corresponding parameter. In these coordinate systems the equations of the sphere are: $|{}^T\rho^3{}_{-1}|^2 \leq {}^T\rho^1{}_1\,{}^T\rho^3{}_3$; $(\mathrm{Re}\,{}^H\rho^3{}_{-1})^2 + (\mathrm{Re}\,{}^H\rho^3{}_1)^2 \leq {}^H\rho^1{}_1\,{}^H\rho^3{}_3$. In (b) the two orthonormal coordinate systems with origin at O correspond to multipole parameter $r_M^{(2)}$ parametrization in the transversity and in helicity frame. In th se coordinate systems the equations of the sphere are: $({}^Tr_2^{(2)})^2 + ({}^Tr_{-2}^{(2)})^2 + ({}^Tr_0^{(2)})^2 \leq \frac{1}{3}$; $({}^Hr_2^{(2)})^2 + ({}^Hr_1^{(2)})^2 + ({}^Hr_0^{(2)})^2 \leq \frac{1}{3}$.

For spin 2, the polarization domain of B-symmetric even polarization states is eight-dimensional. It is convenient to use its projection on the two-dimensional plane of density matrices which are diagonal in any transversity frame. This projection is an isosceles triangle $P_1P_2P_3$ represented in Fig. 5. The angle at the P_1 vertex is

$$\chi = \cos^{-1}(\tfrac{2}{3}) \cong 48.2°.$$

The unpolarized state is projected in the point O ($\rho^2{}_2 = \rho^1{}_1 = \rho^0{}_0 = \frac{1}{5}$). The set of pure states (i.e., rank 1) contains two disconnected parts, a one-dimensional circle whose projection is the point P_2, and a two-dimensional surface whose projection is the segment P_1P_3.

In any helicity quantization the diagonal matrices form another triangle, whose projection in the plane of Fig. 2 is $C_0\,C_1\,C_2$. The point C_0 is the projection of the pure state with ${}^H\rho^0{}_0 = 1$; the points C_1 and C_2 are the projections of the rank-2 states with ${}^H\rho^1{}_1 = \frac{1}{2}$ and ${}^H\rho^2{}_2 = \frac{1}{2}$, respectively.

For a complete study of the spin-2 even polarization domain we refer the reader to Ref. 2a or 2b, and for the study of the spin-$\frac{5}{2}$ even polarization domain to Ref. 2a.

C. Production of Bosons with $j^\eta = 1^+$ When the Natural-Parity Trajectory Dominates

Recently there has been considerable interest[3] in studying the polarization properties of particles with $j^\eta = 1^+$ (A_1, Q mesons) in high-energy reactions. Since the Pomeranchuk trajectory can dominate, such a study provides information concerning the nature of this singularity which is one of the least understood problems in hadronic interactions. As stated in Sec. II C 1, if one studies only the angular distribution $\mathcal{G}(\theta, \varphi)$ of the normal to the decay plane of $1^+ \to 0^- + 0^- + 0^-$, one measures only even multipoles and hence only even polarization. Also such an angular distribution is independent of the decay dynamics.

If we assume that the production of the 1^+ particle is dominated by a Pomeranchuk singularity which can be represented by a single Regge pole with factorizable residues, then from the parity relations (3.14) and (3.15) since $\eta = +1$ and $\sigma_E = +1$,

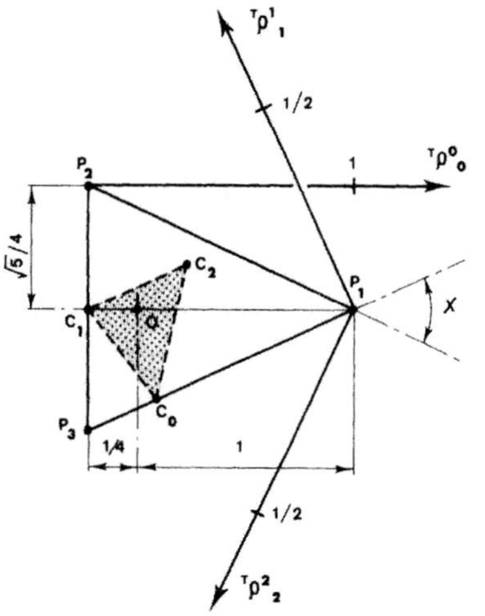

FIG. 5. Projection of the domain of B-symmetric even polarization for spin 2, on the plane of matrices which are diagonal in transversity quantization. The triangle C_0, C_1, C_2 is the projection of the domain of density matrices which are diagonal in any helicity quantization. The three coordinate axes which are drawn correspond to parametrization of the density-matrix element in the transversity frame. Note that these three coordinates are not independent (they satisfy ${}^T\rho^0{}_0+2\,{}^T\rho^1{}_1+2\,{}^T\rho^2{}_2=1$). The coordinate $({}^T\rho^0{}_0,\ {}^T\rho^1{}_1,\ {}^T\rho^2{}_2)$ of the points are $O=(\frac{1}{5},\frac{1}{5},\frac{1}{5})$, $P_1=(1,0,0)$, $P_2=(0,\frac{1}{2},0)$, $P_3=(0,0,\frac{1}{2})$, $C_0=(\frac{1}{4},0,\frac{3}{8})$, $C_1=(0,\frac{1}{4},\frac{1}{4})$, and $C_2=(\frac{3}{8},\frac{1}{4},\frac{1}{16})$.

the amplitudes c and d are

$$c=\begin{bmatrix}\gamma_1\\ \\ -\gamma_1\end{bmatrix},\quad d=[\gamma_0]\,. \tag{4.5}$$

In helicity quantization, the density matrix ${}^H\rho$ defined by (3.11) is

$${}^H\rho=\begin{bmatrix}\gamma_1\\ -\gamma_1\\ \gamma_0\end{bmatrix}[\gamma_1,\ -\gamma_1,\ \gamma_0]=\begin{bmatrix}\gamma_1^2 & -\gamma_1^2 & \gamma_1\gamma_0\\ -\gamma_1^2 & \gamma_1^2 & -\gamma_1\gamma_0\\ \gamma_0\gamma_1 & -\gamma_0\gamma_1 & \gamma_0^2\end{bmatrix}. \tag{4.6}$$

We note that the order of indices is $m, m'=1, -1, 0$, and that the trace-1 condition implies

$$\gamma_0^2+2\gamma_1^2=1\,. \tag{4.7}$$

For $j=1$, the unitary transformation $D^{(j)}(\tilde{R})$ has the form (2.45)

$$D^{(1)}(\tilde{R})=\begin{bmatrix}X & iW\\ iW & Y\end{bmatrix}$$

with

$$X=\tfrac{1}{2}\begin{bmatrix}1 & -1\\ -1 & 1\end{bmatrix},\quad Y=0,\quad \text{and}\quad W=\frac{1}{\sqrt{2}}\begin{bmatrix}1\\ 1\end{bmatrix}. \tag{4.8}$$

In transversity quantization, the density matrix ${}^T\rho$ deduced from ${}^H\rho$ by (2.33) is

$${}^T\rho=\begin{bmatrix}\gamma_1-(i/\sqrt{2})\gamma_0\\ -\gamma_1-(i/\sqrt{2})\gamma_0\\ 0\end{bmatrix}[\gamma_1+(i/\sqrt{2})\gamma_0,\ -\gamma_1+(i/\sqrt{2})\gamma_0,\ 0]$$

$$=\begin{bmatrix}\frac{1}{2} & -(\gamma_1-(i/\sqrt{2})\gamma_0)^2 & 0\\ -(\gamma_1+(i/\sqrt{2})\gamma_0)^2 & \frac{1}{2} & 0\\ 0 & 0 & 0\end{bmatrix}. \tag{4.9}$$

Equations (4.6) and (4.9) exhibit clearly the rank 1 of the density matrix. Note that the matrix ${}^T\rho$ has the general form (3.20)

$${}^T\rho=\begin{bmatrix}A & 0\\ 0 & 0\end{bmatrix},$$

where the 2×2 matrix A has rank 1. Furthermore, note that because of (4.7) there is only one free parameter to describe the density matrix.

In transversity quantization the rank-1 condition of A reads

$$|{}^T\rho^1{}_{-1}|={}^T\rho^1{}_1={}^T\rho^{-1}{}_{-1}=\tfrac{1}{2}\,. \tag{4.10}$$

Therefore in Fig. 3 the representative points should lie on the circle which bounds the basis of the cone. Likewise, for the matrix elements of ${}^H\rho$ we have

$$\begin{aligned}&{}^H\rho^1{}_1+{}^H\rho^1{}_{-1}=0\,,\\ &({}^H\rho^1{}_0)^2={}^H\rho^1{}_1\,{}^H\rho^0{}_0\,.\end{aligned} \tag{4.11}$$

Let us assume that t-channel coordinates are used (e.g., the so-called Gottfried-Jackson frame which is the t-helicity quantization frame). If t-channel helicity is conserved, in addition one has

$${}^H\rho^1{}_{-1}=0,\quad \text{i.e.,}\quad \gamma_1=0\quad\text{and}\quad \gamma_0^2=1\,. \tag{4.12}$$

Then all the points should be confined to the point P_2. If s helicity is conserved, the points should be in the transformed of P_2 by the s- and t-dependent rotation of crossing.

In Figs. 6 and 7 we have plotted for the Q meson, results from Ref. 3a for even polarization. In this experiment the odd part of the polarization has also been measured and the results are compatible with zero. We see immediately from these plots that some experimental points are compatible with the exchange of a single trajectory of natural parity and with t-helicity conservation.

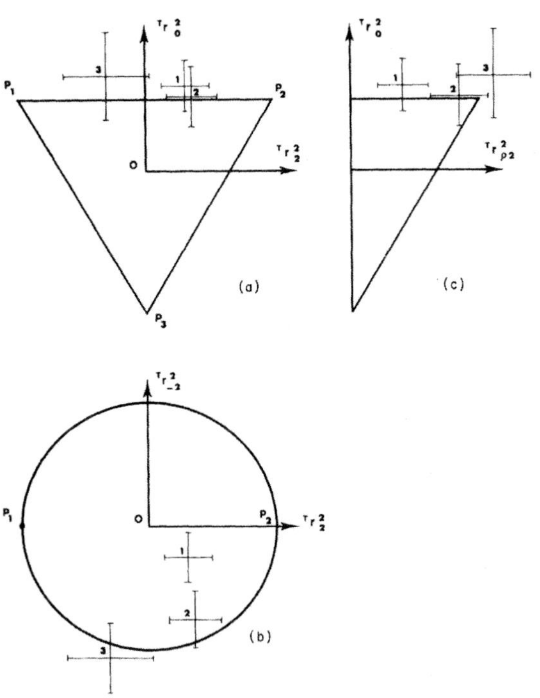

FIG. 6. Plot of experimental data of Q-meson polarization measured in s-channel frame. (a) Projection of the the cone of Fig. 3 on the meridian plane containing P_1 and P_2. (b) Projection of the same cone on a plane perpendicular to its axis. (c) Meridian triangle generating the cone, with $^T r^{(2)}_{\rho 2} = [(^T r^{(2)}_2)^2 + (^T r^{(2)}_{-2})^2]^{1/2}$. The experimental points 1, 2, 3, represented in (a), (b), and (c) are taken from Ref. 3 a. They correspond to a beam momentum of 8.25 GeV/c and successive intervals of momentum transfer. The points barely satisfy positivity conditions [to be inside the triangle (c)]. The three points are compatible with the prediction of the exchange of natural-parity trajectories [to be in the circle P_1P_2 projected on (a) and (b)], and the points 2 and 3 with the predictions of a single-trajectory exchange of natural parity (to be on the boundary of the circle). They are not compatible with the prediction of s-helicity conservation [to be on the straight line P_2O of (a) and (b)].

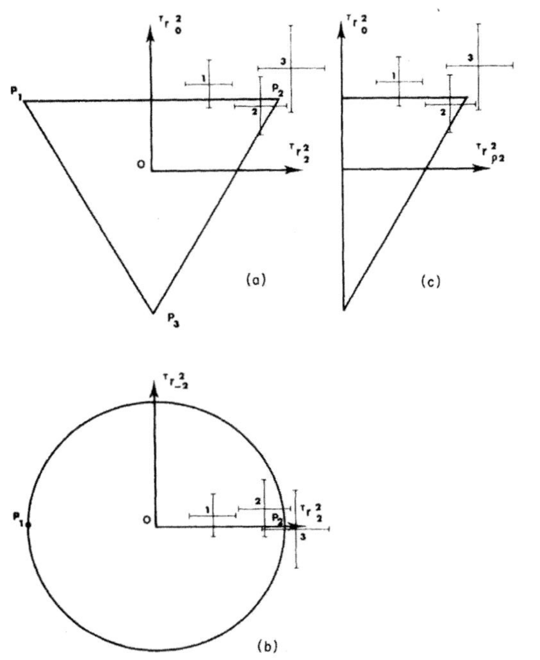

FIG. 7. Plot of the same experimental data as in Fig. 6, measured in the t-channel frame. For the comparison with Fig. 6, recall that both parts (c) have to be identical, and that the points in both parts (b) are rotated from each other by twice the crossing angle χ, which is (Ref. 3 a) $0° < \chi < 30°$ for point 1, $30° < \chi < 60°$ for point 2, and $45° < \chi < 70°$ for point 3. Points 2 and 3 are compatible with the prediction of the exchange of a single natural-parity trajectory (to be on the boundary of circle P_1P_2) and moreover with the prediction of t-helicity conservation (to be on the straight line P_2O).

D. Production of a Boson with $j^{\eta} = 1^-$ When the Natural-Parity Trajectory Dominates

In this case, since $\eta = -1$ and $\sigma_E = +1$, the parity relations (3.14) and (3.15) imply for the amplitudes c and d

$$c = \begin{bmatrix} \gamma_1 \\ \gamma_1 \end{bmatrix}, \quad d = [0]. \tag{4.13}$$

So, in helicity quantization, the density matrix $^H\rho$ defined by (3.11) is

$$^H\rho = \begin{bmatrix} \gamma_1 \\ \gamma_1 \\ 0 \end{bmatrix} [\gamma_1, \ \gamma_1, \ 0] = \begin{bmatrix} \gamma_1^2 & \gamma_1^2 & 0 \\ \gamma_1^2 & \gamma_1^2 & 0 \\ 0 & 0 & 0 \end{bmatrix}. \tag{4.14}$$

The trace-1 condition implies $\gamma_1^2 = \frac{1}{2}$. The whole polarization is then fixed. This is the pure polarization state represented by the vertex P_3 of the cone:

$$^H\rho^1{}_0 = 0, \quad ^H\rho^1{}_1 = {}^H\rho^{-1}{}_{-1} = {}^H\rho^1{}_{-1} = \tfrac{1}{2}. \tag{4.15}$$

In this case there is no possible helicity conservation in any channel.

In transversity quantization the density matrix has the general form (3.21)

$$^T\rho = \begin{bmatrix} 0 & 0 \\ 0 & B \end{bmatrix} = \begin{bmatrix} 0 & 0 & 0 \\ 0 & 0 & 0 \\ 0 & 0 & 1 \end{bmatrix}, \tag{4.16}$$

i.e., the only nonvanishing element is

$$^T\rho^0{}_0 = 1. \quad (4.17)$$

If unnatural-parity trajectory dominates it is clear that all the calculations made for $1^{\pm}$ (with $\sigma_E = +1$) becomes true for $1^{\mp}$ (with $\sigma_E = -1$).

E. Production of a Boson with $j^\eta = 2^+$ When the Natural-Parity Trajectory Dominates

From the parity relations (3.14) and (3.15), since $\eta = +1$ and $\sigma_E = +1$, the amplitudes c and d are

$$c = \begin{bmatrix} \gamma_2 \\ 0 \\ -\gamma_2 \end{bmatrix}, \quad d = \begin{bmatrix} \gamma_1 \\ \gamma_1 \end{bmatrix}, \quad (4.18)$$

so that the matrix $^H\rho$ is

$$^H\rho = \begin{bmatrix} \gamma_2 \\ 0 \\ -\gamma_2 \\ \gamma_1 \\ \gamma_1 \end{bmatrix} [\gamma_2,\ 0,\ -\gamma_2,\ \gamma_1,\ \gamma_1]$$

$$= \begin{bmatrix} \gamma_2^2 & 0 & -\gamma_2^2 & \gamma_1\gamma_2 & \gamma_1\gamma_2 \\ 0 & 0 & 0 & 0 & 0 \\ -\gamma_2^2 & 0 & \gamma_2^2 & -\gamma_1\gamma_2 & -\gamma_1\gamma_2 \\ \gamma_1\gamma_2 & 0 & -\gamma_1\gamma_2 & \gamma_1^2 & \gamma_1^2 \\ \gamma_1\gamma_2 & 0 & -\gamma_1\gamma_2 & \gamma_1^2 & \gamma_1^2 \end{bmatrix} \quad (4.19)$$

with the trace-1 condition

$$\gamma_1^2 + \gamma_2^2 = \tfrac{1}{2}. \quad (4.20)$$

For $j = 2$, the unitary transformation $D^{(2)}(\tilde{R})$ has the form (2.45) with

$$X = \tfrac{1}{4}\begin{bmatrix} 1 & -\sqrt{6} & 1 \\ -\sqrt{6} & -2 & -\sqrt{6} \\ 1 & -\sqrt{6} & 1 \end{bmatrix},$$

$$Y = -\tfrac{1}{2}\begin{bmatrix} 1 & 1 \\ 1 & 1 \end{bmatrix}, \quad (4.21)$$

$$W = \tfrac{1}{2}\begin{bmatrix} 1 & -1 \\ 0 & 0 \\ -1 & 1 \end{bmatrix}.$$

In transversity quantization, the matrix $^T\rho$ deduced from $^H\rho$ by (2.33) is

$$^T\rho = \begin{bmatrix} 0 \\ 0 \\ 0 \\ -\gamma_1 - i\gamma_2 \\ -\gamma_1 + i\gamma_2 \end{bmatrix} [0,\ 0,\ 0,\ -\gamma_1 + i\gamma_2,\ -\gamma_1 - i\gamma_2], \quad (4.22)$$

i.e., $^T\rho$ has the general form (3.21)

$$^T\rho = \begin{bmatrix} 0 & 0 \\ 0 & B \end{bmatrix},$$

where the 2×2 matrix B has rank 1:

$$B = \begin{bmatrix} \tfrac{1}{2} & (\gamma_1 + i\gamma_2)^2 \\ (\gamma_1 - i\gamma_2)^2 & \tfrac{1}{2} \end{bmatrix}. \quad (4.23)$$

From (4.19), (4.20), and (4.23) one can see immediately the whole structure of the density matrix when a single natural-parity trajectory is exchanged. In helicity quantization we get

$$\begin{aligned} &^H\rho^m{}_n + (-)^n\, {}^H\rho^m{}_{-n} = 0 \ \forall\ m, n, \\ &({}^H\rho^2{}_1)^2 = {}^H\rho^2{}_2\, \rho^1{}_1, \end{aligned} \quad (4.24)$$

and in transversity quantization

$$\begin{aligned} &^T\rho^1{}_1 = {}^T\rho^{-1}{}_{-1} = |{}^T\rho^1{}_{-1}| = \tfrac{1}{2}, \\ &^T\rho^m{}_n = 0 \ \text{ for } m \text{ or } n = \pm 2, 0. \end{aligned} \quad (4.25)$$

Equation (4.24) or (4.25) gives the circle of pure state which is projected on point P_2 of Fig. 5. In this case there is no possible helicity conservation in any channel.

F. Production of a Boson with $j^\eta = 2^-$ When the Unnatural-Parity Trajectory Dominates

In this case, since $\eta = -1$ and $\sigma_E = +1$, the parity relations (3.14) and (3.15) imply for the amplitudes c and d,

$$c = \begin{bmatrix} \gamma_2 \\ \gamma_0 \\ \gamma_2 \end{bmatrix}, \quad d = \begin{bmatrix} \gamma_1 \\ -\gamma_1 \end{bmatrix}, \quad (4.26)$$

so that the matrix H_ρ is

$$H_\rho = \begin{bmatrix} \gamma_2 \\ \gamma_0 \\ \gamma_2 \\ \gamma_1 \\ -\gamma_1 \end{bmatrix} [\gamma_2, \gamma_0, \gamma_2, \gamma_1, -\gamma_1] = \begin{bmatrix} \gamma_2^2 & \gamma_0\gamma_2 & \gamma_2^2 & \gamma_1\gamma_2 & -\gamma_1\gamma_2 \\ \gamma_0\gamma_2 & \gamma_0^2 & \gamma_0\gamma_2 & \gamma_0\gamma_1 & -\gamma_0\gamma_1 \\ \gamma_2^2 & \gamma_0\gamma_2 & \gamma_2^2 & \gamma_1\gamma_2 & -\gamma_1\gamma_2 \\ \gamma_1\gamma_2 & \gamma_0\gamma_1 & \gamma_1\gamma_2 & \gamma_1^2 & -\gamma_1^2 \\ -\gamma_1\gamma_2 & -\gamma_0\gamma_1 & -\gamma_1\gamma_2 & -\gamma_1^2 & \gamma_1^2 \end{bmatrix} \tag{4.27}$$

with the trace-1 condition

$$\gamma_0^2 + 2(\gamma_1^2 + \gamma_2^2) = 1 . \tag{4.28}$$

In transversity quantization, the matrix ${}^T\rho$ has the general form (3.20)

$${}^T\rho = \begin{bmatrix} A & 0 \\ 0 & 0 \end{bmatrix} \quad \text{with rank } A = 1,$$

where the 3×3 matrix A is given by

$$A = \tfrac{1}{16} \begin{bmatrix} 2\gamma_2 - \sqrt{6}\gamma_0 - 4i\gamma_1 \\ -2\gamma_0 - 2\sqrt{6}\gamma_2 \\ 2\gamma_2 - \sqrt{6}\gamma_0 + 4i\gamma_1 \end{bmatrix} [2\gamma_2 - \sqrt{6}\gamma_0 + 4i\gamma_1,\ -2\sqrt{6}\gamma_2 - 2\gamma_0,\ 2\gamma_2 - \sqrt{6}\gamma_0 - 4i\gamma_1] . \tag{4.29}$$

From (4.27), (4.28), and (4.29) one can easily get all the relations between spin-density-matrix elements. For instance in helicity quantization one gets

$$\begin{aligned} &{}^H\rho^m_n - (-)^n\, {}^H\rho^m_{-n} = 0 \quad \forall\, m, n, \\ &({}^H\rho^2_0)^2 = {}^H\rho^2_2\, {}^H\rho^0_0 , \\ &({}^H\rho^2_1)^2 = {}^H\rho^2_2\, {}^H\rho^1_1 , \\ &({}^H\rho^1_0)^2 = {}^H\rho^1_1\, {}^H\rho^0_0 , \end{aligned} \tag{4.30}$$

and in transversity quantization,

$$\begin{aligned} &|{}^T\rho^2_0|^2 = {}^T\rho^2_2\, {}^T\rho^0_0 , \\ &|{}^T\rho^2_{-2}|^2 = ({}^T\rho^2_2)^2 , \\ &\arg {}^T\rho^2_{-2} = 2 \arg {}^T\rho^2_0 , \\ &{}^T\rho^m_n = 0 \ \text{ for } m \text{ or } n = \pm 1 . \end{aligned} \tag{4.31}$$

Equations (4.30) or (4.31) give the surface of pure states which is projected on the segment P_1P_3 of Fig. 5. Only one point of this surface is compatible with helicity conservation in any channel. It is projected on C_0 and has as coordinates in the transversity and helicity quantization corresponding to this channel:

$$\begin{aligned} &{}^T\rho^2_2 = {}^T\rho^2_{-2} = \sqrt{\tfrac{3}{2}}\, {}^T\rho^2_0 = \tfrac{3}{8} , \\ &{}^H\rho^0_0 = 1, \quad \text{other } {}^H\rho^m_n = 0 . \end{aligned}$$

As in the spin-1 case, it is clear that if unnatural trajectory dominates all results for $2^\pm$ (with $\sigma_B = +1$) are true for $2^\mp$ (with $\sigma_B = -1$).

G. Production of a Fermion Isobar

Let us consider B-symmetric reactions of the type

$$j_1 + \tfrac{1}{2} \to j_3 + j ,$$

such as, for example, N^* production from a nucleon target.

As we have seen in Sec. III A 1, if the reaction is dominated by the exchange of a simple trajectory, the density matrix $\rho(j)$ has the following structure:

(i) $\rho(j)$ has only even polarization, i.e., in the separation order of indices, the density matrices ${}^T\rho$ and ${}^H\rho$ are

$${}^T\rho = \begin{bmatrix} A & 0 \\ 0 & A^T \end{bmatrix}, \quad {}^H\rho = \begin{bmatrix} C & D \\ -D & C \end{bmatrix}, \tag{4.32}$$

with

$$A = V(C + e^{i\pi j} D) V . \tag{4.33}$$

(ii) Rank $\rho(j) = 2$, hence

$$\text{rank } A = 1 , \tag{4.34}$$

and from (4.33)

$$\text{rank } (C + e^{i\pi j} D) = 1 . \tag{4.35}$$

Let us now apply these results to spin $\frac{3}{2}$ and $\frac{5}{2}$. Note that in these applications the matrix elements are denoted by ρ^{2m}_{2n} instead of ρ^m_n.

1. *Spin* $\frac{3}{2}$

In this case the matrix A is a 2×2 matrix. Hence the rank condition (4.34) reads $\det A = 0$, i.e., the matrix elements of ${}^T\rho$ satisfy

$$|{}^T\rho^3{}_{-1}|^2 = {}^T\rho^3{}_3\, {}^T\rho^{-1}{}_{-1} \tag{4.36}$$

with

$${}^T\rho^3{}_3 + {}^T\rho^{-1}{}_{-1} = \tfrac{1}{2}.$$

The corresponding condition on the density-matrix elements of ${}^H\rho$ is obtained from the rank condition (4.35) which reads $\det(C - iD) = 0$, i.e.,

$$({}^H\rho^3{}_1)^2 + ({}^H\rho^3{}_{-1})^2 = {}^H\rho^3{}_3\, {}^H\rho^{-1}{}_{-1} \tag{4.37}$$

with

$${}^H\rho^m{}_n = \text{real} \quad \forall m, n, \quad {}^H\rho^3{}_3 + {}^H\rho^{-1}{}_{-1} = \tfrac{1}{2}.$$

This result has been already noted by Ringland and Thews.[8]

Relations (4.36) or (4.37) mean that the point which represents the polarization density matrix (see Fig. 4) must be on the sphere, i.e., on the boundary of the polarization domain.

2. *Spin* $\frac{5}{2}$

In this case the matrix A is a 3×3 matrix. By writing the rank-1 condition for the 3×3 matrix A, we get in transversity quantization

$$\begin{aligned} &|{}^T\rho^5{}_{-3}|^2 = {}^T\rho^5{}_5\, {}^T\rho^{-3}{}_{-3},\\ &|{}^T\rho^5{}_1|^2 = {}^T\rho^5{}_5\, {}^T\rho^1{}_1,\\ &|{}^T\rho^1{}_{-3}|^2 = {}^T\rho^{-3}{}_{-3}\, {}^T\rho^1{}_1,\\ &{}^T\rho^5{}_1\, {}^T\rho^1{}_{-3} = {}^T\rho^1{}_1\, {}^T\rho^5{}_{-3}, \end{aligned} \tag{4.38}$$

with

$${}^T\rho^5{}_5 + {}^T\rho^1{}_1 + {}^T\rho^{-3}{}_{-3} = \tfrac{1}{2}.$$

In helicity quantization the rank conditions (3.28) and (3.29) give the relations

$$\begin{aligned} &({}^H\rho^5{}_3)^2 + ({}^H\rho^5{}_{-3})^2 = {}^H\rho^5{}_5\, {}^H\rho^{-3}{}_{-3},\\ &({}^H\rho^5{}_1)^2 + ({}^H\rho^5{}_{-1})^2 = {}^H\rho^5{}_5\, {}^H\rho^1{}_1,\\ &({}^H\rho^1{}_3)^2 + ({}^H\rho^1{}_{-3})^2 = {}^H\rho^{-3}{}_{-3}\, {}^H\rho^1{}_1,\\ &{}^H\rho^5{}_1\, {}^H\rho^1{}_{-3} = {}^H\rho^5{}_{-3}\, {}^H\rho^1{}_1 + {}^H\rho^5{}_{-1}\, {}^H\rho^1{}_3,\\ &{}^H\rho^5{}_{-1}\, {}^H\rho^1{}_{-3} + {}^H\rho^5{}_1\, {}^H\rho^1{}_3 = {}^H\rho^5{}_3\, {}^H\rho^1{}_1, \end{aligned} \tag{4.39}$$

with

$${}^H\rho^m{}_n = \text{real} \quad \forall m, n \quad \text{and} \quad {}^H\rho^5{}_5 + {}^H\rho^1{}_1 + {}^H\rho^{-3}{}_{-3} = \tfrac{1}{2}.$$

The first three relations are those of Ringland and Thews.[8] The number of relations increases with the spin value but the corresponding relations can be computed in the same way for any half-integer spin.

V. CONCLUSION

We have studied two-body (or quasi-two-body) parity-conserving reactions

$$1 + 2 \to 3 + 4$$

with unpolarized beam and target.

From the general consideration of angular momentum and parity conservation we have derived the structure of the density matrices of the final particles 3 and 4. We have emphasized the fact that the structure of the density matrices is more simple in transversity quantization than in helicity quantization. We have also noticed that it is very convenient to choose the separation order for the lines and columns instead of the usual order. All these results on the strucutre of the density matrices have been summarized in Table I.

We have studied the predictions of a single Regge trajectory exchange model for the density matrices. Our aim was to derive all the consequences of reality and factorization of the residue functions, and of parity conservation at each vertex. The results have been summarized in Table II.

The particular form of energy dependence of Regge amplitudes is irrelevant for our conclusions thus, more generally, other models can yield some of these conclusions. For instance any model in which all helicity amplitudes have the same phase [e.g., exchange of several particle members of same multiplet of a symmetry group, e.g.,[11] $SU(6)_W$] will also predict a lack of odd polarization in a B-symmetric reaction.

As an example, we have applied these considerations to experimental results in the case of 1^+ particle and, on this example, we have discussed the usefulness of the geometrical plots. When experimental results are plotted in this way, one can see clearly to what extent the results are meaningful (i.e., if they represent the properties of a particle of given mass, spin, and parity), and to what extent they are consistent with the model in which a single Regge-pole exchange mechanism dominates. With experiments in progress for the determination of density matrices for high spin isobars, it should be clear that such considerations will play an important role.

PROPERTIES OF POLARIZATION DENSITY MATRIX...

ACKNOWLEDGMENTS

All the authors would like to thank the Institut des Hautes Etudes Scientifiques where this work has been done. They also thank Hewlett-Packard France for their help with the desk calculator and plotter used for the figures. M. G. Doncel is grateful to the Grupo Interuniversitario de Física Teórica, and P. Mery to the Commission des Grands Accélérateurs, for partial support.

---

*Permanent address: Departamento de Física Teórica; Universidad Autônoma de Barcelona Cerdanyola (Barcelona), Spain.

†Permanent address: Centre de Physique Théorique, 31 chemin J. Aiguier 13 – Marseille 9°, France.

‡Permanent address: Laboratoire de Physique Théorique, Université de Bordeaux I 33 – Talence, France.

‖Permanent address: Department of Physics, Syracuse University, Syracuse, N. Y. 13210.

[1]M. G. Doncel, L. Michel, and P. Minnaert, Nucl. Phys. B38, 477 (1972).

[2]M. G. Doncel, L. Michel, and P. Minnaert: (a) See "Polarization Density Matrix," issues 1, 2, and 3, to be published in a forthcoming book. They are available as P. T. B. Reports No. 35, No. 37, and No. 44 (unpublished) at Laboratoire de Physique Théorique, Université de Bordeaux I, 33 – Gradignan. (b) See "Matrices Densité de Polarization," in *Lectures at the 1970 Summer School of Gif sur Yvette,* edited by R. Salmeron (Laboratoire de Physique, Ecole Polytechnique, Paris).

[3](a) B. Buschbeck, A. Fröhlich, M. Markytan, G. Otter, P. Schmid, A. Apostolakis, P. Michaelidis, E. Solomos, A. Stergiou, P. Theocharopoulos, T. Kalogeropoulos, E. Simopoulou, P. Tsilimigras, A. Vayaki-Seraphimidou, E. Zevgolatakos, C. Brankin, J. R. Fry, R. Matthews, H. Muirhead, and C. J. Onions, Nucl. Phys. B35, 511 (1971). (b) G. Ascoli, D. V. Brockway, L. Eisenstein, M. L. Ioffredo, U. E. Kruse, P. F. Schultz, C. Caso, G. Tomasini, P. Von Handel, P. Schilling, G. Costa, S. Ratti, P. Daronian, L. Mosca, A. E. Brenner, W. C. Harrison, D. Heyda, W. H. Johnson, Jr., J. K. Kim, M. E. Law, J. E. Mueller, B. M. Salzberg, L. K. Sisterson, T. F. Johnston, J. D. Prentice, N. R. Steenberg, T. S. Yoon, and J. T. Carroll, A. R. Erwin, R. Morse, B. Y. Oh, W. Robertson, and W. D. Walker, Phys. Rev. Letters 26, 929 (1971). (c) Aachen-Berlin-Bonn-CERN-Cracow-Heidelberg-London-Vienna Collaboration, Phys. Letters 34B, 160 (1971).

[4]E. de Rafael, Ann. Inst. Henri Poincaré 5, 83 (1966).

[5]A. R. Edmonds, *Angular Momentum in Quantum Mechanics* (Princeton Univ. Press, Princeton, N. J., 1957).

[6]See, for instance, J. D. Jackson, Rev. Mod. Phys. 42, 12 (1970), for a review and references to original papers.

[7]J. P. Ader, M. Capdeville, G. Cohen-Tannoudji, and Ph. Salin, Nuovo Cimento 56A, 952 (1968).

[8]G. A. Ringland and R. L. Thews, Phys. Rev. 170, 1569 (1968); R. L. Thews, *ibid.* 188, 2264 (1969).

[9]Chan Hong-Mo, K. Kajantie, and G. Ranft, Nuovo Cimento 49A, 157 (1967).

[10]A. Capella and G. Ranft, Nuovo Cimento 55A, 507 (1968); A. Capella, *ibid.* 56A, 701 (1968).

[11]M. G. Doncel, Nuovo Cimento 52A, 617 (1967).

Radially separated monopole solutions in non-Abelian gauge models*

Louis Michel and Lochlainn O'Raifeartaigh†
Institut des Hautes Etudes Scientifiques, Bures-sur-Yvette 91440, France

Kameshwar C. Wali
Physics Department, Syracuse University, Syracuse, New York 13210
(Received 24 March 1977)

We consider a Yang-Mills-Higgs Lagrangian invariant under local gauge transformations belonging to an arbitrary compact group G. The Higgs fields are assumed to belong to a real representation of G. We analyze in detail the conditions imposed on the fields due to the requirement that the static Hamiltonian or the total energy of the system be finite. We then seek static finite-energy solutions for which the radial dependence of the fields is factorized. We show that for the coupled system of nonlinear equations emerging from the Lagrangian the equations for angular functions which carry the internal-symmetry labels and the equations for the radial functions separate into two separate systems of coupled nonlinear equations. We solve the equations for angular functions completely and show that the gauge fields vanish outside a fixed SO(3) subgroup of G and that inside the SO(3) group they reduce to the 't Hooft–Polyakov solution *with unit magnetic charge in appropriate units*. The Higgs fields may belong to any integer representation of this SO(3) group. The static Hamiltonian and consequently the total energy or mass of the monopole depend on the representation of the Higgs field. Thus we obtain in principle a mass formula for the monopoles, the one with the lowest mass corresponding to the 't Hooft–Polyakov case.

I. INTRODUCTION

Non-Abelian gauge theories of the Yang-Mills type have created a great deal of interest recently as possible candidates for a unified theory of strong, weak, and electromagnetic interactions.[1] While the full quantum field aspects of such theories are still far from being thoroughly understood, classical solutions to the field equations have proved to be quite interesting from several points of view.[2] Thus 't Hooft[3] and Polyakov[4] independently found a static, finite-energy solution in a non-Abelian model consisting of a triplet of isovector gauge mesons and a triplet of isovector Higgs mesons with interactions invariant under local gauge group SO(3). The solution has the very interesting feature that it can be thought of as corresponding to a magnetic monopole with magnetic charge of strength $4\pi/e$, which satisfies the famous Dirac quantization condition.[5] Subsequently several authors have attempted to generalize the model in various ways.[6]

In this paper we consider the 't Hooft-Polyakov model generalized to an arbitrary compact gauge group G. In the next section we describe the notation, set up the Lagrangian, and derive the covariant field equations. We consider the most general Higgs potential allowed by renormalization requirements. The conditions for the minima of such a potential and the ensuing mass spectrum of the Higgs mesons are discussed at some length in the Appendix. In the third section we examine in detail the consequences of the assumption that the total energy be finite. This leads to several interesting and strong restrictions on the Higgs and the gauge fields on a sphere at infinity. Of special importance is the integrability condition for the well-known equation[2] which relates the value of the Higgs field at one point on the sphere to that at another point by a gauge transformation. In Sec. IV, we derive the necessary condition for the existence of a solution for this equation and explain its physical significance.

To proceed further, we assume in the remaining sections that the fields can be expressed as products of functions involving separately the radial coordinate r and the polar angles $\omega = (\theta, \psi)$. The internal-symmetry indices are associated with the angular functions. This hypothesis is a natural generalization of the 't Hooft–Polyakov model and it enables us to accomplish a complete separation of the nonlinear coupled system of equations into nonlinear coupled equations for the radial functions and the angular functions. We solve the angular equations and show that the gauge-field angular part is unique; it vanishes outside a fixed-SO(3) subalgebra of the original algebra. Inside the subalgebra, it is the 't Hooft–Polyakov solution. The Higgs fields must also belong to a representation of the fixed SO(3) but, in contrast to the 't Hooft–Polyakov case, they may belong to any integer-valued representation. However, the strength of the magnetic charge is still a unit multiple of $4\pi/e$ as in the 't Hooft–Polyakov model before. The static Hamiltonian and consequently the mass or total energy of the monopole depend on the representation of the Higgs field. We discuss the existence of solutions to the ensuing radial equations and show that the 't Hooft–Polyakov monopole is the one with the lowest mass.

II. THE LAGRANGIAN AND THE FIELD EQUATIONS

We consider a compact Lie group G; its Lie algebra, denoted by $\mathcal{G}$ is assumed to be k dimensional. Let the generators T^α, $\alpha = 1, 2, \ldots, k$ satisfy

$$[T^\alpha, T^\beta] = C^{\alpha\beta\gamma} T^\gamma , \tag{2.1}$$

where $C^{\alpha\beta\gamma}$ are the structure constants of G. If $\underline{a}$ and $\underline{b} \in \mathcal{G}$, then $(\underline{a}, \underline{b})$ given by

$$(\underline{a}, \underline{b}) = -\tfrac{1}{2} a^\alpha b^\beta \operatorname{tr} T^\alpha T^\beta , \quad \underline{a} = a^\alpha T^\alpha , \quad \underline{b} = b^\alpha T^\alpha , \tag{2.2}$$

defines the group-invariant Cartan inner product which is nondegenerate on the semisimple subalgebra of $\mathcal{G}$.

We are interested in a Lagrangian consisting of a set of gauge fields $\underline{A}_\mu(x) = A_\mu^\alpha(x)\tau^\alpha$ and a set of Higgs fields $\Phi(x)$ and invariant under local G transformations.

The gauge fields $\underline{A}_\mu(x)$ belong to the adjoint representation of G given by the antisymmetric matrices $\tau = (\tau^\alpha)$, where

$$(\tau^\beta)^{\alpha\gamma} = C^{\alpha\beta\gamma} . \tag{2.3}$$

The fields $\underline{A}_\mu(x)$ can be looked upon as vector fields valued in $\mathcal{G}$.

We shall assume that the Higgs fields $\Phi(x)$ belong to an arbitrary real representation (not necessarily irreducible) of G. Let the corresponding vector space be denoted by $\mathcal{E}$. There is actually no loss of generality in assuming that the Φ representation is real, since any complex representation can be combined with its conjugate representation to form a real representation and the resultant skew-symmetric invariant does not occur in the Lagrangian. We denote the matrices of the representation to which Φ belongs by $\underline{t}$,

$$(t^\alpha)^{AB} = -(t^\alpha)^{BA} , \tag{2.4}$$

$$\alpha = 1, 2, \ldots, k, \quad A, B = 1, 2, \ldots, K .$$

The covariant derivatives of the Φ and $\underline{A}_\mu$ fields are defined by

$$\underline{D}_\mu \Phi^A = \partial_\mu \Phi^A + e(t^\alpha)^{AB} A_\mu^\alpha \Phi^B , \tag{2.5}$$

$$\underline{D}_\mu A_\nu^\alpha = \partial_\mu A_\nu^\alpha + e C^{\alpha\beta\gamma} A_\mu^\beta A_\nu^\gamma . \tag{2.6}$$

Using the above-described fields, their covariant derivatives, and the metric

$$-g_{00} = g_{11} = g_{22} = g_{33} = 1 , \tag{2.7}$$

we can write the Lagrangian $\mathcal{L}(x)$ invariant under local gauge transformations

$$A_\mu \to A'_\mu = e^{\tau \cdot \chi(x)} A_\mu , \quad \Phi \to \Phi' = e^{t \cdot \chi(x)} \Phi$$

as

$$\mathcal{L}(x) = -\tfrac{1}{4}(\underline{F}_{\mu\nu}, \underline{F}_{\mu\nu}) - \tfrac{1}{2}(\underline{D}_\mu \Phi, \underline{D}_\mu \Phi) - V(\Phi) , \tag{2.8}$$

where

$$\underline{F}_{\mu\nu} = \frac{\partial}{\partial x^\mu} \underline{A}_\nu - \frac{\partial}{\partial x^\nu} \underline{A}_\mu + e[\underline{A}_\mu, \underline{A}_\nu] . \tag{2.9}$$

We note two other useful forms for $\underline{F}_{\mu\nu}$,

$$\underline{F}_{\mu\nu} = \underline{D}_\mu \underline{A}_\nu - \frac{\partial}{\partial x^\nu} \underline{A}_\mu = \frac{1}{e}[\underline{D}_\mu, \underline{D}_\nu] . \tag{2.10}$$

From the Lagrangian (2.8) we can derive the covariant field equations

$$\frac{\partial}{\partial x^\mu} \underline{D}^\mu \Phi^A - e(t^\alpha)^{BA} A_\mu^\alpha \underline{D}^\mu \Phi^B - \frac{\partial V}{\partial \Phi^A} = 0 , \tag{2.11}$$

$$\frac{\partial}{\partial x^\mu} F^{\mu\nu,\alpha} - e C^{\gamma\beta\alpha} F^{\mu\nu,\gamma} A_\mu^\beta - e(\underline{D}^\nu \Phi^A)(t^\alpha)^{AB} \Phi^B = 0 . \tag{2.12}$$

If we define

$$\underline{J}^\nu(\Phi) = (\Phi, \underline{t}\,\underline{D}^\nu \Phi) \ [\text{i.e.}, J^{\nu,\alpha} = \Phi^A (t^\alpha)^{AB} \underline{D}^\nu \Phi^B] , \tag{2.13}$$

the field equations (2.11) and (2.12) take the form

$$\underline{D}^\mu \underline{D}_\mu \Phi = \frac{\partial V}{\partial \Phi} , \tag{2.14}$$

$$\underline{D}_\mu \underline{F}^{\mu\nu} = -e \underline{J}^\nu(\Phi) . \tag{2.15}$$

Note from (2.10) that $\underline{F}^{\mu\nu}$ also satisfies the Jacobi identity,

$$\underline{D}_\mu \underline{\tilde{F}}^{\mu\nu} = 0 , \tag{2.16}$$

where

$$\underline{\tilde{F}}^{\mu\nu} = \tfrac{1}{2} \epsilon^{\mu\nu\sigma\rho} \underline{F}_{\sigma\rho} .$$

The potential $V(\Phi)$ is a function of Φ, which is bounded below and invariant under G. For classical solutions there are no further restrictions on the functional form of V but, if we require a renormalizable quantum field theory, $V(\Phi)$ is restricted to be a fourth-degree polynomial. In simple models with spontaneous symmetry breaking, one generally considers a $V(\Phi)$ of the form

$$V(\Phi) = \frac{\lambda}{4}(\Phi, \Phi)^2 - \frac{\mu^2}{2}(\Phi, \Phi)^2 . \tag{2.17}$$

The symmetry-breaking direction is then given by an arbitrary vector v such that

$$(v, v)^{1/2} = \mu/\sqrt{\lambda} . \tag{2.18}$$

In general, however, a fourth-degree potential may contain trilinear and quadrilinear invariants which are not isotropic in the group, that is, they do not allow an arbitrary direction for spontaneous symmetry breaking. Full details for such potentials are given in the Appendix. The general form is

$$V(\Phi)=\frac{\alpha}{4}(t(\Phi),\Phi)+\frac{\lambda}{4}(\Phi,\Phi)^2+\frac{\beta\mu}{3}(b(\Phi),\Phi)$$
$$-\frac{\mu^2}{2}(\Phi,\Phi), \qquad (2.19)$$

where $t(\Phi)$ and $b(\Phi)$ denote vectors in $\mathcal{S}$ which are trilinear and bilinear (and symmetric) in Φ. The condition for an extremum of $V(\Phi)$ is then

$$\frac{dV}{d\Phi}=\alpha t(\Phi)+\lambda(\Phi,\Phi)\Phi+\beta\mu b(\Phi)-\mu^2\Phi=0. \qquad (2.20)$$

For any solution $\Phi=\Phi_0$ of (2.20), the mass-squared matrices for the Higgs and gauge fields are given in the usual way by

$$\mathfrak{M}_{AB}=\left(\frac{\partial^2 V}{\partial\Phi^A\partial\Phi^B}\right)_{\Phi=\Phi_0}, \qquad (2.21)$$

$$e^2M_{\alpha\beta}=(t^\alpha\Phi_0,t^\beta\Phi_0), \qquad (2.22)$$

respectively.

III. FINITE-ENERGY CONDITIONS IN THE STATIC CASE

We are interested in the static solutions of the field equations (2.14) and (2.15). In addition we shall assume for the moment that $A_0=0$. When all the time derivatives and A_0 are equal to zero, the equations of motion reduce to

$$\underline{\vec{D}}^2\Phi=\frac{\partial V}{\partial\Phi}, \qquad (3.1)$$

$$\underline{\vec{D}}\times\underline{\vec{F}}=-e\underline{\vec{J}}(\Phi), \qquad (3.2)$$

where

$$\underline{F}_i=\tfrac{1}{2}\epsilon_{ijk}\underline{F}_{jk}.$$

The Hamiltonian or the total energy of the system is

$$\mathcal{H}=\int\mathcal{H}(x)d^3x=\int[\tfrac{1}{2}(\underline{\vec{F}},\underline{\vec{F}})+\tfrac{1}{2}(\underline{\vec{D}}\Phi,\underline{\vec{D}}\Phi)+V(\Phi)]d^3x. \qquad (3.3)$$

The boundary conditions for the solutions of (3.1) and (3.2) are imposed by the requirement that $\mathcal{H}$ be finite, i.e., $\mathcal{H}(x)$ be integrable,

$$r^3\mathcal{H}(x)\underset{r\to\infty}{\longrightarrow}0, \qquad (3.4)$$

and by the spontaneous symmetry-breaking mechanism. The latter requires that all the group invariants in the potential $V(\Phi)$ tend to finite constants for large r. In particular

$$\lim_{r\to\infty}(\Phi(\vec{r}),\Phi(\vec{r}))=c^2<\infty, \quad c>0. \qquad (3.5)$$

Hence we assume

$$\lim_{r\to\infty}\Phi(r,\omega)=c\phi(\omega), \quad (\phi(\omega),\phi(\omega))=1, \qquad (3.6)$$

where $\phi(\omega)$ lies on a group orbit which is determined by the other invariants according to (2.20) with $\Phi_0=c\phi(\omega)$.

We next note that, since each term in the Hamiltonian (3.3) is positive, (3.4) imposes a condition separately on each term. For the pure gauge term we satisfy this condition by assuming in the standard way that

$$\lim_{r\to\infty}re\underline{\vec{A}}(r,\omega)=\underline{\vec{a}}(\omega)<\infty, \qquad (3.7)$$

which implies that

$$\lim_{r\to\infty}r^2e\underline{\vec{F}}(r,\omega)=\underline{\vec{f}}(\omega)<\infty. \qquad (3.8)$$

For the kinetic term for the Higgs field in (3.3), the finite-energy condition (3.4) implies

$$\lim_{r\to\infty}r(r\underline{\vec{D}}\Phi,r\underline{\vec{D}}\Phi)=0, \qquad (3.9)$$

which takes the form

$$\underline{\vec{d}}\phi(\omega)=[\vec{\partial}+\vec{a}^\alpha(\omega)t^\alpha]\phi(\omega)=0, \qquad (3.10)$$

where

$$\vec{\partial}=r\vec{\nabla}.$$

Equation (3.10) implies that on the infinite sphere the values of $\phi(\omega)$ for different ω are related by a group transformation, a result which is perhaps more familiar in the integrated form

$$\phi(\omega)=U(g(\omega,\omega_0))\,\phi(\omega_0). \qquad (3.11)$$

If we take the scalar and vector product of (3.10) by the unit vector $\hat{r}$, respectively, we obtain

$$\hat{r}\cdot\vec{a}^\alpha t^\alpha\phi(\omega)=0 \qquad (3.12)$$

and

$$\underline{\vec{J}}\phi(\omega)=-(\hat{r}\times\underline{\vec{d}})\phi(\omega)=(\underline{\vec{L}}+\underline{\vec{b}})\phi(\omega)=0, \qquad (3.13)$$

where

$$\underline{\vec{L}}=-\hat{r}\times\vec{\partial} \text{ and } \underline{\vec{b}}=-\hat{r}\times\vec{a}.$$

Note $\underline{\vec{L}}$ and $\underline{\vec{J}}$ are the ordinary and covariant angular momentum operators (except that they are anti-Hermitian since we have found it convenient to omit the usual factor i).

With (3.6), the mass-squared matrix (2.22) for the gauge field becomes

$$e^2M_{\alpha\beta}(\omega)=c^2(t^\alpha\phi(\omega),t^\beta\phi(\omega)). \qquad (3.14)$$

Now, although this mass matrix is ω dependent, it is physical because the *covariant angular momentum* acting on it is zero on account of (3.10), that is,

$$J_i M_{\alpha\beta} = L_i M_{\alpha\beta} + a_i^\gamma (C_{\gamma\alpha}^{\alpha'} M_{\alpha'\beta} + C_{\gamma\beta}^{\beta'} M_{\alpha\beta'}) = 0\,. \tag{3.15}$$

In particular the eigenvalues of $M_{\alpha\beta}$ are independent of ω.

Finally we note for later convenience that, taking the scalar product of (3.10) with $t^\beta\phi(\omega)$, we obtain

$$M_{\beta\alpha}\vec{a}^{\,\alpha} = (\phi, t^\beta \vec{\partial}\phi)\,, \tag{3.16}$$

which expresses the massive gauge fields in terms of the Higgs currents. From (3.16) we also have

$$a_i^\beta M_{\beta\alpha} a_j^\alpha = (\partial_i\phi, \partial_j\phi)\,. \tag{3.17}$$

IV. INTEGRABILITY CONDITIONS FOR $\phi(\omega)$

The equation (3.10) which connects the ω dependence of ϕ with its internal-symmetry structure is the most striking consequence of the condition that the energy should be finite. However, such a partial-differential equation has global solutions if and only if its integrability conditions are satisfied. Hence it is interesting and also turns out to be very useful to study these conditions. If we write (3.10) in the form

$$\vec{\partial}\phi(\omega) = -\vec{a}^{\,\alpha} t^\alpha \phi(\omega)\,, \tag{4.1}$$

we see that the integrability conditions for these equations are

$$\begin{aligned}\partial_i(a_j^\alpha t^\alpha\phi) - \partial_j(a_i^\alpha t^\alpha \phi) &= -(\partial_i\partial_j - \partial_j\partial_i)\phi \\ &= \epsilon_{ijk} L_k \phi \\ &= -\epsilon_{ijk} b_k \phi\,.\end{aligned} \tag{4.2}$$

Thus they can be written in the form

$$\underline{f}_{ij}\,\phi = 0\,, \tag{4.3}$$

where

$$\underline{f}_{ij} = \partial_i \underline{a}_j - \partial_j \underline{a}_i + [\underline{a}_i, \underline{a}_j] + \epsilon_{ijk}\underline{b}_k\,. \tag{4.4}$$

The fields $\underline{f}_{ij}$ are the analogs, on the sphere, of the field $\underline{F}_{\mu\nu}$ in Minkowski space. If they vanish, the fields $\underline{a}_j$ can be gauged to zero locally. As a matter of fact, because $\pi_2(G) = 0$ for any compact Lie group, they can be gauged away globally. That is to say, if $\underline{f}_{ij} = 0$, there is a function $S(\omega)$ valued in $\mathcal{G}$ such that

$$\underline{a}_i(\omega) = S^{-1}(\omega)\partial_i S(\omega)\,, \tag{4.5}$$

and for ω in S_2 and G compact, $S(\omega)$ is homotopic to zero. We also note that

$$\underline{f}_{ij} = \lim_{r\to\infty} r^2 \underline{F}_{ij}\,. \tag{4.6}$$

Similarly, if we write (3.13) in the form

$$L_i\phi(\omega) = -b_i\phi(\omega)\,, \tag{4.7}$$

we can derive the integrability condition

$$\underline{h}_{ij}\,\phi = 0\,, \tag{4.8}$$

where

$$\underline{h}_{ij} = [\underline{J}_i, \underline{J}_j] - \epsilon_{ijk}\underline{J}_k \tag{4.9}$$

$$= \underline{f}_{ij} - \epsilon_{ijk}[\underline{J}_k, \hat{r}\cdot\underline{\vec{a}}\,]\,. \tag{4.10}$$

In contrast to $\underline{f}_{ij}$ $\underline{h}_{ij}$ has the remarkable property that

$$\hat{r}_i \underline{h}_{ij} = 0 \tag{4.11}$$

and hence

$$\underline{h}_{ij} = \epsilon_{ijk}\hat{r}_k \underline{f}\,, \tag{4.12}$$

where $\underline{f}$ is the scalar

$$\underline{f} = \tfrac{1}{2}\epsilon_{ijk}\hat{r}_k \underline{h}_{ij} = \tfrac{1}{2}\epsilon_{ijk}\hat{r}_k \underline{f}_{ij}\,. \tag{4.13}$$

Thus $\underline{h}_{ij}$ has the same radial component $\underline{h}_r = \hat{r}\underline{f}$ as $\underline{f}_{ij}$ but, unlike $\underline{f}_{ij}$, it has no transverse components. The reason for this is that Eq. (3.13) is weaker than (3.10) since the former can be obtained by taking the cross product of (3.10) with $\hat{r}$, but not conversely. The result is that the vanishing of $\underline{h}_{ij}$ implies only that the transverse components of $\underline{\vec{a}}(\omega)$ can be gauged to zero.

Note in particular that

$$\underline{f}\,\phi = 0\,. \tag{4.14}$$

Thus $\underline{f}$ is scalar in ordinary space, vector in the adjoint representation of G, and is in the little group of $\phi(\omega)$. From (3.14) we then have

$$M_{\alpha\beta} f^\beta = 0\,. \tag{4.15}$$

Thus $f^\alpha(\omega)$ is massless and is therefore a suitable but generally not unique candidate for defining an electromagnetic direction in the Lie algebra.

In any case Eq. (4.14) will play an important role in the sequel and it is convenient to expand the function $\underline{f}$ into

$$\underline{f} = 2\underline{u} + \underline{v}\,, \tag{4.16}$$

where

$$2\underline{u} = -\vec{\mathrm{L}}\cdot\underline{\vec{a}}, \quad \underline{v} = -\epsilon_{ijk}\hat{r}_i \underline{a}_j \underline{a}_k\,. \tag{4.17}$$

Finally we note that the integrability condition (4.3) shows that there are only three possibilities:

(i) $\underline{f}_{ij} = 0$. In this case, the potentials can be gauged away leading to a trivial solution.

(ii) $\underline{f}_{ij} \neq 0$, $\underline{f}_{ij}\,\phi = 0$. Then there exists a solution which is nontrivial.

(iii) $\underline{f}_{ij} \neq 0$, $\underline{f}_{ij}\,\phi \neq 0$. Then there are no solutions.

V. SEPARABILITY HYPOTHESIS: ANGULAR AND RADIAL EQUATIONS

In this section, we seek solutions to the field equations (3.1) and (3.2) in which $\Phi(\vec{r})$ and $\underline{\vec{A}}(\vec{r})$ have the form

$$\Phi^A(\vec{r}) = \phi(\omega)\, c\frac{S(r)}{r}, \quad e\underline{\vec{A}}(\vec{r}) = \underline{\vec{a}}(\omega)\frac{R(r)}{r} \tag{5.1}$$

with the boundary conditions, when $r \to \infty$,

$$\frac{S(r)}{r} \to 1, \; R(r) \to 1, \text{ and } S(0) = R(0) = 0. \tag{5.2}$$

The above separation of variables requires for its definition the choice of a gauge which we take to be the Landau gauge specified by div $\underline{\vec{A}} = 0$. Equation (5.1) constitutes a straightforward generalization of the Wu-Yang hypothesis[7] in the case of pure Yang-Mills fields and the later adaptation by 't Hooft[3] and Polyakov[4] to the case of Yang-Mills fields supplemented by Higgs fields. The internal-symmetry indices are carried by the angular functions $\phi(\omega)$ and $\underline{\vec{a}}(\omega)$. To begin with they are assumed to be arbitrary but for the usual smoothness properties. In particular, they are not assumed to be spherically symmetric.[8]

The radial separation (5.1) reduces the Landau gauge condition to two separate conditions

$$\hat{r}\cdot\vec{a}(\omega) = 0 = \vec{\partial}\cdot\vec{a}(\omega), \tag{5.3}$$

implying $\vec{a}(\omega)$ has no radial component and it is divergence-free. Thus the Landau gauge is the most convenient and natural gauge to impose the separability hypothesis. Using (5.1) the static field equation (3.1) can be written as

$$\phi(\omega) r^2 S''(r) = S(r)\left[-\underline{\vec{d}}_R{}^2\phi(\omega) + r^2\mu^2 W\left(\frac{S(r)}{r}\right)\phi(\omega)\right], \tag{5.4}$$

where

$$\underline{\vec{d}}_R\phi = \vec{\partial}\phi + R(r)\underline{\vec{a}}\cdot\underline{t}\,\phi = \vec{d}_R\phi + [R(r)-1]\underline{\vec{a}}\cdot\underline{t}\phi \tag{5.5}$$

and

$$W(x) = (x-1)(\sigma^2 x + 1), \tag{5.5'}$$

where $\sigma \geq 1$ is a dimensionless parameter which depends on the assumed form for the Higgs potential (see Appendix). Similarly, using (5.1) the static field equations (3.2) can be written as

$$\underline{\vec{a}}(\omega) r^2 R''(r) = R(r)(\underline{\vec{j}}(R,\underline{\vec{a}})) - e^2c^2S^2(r)\underline{\vec{j}}(R,\phi), \tag{5.6}$$

where

$$\underline{\vec{d}}_R\,\underline{\vec{a}}(\omega) = \vec{\partial}\,\underline{\vec{a}}(\omega) + R(r)\,\underline{\vec{a}}\cdot\underline{\tau}\,\underline{\vec{a}}(\omega)$$

and

$$\begin{aligned}\underline{\vec{j}}(R,\underline{\vec{a}}) &= -\underline{\vec{d}}_R{}^2\underline{\vec{a}}(\omega) - R(r)(\underline{\vec{a}},\underline{\tau}\,\vec{\partial}\,\vec{a}),\\ \underline{\vec{j}}(R,\phi) &= (\phi, \underline{t}\,\underline{\vec{d}}_R\phi).\end{aligned} \tag{5.7}$$

Equations (5.4) and (5.6) are coupled nonlinear equations. It is not obvious that the separability ansatz (5.1) will lead to separate equations for the radial functions $S(r)$ and $R(r)$ and the angular functions $\phi(\omega)$ and $\underline{\vec{a}}(\omega)$. We next prove that this indeed is true and derive the radial and angular equations.

A. Polynomials $q(R)$ and $Q(R)$

By taking the scalar product of (5.4) and (5.6) with $\phi(\omega)$ and $\underline{\vec{a}}(\omega)$, respectively, we obtain

$$r^2S''(r) = S(r)[q(R) + r^2\mu^2 W(S/r)], \tag{5.8}$$

$$\nu^2(\omega) r^2 R''(r) = R(r)\nu^2(\omega)Q(R) + e^2c^2\frac{S^2(r)}{2}\frac{dq(R)}{dR}, \tag{5.9}$$

where we have used $(\phi(\omega), \phi(\omega)) = 1$ and defined

$$\nu^2(\omega) = (\vec{a}(\omega), \vec{a}(\omega)). \tag{5.10}$$

$q(R)$ and $\nu^2 Q(R)$ are quadratic polynomials in $R(r)$,

$$\begin{aligned} q(R) &= -(\phi(\omega), \underline{\vec{d}}_R{}^2\phi(\omega)) \\ &= -[(\phi(\omega), \partial^2\phi(\omega)) + 2R(r)(\phi(\omega), \underline{t}\cdot\underline{\vec{a}}\cdot\vec{\partial}\phi(\omega)) \\ &\quad + R^2(r)(\phi(\omega), (\underline{t}\cdot\vec{a})^2\phi(\omega))], \end{aligned} \tag{5.11}$$

$$\begin{aligned} \nu^2(\omega)Q(R) &= (\underline{\vec{a}}(\omega), \underline{\vec{j}}(R,a)) \\ &= -[(\underline{\vec{a}}(\omega), \vec{\partial}^2\underline{\vec{a}}(\omega)) + 3R(r)(\underline{\vec{a}}(\omega), \underline{\tau}\cdot\underline{\vec{a}}\cdot\vec{\partial}\underline{\vec{a}}(\omega)) \\ &\quad + R^2(r)(\underline{\vec{a}}(\omega), (\underline{\tau}\cdot\underline{\vec{a}})^2\underline{\vec{a}}(\omega))]. \end{aligned} \tag{5.12}$$

Both the polynomials $q(R)$ and $Q(R)$ have apparent ω dependence in the coefficients of the different powers of R. However, consider (5.11); if we take two different values of ω, say ω and ω_0, in (5.8) and subtract one equation from the other, we obtain an equation of the form

$$0 = \sum_{k=1}^{3}[X_k(\omega) - X_k(\omega_0)]R^{k-1}(r). \tag{5.13}$$

Now if $R(r)$ is a constant, $R(0) = 0$ forces $R(r)$ to vanish everywhere leading to the trivial solution $\underline{\vec{A}} = 0$ everywhere. If $R(r)$ is not a constant, since the different powers of R are linearly independent functions, it follows that

$$X_k(\omega) = X_k(\omega_0), \tag{5.14}$$

i.e., the coefficient of different powers of $R(r)$ in $q(R)$ are ω-independent constants. Repeated arguments along similar lines prove that $\nu(\omega)$ is a constant independent of ω and so also are the coefficients in the polynomial $Q(R)$. Thus we obtain

$$(\phi(\omega), \partial^2\phi(\omega)) = -l(l+1), \tag{5.15}$$

$$(\phi(\omega), \underline{t}\cdot\vec{\underline{a}}\cdot\vec{\partial}\phi(\omega)) = m\nu, \tag{5.16}$$

$$(\phi(\omega), (\underline{t}\cdot\vec{\underline{a}})^2\phi(\omega)) = -n\nu^2, \tag{5.17}$$

$$(\vec{\underline{a}}(\omega), \partial^2\vec{\underline{a}}(\omega)) = -L(L+1)\nu^2, \tag{5.18}$$

$$3(\vec{\underline{a}}(\omega), \underline{\tau}\cdot\vec{\underline{a}}\cdot\vec{\partial}\,\vec{\underline{a}}(\omega)) = 2M\nu^3, \tag{5.19}$$

$$(\vec{\underline{a}}(\omega), (\underline{\tau}\cdot\vec{\underline{a}})^2\vec{\underline{a}}(\omega)) = -\nu^4/N. \tag{5.20}$$

In the above equations l, m, n, L, M, N, and ν are constants. We have chosen to write them in the specified form for later convenience.

Using the finite-energy condition (3.10) and Eqs. (5.15)–(5.17), we see that the constants l, m, and n satisfy

$$l(l+1) = m\nu = n\nu^2 \tag{5.21}$$

and hence, from (5.11), that the polynomial $q(R)$ takes the form

$$q(R) = l(l+1)(1-R)^2, \tag{5.22}$$

where there is only one free parameter.

Derivation of an analogous expression for $Q(R)$ is more involved. The key to the derivation is to prove that the $r\to\infty$ limit $\vec{\underline{j}}(1,\vec{\underline{a}})$ of the current $\vec{\underline{j}}(R,\underline{a})$ defined in (5.7) is zero. For this purpose we observe that, from (5.7) and the finite-energy condition (3.10),

$$\vec{j}^{\alpha}(1,\vec{a})t^{\alpha}\phi = 0. \tag{5.23}$$

But then from the definition of $\vec{\underline{j}}(R,\phi)$ in (5.7),

$$\begin{aligned}(\vec{\underline{j}}(1,\vec{\underline{a}}), \vec{\underline{j}}(R,\phi)) &= (\phi, \vec{\underline{j}}(1,\vec{\underline{a}})\cdot\underline{t}\cdot\vec{\underline{d}}_R\phi)\\ &= -(\vec{\underline{j}}(1,a)\cdot\underline{t}\phi, \vec{\underline{d}}_R\phi) = 0.\end{aligned} \tag{5.24}$$

Note that Eq. (5.24) holds for all R in $\vec{\underline{j}}(R,\phi)$. If we now take the scalar product of (5.6) with $\vec{\underline{j}}(1,\vec{a})$ and use (5.24), the $\vec{\underline{j}}(R,\phi)$ term drops out and we obtain

$$(\vec{\underline{j}}(1,\vec{a}), \vec{\underline{a}})r^2R'' = R(\vec{\underline{j}}(1,\vec{a}), \vec{\underline{j}}(R,\vec{\underline{a}})). \tag{5.25}$$

However, if $R(r)\to 1$ as $r\to\infty$, $r^2R''\to 0$ as $r\to\infty$. Consequently, taking the limit as $r\to\infty$, (5.25) yields

$$(\vec{\underline{j}}(1,\vec{\underline{a}}), \vec{\underline{j}}(1,\vec{\underline{a}})) = 0$$

and hence the desired result

$$\vec{\underline{j}}(1,\vec{\underline{a}}) = 0. \tag{5.26}$$

Then taking the limit $r\to\infty$ of (5.12), it follows that

$$Q(1) = 0. \tag{5.27}$$

Equation (5.27) has two consequences. First it shows that $Q(R)$ must have a factor $(1-R)$; also from (5.18)–(5.20) and (5.12),

$$2M\nu = L(L+1) + \nu^2/N$$

and hence

$$Q(R) = L(L+1)(1-R)\left[1 - \frac{\nu^2}{L(L+1)N}R\right] \tag{5.28}$$

This is the required analog of (5.22) for $Q(R)$. However, note that it contains two parameters, L and ν^2/N. Secondly, if we take the $r\to\infty$ limit of (5.9) and use (5.27), we obtain

$$\lim_{r\to\infty}\frac{S^2}{2}\frac{dq}{dR} = \lim_{r\to\infty} S^2(1-R) = 0. \tag{5.29}$$

Since $S\sim r$ as $r\to\infty$, the above equation shows that $(1-R)$ falls off faster than r^{-2} as $r\to\infty$. The asymptotic condition (5.29) will be crucial in what follows.

The physical meaning of the equations $q(1) = Q(1) = 0$ can be seen from the radial equations (5.8) and (5.9). As we shall see in Sec. VIII, because of $q(1) = Q(1) = 0$, the fields $(S-r)$ and $(R-1)$ fall off exponentially with increasing r and hence are massive. On the other hand if $q(1)$ or $Q(1)$ were not zero, then one can easily see that $(S-r)$ and $(R-1)$ would fall off like r^{-1} and r^{-2}, respectively, and hence they are long-range or massless fields.

B. Angular equations

We can extract more information than is contained in (5.15)–(5.20) from (5.4) and (5.5). For this purpose consider the Hilbert space V_ϵ of real functions defined on the sphere, valued in $\mathcal{E}$ with the scalar product (denoted by double bold parentheses)

$$((\psi(\omega), \chi(\omega))) = \frac{1}{4\pi}\int(\psi(\omega), \chi(\omega))d\omega. \tag{5.30}$$

Take the scalar product of (5.4) in V_ϵ by $\Psi(\omega)$ such that

$$((\Psi(\omega), \phi(\omega))) = \frac{1}{4\pi}\int(\Psi(\omega), \phi(\omega))d\omega = 0. \tag{5.31}$$

Then

$$((\Psi(\omega), \vec{\underline{d}}_R{}^2\phi(\omega))) = a + bR + cR^2 = 0, \tag{5.32}$$

where a, b, c are numerical constants. Since $R(r)$ is not a constant,

$$a = b = c = 0$$

for all $\Psi(\omega)$ satisfying (5.31). Hence

$$\begin{aligned}((\Psi(\omega), \partial^2\phi(\omega))) &= ((\Psi(\omega), \underline{t}\cdot\vec{\underline{a}}\cdot\vec{\partial}\phi(\omega)))\\ &= ((\Psi(\omega), (\underline{t}\cdot\vec{\underline{a}})^2\phi(\omega)))\\ &= 0.\end{aligned} \tag{5.33}$$

Equation (5.33) combined with (5.15), (5.16), (5.17) give

$$\partial^2\phi(\omega) = -l(l+1)\phi(\omega)\,, \tag{5.34}$$

$$\underline{t}\cdot\vec{\underline{a}}\cdot\vec{\partial}\phi(\omega) = m\nu\phi(\omega)\,, \tag{5.35}$$

$$(\underline{t}\cdot\vec{\underline{a}})^2\,\phi(\omega) = -n\nu^2\phi(\omega)\,. \tag{5.36}$$

Similarly, consider a Hilbert space V_g of real vector functions defined on the sphere and valued in the algebra $\mathcal{G}$. If $\vec{\underline{\chi}}(\omega)$ are vector functions such that

$$((\vec{\underline{\chi}}(\omega), \vec{\underline{a}}(\omega))) = 0\,, \tag{5.37}$$

taking the scalar product of (5.5) by $\underline{\chi}(\omega)$, we obtain an equation of the form

$$a[1-R(r)] + b[1-R(r)]^2 + c[1-R(r)]^3 + S^2(r)\{g + h[1-R(r)]\} = 0\,, \tag{5.38}$$

where a, b, c, g, and h are numerical constants which can be easily determined from (5.5) and (5.18)–(5.20). Now, from the asymptotic conditions $R\to 1$ and $S^2(1-R)\to 0$, we see that (5.38) can be satisfied for nontrivial R only if

$$a = b = c = g = h = 0\,. \tag{5.39}$$

From identifying a, b, c appropriately, we obtain three equations

$$\partial^2\vec{\underline{a}}(\omega) = -L(L+1)\vec{\underline{a}}(\omega)\,, \tag{5.40}$$

$$2\vec{\underline{a}}\cdot\underline{\tau}\cdot\vec{\partial}\,\vec{\underline{a}}(\omega) + (\vec{\underline{a}}, \underline{\tau}\vec{\partial}\,\vec{\underline{a}}) = 2M\nu\vec{\underline{a}}(\omega)\,, \tag{5.41}$$

$$[\underline{\tau}\cdot\vec{\underline{a}}(\omega)]^2\vec{\underline{a}}(\omega) = -\frac{1}{N}\nu^2\vec{\underline{a}}(\omega)\,, \tag{5.42}$$

which are the counterparts of the corresponding equations (5.34)–(5.36) for $\phi(\omega)$. We also obtain two additional equations relating ϕ and $\vec{\underline{a}}$, viz.,

$$(\phi, \underline{t}\vec{\partial}\phi) = -(\phi, \underline{t}(\underline{t}\cdot\vec{\underline{a}})\phi) = n\vec{\underline{a}}\,, \tag{5.43}$$

where the constant n has been computed from (5.43) and (5.35).

C. Summary of all the equations

Utilizing the relations between the numerical constants to eliminate some of them, we summarize all the equations obtained above for both the radial functions $S(r)$ and $R(r)$ and the angular functions $\phi(\omega)$ and $\vec{\underline{a}}(\omega)$.

Equations for angular functions.

(a) $\vec{\underline{d}}\phi(\omega) = [\vec{\partial}1 + \underline{t}\cdot\vec{\underline{a}}(\omega)]\phi(\omega) = 0\,,\quad (\phi(\omega), \phi(\omega)) = 1\,,$ (5.44a)

(b) $\vec{\underline{d}}^2\vec{\underline{a}}(\omega) = -\vec{\underline{j}}(\vec{\underline{a}})\,,\quad (\vec{\underline{a}}(\omega), \vec{\underline{a}}(\omega)) = \nu^2\,,$ (5.44b)

(c) $\partial^2\phi(\omega) = -l(l+1)\phi(\omega)\,,\quad \partial^2\vec{\underline{a}}(\omega) = -L(L+1)\vec{\underline{a}}(\omega)\,,$ (5.44c)

(d) $(\vec{\underline{a}}\cdot\underline{\tau})^2\vec{\underline{a}} = [\underline{a}_i, [\underline{a}_i, \vec{\underline{a}}]] = -\frac{\nu^2}{N}\vec{\underline{a}}\,,$ (5.44d)

(e) $(\phi, \underline{t}\vec{\partial}\phi) = n\vec{\underline{a}}\,.$ (5.44e)

Equations for radial functions.

$$r^2S''(r) = \{l(l+1)[1-R(r)]^2 + r^2\mu^2 W(S/r)\}S(r)\,, \tag{5.45a}$$

$$r^2R''(r) = [1-R(r)]\{L(L+1)R(r)[1-\epsilon R(r)] - e^2c^2nS^2(r)\}\,, \tag{5.45b}$$

where

$$\epsilon = \frac{l(l+1)}{L(L+1)Nn}\,.$$

VI. DISCUSSION OF THE CONSTANTS IN THE EQUATIONS

Before we discuss the solutions to Eqs. (5.44) and (5.45), it is instructive to analyze them briefly and see the nature and significance of the various constants appearing in them. First of all we note that the radial separation implies that all the equations valid in the $r\to\infty$ limit are valid for all r. This is the case of Eq. (5.44a), which is the basic finite-energy condition (3.10). The analogous equation for $\vec{\underline{a}}(\omega)$ contains a Higgs current term in the $r\to\infty$ limit, but the radial separation, as shown in the preceding section, leads to a stronger condition, namely, Eq. (5.44b) involving only $\vec{\underline{a}}$ fields. The latter equation can be written in the equivalent forms

$$(\underline{d}_i\underline{f}_{ij}) = 0\,,\quad (\underline{d}_i\underline{f}) = 0\,, \tag{6.1}$$

where $\underline{f}_{ij}$ and $\underline{f}$ are defined in (4.4) and (4.13). Equations (6.1) are pure Yang-Mills equations for the angular functions $\vec{\underline{a}}(\omega)$ or equivalently for $\underline{f}(\omega)$ on the unit sphere. However, note that (5.44e) relates $\vec{\underline{a}}(\omega)$ to $(\phi, \underline{t}\vec{\partial}\phi)$.

Equations (5.44c) show that $\phi(\omega)$ and $\vec{\underline{a}}(\omega)$ are spherical harmonics of definite order l and L, respectively. Thus this property true for a system of linear equations is also true in the case of the coupled nonlinear system of equations under study. Therefore, l and L are positive integers [if one of them were zero, we would have the trivial solution $\phi(\omega) = \vec{\underline{a}}(\omega) = \text{const}$]. If l is a positive integer, it then follows from (5.21) that $n>0$. Assuming $\nu>0$, we see from (5.21) that $m>0$. Finally, since $(\underline{a}\cdot\underline{\tau})^2$ is a sum of negative operators (being a sum of squares of antisymmetric operators), it follows from (5.44d) that $N>0$.

The same equation (5.44d) can be used to prove that, for each ω, the $\underline{a}_i(\omega)$ generate an SU(2) subalgebra which we shall denote $\mathrm{SU}(2)_\omega$. For this purpose, consider at each ω a triad of unit vectors $\hat{n}^{(1)}(\omega)$, $\hat{n}^{(2)}(\omega)$, $\hat{r}$ such that

$$\hat{n}^{(1)}\cdot\hat{n}^{(2)} = 0\,,\quad \hat{r}(\omega) = \hat{n}^{(1)}\times\hat{n}^{(2)}\,. \tag{6.2}$$

Then

$$\hat{r}_k \hat{r}_l + \sum_{i=1}^{2} n_k^{(i)} n_l^{(i)} = \delta_{kl} . \tag{6.3}$$

Now, since $N > 0$, we can define

$$\underline{e}^{(i)}(\omega) = \frac{\sqrt{N}}{\nu} \hat{n}^{(i)} \cdot \underline{\vec{a}}(\omega), \quad i = 1, 2 , \tag{6.4}$$

and using (6.3) and $\hat{r} \cdot \underline{\vec{a}}(\omega) = 0$, we find

$$\underline{\vec{a}}(\omega) = \frac{\nu}{\sqrt{N}} \sum_{i=1}^{2} \hat{n}^{(i)} \underline{e}^{(i)} . \tag{6.5}$$

A straightforward computation using (6.2), (6.4), and (5.44d) yields

$$[\underline{e}^{(1)}, [\underline{e}^{(1)}, \underline{e}^{(2)}]] = -\underline{e}^{(2)}$$

and $\qquad$ (6.6)

$$[\underline{e}^{(2)}, [\underline{e}^{(2)}, \underline{e}^{(1)}]] = -\underline{e}^{(1)} .$$

Hence, if we define

$$\underline{e}^{(3)} = [\underline{e}^{(1)}, \underline{e}^{(2)}] = -\frac{N}{\nu^2} \underline{v}(\omega), \tag{6.7}$$

we see that the three $\underline{e}^{(i)}$ form the basis of an $SU(2)_\omega$ Lie subalgebra of $\mathcal{G}$ with the usual structure constants ϵ_{ijk}, i.e.,

$$[\underline{e}^{(i)}, \underline{e}^{(j)}] = \epsilon_{ijk} \underline{e}^{(k)} . \tag{6.8}$$

Now any compact Lie algebra splits canonically into the direct sum of an Abelian algebra and a semisimple algebra. From (5.44d), since we can write $\underline{\vec{a}}(\omega)$ as a commutator

$$\frac{\nu^2}{N} \underline{\vec{a}} = [\underline{v}, \underline{\vec{b}}] , \tag{6.9}$$

the $\underline{\vec{a}}(\omega)$ fields take their values inside the semi-simple $\mathcal{G}_s$ of $\mathcal{G}$. There is only a finite set of SU(2) Lie subalgebras in any $\mathcal{G}_s$, up to a conjugation by the group G. Since we are interested only in the regular solution $\underline{\vec{a}}(\omega)$, we conclude that all $SU(2)_\omega$ for different ω are conjugated into each other by G.

The normalization constant ν^2 [Eq. (5.44b)] depends on the arbitrary scale which one can choose for the invariant orthogonal metric (i.e., the Cartan-Killing bilinear form) on $\mathcal{G}_s$. However, as can easily be verified, the scalar product of $\underline{e}$'s in G space is independent of ν^2, but depends only on N,

$$(\underline{e}^{(1)}, \underline{e}^{(1)}) = (\underline{e}^{(2)}, \underline{e}^{(2)}) = (\underline{e}^{(3)}, \underline{e}^{(3)}) = N/2 . \tag{6.10}$$

Since all the representations of the SU(2) Lie algebras are known, Eq. (6.10) yields

$$\begin{aligned} N &= -\tfrac{1}{2} \operatorname{tr}(\underline{e}^{(3)})^2 \\ &= \sum_{j=0}^{\infty} \tfrac{1}{3} c_j j(j+1)(2j+1) \\ &= \sum_{j=0}^{\infty} \tfrac{1}{2} c_j \binom{2j+2}{3} , \end{aligned} \tag{6.11}$$

where c_j is the multiplicity of the $(2j+1)$-dimensional irreducible representation in the reduction of the adjoint representation $\mathcal{G}$ restricted to a specific $SU(2)_\omega$ subalgebra. Of course $c_1 \geq 1$ and, since $\mathcal{G}$ is real, c_j is even when j is half-integer; then (6.11) shows that

$$N \text{ is an integer} \geq 2 . \tag{6.12}$$

To summarize, at each ω, $\underline{\vec{a}}(\omega)$ generate an $SU(2)_\omega$; for different ω the different $SU(2)_\omega$ are conjugated. N is a purely group-theoretic number which depends only on the embedding of these SU(2) in $\mathcal{G}$. For instance, the values of N for some low-rank groups are the following.

$\mathcal{G}$	SO(3)	SO(4)	SU(3)
N	2	2, 2, 4	3, 12

(6.13)

The only constant that remains to be discussed is n. First of all we note that, from the definition (3.14) of $M_{\alpha\beta}$ and the Eqs. (5.44a) and (5.44e), we have

$$M_{\alpha\beta}(\omega) \vec{a}^{\beta} = c^2 n \vec{a}^{\alpha} , \tag{6.14}$$

which shows that the nonvanishing gauge fields are actually eigenfields of the mass-squared operator with the same mass squared $m^2 = c^2 n$. Next, Eq. (5.36) can be written as

$$-(\underline{e}^{(1)} \cdot \underline{t})^2 \phi - (\underline{e}^{(2)} \cdot \underline{t})^2 \phi = nN\phi \tag{6.15}$$

or also

$$\begin{aligned} \left[-\sum_{i=1}^{2} (\underline{e}^{(i)} \cdot \underline{t})^2 + (\underline{e}^{(3)} \cdot \underline{t})^2 \right] \phi &= [t(t+1) - m_t^2] \phi \\ &= nN\phi , \end{aligned} \tag{6.16}$$

where $2t$, $(t - m_t)$ are non-negative integers and $-t \leq m_t \leq t$. This implies that, if we call t the spin of the irreducible representation of $SU(2)_\omega$ which appears in the space of ϕ, $\phi(\omega)$ is an eigenvector of $(\underline{\vec{a}} \cdot \underline{t})^2$ and

$$2nN = 2[t(t+1) - m_t^2] \text{ is an integer} > 0 , \tag{6.17}$$

barring the trivial case $\phi(\omega) = \underline{\vec{a}}(\omega) = \text{const}$.

VII. SOLUTION OF THE ANGULAR EQUATIONS

In this section we obtain the complete solution of the angular equations (5.44). The first step is to use Eq. (5.44b) to show that the functions $\underline{u}$ and $\underline{v}$ which make up $\underline{f}$ in (4.16) are not independent but satisfy the relationship

$$\underline{f} = \underline{u} = -\underline{v} . \tag{7.1}$$

For this purpose we introduce the identity

$$L_i L_j + \partial_i \partial_j = (\delta_{ij} - \hat{r}_i \hat{r}_j) \vec{L}^2 \tag{7.2}$$

for functions on the sphere. Using this identity we have, from the definition (4.17) of $\underline{u}$, the relationship

$$\underline{\vec{a}}(\omega) = \frac{-2}{L(L+1)} \vec{L}\underline{u}(\omega), \tag{7.3}$$

which shows that $\underline{u}(\omega)$ acts as a potential for $\underline{\vec{a}}(\omega)$. We next note that, since for $\hat{r}\cdot\underline{\vec{a}}=0$

$$\underline{f}_{ij} = \epsilon_{ijk}\hat{r}_k \underline{f}, \tag{7.4}$$

Eq. (5.44b) yields

$$d_i \underline{f} = 0. \tag{7.5}$$

Using the definition $\underline{f} = 2\underline{u} + \underline{v}$, we then have from (7.5)

$$\partial_i \underline{v} + 2[\underline{a}_i, \underline{u}] = -2\partial_i \underline{u} - [\underline{a}_i, \underline{v}]. \tag{7.6}$$

The terms on the right-hand side of (7.6) can be evaluated as follows: From (7.3) we obtain by taking the cross product with $\hat{r}$

$$\partial_i \underline{u} = \frac{L(L+1)}{2} \underline{b}_i, \tag{7.7}$$

and from (5.44d) and the definition (4.17) of $\underline{v}$ we obtain

$$[\underline{a}_i, \underline{v}] = \frac{\nu^2}{N} \underline{b}_i. \tag{7.8}$$

Hence (7.6) becomes

$$\partial_i \underline{v} + 2[\underline{a}_i, \underline{u}] = -[L(L+1) + \nu^2/N]\underline{b}_i. \tag{7.9}$$

If we now take the (noncovariant) divergence of (7.9), recall that $\underline{\vec{a}}$ is divergence-free, and use (7.7) twice, we obtain

$$\vec{\delta}^2 \underline{v} + L(L+1)[\underline{a}_r, \underline{b}_r] = 2[L(L+1) + \nu^2/N]\underline{u}, \tag{7.10}$$

which from the definition (4.17) of $\underline{v}$ reduces to

$$\vec{\delta}^2 \underline{v} - 2L(L+1)\underline{v} = 2[L(L+1) + \nu^2/N]\underline{u}. \tag{7.11}$$

Now from the definition (4.17) of $\underline{u}$ and from (5.44c) we see that $\underline{u}$ is a spherical harmonic of order L. Hence operating on (7.11) with $\vec{\delta}^2 + L(L+1)$, we obtain

$$[\vec{\delta}^2 + L(L+1)][\vec{\delta}^2 - 2L(L+1)]\underline{v} = 0. \tag{7.12}$$

But for nontrivial L ($L \neq 0$) the operator $\vec{\delta}^2 - 2L(L+1)$ is negative definite and therefore nonsingular. Thus (7.12) implies that

$$[\vec{\delta}^2 + L(L+1)]\underline{v} = 0. \tag{7.13}$$

Thus $\underline{v}$ is also a spherical harmonic of order L. If we now insert (7.13) into (7.11) we obtain

$$\underline{v}(\omega) = \lambda \underline{u}(\omega),$$

where (7.14)

$$\lambda = -2[L(L+1) + \nu^2/N]/3L(L+1).$$

Thus $\underline{v}$, and therefore $\underline{f}$, are proportional to $\underline{u}$; indeed (4.16) yields

$$\underline{f} = (2+\lambda)\underline{u}. \tag{7.15}$$

We assume that the integrability condition (4.14) is satisfied nontrivially, i.e., $\underline{f} \neq 0$. Then Eq. (4.15) reads

$$M_{\alpha\beta} u^\beta = 0 = M_{\alpha\beta} v^\beta. \tag{7.16}$$

To determine λ, we consider the equation

$$\underline{\vec{J}} \cdot \underline{\vec{a}} = 2(\underline{u} + \underline{v}) \tag{7.17}$$

and operate on it by the mass matrix M. Now

$$M_{\alpha\beta}(\underline{\vec{J}}\cdot\underline{\vec{a}})^\beta = \underline{\vec{J}}\cdot(M_{\alpha\beta}\vec{a}^\beta) = c^2 n(\underline{\vec{J}}\cdot\vec{a}^\alpha), \tag{7.18}$$

where we have used (3.15) (vanishing of the covariant derivative of M) and (6.14). From (7.16), (7.17), and (7.18), since $c^2 n \neq 0$ for a nontrivial solution, we obtain

$$\underline{u} + \underline{v} = 0 \text{ or } \lambda = -1. \tag{7.19}$$

Using $\lambda = -1$ in (7.14), we also pick up the relation

$$\frac{\nu^2}{N} = \frac{L(L+1)}{2} \tag{7.20}$$

for the constants. This completes the first step in the solution of the angular equations.

The next step is to construct a self-coupling equation for $\underline{f}$. This is not difficult since we have $\underline{f} = \underline{u}$. First from (7.1) and (7.13) we see that

$$\vec{\delta}^2 \underline{f} = -L(L+1)\underline{f}, \tag{7.21}$$

so that $\underline{f}$ also is a spherical harmonic of order L, and from (7.3) we have

$$\underline{a}_i(\omega) = \frac{2}{L(L+1)} L_i \underline{f}, \tag{7.22}$$

but from the angular equation (6.1) we have

$$d_i \underline{f} = \partial_i \underline{f} + [\underline{a}_i, \underline{f}] = 0. \tag{7.23}$$

Inserting (7.22) into this equation, we obtain

$$\partial_i \underline{f} = \frac{2}{L(L+1)} [\underline{f}, L_i \underline{f}], \tag{7.24}$$

which is the required self-coupling equation. Furthermore (6.1), which is similar to (3.10), means that the values of $\underline{f}$ in $\mathcal{G}$ for different ω are obtained from each other by a transformation of G [similar to (3.11)]. Hence

$$(\underline{f}(\omega), \underline{f}(\omega)) = \kappa^2, \tag{7.25}$$

where κ^2 is a constant. To evaluate κ^2 we use (7.22), which gives

$$\nu^2 = (\underline{a}_i, \underline{a}_i) = \left[\frac{2}{L(L+1)}\right]^2 (L_i \underline{f}, L_i \underline{f}) = -\left[\frac{2}{L(L+1)}\right]^2 (\underline{f}, L^2 \underline{f}) .$$

Hence from (7.21) we have

$$\nu^2 = \frac{4}{L(L+1)} (\underline{f}, \underline{f}) = \frac{4\kappa^2}{L(L+1)} \tag{7.26}$$

and from (7.20)

$$\kappa^2 = NL^2(L+1)^2/8 . \tag{7.27}$$

Finally, we note from (7.21) that $\underline{f}$ has the expansion

$$\underline{f}^\alpha(\omega) = \sum_m t_m^\alpha Y_m^L(\omega), \quad \bar{t}_m^\alpha = (-1)^m t_{-m}^\alpha \tag{7.28}$$

in terms of one spherical harmonic Y_m^L. Since $\underline{f}$ is real and the Y_m^L are complex, the expansion coefficients t_m^α are complex, but they satisfy the relation displayed.

To exploit the self-coupling equation (7.24), using the form (7.28) for $\underline{f}(\omega)$ we need the following inversion formula which can be proved using standard spherical-harmonic analysis.

Let $\chi(\omega)$ be a function such that

$$\chi(\omega) = \sum c_{rs} Y_r^L(\omega) Y_s^L(\omega), \quad c_{rs} = c_{sr} . \tag{7.29}$$

Then

$$c_{rs} = \sum_J \left(\!\!\begin{pmatrix} L & L & J \\ r & s & \mu \end{pmatrix}\!\!\right) \int Y_\mu^J(\omega) \chi(\omega) d\omega , \tag{7.30}$$

where

$$\left(\!\!\begin{pmatrix} L & L & J \\ r & s & \mu \end{pmatrix}\!\!\right) = [4\pi(2J+1)]^{1/2} \times \begin{pmatrix} L & L & J \\ r & s & \mu \end{pmatrix} \Big/ \begin{pmatrix} L & L & J \\ 0 & 0 & 0 \end{pmatrix} (2L+1) , \tag{7.31}$$

and $\binom{L\ L\ J}{r\ s\ \mu}$ are the conventional 3-j symbols.[9]

When written out explicitly in terms of the structure constants the third component of the self-coupling equation (7.24) reads

$$\partial_3 f^\alpha(\omega) = \frac{2}{L(L+1)} c_{\beta\gamma}^\alpha f^\beta(\omega) L_3 f^\gamma(\omega) , \tag{7.32}$$

and if we now use the expansion (7.28) we see that it becomes

$$\sum_m t_m^\alpha \partial_3 Y_m^L(\omega) = \sum_{rs} \frac{3-r}{L(L+1)} c_{\beta\gamma}^\alpha t_r^\beta t_s^\gamma Y_r^L(\omega) Y_s^L(\omega) , \tag{7.33}$$

where we have used the antisymmetry of the structure constants in β and γ. Equation (7.33) is then of the form (7.29). Hence if we use the inversion formula we obtain

$$\frac{s-r}{L(L+1)} c_{\beta\gamma}^\alpha t_r^\beta t_s^\gamma = t_m^\alpha \sum_J \left(\!\!\begin{pmatrix} L & L & J \\ r & s & \mu \end{pmatrix}\!\!\right) \int d\omega\, Y_\mu^J(\omega) \partial_3 Y_m^L(\omega) , \tag{7.34}$$

where m is not summed over because $m = -\mu = r + s$ on account of the integration over ω and the additivity of the magnetic quantum number in the 3-j symbols.

Similarly, if we insert the expansion (7.28) into (7.25) we obtain

$$\kappa^2 = \sum_{rs} t_r^\alpha t_s^\alpha Y_r^L(\omega) Y_s^L(\omega) , \tag{7.35}$$

which is also of the form (7.29). Hence, by the inversion formula, we have

$$t_r^\alpha t_s^\alpha = \kappa^2 \left(\!\!\begin{pmatrix} L & L & 0 \\ r & s & 0 \end{pmatrix}\!\!\right) = (-1)^r \delta_{r+s,0} \frac{\sqrt{4\pi}}{2L+1} \kappa^2 . \tag{7.36}$$

We now use (7.36) to analyze (7.34) as follows: Define the matrix

$$P_{\alpha\beta} = \rho \sum_r t_r^\alpha t_{-r}^\beta (-1)^r , \quad \rho = \frac{2L+1}{\sqrt{4\pi}\,\kappa^2} . \tag{7.37}$$

It is easy to see that the matrix $P_{\alpha\beta}$ is a projection operator, since from (7.28) it is real and symmetric and from (7.36) we have

$$P_{\alpha\beta} P_{\beta\gamma} = P_{\alpha\gamma} . \tag{7.38}$$

Furthermore from (7.36) we see that

$$P_{\alpha\beta} t_r^\beta = t_r^\alpha \text{ and } P_{\alpha\beta} f^\beta(\omega) = f^\alpha(\omega) , \tag{7.39}$$

so that the t_r^α and $f^\alpha(\omega)$ are eigenvectors of $P_{\alpha\beta}$ with eigenvalue unity. Finally, since from (7.28) and (7.37), respectively, we have for any vector λ_α

$$f^\alpha(\omega)\lambda_\alpha = 0 \rightarrow t_r^\alpha \lambda_\alpha = 0 \rightarrow P_{\beta\alpha}\lambda_\alpha = 0 , \tag{7.40}$$

we see that $f^\alpha(\omega)$ spans the complete subspace $\mathfrak{F}$ onto which $P_{\alpha\beta}$ projects. But since from (7.36) the trace of $P_{\alpha\beta}$ is $2L+1$, this subspace is $2L+1$ dimensional.

If we now return to (7.34) and operate with the projection operator $(\delta_{\alpha\alpha'} - P_{\alpha\alpha'})$, the right-hand side drops out and we obtain from (7.39)

$$(\delta_{\alpha\alpha'} - P_{\alpha\alpha'}) \frac{s-r}{L(L+1)} c_{\beta\gamma}^{\alpha'} t_r^\beta t_s^\gamma = 0 . \tag{7.41}$$

But since $c^{\alpha}_{\beta\gamma} t^{\beta}_{r} t^{\gamma}_{s}$ vanishes in any case for $r = s$, we then have

$$(\delta_{\alpha\alpha'} - P_{\alpha\alpha'}) c^{\alpha'}_{\beta'\gamma'} P_{\beta'\beta} P_{\gamma'\gamma} = 0 . \qquad (7.42)$$

Now let ${}^{(r)}\mu$, $r = 1, 2$, be any two elements of the Lie algebra lying in the $(2L+1)$-dimensional subspace $\mathfrak{F}$ projected out by $P_{\alpha\beta}$. Then by hypothesis

$${}^{(r)}\mu = {}^{(r)}\mu_{\alpha}\tau_{\alpha} = {}^{(r)}\mu_{\alpha'} P_{\alpha'\alpha}\tau_{\alpha} , \qquad (7.43)$$

and hence

$$[{}^{(1)}\mu, {}^{(2)}\mu] = {}^{(1)}\mu_{\alpha'} {}^{(2)}\mu_{\beta'} P_{\alpha'\alpha} P_{\beta'\beta} c^{\gamma}_{\alpha\beta}\tau_{\gamma} = W^{\gamma}\tau_{\gamma} . \qquad (7.44)$$

But from (7.42) we have

$$P_{\gamma\gamma'} W^{\gamma'} = W^{\gamma} . \qquad (7.45)$$

Thus the commutator also lies in the subspace $\mathfrak{F}$ projected out by $P_{\alpha\beta}$. It follows that the $2L+1$ subspace $\mathfrak{F}$ of the Lie algebra projected out by $P_{\alpha\beta}$ is actually a subalgebra of $\mathcal{G}$.

Every compact Lie algebra is the direct sum of an Abelian Lie algebra and semisimple Lie algebra, the latter being generated by the commutators of the algebra. The Lie algebra $\mathfrak{F}$, which is compact since it is a subalgebra of $\mathcal{G}$, is generated by the fields $\underline{f}(\omega) = -\underline{v}(\omega)$ which are commutators [see Eq. (4.17)]; hence $\mathfrak{F}$ is semisimple.

For the last step of the solution we return to (7.36) which shows that the spatial rotation group acts on $\mathfrak{F}$ by its $(2L+1)$-dimensional irreducible representation:

$$D^{(L)}(R)_{\alpha\beta} f^{\beta}(\omega) = f^{\alpha}(R\omega) . \qquad (7.46)$$

By this action, SO(3) is a group of automorphisms of the Lie algebra $\mathfrak{F}$; indeed, with the use of (7.30) and (7.39), one obtains from (7.36) the explicit expression for the structure constants of $\mathfrak{F}$,

$$c^{\alpha}_{\beta\gamma} = -\frac{(2L+1)^2 L(L+1)}{4\pi\kappa^4} \sum_{rs} \sum_{J\mu} \frac{1}{r-s} \bar{t}^{\beta}_{r} \bar{t}^{\beta}_{s} t^{\alpha}_{m} \left(\begin{pmatrix} L & L & J \\ r & s & \mu \end{pmatrix}\right) \int Y^{j}_{\mu}(\omega) \partial_3 Y^{L}_{m}(\omega) d\omega , \qquad (7.47)$$

and the integration in this equation shows that the structure constants are invariant by the rotation on ω defined in (7.46). For a semisimple Lie group every connected Lie group of automorphisms is a group of inner automorphisms. Hence $\mathfrak{F}$ contains an SO(3) subalgebra which acts irreducibly [by (7.46)] on the $(2L+1)$-dimensional algebra $\mathfrak{F}$; this is impossible unless this SO(3) is $\mathfrak{F}$ itself. This requires L to be unity.

We denote by $SO(3)_I$ the fixed-SO(3) subalgebra of $\mathcal{G}$ generated by the $\underline{f}(\omega)$'s. From (7.1) and (7.3), we see that $SO(3)_I$ is also generated by the gauge fields $\bar{\mathbf{a}}(\omega)$. If there are several nonconjugated SO(3) subalgebras of $\mathcal{G}$ the Higgs mechanism will in general choose among them. Equations (7.24), (7.1), and (7.22) yield, up to an arbitrary orthogonal choice of basis in $SO(3)_I$,

$$f^{\alpha} = -\kappa\hat{r}_{\alpha}, \quad a^{\alpha}_{i} = \kappa\epsilon_{\alpha ij}\hat{r}_{j}, \quad \kappa = (N/2)^{1/2}. \qquad (7.48)$$

Since we had absorbed the charge e in the definition of $\underline{f}(\omega)$ and $\bar{\mathbf{a}}(\omega)$, Eq. (7.48) is exactly the 't Hooft–Polyakov solution for the charge $e_N = e(2/N)^{1/2}$. This charge, which varies with the embedding in G, is the natural charge for $SO(3)_I$, and the monopole strength has the minimum value of $4\pi/e_N$.

Equation (7.48) completely solves the angular Eqs. for $a(\omega)$, and it remains only to consider the Higgs fields. From (5.44a) and (7.48)

$$\partial_i \phi^A = -(t^{\alpha})^{AB} a^{\alpha}_{i} \phi^B = -(t^{\alpha})^{AB} \kappa\epsilon_{\alpha ij}\hat{r}_j \phi^B , \qquad (7.49)$$

which implies that

$$(L_i + \bar{t}_i)\phi(\omega) = 0 , \qquad (7.50)$$

where the $\bar{t}_i$ are the generators of $SO(3)_I$, normalized in the conventional way. Equations (7.50) and (5.44c) show that the SO(3) representation of $\phi(\omega)$ if fixed to be $t(t+1) = l(l+1)$. For given $l = t$, it is easy to construct explicit solutions for $\phi(\omega)$ in terms of linear combinations of real and imaginary parts of appropriate spherical harmonics with real coefficients.

The above results show that the condition of separation of variables is so strong that it reduces the solution for general G to that for $G = SO(3)$, and that for SO(3) the solution differs from the original 't Hooft–Polyakov ansatz only in that the Higgs field can belong to an arbitrary integral representation of SO(3) [and then only if the spin and isospin compensate as in (7.50)]. Thus the 't Hooft–Polyakov solution for $G = SO(3)$ is essentially the only solution with separated variables for arbitrary G.

VIII. RADIAL EQUATIONS

Finally we come to the discussion of the radial equations. Using (7.50) to determine the constants and replacing $l(l+1)$ by its gauge-invariant counterpart $t(t+1)$, the radial equations (5.45a), (5.45b) are easily seen to reduce to

$$r^2 H'' = H[(t+1)K^2 + \mu^2(S - r)(\sigma^2 S + r)], \quad H = e_N c S ,$$
$$r^2 K'' = K[(K^2 - 1) + \tfrac{1}{2}t(t+1)H^2], \quad K = 1 - R . \qquad (8.1)$$

The boundary conditions for these equations are

$$K(0) = 1, \quad H(0) = 0$$

and (8.2)

$$K(r) \to 0, \quad S(r) \sim r \quad \text{as } r \to \infty .$$

Inserting the conditions at $r \to \infty$ in the radial equations we easily see that $K(r)$ and $S(r)$ take the asymptotic forms

$$K(r) \sim \text{const} \times \exp\{-e_N c[\tfrac{1}{2}t(t+1)]^{1/2} r\},$$
$$S(r) \sim r + \text{const} \times \exp[-\mu(1+\sigma^2)^{1/2} r]. \tag{8.3}$$

Thus $K(r)$ and $S(r) - r$ describe fields of mass $e_N c[t(t+1)/2]^{1/2}$ and $\mu(1+\sigma^2)^{1/2}$, respectively. This is exactly what we should expect since, as we have shown in the Appendix, $\mu(1+\sigma^2)^{1/2}$ is the mass acquired by the Higgs field after the spontaneous symmetry breaking and, from (2.22), $e_N c[t(t+1)/2]^{1/2}$ is the mass acquired by the gauge field through the Higgs mechanism.

It is worth noting that the radial equations are derivable directly from the Hamiltonian (3.3), which now takes the form

$$\mathcal{H} = \frac{4}{e_N^2} \int_0^\infty \frac{dr}{r^2} \left\{ (rK')^2 + \tfrac{1}{2}(rH' - H)^2 + \tfrac{1}{2}(1-K^2)^2 + \frac{t(t+1)}{2} H^2K^2 + \frac{\mu^4}{4} H^2[\sigma^2 S^2 + \tfrac{4}{3}(1-\sigma^2) rS - 2r^2] \right\}. \tag{8.4}$$

Now the existence of nontrivial solutions that minimize $\mathcal{H}$ [equivalently, solutions to the radial equations (8.1)] has been proved by Tyupkin *et al.*,[10] in the case when $t = 1$. It is easy to verify that the proof can be extended to any $t > 0$. Thus finite-energy solutions to the radial equations (8.1) certainly exist.

An interesting feature of the solutions for general t is that they provide a *mass formula* $m(t) = \mathcal{H}$ on account of the dependence on t in (8.4). Differentiating $\mathcal{H}$ with respect to t (which is isospin, not time), we obtain

$$\frac{d\mathcal{H}}{dt} = \int d^3x \frac{\delta\mathcal{H}}{\delta\psi_s} \frac{\partial\psi_s}{\partial t} + \frac{\partial\mathcal{H}}{\partial t}, \quad \psi_s = (H, K)_{\text{solution}} \tag{8.5}$$

and, since the first term on the right-hand side of (8.5) vanishes on account of the field equations, we have

$$\frac{d\mathcal{H}}{dt} = \frac{\partial\mathcal{H}}{\partial t} = \left(\frac{2t+1}{2}\right) \int_0^\infty \frac{dr}{r^2} H^2K^2 > 0 .$$

Thus the mass is a monotonically increasing function of the isospin and the solution with $t = 1$ is the lowest-lying nontrivial solution.

In conclusion we note that the above formalism can be extended to include dyons by making the ansatz

$$eA_0^\alpha = \hat{r}^\alpha \left(\frac{J(r)}{r}\right) . \tag{8.6}$$

Since A_0^α plays the role of a second Higgs field, which lies in the adjoint representation and which has no potential and no interaction[11] with $\phi(\omega)$, one sees by analogy with ϕ that the only effect of A_0^α on the field equations is to add to the Hamiltonian density a term of the form

$$\frac{1}{e^2 r^2} [\tfrac{1}{2}(rJ' - J)^2 + \tfrac{1}{2} J^2K^2] . \tag{8.7}$$

The corresponding equations of motion are then the generalization of the Julia and Zee equations[12] for the special case $t = 1$.

Note added in proof. We are grateful to John Rawnsley for pointing out that since the value of $m(t)$ is less than the value of (8.4) for any trial functions and the t dependence drops out for nonoverlapping trial functions H and K, $m(t)$, although it increases monotonically, must tend to a finite limit $m(\infty)$ as $t \to \infty$. A more detailed discussion of the radial equations will be given elsewhere.

ACKNOWLEDGMENTS

This work was begun in the Institut des Hautes Etudes Scientifiques, Bures-sur-Yvette, continued at Los Alamos, and finally finished at Syracuse. The authors are grateful to several colleagues at these different places for many helpful discussions and hospitality. One of the authors (K.C.W.) would also like to thank George Campbell for a careful and critical reading of the manuscript.

APPENDIX

In this appendix we study the symmetry-breaking minima of G-invariant polynomials, where G is a compact Lie group acting linearly on a finite-dimensional real vector space $\mathcal{E}$:

$$G \ni g \to \Delta(g) \in \mathcal{L}(\mathcal{E}) \tag{A1}$$

where $\mathcal{L}(\mathcal{E})$ is the space of linear operators on $\mathcal{E}$.

For this representation we only assume[13] that it does not contain the trivial representation, i.e., there is no nonzero vector in $\mathcal{E}$ invariant by G. Since G is compact, this representation is equivalent to an orthogonal representation; hence, without loss of generality, we assume

$$\Delta(g^{-1}) = \Delta(g)^T . \tag{A2}$$

We denote by Φ an arbitrary element of $\mathcal{E}$ and by $p^{(n)}(\Phi)$, a G-invariant nth-degree homogeneous polynomial on $\mathcal{E}$. It satisfies

$$p^{(n)}(\Delta(g)\Phi) = p^{(n)}(\Phi), \quad p^{(n)}(\lambda\Phi) = \lambda^n p^{(n)}(\Phi) . \tag{A3}$$

Any linear combination of two such polynomials for a given n is again a $p^{(n)}$. So the most general G-invariant polynomial potential $V(\Phi)$ is of the form[14]

$$V(\Phi) = \sum_{n=0}^{N} p^{(n)}(\Phi) . \tag{A4}$$

Since we are interested only in the extrema of V, the constant term is irrelevant. From our assumption of no invariant vectors on $\mathcal{E}$, V has no linear term, and its gradient $dV/d\Phi$ vanishes at the origin. To obtain symmetry breaking we do not want the extremum at the origin to be a minimum. Hence the quadratic term on at least one G-invariant subspace of $\mathcal{E}$ has to be of the form $-k(\Phi,\Phi)$, $k>0$, where (Φ,Φ) is a G-invariant, strictly positive orthogonal scalar product on $\mathcal{E}$. We assume that, on the whole space $\mathcal{E}$,

$$p^{(2)}=-\tfrac{1}{2}\mu^2(\Phi,\Phi),\quad (\Phi,\Phi)\geq 0,\quad (\Phi,\Phi)=0\rightarrow\Phi=0 \tag{A5}$$

and leave to the reader the straightforward generalization. The notation $\mu^2/2$ for the coefficient of the quadratic term is traditional in the literature on this subject, μ giving the mass scale. Furthermore, V has to be bounded below in order to have a lowest, finite-value minimum; this requires

$$p^{(N)}(\Phi)\geq 0 \text{ for every } \Phi\,, \tag{A6}$$

which implies that N is even. The lowest possible value of N is therefore 4. It happens that this is the only possible value for the quantum theory to be renormalizable. However, no such limitation exists for a purely classical field theory.

Given a $p^{(n)}(\Phi)$, one obtains by polarization an n-multilinear form

$$l^{(n)}(\Phi_1,\ldots,\Phi_n)=\frac{1}{n!}\sum_{k=0}^{n-1}(-1)^k\sum_{(k)}p^{(n)}\Big(\sum_k\Phi_i\Big)\,, \tag{A7}$$

where $\sum_k$ means the sum over $n-k$ different vectors among the n variables of $l^{(n)}$ and the first $\sum_{(k)}$ means the sum over all possible terms. For example, for $n=3$,

$$l^{(3)}(\Phi_1,\Phi_2,\Phi_3)=\frac{1}{3!}\,[\,p^{(3)}(\Phi_1+\Phi_2+\Phi_3)-p^{(3)}(\Phi_1+\Phi_2)-p^{(3)}(\Phi_2+\Phi_3)-p^{(3)}(\Phi_3+\Phi_1)+p^{(3)}(\Phi_1)+p^{(3)}(\Phi_2)+p^{3}(\Phi_3)]\,. \tag{A8}$$

$l^{(n)}$ is completely symmetrical in its n variables and is G invariant. We can also define from $p^{(n)}$ a vector $t^{(n-1)}$ and a linear operator $T^{(n-2)}$ depending linearly and completely symmetrically on, respectively, $(n-1)$ and $(n-2)$ variables by

$$l^{(n)}(\Phi_1,\ldots,\Phi_n)=(\Phi_n,t^{(n-1)}(\Phi_1,\ldots,\Phi_{n-1}))=(\Phi_n,T^{(n-2)}(\Phi_1,\ldots,\Phi_{n-2})\Phi_{n-1})\,, \tag{A9}$$

i.e.,

$$T^{(n-2)}(\Phi_1,\ldots,\Phi_{n-2})\Phi_{n-1}=t^{(n-1)}(\Phi_1,\ldots,\Phi_{n-1})\,. \tag{A9'}$$

Note that T is a symmetric operator and that t and T are covariant under G:

$$(T^T=T)\,, \tag{A10}$$

$$\Delta(g)t^{(n-1)}(\Phi_1,\ldots,\Phi_{n-1})=t^{(n-1)}(\Delta(g)\Phi_1,\ldots,\Delta(g)\Phi_{n-1})\,, \tag{A10'}$$

$$\Delta(g)T^{(n-2)}(\Phi_1,\ldots,\Phi_{n-2})\Delta(g)^T=T^{(n-2)}(\Delta(g)\Phi_1,\ldots,\Delta(g)\Phi_{n-2})\,. \tag{A10''}$$

We finally remark that the gradient $dp^{(n)}/d\Phi$ and the Hessian $d^2p^{(n)}/d\Phi^2$ of $p^{(n)}$ are given by

$$\frac{dp^{(n)}}{d\Phi}=nt^{(n-1)}(\Phi,\ldots,\Phi)\,,\quad \frac{d^2p^{(n)}}{d\Phi^2}=n(n-1)T^{(n-2)}(\Phi,\ldots,\Phi)\,. \tag{A11}$$

Consider the simplest examples

$$p^{(2)}=(\Phi,\Phi)\,,\quad t^{(1)}=2\Phi,\quad T^{(0)}=2I\quad (I=\text{identity matrix})\,, \tag{A12}$$

$$p^{(4)}=(\Phi,\Phi)^2\,,\quad t^{(3)}=\tfrac{1}{3}[(\Phi_1,\Phi_2)\Phi_3+(\Phi_2,\Phi_3)\Phi_1+(\Phi_3,\Phi_1)\Phi_2]\,, \tag{A13}$$

$$T^{(2)}(\Phi_1,\Phi_2)=\tfrac{1}{3}[(\Phi_1,\Phi_2)I+|\Phi_1\rangle\langle\Phi_2|+|\Phi_2\rangle\langle\Phi_1|]\,, \tag{A13'}$$

where $|\Phi_1\rangle\langle\Phi_2|$ is the dyadic operator whose action on Φ_3 is $\Phi_1(\Phi_2,\Phi_3)$. Hence

$$\frac{d}{d\Phi}(\Phi,\Phi)^2=4\Phi(\Phi,\Phi)\,,$$
$$\frac{d^2}{d\Phi^2}(\Phi,\Phi)^2=\tfrac{4}{3}(\Phi,\Phi)(I+2P_\Phi)\,, \tag{A14}$$

where P_Φ is the orthogonal projector on Φ, i.e., $P_\Phi=(\Phi,\Phi)^{-1}|\Phi\rangle\langle\Phi|$.

If we denote the general third- and fourth-degree invariants by $\theta(\Phi)$ and $\omega(\Phi)$, we can write them as

$$\theta(\Phi)=(b(\Phi),\Phi),\quad \omega(\Phi)=(t(\Phi),\Phi)\,, \tag{A15}$$

where for simplicity we have set the bilinear vector[15] $t^{(2)}(\Phi,\Phi)=b(\Phi)$ and the trilinear vector $t^{(3)}(\Phi,\Phi,\Phi)=t(\Phi)$. With this notational simplification the most general inhomogeneous fourth-degree Higgs potential has the form

$$V(\Phi)=\tfrac{1}{4}[\lambda(\Phi,\Phi)^2+\alpha\omega(\Phi)]+\frac{\beta}{3}\mu\theta(\Phi)-\frac{\mu^2}{2}(\Phi,\Phi), \tag{A16}$$

the invariant $\omega(\phi)$ is in general linearly independent from $(\Phi,\Phi)^2$, and further it is assumed to be positive since, if necessary, we can always add a term $k(\Phi,\Phi)^2$, $k>0$ to $\omega(\Phi)$ and call the new term $\omega(\Phi)$. Let χ_m and χ_M be, respectively, the minimum and maximum of $\omega(\Phi)$ on the unit sphere $(\Phi,\Phi)=1$ $(0\leq\chi_m<\chi_M)$. The condition that V is bounded below gives

$$0<\lambda+\alpha\chi \text{ for } \chi_m\leq\chi<\chi_M. \tag{A17}$$

From (A16), we obtain

$$\frac{dV}{d\Phi}=\alpha t(\Phi)+\beta\mu b(\Phi)+[\lambda(\Phi,\Phi)-\mu^2]\Phi \tag{A18}$$

$$=\{\alpha T(\Phi,\Phi)+\beta\mu\mathfrak{D}(\Phi)+[\lambda(\Phi,\Phi)-\mu^2]I\}\Phi, \tag{A19}$$

$$\frac{d^2V}{d\Phi^2}=3\alpha T(\Phi,\Phi)+2\beta\mu\mathfrak{D}(\Phi)+[\lambda(\Phi,\Phi)-\mu^2]I+2\lambda(\Phi,\Phi)P_\Phi, \tag{A20}$$

and $V(\Phi)$ is minimum for the Φ which satisfies

$$\left.\frac{dV}{d\Phi}\right|_{\phi=\phi_0}=0,\quad \left.\frac{d^2V}{d\Phi^2}\right|_{\phi=\phi_0}>0. \tag{A21}$$

Furthermore the spectrum of $d^2V/d\Phi^2|_{\phi=\phi_0}$ gives the squares of the masses for the Higgs fields Φ; however, the Φ's in the kernel of $d^2V/d\Phi^2$, i.e., those of the form $t(a)\Phi_0$ for all $a\in\mathcal{G}$ (they form the tangent plane to the orbit $G\cdot\Phi$) constitute the zero-mass Goldstone bosons. The gradients of the latter fields $t^\alpha\nabla_\mu\Phi$ become part of the gauge-field components A_μ^α which then acquires a mass in the Higgs phenomenon.

Let us first study the simple case $\beta=0$, which is the most general case when there are no third-degree invariants[16] on $\mathcal{E}$.

Let Φ_0 be a vector of $\mathcal{E}$ which minimizes V; we write

$$\Phi_0=c\phi \text{ with } (\phi,\phi)=1,\quad c>0. \tag{A22}$$

Then Eq. (A18) yields

$$c^2=\frac{\mu^2}{\lambda+\alpha\chi_0} \text{ with } \chi_0=\omega(\phi), \tag{A23}$$

which is consistent with (A17). We also obtain from (A16) the value of the minimum for the potential $V_{\min}$,

$$V_{\min}=-\frac{\mu^4}{4}\frac{1}{\lambda+\alpha\chi_0}. \tag{A24}$$

This shows that

$$\text{if } \alpha>0,\ \chi_0=\chi_m;\ \text{if } \alpha<0,\ \chi_0=\chi_M. \tag{A25}$$

The case $\alpha=0$ [which occurs when G is the orthogonal group $O(n)$ and $\mathcal{E}$ is the n-dimensional space] is well known. In any case, the mass square in the radial direction is

$$\left(\phi,\frac{d^2V}{d\phi^2}\phi\right)=3(\lambda+\alpha\chi_0)c^2-\mu^2=2\mu^2. \tag{A26}$$

Finally, we need to compute the polynomial $W(\xi)$, used in Eq. (5.5′) of the text, which is defined by

$$\frac{dV}{d\phi}(\xi c\phi)=W(\xi)\mu^2\phi. \tag{A27}$$

We find from (A18) and (A23) that

$$W(\xi)=\xi^2-1. \tag{A27′}$$

Returning to the general case (A16) when both α and β are nonvanishing, we assume that the minimum is attained when the three vectors $t(\Phi)$, $b(\Phi)$, and Φ are collinear. More generally it might happen that $t(\Phi,\Phi,\Phi)$, $b(\Phi)$, and Φ form a two-plane and then there are two independent conditions for the minimum. We shall not consider the latter case here. The lower bound of V is therefore reached on the orbit of $\Phi_0=c\phi$ with

$$(\phi,\phi)=1,\quad b(\phi)=\phi_v\phi=\eta\phi,$$
$$t(\phi)=\chi_0\phi \text{ with } \chi_m\leq\chi_0<\chi_M; \tag{A28}$$

the extremum condition (A18) then gives

$$(\lambda+\alpha\chi_0)c^2+\beta\mu\eta c-\mu^2=0.$$

As we expect for a fourth-degree polynomial with a maximum at the origin, there are two solutions for c; that which corresponds to the lowest radial minimum and therefore to the lowest bound for V is (with the convention $c>0$)

$$c=\frac{\mu}{(\lambda+\alpha\chi)^{1/2}}[(1+\rho^2)^{1/2}+\rho],$$
$$\rho=\frac{-\beta\eta}{2(\lambda+\alpha\chi_0)^{1/2}}\geq 0. \tag{A29}$$

Then $W(\xi)$, as defined in (A27) and used in Eq. (5.5′), can be written as

$$W(\xi)=(\xi-1)(\sigma^2\xi+1) \text{ with } \sigma=(1+\rho^2)^{1/2}+\rho\geq 1. \tag{A30}$$

Note that

$$c=\frac{\mu}{(\lambda+\alpha\chi_0)^{1/2}}\sigma \tag{A31}$$

gives a generalization of (A23) when $\beta\neq 0$. The generalizations of (A24) and (A26) are easily found to be

$$V_{\min} = -\frac{\mu^4}{\lambda + \alpha\chi_0}\,\frac{4(1+\rho\sigma)^2 - 1}{3} = -\frac{\mu^4}{\lambda + \alpha\chi_0}\,\frac{\sigma^2(\sigma^2+2)}{3} \tag{A32}$$

and the "radial" mass squared

$$\left(\phi, \left.\frac{d^2V}{d\phi^2}\right|_{\phi_0} \phi\right) = 2\mu^2(1+\rho\sigma) = \mu^2(1+\sigma^2), \tag{A33}$$

where we have used the identity $\sigma^2 = 1 + 2\rho\sigma$.

---

*Work supported in part by the U. S. Energy Research and Development Administration under Contract No. EY-76-S-02-3533.

†On leave of absence from Dublin Institute for Advanced Study, Dublin 4, Ireland.

[1]See, for instance, for the general approach, E. S. Abers and B. W. Lee, Phys. Rep. 9C, 1 (1973); also S. Weinberg, Rev. Mod. Phys. 46, 255 (1974).

[2]For an excellent review, see S. Coleman, in lectures at the 1975 International School of Subnuclear Physics, "Ettore Majorana" [Harvard report, 1975 (unpublished)].

[3]G. 't Hooft, Nucl. Phys. B79, 276 (1974).

[4]A. M. Polyakov, Zh. Eksp. Teor. Fiz. Pis'ma Red. 20, 430 (1974) [JETP—Lett. 20, 194 (1974)].

[5]P. A. M. Dirac, Proc. R. Soc. London A133, 60 (1934); Phys. Rev. 74, 817 (1948).

[6]Here we mean solutions throughout all three-space. Generalizations of Refs. 3 and 4 using topological considerations on the sphere at infinity have been considered elsewhere. A partial list, containing other references, is as follows: E. Corrigan, D. B. Fairlie, J. Nuyts, and D. I. Olive, Nucl. Phys. B106, 475 (1976); E. Corrigan and D. I. Olive, *ibid.* B110, 237 (1976); A. Goldhaber and D. Wilkinson, *ibid.* B114, 317 (1976); F. Englert and P. Windy, Univ. Libre, Bruxelles report, 1976 (unpublished); A. Chakrabarti, Ann. Inst. Henri Poincaré 13, 235 (1975); R. Shankar, Phys. Rev. D 14. 1107 (1976).

[7]T. T. Wu and C. N. Yang, in *Properties of Matter Under Unusual Conditions,* edited by H. Mark and S. Fernbach (Interscience, New York, 1969), p. 349.

[8]E. J. Weinberg and A. H. Guth, Phys. Rev. D 14, 1660 (1976). See also L. O'Raifeartaigh, Lett. Nuovo Cimento 18, 205 (1977).

[9]Note that the summation is over μ and even J only. The symbol

$$\left(\begin{pmatrix} L & L & J \\ r & s & \mu \end{pmatrix}\right)$$

is not defined for odd J. For odd J, our equations turn out to be identities $0=0$.

[10]Yu. S. Tyupkin, V. A. Fateev, and A. S. Shvarts, Teor. Mat. Fiz. 26, 397 (1975) [Theor. Math. Phys. 26, 270 (1976)].

[11]This is because $f^\alpha = -\kappa\hat{r}^\alpha$ is in the little group of $\phi(\omega)$.

[12]B. Julia and A. Zee, Phys. Rev. D 11, 2227 (1975). See also M. K. Prasad and C. S. Sommerfield, Phys. Rev. Lett. 35, 760 (1975).

[13]If there were nonzero G-invariant vectors on $\mathcal{E}$, they could be minima of V and there would be no symmetry breaking.

[14]The vector Φ itself is a scalar field which depends on the space-time point x. However, our problem is to vary V as a function of Φ and every minimum of $V(\Phi)$ is a minimum of $V(\Phi(x))$ stable against variations of $\Phi(x)$ at each x.

[15]The vector bilinear in (Φ_1, Φ_2) is also denoted by $\Phi_1 \vee \Phi_2 = \Phi_2 \vee \Phi_1$. The corresponding operator linear on Φ is denoted by $\mathfrak{D}(\Phi)$. See L. Michel and L. Radicati, Ann. Inst. Henri Poincaré 18, 85 (1973), for a systematic study of these algebras and their relation, in the case of SU(3), to Gell-Mann's d algebra [M. Gell-Mann, Phys. Rev. 125, 1097 (1962)]; also L. C. Biedenharn, J. Math. Phys. 4, 436 (1963).

[16]This is, for instance, the case for compact Lie groups which contain in their representations $-I$ on $\mathcal{E}$ and for all simple compact G which are not SU(n) when $\mathcal{E}$ is the adjoint representation. This is also always the case in another physical situation, namely, in the Landau theory of second-order phase transitions.

Multigenerational Flavor-Color-Hypercolor Unification

Aharon Davidson[(a)] and Kameshwar C. Wali
Physics Department, Syracuse University, Syracuse, New York 13210

and

Philip D. Mannheim
Department of Physics, University of Connecticut, Storrs, Connecticut 06268
(Received 7 July 1980)

A vertical-horizontal symmetric $[SO(10)]_V \otimes [SO(10)]_H$ flavor unifying scheme characterized by a natural generation structure is presented. No-go theorems associated with SO(10) embeddings are trivially avoided. All the fermions are assigned to a single irreducible representation, and the intergenerational structure is dictated by the presence of a hypercolor subgroup for which SU(4) is the only tenable candidate. The theory predicts exactly four generations and suggests a color (Pati-Salam) -hypercolor symmetry at medium energies.

PACS numbers: 11.30.Ly, 12.20.Hx, 14.60.-z, 14.80.Dq

The old e-μ puzzle, namely the mystery associated with the question: "Why does Nature repeat herself?", has been significantly enlarged in recent years to encompass all the presently known families of quarks and leptons. The proliferation of fermion representations, also known as the superfluous replication, is unacceptable in a fundamental theory, but as yet does not have any satisfactory explanation. Currently popular unifying schemes such as[1] SU(5) and[2] SO(10) are actually single-generation theories, and hence are not capable of dealing with such a problem. Nonetheless, the relative successes of these theories, and especially the fact that they offer a striking correlation between the structure of the electro/nuclear interactions and the content of the fundamental sets of fermions, strongly indicate that the single-generation group factor does play a crucial role in unification.

The above considerations have led recently to an investigation[3] of flavor-chiral grand unifying theories based on semisimple gauge groups. Such theories have the following characteristics: (i) A discrete vertical-horizontal symmetry ensures a single gauge coupling constant, (ii) the single-generation group is *not* an accidental consequence of the symmetry breaking process, (iii) a pure generation structure exists at all energy regimes, (iv) the intergenerational structure (including the total number of generations) is dictated by the underlying single-generation group, and (v) the fact that each generation has to have a vanishing anomaly finds a simple explanation.[4] As a prototype of such theories, we have already discussed[3] the minimal $[SU(5)]_V \otimes [SU(5)]_H$ flavor unifying scheme. This model contains attractive properties which are absent from models[5] based on simple SU($N > 5$) groups, but still suffers from some conventional diseases: The anomaly cancellation must be imposed; the fermions do not furnish a single representation[6]; a complicated Higgs system is needed to generate the fermionic mass spectrum; and the scheme is too small to incorporate the hypercolor[7] alternative. All these typical problems are automatically resolved within an $[SO(10)]_V \otimes [SO(10)]_H$ theory which we hereby present. It is a very tight scheme where the intergenerational structure is fully controlled by the size of an associated hypercolor group, with SU(4) being the only tenable candidate for the latter. The theory predicts four generations which interact horizontally via a "nearest-neighbor interaction" as in Harari[8] and Fritzsch.[8] Moreover, we provide simple solutions to the two no-go theorems[9,10] associated with the SO(10) embedding attempts.

As far as the single-generation group $[SO(10)]_V$ is concerned, the only representations which do not introduce bizzare fermions[11] under $[SU(3)]_C$ are the spinorial $\underline{16}$ and the vectorial $\underline{10}$. Thus, within the framework of the flavor-chiral and V $\leftrightarrow$ H symmetric $[SO(10)]_V \otimes [SO(10)]_H$ theory, the set ψ_L of *all* fundamental left-handed fermions is given by either

$$\psi_L{}^{(1)} = (\underline{16}; \underline{16}) \tag{1}$$

$$\psi_L{}^{(2)} = [(\underline{16}; \underline{10}) + (\underline{10}; \underline{16})]. \tag{2}$$

Note that $\psi_L{}^{(2)}$ is irreducible because of the discrete V-H symmetry, so that the $(\underline{16}; \underline{10})$ cannot appear in the theory without its $(\underline{10}; \underline{16})$ partner and vice versa. This is why we do not consider the complex $(\underline{16}; \underline{16}^*)$ which is necessarily accompanied by the $(\underline{16}^*;\underline{16})$ resulting in a real representation. Also we note that single-generation representations such as $[(\underline{16};\underline{1})+(\underline{1};\underline{16})]$ are of no interest at a multigeneration level. At this stage, several remarks concerning the structure of the fermionic representation are in order:

(1) Simple-minded SO($N>10$) embeddings of SO(10) face the severe difficulty[10] that the SO(10) spinors always appear with an equal number of "conjugate" spinors, leaving us (after the implementation of Georgi's mechanism[12]) with no ordinary-mass generations at all. To overcome this problem, Gell-Mann, Ramond, and Slansky[13] have demonstrated the need for hypercolor, showing how the hypercolor-free fermions alone possess the desirable SO(10) structure in a class of SO($10+4k$) models. We therefore emphasize that such a problem is trivially avoided in our $[SO(10)]_V\otimes[SO(10)]_H$ theory, simply because $[SO(10)]_V$ is *not* a subgroup of some higher orthogonal group. The hypercolor interactions on the other hand do play a major role in the present theory as well.

(2) The fundamental $\underline{10}$ is the *smallest* representation of SO(10) but is not usually considered since by itself it is a real representation. Within $[SO(10)]_V$ it corresponds to an extraordinary family recently discussed by Georgi and Glashow[14] in conjunction with topless (t-free) models which incorporate a generalized Glashow-Iliopoulos-Maiani (GIM) mechanism. As we shall see later, only $\psi_L{}^{(2)}$ leads to a realistic low-energy model with or without extraordinary families, so that the $\underline{10}$ plays a key role in our theory.

(3) The $[SO(10)]_V\otimes[SO(10)]_H$ theory is asymptotically free. The overall renormalization-group β function is

$$\beta=-(g^3/42\pi^2)(22-D-\tfrac{1}{2}D')+O(g^5), \tag{3}$$

where the fermionic contributions[15] D and D' are such that

$$D=16,\quad D'=0 \text{ for } \psi_L{}^{(1)}; \tag{4}$$

$$D=10,\quad D'=16 \text{ for } \psi_L{}^{(2)}. \tag{5}$$

The *negative* β function (for both cases) shows how Frampton's no-go theorem[9] which forbids simple grand unification involving superstrong interactions[16] does not hold for the semisimple extensions.

Next we turn our attention to the structure of the symmetry breaking. $[SO(10)]_V$ is expected to be broken primarily either to SU(5)⊗U(1) according to Georgi and Glashow[1] or to the Pati-Salam[17] subgroup SU(4)⊗SU(2)⊗SU(2), and then to $[SU(3)]_C\otimes[U(1)]_{em}$ via the standard Weinberg-Salam stage. Simultaneously, $[SO(10)]_H$ gets broken too, with its surviving unbroken subgroup being interpreted as the hypercolor group. The main question is what can this hypercolor group be? Following standard wisdom, it must be bigger than SU(3) [if it is semisimple, each simple group factor has to be bigger than SU(3)] in order that the hypercolor interaction be superstrong. On the other hand, it is obviously smaller than SO(10). In fact, it must be even smaller than SO(8) to allow for more than two ordinary generations. The above arguments leave SO(5), SO(6) ≈ SU(4), and SO(7) as the only candidates for the hypercolor group. SU(4) emerges thus as the only *unitary* candidate, and also as the only candidate which admits complex hypercolor fermionic representations. The choice of SU(4) is also favored by the attractive possibility that the V↔H discrete symmetry survives the primary stage of symmetry breaking. This way, the SU(4)-hypercolor becomes the horizontal analog of the vertical Pati-Salam SU(4).

A priori, consistent with the emergence of an SU(4) hypercolor group, there exist two independent alternatives concerning the horizontal assignments of the fundamental fermions. These alternatives correspond to the two different ways of picking up an SU(4) subgroup (rather than any other unitary subgroup) of SO($2N$). Physically, as we shall see, they give rise to models with or without the t quark, respectively, and are therefore both of current interest. The final choice will hopefully be dictated by the dynamical symmetry breaking in conformity with future experiments. The two alternatives are the following:

Case I: $[SO(10)]_H\to SO(6)\otimes SO(4)$.—This alternative, which uses the fact that the algebra of SO(6) is equivalent to that of SU(4), fixes SO(4) ≈ SU(2)⊗SU(2) as the horizontal component which distinguishes the "repeated" fermion structures. The relevant SO(10) representations decompose according to

$$\underline{16}=(\underline{4},\underline{2})+(\underline{4}^*,\underline{2}'), \tag{6}$$

$$\underline{10}=(\underline{6},\underline{1})+(\underline{1},\underline{4}). \tag{7}$$

Thus the spinorial 16 does not contain any hypercolor singlet at all, while the vectorial $\underline{10}$ has exactly four such objects. Consequently, $\psi_L{}^{(1)}$ is not capable of providing a low-energy generation structure, and hence it can be ruled out. $\psi_L{}^{(2)}$, on the other hand, offers *four* ordinary generations {an ordinary generation is defined as a hypercolor-free $\underline{16}$ of $[SO(10)_V]$} originating *only* from its $(\underline{16};\underline{10})$ part. Notice especially that, apart from the corresponding mass-scale $\gtrsim (1/\alpha)m(W^{\pm})$,[18] we do not have any additional left-over degrees of freedom with respect to the horizontal interactions. The four predicted generations turn out to interact horizontally via the fundamental $\underline{4}$ of SO(4), which means, as stated before, a "nearest-neighbor interaction".[8]

Case II: $SO(10)_H \to SO(8)\otimes SO(2)$.—As discussed earlier, SO(8) is too big to be considered as a candidate for the hypercolor group. However, as an SO(2*N*) group, its representations can be expressed in an SU(*N*) basis,[19] e.g., the $\underline{16}$ of $[SO(10)]_V$ decomposes into the familiar $\underline{1}+\underline{5}^*+\underline{10}$ combination under the Georgi-Glashow SU(5). The two inequivalent spinors $\underline{8}$ and $\underline{8}'$ of SO(8) have the SU(4) contents $(\underline{4}+\underline{4}^*)$ and $(\underline{1}+\underline{6}+\underline{1})$, respectively, while the vectorial $\underline{8}''$ decomposes into $(\underline{4}+\underline{4}^*)$. Now, decomposing according to SO(8) $\otimes$SO(2) and SU(4)$\otimes$SO(2), we have

$$\underline{16}=(\underline{8},\underline{1})+(\underline{8}',\underline{1})=(\underline{4}+\underline{4}^*,\underline{1})+(\underline{1}+\underline{6}+\underline{1},\underline{1}), \quad (8)$$

$$\underline{10}=(\underline{8}'',\underline{1})+(\underline{1},\underline{2})=(\underline{4}+\underline{4}^*,\underline{1})+(\underline{1},\underline{2}). \quad (9)$$

Altogether, the spinor and the vector representations of $[SO(10)]_H$ contain this time two SU(4)-hypercolor singlets each. In other words, $\psi_L^{(1)}$ has only two ordinary generations and is ruled out once again on experimental grounds, while $\psi_L^{(2)}$ contains an addition to two ordinary generations and also to two extraordinary ones. Here we refer to an extraordinary generation such as the hypercolor-free $\underline{10}$ of $[SO(10)]_V$. Such a family is just the Georgi-Glashow $\underline{5}+\underline{5}^*$ combination[14] that contains no $|Q|=\frac{2}{3}$ fermions. Note that the residual horizontal symmetry in this case[20] is U(1)$\otimes$U′(1) which is *not* a subgroup of SO(4).

As far as the fermionic content of the theory is concerned, both symmetry-breaking alternatives require $\psi_L^{(2)}$ to be the correct fermionic set. Only this choice allows us to make contact with the fact that the *b* quark does exist. However, it is still not clear which of the two symmetry-breaking alternatives is the physical one since firstly the *t* quark has not yet been found and secondly the decay modes of the *b* quark have yet to be established. For the moment, however, we note that case I is favored because of the following theoretical reasons: (i) The four ordinary generations predicted in case I are absolutely equivalent from the $[SO(10)]_V$ point of view, and thus automatically obey the GIM mechanism. This may not be true for case II where the four $|Q|=\frac{1}{3}$ quarks do not have the same Weinberg-Salam assignments. (ii) The possibility of having an effective V$\leftrightarrow$H symmetric $[SU(4)\otimes SU(2)\otimes SU(2)]_V \otimes[SU(4)\otimes SU(2)\otimes SU(2)]_H$ theory at medium energies is a very attractive idea. SU(4)-hypercolor becomes the horizontal analog of the vertical Pati-Salam SU(4), while the various flavors are listed (within a single generation) and distinguished (i.e., horizontally) by identical SU(2)$\otimes$SU(2) group factors! In fact, recalling the V$\leftrightarrow$H symmetry, the emergence of SU(4)-hypercolor on the horizontal side may indicate that $[SO(10)]_V$ descends into SU(3)$\otimes$SU(2)$\otimes$U(1) via the Pati-Salam stage rather than via the Georgi-Glashow SU(5) alternative. Only case I can support such an idea.

We conclude by observing that a grand unifying theory based on the semisimple group structure $[SO(10)]_V\otimes[SO(10)]_H$ has many interesting and desirable features. With the single-generation structure still being described by the usual SO(10) theory, a natural multigeneration structure with exactly four generations emerges. The option of having or not having the *t* quark exists, to be decided by further study of the symmetry-breaking mechanism and by experiments. The anomaly cancellation is automatic and asymptotic freedom is guaranteed for the chosen irreducible fermionic representation. The many difficulties discussed in the literature, such as the "conjugate" spinor problem, in embedding SO(10) in a bigger Lie group are easily avoided. The theory suggests symmetry breaking with SU(4) as the only possible subgroup for hypercolor. This in turn points to an interesting symmetry between the color of Pati and Salam and hypercolor.

This work was supported in part by the U. S. Department of Energy under Contracts No. DE-AS02-79ER03533 and No. DE-AC02-79ER10336.A.

---

[a]Present address: Department of Nuclear Physics, The Weizman Institute of Science, Rehovot, Israel.

[1]H. Georgi and S. L. Glashow, Phys. Rev. Lett. 32, 438 (1974).

[2]H. Georgi, in *Particles and Fields* (American Institute of Physics, New York, 1975), p. 575; H. Fritzsch and P. Minkowski, Ann. Phys. (N.Y.) 93, 193 (1975).

[3]A. Davidson, Phys. Lett. 90B, 87 (1980); A. Davidson, Phys. Lett. 93B, 183 (1980); K. C. Wali, in Proceedings of the First Workshop on Grand Unification, New Hampshire, 1980 (unpublished). Other semisim-

ple grand-unifying alternatives are the following: $[SU(4)]^4$, J. C. Pati and A. Salam, Phys. Lett. 58B, 333 (1975); $[SU(8)]_L \otimes [SU(8)]_R$, N. G. Deshpande and P. D. Mannheim, Phys. Lett. 94B, 359 (1980). In contrast with our scheme, however, these are left-right-symmetric theories.

[4]For the sake of renormalizability, only the overall anomaly has to vanish.

[5]SU(7), P. H. Frampton, Phys. Lett. 88B, 298 (1979); SU(8), J. Chakrabarti, M. Popovic, and R. N. Mohapatra, Phys. Rev. D 21, 3212 (1980); SU(9), P. H. Frampton and S. Nandi, Phys. Rev. Lett. 43, 1460 (1979); SU(11), H. Georgi, Nucl. Phys. B156, 126 (1979).

[6]To avoid this particular difficulty within an $[SU(N)]_V \otimes [SU(N)]_H$ theory, one must choose $N = 7$. See A. Davidson and K. C. Wali, in Proceedings of the Twentieth International Conference on High Energy Physics, Madison, Wisconsin, 1980 (to be published).

[7]L. Susskind, Phys. Rev. D 20, 2619 (1979).

[8]H. Harari, SLAC Report No. SLAC-PUB-2363 (unpublished); H. Fritzsch, CERN Report No. TH-2672-CERN (unpublished). By "nearest-neighbor interaction" we mean that the current $\bar{\psi}_i \gamma \psi_j$, with i and j being generation indices, couples to a horizontal gauge boson only if $|i-j| \leq 1$.

[9]P. H. Frampton, Phys. Rev. Lett. 43, 1912 (1979), and 44, 299(E) (1980).

[10]See, for example, S. L. Glashow, Harvard University Report No. HUTP-79/A059, 1979 (unpublished); F. Wilczek, in *Proceedings of the Ninth International Symposium on Lepton and Photon Interactions at High Energies, Batavia, Illinois, 1979*, edited by T. B. W. Kirk and H. D. I. Abarbanel (Fermilab, Batavia, Ill., 1979).

[11]M. Gell-Mann, P. Ramond, and R. Slansky, Rev. Mod. Phys. 50, 721 (1978).

[12]H. Georgi, Nucl. Phys. B156, 126 (1979).

[13]M. Gell-Mann, P. Ramond, and R. Slansky, in *Supergravity*, edited by P. Van Nieuwenhuizen and D. Z. Freedman (North-Holland, Amsterdam, 1979).

[14]H. Georgi and S. L. Glashow, Nucl. Phys. B167, 173 (1980).

[15]To calculate those contributions, see, for example, S. Okubo, J. Math. Phys. 18, 2382 (1977).

[16]The SU(7) model of E. Farhi and L. Susskind, Phys. Rev. D 20, 3404 (1979), involves an SU(2)-hypercolor group.

[17]J. C. Pati and A. Salam, Phys. Rev. D 10, 275 (1974).

[18]The lower bound arises from the K_L-K_S mass difference.

[19]See, for example, R. N. Mohapatra and B. Sakita, Phys. Rev. D 21, 1062 (1980).

[20]The idea of using the horizontal symmetry $[U(1)]^N$ for 2^N generations has been discussed by A. Davidson and K. C. Wali, Phys. Rev. Lett. 43, 92 (1979).

Hypercolor, extended hypercolor, and the generation problem

Aharon Davidson
Department of Nuclear Physics, Weizmann Institute of Science, Rehovot, Israel

Philip D. Mannheim
Department of Physics, University of Connecticut, Storrs, Connecticut 06268

Kameshwar C. Wali
Physics Department, Syracuse University, Syracuse, New York 13210
(Received 20 August 1981)

We analyze some of the detailed consequences of the recently proposed grand unified theory based on $SO(10)_V \times SO(10)_H$ subject to a discrete $V \leftrightarrow H$ symmetry. The model attempts to unify flavor, color, and hypercolor, and provide a multigenerational grand unified theory. It predicts SU(4) as the unique unitary group for hypercolor. With the fundamental fermions belonging to the $(\underline{16},\underline{10}) \oplus (\underline{10},\underline{16})$ irreducible representation of $SO(10)_V \times SO(10)_H$, there are exactly four generations of ordinary fermions (hypercolor singlets). The dynamical symmetry breaking which gives masses to the vector bosons of the standard electroweak theory is accomplished through the condensates of a single generation of hyperfermions belonging to the sextet of SU(4). We show that the Weinberg relation $M_W = M_Z \cos\theta$ will be satisfied provided the hypercolor dynamics satisfies certain constraints. Present in the model are the much desired extended-hypercolor gauge bosons whose radiative transitions between the ordinary fermions and the hyperfermions give rise to both masses and mixing of the ordinary fermions; hence the generalized Cabibbo angles are in principle calculable. We also analyze the pseudo-Goldstone bosons and the rare decay modes of the K and D mesons. With the hyperfermions in the TeV range and the extended gauge bosons in the 1000-TeV range, we show that there is no conflict in the model with any known experimental bounds on such decays. We establish a hypercolor—Pati-Salam-color symmetry at medium energies.

I. INTRODUCTION

In this paper we combine two currently popular ideas in particle theory, namely grand unification of the fundamental strong, electromagnetic, and weak interactions, and the possible existence of a new superstrong hypercolor gauge interaction. We shall be particularly interested in hypercolor not only as a possible origin for a superstrong force which could lead to the dynamical symmetry breaking of unified theories, but also as an extra group-theoretical degree of freedom which will provide us with a new group-theoretical structure to both label and relate the various generations of fundamental fermions. Thus we shall seek to enlarge grand unifying theories to include hypercolor interactions as well.

With regard to grand unification we remark that while there is as of yet no theoretical unanimity or experimental preference regarding the choice of grand unifying group there is a lot of support in the current literature for grand unification based on the group SU(5).[1] This support derives from the fact that SU(5) involves the smallest possible number of additional vector bosons beyond those of $SU(3)_C \times SU(2)_L \times U(1)$, thus making SU(5) the simplest choice. Its immediate shortcoming is that it puts the fundamental fermions into many differing $\underline{5}^*$ and $\underline{10}$ representations. It is possible to partially remedy this by embedding the theory[2] in SO(10) provided some so far unobserved right-handed neutrinos which are to be singlets under SU(5) are introduced. The $\underline{16}$ representation of SO(10) then nicely accommodates 16 two-component fermions which transform as $\underline{5}^* \oplus \underline{10} \oplus \underline{1}$ under SU(5). Each set of 16 fermions of this type is known as a *generation* or *family*. Thus SO(10) puts all the members of a given generation into a single irreducible representation. However it still treats the different generations as separate irreducible representations under the group. Consequently SO(10), like its SU(5) subgroup, is strictly a single-generation grand unifying theory which can thus not adequately address questions such as how many generations there are or in what way they mix with each other. With regard to this generation problem two points of view can be adopted. One is to say that the very fact that SO(10) [or SU(5)] is a single-generation theory could be an indication that it is simply the wrong group for grand unification altogether; while the other is to take SO(10) as a good starting point with the generation structure then hopefully emerging after SO(10) is itself embedded into a yet larger group again. In the absence of an equally popular alternative to SO(10) we shall thus only explore the latter alternative in this paper, though we believe that the former alternative merits further study.

With regard to hypercolor we recall that the primary motivation for hypercolor was to understand the origin for a force which could be strong enough to produce a typical weak-interaction mass scale of the order of 100 GeV.[3,4] It was argued that hypercolor might act as a scaled-up version of quantum chromodynamics (QCD) which is itself thought of as being responsible for producing a scale of the order of 100 MeV for the usual strong interactions by breaking the strong-interaction chiral symmetry dynamically. Using this as-yet-poorly-understood analogy hypercolor would then be expected to break the weak-interaction flavor symmetry dynamically and provide a mass scale for the intermediate vector bosons of the $SU(2)_L \times U(1)$ theory. Indeed it was noted that the Weinberg mixing formula $M_W = M_Z \cos\theta$ would follow naturally (i.e., group theoretically) if the hypercolor force caused hypercolor-carrying fermions (hyperfermions) to form fermion bilinear condensates which acquired nonvanishing vacuum expectation values.

While these features of the hypercolor scheme are very attractive, the scheme nonetheless fails in one significant respect. In the conventional Weinberg-Salam theory a fundamental Higgs field gives masses to both the intermediate vector bosons and the usual fermions (now the hypersinglet fermions of the hypercolor scheme). In the dynamical hypercolor scheme the hyperfermion condensates which are to replace the fundamental Higgs fields give 100-GeV masses to the intermediate vector bosons and TeV masses to the hyperfermions, leaving the usual fermions massless. While this already constitutes a considerable though acceptable departure from the structure of the conventional Weinberg-Salam theory, it also leaves open the question of where do the usual fermions in fact get their masses from. To this end yet another even stronger interaction has been proposed, extended hypercolor,[5] which would then serve to mediate processes in which the now massive hyperfermions could then contribute radiatively to the self-energies of the usual fermions. The hypercolor theory as currently formulated is thus still somewhat unsatisfying in that it appeals to a still poorly understood dynamics, gives no indication of what the hypercolor group actually is [save that it be larger than SU(3) so that renormalization-group effects would make it stronger than QCD in the TeV region], and needs to be augmented by an additional extended hypercolor interaction. Thus further guiding principles are needed in order to make hypercolor a more restricted and hence more predictive theory. In this paper we shall provide such principles by using the group structure of hypercolor and extended hypercolor in order to solve the generation problem. We believe that this now puts the whole hypercolor idea on a more secure footing.

Our aim is thus to enlarge the SO(10) grand unifying theory to include hypercolor interactions as well. The immediate possibility of course would be to embed SO(10) into $SO(10+4n)$ groups. However this choice has been shown to be afflicted with serious phenomenological difficulties since it has to introduce unobserved conjugate $\underline{16}^*$ spinor representations and can only admit of two separate generations of usual fermions.[6] Motivated by our recent work in semisimple grand unification[7] we have instead proposed to embed SO(10) into the semisimple $SO(10)_V \times SO(10)_H$ group.[8] Here $SO(10)_V$ is our initial "vertical" single generation SO(10) while $SO(10)_H$ is a new "horizontal" SO(10) symmetry which as we will see will connect the various generations. Further we will identify $SO(10)_H$ as the complete hypercolor group, so that $SO(10)_H$ will contain both hypercolor-carrying and flavor-carrying (generation-changing) gauge bosons. Thus one single group structure restricts the hypercolor- and the flavor-changing interactions simultaneously. There is of course initially some freedom in specifying the horizontal group factor. However the simple imposition of a discrete vertical-horizontal symmetry, which we make, then unambiguously forces the horizontal group to be SO(10) just like its vertical counterpart, while also fixing the otherwise unknown horizontal coupling g_H to be equal to the vertical g_V at the grand unification mass scale. Moreover, as we will see below, there will also be vestiges of the discrete symmetry in the low-energy structure of our theory following the symmetry breaking. This will then lead to an attractive approximate low-energy color-hypercolor symmetry.

While we have now fixed the horizontal symmetry, our model is not yet completely specified since we still have to classify the fermions according to an appropriate irreducible representation of $SO(10)_V \times SO(10)_H$ and still have to identify which subgroup of $SO(10)_H$ is to serve as the hypercolor group. Whatever generators of $SO(10)_H$ are then left over will serve to count generations. Thus because of the embedding into an extended hypercolor $SO(10)_H$ group the size of the hypercolor group and the number of independent generations of fermions are simultaneously fixed.

In classifying the fermions the obvious embeddings of the $\underline{16}$ of $SO(10)_V$ would be in $(\underline{16},\underline{16}^*)$ or $(\underline{16},\underline{16})$ under $SO(10)_V \times SO(10)_H$. Because of the discrete symmetry however the $(\underline{16},\underline{16}^*)$ would have to be accompanied by a $(\underline{16}^*,\underline{16})$ to thus give a real representation which is not appropriate for fermions. The $(\underline{16},\underline{16})$ would be acceptable on this score but was found not to be acceptable phenomenologically as it could only lead to two ordinary generations of usual fermions.[8] Because of this we instead proposed to classify the fermions according to the $(\underline{16},\underline{10})$ representation which would then be accompanied by a $(\underline{10},\underline{16})$ representation to maintain irreducibility under the discrete symmetry. The $\underline{10}$ would not ordinarily be considered in grand unified models since by itself it would be real. However when combined with the $\underline{16}$ it leads to a complex $(\underline{16},\underline{10}) \oplus (\underline{10},\underline{16})$ representation which now is appropriate for fermions. Given this classification of fermions it was found that there was only one choice for the hypercolor subgroup which could lead to both a unitary hypercolor group and to more than two ordinary generations of fermions. Specifically, we decomposed $SO(10)_H$ according to $SO(6)_H \times SO(4)_H$ to obtain

$$\begin{aligned} \underline{16} &= (\underline{4},\underline{2}) + (\underline{4}^*,\underline{2}') , \\ \underline{10} &= (\underline{6},\underline{1}) + (\underline{1},\underline{4}) \end{aligned} \tag{1.1}$$

with the $(\underline{16},\underline{10}) \oplus (\underline{10},\underline{16})$ then containing altogether four generations of fermions [i.e., fermions which transform as the $\underline{16}$ under $SO(10)_V$] which were $SO(6)_H$ singlets. Thus identifying $SO(6)_H$ as the hypercolor group enables us to obtain the previously *ad hoc* choice of SU(4) uniquely as the hypercolor symmetry while yielding at the same time exactly four generations of usual fermions. As we noted in Ref. 8, the emergence of four generations occurred only because we identified SU(4) with the $SO(6)_H$ subgroup of $SO(10)_H$ using the SU(4), SO(6) homomorphy. Had we instead identified SU(4) through the $SO(8)_H$ subgroup of $SO(10)_H$ [using the connection between SU(N) and SO($2N$) groups] we would only have obtained two generations. Thus in our model the hypercolor group is special because it is both a unitary and an orthogonal group, a situation which only occurs for SU(4). It is this feature of our model which makes SU(4) special and not merely the fact that because SU(4) is larger than SU(3) it becomes strong at a higher energy than QCD.

Having chosen to identify $SO(6)_H$ as the hypercolor subgroup and $SO(4)_H$ as the generation-counting subgroup of $SO(10)_H$, we can decompose the adjoint representation of $SO(10)_H$ according to $SO(6)_H \times SO(4)_H$ to obtain

$$\underline{45} = (\underline{15},\underline{1}) + (\underline{1},\underline{6}) + (\underline{6},\underline{4}) \; . \tag{1.2}$$

The $(\underline{15},\underline{1})$ and $(\underline{1},\underline{6})$ gauge bosons are the generators of $SO(6)_H$ and $SO(4)_H$, respectively, while the additional $(\underline{6},\underline{4})$ gauge bosons carry both hypercolor and flavor. These latter gauge bosons couple the hyperfermions to the usual fermions, and thus enable the usual fermions to acquire radiative masses.[9] Our $SO(10)_V \times SO(10)_H$ model thus automatically contains the previously *ad hoc* extended-hypercolor gauge bosons of Ref. 5. Moreover, since the different generations of fermions all belong to one irreducible representation of $SO(10)_V \times SO(10)_H$, the spontaneous breakdown of the model leads not merely to fermion masses but also to fermion mass mixing. Thus we naturally obtain Cabibbo mixing in our model[9] by taking advantage of the correlation between hypercolor and the generation structure which our model possesses. Hence we exploit the group-theoretical structure of hypercolor in a nontrivial manner to make the hypercolor concept highly predictive.

In this present work we shall report on all the various interesting aspects of our theory, to complete our analysis of the central features of the model. This paper is organized as follows. In Sec. II we present the general group-theoretical structure, and the classification of all the fermions and gauge bosons of the model. We identify the relevant hyperfermion condensates needed for the dynamical symmetry breaking. In Sec. III we show how the Weinberg mixing relation $M_W = M_Z \cos\theta$ arises in the model. Since the weak-interaction local flavor group is the chiral $SU(2)_L \times SU(2)_R \times U(1)$ group the analysis needed to obtain this mixing relation dynamically is substantially different from that given in Refs. 3 and 4 where only the Weinberg-Salam weak interactions are local. In Sec. IV we discuss the generation of the light-fermion self-energies and of Cabibbo mixing. In Sec. V we identify the Goldstone and pseudo-Goldstone bosons of our model, and in Sec. VI we show how the model successfully meets the constraints due to rare flavor-changing decay processes. In Sec. VII we catalog all the various symmetry-breaking scales, and in Sec. VIII we present our conclusions. Finally for the sake of completeness we discuss briefly, in an appendix, the other alternatives for the choice of hypercolor and generation-counting subgroups of $SO(10)_H$. Some of these alternatives are theoretically permissible, but not as attractive as the one we have discussed in detail in this paper.

II. STRUCTURE OF THE MODEL

Our model is an $SO(10)_V \times SO(10)_H$ gauge theory in which the fermions belong to a left-handed $(\underline{16},\underline{10}) \oplus (\underline{10},\underline{16})$ representation and the gauge bosons to the $(\underline{45},\underline{1}) \oplus (\underline{1},\underline{45})$ representation. It is useful to decompose each SO(10) according to $SU(4)\times SU(2)\times SU(2)$. Under this decomposition

$$\begin{aligned} \underline{10} &= (\underline{6},\underline{1},\underline{1}) + (\underline{1},\underline{2},\underline{2}) , \\ \underline{16} &= (\underline{4},\underline{2},\underline{1}) + (\underline{4}^*,\underline{1},\underline{2}) , \\ \underline{45} &= (\underline{15},\underline{1},\underline{1}) + (\underline{6},\underline{2},\underline{2}) + (\underline{1},\underline{3},\underline{1}) + (\underline{1},\underline{1},\underline{3}) \end{aligned} \tag{2.1}$$

with the complete classification of the fermions and gauge bosons being given in Table I. Physically on the vertical side we identify the two SU(2) groups as $SU(2)_L \times SU(2)_R$ and the SU(4) as the Pati-Salam vector color group which extends QCD to include the leptons as a fourth color.[10] Thus for the usual fermions of the first generation we identify

$$\begin{aligned} &\begin{bmatrix} \nu_e \\ u_R \\ u_G \\ u_B \end{bmatrix}_L , \begin{bmatrix} e \\ d_R \\ d_G \\ d_B \end{bmatrix}_L \in (\underline{4},\underline{2},\underline{1};\underline{1},\underline{2},\underline{2}) , \\ &\begin{bmatrix} \nu_e^C \\ u_R^C \\ u_G^C \\ u_B^C \end{bmatrix}_L , \begin{bmatrix} e^C \\ d_R^C \\ d_G^C \\ d_B^C \end{bmatrix}_L \in (\underline{4}^*,\underline{1},\underline{2};\underline{1},\underline{2},\underline{2}) , \end{aligned} \tag{2.2}$$

where R, G, and B denote the three QCD colors red, green, and blue. The 15 generators of SU(4) are then the 8 $SU(3)_C$ generators, 6 generators which mix quarks and leptons and one extra diagonal generator $Y = (B - L)/2$ with current

TABLE I. Classification of the fermions and gauge bosons.

	$[SU(4)\times SU(2)\times SU(2)]_V$			$\times$ $[SU(4)\times SU(2)\times SU(2)]_H$		
$(\underline{16},\underline{10})$	4	2	1	6	1	1
	4*	1	2	6	1	1
	4	2	1	1	2	2
	4*	1	2	1	2	2
$(\underline{10},\underline{16})$	6	1	1	4	2	1
	6	1	1	4*	1	2
	1	2	2	4	2	1
	1	2	2	4*	1	2
$(\underline{45},\underline{1})$	15	1	1	1	1	1
	6	2	2	1	1	1
	1	3	1	1	1	1
	1	1	3	1	1	1
$(\underline{1},\underline{45})$	1	1	1	15	1	1
	1	1	1	6	2	2
	1	1	1	1	3	1
	1	1	1	1	1	3

$$Y_\lambda = \sum_{i=R,G,B} \tfrac{1}{6}(\bar{u}_i\gamma_\lambda u_i + \bar{d}_i\gamma_\lambda d_i) - \tfrac{1}{2}(\bar{\nu}_e\gamma_\lambda\nu_e + \bar{e}\gamma_\lambda e)\ . \tag{2.3}$$

Denoting by T_L^3 and T_R^3 the third components of $SU(2)_L$ and $SU(2)_R$, respectively, we identify the electric-charge generator as

$$Q = T_L^3 + T_R^3 + \tfrac{1}{2}(B-L)\ , \tag{2.4}$$

so that

$$Y_{WS} = T_R^3 + \tfrac{1}{2}(B-L) \tag{2.5}$$

is the $U(1)_{WS}$ generator of the Weinberg-Salam group $SU(2)_L \times U(1)_{WS}$. With regard to the vertical gauge bosons we note also that the $(\underline{6},\underline{2},\underline{2};\underline{1},\underline{1},\underline{1})$ gauge bosons are the leptoquark bosons which lead to proton decay and hence possess masses of the order of the grand unification 10^{15}-GeV mass scale.

On the horizontal side we identify the two SU(2) groups together as an SO(4) which counts and links the generations. The SU(4) group is the hypercolor group. Thus the $(\underline{16},\underline{10})$ contains the following particles:

$$\begin{aligned} \text{hyperfermions} &\sim (\underline{4},\underline{2},\underline{1};\underline{6},\underline{1},\underline{1}) + (\underline{4}^*,\underline{1},\underline{2};\underline{6},\underline{1},\underline{1})\ , \\ \text{usual fermions} &\sim (\underline{4},\underline{2},\underline{1};\underline{1},\underline{2},\underline{2}) + (\underline{4}^*,\underline{1},\underline{2};\underline{1},\underline{2},\underline{2})\ , \end{aligned} \tag{2.6}$$

i.e., hyperfermions in the $\underline{6}$ of SU(4) and four generations of usual fermions (i.e., hypercolor singlets). In the following we shall denote the hyperfermions, whose $SO(10)_V$ content is the same as that of the usual fermions, by N, U_i, E, D_i. It is important to note that the hyperfermions are generation singlets. Thus, as we shall see in the following, hyperfermion condensation will not lead to strangeness-changing processes. Further, the $(\underline{10},\underline{16})$, whose presence is required by the discrete symmetry, contains no hypercolor singlets at all, and hence leads to no other observable particles, assuming of course that hypercolor confines all its hypercolor-nonsinglet particles.

With regard to the horizontal gauge bosons we note that the $(\underline{1},\underline{1},\underline{1};\underline{6},\underline{2},\underline{2})$ gauge bosons carry both hypercolor and flavor and hence connect a hyperfermion with a usual fermion to mediate the self-energy diagram of Fig. 1. The model thus naturally possesses the extended hypercolor gauge bosons which had previously been introduced by hand. The $(\underline{1},\underline{1},\underline{1};\underline{1},\underline{3},\underline{1})$ and $(\underline{1},\underline{1},\underline{1};\underline{1},\underline{1},\underline{3})$ gauge bosons lead to flavor-changing processes in the tree approximation. These will readily be seen to be easily suppressible in the following.

The symmetry breaking of the model is achieved first by fundamental Higgs fields (which might be an expression of our ignorance of yet stronger and stronger forces), and then in the TeV region by dynamical symmetry breaking. The force responsible for dynamical symmetry breaking is the exchange of the $(\underline{1},\underline{1},\underline{1};\underline{15},\underline{1},\underline{1})$ hypergluons. In our left-handed theory they can only produce Majorana-type masses $\psi_L C\psi_L$ which are symmetric in the internal $SO(10)_V \times SO(10)_H$ symmetry indices. When viewed from the classification according to the vertical $SU(2)_L \times SU(2)_R$ subgroup this mass term will include both Majorana and Dirac masses. Since hypergluon exchange is not to lead to dynamical symmetry breaking of hypercolor itself there can be no condensation of a $(\underline{16},\underline{10})$ fermion with a $(\underline{10},\underline{16})$ fermion. Further, bilinears built out of a pair of $(\underline{10},\underline{16})$ fermions will play no significant role in the following. (See however Sec. VII.) The key condensates are contained in a pair of $(\underline{16},\underline{10})$ bilinears and are

$$\begin{aligned} &(\underline{4},\underline{2},\underline{1};\underline{6},\underline{1},\underline{1}) \times (\underline{4},\underline{2},\underline{1};\underline{6},\underline{1},\underline{1}) \to (\underline{10},\underline{3},\underline{1};\underline{1},\underline{1},\underline{1}) + (\underline{6},\underline{1},\underline{1};\underline{1},\underline{1},\underline{1})\ , \\ &(\underline{4}^*,\underline{1},\underline{2};\underline{6},\underline{1},\underline{1}) \times (\underline{4}^*,\underline{1},\underline{2};\underline{6},\underline{1},\underline{1}) \to (\underline{10}^*,\underline{1},\underline{3};\underline{1},\underline{1},\underline{1}) + (\underline{6},\underline{1},\underline{1};\underline{1},\underline{1},\underline{1})\ , \\ &(\underline{4}^*,\underline{1},\underline{2};\underline{6},\underline{1},\underline{1}) \times (\underline{4},\underline{2},\underline{1};\underline{6},\underline{1},\underline{1}) \to (\underline{15},\underline{2},\underline{2};\underline{1},\underline{1},\underline{1}) + (\underline{1},\underline{2},\underline{2};\underline{1},\underline{1},\underline{1})\ . \end{aligned} \tag{2.7}$$

On the vertical side we decompose SU(4) according to $SU(3)_C$ to obtain

$$\begin{aligned} \underline{10} &= \underline{6} + \underline{3} + \underline{1}\ , \\ \underline{6} &= \underline{3} + \underline{3}^*\ , \\ \underline{15} &= \underline{1} + \underline{3} + \underline{3}^* + \underline{8}\ . \end{aligned} \tag{2.8}$$

Thus the only fermion bilinears which can leave $SU(3)_C$ unbroken are

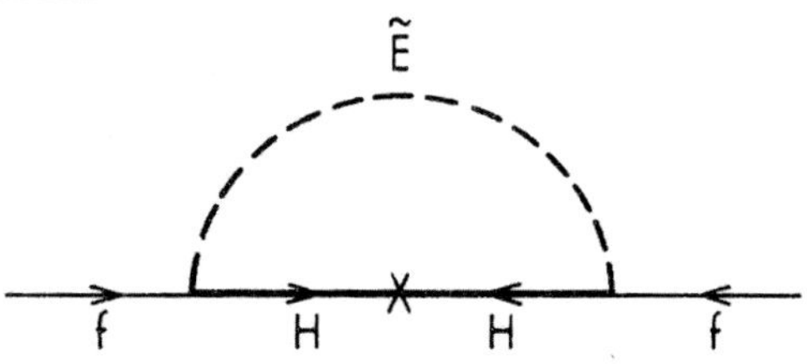

FIG. 1. Radiative graph that generates the light-fermion masses.

$$\Delta_L = (\underline{10},\underline{3},\underline{1};\underline{1},\underline{1},\underline{1}) ,$$
$$\Delta_R = (\underline{10}^*,\underline{1},\underline{3};\underline{1},\underline{1},\underline{1}) , \quad (2.9)$$
$$X = (\underline{15},\underline{2},\underline{2};\underline{1},\underline{1},\underline{1}) + (\underline{1},\underline{2},\underline{2};\underline{1},\underline{1},\underline{1}) .$$

With electric charge to also be unbroken we recognize Δ_L and Δ_R as Majorana masses for the hyperneutrinos N, while X is a Dirac mass for the hyperfermions N, U_i, E, D_i. Since the hyperfermions have the same $SO(10)_V$ properties as the usual fermions we note that the hypercolor-singlet condensates Δ_L, Δ_R, and X have the same group-theoretical content as condensates built out of the regular fermions. Now precisely such condensates (at the usual fermion level) were introduced by one of us recently to obtain the Weinberg-Salam theory by breaking chiral theories dynamically.[11] Thus all of the group theory of those works will carry over into the present theory with the bonus that now a new hypercolor dynamics has been introduced to provide a basis for the dynamical symmetry breaking in the first place. Thus Δ_L and Δ_R which were previously introduced as neutrino-pairing terms are now reinterpreted as hyperneutrino-pairing terms with all the group-theoretical structure remaining intact. Since that group theory is somewhat different than that discussed by Susskind and by Weinberg in their attempt to obtain $M_W = M_Z\cos\theta$ dynamically we shall now discuss the $M_W = M_Z\cos\theta$ problem in our model in some detail.

III. DERIVATION OF THE WEINBERG MIXING RELATION

In their original analyses Susskind[3] and Weinberg[4] identified a possible origin for the Weinberg mixing relation

$$\rho = \frac{M_W}{M_Z\cos\theta} = 1 \quad (3.1)$$

in a dynamically broken Weinberg-Salam theory of the weak interactions. They discussed the situation in which the condensate potential possesses a global chiral flavor $SU(2)_L \times SU(2)_R \times U(1)_{L+R}$ invariance [the $U(1)_{L+R}$ generator is the $(B-L)/2$ current of Eq. (2.3)] with its Weinberg-Salam $SU(2)_L \times U(1)_{WS}$ subgroup being given a local extension [the $U(1)_{WS}$ generator is that of Eq. (2.5)]. The fermion Dirac mass term (to be labeled X) transforms as the $(\underline{2},\underline{2}^*,\underline{0}) \oplus (\underline{2}^*,\underline{2},\underline{0})$ representation of the chiral group. If this mass term breaks diagonally down to a residual $SU(2)_{L+R} \times U(1)_{L+R}$ subgroup [i.e., if $\langle \bar{U}U \rangle = \langle \bar{D}D \rangle$], then the dynamical potential produces just the right three Goldstone bosons required to give masses to three of the four intermediate vector bosons. Since the $SU(2)_L$ Weinberg-Salam gauge bosons transform as a triplet under this residual unbroken $SU(2)_{L+R}$ subgroup they acquire degenerate masses by the Higgs mechanism. This then yields Eq. (3.1) after the mixing with the $U(1)_{WS}$ gauge boson. The residual global $SU(2)_{L+R}$ symmetry of the condensate potential thus provides a group-theoretical origin for the Weinberg mixing relation.[3,4]

The above standard discussion is deficient in three respects. The first is that the same residual $SU(2)_{L+R}$ symmetry which enforces Eq. (3.1) also entails an isospin invariance for the up and down hyperquarks to yield $M_U = M_D$ and hence (as we shall see in Sec. IV) $m_u = m_d$, a mass relation which is known to be violated experimentally. The second difficulty is that any attempt to split M_U and M_D would not just spoil Eq. (3.1) but would also reduce the residual symmetry of the condensate potential to $T^3_{L+R} \times U(1)_{L+R}$. Then the potential would generate more Goldstone bosons than could be removed by the Weinberg-Salam gauge bosons. Finally, the above analysis does not by itself even yield the usual Weinberg-Salam phenomenology at all in the event that the full $SU(2)_L \times SU(2)_R \times U(1)_{L+R}$ group is given a local extension, since then both the fermion mass term and the gauge-boson sector would be parity conserving. The grand unified theory under discussion contains the full $SU(2)_L \times SU(2)_R \times U(1)_{L+R}$ as a local vertical subgroup of $SO(10)_V$. Thus all of the above problems need to be addressed on our model. It turns out that the general resolution of all of these difficulties has been proposed recently.[11,12] Since the results of this analysis are applicable to our model we shall briefly review the important aspects.

The key point of Ref. 12 was to note that while a residual symmetry yields both $\rho = 1$ and $M_U = M_D$, we cannot immediately say by how much these relations are not satisfied when the symmetry is only approximate, this being a model-dependent statement. Reference 12 then constructed an explicit model in which Eq. (3.1) was found to be approximately valid while M_U and M_D were free to be very different from each other. For our purposes here we note that the model considered in Ref. 12 was none other than a local chiral $SU(2)_L \times SU(2)_R \times U(1)_{L+R}$ weak-interaction theory.

With regard to this local chiral theory we have

to introduce some new additional symmetry breaking which first breaks the local chiral flavor group down to $SU(2)_L \times U(1)_{WS}$. Now precisely such a breaking has recently been introduced through the idea of neutrino pairing[11] in which a Higgs field which transforms as a pair of right-handed neutrinos acquires a vacuum expectation value. This neutrino pairing term (to be labeled Δ_R) transforms as the $(\underline{1},\underline{3},-\underline{1})$ representation of $SU(2)_L \times SU(2)_R \times U(1)_{L+R}$ and hence breaks both $SU(2)_R$ and $U(1)_{L+R}$ (since it breaks lepton number), while leaving $SU(2)_L$ and Q unbroken. Hence according to Eqs. (2.4) and (2.5) it leaves $SU(2)_L \times U(1)_{WS}$ unbroken, and thus exactly breaks the local chiral theory down to the Weinberg-Salam model. Moreover, since the neutrino-pairing term transforms as a difermion, another fermion bilinear, this breaking can also be achieved dynamically.

This neutrino-pairing phenomenon has been studied in detail in Ref. 11 not in the dynamical-symmetry-breaking situation itself, but rather in a fundamental-Higgs-field theory. The model considered in Ref. 11 contained Higgs fields which transform as all the available fermion bilinears, namely X and Δ_R given above and a left-handed counterpart Δ_L which transforms as the $(\underline{3},\underline{1},-\underline{1})$ representation. In Ref. 11 (and also Ref. 13) a tree-approximation minimum to the Higgs potential was found in which

$$\Delta_R \gg X \gg \Delta_L \qquad (3.2)$$

and in which the fermion masses were split by an arbitrarily large amount. Explicit counting shows that because there are now seven gauge bosons we now have just the right number of gauge bosons to remove the six Goldstone bosons generated by the breaking pattern of Eq. (3.2) while simultaneously leaving the photon massless. Thus in a local chiral theory we can split the fermion masses without having any superfluous Goldstone bosons. Finally, evaluating the gauge-boson mass matrix we find all the usual Weinberg-Salam phenomenology with Eq. (3.1) being accurate to order

$$\frac{M_W}{M_Z \cos\theta} = 1 + O\left[\frac{X^2}{\Delta_R{}^2}\right] + O\left[\frac{\Delta_L{}^2}{X^2}\right]. \qquad (3.3)$$

Thus in the tree approximation there is no residual symmetry at all and yet ρ only deviates from unity by a small amount dependent on the relative strengths of the left- and right-handed currents of the model.

Having now obtained a good limit in which ρ is close to unity and in which the fermions are far from degenerate, Ref. 12 then studied the radiative corrections to the tree approximation. It was found that there was sufficient approximate residual symmetry in the theory so that Eq. (3.3) continued to hold while the fermions stayed far from degenerate, no matter how big the coupling constants were. Thus the Weinberg mixing relation can be obtained in a local chiral weak-interaction theory to any required degree of accuracy with the fermion masses being unconstrained.

Returning now to the $SO(10)_V \times SO(10)_H$ model, we observe that the hyperfermion Dirac mass term X of Eq. (2.9) contains exactly one $(\underline{2},\underline{2}^*,\underline{0}) \oplus (\underline{2}^*,\underline{2},0)$ piece under $SU(2)_L \times SU(2)_R \times U(1)_{L+R}$ which is an $SU(3)_C$ singlet, while the hyperfermion Majorana mass terms Δ_L and Δ_R of Eq. (2.9) contain exactly one $(\underline{3},\underline{1},-\underline{1}) \oplus (\underline{1},\underline{3},-\underline{1})$ piece under $SU(2)_L \times SU(2)_R \times U(1)_{L+R}$ which is an $SU(3)_C$ singlet. Thus our $SO(10)_V \times SO(10)_H$ model contains just the right condensates required for the analyses of Refs. 11 and 12, this therefore being a highly nontrivial feature of our theory. Since dynamical symmetry breaking is equivalent in content to a fundamental theory plus its radiative corrections the analysis of Ref. 12 will also hold dynamically. Thus if the hypercolor interactions of the horizontal sector of our theory are able to produce a hyperfermion bilinear condensate breaking pattern of the form of Eq. (3.2) dynamically (which of course we can make no comment on without a detailed knowledge of the dynamics of the actual breaking mechanism), then all the usual Weinberg-Salam phenomenology (and especially $\rho \simeq 1$ and m_u/m_d far from one) will obtain dynamically in the vertical sector of the theory.

We would like to make three additional remarks about our breaking pattern. First, we note that Eq. (3.2) has a clear experimental signal in that the right-handed and left-handed neutrinos acquire very different Majorana masses, with this disparity between the left- and right-handed sectors of the theory being due to the way we have spontaneously broken parity in Eq. (3.2). Further, since Δ_R and X both arise dynamically via the exchange of the same hypergluons we would not expect their vacuum expectation values to be overwhelmingly different. Hence we may anticipate that the right-handed neutrinos will acquire GeV region masses. Thus in models in which local chiral theories are broken dynamically by hypercolor interactions there will be relatively light Majorana neutrinos, so that such models can be tested at normal energies way below the 10^{15}-GeV grand unification mass scale. Moreover, our neutrino mass spectrum stands in sharp contrast to that obtained in the standard treatment of the $SO(10)_V$ model. There $SO(10)_V$ is first broken down to SU(5) by a right-handed neutrino Majorana mass, which has to be of order 10^{15} GeV.[14] As we shall see below in more detail, in our model we break $SO(10)_V$ down to $SU(4)\times SU(2)_L\times SU(2)_R$ instead and never go via SU(5) at all. Hence in our scheme the right-handed neutrinos are decoupled from the 10^{15}-GeV region physics altogether and are thus able to be relatively light particles.

Our second remark concerns a subtle interplay between parity nonconservation and hypercolor conservation in our model. As we have just seen, in order to break $SU(2)_L\times SU(2)_R\times U(1)_{L+R}$ dynamically we have to break with Δ_R, a dihyperfermion. On the other hand, we do not want to break hypercolor itself, so that the dihyperfermion must be a hypercolor singlet. Since the product of two fundamentals of SU(4) contains no singlet, we see that we must not put the hyperfermions into the $\underline{4}$ of SU(4) of hypercolor, the "obvious" choice. Indeed, in our model we did not put the hyperfermions into the $\underline{4}$ but rather into the $\underline{6}$ [as they go into the $\underline{10}$ of $SO(10)_H$], so that Δ_R does nicely contain a hypercolor-singlet piece. Now our original choice of the $\underline{10}$ was determined for completely different phenomenological reasons [the classification according to the $(\underline{16},\underline{10})\oplus(\underline{10},\underline{16})$ representation is the only possible one for the fermions in our model which is complex, which contains no particles with bizarre quantum numbers, and which possesses the usual $SO(10)_V$ structure. See Ref. 8]; thus we see that the fact that Δ_R is a hypercolor singlet is then a highly nontrivial property of the model. Moreover, we will see below that the radiative corrections due to extended hypercolor are also only nonvanishing because the fermions are classified according to the $\underline{10}$ of $SO(10)_H$. Thus the model provides strong support for the use of the vector representation of orthogonal groups in grand unified theories, rather than just of the overwhelmingly popular spinor representation.

Our third and final remark concerns a subtle interplay between electric charge conservation and color conservation in the model. As well as serving to break $SU(2)_L\times SU(2)_R$ on the vertical side the difermion condensates Δ_R and Δ_L also break the vertical $SU(4)_V$ color group of Pati and Salam,[10] as they both transform according to the $\underline{10}$ of $SU(4)_V$. Now we note that the $\underline{10}$ of $SU(4)_V$ only contains one $SU(3)_C$ singlet. Thus there is only one breaking pattern which does not lead to QCD breaking. However, within each $\underline{10}$ of $SU(4)_V$ the $SU(3)_C$ singlet is the unique piece which is electrically neutral. Thus if the difermions do not break electric charge then only the dihyperneutrinos can acquire an expectation value, so that QCD is then necessarily not broken. Thus we correlate the conservation of electric charge with the lack of spontaneous breaking of QCD which is very interesting. Finally, to summarize we note that if hypercolor exchange is clever enough to break according to Eq. (3.2) we will obtain the relation $M_W=M_z\cos\theta$ in our theory to any required degree of accuracy of order $\chi^2/\Delta_R{}^2$. In this way the hypercolor dynamics produces all the usual $SU(3)_C\times SU(2)_L\times U(1)_{WS}$ phenomenology.

IV. THE LIGHT-FERMION SELF-ENERGIES AND THE CABIBBO ANGLE

We turn now to a discussion of the mass-generation mechanism for the usual fermions in our model. As we have seen, the hyperfermion condensates are responsible for the dynamical symmetry breaking. They give masses to the Weinberg-Salam gauge bosons by giving Dirac masses M_U, M_D, M_E, and M_N (to be referred to collectively as M_{HF}) to the hyperfermions. Specifically, these condensates produce a triplet of hyperpion Goldstone bosons with decay constant F. These hyperpions then become the longitudinal components of the gauge bosons and give them masses of order eF by the Higgs mechanism so that F is typically of order 125 GeV. Further, these hyperpions give a mass of order κF to the hyperfermions via the Goldberger-Treiman relation, where κ is the coupling of the hyperpions to the hyperfermions. If κ takes the same value here as it does for the coupling of ordinary pions to ordinary fermions we would expect M_{HF} to be of order 1 TeV, so that 1 TeV is the typical mass scale of hypercolor theories.

While the hyperfermion condensates give masses to the hyperfermions, we note that in a pure hypercolor theory these condensates do not give masses to the usual fermions, since these fermions carry no hypercolor and hence undergo no hyperstrong interaction in the first place. However, in our extended hypercolor model, the usual fermions can get masses through the radiative corrections of Fig. 1 which involve the hyperfermions and the additional $(\underline{1},\underline{1},\underline{1};\underline{6},\underline{2},\underline{2})$ extended hypergluons of the model. Specifically these extended hypergluons (to be labeled $E_{\alpha i}$, $\alpha=1,\ldots,6$, $i=1,\ldots,4$) can couple a $(\underline{4},\underline{2},\underline{1};\underline{6},\underline{1},\underline{1})$ hyperfermion H_α to a usual

$(\underline{4},\underline{2},\underline{1};\underline{1},\underline{2},\underline{2})$ fermion f_i with an interaction Lagrangian $\mathscr{L}$ of the form [suppressing all $SO(10)_V$ indices]

$$\mathscr{L}=g_H\sum_{\alpha i}E^{\lambda}_{\alpha i}(\bar{f}_i\gamma_\lambda H_\alpha-\bar{H}_\alpha\gamma_\lambda f_i) \tag{4.1}$$

since the extended hypergluons carry both hypercolor and a generation index. We thus recognize $\mathscr{L}$ as the required interaction of Ref. 5, with the intermediate hyperfermions in the graph of Fig. 1 then providing a mass scale for the usual fermions. Since the dynamical symmetry breaking is only an effective TeV region phenomenon we evaluate the graph of Fig. 1 by cutting off the loop integration at $M_{\rm HF}$. Momentarily ignoring $SO(10)_H$ indices we thus expect light-fermion self-energies m of order

$$m=g_H{}^2\frac{M_{\rm HF}{}^3}{M_{\rm EHG}{}^2} \tag{4.2}$$

up to kinematic factors. Here $M_{\rm EHG}$ denotes the mass of an extended hypergluon. From Eq. (4.2) we thus typically expect $M_{\rm EHG}$ to be in the 1000-TeV region. Thus given the interaction of Eq. (4.1) the usual fermons acquire their masses by radiative corrections in a straightforward manner in our model.

We would like to stress the group-theoretical structure of our model which permits the interaction $\mathscr{L}$ to exist in the first place. On the vertical side we recall that the extended hypergluons are singlets. Consequently the hyperfermions and the usual fermions must both transform the same way under $SO(10)_V$ in order to couple in $\mathscr{L}$, and this exactly occurs in our model as they both transform as the $\underline{16}$ of $SO(10)_V$. Further, on the horizontal side we note that the extended hypergluons transform according to the vector representations of $SO(6)_H$ and $SO(4)_H$, and hence can only couple H_α to f_i if the fermions also transform according to a horizontal vector representation. Moreover, since the extended hypergluons have to belong to the adjoint representation of $SO(10)_H$ their classification according to the $SO(6)_H$ hypercolor vector representation is unique. With the usual fermions necessarily being hypercolor singlets, hypercolor conservation then forces the hyperfermions to also be in the vector representation of $SO(6)_H$ if they are to couple in $\mathscr{L}$ at all. Thus the very existence of the interaction of Eq. (4.1) is seen to be a highly nontrivial property of the model since we had already been obliged to introduce the vector $\underline{10}$ in the classification of the fermions according to the $(\underline{16},\underline{10})\oplus(\underline{10},\underline{16})$ representation of $SO(10)_V\times SO(10)_H$ for the completely different set of reasons previously outlined.

When $SO(10)_V\times SO(10)_H$ is spontaneously broken the $SO(4)_H$ group will necessarily have to be broken since there are no known massless gauge bosons associated with it. This will then both give masses to and mix the $E_{\alpha i}$ gauge bosons into new mass eigenstates

$$\tilde{E}_{\alpha i}=\sum_j A_{ij}E_{\alpha j} \tag{4.3}$$

so that the gauge-boson squared-mass matrix B can be written in terms of its eigenvalues $(\tilde{M}_k)^2$ as follows:

$$\sum_{\alpha ij}E_{\alpha i}(B)_{ij}E_{\alpha j}=\sum_{\alpha k}\tilde{E}_{\alpha k}(\tilde{M}_k)^2\tilde{E}_{\alpha k}\ . \tag{4.4}$$

Thus

$$(B)_{ij}=\sum_k(A^{-1})_{ik}(\tilde{M}_k)^2A_{kj}\ . \tag{4.5}$$

With this diagonalization the interaction of Eq. (4.1) is replaced by

$$\mathscr{L}=g_H\sum_{\alpha ij}(A^{-1})_{ij}\tilde{E}^{\lambda}_{\alpha j}(\bar{f}_i\gamma_\lambda H_\alpha-\bar{H}_\alpha\gamma_\lambda f_i)\ . \tag{4.6}$$

In terms of a new basis for the usual fermions

$$g_i=\sum_j A_{ij}f_j\ , \tag{4.7}$$

we can completely rediagonalize $\mathscr{L}$ as

$$\mathscr{L}=g_H\sum_{\alpha i}\tilde{E}^{\lambda}_{\alpha i}(\bar{g}_i\gamma_\lambda H_\alpha-\bar{H}_\alpha\gamma_\lambda g_i)\ . \tag{4.8}$$

From Fig. 1 the interaction of Eq. (4.8) leads to a light-fermion mass matrix

$$\sum_{ij}\bar{f}_i m_{ij}f_j=g_H{}^2M_{\rm HF}{}^3\sum_{ijk}\bar{f}_i\frac{(A^{-1})_{ik}A_{kj}}{\tilde{M}_k{}^2}f_j\ . \tag{4.9}$$

From Eq. (4.5) we see that

$$[m]_{ij}=g_H{}^2M_{\rm HF}{}^3[(B)^{-1}]_{ij} \tag{4.10}$$

to thus relate the fermion mass matrix to the gauge-boson squared-mass matrix. This is a particularly useful relation since the gauge-boson squared-mass matrix B is given directly by the symmetry-breaking mechanism. In terms of the basis of Eq. (4.7) we can also write the fermion mass matrix as

$$\sum_{ij} \bar{f}_i m_{ij} f_j = \sum_k \bar{g}_k \tilde{m}_k g_k \,, \tag{4.11}$$

with the mass eigenvalues $\tilde{m}_k$ satisfying

$$\tilde{m}_k = g_H{}^2 \frac{M_{\rm HF}{}^3}{\tilde{M}_k{}^2} \,. \tag{4.12}$$

Thus we see that in our model the spontaneous breakdown of $SO(4)_H$ leads not only to light-fermion masses but also to fermion mass mixing, and necessarily so in fact since the four generations are all in the same irreducible representation of $SO(10)_H$ to thus share a common group theory. The light fermions mix through the same angles as do the gauge bosons according to Eq. (4.7) while the diagonal light-fermion masses stand in the same ratios as the gauge-boson inverse-squared masses according to Eq. (4.12). The model thus provides an in-principle mechanism to calculate intergenerational mass splittings and the generalized Cabibbo angles which mix the generations.

We shall refer to the eigenstates of Eq. (4.7) by the sets of generations (u,d,e,ν_e), (c,s,μ,ν_μ), (t,b,τ,ν_τ), and (h,l,σ,ν_σ). The masses of Eq. (4.12) are the so-called *weak-interaction current masses* and writing out Eq. (4.12) in detail, we see that the masses satisfy

$$\begin{aligned} g_H{}^2 M_U{}^3 &= m_u \tilde{M}_1{}^2 = m_c \tilde{M}_2{}^2 = m_t \tilde{M}_3{}^2 = m_h \tilde{M}_4{}^2 \,, \\ g_H{}^2 M_D{}^3 &= m_d \tilde{M}_1{}^2 = m_s \tilde{M}_2{}^2 = m_b \tilde{M}_3{}^2 = m_l \tilde{M}_4{}^2 \,, \\ g_H{}^2 M_E{}^3 &= m_e \tilde{M}_1{}^2 = m_\mu \tilde{M}_2{}^2 = m_\tau \tilde{M}_3{}^2 = m_\sigma \tilde{M}_4{}^2 \,, \\ g_H{}^2 M_N{}^3 &= m(\nu_e) \tilde{M}_1{}^2 = m(\nu_\mu) \tilde{M}_2{}^2 = m(\nu_\tau) \tilde{M}_3{}^2 \\ &= m(\nu_\sigma) \tilde{M}_4{}^2 \,, \end{aligned} \tag{4.13}$$

with each generation having its own associated mass scale $\tilde{M}_k{}^2$. Hence the generations will be reasonably separated in mass.

Typical mass scales can be inferred from the lepton masses. With M_E of order 1 TeV and g_H of order unity we expect $\tilde{M}_1$ to be of order 1000 TeV, $\tilde{M}_2$ of order 100 TeV, $\tilde{M}_3$ of order 10 TeV, which suggests $\tilde{M}_4$ is of order 1 TeV. With m_u and m_d thought to be in the few-MeV region, we find M_D and M_U to both be of order 1 TeV (but not equal to each other as we discussed in Sec. III). Consequently we also take M_N to be of order 1 TeV. With these mass values we can now estimate the other fermion masses. From Eq. (4.13) we obtain many mass formulas. The most interesting ones involve the charged fermions of the three so far observed generations, viz.,

$$\begin{aligned} \frac{\tilde{M}_2{}^2}{\tilde{M}_1{}^2} &= \frac{m_u}{m_c} = \frac{m_d}{m_s} = \frac{m_e}{m_\mu} \,, \\ \frac{\tilde{M}_3{}^2}{\tilde{M}_1{}^2} &= \frac{m_u}{m_t} = \frac{m_d}{m_b} = \frac{m_e}{m_\tau} \,. \end{aligned} \tag{4.14}$$

We will discuss the validity of these relations below after we discuss Cabibbo mixing in detail.

In the neutral fermion sector the right-handed hyperneutrino acquires a Majorana mass $\Delta_R(N)$ from the Δ_R condensate. [For simplicity we set $\Delta_L = 0$ in Eq. (3.2)]. Through the radiative corrections of Fig. 1 the four ordinary neutrinos then also acquire right-handed Majorana masses which are diagonal in the same generation-space basis as the Dirac masses of Eq. (4.13) and are given by

$$\begin{aligned} g_H{}^2[\Delta_R(N)]^3 &= \Delta_R(\nu_e)\tilde{M}_1{}^2 = \Delta_R(\nu_\mu)\tilde{M}_2{}^2 \\ &= \Delta_R(\nu_\tau)\tilde{M}_3{}^2 = \Delta_R(\nu_\sigma)\tilde{M}_4{}^2 \,, \end{aligned} \tag{4.15}$$

For each ordinary neutrino its Dirac and right-handed Majorana masses will mix in its mass matrix to induce a left-handed Majorana mass. As noted in Sec. III, $\Delta_R(\nu_e)$ is in the GeV region [since $\Delta_R(N)$ is in the TeV region] and so these final observable left-handed Majorana masses such as $\Delta_L(\nu_e)$ will typically be of order 1 eV (see, e.g., Ref. 13). This is many orders of magnitude larger than the value of perhaps 10^{-5} eV expected in the usual treatment of the $SO(10)_V$ theory where $\Delta_R(\nu_e)$ is in the 10^{15}-GeV region.[14] Hence our model differs from the standard $SO(10)_V$ picture in its predictions for both $\Delta_R(\nu_e)$ and $\Delta_L(\nu_e)$.

We turn now to a discussion of Cabibbo mixing in our model. As we have already noted the very fact that $SO(4)_H$ is spontaneously broken forces the fermions to mix, and indeed if we had a detailed knowledge of the gauge-boson squared-mass matrix B we could determine all the mixing angles directly from Eq. (4.10). In the absence of such knowledge we must instead make some assumptions, and though we shall make what we regard as reasonable assumptions our analysis should only be taken as illustrative of the mixing phenomenon. To simplify we shall restrict ourselves to the first two generations of quarks by taking $\tilde{E}_{\alpha 1}$ and $\tilde{E}_{\alpha 2}$ to be heavier than $\tilde{E}_{\alpha 3}$ and $\tilde{E}_{\alpha 4}$. In this case $E_{\alpha 1}$ and $E_{\alpha 2}$ mix through an angle ϕ so that f_1 and f_2 also mix through the same angle ϕ according to Eq. (4.7).

To determine the gauge-boson mixing angle ϕ we need to make a model for the gauge-boson mass matrix. The simplest possibility is to break $SO(4)_H$ in the quartet representation by fundamental Higgs fields in the requisite 1000-TeV region. (The dynamical hyperfermion condensates only act at the 1-TeV level.) Since we are dealing with two generations we shall need two quartets which we label ϕ_i^a and ϕ_i^b ($i=1,\ldots,4$). To implement the breaking we note that the extended hypergluons are minimally coupled to the $\underline{10}$ of $SO(10)_H$ which contains the $\underline{4}$ of $SO(4)_H$, so that the gauge-boson mass matrix is obtained as

$$\mathcal{L}_{\text{mass}}=g_H{}^2\sum_\alpha\left\{\left|\sum_i E_{\alpha i}\phi_i^a\right|^2+\left|\sum_i E_{\alpha i}\phi_i^b\right|^2\right\}. \tag{4.16}$$

For the two quartets the most general breaking pattern in the $E_{\alpha 1}$, $E_{\alpha 2}$ sector of interest is

$$\begin{aligned}\phi^a&=a(1,0,0,0)\,,\\ \phi^b&=b(\sin\gamma,\cos\gamma,0,0)\,.\end{aligned} \tag{4.17}$$

In terms of the parameters a, b, and γ we can determine the eigenvalues and eigenvectors of the gauge-boson mass matrix as

$$\widetilde{M}_1{}^2=\frac{g_H{}^2}{2}[(a^2+b^2)+\sqrt{A}\,]\,, \tag{4.18a}$$

$$\widetilde{M}_2{}^2=\frac{g_H{}^2}{2}[(a^2+b^2)-\sqrt{A}\,]\,, \tag{4.18b}$$

$$\tan\phi=\frac{\sqrt{A}-a^2+b^2\cos2\gamma}{b^2\sin2\gamma}\,, \tag{4.18c}$$

where $A=a^4+b^4-2a^2b^2\cos2\gamma$. From Eqs. (4.18) we can obtain the useful relation

$$\tan^2\phi=\frac{\widetilde{M}_2{}^2}{\widetilde{M}_1{}^2}(1+Y)\,, \tag{4.19}$$

where

$$Y=\frac{(a^2-b^2)[(a^2-b^2)-\sqrt{A}\,]}{2a^2b^2\sin^2\gamma}\,. \tag{4.20}$$

Finally, recalling that ϕ is also the d-s mixing angle ϕ_{ds} we find from Eq. (4.14) that

$$\tan^2\phi_{ds}=\frac{m_d}{m_s}(1+Y)\,. \tag{4.21}$$

The d-s mixing angle is thus expressed completely in terms of the parameters of the gauge-boson mass matrix.

With the parameter Y being completely unknown we cannot use Eq. (4.21) to extract out a numerical value for ϕ_{ds}. However we now note some unusual features of the structure of the mixing matrices. From Eq. (4.14) we observe that $\widetilde{M}_2{}^2/\widetilde{M}_1{}^2$ is a very small number since $m_e \ll m_\mu$. Hence from Eqs. (4.18) we see that

$$\frac{a^2b^2\cos^2\gamma}{(a^2+b^2)^2}\ll1\,, \tag{4.22}$$

$$Y=\frac{b^2-a^2}{a^2\sin^2\gamma}\,. \tag{4.23}$$

Now there are four ways of satisfying the constraint of Eq. (4.22) and they lead to differing values of Y. Specifically they are

$$\begin{aligned}&\text{(i) } a\sim b\sin\gamma\gg b\cos\gamma;\ Y\sim0\,,\\ &\text{(ii) } a\gg b\sin\gamma\sim b\cos\gamma;\ Y\sim1\,,\\ &\text{(iii) } a\gg b\sin\gamma\gg b\cos\gamma;\ Y\sim1\,,\\ &\text{(iv) } a\gg b\cos\gamma\gg b\sin\gamma;\ Y\gg1\end{aligned} \tag{4.24}$$

(cases in which $b\gg a$ are equivalent). All of cases (i) through (iii) lead to

$$\tan^2\phi_{ds}=O\left[\frac{m_d}{m_s}\right]\ll1 \tag{4.25}$$

while only case (iv) can make ϕ_{ds} large. Without further knowledge of the structure of the Higgs potential which produces the set of expectation values of Eq. (4.17) we cannot make any further statement. However, we note that case (iv) [and also case (iii)] requires that there be a hierarchy of two separate scales of breaking, while cases (i) and (ii) only require one such hierarchy. Since all of the Higgs expectation values are to arise from one single potential we regard it as perhaps unlikely that case (iv) could emerge. Hence, we can reasonably expect Y to be of order one, so that Eq. (4.25) is a realistic expectation for our model.

While Eq. (4.21) is an encouraging result we note that since the hyperfermions are generation singlets an analogous relation also exists in the u-c sector where the mixing angle satisfies

$$\tan^2\phi_{uc}=\frac{m_u}{m_c}(1+Y) \tag{4.26}$$

according to Eqs. (4.19) and (4.14). Thus from Eq. (4.14) we obtain

$$\phi_{ds}=\phi_{uc}\,. \tag{4.27}$$

Thus the extended-hypercolor contributions to the weak-interaction current mass matrix mixing are the same in the d-s and u-c sectors. Consequently an identification of ϕ_{ds}-ϕ_{uc} with the observed Cabibbo angle would lead us to conclude that there would be no Cabibbo mixing in our model. However as we noted in Ref. 9, $(\phi_{ds}-\phi_{uc})$ as determined from Eqs. (4.21) and (4.26) is not necessarily the observed Cabibbo angle since further mixing contributions can also come from strong-interaction chiral-symmetry breaking.

Specifically with four quarks the QCD Lagrangian possesses a global chiral flavor $SU(4)_L \times SU(4)_R$ symmetry. This symmetry is spontaneously broken by the color dynamics and gives Goldberger-Treiman masses m_u^0, m_d^0, m_s^0, and m_c^0 to the quarks to contribute an extra diagonal piece to the quark mass matrix. In the event that QCD spontaneously breaks $SU(4)_L \times SU(4)_R$ down beyond $SU(4)_{L+R}$ not all of these strong-interaction masses are degenerate, so that the actual observable quark mixing angles are only obtained after a simultaneous diagonalization of both the weak and strong contributions. In the d-s sector for instance we can express the full mass matrix $[m'_{ij}]$ in our initial f_i, f_j basis as

$$\sum_{ij} \bar{f}_i m'_{ij} f_j = \sum_{ij} \bar{f}_i m_{ij} f_j + \bar{f}_1 f_1 m_d^0 + \bar{f}_2 f_2 m_s^0 \tag{4.28}$$

according to Eq. (4.9). Expressing m_{ij} in terms of its eigenvectors and eigenvalues yields

$$[m'_{ij}] = \begin{bmatrix} m_d^0 + m_d \cos^2\phi + m_s \sin^2\phi & \sin\phi\cos\phi(m_d - m_s) \\ \sin\phi\cos\phi(m_d - m_s) & m_s^0 + m_d \sin^2\phi + m_s \cos^2\phi \end{bmatrix}, \tag{4.29}$$

where ϕ is still the extended-hypergluon mixing angle and m_d and m_s are still given by Eq. (4.13). The mixing angle associated with $[m'_{ij}]$ is readily calculated to be

$$\tan 2\theta_{ds} = \frac{\sin 2\phi}{(k_{ds} + \cos 2\phi)}, \tag{4.30}$$

where

$$k_{ds} = \frac{m_s^0 - m_d^0}{(m_s - m_d)}. \tag{4.31}$$

Similarly in the u-c sector we obtain

$$\tan 2\theta_{uc} = \frac{\sin 2\phi}{(k_{uc} + \cos 2\phi)}, \tag{4.32}$$

where

$$k_{uc} = \frac{m_c^0 - m_u^0}{(m_c - m_u)}. \tag{4.33}$$

Thus, finally, the observable Cabibbo angle is given as

$$\begin{aligned} \theta_C &= \theta_{ds} - \theta_{uc} \\ &= \tfrac{1}{2}\tan^{-1}\left[\frac{\sin 2\phi}{k_{ds} + \cos 2\phi}\right] \\ &\quad - \tfrac{1}{2}\tan^{-1}\left[\frac{\sin 2\phi}{k_{uc} + \cos 2\phi}\right], \end{aligned} \tag{4.34}$$

where ϕ is given in Eq. (4.19).

While the above discussion of the influence of QCD on θ_C is straightforward we note that such an analysis is not customarily made in the literature as it is always assumed that m_u^0, m_d^0, m_s^0, and m_c^0 are all equal; i.e., it is always assumed that QCD only breaks $SU(4)_L \times SU(4)_R$ down to $SU(4)_{L+R}$. As far as we are aware such an assumption has not been justified in the literature and in fact demands further study as a problem in its own right even independent of our work here, especially since it affects what we even mean by the Cabibbo angle. Further, the possible breaking of the strong-interaction chiral symmetry beyond $SU(4)_{L+R}$ is directly amenable to experimental testing. As we noted in Ref. 9 there would have to exist some observable scalar Goldstone bosons in addition to the familiar 15 pseudoscalar Goldstone bosons such as the pion and we advocate a vigorous search for such particles. On the phenomenological side we also take note of the detailed study of Das and Deshpande[15] using a partial conservation of the axial-vector current (PCAC) analysis. They found that there are good indications that $SU(4)_{L+R}$ is in fact broken spontaneously, and moreover quite a lot, although $SU(3)_{L+R}$ apparently survives intact as an unbroken symmetry. Hence we can effectively take m_s^0 equal to m_d^0 and m_c^0 very large. Then in the case where Y is zero we obtain finally from Eq. (4.34)

$$\tan^2\theta_C = \frac{m_d}{m_s}, \tag{4.35}$$

where we recall that m_d and m_s are the weak-interaction contributions to the quark masses. We are not of course asserting that Eq. (4.35) is strictly obeyed, by only that it constitutes a reasonable estimate of the smallness of the Cabibbo angle.

Since the quark mass pattern that we have used is somewhat different than the conventional one we would like to comment on the general determinations of the quark mass parameters. In the days before the discovery of charm the chiral symmetry of the strong interaction was taken to be $SU(3)_L \times SU(3)_R$ and the conventional PCAC analysis indicated that it was broken down spontaneously only to $SU(3)_{L+R}$ so that m_u^0, m_d^0, and m_s^0 are all equal, with a typical value of 300 MeV then usually being taken for these masses. The same PCAC analysis was also used to estimate the weak-interaction current masses m_u, m_d, and m_s with typical values of 5, 10, and 200 MeV, respectively, being found for them. The reliability of these numbers is still open to question since apart from anything else there is quite a big continuation to the kaon mass shell. The full constituent masses of the quarks which make up the hadrons are then given by combining m_i^0 and m_i $(i = u,d,s)$ perhaps geometrically. With the discovery of charm the above picture had to be extended to $SU(4)_L \times SU(4)_R$ and it became necessary to know how to distribute the 1.5 GeV constituent charm mass between m_c^0 and m_c. The conventional so-far-unjustified approach is to take m_c^0 simply equal to m_u^0, m_d^0, and m_s^0 [so that $SU(4)_{L+R}$ remains unbroken] while taking m_c very large. Instead we are suggesting that rather it is m_c^0 that is large and that m_c might be much smaller than previously anticipated. Further, if the QCD effects are not $SU(4)_{L+R}$ flavor invariant, then QCD itself will also make radiative contributions to m_u, m_d, m_s, and m_c. With m_c^0 large we might thus expect m_c to undergo a big renormalization so that the m_c as measured in PCAC for D mesons may no longer satisfy the relation $m_u/m_c = m_d/m_s$ of Eq. (4.14). Moreover, there are even further potential renormalizations of the weak masses in our model. It is possible that the fourth generation of fermions might have masses as heavy as the Weinberg-Salam gauge bosons or even some of the the six flavor-changing $SO(4)_H$ generators. They would then make radiative corrections to the self-energies of the light fermions to cause some further deviations from Eq. (4.14) (so that for instance m_d/m_s would no longer be equal to m_e/m_μ). Thus we should only take Eq. (4.14) as a guide to the light-fermion masses, and not as exact mass relations.

To conclude this section we remark again that by exploiting the group-theoretical correlation between hypercolor and the generation problem we have been able to construct a realistic theory of extended hypercolor which gives a reasonable estimation of the Cabibbo angle. We would also like to note that the pattern of masses and mixings we have discussed in detail is assuming that the extended-hypercolor mechanism is the only source. One might also use an appropriate Higgs system instead or in conjunction with the extended hypercolor within the framework of our model.

V. THE GOLDSTONE AND PSEUDO-GOLDSTONE BOSONS OF THE MODEL

We turn now to the classification of the various Goldstone and pseudo-Goldstone bosons associated with the hypercolor dynamics of our model. The Lagrangian of the coupling of the $(\underline{4},\underline{2},\underline{1};\underline{6},\underline{1},\underline{1})$ and $(\underline{4}^*,\underline{1},\underline{2};\underline{6},\underline{1},\underline{1})$ hyperfermions to the $(\underline{1},\underline{1},\underline{1};\underline{15},\underline{1},\underline{1})$ hypergluons which is effective in the TeV region possesses an $SU(8)_L \times SU(8)_R \times U(1)_{L+R}$ global chiral symmetry in the U_i,D_i,E,N space. [Technically there is also an axial $U(1)_{L-R}$ symmetry, but this is destroyed by instantons.] Each SU(8) contains seven $SU(3)_C$ color-singlet currents, four color octets, and eight color triplets. Thus $SU(8)_L \times SU(8)_R \times U(1)_{L+R}$ contains 15 color-singlet currents altogether. We shall not concern ourselves with the color-nonsinglet piece in the following since any associated Goldstone or pseudo-Goldstone bosons will be confined.

We break the global chiral symmetry with the Δ_R, Δ_L, and χ condensates of Eq. (2.9). The right-handed Δ_R breaks as $N_R C N_R$ and thus spontaneously breaks three right-handed currents, $\bar{N}_R\gamma_\lambda N_R$, $\bar{N}_R\gamma_\lambda E_R$, and $\bar{E}_R\gamma_\lambda N_R$ which are all color singlets. Similarly, the left-handed Δ_L breaks the three color-singlet currents $\bar{N}_L\gamma_\lambda N_L$, $\bar{N}_L\gamma_\lambda E_L$, and $\bar{E}_L\gamma_\lambda N_L$. The Dirac mass χ gives different masses M_U, M_D, M_E, and M_N to the hyperfermions. It thus breaks all 63 axial-vector currents and also breaks 44 vector currents leaving unbroken only the 20 vector currents $\bar{N}\gamma_\lambda N$, $\bar{E}\gamma_\lambda E$, $\bar{U}_i\gamma_\lambda U_j$, and $\bar{D}_i\gamma_\lambda D_j$ $(i,j = R,G,B)$. Of these last 20 there are 4 color singlets, $\bar{N}\gamma_\lambda N$, $\bar{E}\gamma_\lambda E$, $\sum \bar{U}_i\gamma_\lambda U_i$, and $\sum \bar{D}_i\gamma_\lambda D_i$. Thus χ breaks 11 singlet currents altogether, of which 4 are vector ($\bar{N}\gamma_\lambda E$, $\bar{E}\gamma_\lambda N$, $\sum \bar{U}_i\gamma_\lambda D_i$, and $\sum \bar{D}_i\gamma_\lambda U_i$) and 7 are axial-vector [$\bar{N}\gamma_\lambda\gamma_5 N$, $\bar{E}\gamma_\lambda\gamma_5 E$, $\bar{N}\gamma_\lambda\gamma_5 E$,

$\bar{E}\gamma_\lambda\gamma_5 N$, $\sum \bar{U}_i\gamma_\lambda\gamma_5 U_i$, $\sum \bar{U}_i\gamma_\lambda\gamma_5 D_i$, $\sum \bar{D}_i\gamma_\lambda\gamma_5 U_i$, and $\sum \bar{D}_i\gamma_\lambda\gamma_5 D_i$, then minus the $U(1)_{L-R}$ current]. Consequently Δ_R, Δ_L, and χ together break a grand total of 12 linearly independent color-singlet chiral currents. The dynamical potential thus produces 12 light spin-zero particles.

In our grand unified model we give a local extension to the 45 gauge bosons associated with the $SO(10)_V$ subgroup of the chiral $SU(8)_L \times SU(8)_R \times U(1)_{L+R}$. Within the $SO(10)_V$ there are altogether 7 color-singlet currents which form the $SU(2)_L \times SU(2)_R \times U(1)_{L+R}$ group studied in detail in Sec. III. Since one of its gauge bosons is to be the photon we see that only 6 color-singlet gauge bosons acquire masses by the Higgs mechanism. Hence, out of our initial 12 spin-zero particles above we see that 6 are removed by the Higgs mechanism while the other 6 remain in the spectrum. These latter 6 bosons can only acquire masses by radiative corrections involving the gauge bosons of the standard model and become relatively light pseudo-Goldstone particles with masses of order eM_W, i.e., of order perhaps 20 GeV or less. Such pseudo-Goldstone particles are typical of minimally gauged theories such as $SO(10)_V$ where the gauge sector has a lower symmetry than that of the potential. In maximally gauged theories on the other hand the full $SU(8)_L \times SU(8)_R \times U(1)_{L+R}$ group would be gauged with all twelve spin-zero particles then being removed by the Higgs mechanism to leave no observable pseudo-Goldstone particles at all. Thus we do not regard the pseudo-Goldstone bosons of our $SO(10)_V \times SO(10)_H$ model with too much concern, as they can all be removed if necessary by embedding the grand unified theory in an appropriately large-enough gauge group.

VI. CONSTRAINTS FROM RARE DECAYS

In this section we analyze the flavor-changing rare decay processes which occur in our model due to the generation mixing. The various possibilities we consider are tree-level single-$SO(4)_H$-gauge-boson exchange, box-diagram extended-hypergluon exchange, pseudo-Goldstone-boson exchange, and tree-level Pati-Salam-$SU(4)_V$-gauge-boson exchange. As we shall see a mass scale in the 1000-TeV region will suffice to provide sufficient suppression to keep all rare decay processes within current experimental bounds.

The single-gauge-boson exchanges yield tree graphs such as that of Fig. 2 with an effective coupling

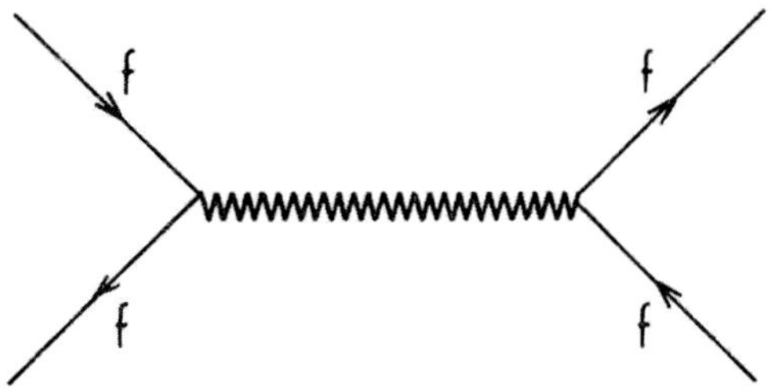

FIG. 2. A single-gauge-boson-exchange graph. The exchanged gauge bosons belong to the adjoint representation of $SO(4)_H$.

$$V_T = \frac{g_H{}^2}{M^2}, \tag{6.1}$$

ignoring quantum numbers for the moment. The exchange of a pair of $(\underline{1},\underline{1},\underline{1};\underline{6},\underline{2},\underline{2})$ extended hypergluons gives the box graph of Fig. 3. Here the exchanged fermions are the same hyperfermions as those of Fig. 1. We evaluate the box graph by cutting off the internal momentum at $M_{\rm HF}$ as in Fig. 1 to obtain

$$V_{\rm box} \sim g_H{}^4 \frac{M_{\rm HF}{}^2}{M_{\rm EHG}{}^4}, \tag{6.2}$$

which we can reexpress as

$$V_{\rm box} \sim \frac{m^2}{M_{\rm HF}{}^4}, \tag{6.3}$$

using Eq. (4.2). Thus $V_{\rm box}$ is a number whose magnitude can be estimated independently of g_H and $M_{\rm EHG}$. To evaluate the contributions due to the pseudo-Goldstone bosons, to be labeled P, we note that they only couple to the ordinary fermions via the hyperfermions and the extended hypergluons to give the effective Yukawa coupling of Fig. 4, viz.,

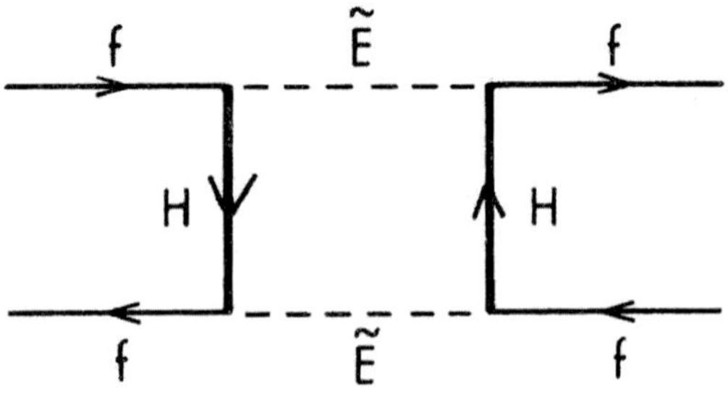

FIG. 3. Box diagram involving the exchange of hyperfermions and extended hypergluons.

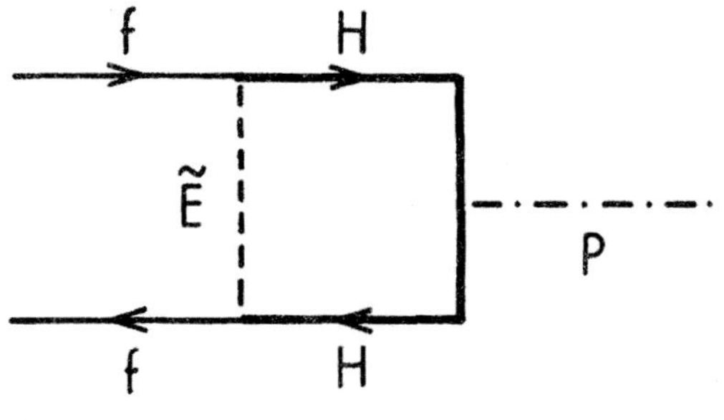

FIG. 4. Pseudo-Goldstone-boson effective coupling.

$$V_Y \sim \kappa g_H^2 \frac{M_{\rm HF}^2}{M_{\rm EHG}^2} , \tag{6.4}$$

where κ, the pseudo-Goldstone-boson—hyperfermion coupling constant, was introduced in Sec. IV. Using Eq. (4.2) and the hypercolor Goldberger-Treiman relation we can reexpress Eq. (6.4) as

$$V_Y \sim \frac{m}{F} , \tag{6.5}$$

where F is the hyperpion decay constant. Again we see that the strength of the interaction can be estimated. Finally pseudo-Goldstone-boson exchange leads to the graph of Fig. 5 with an effective strength

$$V_P \sim \frac{m^2}{F^2 M_P^2} , \tag{6.6}$$

where M_P is the pseudo-Goldstone-boson mass. To evaluate the effects due to all these processes we must now put in all the appropriate quantum numbers.

We discuss first the tree graphs due to the exchange of the $SO(4)_H$ generators, the $(\underline{1},\underline{1},\underline{1};\underline{1},\underline{3},\underline{1})$ and $(\underline{1},\underline{1},\underline{1};\underline{1},\underline{1},\underline{3})$ gauge bosons. To match our previous notation of Eq. (4.1) we shall label these gauge bosons E_{ij} $(i,j,=1,\ldots,4,\ i<j)$. These gauge bosons connect fermions of different generations to each other and are thus flavor changing even in the symmetry limit. In terms of our original $(\underline{4},\underline{2},\underline{1};\underline{1},\underline{2},\underline{2})$ fermion basis f_i we have couplings of the form

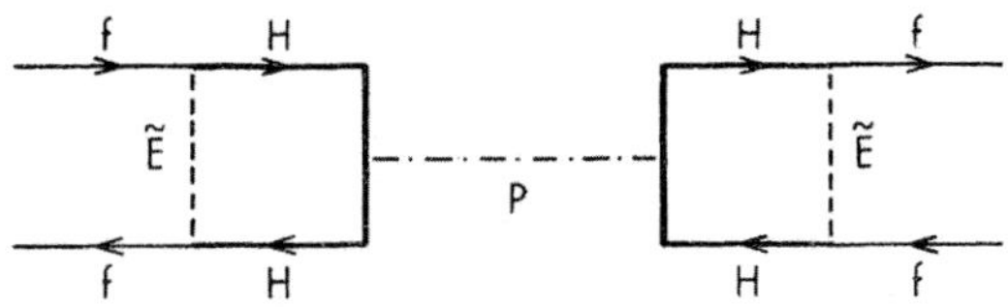

FIG. 5. Pseudo-Goldstone-boson exchange.

$$\mathscr{L} = g_H \sum_{i,j} E_{ij}^{\lambda} (\bar{f}_i \gamma_\lambda f_j - \bar{f}_j \gamma_\lambda f_i) . \tag{6.7}$$

We must now change the fermion basis to that of the eigenstates of the fermion mass matrix. We introduce the notation $d',s',u',c',\ldots$ to denote the exact mass eigenstates of the full $[m'_{ij}]$ of Eq. (4.28) which includes both extended-hypergluon and QCD contributions. In terms of the convenient $d,s,\ldots$ basis which diagonalizes the extended-hypergluon-induced mass matrix $[m_{ij}]$ of Eq. (4.9) alone we can write

$$d' = d \cos\alpha + s \sin\alpha ,$$
$$s' = -d \sin\alpha + s \cos\alpha \tag{6.8}$$

and

$$u' = u \cos\beta + c \sin\beta ,$$
$$c' = -u \sin\beta + c \cos\beta . \tag{6.9}$$

Here

$$\tan 2\alpha = \frac{\sin 2\phi}{(k_{ds}^{-1} + \cos 2\phi)} ,$$
$$\tan 2\beta = \frac{\sin 2\phi}{(k_{uc}^{-1} + \cos 2\phi)} , \tag{6.10}$$

and k_{ds} and k_{uc} are defined in Eqs. (4.31) and (4.33), respectively. [Eq. (6.10) differs from Eq. (4.30) because we have expressed d' in terms of d, and not in terms of f_i as previously.] Thus we can express Eq. (6.7) in terms of mass eigenstates.

As far as the first two generations of fermions are concerned we note that the antisymmetric vertices of Eq. (6.7) are left invariant by the Cabibbo rotations used to diagonalize the fermion mass matrix. Thus Eq. (6.7) leads to flavor-changing vertices of interest of the form

$$\mathscr{L} = g_H E_{12}^{\lambda} [(\bar{s}'\gamma_\lambda d' - \bar{d}'\gamma_\lambda s') + \bar{\mu}\gamma_\lambda e - \bar{e}\gamma_\lambda \mu$$
$$+ (\bar{c}'\gamma_\lambda u' - \bar{u}'\gamma_\lambda c')$$
$$+ \bar{\nu}_\mu \gamma_\lambda \nu_e - \bar{\nu}_e \gamma_\lambda \nu_\mu] \tag{6.11}$$

[the lepton basis is that of Eq. (4.13)]. Thus Fig. 2 leads to typical effective interactions such as

$$\mathscr{L}_T = \frac{g_H^2}{M_{12}^2} (\bar{s}'\gamma_\lambda d' - \bar{d}'\gamma_\lambda s')_{(1)}$$
$$\times (\bar{s}'\gamma_\lambda d' - \bar{d}'\gamma_\lambda s')_{(2)} , \tag{6.12}$$

where M_{12} is the mass of E_{12} and (1) and (2) denote the two vertices of Fig. 2. In the direct channel [where we construct a K meson by contracting $\bar{s}'$ in (1) with d' in (1)] the couplings are pure vector, and hence the pseudoscalar mesons cannot couple to the quarks at all. Thus there are no direct-channel contributions to both the K_S-K_L and D_S-D_L mass differences or to K and D decays into lepton pairs at all due to Fig. 2. There is, however, a vector $K\pi$ vertex so that $K \to \pi\mu e$ does occur in the direct channel. This process is then sufficiently suppressed if $M_{12} > 5$ TeV. Stronger constraints are found from the cross channel where we can construct a K meson by contracting $\bar{s}'$ in (1) with d' in (2). Unlike the direct channel this channel does possess pseudoscalar projections. The effective strangeness-changing vertices of interest are of the form

$$\mathscr{L}_T = \frac{g_H^2}{M_{12}^2}(\bar{s}'_{(1)}\gamma_5 d'_{(2)}\bar{s}'_{(2)}\gamma_5 d'_{(1)} + \bar{d}'_{(1)}\gamma_5 s'_{(2)}\bar{d}'_{(2)}\gamma_5 s'_{(1)}) \tag{6.13}$$

$$= \frac{g_H^2}{M_{12}^2}(K_0^2 + \bar{K}_0^2) = \frac{g_H^2}{M_{12}^2}(K_S^2 - K_L^2) \ .$$

This yields a K_S-K_L mass difference ΔM_K of the form

$$\frac{\Delta M_K}{M_K f_K^2} = \frac{g_H^2}{M_{12}^2} \ , \tag{6.14}$$

where f_K is the usual kaon decay constant. Experimentally the left-hand side of Eq. (6.14) is less than 10^{-6} TeV^{-2}. Thus with g_H of order 1 we require $M_{12} \geq 1000$ TeV. Similarly the tree graph contributes to the D_S-D_L mass difference. Though this has yet to be measured it has been estimated theoretically[16] to satisfy

$$\frac{\Delta M_D}{M_D f_D^2} = 10^{-4}\text{TeV}^{-2} \tag{6.15}$$

and is thus well suppressed by a 1000-TeV mass for E_{12}. Since there are no lepton-pair decays possible in the crossed channel we see that this one scale suppresses all flavor-changing processes due to the $SO(4)_H$ horizontal gauge bosons.

The value we have found for M_{12} is very encouraging since it is of the same order as $\widetilde{M}_1$. It is thus very tempting to generate both these masses by the same Higgs mechanism. Indeed we can give all six $SO(4)_H$ generators a mass by breaking according to four $SO(4)_H$ quartets. At the same time this would give masses to all 24 of the $(\underline{1},\underline{1},\underline{1};\underline{6},\underline{2},\underline{2})$ extended hypergluons by breaking $SO(10)_H$ to $SO(6)_H$ in one step. Specifically with four quartets ϕ_i^a, ϕ_i^b, ϕ_i^c, and ϕ_i^d we obtain a mass term [analogous to Eq. (4.16)]

$$\mathscr{L}_{\text{mass}} = g_H^2 \sum_{a,\alpha} \left[\left| \sum_i E_{\alpha i}\phi_i^a \right|^2 \right] + g_H^2 \sum_{a,i} \left[\left| \sum_j E_{ij}\phi_j^a \right|^2 \right] . \tag{6.16}$$

For simplicity we shall ignore mixings with the third and fourth generations by taking

$$\begin{aligned} \phi^a &= a(1,0,0,0) \ , \\ \phi^b &= b(\sin\gamma,\cos\gamma,0,0) \ , \\ \phi^c &= c(0,0,1,0) \ , \\ \phi^d &= d(0,0,0,1) \ . \end{aligned} \tag{6.17}$$

This yields [using Eqs. (4.18)]

$$\begin{aligned} \mathscr{L}_{\text{mass}} = {} & \sum_\alpha (\widetilde{M}_1^2 \widetilde{E}_{\alpha 1}^2 + \widetilde{M}_2^2 \widetilde{E}_{\alpha 2}^2 + g_H^2 c^2 E_{\alpha 3}^2 + g_H^2 d^2 E_{\alpha 4}^2) + g_H^2(a^2+b^2)E_{12}^2 + g_H^2(c^2+d^2)E_{34}^2 \\ & + g_H^2[a^2 E_{13}^2 + c^2(E_{13}^2 + E_{23}^2) + b^2(E_{13}\sin\gamma + E_{23}\cos\gamma)^2] \\ & + g_H^2[a^2 E_{14}^2 + d^2(E_{14}^2 + E_{24}^2) + b^2(E_{14}\sin\gamma + E_{24}\cos\gamma)^2] \ . \end{aligned} \tag{6.18}$$

Thus we find that

$$M_{12}^2 = g_H^2(a^2+b^2) = \widetilde{M}_1^2 + \widetilde{M}_2^2 \tag{6.19}$$

according to Eq. (4.18), so that indeed M_{12} can naturally be 1000 TeV in our model. Finally using Eq. (4.13) we can reexpress the K_S-K_L mass difference as

$$\frac{\Delta M_K}{M_K f_K^2} = \frac{m_e}{M_E^3} \tag{6.20}$$

to order m_e/m_μ, so we see that in our model it is m_e being small which suppresses the tree-graph contribution to ΔM_K. In Sec. VII we shall study further this interesting connection between the various mass scales.

The second type of rare decay processes that we need to consider are the extended-hypergluon box graphs of Fig. 3. Some simplification comes by noting that since $SO(10)_V \times SO(10)_H$ is an orthogonal group the coupling at each vertex in Fig. 3 is antisymmetric in the fermion indices. Thus after contracting the internal fermion lines the graphs will be symmetric in the external fermion lines. For instance contracting $(\bar{d}D-\bar{D}d)$ with $(\bar{s}D-\bar{D}s)$ yields $(\bar{s}d+\bar{d}s)$. Thus the box diagrams only couple to K_S in the direct channel. Analogous to our discussion of the crossed-channel structure of Fig. 2 we thus see that there are no extended-hypergluon box-diagram contributions to $K_L \to \mu e$ at all, although there are both direct- and crossed-channel contributions to the K_S-K_L mass difference.

To evaluate these contributions explicitly we find it convenient to first classify the graphs according to the eigenstates of $[m_{ij}]$ of Eq. (4.9) since this is the basis in which the extended-hypergluon mass matrix is diagonal, and then only at the end make the QCD-induced rotation to the final quark basis of $[m'_{ij}]$ of Eq. (4.28). Using Eq. (4.8) we find the box graphs yield an effective action of interest,

$$\mathscr{L}_{\text{box}} = \frac{4m_d^2}{M_D^4}(\bar{d}\gamma_5 d)^2 + \frac{4m_s^2}{M_D^4}(\bar{s}\gamma_5 s)^2 + \frac{m_d m_s}{M_D^4}(\bar{d}\gamma_5 s+\bar{s}\gamma_5 d)^2$$
$$+\frac{4m_d m_e}{M_D^2 M_E^2}(\bar{d}\gamma_5 d)(\bar{e}\gamma_5 e) + \frac{4m_s m_\mu}{M_D^2 M_E^2}(\bar{s}\gamma_5 s)(\bar{\mu}\gamma_5 \mu) + \frac{m_d m_\mu}{M_D^2 M_E^2}(\bar{d}\gamma_5 s+\bar{s}\gamma_5 d)(\bar{e}\gamma_5\mu+\bar{\mu}\gamma_5 e) \tag{6.21}$$

with an analogous expression in the charm sector. Changing now to the final d',s' basis of Eq. (6.8) yields

$$\mathscr{L}_{\text{box}} = \left[\frac{(m_d^2+m_s^2)}{M_D^4}\sin^2 2\alpha + \frac{m_d m_s}{M_D^4}\cos^2 2\alpha\right](\bar{d}'\gamma_5 s'+\bar{s}'\gamma_5 d')^2$$
$$-\frac{2\sin 2\alpha}{M_D^2 M_E^2}(\bar{d}'\gamma_5 s'+\bar{s}'\gamma_5 d')(m_d m_e \bar{e}\gamma_5 e - m_s m_\mu \bar{\mu}\gamma_5\mu)$$
$$+\frac{m_d m_\mu}{M_D^2 M_E^2}\cos 2\alpha(\bar{d}'\gamma_5 s'+\bar{s}'\gamma_5 d')(\bar{e}\gamma_5\mu+\bar{\mu}\gamma_5 e)\ . \tag{6.22}$$

We note that since the rotation of Eq. (6.8) is real the above interactions still only involve K_S in the direct channel. For the K_S-K_L mass difference Eq. (6.22) yields

$$\frac{\Delta M_K}{M_K f_K^2} = \frac{(m_d^2+m_s^2)\sin^2 2\alpha + m_d m_s \cos^2 2\alpha}{M_D^4}\ . \tag{6.23}$$

With $m_s = 200$ MeV and $M_D = 1$ TeV the box-graph contribution is then at least an order of magnitude smaller than the experimental value. Further, Eq. (6.22) leads to a muon-number-violating $K_S \to \mu e$ decay. However, the branching ratio is found to be of order 10^{-22} and is thus essentially unobservable. The analogous charm-changing effective interactions are of the form

$$\mathscr{L}_{\text{box}} = \left[\frac{(m_u^2+m_c^2)}{M_U^4}\sin^2 2\beta + \frac{m_u m_c}{M_U^4}\cos^2 2\beta\right](\bar{u}'\gamma_5 c'+\bar{c}'\gamma_5 u')^2$$
$$-\frac{2\sin 2\beta}{M_U^2 M_E^2}(\bar{u}'\gamma_5 c'+\bar{c}'\gamma_5 u')(m_u m_e \bar{e}\gamma_5 e - m_c m_\mu \bar{\mu}\gamma_5\mu)$$
$$+\frac{m_c m_\mu}{M_U^2 M_E^2}\cos 2\beta(\bar{u}'\gamma_5 c'+\bar{c}'\gamma_5 u')(\bar{e}\gamma_5\mu+\bar{\mu}\gamma_5 e) \tag{6.24}$$

so that the D_S-D_L mass difference is well below the estimate of Ref. 16. Thus the box diagrams are also sufficiently suppressed.

The third type of decay processes that we need to consider are the pseudo-Goldstone-boson exchanges of

Fig. 5. The Yukawa couplings of Fig. 4 give an effective interaction of interest of the form

$$\mathscr{L}_Y=\frac{P}{F}(m_d\bar{d}\gamma_5 d+m_s\bar{s}\gamma_5 s+m_e\bar{e}\gamma_5 e+m_\mu\bar{\mu}\gamma_5\mu+m_u\bar{u}\gamma_5 u+m_c\bar{c}\gamma_5 c) \tag{6.25}$$

which is flavor diagonal in the $d,s,\ldots$ basis since the hyperfermion condensates are flavor singlets. Thus at this stage there are still no flavor-changing processes associated with Fig. 5. Rotating to the final $d',s',\ldots$ basis of Eqs. (6.8) and (6.9) however yields a flavor-changing interaction of interest,

$$\mathscr{L}_Y=\frac{P}{F}\{(m_s-m_d)\sin\alpha\cos\alpha(\bar{d}'\gamma_5 s'+\bar{s}'\gamma_5 d')+(m_c-m_u)\sin\beta\cos\beta(\bar{u}'\gamma_5 c'+\bar{c}'\gamma_5 u')$$
$$+m_e\bar{e}\gamma_5 e+m_\mu\bar{\mu}\gamma_5\mu\}\ . \tag{6.26}$$

Thus we obtain

$$\frac{\Delta M_K}{M_K f_K{}^2}=\frac{(m_s-m_d)^2\sin^2\alpha\cos^2\alpha}{F^2 M_P{}^2}\ ,$$
$$\frac{\Delta M_D}{M_D f_D{}^2}=\frac{(m_c-m_u)^2\sin^2\beta\cos^2\beta}{F^2 M_P{}^2}\ . \tag{6.27}$$

The K_S-K_L mass difference will thus only be suppressed enough if

$$M_P>\sin 2\alpha \tag{6.28}$$

in TeV mass units. Thus even with α as small as θ_C, M_P would have to be at least 500 GeV. However P is a pseudo-Goldstone boson and is thus perhaps as light as a few GeV. Thus the only way to suppress the P exchange enough is to take $\alpha=0$, i.e., to set $k_{ds}=0$. However we have exactly argued in Sec. IV that k_{ds} is in fact zero since $SU(3)_{L+R}$ is left unbroken after QCD breaks the global chiral $SU(4)_L\times SU(4)_R$ symmetry of the strong interactions. Thus the symmetry relation $m_s^0=m_d^0$ completely suppresses pseudo-Goldstone-boson-exchange contributions to the K_S-K_L mass difference. Then with $\alpha=0$ $\mathscr{L}_Y$ leads to no K_S leptonic decays either. Thus the global $SU(3)_{L+R}$ invariance of the strong interaction suppresses all strangeness-changing processes due to Fig. 5 in our model.

Since $SU(4)_{L+R}$ is broken by m_c^0 being unequal to m_u^0, m_d^0, and m_s^0, there will instead be charm-changing processes. Indeed with m_c^0 very large, $k_{uc}{}^{-1}=0$ so that $\beta=\phi$. According to Eq. (4.34) ϕ is θ_C in this limit, and so we take $\beta=\theta_C$. Equation (6.26) then yields

$$\frac{\Delta M_D}{M_D f_D{}^2}=\frac{(m_c-m_u)^2\sin^2\theta_C\cos^2\theta_C}{F^2 M_P{}^2}\ . \tag{6.29}$$

Taking $m_c=m_u m_s/m_d\simeq 100$ MeV then requires M_P to be of order at least 20 GeV, which is just our previous estimate of Sec. V. Given the uncertainties in estimating ΔM_D and M_P theoretically the pseudo-Goldstone-boson exchanges do not appear to pose any difficulties at present. Of course experimental improvements in the charm sector (there are currently no bounds on leptonic decay modes) would considerably tighten the current freedom in the model.

The fourth and final class of exchanges that we must consider are the tree-level Pati-Salam-$SU(4)_V$-gauge-boson[10] exchanges. As we noted in Sec. II this $SU(4)_V$ group contains an $SU(3)_C\times(B-L)$ subgroup and six other generators (to be denoted generically as leptoquarks [LQ]) which transform as a $\underline{3}\oplus\underline{3}^*$ under $SU(3)_C$ and hence couple quarks to leptons. Additionally the $SU(4)_V$ Lagrangian also possesses a global $3B+L$ fermion-number invariance. Thus $SU(4)_V$ breaking can potentially lead to proton decay or lepton pair rare decays. With regard first to proton decay we note that in the symmetry limit the [LQ]-exchange diagrams lead only to elastic $\bar{q}+l\to\bar{q}+l$ processes and not to $\bar{q}+l\to q+q$. Thus there will be no proton decay unless baryon number is explicitly spontaneously broken at the $SU(4)_V$ level. Since breaking $SU(4)_V$ in the $\underline{10}$ (i.e., as $N_R C N_R$ as in Sec. III) breaks $SU(4)_V$ down to $SU(3)_C$ while only breaking $(B-L)$ and $(3B+L)$ in the right-handed neutrino sector, and since breaking $SU(4)_V$ in the $\underline{15}$ (i.e., as $\bar{N}N$) breaks $SU(4)_V$ down to $SU(3)_C\times(B-L)$ while not breaking the global $3B+L$ symmetry at all, we see that we can break down to the standard model without needing to break baryon number at the $SU(4)_V$ level at all. Hence we can, and will, give masses to the [LQ] gauge bosons which are much less than the 10^{15}-GeV grand unification mass scale.

With regard to flavor-changing processes we note that the $SU(4)_V$ generators are $SO(4)_H$ singlets and thus even in the symmetry limit connect quark-lepton combinations of different generations.

Specifically, in terms of our original $(\underline{4},\underline{2},\underline{1};\underline{1},\underline{2},\underline{2})$ fermion basis f_i we have couplings of the form

$$\mathscr{L}=[\mathrm{LQ}]\sum_i \bar{f}_i(q)f_i(l)\ , \qquad (6.30)$$

where $f_i(q)$ denotes a quark in f_i, and $f_i(l)$ a lepton. When the $\mathrm{SO}(4)_H$ symmetry is broken the fermions mix. As we noted in Eq. (4.7) the extended-hypercolor-induced mixings are the same for quarks and leptons. Hence in terms of the basis of Eq. (4.13) the interaction of Eq. (6.30) yields flavor-changing vertices of the form

$$\mathscr{L}=[\mathrm{LQ}](\bar{d}e+\bar{s}\mu+\bar{b}\tau+\bar{l}\sigma+\bar{u}\nu_e+\bar{c}\nu_\mu+\bar{t}\nu_\tau+\bar{h}\nu_\sigma)\ . \qquad (6.31)$$

Thus in the crossed channel [LQ] exchange leads to both K_S and K_L decays into μe. The current bound of 2×10^{-9} on the $K_L\rightarrow\mu e$ branching ratio then requires that the [LQ] masses be at least of order 50 TeV. Thus a mass scale of only 50 TeV suppresses all $\mathrm{SU}(4)_V$-induced rare decays, enabling $\mathrm{SU}(4)_V$ to be a relatively low-lying symmetry. Also, and most intriguingly, we see that this leptoquark mass scale is again typical of the horizontal mass scales we have just found. Consequently, we can anticipate that all of these gauge bosons acquire their masses from a common origin. As we shall show in detail in Sec. VII, the leptoquark gauge bosons can also get their masses from the $\mathrm{SO}(4)_H$ quartets of Higgs fields of Eq. (6.17). In such a situation we find that the [LQ] mass is at most of order of $\tilde{M}_1$. Then we obtain a *lower* bound on the muon-number-violating $K_L\rightarrow\mu e$ decay

$$\frac{\Gamma(K_L\rightarrow\mu e)}{\Gamma(K_L\rightarrow \text{all})}>10^{-18}\ . \qquad (6.32)$$

While this number is too small to be practical its significance (and also that of the box-diagram rate for $K_S\rightarrow\mu e$ presented earlier) lies in the fact that it is a lower rather than an upper bound on a rare decay. Our ability to bound $K_L\rightarrow\mu e$ from below is due to the fact that the extended hypergluon masses are bounded from above by the observed lepton masses. It is amusing to note that the present lower bound on $K_L\rightarrow\mu e$ obtained here is of the same order of magnitude as that obtained in an $\mathrm{SU}(4)_L\times\mathrm{U}(1)$ weak interaction model where the gauge-boson masses were bounded from above by the observed rate for *CP* violation.[17]

Finally, to conclude this section we remark again that a single 1000-TeV mass scale is sufficient to suppress all the rare decay processes which occur in our model, so that these decays do not present us with any problems at all.

VII. THE MASS SCALES OF THE MODEL

Out model possesses a variety of scales and in this section we classify all of them. With regard first to $\mathrm{SO}(10)_V$ we recall that, like its SU(5) subgroup, this group is baryon-number violating in the Lagrangian, i.e., in the symmetry limit, through the exchange of the $(\underline{6},\underline{2},\underline{2};\underline{1},\underline{1},\underline{1})$ gauge bosons. Consequently they must acquire masses of the order of 10^{15} GeV to keep proton decay within experimental bounds. This value for the grand unification mass scale also leads to an acceptable renormalization of the Weinberg angle. Hence we first break $\mathrm{SO}(10)_V$ by some fundamental Higgs field at 10^{15} GeV. This Higgs breaking is required to leave $\mathrm{SU}(4)_V\times\mathrm{SU}(2)_L\times\mathrm{SU}(2)_R$ unbroken [or possibly $\mathrm{SU}(4)_V\times\mathrm{SU}(2)_L\times T_R^3$ only since the charged gauge bosons of $\mathrm{SU}(2)_R$ play no role at all in the subsequent breaking to the standard-model phenomenology given in Sec. III]. The easiest way to break $\mathrm{SO}(10)_V$ down to $\mathrm{SU}(4)_V\times\mathrm{SU}(2)_L\times\mathrm{SU}(2)_R$ is with a single $\underline{54}$ representation of SO(10), as it contains exactly one singlet under the subgroup. Moreover, since

$$\begin{aligned}\underline{10}\times\underline{10}&=\underline{1}+\underline{45}+\underline{54}\ ,\\ \underline{16}\times\underline{10}&=\underline{16}+\underline{144}\ ,\\ \underline{16}\times\underline{16}&=\underline{10}+\underline{120}+\underline{126}\ ,\\ \underline{16}\times\underline{16}^*&=\underline{1}+\underline{45}+\underline{210}\ ,\end{aligned} \qquad (7.1)$$

we see that breaking in the $(\underline{54},\underline{1})\oplus(\underline{1},\underline{54})$ representation of $\mathrm{SO}(10)_V\times\mathrm{SO}(10)_H$ will not give a mass to any of the fermions of our model, since they are only allowed to have Majorana masses at the $\mathrm{SO}(10)_V\times\mathrm{SO}(10)_H$ level. On the $\mathrm{SO}(10)_H$ side we note that no gauge boson can get a 10^{15} GeV mass according to our analysis of Sec. IV and thus the discrete symmetry between $\mathrm{SO}(10)_V\times\mathrm{SO}(10)_H$ is broken spontaneously at the 10^{15}-GeV level. Thus in the primary stage of breaking we break $\mathrm{SO}(10)_V\times\mathrm{SO}(10)_H$ according to the $(\underline{54},\underline{1})\oplus(\underline{1},\underline{54})$ representation to reduce the symmetry to $\mathrm{SU}(4)_V\times\mathrm{SU}(2)_L\times\mathrm{SU}(2)_R\times\mathrm{SO}(10)_H$.

Before we discuss the second stage of breaking at 1000 TeV we take note of some constraints on the model due to asymptotic freedom requirements. The standard renormalization-group analysis gives a lowest-order β function of the form

$$\beta = \frac{-g^3}{12\pi^2}\left[\frac{11C_2}{4} - \frac{T(R)}{2} - \frac{T(R)}{8}\right], \qquad (7.2)$$

where C_2 is the contribution of the adjoint, $T(R)/2$ that of each irreducible representation of two-component spinors, and $T(R)/8$ that of each real irreducible representation of scalars. For the adjoint representation of SU(N) the value of C_2 is N, for the fundamental $T(R)$ is $\frac{1}{2}$, and for the symmetric and antisymmetric second-rank tensors $T(R)$ is $(N+2)/2$ and $(N-2)/2$, respectively. We calculate first the value of β for $\mathrm{SO(10)}_V \times \mathrm{SO(10)}_H$. An easy way to do this is to decompose SO(10) according to SU(5), viz.,

$$\underline{10} = \underline{5} + \underline{5}^*, \quad \underline{45} = \underline{1} + \underline{24} + \underline{10} + \underline{10}^*,$$
$$\underline{16} = \underline{1} + \underline{5}^* + \underline{10}, \quad \underline{54} = \underline{15} + \underline{15}^* + \underline{24}, \qquad (7.3)$$

to give $C_2(45)=8$, $T(16)=2$, $T(10)=1$, and $T(54)=12$. Thus for the adjoint gauge bosons, the $(\underline{16},\underline{10})$ and the $(\underline{10},\underline{16})$ fermions, and the $(\underline{54},\underline{1}) \oplus (\underline{1},\underline{54})$ scalars we obtain for either $\mathrm{SO(10)}_V$ or $\mathrm{SO(10)}_H$,

$$\beta_{\mathrm{GUT}} = \frac{-g^3}{12\pi^2}(22 - 10 - 8 - \tfrac{3}{2}) < 0 . \qquad (7.4)$$

Thus as we noted in Ref. 8,[18] our $\mathrm{SO(10)}_V \times \mathrm{SO(10)}_H$ model is asymptotically free despite its large fermionic content.

We now study the renormalization for the subgroups. For $\mathrm{SO(6)}_H$ the same analysis gives [ignoring now the superheavy $(\underline{54},\underline{1}) \oplus (\underline{1},\underline{54})$ scalars as they play no role at low energies]

$$\beta_{\mathrm{hyp}} = \frac{-g^3}{12\pi^2}(11 - 18) \qquad (7.5)$$

which is positive. The reason for this change of sign is that we have reduced the number of gauge bosons but not the number of fermions compared with Eq. (7.4). Thus the only way to retain asymptotic freedom for the hypercolor group is to freeze out some of the fermionic degrees of freedom. We thus give the $(\underline{10},\underline{16})$ fermions (which have so far played no role in this paper) large masses so that after they condense out only the $(\underline{16},\underline{10})$ fermions survive. These latter fermions alone then yield

$$\beta_{\mathrm{hyp}} = \frac{-g^3}{12\pi^2}(11 - 8) \qquad (7.6)$$

which is now nicely asymptotically free.

In order to give the $(\underline{10},\underline{16})$ fermions a mass we note that the product of a pair of $(\underline{10},\underline{16})$ representations contains the following hypercolor singlets:

$$\begin{aligned}
&(\underline{1},\underline{1},\underline{1};\underline{1},\underline{2},\underline{2}), \quad (\underline{1},\underline{1},\underline{1};\underline{1},\underline{2},\underline{2}), \\
&(\underline{15},\underline{1},\underline{1};\underline{1},\underline{2},\underline{2}), \quad (\underline{1},\underline{3},\underline{1};\underline{1},\underline{2},\underline{2}), \\
&(\underline{20},\underline{1},\underline{1};\underline{1},\underline{2},\underline{2}), \quad (\underline{1},\underline{1},\underline{3};\underline{1},\underline{2},\underline{2}), \\
&(\underline{6},\underline{2},\underline{2};\underline{1},\underline{2},\underline{2}), \quad (\underline{1},\underline{3},\underline{3};\underline{1},\underline{2},\underline{2}), \\
&(\underline{6},\underline{2},\underline{2};\underline{1},\underline{2},\underline{2}) .
\end{aligned} \qquad (7.7)$$

Out of these a set of four is of particular interest, viz.,

$$\begin{aligned}
&(\underline{1},\underline{1},\underline{1};\underline{1},\underline{2},\underline{2}), \quad (\underline{1},\underline{1},\underline{1};\underline{1},\underline{2},\underline{2}), \\
&(\underline{15},\underline{1},\underline{1};\underline{1},\underline{2},\underline{2}), \quad (\underline{1},\underline{1},\underline{3};\underline{1},\underline{2},\underline{2}) .
\end{aligned} \qquad (7.8)$$

This set of four is just sufficient to break $\mathrm{SO(10)}_H$ all the way down to $\mathrm{SO(6)}_H$ on the horizontal side, while breaking $\mathrm{SU(4)}_V \times \mathrm{SU(2)}_L \times \mathrm{SU(2)}_R$ down to $\mathrm{SU(3)}_C \times (B-L) \times \mathrm{SU(2)}_L \times T_R^3$ on the vertical side as required in Sec. VI. Further with these same four representations we can also give masses to all of the $(\underline{10},\underline{16})$ fermions. While the fourth representation in Eq. (7.8) gives masses to the charged gauge bosons of $\mathrm{SU(2)}_R$ it leaves the neutral sector alone, which, as we already noted, is sufficient for subsequently obtaining the usual Weinberg-Salam phenomenology. It is thus remarkable that the set of representations of Eq. (7.7) contains just the right set of fields with just the right quantum numbers to yield the set of Eq. (6.17), which is precisely the set we used to be able to control all the rare decays with one common scale. Thus with four fields which transform as in Eq. (7.8) we can simultaneously give common 1000-TeV region masses to the 160 $(\underline{10},\underline{16})$ fermions, the 24 extended hypergluons, the 6 $\mathrm{SO(4)}_H$ generators, the 2 charged gauge bosons of $\mathrm{SU(2)}_R$ and the 6 leptoquark [LQ] gauge bosons of the Pati-Salam $\mathrm{SU(4)}_V$ group. While we shall use Higgs fields to explicitly do this it is extremely tempting to speculate that the $(\underline{10},\underline{16})$ fermions undergo dynamical symmetry breaking through some new 1000-TeV dynamics—or perhaps, more economically, through hypercolor exchange itself. [This is not so unreasonable since we have seen that some of the extended hypergluons associated with the third and fourth generations have masses in the 1-to-10-TeV region. Also the group-theory factors associated with hypercolor exchanges are quite different for the $\underline{6}$ and $\underline{4}$ hypercolor representations so that the $(\underline{10},\underline{16})$ condensates could anyway have different scales than the $(\underline{16},\underline{10})$ condensates even while both are bound by the same hypergluons.] We feel that the mechanism for giving the ($\underline{10}$,$\underline{16}$) fermions their masses needs to be explored in detail in the future, as we appear to have uncovered a new scale at 1000 TeV characteristic of unified theories. Thus the desert is beginning to bloom.

With the ($\underline{10}$,$\underline{16}$) fermions now frozen out at 1000 TeV the hypercolor group is asymptotically free at lower energies so that the ($\underline{16}$,$\underline{10}$) fermions can condense and dynamically break $SU(3)_C \times (B-L) \times SU(2)_L \times T_R^3 \times SO(6)_H$ down to $SU(3)_C \times U(1)_{EM} \times SO(6)_H$ as we discussed previously in Sec. III. As we continue down in energy from 1000 TeV there would at first appear to be a possibility that QCD could break its global symmetry before $SO(6)_H$ does. This is nicely avoided on our model however, since we note that the complete ($\underline{16}$,$\underline{10}$) contribution to the $SU(3)_C$ renormalization is

$$\beta_C = \frac{-g^3}{12\pi^2}\left(\tfrac{33}{4} - 10\right) \tag{7.9}$$

which is positive, not negative. Thus as we continue down below 1000 TeV only $\beta_{\rm hyp}$ is negative so that $SO(6)_H$ causes dynamical symmetry breaking first in the TeV region. After this happens the hyperfermions freeze out leaving only the four generations of ordinary fermions. Their contribution to QCD alone gives

$$\beta_C = \frac{-g^3}{12\pi^2}\left(\tfrac{33}{4} - 4\right) \tag{7.10}$$

which is now negative again, so that below the hypercolor threshold only the ordinary fermions are operative allowing QCD to finally spontaneously break the strong-interaction chiral symmetry in the GeV region.

Having now identified the various breaking scales we can now calculate the observable low-energy parameters of our model. Unlike the usual $SO(10)_V$ extrapolation from the grand unified mass scale M down to an ordinary mass scale m we note that in our theory we have to go through various dynamical thresholds to get to the low-energy region. At these thresholds highly nonperturbative effects are taking place which we do not know how to calculate. Thus without a detailed theory of dynamical symmetry breaking we are actually unable to do anything. We note that the usual renormalization-group analysis actually provides two different types of information, namely, explicit low-energy values, $g_3(m)$, $g_2(m)$, and $g_1(m)$ for the $SU(3)_C \times SU(2)_L \times U(1)_{WS}$ coupling constants, and second, a low-energy value for $\sin^2\theta_W(m)$. Because we go through all the dynamical thresholds and freeze out fermions at different energies we do not believe that the standard renormalization-group extrapolation for $g_3(m)$, $g_2(m)$, and $g_1(m)$ is reliable in our model and hence we shall make no attempt to evaluate these quantities. However we recall that $\sin^2\theta_W(m)$ only depends on the difference $[g_1(m)^{-2} - g_2(m)^{-2}]$. In this difference all fermionic contributions to the appropriate β functions drop out and so this difference is presumably not sensitive to threshold effects. Consequently the standard renormalization-group analysis should hold in our model for $\sin^2\theta_W(m)$.

In explicitly evaluating the renormalization of the Weinberg angle we note first that the ($\underline{16}$,$\underline{10}$) and ($\underline{10}$,$\underline{16}$) fermions contribute in the standard manner to yield $\sin^2\theta_W(M) = \frac{3}{8}$ at the grand unified mass scale. The low-energy value for $\sin^2\theta_W(m)$ depends on which subgroup of $SO(10)_V$ is light. If only $SU(2)_L \times U(1)_{WS}$ is light we obtain the standard relation

$$\sin^2\theta_W(m) = \tfrac{3}{8} - \frac{55e^2(m)}{96\pi^2}\, ln\frac{M}{m} \; . \tag{7.11}$$

Now , it was noted in the third reference cited in Ref. 7 that this same relation is obtained if $SU(2)_L \times T_R^3 \times U(1)_{L+R}$ is light, with the relation

$$\sin^2\theta_W(m) = \tfrac{3}{8} - \frac{11e^2(m)}{48\pi^2}\, ln\frac{M}{m} \tag{7.12}$$

obtaining if the full chiral $SU(2)_L \times SU(2)_R \times U(1)_{L+R}$ is light. With the standard low-energy values for $\sin^2\theta_W(m)$ and $e^2(m)$ we find that Eq. (7.11) requires $M \sim 10^{15}$ GeV while Eq. (7.12) leads to a completely unacceptable value for M. Now we recall that in our analysis of the mass scales of our model given above we precisely found that $SU(2)_R$ should not be a light subgroup but rather only T_R^3. Thus that analysis dovetails nicely with the analysis of $\sin^2\theta_W(m)$ leading us to Eq. (7.11) and grand unification in the 10^{15}-GeV region. Finally, we note that this value for M leads to a proton lifetime in the standard 10^{30} yr region. However, because we are unable to extrapolate $g_3(m)$, $g_2(m)$, and $g_1(m)$ back to the grand unified scale we cannot give an exact value for the lifetime, but only an order-of-magnitude estimate. Hence in our model even though $\sin^2\theta_W(m)$ is not sensitive to the details of the dynamical symmetry breaking, the proton lifetime is, and thus remains a little uncertain.

Thus we identify the three main scales of our model, a primary superheavy breaking at 10^{15} GeV and then two sequential low-lying breakings at

1000 TeV and then at 1 TeV. For convenience we display the complete pattern in Fig. 6. We note that there emerges a very attractive approximate low-energy $SU(4)_V \times SU(4)_H$ symmetry which is broken in the 1000-TeV region only and is otherwise good up to the grand unification mass scale. Thus the hypercolor group emerges as an approximate horizontal counterpart of the Pati-Salam color group.

VIII. CONCLUDING REMARKS

With quarks and leptons as the fundamental fermionic entities, the occurrence of repeated fermionic generations poses a serious and difficult problem for the program of grand unification of the strong, weak, and electromagnetic interactions. Attempts in the literature[19] to construct a multigenerational grand unified theory based on a simple group and satisfying some physically reasonable constraints have proved quite unsatisfactory. The problem is made more acute because of a recent demonstration by Tosa and Okubo[20] who show under very general assumptions that SU(5) and SO(10) are almost unique as unitary and orthogonal group candidates for a grand unified theory. But both of these lead to a single generation of the basic fermions. Accommodating more than one generation in such theories is accompanied by a large number of free, adjustable parameters. If simple groups cannot provide the needed framework, the next simplest alternative is a semisimple group structure with a discrete symmetry imposed so as to ensure a single coupling constant. From this point of view we have studied in this paper a grand unified theory based on $SO(10)_V \times SO(10)_H$.

The semisimple group structure is forced upon us for another reason. The problems in theories in which fundamental scalars bring about the spontaneous symmetry breaking have been discussed so

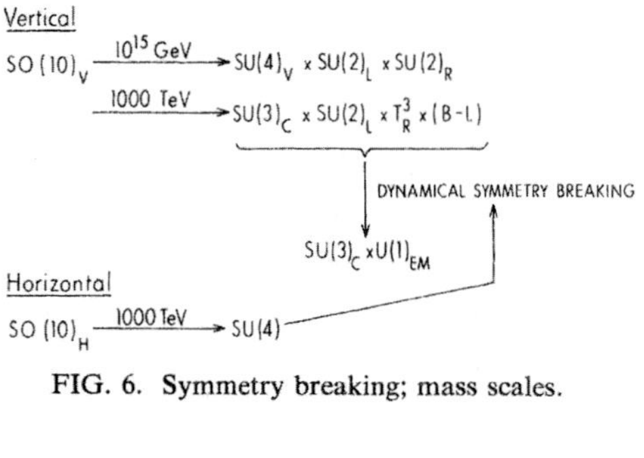

FIG. 6. Symmetry breaking; mass scales.

often in the literature that it is not necessary to repeat them here. The alternative to the Higgs mechanism is the so-called dynamical symmetry breaking which requires forces that become strong in the TeV region. Since the QCD forces described by SU(3) color symmetry become strong in the GeV region, according to the prevailing ideas on how renormalization-group equations govern the strength of the coupling constant, the required superstrong group must be larger than or equal to SU(4). This requirement along with some others which have become standard rules of the game cannot be met within the framework of a simple group.[21] As we have seen they are met in our semisimple grand unified model, however.

The model discussed in this paper has indeed some very attractive features from a group-theoretic point of view. The "vertical" $SO(10)_V$ describes each family of ordinary fermions and hence incorporates automatically all the good features of the grand unified theory based on $SO(10)_V$.[2] The "horizontal" $SO(10)_H$ contains hypercolor, extended hypercolor, and horizontal interactions. The irreducible representation—an unusual combination of spinor and vector representations, namely $(\underline{16},\underline{10}) \oplus (\underline{10},\underline{16})$—is almost uniquely selected out. The ensuing particle spectrum and quantum numbers allow the existence of all the needed interactions, unlike other hypercolor models where one has to introduce such interactions from the outside. Further, the model satisfies the criteria for renormalizability and asymptotic freedom. It has a nontrivial generation structure of four conventional fermionic families.

The important relation $M_W = M_Z \cos\theta$ can be satisfied to any desired degree of accuracy provided the hypercolor dynamics satisfies certain criteria. These criteria are not unique to our model; they are shared by other theories. The masses of the ordinary quarks and their mixings come out satisfactorily, although much more work is necessary to establish them on a quantitative basis. In this regard we take an unconventional view concerning the mass matrix of the usual fermions in the low-energy region. It appears that in the literature one either neglects or implicitly assumes that strong forces preserve an $SU(N)_{L+R}$ symmetry while breaking the $SU(N)_L \times SU(N)_R$ global chiral flavor symmetry when considering the mass matrix. The entire mass matrix is then the current quark mass matrix generated either through Higgs or an extended hypercolor mechanism. We propose that this is not necessarily the correct procedure. The QCD forces could break part of the vector flavor symmetry as well. On that basis, our results concerning masses and mixing angles are encouraging. Also our analysis of the rare decay modes shows that no serious difficulties are encountered there. This is due to a fortuitous combination of group-theoretic features of the model, which constrain the effective exchanges of single bosons in their couplings to K and D mesons. There are definite predictions concerning the rare decays of the charmed particles. When experimental information concerning these decays becomes available the model can be further tightened.

Perhaps the most unsatisfactory feature of the model is the use of more than one primary Higgs mechanism to cause the initial symmetry breaking. The discrete symmetry which fixes the group structure and the fermionic representation has to be broken right away. Nonetheless, an attractive feature emerges, namely that there is a surviving correspondence between Pati-Salam SU(4) color in the vertical sector and the SU(4) hypercolor in the horizontal sector. In the TeV region the dynamical symmetry breaking takes over with no further need for fundamental scalars. Clearly a great deal of further work is necessary to clarify and establish many features of the model. It does however contain many realistic features and has a rich structure which merits further study.

ACKNOWLEDGMENTS

This work was supported in part by the U.S. Department of Energy under Contracts Nos. DE-AC02-79ER10336.A (P.D.M.) and DE-AC02-76ER03533 (K.C.W.).

APPENDIX

In this paper we have considered in great detail the decomposition $SO(10)_H \to SO(6)\times SO(4)$, identifying the $SO(6) \sim SU(4)$ subgroup as the candidate for the hypercolor group. In this appendix we briefly consider other possible subgroups which could serve as candidates for hypercolor and analyze the hypercolor content of the $(\underline{16},\underline{10}) \oplus (\underline{10},\underline{16})$ and $(\underline{16},\underline{16})$ representations for each case.

(1) $SO(10)_H \to SO(8)\times SO(2)$. Under this decomposition

$$\begin{aligned} \underline{16} &= (\underline{8},\underline{1}) + (\underline{8}',\underline{1}) , \\ \underline{10} &= (\underline{8},\underline{1}) + (\underline{1},\underline{2}) . \end{aligned} \tag{A1}$$

If we identify SO(8) itself as the hypercolor group, it would imply that the $(\underline{16},\underline{10}) \oplus (\underline{10},\underline{16})$ representation contains two ordinary generations, while the $(\underline{16},\underline{16})$ representation contains no ordinary fermions at all. This choice is thus excluded phenomenologically.

(2) $SO(10)_H \to SO(7)\times SO(3)$. Here

$$\begin{aligned} \underline{16} &= (\underline{8},\underline{2}) , \\ \underline{10} &= (\underline{7},\underline{1}) + (\underline{1},\underline{3}) . \end{aligned} \tag{A2}$$

With SO(7) as the hypercolor subgroup, we see that the $(\underline{16},\underline{16})$ representation contains no hypercolor singlets. The $(\underline{16},\underline{10}) \oplus (\underline{10},\underline{16})$ representation contains *three hypercolor-singlet generations* which transform like the vector representation of SU(2). This is then the horizontal flavor-chiral version of the Wilczek and Zee model.[22] The hypercolor β function contribution due to the $(\underline{16};\underline{7},\underline{1})$ fermions is

$$\beta = -\frac{g^3}{12\pi^2}\left(\frac{55}{4} - 8\right) . \tag{A3}$$

[In SO(N), $C_2 = N-2$ for the adjoint, and $T(R)=1$ for the vector representation.] The decomposition is thus asymptotically free and therefore provides a viable phenomenological alternative, and is worthy of further study. We did not pursue it here only because the hypercolor group is orthogonal rather than unitary. Though there is no compelling reason to use a unitary hypercolor group, we note that only unitary groups have a nontrivial topological structure, which could provide an origin for quantum number confinement.

(3) $SO(10)_H \to SO(5)\times SO(5)$. Here

$$\begin{aligned} \underline{16} &= (\underline{4},\underline{4}) , \\ \underline{10} &= (\underline{5},\underline{1}) + (\underline{1},\underline{5}) . \end{aligned} \tag{A4}$$

With SO(5) as the hypercolor group, we see that the $(\underline{16},\underline{16})$ representation contains no ordinary fermions; the $(\underline{16},\underline{10}) \oplus (\underline{10},\underline{16})$ representation contains five generations. The hypercolor β function due to the $(\underline{16};\underline{5},\underline{1})$ fermions is given by

$$\beta = -\frac{g^3}{12\pi^2}\left(\frac{33}{4} - 8\right) . \tag{A5}$$

Though β is negative, it is so small that it is unlikely to lead to dynamical symmetry breaking in the TeV region.

(4) $SO(10)_H \to SU(4)\times SO(2)$. Here we identify SU(4) as an SO(8) subgroup using the standard connection between SO($2N$) and SU(N) groups. Then

$$\underline{16}=(\underline{4}+\underline{4}^*,\underline{1})+(\underline{1}+\underline{6}+\underline{1},\underline{1}) ,$$
$$\underline{10}=(\underline{4}+\underline{4}^*,\underline{1})+(\underline{1},\underline{2}) . \tag{A6}$$

The $(\underline{16},\underline{16})$ representation then provides only two generations of ordinary fermions and hence can be set aside phenomenologically. The $(\underline{16},\underline{10})$ representation also contains only two generations; however, the $(\underline{10},\underline{16})$ representation contains some hypercolor singlets. Thus the representation $(\underline{16},\underline{10}) \oplus (\underline{10},\underline{16})$ contains altogether two ordinary $[SO(10)_V$ spinor] and two extraordinary $[SO(10)_V$ vector] fermionic generations. It leads to the topless model of Georgi and Glashow.[23] If this decomposition is to be viable experimentally (to accommodate the b quark), both the $(\underline{16},\underline{10})$ and $(\underline{10},\underline{16})$ representations of the fermions must be light. But then, however, the hypercolor β function is positive [see Eq. (7.5)]. Further, currently available experimental information favors b-quark assignment to an ordinary family. On these grounds one may reject this possibility.

(5) $SO(10)_H \rightarrow SU(3)\times SO(4)$. As in the previous case, here we identify SU(3) through its connection with SO(6). Then

$$\underline{16}=(\underline{3}+\underline{1},\underline{2})+(\underline{3}^*+\underline{1},\underline{2}') ,$$
$$\underline{10}=(\underline{3}+\underline{3}^*,\underline{1})+(\underline{1},\underline{4}) . \tag{A7}$$

Now the $(\underline{16},\underline{16})$ representation contains four generations as in the preferred case discussed in the paper. However, the hypercolor β function due to the $(\underline{16},\underline{16})$ fermions is

$$\beta=-\frac{g^3}{12\pi^2}(\tfrac{33}{4}-16) , \tag{A8}$$

which is not asymptotically free and so must be rejected. The $(\underline{16},\underline{10})$ representation contains four generations, while the $(\underline{10},\underline{16})$ representation this time also contains four generations. With altogether eight generations of light fermions, the hypercolor group is not asymptotically free as previously note in Sec. VII. With only the $(\underline{16},\underline{10})$ fermions light,

$$\beta=-\frac{g^3}{12\pi^2}(\tfrac{33}{4}-8) . \tag{A9}$$

This is again unrealistically small and so must be rejected.

Thus we see that there are only two choices allowed if we take into account the requirements of asymptotic freedom and consider it as established that more than two generations exist. These are case (2) and the case considered in the main body of the paper. The latter is the only case which gives *a unitary hypercolor group.*

[1]H. Georgi and S. L. Glashow, Phys. Rev. Lett. 32, 438 (1974).

[2]H. Georgi, in *Particles and Fields—1975,* proceedings of the Meeting of the Division of Particles and Fields, Williamsburg, Virginia, edited by Carl Carlson (AIP, New York, 1975), p. 575; H. Fritzsch and P. Minkowski, Ann. Phys. (N.Y.) 93, 193 (1975).

[3]L. Susskind, Phys. Rev. D 20, 2619 (1979).

[4]S. Weinberg, Phys. Rev. D 13, 974 (1976); 19, 1277 (1979).

[5]S. Dimopoulos and L. Susskind, Nucl. Phys. B155, 237 (1979).

[6]M. Gell-Mann, P. Ramond, and R. Slansky, in *Supergravity,* edited by P. Van Nieuwenhuizen and D. Z. Freedman (North-Holland, Amsterdam, 1979).

[7]A. Davidson and K. C. Wali, Phys. Rev. D 23, 477 (1981); N. G. Deshpande and P. D. Mannheim, Phys. Lett. 94B, 355 (1980); Phys. Rev. D 24, 2923 (1981).

[8]A. Davidson, K. C. Wali, and P. D. Mannheim, Phys. Rev. Lett. 45, 1135 (1980).

[9]A. Davidson, P. D. Mannheim, and K. C. Wali, Phys. Rev. Lett. 47, 149 (1981); 47, 620(E) (1981).

[10]J. C. Pati and A. Salam, Phys. Rev. D 10, 275 (1974).

[11]P. D. Mannheim, Phys. Lett. 85B, 253 (1979); Phys. Rev. D 22, 1729 (1980).

[12]P. D. Mannheim, University of Connecticut report, 1981 (unpublished).

[13]R. N. Mohapatra and G. Senjanovic, Phys. Rev. D 23, 165 (1981).

[14]See, e.g., E. Witten, in *First Workshop on Grand Unification, Durham, New Hamsphire, 1980,* edited by P. H. Frampton, S. L. Glashow, and A. Yildiz (Math Sci. Press, Brookline, Massachusetts, 1980).

[15]K. P. Das and N. G. Deshpande, Phys. Rev. D 19, 3387 (1979).

[16]S. Dimopoulos and J. Ellis, Nucl. Phys. B182, 505 (1981).

[17]N. G. Deshpande, R. C. Hwa, and P. D. Mannheim, Phys. Rev. D 19, 2703 (1979).

[18]The overall numerical factor $12\pi^2$ was unfortunately incorrectly given in Ref. 8.

[19]See for instance, H. Georgi, Nucl. Phys. B156, 126 (1979); P. H. Frampton, Phys. Lett. 89B, 352 (1980); J. Chakrabarti, M. Popovic, and R. N. Mohapatra, Phys. Rev. D 21, 3212 (1980).

[20]Y. Tosa and S. Okubo, Phys. Rev. D 23, 2486 (1981); 23, 3058 (1981).

[21]P. H. Frampton, Phys. Rev. Lett. 43, 1912 (1979); 44, 299(E) (1980).

[22]F. Wilczek and A. Zee, Phys. Rev. Lett. 42, 421 (1979).

[23]H. Georgi and S. L. Glashow, Nucl. Phys. B167, 173 (1980).

ABSOLUTE MINIMA OF A SO(10) INVARIANT HIGGS POTENTIAL

Ömer KAYMAKCALAN, Louis MICHEL[1] and Kameshwar C. WALI

Physics Department, Syracuse University, Syracuse, New York 13210, USA

W.D. McGLINN

Physics Department, University of Notre Dame, Notre Dame, Indiana 46556, USA

Lochlainn O'RAIFEARTAIGH

Dublin Institute for Advanced Studies, Dublin 4, Ireland

Received 12 June 1985
(Revised 15 October 1985)

The minima of a SO(10) invariant Higgs potential with the Higgs field transforming as a reducible $45 \oplus 54$ representation are studied. It is shown that there exist absolute minima of the potential in an open region of the full eleven-parameter space that characterizes the potential. The isotropy algebras of some of these absolute minima *contain* the physically interesting $\mathcal{H}_0 = SU(3) \times SU(2) \times U(1)$, but none of them correspond exactly to $\mathcal{H}_0$. There is a solution whose isotropy algebra is $\mathcal{H} = SU(3) \times SU(2) \times U(1)$, but it is not the desired physical isotropy algebra as it does not lead to the familiar quantum numbers of 16, 16 representations for the fermions.

1. Introduction

Spontaneous symmetry breaking (SSB) plays a crucial role in theories that attempt to unify all the fundamental, elementary particle interactions. At the present time, the Higgs mechanism seems to be indispensable in characterizing SSB, and therefore in spite of many of its shortcomings* which are discussed in the literature, it is necessary to study the mechanism in all its detail if we want to extract experimentally testable consequences from unified theories which exhibit their full symmetries only at very high energies.

[1] Permanent address: Institut des Hautes Études Scientifiques 91440, Bures-sur-Yvette, France.

* Unlike the gauge interactions, the Higgs interactions are not determined from the gauge principles alone. There is considerable arbitrariness in the choice of Higgs multiplets and the form of the Higgs potential. The gauge hierarchy and the problem of fine tuning are also problems inherent in the conventional grand unified theories which utilise Higgs mechanism for spontaneous symmetry breaking.

Simple in principle, SSB à la Higgs presents technical problems which are difficult to implement in practice. In a given unified gauge theory, SSB may proceed through several stages to the final one. The symmetry breaking pattern or chain is determined, firstly, by an appropriate choice of a combination of irreducible representations (IRPs) of Higgs fields, and secondly by a pattern of vacuum expectation values (VEV's) for these fields which minimizes a chosen Higgs potential. Given a set of Higgs representations and the requirement that a subalgebra such as $\mathcal{K}_0 = SU(3) \times SU(2) \times U(1)$ remains unbroken, one can deduce one or more symmetry breaking chains purely from group theory. Then the important question arises, whether any of the chains can be realized in practice by the chosen potential which involves, in general, several parameters. Minimization has to provide a domain in multiparameter space, in which a desired symmetry breaking pattern is realizable. Further, it needs to be established that the minimum of the potential is the absolute minimum and the solution is stable against small changes in parameter space.

A complete analysis along the above lines has been carried out only in a few instances* [1]. Model builders generally overlook the technical problems, and content themselves by assuming that a desired symmetry breaking chain is attainable, and proceed to draw some general consequences. While this may be good as a starting point, detailed quantitative predictions of a given model cannot be deduced in this way. A proliferation of models is the result, leaving one model indistinguishable from the other.

The main purpose of our paper is to elucidate some of the above discussed technical problems by analysing in detail a non-trivial problem involving a reducible combination of $\underline{45} \oplus \underline{54}$ Higgs representation in SO(10) grand unified theory. This combination of two IRPs, the adjoint and a symmetric traceless second rank tensor, forming a general second rank tensor field is a possible candidate for SSB in some SO(10) models [2].

In the next section, we discuss some general aspects of the isotropy groups of $SO(n)$ and $SO(n_1) \times SO(n_2)$ groups. In sect. 3, we study Lie algebras of the isotropy groups of the $(\underline{45} \oplus \underline{54})$ representation of SO(10) which contain $\mathcal{K}_0 = SU(3) \times SU(2) \times U(1)$. Sect. 4 is devoted to the discussion of the actual Higgs potential from these representations, a study of its minima, symmetry breaking patterns and the stability of the solutions. The final section contains a discussion of the results and some concluding remarks.

The second rank tensor Higgs field which we are considering in this paper is a representation of SO(10) and of course a representation of $\widetilde{SO}(10)$, the covering group of SO(10). The spinor representations used for the fermions are not representations of SO(10) but only of $\widetilde{SO}(10)$. In the body of the text, we consider either the isotropy groups as subgroups of SO(10) or only the Lie algebras of the isotropy

* See ref. [1]. Barring a few exceptions, the analysis carried out is generally for a single irreducible representation or reducible representation but confined to local minima.

TABLE 1
Some IRPs of SO(n) and their properties

Dynkin label	Dimension	Elements of carrier vector space $\mathcal{E}$	Faithful representations of: n odd	n even
(0,0,0,…,0,0)	1	Real numbers	none	none
(1,0,0,…,0,0)	n	n = component vectors	SO($2l-1$)	SO($2l$)
(2,0,0,…,0,0)	$\frac{1}{2}n(n+1)$	Symmetric traceless $n\times n$ matrices	SO($2l+1$)	$\widetilde{\mathrm{SO}}(2l)$
(0,1,0,…,0,0)	$\frac{1}{2}n(n-1)$	Anti-symmetric $n\times n$ matrices	SO($2l+1$)	$\widetilde{\mathrm{SO}}(2l)$

$\widetilde{\mathrm{SO}}(2l)$ in the last column denotes the quotient SO($2l$)/Z_2 where Z_2 is the center of SO($2l$) consisting of two elements 1 and −1. Thus $\widetilde{\mathrm{SO}}(2l)$ is the adjoint of group of SO($2l$).

groups. Because the $\underline{16}$ spinor representation is used in physics, the global gauge group is $\widetilde{\mathrm{SO}}(10)$ and the isotropy groups as subgroups of $\widetilde{\mathrm{SO}}(10)$ should be identified. This is done in appendix A. In appendix B, an alternate method is used to exhibit one of the solutions obtained in sect. 4 in a different region of parameter space of the Higgs potential.

2. Isotropy groups of some irreducible representations of SO(n) and SO(n_1) ⊕ SO(n_2)

Our purpose in this and the next section is to lead to table 2 which lists the possible solutions to the symmetry breaking problem purely from group theoretic consideration in the case of the reducible combination of the two irreducible representations $\underline{54} \oplus \underline{45}$. The possible solutions are required to contain $\mathcal{K}_0$ = SU(3) × SU(2) × U(1) as a surviving symmetry subalgebra.

Table 1 contains the irreducible representations of SO(n) of interest to us in this paper along with some pertinent facts. The last representations in table 1 are the adjoint representations defined by the action of the group on the vector space $\mathcal{E}$ of its Lie algebra.

The isotropy group (or the little group) G_x of $x \in \mathcal{E}$ is the *largest* subgroup of G which leaves x fixed. The set Gx of G-transforms of x is called the orbit of x. Isotropy groups of an orbit form a conjugation class of subgroups of G, which characterizes completely the type of the orbit. The union of orbits of the same type is called a *stratum*.

There is a natural order (by inclusion up to a conjugation) on the conjugation classes of closed subgroups of a compact group. This induces a partial ordering of the subgroups themselves. Hence the isotropy groups which are closed subgroups of the group G are partially ordered. It is a non-trivial theorem which states that there exists a unique *minimal isotropy group*, and that the corresponding stratum (called the generic stratum) is open dense. Excluding the group itself, there are in general several conjugation classes of *maximal isotropy groups* whose strata are closed.

In this section, we shall establish the list of isotropy groups appearing in the representations of SO(n) that we study and specify the minimal and maximal ones. Of course, the isotropy groups (or algebras) of some of these representations are discussed in the literature [1].

2.1 ISOTROPY GROUPS OF THE VECTOR REPRESENTATION

Any n-dimensional vector can be transformed by SO(n) transformations into any other vector of the same length. Hence it is sufficient to consider the first basis vector with components $(1,0,0,\ldots,0)$ whose isotropy group is SO($n-1$). Outside the origin (null vector), there is only one stratum. Thus, SO($n-1$) is both a minimal and maximal isotropy group.

More generally, for a reducible representation which transforms like a product of k-vector representations, the isotropy groups are SO($n-m$) for $1 \leqslant m \leqslant \min(k, n-2)$. Thus the maximal isotropy group is SO($n-1$) and the minimal isotropy group is SO($\max[n-k, 2]$).

2.2. ISOTROPY GROUPS OF REPRESENTATION (2,0,0,...,0,0)

The elements of the carrier vector space for this representation are the $n \times n$, real symmetric, traceless matrices. The action of G ≡ SO(n) on such a matrix S is given by

$$S \to gSg^{-1} = gSg^{\mathrm{T}},$$

where

$$g \subset \mathrm{SO}(n), \qquad S = S^{\mathrm{T}} = \bar{S}, \qquad \operatorname{tr} S = 0. \tag{2.1}$$

By this action, S can be diagonalized and its diagonal elements for $S \neq 0$ can be made to satisfy

$$\sigma_1 \geqslant \sigma_2 \geqslant \cdots \geqslant \sigma_{n-1} \geqslant \sigma_n, \qquad \sum \sigma_i = 0. \tag{2.2}$$

The isotropy group of such a diagonal matrix is the subgroup of orthogonal matrices that commute with it.

Now if all the σ's are distinct, the orthogonal matrices that belong to the isotropy group must themselves be diagonal: the diagonal elements are ± 1 with an even number of -1 in order that the special orthogonal matrix has determinant $+1$. It then follows that the *minimal* isotropy group is Z_2^{n-1}. (Z_2 is the two-element group).

More generally, if $p_1, p_2, \ldots$ are the multiplicities of the eigenvalues $\sigma_1, \sigma_2, \ldots$ so that

$$\sum p_i = n,$$

the isotropy group is

$$G_S = S\Big(\prod_i O(p_i)\Big), \tag{2.4}$$

where the S in front of the parenthesis specifies that the product of the det G_i, $G_i \in O(p_i)$ must be unity. Note $O(1) = Z_2$.

The maximal isotropy groups are those of the matrices with only two distinct eigenvalues,

$$S(O(p_1) \times O(p_2)); \qquad p_1 + p_2 = n, \qquad p_1 \geqslant p_2 > 0. \tag{2.5}$$

2.3. ISOTROPY GROUPS OF THE ADJOINT REPRESENTATIONS (0, 1, 0, ..., 0, 0)

The elements of the carrier vector space in this case are $n \times n$, real antisymmetric matrices L,

$$-L^{\mathrm{T}} = L = \bar{L}, \qquad \mathrm{ad}(G)L \to GLG^{-1} = GLG^{\mathrm{T}}, \tag{2.6}$$

where ad(G) denotes the group action. Any element of G is of the form $e^{\theta l}$. Equivalently, and more in keeping with the physicists' convention, we can consider the elements of the vector space as $n \times n$ antisymmetrical, hermitian matrices A,

$$A = A^{\dagger} = -A^{\mathrm{T}}, \tag{2.7}$$

with the Lie product law $A_1 \wedge A_2 \equiv \frac{1}{2}i[A_1, A_2]$. The action of the group is still given by (2.6). We note that if α_i is an eigenvalue of A, so is $-\alpha_i$. Hence if n is odd, there must be at least one eigenvalue zero.

Using the transformation property (2.6), we can bring any A to the form A_0, where

$$A_0 = \begin{bmatrix} \alpha_1 \tau_2 & & 0 \\ & \alpha_2 \tau_2 & \\ 0 & & \ddots \end{bmatrix}, \qquad \alpha_1 \geqslant \alpha_2 \geqslant \cdots \geqslant \alpha_l \geqslant 0, \qquad l = \left[\tfrac{1}{2}n\right], \tag{2.8}$$

where l, the largest integer $< \frac{1}{2}n$, is the rank of SO(n), and τ_2 is the 2×2 Pauli matrix.

The set of $\pm A_0$ form a maximal subalgebra of the Lie algebra, i.e., a Cartan subalgebra with the corresponding Cartan subgroup consisting of matrices of the form $e^{i\theta A_0}$,

$$e^{i\theta A_0} = \begin{bmatrix} I_2 - i\tau_2 \sin\theta_1 & & 0 \\ & I_2 - i\tau_2 \sin\theta_2 & \\ & & \ddots \\ 0 & & \end{bmatrix}, \qquad \theta_i = \alpha_i \theta, \tag{2.9}$$

where I_2 is a 2×2 unit matrix.

The matrices in (2.9) are isomorphic to $SO(2)^l \sim U(1)^l$. When the α's are non-degenerate, $SO(2)^l$ is the connected isotropy group of A_0 and the non-connected minimal isotropy group is $S(O(2)^l)$.

We now consider the case when k of the α's are equal. The corresponding $2k \times 2k$ matrix can be written $\alpha I_k \otimes \tau_2$ where I_k is $k \times k$ unit matrix.

Lemma. In the adjoint representation of $SO(2k)$, the isotropy group of $I_k \otimes \tau_2$ is isomorphic to $U(k)$.

Proof. Let u be a $k \times k$ unitary matrix and $\bar{u}$ its complex conjugate. Then there is a unique decomposition,

$$u = x + iy, \qquad x = \frac{u + \bar{u}}{2}, \qquad y = \frac{u - \bar{u}}{2i}, \tag{2.10}$$

where x and y are $k \times k$ real matrices.

Define

$$U = \begin{bmatrix} u & 0 \\ 0 & \bar{u} \end{bmatrix} = I_2 \otimes x + i\tau_3 \otimes y, \tag{2.11}$$

$$R = \sqrt{\tfrac{1}{2}}\,(I_2 - i\tau_1) \otimes I_k. \tag{2.12}$$

Then

$$V = RUR^{-1} = I_2 \otimes x - i\tau_2 \otimes y$$

$$= \begin{bmatrix} x & -y \\ y & x \end{bmatrix} \in SO(2k). \tag{2.13}$$

Permutations of rows and columns carries $V \to V' = x \otimes I_2 - iy \otimes \tau_2$, the set of matrices forming the isotropy group of $I_k \otimes \tau_2$.

The $2k$-dimensional real orthogonal representation of $U(k)$ is irreducible on the reals. It yields an inclusion $U(k) < SO(2k) < SO(n)$, $2k \leqslant n$ of $U(k)$ as subgroup of $SO(n)$. Moreover, $U(k)$ is a maximal subgroup of $SO(2k)$. Its center is made of the matrices of the form $I_k \otimes (I_2 \cos\theta - i\tau_2 \sin\theta)$. This real representation of $U(k) \subset$

SO($2k$) is the isotropy group of $I_k \otimes \tau_2$. When n is odd, the matrix A_0 in (2.8) has at least one zero eigenvalue. If q is the multiplicity of the zero-diagonal elements of A_0, the full SO(q) commutes with this zero $q \times q$ block. Then the connected isotropy groups of the adjoint representation of SO(n) are isomorphic to

$$\left(\prod_i \mathrm{U}(k_i)\right) \times \mathrm{SO}(q), \qquad q + \sum_i 2k_i = n. \tag{2.14}$$

The maximal isotropy subgroups are

$$\mathrm{U}(k) \times \mathrm{SO}(q), \quad 2k + q = n.$$

2.4. ISOTROPY GROUPS OF THE $n_1 \times n_2$ DIMENSIONAL IRREDUCIBLE REPRESENTATIONS OF SO(n_1) × SO(n_2)

We shall need the isotropy groups of the above representations in the next section. The carrier space consists of $n_1 \times n_2$ rectangular real matrices B. The action of the group $G = \mathrm{SO}(n_1) \times \mathrm{SO}(n_2)$ is given by

$$B \to g_1 B g_2^{-1} = g_1 B g_2^{\mathrm{T}}, \tag{2.15}$$

where $g_1 \in \mathrm{SO}(n_1)$ and $g_2 \in \mathrm{SO}(n_2)$. Then it follows that the real symmetric square matrices BB^{T} and $B^{\mathrm{T}}B$ are transformed according to

$$BB^{\mathrm{T}} \to g_1 BB^{\mathrm{T}} g_1^{\mathrm{T}}, \quad B^{\mathrm{T}}B \to g_2 B^{\mathrm{T}} B g_2^{\mathrm{T}}. \tag{2.16}$$

The isotropy group in this case is determined by the rank of B, which can be defined by the dimension of its range. The row rank and the column rank are the same, and rank B = rank BB^{T} = rank $B^{\mathrm{T}}B$. These ranks are invariant under the action (2.15). By chosing an appropriate pair g_1, g_2 we can, by the action (2.15), cast the matrix B in the form [3]

$$\begin{bmatrix} \begin{matrix} \lambda_1 & & 0 \\ & \lambda_2 & \\ & & \ddots & \\ 0 & & & \lambda_r \end{matrix} & 0_{r,n_2-r} \\ 0_{n_1-r,r} & 0_{n_1-r,n_2-r} \end{bmatrix}. \tag{2.17}$$

where $0_{i,j}$ is a $i \times j$ matrix with all zero entries. Hence under the action (2.15), the isotropy group of the matrix (2.17) is

$$S\left[\prod_i \mathrm{O}(p_i)\right] \times \mathrm{SO}(n_1 - r) \times \mathrm{SO}(n_2 - r), \tag{2.18}$$

where p_i is the degeneracy of λ_i and $\Sigma p_i = r$. ($O(p_i)$ occurs rather than $SO(p_i)$ because in the subspace of dimension p_i, one can multiply by -1 from the right and left.)

3. Isotropy Lie algebras of the $\underline{45} \oplus \underline{54}$ representation of SO(10), which contain $\mathcal{H} = \overline{SU(3) \times} SU(2) \times U(1)$

Given a direct sum $\mathcal{E} = \mathcal{E}_1 \oplus \mathcal{E}_2$ of two inequivalent, irreducible representations of a group G, the isotropy groups of $\mathcal{E}$ are the intersections of those of $\mathcal{E}_1$ and $\mathcal{E}_2$. More precisely, if

$$x = x_1 \oplus x_2, \quad x_i \subset \mathcal{E}_i,$$

then the isotropy group of G_x is

$$G_x = G_{x_1} \cap G_{x_2}. \tag{3.1}$$

In this section, we investigate the intersection of the isotropy groups of the 45 and 54 representations of SO(10) which have Lie algebras containing $\mathcal{H}_0 - \overline{SU(3) \times} SU(2) \times U(1)$. In fact we consider only the Lie algebras, neglecting the non-connected components of the isotropy groups and their topology.

We note that $\mathcal{H}_0$ is of rank 4, and hence any algebra that contains it has to be of at least rank 4. If we recall the isomorphisms of classical Lie algebras,

$$SO(2) \sim U(1), \quad SO(3) \sim SU(2), \quad SO(4) \sim SU(2) \times SU(2), \quad SO(6) \sim SU(4),$$

and also, as seen in subsect. 2.2, that the isotropy Lie algebras of 54 of SO(10) are of the form $S(\Pi O(p_i))$, $\Sigma p_i = n$, it follows that the smallest isotropy Lie algebra which contains $\mathcal{H}_0$ is

$$SO(6) \otimes SO(3), \tag{3.2}$$

which itself is a subalgebra of the three maximal isotropy Lie algebras,

$$SO(9), \qquad SO(7) \times SO(3), \qquad SO(6) \times SO(4). \tag{3.3}$$

Similarly, the smallest isotropy Lie algebra of the SO(10) adjoint representation which contains $\mathcal{H}_0$, according to subsect. 2.3, is $U(3) \times U(2)$. This algebra is itself a subalgebra of the maximal isotropy Lie algebras,

$$U(5), \qquad U(3) \times SO(4), \qquad U(2) \times SO(6) \tag{3.4}$$

of the adjoint representation.

We can now discuss the isotropy algebras of the $\underline{45} \oplus \underline{54}$ representation which contains $\mathcal{H}_0$.

The strategy is to start from an isotropy group G_S of the 54 representation containing a subgroup whose Lie algebra is $\mathcal{K}_0$. The isotropy groups of the action of G_S in the 45 representation are isotropy groups on the $54 \oplus 45$ representation. We keep those whose Lie algebra contains $\mathcal{K}_0$. By varying G_S we exhaust the possible isotropy groups.

Consider a maximal isotropy algebra $\mathcal{S} = SO(p) \times SO(q)$, $p + q = 10$ of the $\underline{54}$ representation of SO(10). It commutes with the matrix

$$S_{pq} = \begin{bmatrix} qI_p & 0 \\ 0 & -pI_q \end{bmatrix}. \tag{3.5}$$

Now to study the action on $\mathcal{A}$, the space of 10×10 antisymmetrical hermitian matrices, we decompose $A \in \mathcal{A}$ into a block structure,

$$A = \begin{bmatrix} A_p & iB_{pq} \\ -iB^{\mathrm{T}}_{pq} & A_q \end{bmatrix},$$

where $A^{\dagger}_p = A_p = -A^{\mathrm{T}}_p$; $A^{\dagger}_q = A_q = -A^{\mathrm{T}}_q$ and B_{pq} are real $p \times q$ matrices.

Observe that $SO(p) \times SO(q)$ action on $\mathcal{A}$ corresponds to a decomposition of A into irreducible subspaces

$$\mathcal{A} = \mathcal{A}_p \oplus \mathcal{A}_q \oplus \mathcal{B},$$

where $\mathcal{A}_p$, $\mathcal{A}_q$ are the carrier spaces for the adjoint representations of $SU(p)$ and $SU(q)$ respectively and $\mathcal{B}$ carries the pq dimensional representation of $SO(p) \times SO(q)$. The isotropy groups of B are given in subsect. 2.4, and, as shown there, the maximal isotropy group of B (other than $SO(p) \times SO(q)$) is $SO(p-1) \times SO(q-1)$. If p, q have values listed in (3.2) and (3.3), then the isotropy algebras of B do not contain $\mathcal{K}$. Thus a *necessary condition for the isotropy group $G_S \cap G_A$ to have an algebra which contains $\mathcal{K}$ is that B in (3.6) vanish, that is S and A commute*. This is an important result for our consideration of the minima of the Higgs potential.

To complete the classification of isotropy algebras containing $\mathcal{K}_0$, we have only to consider a pair of matrices S, A corresponding to (3.2), (3.3) and (3.4), and select the commuting pairs, note their isotropy algebras and compute their intersection. The results are given in table 2. For instance, consider

$$S_{64} = \left[\begin{array}{c|c} 4I_6 & \\ \hline & -6I_4 \end{array}\right], \quad A_3 = \left[\begin{array}{c|c} I_3 \times \tau_2 & 0 \\ \hline 0 & 0 \end{array}\right],$$

which commute. Their individual isotropy algebras are $S(O(6) \times O(4))$ and $U(3) \times SO(4)$ respectively. Their intersection $U(3) \times SO(4)$ contains $\mathcal{K}_0$.

Table 2 indicates there are seven possible solutions. In the next section we prove that at least four, marked A, B, C, and D, can be realized by an explicit Higgs

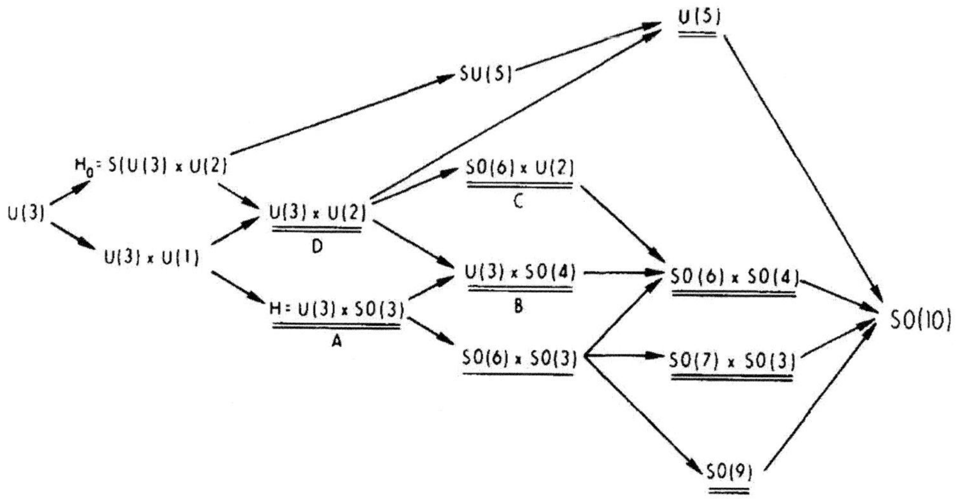

Fig. 1. The diagram gives the partial ordering of the conjugation classes of the SO(10) subgroups containing either of the two conjugation classes of H_0 and H which have the same Lie algebra $SU(3) \times SU(2) \times U(1)$. Isotropy groups of the $45 \oplus 54$ are underlined. Those which can be isotropy groups of a stable minimum of a Higgs potential have a double underline.

potential. However, we have to distinguish between isomorphic subalgebras which are not conjugated in SO(10). Indeed there are two distinct conjugation classes for $\mathcal{H}_0$. This can be seen in fig. 1 which shows the ordering of conjugation classes of some "large" subgroups of SO(10). Note that the subalgebra $\mathcal{H}$ of table 2 is a subalgebra of $SO(4) \times U(3)$ and not a subalgebra of SU(5) which is contained in U(5). The corresponding subgroup H in fig. 1 is not conjugated to the physical H_0. The latter is not an isotropy group - it is the intersection of $U(3) \times SO(4)$ and SU(5).

4. Higgs potentials on the $\underline{54} \oplus \underline{45}$ representations of SO(10) and their minima

We shall begin this section by a review of some formalism useful in analyzing Higgs potentials consisting of the reducible combination of two irreducible representations of the type $\underline{54}$ and $\underline{45}$. Although for the sake of definiteness, we concentrate on SO(10), the formalism should be useful for combinations of such representations (that is, those characterized by (2,0,0,...,0) and (0,1,0,...,0) Dynkin labels) for $SO(2n)$ groups.

4.1. GEOMETRICAL ASPECTS

Consider the real vector space $\mathcal{H}$ of 10×10 hermitian matrices which can be thought of as the Lie algebra of U(10) with the Lie product defined as usual by the commutator. For SU(10), such matrices provide a reducible sum of the singlet and the adjoint representation. Restricted to SO(10), they split into irreducible representations, $\underline{54}$ (symmetrical matrices), and $\underline{45}$ (anti-symmetrical matrices). A natural invariant scalar product for $X, Y \in \mathcal{H}$ is given by

$$(X, Y) = \operatorname{tr} XY = \operatorname{tr} YX = (Y, X). \tag{4.1}$$

If we define the linear map $\mathcal{H} \xrightarrow{F} \mathcal{L}(\mathcal{H})$, where

$$F_X Y = X_\wedge Y \equiv \tfrac{1}{2} i [X, Y], \tag{4.2}$$

then, according to (4.1), F_X is an antisymmetrical operator:

$$F_X^{\mathrm{T}} = -F_X, \quad \mathrm{T} \equiv \text{Transpose},$$

since for every $Y, Z \in \mathcal{H}$,

$$(Y, F_X Z) = (Y, X_\wedge Z) = -(F_X Y, Z) = -(X_\wedge Y, Z). \tag{4.3}$$

We can also introduce a symmetric non-associative SU(10) covariant algebra [4] on $\mathcal{H}$ by the use of a symmetric product law

$$X_\vee Y = Y_\vee X = \tfrac{1}{2}(XY + YX) - \tfrac{1}{10}(X, Y) I, \tag{4.4}$$

and define a linear map $\mathcal{H} \xrightarrow{D} \mathcal{L}(\mathcal{H})$, where

$$D_X Y = X_\vee Y, \qquad X, Y \subset \mathcal{H}.$$

D_X is a symmetrical operator,

$$D_X^{\mathrm{T}} = D_X; \qquad (Y, D_X Z) = (Y, X_\vee Z) = (D_X Y, Z) = (X_\vee Y, Z). \tag{4.5}$$

Let $\mathcal{S}$ and $\mathcal{A}$ be the vector spaces of the $\underline{54}$ and $\underline{45}$ representations of SO(10). Recall that their carrier spaces consist of hermitian real symmetric and hermitian, purely imaginary antisymmetric matrices respectively. Then it is straightforward to verify that

$$\mathcal{A}_\wedge \mathcal{A} = \mathcal{A}, \qquad \mathcal{S}_\wedge \mathcal{S} = \mathcal{A}, \qquad \mathcal{A}_\wedge \mathcal{S} = \mathcal{S} = \mathcal{S}_\wedge \mathcal{A}. \tag{4.6}$$

$$\mathcal{S}_\vee \mathcal{S} = \mathcal{S}, \qquad \mathcal{A}_\vee \mathcal{A} = \mathcal{S}, \qquad \mathcal{A}_\vee \mathcal{S} = \mathcal{A} = \mathcal{S}_\vee \mathcal{A}. \tag{4.7}$$

Finally, let $\mathcal{P}(\omega)$ be a real valued polynomial function defined on $\mathcal{H}$. Its *differential* $\mathrm{d}\mathcal{P}_X$ at X is defined to be the linear form,

$$\mathrm{d}\mathcal{P}_X(Y) = \lim_{\theta \to 0} \frac{1}{\theta}(\mathcal{P}(X + \theta Y) - \mathcal{P}(X)). \tag{4.8}$$

Then through the use of the scalar product, (4.1), we can define the gradient vector of $\mathcal{V}$, $\mathrm{d}\mathcal{V}(X)/\mathrm{d}X$ at X, as given by

$$\mathrm{d}\mathcal{V}_X(Y)=\left(\frac{\mathrm{d}\mathcal{V}(X)}{\mathrm{d}X},Y\right). \tag{4.9}$$

Similarly, we can define the second differential,

$$\mathrm{d}^2\mathcal{V}_X(Y,Z)=\mathrm{d}(\mathrm{d}\mathcal{V}_X(Y))(Z)=\mathrm{d}(\mathrm{d}\mathcal{V}_X(Z))(Y) \tag{4.10}$$

and the hessian $\mathrm{d}^2\mathcal{V}_X(Y,Z)/\mathrm{d}X^2$ as the symmetric operator given by

$$\mathrm{d}^2\mathcal{V}_X(Y,Z)=\left(Y,\frac{\mathrm{d}^2\mathcal{V}_X}{\mathrm{d}X^2}Z\right)=\left(\frac{\mathrm{d}^2\mathcal{V}_X}{\mathrm{d}X^2}Y,Z\right). \tag{4.10}$$

4.2. MOST GENERAL HIGGS POTENTIAL ON $\mathcal{S}\oplus\mathcal{A}$: ITS GRADIENT, AND ITS HESSIAN

A Higgs potential on $\mathcal{S}\oplus\mathcal{A}$ is a SO(10) invariant, real valued, inhomogeneous fourth degree polynomial $\mathcal{V}(X)$ bounded from below. We can write this as

$$\mathcal{V}(X)=\mathcal{V}_{\mathcal{S}}(S)+\mathcal{V}_{\mathcal{A}}(A)+\mathcal{V}_I(S,A), \tag{4.11}$$

where $X\in\mathcal{S}\oplus\mathcal{A}$, and

$$\mathcal{V}_{\mathcal{S}}=\tfrac{1}{4}\lambda_s(S,S)^2+\tfrac{1}{4}\alpha_s(S\vee S,S\vee S)+\tfrac{1}{3}\beta\mu_s(S\vee S,S)-\tfrac{1}{2}\mu_s^2(S,S), \tag{4.12}$$

$$\mathcal{V}_{\mathcal{A}}=\tfrac{1}{4}\lambda_a(A,A)^2+\tfrac{1}{4}\alpha_a(A\vee A,A\vee A)-\tfrac{1}{2}\mu_a^2(A,A), \tag{4.13}$$

$$\mathcal{V}_I=\tfrac{1}{2}\gamma_0(S\wedge A,S\wedge A)+\tfrac{1}{2}\gamma_1(S\vee S,A\vee A)+\tfrac{1}{2}\gamma_2(S,S)(A,A)+\gamma_3\mu_s(S,A\vee A), \tag{4.14}$$

where $S\in\mathcal{S}$ and $A\in\mathcal{A}$.

Only one third-degree term is possible in (4.14) since $(S,A\vee A)=(S\vee A,A)$. There is some ambiguity in the form of the fourth-degree terms because of the identity,

$$(S\vee A,S\vee A)+(S\wedge A,S\wedge A)-(S\vee S,A\vee A)-\tfrac{1}{10}(S,S)(A,A)=0, \tag{4.15}$$

which follows from the definitions of the products in (4.2) and (4.4). We have used this identity to write the quartic terms in (4.14).

The potential given in (4.11)–(4.14) contains eleven parameters*. In what follows, we shall assume, in order to include the possibility of no symmetry breaking (i.e.

* Since without any loss of generality we can rescale the parameters by dividing throughout by λ's, there are nine essential parameters.

zero is a minimum) in one or both the IRPs.

$$\mu_s^2, \mu_a^2 > 0, \tag{4.16}$$

and observe that to be bounded from below, the sums of all the quartic terms be non-negative for *any non-vanishing* $X \in \mathcal{S} \oplus \mathcal{A}$. From (4.4), we note that

$$S_{55} \vee S_{55} = 0, \qquad A_5 \vee A_5 = 0, \tag{4.17}$$

with $A_5 = I_5 \times \tau_2$. Since the quartic terms in $\frac{1}{4}\alpha_s\,(S \vee S, S \vee S)$ or $\frac{1}{4}\alpha_a\,(A \vee A, A \vee A)$ can vanish, the quartic terms $\frac{1}{4}\lambda_s(S_{55}, S_{55})^2, \frac{1}{4}\lambda_a(A_5, A_5)^2$ have to be positive in order to satisfy the boundedness requirement. This leads us to conclude

$$\lambda_s > 0, \qquad \lambda_a > 0 \tag{4.18}$$

and without any loss of generality to assume $\lambda_s = \lambda_a = 1$, since we can easily rescale the parameters in $\mathcal{V}$. Hence $\mathcal{V}$ depends on *nine* essential parameters.

In order that $\mathcal{V}(X)$ is minimum at X,

$$\frac{d\mathcal{V}}{dX}(X) = 0, \quad \frac{d^2\mathcal{V}}{dX^2}(X) \geqslant 0, \tag{4.19}$$

where by the inequality in (4.19) we mean for every $Z \in \mathcal{S} \oplus \mathcal{A}$,

$$\left(Z, \frac{d^2\mathcal{V}}{dX^2}(X)Z\right) \geqslant 0. \tag{4.20}$$

The vanishing of the scalar product in (4.20) will occur on the tangent plane to the SO(10) orbit at the minimum X. If the expectation value of the hessian would vanish on a larger subspace, it would mean that there are pseudo-Goldstone bosons in the theory. We shall return to this interesting point in subsect. 4.3.

From eqs. (4.9) and (4.10), it is straightforward to compute the gradient and the hessian of the Higgs potential.

Gradients:
$$\frac{d\mathcal{V}}{dX} = \frac{d\mathcal{V}_S}{dX} + \frac{d\mathcal{V}_A}{dX} + \frac{d\mathcal{V}_I}{dX}, \qquad X = \mathcal{S} \oplus \mathcal{A}. \tag{4.21}$$

$$\frac{d\mathcal{V}_S}{dS} = \left(\alpha_s D_S^2 + \beta\mu_s D_S + \lambda_s(S, S) - \mu_s^2\right)S, \qquad \frac{d\mathcal{V}_S}{dA} = 0. \tag{4.22}$$

$$\frac{d\mathcal{V}_A}{dA} = \left(\alpha_a D_A^2 + \lambda_A(A, A) - \mu_a^2\right)A, \quad \frac{d\mathcal{V}_A}{dS} = 0. \tag{4.23}$$

$$\frac{d\mathcal{V}_I}{dS} = -\gamma_0 F_A^2 + \gamma_1 D_{A,A}\hat{S} + \gamma_2(A, A)S + \gamma_3 A \vee A, \tag{4.24}$$

$$\frac{d\mathcal{V}_I}{dA} = \left(-\gamma_0 F_S^2 + \gamma_1 D_{S,S} + \gamma_2(S, S) + 2\gamma_3 D_S\right)A. \tag{4.25}$$

To write the hessians, it is convenient to introduce the orthogonal projection P_X of a vector Y along another vector X,

$$P_X = \frac{(X, Y)}{(X, X)} X$$

and the dyadic operator $|X\rangle\langle Y|Z = (Y, Z)X$. Then,

Hessians:

$$\frac{\mathrm{d}^2 \tilde{V}_S}{\mathrm{d}S^2} = 2\alpha_s D_S^2 + \alpha_s D_{S \vee S} + 2\beta\mu_s D_S + \left(\gamma_s (S, S) - \mu_s^2\right) I + 2\lambda_s (S, S) P_S , \tag{4.26}$$

$$\frac{\mathrm{d}^2 \tilde{V}_A}{\mathrm{d}A^2} = 2\alpha_a D_A^2 + \alpha_a D_{A \wedge A} + \left(\lambda_a (A, A) - \mu_a^2\right) I + 2\lambda_a (A, A) P_A , \tag{4.27}$$

$$\frac{\mathrm{d}^2 \tilde{V}_I}{\mathrm{d}S^2} = -\gamma_0 F_A^2 + \gamma_1 D_{A \wedge A} + \gamma_2 (A, A) I , \tag{4.28}$$

$$\frac{\mathrm{d}^2 \tilde{V}_I}{\mathrm{d}A^2} = -\gamma_0 F_S^2 + \gamma_1 D_{S \vee S} + \gamma_2 (S, S) I + 2\gamma_3 D_S , \tag{4.29}$$

$$\frac{\mathrm{d}^2 \tilde{V}_I}{\mathrm{d}S\,\mathrm{d}A} = \gamma_0 F_{A \wedge S} + \gamma_0 F_A F_S + 2\gamma_1 D_S D_A + 2\gamma_2 S\rangle\langle A + 2\gamma_3 D_A = \left[\frac{\mathrm{d}^2 \tilde{V}_I}{\mathrm{d}A\,\mathrm{d}S}\right]^{\mathrm{T}} . \tag{4.30}$$

TABLE 2

Isotropy algebras of the $\underline{54}$ and $\underline{45}$ and their intersections which contain $\mathfrak{k}$

	54 / 45	(i) SO(9)	(ii) SO(3) × SO(7)	(iii) SO(4) × SO(6)	(iv) SO(3) × SO(6)
a	U(5)			U(2) × U(3) (E)	
a'	SO(4) × U(3)	$\mathfrak{k}$	$\mathfrak{k}$	SO(4) × U(3) (B)	$\mathfrak{k}$ (A)
a''	U(2) × SO(6)			U(2) × SO(6) (C)	
a'''	U(3) × U(2)			U(3) × U(2) (D)	

Cases marked A, B, C, D are realizable by the absolute minima of a Higgs potential as explained in the text.

We now have the mechanism to investigate which of the seven cases of isotropy groups of table 2 can be the isotropy group of the minimum of a Higgs potential.

4.3. MINIMA OF THE GENERAL HIGGS POTENTIAL

The minima of $\mathcal{V}_S$ and $\mathcal{V}_A$ separately have been investigated in the literature [5]. For the sake of completeness and for our later use, we shall discuss briefly the results. In the case of $\mathcal{V}_A$, the required $\tilde{A}$ which satisfies (4.23) is of the form

$$\tilde{A} = \xi_k \hat{A}_k , \qquad \hat{A}_k = \frac{1}{\sqrt{2k}} \begin{bmatrix} I_k \times \tau_2 & 0 \\ 0 & 0 \end{bmatrix} , \tag{4.31}$$

where

$$(\hat{A}_k , \hat{A}_k) = 1 , \qquad (\tilde{A}, \tilde{A}) = \xi_k^2 , \qquad 1 \leqslant k \leqslant [\tfrac{1}{2}n] ,$$

$$\left(\alpha_a + \left(\lambda_a - \frac{\alpha_a}{n} \right) \right) \xi_k^2 = \mu_a^2 . \tag{4.32}$$

Therefore, apart from $\tilde{A} = 0$, there are $[\frac{1}{2}n]$ orbits of extrema labelled by k with the corresponding values of $\mathcal{V}_A$ given by

$$\mathcal{V}_A = - \frac{\mu_a^4}{4\lambda_a} \frac{1}{1 + (\alpha_a/\lambda_a)(n - 2k)/2nk} . \tag{4.33}$$

The *absolute minimum* of $\mathcal{V}_A$ depends on whether $\alpha_A \gtrless 0$, because once the length of A is fixed, the variation of $\mathcal{V}_A$ is due to the term $\alpha_a(A \vee A, A \vee A)$. Now it is easy to verify that

$$(A_k \vee A_k , A_k \vee A_k) = \frac{n - 2k}{2nk} . \tag{4.34}$$

so that:

(i) If $\alpha_a > 0$, the potential minimum is obtained for $k = [\frac{1}{2}n]$, and

$$\mathcal{V}_{A,\min} = - \frac{\mu_a^4}{4\lambda_a} \begin{cases} 1 , & n \text{ even} \\ \dfrac{1}{1 + (\alpha_a/\lambda_a)(n(n-1))^{-1}} , & n \text{ odd} . \end{cases} \tag{4.35}$$

The isotopy group in either case is $U([\frac{1}{2}n])$.

(ii) If $\alpha_a < 0$, the potential minimum is obtained for $k = 1$, and

$$\mathcal{V}_{A,\min} = - \frac{\mu_a^4}{4\lambda_a} \frac{1}{1 + (\alpha_a/\lambda_a)(n - 2)/2n} , \tag{4.36}$$

independent of whether n is even or odd. The isotropy group is $SO(n - 2) \times SO(2)$.

The case of $\mathcal{V}_S$ is slightly more complicated because of the presence of the cubic term. As shown by several authors, the extrema of $\mathcal{V}_S$ occur for matrices $\hat{S}$ with only two distinct eigenvalues (instead of the allowed three from the cubic equation for $\mathrm{d}\mathcal{V}_S/\mathrm{d}S = 0$), that is S is of the form

$$\tilde{S} = \xi_p \hat{S}_p, \qquad \hat{S}_p = \frac{S_{pq}}{\sqrt{npq}}, \qquad p + q = n, \qquad 1 \leqslant p \leqslant n - 1, \qquad (4.37)$$

where

$$S_{pq} = \begin{bmatrix} qI_p & 0 \\ 0 & -pI_q \end{bmatrix}.$$

Note that

$$\hat{S}_{n-p} = -\hat{S}_p, \quad \hat{S}_p \vee \hat{S}_p = \eta_p \hat{S}_p,$$

where

$$\eta_p = \frac{q-p}{\sqrt{npq}}. \qquad (4.38)$$

The extremum equation $(\mathrm{d}\mathcal{V}_S/\mathrm{d}S) = 0$ leads to the condition

$$\sqrt{\lambda_s + \alpha_s \eta_p^2}\, \xi_p^2 - 2\rho\mu_s \xi_p - \frac{\mu_s^2}{\sqrt{\lambda_s + \alpha_s \eta_p^2}} = 0, \qquad (4.39)$$

with the roots

$$\xi_p = \frac{\mu_s \sigma_\epsilon}{\sqrt{\lambda_s + \alpha_s \eta_p^2}}. \qquad (4.40)$$

where

$$\sigma_\epsilon = \rho + \epsilon\sqrt{1 + \rho^2}, \qquad \epsilon = \pm 1, \quad \rho = -\beta\eta_p / 2\sqrt{\lambda_s + \alpha_s \eta_p^2}.$$

The extremum values of the potential $\mathcal{V}_S(\tilde{S})$ are

$$\mathcal{V}_S(\tilde{S}) = -\frac{\mu_s^4}{12} \frac{\sigma_\epsilon^2(\sigma_\epsilon^2 + 2)}{\lambda_s + \alpha_s \eta_p^2}. \qquad (4.41)$$

and for a given p, $\epsilon = \operatorname{sign} \rho$ gives the lowest value for $\mathcal{V}_S$ in (4.41). Note that (i) the positivity condition on the quartic terms requires $\lambda_s > 0$, $\lambda_s + (n-2)^2/(n(n-1))\alpha_s > 0$, (ii) the value of the potential is independent of the sign of β; so is the isotropy group, and (iii) interchange of $p \to q = n - p$ introduces only redundant solutions.

Hence, we need consider only such values of p that make $\eta_p \leqslant 0$ and then $\varepsilon = \operatorname{sign} p = \operatorname{sign}(-\beta)$.

Although the above results are known in the literature, we shall study the p-dependence of the minima of $\tilde{U}_S$ in some detail, as we shall find the method useful in our later discussion. Since $\tilde{U}_S/\lambda_S$ depends only on α_s/λ_s, β/λ_s, and $\mu_s/\sqrt{\lambda_s}$, we can choose $\lambda_s = 1$ and study the function

$$\tilde{U}_p = \frac{\sigma_\varepsilon^2(\sigma_\varepsilon^2 + 2)}{\lambda_s + \alpha_a \eta_p^2}\,,$$

as a function of α_s and $\beta > 0$ for different values of p. For $N = 10$, with the restriction $\eta_p \leqslant 0$, we need to consider 5 functions $\tilde{U}_p$, $p = 5, \ldots, 9$. For fixed values of α_s, β, the minimum of the potential is determined by the maximum of the 5 values $\tilde{U}_p$. We can argue

(i) $|\eta_i| > |\eta_j|$ for $i > j$;

(ii) $\sigma_\varepsilon^2\,(\sigma_\varepsilon^2 + 2)$ is a positive function which increases with increasing p;

(iii) $\omega_i > \omega_j$ for $i > j$;

(iv) $\tilde{U}_5(\beta) = 3$, and when $\alpha_5 > 0$

$$3 = U_5(0) > U_6(0) > \tilde{U}_7(0) > U_8(0) > \tilde{U}_9 > 0\,.$$

(i), (ii), and (iii) imply that for $\alpha_s < 0$ the maximum $\tilde{U}_p$ is $\tilde{U}_9$. Thus for $\alpha_s < 0$, the isotropy group of the minimum is O(1). Further (iv) implies that the isotropy group in the positive α_s axis is SO(5) × SO(5). The curves of separation of the different basins in the α_s, β plane corresponding to the largest value of the five functions $\tilde{U}_p(\alpha_s, \beta)$ are among the curves Γ_{ij} defined by $\tilde{U}_i(\alpha_s, \beta) = \tilde{U}_j(\alpha_s, \beta)$. When the value $\tilde{U}_i$ on this curve is the largest value of all $\tilde{U}_p$ it is a basin boundary. At $\alpha = 0$ $\beta = 0$, $\tilde{U}_p = 3$ for all five functions so the curves Γ_{ij} exist and pass through the origin. We can compute the tangent at the origin for the curve Γ_{ij}

$$\left.\frac{d\alpha}{d\beta}\right|_{\substack{\alpha=0\\ \beta=0}} = \frac{4}{3}\,\frac{1}{\eta_i + \eta_j} > 0\,.$$

Since the slope of Γ_{5i} decreases with decreasing i, $\tilde{U}_5 = \tilde{U}_6$ with $d\alpha/d\beta = 1/|\eta_6|$ is the boundary of the basin $p = 5$ and $p = 4$. Similarly the boundary between 6 and 7 has the slope $d\alpha/d\beta = \frac{4}{3}\cdot 1/(|\eta_6| + |\eta_7|)$, between 7 and 8 the slope is $d\alpha/d\beta = \frac{4}{3}\cdot 1/(|\eta_7| + |\eta_8|)$ and finally between 8 and 9 the slope is $d\alpha/d\beta = \frac{4}{3}\cdot 1/(|\eta_8| + |\eta_9|)$. All these slopes refer to the boundary of the basins into the positive α, β quadrant. We have already noted that the basin for $\alpha < 0$ is that of $p = 1$ and that $\beta \to -\beta$ does not change the basin. The full characterization of the basins, which has been generated by computer calculation of the curves Γ_{ij}, is pictured in fig. 2.

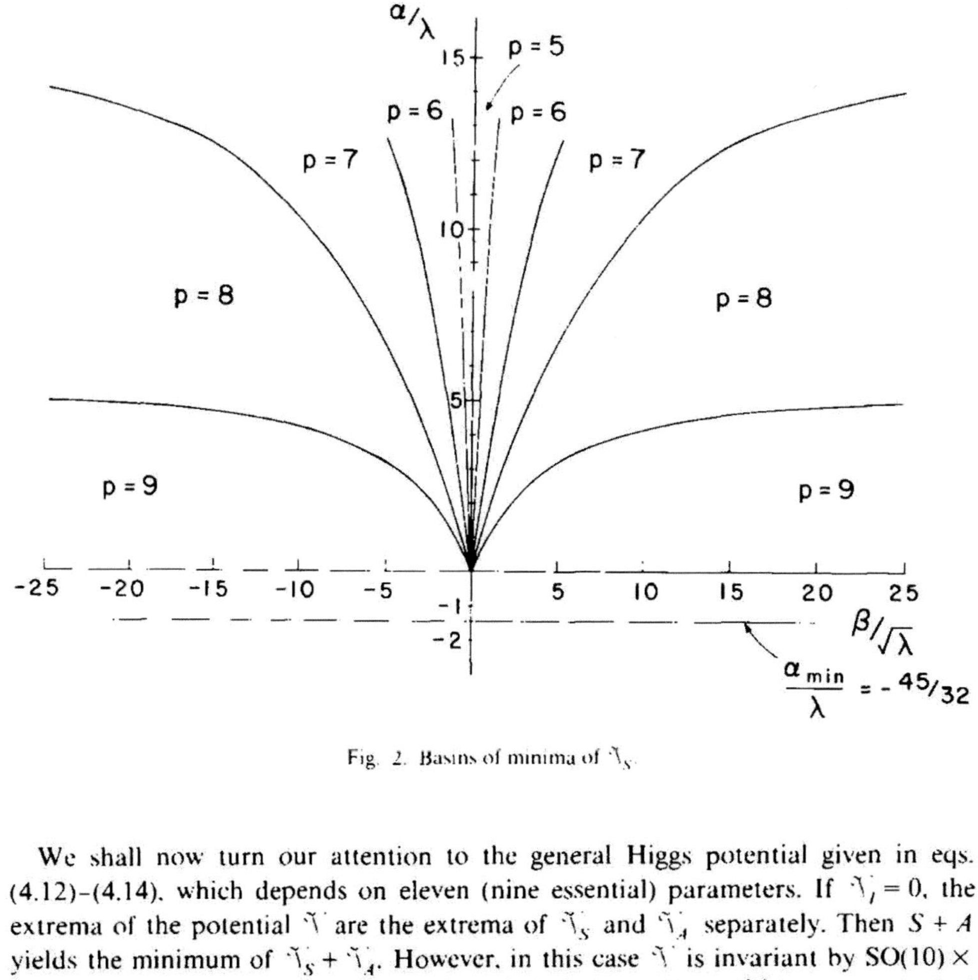

Fig. 2. Basins of minima of $\mathcal{V}_S$.

We shall now turn our attention to the general Higgs potential given in eqs. (4.12)–(4.14), which depends on eleven (nine essential) parameters. If $\mathcal{V}_I = 0$, the extrema of the potential $\mathcal{V}$ are the extrema of $\mathcal{V}_S$ and $\mathcal{V}_A$ separately. Then $S + A$ yields the minimum of $\mathcal{V}_S + \mathcal{V}_A$. However, in this case $\mathcal{V}$ is invariant by SO(10) × SO(10) acting on the space $\underline{45} \oplus \underline{54}$. The orbit of the minimum $\mathcal{V}$ is the topological product of the orbit of the minimum of $\mathcal{V}_S$ and $\mathcal{V}_A$. The "physical" symmetry group SO(10) is the diagonal subgroup of SO(10) × SO(10), the orbits of the latter being a union of an infinity of SO(10) orbits. Hence, there is a degeneracy leading to pseudo-Goldstone bosons. Adding $\mathcal{V}_I$ will partially resolve the degeneracy. Consider, for instance,

$$\mathcal{V} = \mathcal{V}_S + \mathcal{V}_S + \tfrac{1}{2}\gamma_0 (S \wedge A, S \wedge A), \qquad \gamma_0 > 0. \tag{4.42}$$

This interaction term is $\geqslant 0$, and hence on the SO(10) × SO(10) orbit of the minima of $\mathcal{V}_S + \mathcal{V}_A$, the SO(10) orbit of the minima $\mathcal{V}$ will occur when $S \wedge A = 0$ or $SA = AS$. This shows that the case marked E in table 2 is obtained with $\gamma_0 > 0$, $\alpha_A > 0$, and α_s and β in the basin 6 of fig. 2. Further it can be shown [6] that if this case is perturbed by small γ_1, γ_2, and γ_3 terms, the isotropy group remains unchanged.

The question we would like to answer is which of the cases allowed by group algebraic considerations can actually be realized as having absolute minima of an explicit Higgs potential *in an open subspace of the eleven parameter space*. In appendix B, we use a non-perturbative method to obtain the solution marked D in table 2. Here, by considering perturbations of the potential around a point where $\mu_a^2 = \gamma_1 = \gamma_2 = \gamma_3 = 0$, we show that solutions marked A, B, C, and D in the table are realizable. Henceforth, we shall consider μ_a, $\gamma_1, \gamma_2, \gamma_3$ as small parameters.

With $\mu_a^2 = \gamma_1 = \gamma_2 = \gamma_3 = 0$, A appears in the potential as follows:

$$\tfrac{1}{4}\lambda_a (A, A)^2 + \tfrac{1}{4}\alpha_a (A \vee A, A \vee A) + \tfrac{1}{2}\gamma_0 (S \wedge A, S \wedge A),$$

with the condition that the sum of the quartic terms > 0 if $A \neq 0$. The minimum occurs at $A = 0$ independent of S. Hence, the minimum of general $\mathcal{V}$ in this case is given by the minimum of $\mathcal{V}_S$ studied in the previous subsection. It is on the orbit $S = \xi_p S_p$ for a fixed p chosen by the relative values of α_s/λ_s, $\beta/\sqrt{\lambda_s}$ and μ_s^2.

When we perturb this solution with small μ_a^2, γ_1, γ_2, and γ_3, because of the condition $SA = AS$, we are led to consider the perturbed A, A_p, in the form

$$A_p = \begin{bmatrix} A' & 0 \\ 0 & A'' \end{bmatrix}, \tag{4.43}$$

where A' is $p \times p$, A'' is $(n-p)\times(n-p)$ anti-symmetrical hermitian matrices. If the value of S at the unperturbed minimum is denoted by S_0 and the perturbed minimum $S_0 + \delta S$, in the lowest non-vanishing order the condition for extremum of eq. (4.11) becomes

$$\frac{\partial \mathcal{V}}{\partial S}\bigg|_{\substack{S_0+\delta S=\\ A_p}} \frac{\partial^2 \mathcal{V}_S}{\partial S^2}\bigg|_{S_0} \delta S + \gamma_1 D_{A_p \vee A_p} S_0 + \gamma_2 (A_p, A_p) S_0 + \gamma_3 (A_p \vee A_p) = 0, \tag{4.44a}$$

$$\frac{\partial \mathcal{V}}{\partial A} = \left[\alpha_a D_{A_p}^2 + \lambda_a (A_p, A_p) - \mu_a^2\right] A_p$$

$$+ \gamma_1 D_{S_0} S_0 A_p + \gamma_2 (S_0, S_0) A_p + 2\gamma_3 D_{S_0} A_p = 0. \tag{4.44b}$$

From (4.37), we know that S_0 is of the form $\xi_p \hat{S}_p$. Using this and the form for A_p given in (4.43), in eq. (4.44b), we can show that

$$A' = \xi_{k'} A_{k'}, \qquad 1 \leqslant k' \leqslant \tfrac{1}{2}p,$$

$$A'' = \xi_{k''} A_{k''}, \qquad 1 \leqslant k'' \leqslant \tfrac{1}{2}(n-p). \tag{4.45}$$

where $\xi_{k'}$ and $\xi_{k''}$ satisfy the equations,

$$\xi_{k'}\left(\frac{\alpha_a\xi_{k'}^2}{2k} + (\lambda_a - \tfrac{1}{10}\alpha_a)(\xi_{k'}^2 + \xi_{k''}^2)\right) = \mu_{a'}^2\xi_{k'}, \tag{4.46}$$

$$\xi_{k''}\left(\frac{\alpha\xi_{k''}^2}{2k''} + (\lambda_a - \tfrac{1}{10}\alpha_a)(\xi_{k'}^2 + \xi_{k''}^2)\right) = \mu_{a''}^2\xi_{k''}, \tag{4.47}$$

with

$$\mu_{a'}^2 = \mu_a^2 - \gamma_2\xi_p^2 + \frac{\xi_p}{\sqrt{10p(10-p)}}(\gamma_1\xi_p\eta_p + 2\gamma_3)(n-p), \tag{4.48}$$

$$\mu_{a''}^2 = \mu_a^2 - \gamma_2\xi_p^2 + \frac{\xi_p}{\sqrt{10p(10-p)}}(\gamma_1\xi_p\eta_p + 2\gamma_3)(-p). \tag{4.49}$$

Then the part of $\mathcal{V}$ involving A takes the form

$$\mathcal{V}_A + \mathcal{V}_I = \frac{\alpha_a}{4}\frac{\xi_{k'}^4}{2k'} + \frac{\alpha_a}{4}\frac{\xi_{k''}^4}{2k''} + \tfrac{1}{4}(\lambda_a - \tfrac{1}{10}\alpha_a)(\xi_{k'}^2 + \xi_{k''}^2)^2$$
$$- \tfrac{1}{2}\mu_{a'}^2\xi_{k'}^2 - \tfrac{1}{2}\mu_{a''}^2\xi_{k''}^2. \tag{4.50}$$

We will impose the restriction $\alpha_a > 0$ and $(\lambda_a - \frac{1}{4}\alpha_a) > 0$. This is still an open region in α_a, λ_a parameter space. From (4.45), (4.48), and (4.49), we see that if $\mu_{a'}^2$, $\mu_{a''}^2 < 0$, the minimum occurs at $\xi_{k'} = \xi_{k''} = 0$. We are not interested in this case as that implies $A_p = 0$. Thus we should consider three cases:

$$\text{(i)}\ \mu_{a'}^2 > 0,\quad \mu_{a''}^2 > 0,\quad \text{(ii)}\ \mu_{a'}^2 > 0,\quad \mu_{a''}^2 < 0,\quad \text{(iii)}\ \mu_{a'}^2 < 0,\quad \mu_{a''}^2 > 0.$$

The cases (ii) and (iii) are easily handled. For (ii), $\mathcal{V}_A + \mathcal{V}_I$ is extremum at $\xi_{k''} = 0$ and $\xi_{k'}$, given by

$$\alpha_a\frac{\xi_{k'}}{2k'} + (\alpha_a - \tfrac{1}{10}\alpha_a)\xi_{k'}^2 = \mu_{a'}^2, \tag{4.51}$$

which has the same form as (4.32) with μ_a replaced by $\mu_{a'}$. The minimum will occur with $k' = [\frac{1}{2}p]$. Similarly for case (iii), $k'' = [\frac{1}{2}(n-p)]$.

Case (i) is more complicated. The value of the potential $\mathcal{V}_A + \mathcal{V}_I$ at an extremum satisfying (4.52) and (4.53) is given by

$$\mathcal{V}_A + \mathcal{V}_I = -\frac{(\mu_{a'}^2 - \mu_{a''}^2)}{4}\,\frac{k'a + k''b + k'k''c}{c(k' + k'') + 1}, \tag{4.52}$$

where

$$a = \frac{\mu^4_{a'}}{(\mu^2_{a'} - \mu^2_{a''})^2} > 0\,,$$

$$b = \frac{\mu^4_{a''}}{(\mu^2_{a''} - \mu^2_{a'})^2} > 0\,,$$

$$c = \frac{2(\lambda_a - \frac{1}{4}\alpha_a)}{\alpha_a} > 0\,. \tag{4.53}$$

The constants a, b, and c can take on any positive value with either a or b (or both) greater than one. The minimum of the potential is achieved with values k', k'' such that the expression in the second term in (4.52) is maximum.

We are now in a position to investigate which of the solutions of table 2 can be obtained by this perturbative technique.

(i) $p = 1$, $q = 9$, which corresponds to SO(9) case in table 2, $A' = 0$. The isotropy algebra of the perturbed A_p is U(4) × SO(2) which does not contain $\mathcal{H}$. Thus, although there is the isotropy algebra intersection H in 54 ⊕ 45, it cannot be obtained by a Higgs potential by the perturbative techniques we have investigated.

(ii) $p = 3$, $q = 7$: if $\mu^2_{a'} > 0$, $\mu^2_{a''} < 0$ (case (ii)), $k' = 1$, $A'' = 0$. The isotropy algebra of A_p is U(1) × SO(8) which does not contain $\mathcal{H}$. Likewise, if $\mu^2_{a'}$, $\mu^2_{a''} > 0$ (case (i)), the largest possible isotropy algebra of A_p is U(1) × U(3) × SO(2) which also does not contain $\mathcal{H}$.

This leaves the interesting case (iii) when $\mu^2_{a'} < 0$, $\mu^2_{a''} > 0$. Then $A' = 0$, $k'' = 3$. The isotropy algebra of A_p is SO(4) × U(3), and

$$A_p \wedge A_p \propto \begin{bmatrix} 6I_4 & 0 \\ 0 & 4I_6 \end{bmatrix}.$$

When we use this in eq. (4.44a), we find that the perturbed

$$(S_0 + \delta S) \propto \begin{bmatrix} aI_3 & & \\ & bI_1 & \\ & & cI_6 \end{bmatrix}$$

which has the isotropy algebra SO(3) × SO(6). The combined effect produces the solution marked A in the table. Notice that the isotropy algebra of the minimum is precisely $\mathcal{H}$. It is interesting to note that it *cannot* be obtained from an actual minimum of $\mathcal{V}_S$. As shown in the previous section SO(3) × SO(6) is not the isotropy group of the Higgs minimum of $\mathcal{V}_S$. Further, as noted before, this $\mathcal{K}$ is not conjugate to the Lie algebra of the physical group H_0.

(iii) $p = 4$, $q = 6$; U(5) will be the isotropy group of A_p, if $\mu^2_{a'} = \mu^2_{a''}$ which is a closed subset of parameter space which has no open subset. We have considered this case at the beginning of this section. If $\mu^2_{a'} < 0, \mu^2_{a''} > 0$ (case (iii)), the isotropy algebra of A_p $(k' = 0,\ k'' = 3)$ is SO(4) × U(3). $A_p A_p \propto S_0$ and hence from eq. (4.44a) the isotropy algebra of $(S_0 + \delta S)$ remains unchanged. Thus we obtain the solution B in the table. If $\mu^2_{a'} > 0,\ \mu^2_{a''} < 0$ (case (ii)), the isotropy algebra of A_p $(k' = 2,\ a'' = 0)$ is U(2) × SO(6). Again $A_p A_p \propto S_0$ and the isotropy algebra of the perturbed S remains unchanged. Solution C is obtained. Finally, when $\mu^2_{a'}, \mu^2_{a''} > 0$ (case (i)), a minimum with an isotropy algebra U(2) × U(3) of A_p can be obtained in an open region in parameter space. For instance, in a region where c is much larger than a and b, the minimum occurs for $k' = 2$ and $k'' = 3$. Again for this case, $A_p A_p \propto S_0$ and the isotropy algebra of S remains unchanged. The solution D is possible this way.

5. Concluding remarks

We have considered in this paper a reducible direct sum of two irreducible representations (54 + 45) of SO(10) as the Higgs multiplet responsible for spontaneous symmetry breaking. By looking at the separate isotropy groups of 54 and 45 that contain the desired surviving symmetry algebra $\mathcal{K}$ = SU(3) × SU(2) × U(1) and their intersections that contain $\mathcal{K}$, we can limit the possible symmetry breaking patterns from group algebraic considerations.

This in general is not enough for there to exist a Higgs potential with an absolute minimum having such isotropy algebras. Hence, we have examined the most general, renormalizable Higgs potentials, first finding a general criterion for an extremum to contain $\mathcal{K}$ $(AS = SA)$. This Higgs potential contains eleven (nine essential) parameters. We require that for an open subspace in this eleven parameter space, the absolute minima have the same isotropy algebra. We consider this a necessary requirement which is not emphasized sufficiently in the literature. An absolute minimum for a specific set of ad hoc parameters is not sufficient, for radiative corrections generally alter the value of the parameters which in turn could change the isotropy group of the minimum*.

* After the completion of this work, a paper by Babu and Ma [7] came to our attention. In this paper, the authors discuss the Higgs potential containing the 45 ⊕ 54 ⊕ 16 ⊕ $\overline{16}$, but the "minimum" of the Higgs potential is not shown to be an absolute minimum. It is also not clear to us that in fact it is a local minimum since the most general variation of the Higgs field is not considered. Further, since restrictions on the parameters of the potential (such as the vanishing of the cubic term of the 54) are imposed, the stability of the solution under small changes of the potential is not assured.

By using perturbative techniques, we have found near a point where the absolute minimum has the $\underline{45}$ field $\tilde{A} = 0$, open regions in the full eleven parameter space where the absolute minimum of the SO(10) invariant potential has isotropy algebras,

$SO(4) \times U(3)$, solution marked B in table 2.

$SU(2) \times SO(6)$, solution marked C in table 2.

$U(3) \times U(2)$, solution marked D in table 2,

and finally, a solution whose isotropy algebra $\mathcal{H} = SU(3) \times SU(2) \times U(1)$. However, this last case, is not physical. It does not lead to the familiar assignments of quantum numbers to the fermions belonging to $\underline{16}$ and $\overline{16}$ spinor representations.

Finally, although we have concentrated our attention on a particular reducible representation, we think that the methods and formalism presented are useful in other cases. Of course, the physical consequences of the SO(10) model that emerges from these minima is another matter and that needs further investigation.

This work was supported in part by the US Department of Energy under contract No. DE-AC02-76ER03533. W. D. McGlinn would like to thank the Syracuse High Energy Theory Group for its hospitality during his stay when this work was started. He would also like to thank R. Thornburg for useful discussions.

This paper is dedicated to the memory of Ömer Kaymakcalan whose untimely death on February 22, 1985, due to leukemia, ended a brilliant career.

Appendix A

The purpose of this appendix is to clarify the association between Lie algebras and the corresponding Lie groups*. Generally there exist several non-isomorphic, compact Lie groups G_i which have the same Lie algebra $\mathcal{G}$. The question arises as to which of these groups is the physically chosen group defined by the global transformations. The physical consequences of the choice of G are of two types:

(i) the different G_i's have different sets of representations and hence generally different sets of internal quantum numbers that label the states of the irreducible representations.

(ii) The different G_i's have different first homotopy groups; this is relevant to the nature of the monopole charges.

* For instance, see ref. [8]. They discuss the physical significance of such association and go on to deduce the global isotropy groups for the case of the Higgs field in the adjoint representation with a resulting isotropy algebra of $U(1) \times K$, where K is semi-simple. They assume the unbroken group $\tilde{G}$ is simple and simply connected, which obtains if the matter fields and their products contain all irreducible representations. Here we mainly consider $\tilde{G} = \widetilde{SO}(10)$ and deduce the isotropy groups when the Higgs fields transform as *a reducible* $\underline{45} + \underline{54}$.

Here we will consider only the first type of consequences.

For every simple compact Lie algebra $\mathcal{G}$, there correspond (a) a unique, simply connected Lie group $\tilde{G}$ whose center $C(\tilde{G})$ is finite; $\tilde{G}$ is called the covering group, (b) a unique group $\underset{\sim}{G}$ without a center; $\underset{\sim}{G}$ is called the adjoint group, and (c) other connected groups of the type $\tilde{G}/F$ where F is a subgroup of $C(\tilde{G})$.

Although our primary concern in this paper is the Lie algebra of SO(10) and the corresponding Lie groups, let us first consider the standard model of electroweak and strong interactions which is often stated as based on the *gauge group* $SU(3)_C \times SU(2) \times U(1)$. For this purpose, we observe that every compact Lie group G can be written as a quotient of the type

$$G = N/F; \qquad N = (\chi_1 \times \chi_2 \times \cdots \chi_p) \times U(1)^q,$$

where χ_i are simple, simply connected compact Lie groups and F is a finite subgroup of the center of N. For instance,

$$U(l) = (SU(l) \times U(1))/Z_l, \tag{A.1}$$

where

$$Z_l = \{(I_l e^{2\pi ik/l}, e^{-2\pi ik/l}), 0 \leqslant k \leqslant l\}.$$

Now in the standard Dynkin diagram of $SU(l)$ with Dynkin labels $(b_1, b_2, \ldots, b_{l-1})$, the center element $\phi_k = I_l e^{2\pi ik/l}$ of $SU(l)$ is represented by $I\phi_k^\beta$ with

$$\beta = \sum_{j=1}^{l-1} jb_j. \tag{A.2}$$

and similarly the element $e^{i\phi}$ of U(1) is represented by $e^{im\phi}$ in the representation of U(1) labelled in m. Since the IRP's of $U(l)$ must contain Z_l in their kernel,

$$e^{2\pi i(\beta - m)k/l} = 1, \quad 0 \leqslant k \leqslant l. \tag{A.3}$$

Hence,

$$\beta = m(\text{mod } l). \tag{A.4}$$

From the assignments of the quantum numbers in the SU(2) × U(1) theory of electroweak interactions we can easily deduce that the relation (A.4) with $l = 2$ and β, m corresponding to SU(2) and U(1) are satisfied. Hence, *the gauge group* is U(2) and not SU(2) × U(1). Similarly, the strictly conserved gauge symmetry group of chromodynamics and electrodynamics is U(3) and not SU(3) × U(1) to explain the fractional electrical charges of the quarks; the exact gauge group of the low-energy physics is not SU(3) × SU(2) × U(1), but

$$H = (SU(3) \times SU(2) \times U(1))/Z_6. \tag{A.5}$$

where Z_6 is the center of SU(3) × SU(2) × U(1). Indeed it is the gauge symmetry of H that can be enlarged to SU(5), since it is H and not SU(3) × SU(2) × U(1) which is a subgroup (maximal isotropy group of the adjoint) of SU(5).

Returning to the case of SO(10), its universal covering group $\widetilde{SO}(10)$ (spinor 10) is the unique simply connected Lie group with the finite center consisting of four elements $Z_4 = \{1, \zeta, \zeta^2, \zeta^4\}$ and of course the same Lie algebra as SO(10). Further $\overline{SO}(10) = \widetilde{SO}(10)/Z_4$ and $SO(10) = \widetilde{SO}(10)/Z_2$ with $Z_2 = \{1, \zeta^2\}$. The distinction between SO(10) and $\widetilde{SO}(10)$ becomes necessary when fermions belonging to the spinor representations characterized by Dynkin labels (0,0,0,1,0) and (0,0,0,0,1) are included in the theory. These are not representations of SO(10) or $\overline{SO}(10)$, but only of $\widetilde{SO}(10)$. With this in mind, we discuss the isotropy groups of the minima A, B, C, D of fig. 1.

We begin by noting the following results,

$$C(\widetilde{SO}(N)) = \begin{cases} Z_2 & \text{for } n \text{ odd} \\ Z_4 & \text{for } n = 2 \bmod(4) \\ Z_2 \times Z_2 & \text{for } n = 0 \bmod(4) . \end{cases} \tag{A.6}$$

Let ε be the element of the center of SO(n) which generates the group $Z_2(\varepsilon)$ so that $SO(n) = \widetilde{SO}(n)/Z_2(\varepsilon)$. Then, the connected subgroups of the maximal isotropy groups of the symmetric tensor representation of $\widetilde{SO}(n)$ are

$$(\widetilde{SO}(p_1) \times \widetilde{SO}(p_2))/Z_2(\varepsilon_1, \varepsilon_2) , \tag{A.7}$$

where $p_1 + p_2 = n$. Now in the case of $n = 10$, the isotropy group of interest K has $p_1 = 6$ and $p_2 = 4$. The isomorphisms $\widetilde{SO}(6) \sim SU(4), \widetilde{SO}(4) \sim SU(2) \times SU(2)$ show that K given by

$$K = (SU(4) \times SU(2) \times SU(2))/Z_2(-I_4, -I_2, -I) \tag{A.8}$$

is a subgroup of $\widetilde{SO}(10)$. Then the isotropy groups of the minima B, C, D, which are subgroups of K, can be shown to be

$$\begin{aligned} &B: \quad [SU(2) \times SU(2) \times SU(3) \times U(1)]/Z_6(\omega); \quad \omega = (-I_2, -I_2, e^{2\pi i/3} I_3, e^{\pi i/3}) , \\ &C: \quad [SU(2) \times U(1) \times SU(4)]/Z_2(\omega); \quad \omega = (-I_2, -1, -I_4) , \\ &D: \quad [SU(2) \times U(1) \times SU(3) \times U(1)]/Z_6(\omega), \quad \omega = (-I_2, -1, e^{2\pi i/3} I_3, e^{\pi i/3}) , \end{aligned} \tag{A.9}$$

by noting that the coverings of U(2) in SO(4), U(3) in SO(6) are SU(2) × U(1) in SU(2) × SU(2) and U(3) in SU(4) respectively.

We leave it to the reader to check that the isotropy group of solution A (which is a subgroup of that of B) is isomorphic to the group H = [SU(3) × SU(2) × U(1)]/Z_6. However, as we emphasized in the text, it is not conjugated in $\widetilde{\mathrm{SO}}(10)$ to the physical H_0 (see fig. 1).

Appendix B

It is interesting that the most general potential in (4.11) can be expressed in the form

$$V = a\left((S,S) - m_s^2\right)^2 + b\left((A,A) - m_a^2\right)^2 + \frac{c}{m_a^2 m_s^2}\left(m_a^2(S,S) - m_s^2(A,A)\right)^2$$

$$+ \sigma(S_\wedge A, S_\wedge A) + f(F,F) + g(G,G), \qquad \text{(B.1)}$$

where

$$F = \cos\alpha S_\vee S + m_s \sin\alpha S, \qquad \text{(B.2)}$$

$$G = \cos\beta A_\vee A + \sin\beta(\cos\gamma S_\vee S + m_s \sin\gamma S). \qquad \text{(B.3)}$$

Of course there is a non-trivial connection between the parameters of (B.1) and those of (4.11). If the coefficients a, b, c, σ, f and g are chosen positive, V is a sum of positive semi-definite terms. One could them exhibit an absolute minimum of V by restricting the other parameters such that the individual terms are zero for some value of the fields. The stability of the isotropy algebra of this minimum is implied if there exists no pseudo-Goldstone modes at this minimum*.

The absolute minimum value of V is zero if

$$(\tilde{S}, \tilde{S}) = m_s^2, \qquad \text{(B.4)}$$

$$(\tilde{A}, \tilde{A}) = m_a^2, \qquad \text{(B.5)}$$

$$[\tilde{S}, \tilde{A}] = 0, \qquad \text{(B.6)}$$

$$F = 0, \qquad \text{(B.7)}$$

$$G = 0. \qquad \text{(B.8)}$$

* The isotropy group of the minimum of an invariant potential is unchanged for any small invariant perturbation of the *potential* if there are no pseudo-Goldstone modes. This result which is perhaps well known is discussed in ref. [6].

Eqs. (B.4) and (B.7) imply

$$\tilde{S} = \frac{m_s}{\sqrt{npq}} \begin{bmatrix} pI_q & 0 \\ 0 & -qI_p \end{bmatrix}, \tag{B.9}$$

$$\tan\alpha = -\frac{p-q}{\sqrt{npq}}, \tag{B.10}$$

with $p+q=n$.

(B.6) implies $\tilde{A}$ can be written

$$\tilde{A} = \begin{bmatrix} A' & 0 \\ 0 & A'' \end{bmatrix} \tag{B.11}$$

and then from (B.8).

$$\begin{bmatrix} A'^2 & 0 \\ 0 & A''^2 \end{bmatrix} = \begin{bmatrix} a'I_q & 0 \\ 0 & a''I_p \end{bmatrix}, \tag{B.12}$$

where

$$a' = \frac{m_a^2}{n} - p\frac{\tan\beta\sin(\gamma-\alpha)}{\cos\alpha\sqrt{npq}} m_s^2,$$

$$a'' = \frac{m_a^2}{n} + q\frac{\tan\beta}{\cos\alpha}\frac{\sin(\gamma-\alpha)}{\sqrt{npq}} m_s^2.$$

If $a'=0$, $a''\neq 0$, then $A'=0$ and $A''=\alpha I_{p/2}\times\tau_2$. The isotropy algebra is $SO(q)\times U(p)$.

If a', $a''\neq 0$, then $A'=\alpha I_{q/2}\times\tau_2$ and $A''=\alpha I_{p/2}\times\tau_2$. The isotropy algebra is $U(q)\times U(p)$.

The "sum of squares" formulation of the potential (B.1) is well suited to investigate pseudo-Goldstone modes. The potential is of the form

$$V = \sum_i C_i(E_i, E_i), \qquad C_i > 0,$$

$$\delta V|_{V=0} = \sum_i C_i(\delta E_i, \delta E_i). \tag{B.13}$$

Thus Goldstone and pseudo-Goldstone modes satisfy

$$\delta E_i = 0 \qquad \text{for all } i. \tag{B.14}$$

With the use of (B.14), one can show that in the case a', $a''\neq 0$, there are no pseudo-Goldstone modes. Hence, one obtains a stable solution corresponding to the ones marked D when $a' \neq a''$, and E when $a' = a''$, in table 2. However, when a' or a'' vanishes, pseudo-Goldstone modes exist*. The isotropy algebras in this case correspond to the solutions marked B and C on the table. This implies that, stable absolute minima with the specified isotropy groups cannot be obtained by this technique.

References

[1] J.S. Kim, Nucl. Phys. B196 (1982) 285;
Dan-di Wu, Nucl. Phys. B199 (1982) 523;
H. Ruegg, Phys. Rev. D22 (1980) 2040;
F. Buccella, L. Cocco and C. Wetterich, Nucl. Phys. B243 (1984) 273

[2] H. Georgi, in Particles and fields (AIP Press, NU, 1975) p. 575;
H. Fritzsch and P. Minkowski, Ann. of Phys. (NY) 93 (1975) 193

[3] G. Birkhoff and S. Maclane, A survey of modern algebra (Macmillan, NY, 1953), p. 234

[4] L.C. Biedenharn, J. Math. Phys. 4, (1963) 436;
L. Michel and L. Radicatti, Ann. Inst. Henry Poincaré 18 (1973) 85

[5] L.F. Li, Phys. Rev. D9 (1974) 1725;
D. Ruegg, in ref. [2]

[6] R. Thornburg and W.D. McGlinn, An absolute minimum and mass eigenstates of a SO(n)-invariant potential with a reducible second rank tensor higgs field, Notre Dame preprint

[7] K.S. Babu and E. Ma, Phys. Rev. D31 (1985) 2316

[8] P. Goddard and D. Olive, Nuc. Phys. B191 (1981) 511

* We are indebted to R. Thornburg for pointing this out to us.

12. Introduction to Grand Unified Theories

K. C. WALI
Physics Department, Syracuse University, Syracuse, NY 13244, U.S.A.

1. Grand Unification – A Survey of Basic Ideas

1.1. Introduction

Grand unified theories (GUTs) postulate that the description of the elementary particles and their interactions will simplify enormously at some very high energy E,

$$E > M_{\text{GUM}} \text{(grand unification mass)}.$$

The fundamental interactions which we distinguish and recognize as weak, electromagnetic, and strong at low or laboratory energies, will become the manifestations of one gauge interaction above M_{GUM}. This unified interaction will operate between a set of basic constituents of all matter. In its most pristine form, such a unified theory is described by a Lagrangian $\mathscr{L}$, where

$$\mathscr{L} = -\tfrac{1}{4}\mathbf{G}_{\mu\nu}\mathbf{G}^{\mu\nu} - \tfrac{1}{2}\bar{F}\not{D}F, \tag{1}$$

with $\mathbf{F}$ being a set of fermions belonging to an irreducible representation of a grand unification group G. $\mathbf{G}_{\mu\nu}$ represents the field tensor corresponding to the gauge bosons belonging to the adjoint representation of G and D is the covariant derivative such that $\mathscr{L}$ is invariant under a set of local, non-Abelian gauge transformations on F and $\mathbf{G}_{\mu\nu}$.

The simple-looking Lagrangian (1) ideally contains the secret to everything – the unfolding of the universe from the instant of its creation to its present state. Unfortunately, we do not have the necessary calculational tools. In the absence of the latter, it becomes necessary to add additional terms to $\mathscr{L}$ in (1) to make it amenable to practical calculations. It is clear that the full symmetry of G does not manifest itself in the low energy region. The symmetry of G has to be broken. This symmetry breaking could, in principle, arise from the dynamics described by the Lagrangian (1) (dynamical symmetry breaking), but in the absence of reliable calculational techniques, at least as a practical recourse, it becomes necessary to introduce fundamental scalars (Higgs mesons) and add $\mathscr{L}_\phi$

to $\mathscr{L}$, where

$$\mathscr{L}_\phi = -|D\phi|^2 + h\bar{F}\phi F - V(\Phi), \tag{2}$$

in which the first term represents the kinetic energy of the scalar fields, including the interactions of the scalar mesons with the gauge bosons. The second term represents the fermion-Higgs couplings and the third term, the Higgs potential, contains the self-interactions of the ϕ's. Initially, $(\mathscr{L} + \mathscr{L}_\phi)$ is constructed so that it is gauge invariant under local gauge transformations belonging to G, but by giving nonvanishing vacuum expectation values to appropriate ϕ's, the symmetry of G is broken in one or several stages to $SU(3)_C \times SU(2) \times U(1)$ and eventually to $SU(3)_C \times U(1)_{EM}$. The gauge bosons other than those that belong to the residual symmetry, acquire mass at the expense of the would-be Goldstone bosons and their effects at low energies generally become suppressed by inverse powers of their masses. Fermions and the surviving Higgs mesons can also acquire masses in the process of symmetry breaking. However, the implementation of symmetry breaking this way is the most troublesome feature of the theory, as it introduces considerable arbitrariness. Given the fermionic content, gauge invariance principles establish uniquely the gauge boson content and the gauge interactions in the fundamental Lagrangian $\mathscr{L}$. This cannot be said of the Higgs part of the Lagrangian. The choice of the Higgs representations, the form of the Higgs potential, and the fermion-Higgs coupling parameters are not fixed by gauge principles alone. Consequently, the particle spectrum and the particle masses are not completely predictable.

In addition, at the outset, one may raise several questions of a fundamental nature. Are the basic fermions F, quarks and leptons as we know them in the low energy region, or are they ultimately composites of even more 'elementary' constituents called 'preons'? If the scalars are fundamental, should one regard them in a symmetric relation with the fundamental fermions as in supersymmetric models, or should they be regarded as composites or condensates of the fundamental fermions as in dynamical symmetry breaking approach? At present, we believe there are three families or generations of fermions (quarks and leptons) which have identical $SU(3)_C \times SU(2) \times U(1)$ quantum numbers. What is the origin of this degenerate fermionic family structure?

In spite of these fundamental questions, we shall assume for the most part of this article that quarks and leptons are the fundamental constituents and explore some general consequences of GUT ideas in this section. Subsequently, we shall focus our attention on a specific GUT model based on $G \equiv SU(5)$ to illustrate some of the calculational details leading to quantitative results.

1.2. Motivation for Grand Unification

The strong motivation for grand unification arises mainly from the remarkably successful theory of Glashow, Salam and Weinberg, which describes electromagnetic and weak interactions (electro weak) in a 'unified' manner. This theory, for energies $E > 100\,\text{GeV}$, has as its starting point a Lagrangian of the form (1) in which the relevant group $G_{\text{EW}} = \text{SU}(2) \times \text{U}(1)$. The left-handed (right-handed) components of known quarks and leptons transform as doublets (singlets) under the 'weak iso-spin' group SU(2) of G_{EW}; they carry a 'weak hyper-charge' corresponding to the U(1) of G_{EW}. The four gauge bosons of G_{EW} are also massless to begin with.

A single complex doublet of Higgs scalars is used to break the symmetries of G_{EW} (retaining the renormalizability of the theory). When the neutral component of the Higgs doublet is given a nonvanishing vacuum expectation value, three of the four gauge bosons and the fermions acquire mass. The remaining massless gauge boson can be identified with the photon. The chosen mass scale at which the symmetry breaking occurs, distinguishes weak and electromagnetic interactions at low energies. The weak interactions are described by effective four-fermion, V-A, charged and neutral current interactions. The electromagnetic interactions are described in the usual fashion by the surviving $\text{U}(1)_{\text{EM}}$ gauge symmetry. Combining this with quantum chromodynamics (QCD) which attributes strong interactions to the gauge bosons of an exact SU(3)-color ($\text{SU}(3)_{\text{C}}$) symmetry group, we have the completed picture of elementary interactions at low energies ($E < 100\,\text{GeV}$)

$$\text{SU}(3)_{\text{C}} \times \text{SU}(2) \times \text{U}(1) \xrightarrow[100\,\text{GeV}]{M_{\text{EW}}} \text{SU}(3)_{\text{C}} \times \text{U}(1)_{\text{EM}}.$$

The enormous simplicity and success of the above-described theory, leads one naturally to ask, as Pati and Salam first did: can one obtain a unified description of quarks and leptons and the three low energy interactions? It is then straightforward to seek a group G which, in one or several stages of symmetry breaking, leads to

$$G \xrightarrow{M_G} G_1 \to G_2 \cdots \to G_{\text{EW}} \xrightarrow{M_{\text{EW}}} \text{SU}(3)_{\text{C}} \times \text{U}(1)_{\text{EM}}.$$

Such an idea, besides the aesthetic appeal of unification, will help to reduce the number of parameters (more than 25 of them) required to describe the new physics in the low energy region associated with the old and new generations of quarks and leptons. The embedding of $\text{U}(1)_{\text{EM}}$ in a simple or semi-simple group, could explain the long-standing puzzle of charge quantization. Quark-lepton unification could counter questions like why is $|Q_{\text{p}} + Q_{\text{e}}| < 10^{-21}$?

The known low energy interactions will all have a common strength g at $E \geqslant M_G$. Hence, the first requisite of a grand unified theory is to explain the differing strengths and other distinguishing features of the observed strong, weak and electromagnetic interactions in the low energy region. The non-Abelian nature of the theory makes the coupling strength g energy dependent and, at $E < M_G$, g evolves via renormalization group equations in different ways in SU(3), SU(2) and U(1) sectors (depending on the spectrum in each sector) giving the low energy values g_3, g_2, g_1, respectively. We can, of course, consider this in the reverse order. Knowing g_3, g_2 and g_1 in the low energy region, we can study their evolution as a function of energy. If no additions to the low energy particle spectrum are postulated, one finds that g_3, g_2 and g_1 merge to a common value at $M_G \simeq 10^{14} - 10^{15}$ GeV.

This mass scale is relevant from another consideration. The quark-lepton unification *implies* in general that quarks and leptons be put in one multiplet. The unified NEW (nuclear-electro-weak) interactions will then cause quark-lepton transitions leading to baryon and lepton number violations. The most dramatic consequence of this is the possibility of observing proton decay! Viewed as an effective four-fermion process, we can associate a mass scale with proton decay, a mass scale which makes the decay rate consistent with the known experimental bounds, but still large enough to be observable. In the simplest possible model, this mass scale turns out to be 10^{15} GeV, roughly the same scale at which the low energy coupling constants merge together. This remarkable coincidence is one of the reasons why grand unification appears to be more than just a wild speculation.

1.3. General Consequences of Grand Unification Ideas

From the previous subsection, it should be clear that the most natural and important consequence of grand unification ideas is *the possible violation* of the known low energy conservation laws such as *the baryon number* and *lepton number*. Another consequence of grand unification is the existence of exotic species of particles at very high energies. The surviving heavy Higgs scalars is one such species, the non-Abelian magnetic monopole is another. Since $U(1)_{EM}$ is embedded in a simple or semi-simple group G, electric charge is quantized. The breakdown of G then inevitably produces magnetic monopoles. All these are general consequences of grand unification which are true in any model. They are of relevance to both terrestrial and astrophysical measurements, some of which we summarize in Table I.

Since baryon and lepton number violations are such important consequences of grand unification, we shall consider a few details regarding them, concerning ourselves with some model independent questions such as the selection rules, mass scales and the underlying mechanisms. Since the interactions responsible for these violations have a mass scale $M \gg M_{EW} \approx 10^2$, it is convenient to write the transition matrix elements for $\Delta B \neq 0$, $\Delta L \neq 0$ processes in terms of an effective Lagrangian composed of four fermion-type operators involving quarks, leptons and Higgs fields. Because of the mass scales involved, it is also convenient to use the transformation properties of these fields under $SU(3)_C \times SU(2) \times U(1)$ and construct Lorentz and gauge invariant products of the fields making up local operators of different dimensions. The breaking of $SU(2) \times U(1) \rightarrow U(1)_{EM}$ can afterwards be treated as perturbative corrections. Thus, one can write an effective

Table I. General consequences and their significance to terrestrial and astrophysical measurements

Consequences of grand unification	Significance: Terrestrial	Significance: Astrophysical
Baryon-number violation	Proton decay, N-$\bar{\text{N}}$ oscillations. H-H transitions.	Baryon asymmetry of the universe n_B/n_γ.
Lepton-number violation	Neutrinos with mass, Majorana neutrinos, neutrino oscillations, $\nu_e \leftrightarrow \nu_\mu \leftrightarrow \nu_\tau$.	Missing mass in galaxies, clusters, formation of galaxies.
Monopoles	Existence of monopoles, monopole-induced proton decay.	Monopole problem.
Phase transitions	–	Inflationary universe, solution of horizon and flatness problems.

Lagrangian $\mathscr{L}_{\text{eff}}$,

$$\mathscr{L}_{\text{eff}} \approx \sum_i c_i M^{4-d} O_d^{(i)}, \tag{3}$$

where c is a dimensionless constant which will be related in specific models to the gauge or Higgs couplings via group theory and renormalization group equations. M is the mass scale relevant to the transition. O_d is a polynomial in quark, lepton and Higgs fields of dimension d. Each O_d has characteristic selection rules on $\Delta B, \Delta L$ and represents a specific transition. Thus, we can classify O_d's by d, ΔB, ΔL. A few examples of such operators are shown in Table II.

It is worth noting at this stage that to calculate real physical processes (proton lifetime, N-$\bar{\text{N}}$ transition time) from $\mathscr{L}_{\text{eff}}$ in (3), we need to make further dynamical assumptions and approximations. The characteristic mass scales given in Table II can, at best, be described as estimates. First of all, we note that for a fixed value of the mass scale, if we have several O_d's which contribute to a process, the operator with the smallest dimension dominates, since it suffers the least suppression by inverse powers of the mass. Conversely, if O_d's of different dimensions contribute comparably to a process, the associated mass scales will be different, M decreasing as the dimension increases. The values of the mass scales given in Table II take these facts into consideration. They are also constrained by existing limits on $\Delta \text{B} \neq 0$ processes. Further, one assumes, prompted by the desire to have testable predictions, that although unobserved so far, these

Table II. Examples of operators for $\Delta B \neq 0$ processes

Dim. d	Sel. Rule	Fields in O_d	Observable process	Mass scale
6	$\Delta B = \Delta L = -1$ $\Delta S = 0, 1$	qqql	$p \to e^+ \pi^0$ $n \to e^+ \pi^-$	10^{14}–10^{15} GeV
7	$\Delta B = -L = -1$ $\Delta S = 1$	$qqql^c\phi$	$n \to e^- K^+$	10^{10}–10^{11} GeV
11	$\Delta B = -\frac{\Delta L}{3} = \frac{1}{3}$	$qqqlll\phi\phi$	$p \to e^+ \overline{\nu\nu}$	10^4–10^5 GeV
9	$\Delta B = -2$	qqqqqq	$n \leftrightarrow \bar{n}$	10^5–10^6 GeV
⋮	⋮	⋮	⋮	⋮

$d = 6$ – lowest dimensional operator for $\Delta B \neq 0$.

processes will be observable in the present or next generation of experiments. In other words, the mass scales given in Table II are only lower bounds and not rigid constraints on the model. In addition, there are, in practice other customary approximations concerning the hadronic wave functions, questions regarding the relative significance or insignificance of specific diagrams and different methods of calculations. Because of these reasons, predictions of the theory carry some uncertainty in quantitative details. For example, there is an uncertainty and variation among calculations by different groups of about two orders of magnitude for the proton lifetime. The current theoretical and experimental values of τ_p and $\tau_{n\bar{n}}$ are

	Experiment	*Theory*
τ_p	7×10^{30} years (Kolar gold field)	10^{29} to 10^{33} years
$\tau_{n\bar{n}}$	$>3 \times 10^7$ seconds (From stability of nuclear matter)	10^7 seconds to anything

Similar considerations hold in purely leptonic processes with $\Delta L \neq 0$. On the experimental side, there are some positive indications about neutrino masses and neutrino flavor oscillations. Other processes with $\Delta L \neq 0$ which will play an important role in the future are the $\mu \to e\gamma$ transition and neutrinoless double β-decay.

1.4. *Astrophysics and Grand Unification*

The important consequences of grand unification we have discussed so far, are the low energy relics of the spontaneously broken grand unification symmetry.

To obtain a full display of the NEW interactions implied by grand unification, we need energies $>10^{15}$ GeV which are beyond present-day or future terrestrial accelerators. Astrophysics and early cosmology are the natural arena for testing grand unified theories. In the standard Friedmann–Robertson–Walker model of the Universe, we start with a hot dense phase with temperatures exceeding 10^{16} GeV in the first 10^{-35} sec after the big bang. In its very early stages, our Universe was like a giant accelerator, the high temperatures and densities being maintained for a relatively longer time than would be possible in terrestrial accelerators. Under those conditions, there should have been copious production of all the particles that we know and those we do not know – all the particles including the superheavy ones that grand unified theories predict. Forces of grand unification were in full operation. One can trace the effects of these forces through the subsequent adiabatic cooling of the universe down to the present epoch and compare with contemporary astrophysical measurements.

The most exciting possibility in this direction is the resolution of an old puzzle: why is there very little antimatter in the Universe, in particular why is the net baryon density $n_B \sim 10^{-8} n_\gamma$, where n_γ is the number density of photons? One could postulate this baryon asymmetry as an initial condition for the Universe, but it would be more satisfying if we could start with an initial state of zero asymmetry and produce the observed n_B/n_γ in the course of the evolution of the Universe. As noted long ago by Sakharov, this requires (a) $\Delta B \neq 0$ processes, (b) CP-violation, and (c) a nonequilibrium state of the Universe when the $\Delta B \neq 0$ processes are being frozen out. The last requirement can be obtained in an expanding Universe. In a grand unified version of the expanding Universe, we can thus find a natural explanation for baryon asymmetry. In most grand unified theories, we can produce, by a judicious choice of parameters, the number $\sim 10^{-8}$ for n_B/n_γ but only just. We can by no means claim that currently available grand unified models provide a quantitatively perfect solution of the problem.

The interaction with astrophysics provides several constraints on theories and parameters in particle physics. Table III summarizes some of these constraints.

1.5. Theoretical Problems

Unification leads to theories with great predictive potential, since the number of arbitrary parameters is reduced. But as we have noted earlier, this advantage is partially sabotaged by the Higgs sector of the theory. There is arbitrariness in the choice of the Higgs representation and the form and parameters of the Higgs potential. Besides, we may have problems of gauge hierarchy and fine tuning. (This is a strictly technical problem of perturbative computation but, nevertheless, quite worrisome.) There have been several attempts at a more satisfactory description of symmetry breaking. The dynamical breakdown of symmetry is one of them, the use of extra symmetries like supersymmetry is another. Studies of dynamical symmetry breaking are patterned after the

Table III. Examples of astrophysical constraints

Astrophysical measurement	Mechanism	Constraints on particle theory
Energy loss from stars	Bremstrahlung and photo-production of light particles like axion, Majoron, ... rate $\leqslant$ rate of loss due to photons	Vacuum exp. values of Higgs fields, Number of neutrino flavors $N_\nu \leqslant 760$ (red giants) $\leqslant 23$ (carbon burning stars) $\leqslant 50$ (neutron stars)
Deceleration parameter expansion rate	Mass density of the universe $\rho \leqslant \rho_c$, Mass density due to stable and long lived particles $\leqslant \rho_c$.	Scale invariant densities of grand unified monopoles, background sea of neutrinos vacuum walls, spontaneous super-symmetry breaking scale, gravitino masses.
He abundance	Part. species at $T = 1$ MeV, Expansion rate, Freeze out temp., n/p ratio	Light neutrino flavors $\leqslant 4$
Deuterium abundance	Decay of light Higgs particles, axions affects entropy per baryon.	Masses of light Higgs particles, axions.
Estimate of dark matter	Dark matter due to neutrinos? Axions?	Mass of neutrinos
Baryon asymmetry	Baryon-number viola-tion, C, CP viola-tion, departure from equilibrium necessary.	Such interactions exist in grand unified theories. Gives con-straints on parameters of the theory.

breakdown of chiral symmetries in QCD. If we ignore the current algebra masses of the quarks, QCD has $SU(N_F)_L \times SU(N_F)_R \times U(1)_{L+R}$ chiral symmetry. The strong binding forces of QCD produce $q\bar{q}$ condensates which break this chiral symmetry to $SU(N)_{L+R} \times U(1)_{L+R}$. This can be thought of as effective Higgs scalars being formed as $q\bar{q}$ bound states, which then acquire nonzero vacuum values. Since the weak group $SU(2)_L \times U(1)$ is a nondiagonal subgroup of the chiral group, this process gives a dynamical breakdown of $SU(2)_L \times U(1)$, resulting in W-bosons with a mass ~ 30 MeV instead of the desired ~ 100 GeV. In technicolor models, therefore, one repeats this scenario at TeV energies so that the W-boson masses come out right. For this purpose, one needs extra fermions (technifermions) to provide the effective scalars. One also needs massive 'extended technicolor' (ETC) gauge bosons to provide masses to ordinary quarks and leptons. We may thus need scalars again to make the ETC bosons massive. This perhaps can also be done dynamically, but one ends up with a baroque theory of very little predictive power. The ETC bosons also lead to flavor-changing neutral currents (FCNC) and this has made technicolor models unpopular. Nevertheless, realistic models are possible where FCNC are suppressed to acceptable levels. Technicolor, it should be noted, provides a simple solution to the fine tuning problem, although there are still unsatisfactory points when one gets into details.

Supersymmetry provides another resolution of the problem. In technicolor models, there are no quadratic divergences and, hence, there are no radiative corrections to W-boson masses which are of the grand unification scale in magnitude. The SUSY answer to hierarchy is to cancel the unpleasant quadratic divergences by contributions from the supersymmetric partners. An example is the mixing of M_G and M_W by gauge boson loops which is cancelled by the gaugino loop contribution. Thus, the vastly different values of M_G and M_W can be naturally preserved in perturbation theory. Since we need to break supersymmetry (based on phenomenological considerations), there will be some mixing of the two scales, but this can be controlled by the scale of SUSY breaking.

There is a second level of the hierarchy problem. Beyond the question of keeping M_G and M_W separate once they are chosen at the tree level, one might ask: how can we, in an intrinsic sense, understand why $M_G \gg M_W$? SUSY models might provide an answer through the inverted hierarchy scheme of Witten [5].

Some of the other facts about SUSY models are: (a) $N = 4$ SUSY Yang–Mills theory is finite to all orders in perturbation theory (except gauge-dependent parameters); (b) in supergravity models one may be able to predict M_W from $M_{\rm Planck}$ and M_G; (c) $N = 8$ super-gravity may be finite up to 7-loop order.

These provide powerful incentives to make realistic SUSY models. On the experimental side, the most significant fact about SUSY models is the incredibly rich particle spectrum. If SUSY is broken using an extra U(1) symmetry, the scalar partners of quarks and leptons should have masses $< 40\,\text{GeV}$. For radiative mass generation mechanisms with no extra U(1), these masses should be $\lesssim M_W$. The next generation of accelerators should thus provide a definitive answer on SUSY models.

1.6. Concluding Remarks

Grand unified theories are an important advance in the quest for a unified description of elementary particle interactions. Beyond the appeal of unification, they have also raised the possibility of interesting and new phenomena, e.g. proton decay, $N\bar{N}$ oscillations, neutrino masses. Ideas for understanding old puzzles like the baryon asymmetry of the Universe, have also been generated.

While our experience with successful gauge theories, like electrodynamics and the standard electro-weak theory, gives confidence in the use of gauge principles as a unique basis for constructing grand unified theories, choice of gauge group, representations, symmetry breaking mechanism, etc. have a great deal of flexibility. And the very idea that the unification (if it occurs at all) must occur at very high energies unreachable by terrestrial accelerators, makes it very difficult to provide detailed tests on the basis of which grand unification models can be discriminated. Proton decay experiments currently under progress – if they establish proton decay conclusively – may come to being the closest in providing strong support for the underlying ideas of unification. Precise data on the various branching ratios, observation of $N\bar{N}$ oscillations, rare decay processes, will certainly help to make further progress.

In view of the high energies of the order of 10^{15} GeV or greater needed to directly test the grand unification ideas, one may look to astrophysics and our ideas of the very early Universe. Astrophysical measurements do provide constraints on models of grand unification, but these involve matching numbers which depend on cosmological and astrophysical models with the predictions of models of unification. Astrophysics can only give us guidelines. We need terrestrial experiments to sort out the various models. The SU(5) model, because of its minimality, has been analyzed to the fullest extent. If it is not the right model, it can be ruled out most easily because of the least ambiguity in its predictions.

There are also new theoretical directions towards a fundamental theory of elementary particle interactions. The proliferation of the number of degrees of freedom that we should consider, viz. color, flavor, technicolor, etc., and the perplexing mass spectrum of quarks and leptons, have prompted investigations of composite models of quarks, leptons and gauge bosons. However, most of these 'preon' models do not have any verifiable predictions. The question of compositeness for quarks and leptons can be discussed in a serious way only after high energy regimes become experimentally accessible.

We have discussed supersymmetric grand unification only briefly. Supersymmetry is, in some sense, the maximal symmetry of the S-matrix and, as such, would be a nice starting point for a theory. Furthermore, as we have mentioned, this may be the way to incorporate several desirable features like hierarchy and finiteness in our theories. The full potential of SUSY models of grand unification and their embedding in supergravity theories is only now being revealed by current investigations.

Kaluza–Klein-type theories are another direction of investigation. One starts with dynamics in a higher-dimensional (>4) spacetime and comes down to ordinary spacetime by 'dimensional-reduction', i.e. the extra dimensions are frozen out by subsidiary conditions. Ordinary 'internal' degrees of freedom correspond to the 'frozen' degrees of freedom which are the relics of spacetime symmetries associated with these extra dimensions. However, general considerations seem to indicate that it is not possible to obtain chiral fermions in four-dimensional space in this way. Currently, it is believed, most fervently in some quarters, that the best way to implement such ideas is through *superstring* theories. In the latter theories, elementary particles are one-dimensional objects in 10- or 26-dimensional spacetime. An effective field theory resulting from such a starting point only has no anomalies for two internal symmetry groups, namely, SO(32) and $E_8 \times E_8$. The resulting, renormalizable field theory includes gravity, has supersymmetry and is believed to be capable of predicting the number of generations. We have to wait and see whether this highly ambitious approach leads to the ultimate solution we are seeking in our understanding of elementary particles and their interactions.

2. Grand Unified Theory Based on $G \equiv SU(5)$ (Minimal SU(5))

In this section, we shall study a grand unified theory based on the gauge group SU(5). Generalization to any SU(n) should be apparent. By decomposing SU(5) with respect to $SU(3)_C \times SU(2)_L \times U(1)_W$, we see how the particle content and various interactions appear in the theory. In this exact SU(5) limit, all the interactions are governed by one gauge coupling strength. In addition to the known strong (QCD) and electroweak interactions ($SU(2) \times U(1)$), new interactions appear which are due to the additional gauge bosons (lepto-quark) contained in SU(5) and, as their name implies, they can effect transitions between quarks and leptons. They are responsible for proton decay.

2.1. Background; the Standard Model

Let us first recall the salient features of the standard model of 'low energy' strong and electroweak interactions.

(1) It is a non-Abelian gauge theory based on the semi-simple Lie algebraic structure H,

$$H = SU(3)_C \times SU(2)_L \times U(1)_W,$$

where SU(3)-color, $SU(3)_C$, is an exact symmetry describing the strong interactions. The electroweak interactions are governed by $SU(2)_L \times U(1)_W$ non-Abelian gauge theory which is also exact above energies $E > M_W \sim 100$ GeV but, for $E < M_W$, the theory breaks spontaneously to the usual quantum electrodynamics of $U(1)_{EM}$ and the V–A theory of weak interactions.

(2) A family or generation of quarks and leptons consists of 15 helicity objects. Quarks (leptons) transform according to **3** (**1**) representation of $SU(3)_C$. Both quarks and leptons transform as doublets or singlets under left-handed $SU(2)_L$

and their electric charge Q obeys the relation

$Q = T_3$	$+Y$
$\pm\frac{1}{2}, 0$	Weak
Weak	hypercharge
isospin	$U(1)_W$
$SU(2)_L$	

At present, we think that there are three families of quarks and leptons. These, along with their quantum numbers, are shown in Table IV.

Table IV. The three families of quarks and leptons. The subscript i denotes the color index and L the left-handedness of the particles. The superscript c denotes the charge-conjugate field.

$SU(3)_C$	SU(2)	U(1)			
(3	2	$\frac{1}{6}$)	$\begin{pmatrix} u_i \\ d_i \end{pmatrix}_L$	$\begin{pmatrix} c_i \\ s_i \end{pmatrix}_L$	$\begin{pmatrix} t_i \\ b_i \end{pmatrix}_L$
($\bar{3}$	1	$-\frac{2}{3}$)	$(u_i^c)_L$	$(c_i^c)_L$	$(t_i^c)_L$
($\bar{3}$	1	$\frac{1}{3}$)	$(d_i^c)_L$	$(s_i^c)_L$	$(b_i^c)_L$
(1	2	$-\frac{1}{2}$)	$\begin{pmatrix} \nu_e \\ e^- \end{pmatrix}_L$	$\begin{pmatrix} \nu_\mu \\ \mu^- \end{pmatrix}_L$	$\begin{pmatrix} \nu_\tau \\ \tau^- \end{pmatrix}_L$
(1	1	1)	$(e^+)_L$	$(\mu^+)_L$	$(\tau^+)_L$

In gauge theories with spontaneous symmetry breaking, it is convenient to use two-component Weyl spinors, instead of the conventional four-component Dirac spinors, to describe the fermions, since they are all massless at the beginning and acquire mass only through spontaneous symmetry breaking. Further, it is also more convenient to use only the left- or right-handed components. Here, we shall only use the left-handed component of the corresponding charge-conjugate field to denote the right-handed components. (For details see the Appendix.)

(3) The quantum numbers satisfy certain 'anomaly constraints' which are necessary in order that the theory is renormalizable. In the case of the standard model these constraints are

$$\text{(a)}\quad \mathrm{Tr}'Y = 0, \qquad \text{(b)}\quad \mathrm{Tr}''Y = 0, \qquad \text{(c)}\quad \mathrm{Tr}\,Y^3 = 0,$$

where the single prime denotes the sum over all the colored states, the double prime denotes the sum over the weak iso-spin states and, in (c), we have to sum over both color and iso-spin states. Thus,

$$\text{(a)}\quad \underbrace{3\cdot\tfrac{1}{6}\cdot 2 - 3\cdot\tfrac{2}{3} + 3\cdot\tfrac{1}{3}}_{\text{Contribution from quarks}} = \underbrace{2\cdot(-\tfrac{1}{2}) + 1}_{\text{Contribution from Leptons}} = 0$$

(b) $\underbrace{2\cdot\frac{1}{6}-\frac{2}{3}+\frac{1}{3}}_{\text{Quarks}}=\underbrace{2\cdot(-\frac{1}{2})+1}_{\text{Leptons}}=0$

(c) $\underbrace{6(\frac{1}{6})^3+3(-\frac{2}{3})^3+3(\frac{1}{3})^3}_{\text{Quarks}}+\underbrace{2(-\frac{1}{2})^3+1}_{\text{Leptons}}=0$

Note that each family is anomaly-free and, while the first two relations (a) and (b) are satisfied for quarks and leptons separately, (c) is satisfied iff both quarks and leptons are included in the family. Thus, anomaly constraints establish a link between quarks and leptons. The existence of the (ν_τ, τ) leptons demands the existence of the pair (t, b) quarks.

(4) There are 12 gauge bosons in the theory: eight gluons and four vector bosons out of which three acquire masses and become the intermediate vector bosons $W^\pm$, Z^0. The fourth remains massless to be the photon.

(5) The spontaneous symmetry breaking is implemented by introducing a complex Higgs doublet $\binom{\phi^+}{\phi^0}$.

The standard model based on the above considerations has proved immensely successful in the low energy regime accessible so far. Why, then, go further and 'unify'? The reasons are manyfold:

(1) The theory, as it stands, contains too many arbitrary parameters. There are three distinct coupling parameters corresponding to the three gauge groups $SU(3)_C$, $SU(2)_L$, and $U(1)_W$. The weak mixing angle θ_W is arbitrary in the theory, and so are the quark and lepton masses, the generalized Cabibbo-type mixing angles, the CP-violating phase, the parameters in the Higgs potential and so on.

(2) The phenomenon of 'superfluous replication', that is, the occurrence of families with identical electroweak quantum numbers, is unaccounted for. It suggests additional symmetries and/or selection rules.

(3) The quark-lepton correspondence (i.e. the identical classification of quarks and leptons under $SU(2)_L$ and $U(1)_W$) and their linking in the anomaly cancellation constraints, suggest the possibility of further closer correspondence between quarks and leptons. They may be parts of a single representation of a higher symmetry group.

(4) In non-Abelian gauge theories, the coupling parameters are functions of energy or mass scale. The strong coupling parameter of QCD decreases with energy, whereas the SU(2) coupling parameter increases with energy. It is only natural, then, to imagine that, at some high energy, these coupling parameters attain the same strength and one may have a single parameter instead of three describing the various interactions.

(5) In the standard model, there is no explanation for the charge quantization (i.e., only the discrete values for electrical charges, $\pm\frac{2}{3}$, $\pm\frac{1}{3}$, $\pm 1, 0$ in units of e). Also, no explanation for such relations as Q (electron) $= -Q$ (proton),

$$Q(\nu_e) - Q(e^-) = Q(u) - Q(d), \qquad Q(e) = \tfrac{1}{3}Q(d), \quad \text{etc.}$$

These and other shortcomings of the standard model lead us to consider further the higher symmetries which unify all three interactions as a first step. A complete unification, of course, would be required, including gravity. We, therefore, seek a group G,

$$G \supset H \equiv SU(3)_C \times SU(2)_L \times U(1)_W, \tag{1}$$

preferably simple so that if it describes all the three interactions in a unified manner, there will be single coupling parameter g corresponding to the gauge group G. Further, since H has rank 4, the rank of G is prescribed to be at least 4. If $G = SU(N)$, the smallest group with rank 4 is SU(5).

2.2. *SU(5): Group Theory Aspects*

The fundamental representation of SU(5) consists of 5 × 5 unitary, unimodular matrices U,

$$U^\dagger U = 1, \qquad \det U = 1. \tag{2}$$

A general SU(5) transformation can be written in the form

$$U = \exp\left(-i \sum_{i=1}^{24} \theta_i T_i \right), \tag{3}$$

where T_i, called the generators, are 5 × 5, Hermitian, traceless matrices. θ_i are real, continuously variable parameters. The rank of SU(5) is 4 and, hence, there are four generators which can be simultaneously diagonalized.

A convenient basis to represent the generators is provided by the non-Hermitian matrices T^a_b, $a, b = 1, \ldots, 5$,

$$(T^a_b)_{cd} = \delta^c_b \delta^a_d - \tfrac{1}{5} \delta^a_b \delta^c_d. \tag{4}$$

These matrices satisfy the commutation relations

$$[T^a_b, T^c_d] = \delta^a_d T^c_b - \delta^c_b T^a_d \tag{5}$$

and they correspond to the usual raising and lowering operators. The Hermitian generators are linear combinations $\frac{1}{2}(T^a_b + T^b_a)$, $i/2(T^a_b - T^b_a)$.

It is evident from (4), that the nondiagonal matrices have only one nonvanishing element $(T^a_b)_{ba} = 1$ and zero everywhere else. There are five diagonal generators, T^a_a

$$(T^a_a)_{aa} = \tfrac{4}{5}; \qquad (T^a_a)_{bb} = -\tfrac{1}{5}, \quad a \neq b, \tag{6}$$

but $\Sigma^5_{a=1} T^a_a = O$ and, hence, there are only four independent, diagonal generators, as there should be, since the rank of the group is 4. Therefore, a more convenient choice for the independent diagonal operators is the following:

$$\tfrac{1}{2}\,\mathrm{diag}(1,\ -1,\ 0,\ 0,0),\ \frac{1}{2\sqrt{3}}\,\mathrm{diag}(1,\ 1,\ -2,\ 0,\ 0),\ \frac{1}{2\sqrt{6}}(1,\ 1,\ 1,\ -3,\ 0),$$
$$\frac{1}{2\sqrt{10}}(1,\ 1,\ 1,\ 1,\ -4). \tag{7}$$

The normalization of the generators is arbitrary, but we shall stick to the convention

$$\mathrm{Tr}(T_a T_b) = \tfrac{1}{2}\delta_{ab}. \tag{8}$$

Now our task is to label the generators so that the embedding of SU(3) × SU(2) × U(1) is self-evident. To this end, let us denote the SU(5) generators as

$$T_a^b,\ a,\ b = 1, \ldots, 5. \tag{9}$$

Then define the subset of the first eight generators as SU(3) generators. Assign them the Greek labels $\alpha, \beta, \ldots$ which are assumed to take only the values 1, 2, 3, so that

$$T_\alpha^\beta - \tfrac{1}{3}T_\gamma^\gamma\,\delta_\alpha^\beta \tag{10}$$

generate the SU(3) subgroup. Similarly, by prescribing, r, s to take only the values 4, 5, define the generators

$$T_r^s - \tfrac{1}{2}\delta_r^s\,T_{r'}^{r'} \tag{11}$$

which will generate the SU(2) group. The U(1) group can be identified as

$$-\tfrac{1}{3}T_\alpha^\alpha + \tfrac{1}{2}T_r^r. \tag{12}$$

These make twelve of the 24 SU(5) generators. The remaining twelve SU(5) generators are T_α^r and $T_r^\alpha(T_1^4, T_1^5, \ldots;\ T_4^1, T_5^1, \ldots)$.

The two $\mathrm{SU(3)_C}$ diagonal operators T_3^C, T_8^C can be identified as

$$T_3^C = \tfrac{1}{2}\begin{pmatrix}1&&&&\\&-1&&&\\&&0&&\\&&&0&\\&&&&0\end{pmatrix} = \tfrac{1}{2}(T_1^1 - T_2^2),$$

$$T_8^C = \frac{1}{2\sqrt{3}}\begin{pmatrix}1&&&&\\&1&&&\\&&-2&&\\&&&0&\\&&&&0\end{pmatrix} = \frac{1}{2\sqrt{3}}(T_1^1 + T_2^2 - T_3^3). \tag{13}$$

The SU(2) diagonal operator which we shall designate as T_3^W (third component of weak iso-spin),

$$T_3^W = \frac{1}{2}\begin{pmatrix} 0 & & & & \\ & 0 & & & \\ & & 0 & & \\ & & & 1 & \\ & & & & -1 \end{pmatrix} = \frac{1}{2}(T_4^4 - T_5^5). \tag{14}$$

The U(1) generator which we shall identify with T_Y^W given by (12) and normalized according to (8) is

$$T_Y^W = \sqrt{\frac{3}{5}}(-\tfrac{1}{3}T_\alpha^\alpha + \tfrac{1}{2}T_r^r) = \sqrt{\frac{3}{5}}\begin{pmatrix} -\frac{1}{3} & & & & \\ & -\frac{1}{3} & & & \\ & & -\frac{1}{3} & & \\ & & & \frac{1}{2} & \\ & & & & \frac{1}{2} \end{pmatrix}. \tag{15}$$

Let the electric charge operator Q be defined so that, in the fundamental representation, it is

$$Q = \begin{pmatrix} -\frac{1}{3} & & & & \\ & -\frac{1}{3} & & & \\ & & -\frac{1}{3} & & \\ & & & 1 & \\ & & & & 0 \end{pmatrix}, \tag{16}$$

so that

$$Q = T_3^W + \sqrt{\frac{5}{3}}\, T_Y^W \tag{17}$$

2.3. *Particles; Their Representations*

2.3.1. Fermions. Suppose we define the fundamental representation **5** of SU(5), decomposed with respect to the subgroup SU(3) × SU(2) × U(1), as follows:

$$\mathbf{5} = (\mathbf{3}, \mathbf{1}, -\tfrac{1}{3}) + (\mathbf{1}, \mathbf{2}, \tfrac{1}{2}), \tag{18}$$

(That the SU(3)-triplet must be SU(2)-singlet and the SU(3)-singlet must be SU(2)- doublet is obvious. The U(1) quantum numbers follow from (15) and the charge operator *defined* in (16) and (17).)

Then, the **5***-representation contains

$$5^* = (\mathbf{3}^*, \mathbf{1}, \tfrac{1}{3}) + (\mathbf{1}, \mathbf{2}, -\tfrac{1}{2}), \tag{19}$$

leading to charges of 1/3 for the SU(3)-triplet and charges 0 and −1 for the SU(3)-singlet, SU(2)-doublet components.

The quarks $(d_i^c)_L$ and the leptons $(\nu_e, e^-)_L$ of the 15-plet family have exactly the quantum numbers contained in 5*. Hence, we can fit them into the **5*** representation.

The anti-symmetrized direct product $\mathbf{5} \times \mathbf{5}$ gives rise to a **10** representation of SU(5). Its decomposition with respect to SU(3) × SU(2) × U(1) yields

$$\mathbf{10} = (\mathbf{3}^*, \mathbf{1}, -\tfrac{2}{3}) + (\mathbf{3}, \mathbf{2}, \tfrac{1}{6})_A + (1, 1, 1)$$

$$Q = -\tfrac{2}{3} \quad Q = \begin{pmatrix} \tfrac{2}{3} \\ -\tfrac{1}{3} \end{pmatrix}, \quad Q = 1. \tag{20}$$

The three $(u_i^c)_L$ quarks, the six $(u_i, d_i)_L$ quarks, and the lepton $(e^+)_L$, altogether ten, exhaust the quantum numbers of the **10**. Thus, the **15**-plet family can be made to belong to a reducible combination of **5*** and **10** representations of SU(5),

$$\mathbf{15} = \mathbf{5}^* + \mathbf{10} \tag{21}$$

That each family belongs to a *reducible* and *not to an irreducible* representation is considered as one of the unsatisfactory features of SU(5). But it is not a very serious objection at this stage. On the contrary, what is worth noting is that the combination is *anomaly free* and, hence, it can lead to a renormalizable gauge theory.

In tensor notation, let ψ^a be fields that transform according to 5 representation, and $\chi_a = (\psi^a)^\dagger$ transform according to 5*. Higher representations transform as the tensor products of the above representations. Denoting the anti-symmetric tensor belonging to **10** by ψ_L^{ab}, we have for the 15-plet family of fermions,

$$\mathbf{5}^*: \chi_L = \begin{bmatrix} d_1^c \\ d_2^c \\ d_3^c \\ e^- \\ -\nu_e \end{bmatrix}_L, \tag{22}$$

$$10: \psi_L = \frac{1}{\sqrt{2}} \begin{bmatrix} 0 & u_3^c & -u_2^c & -u^1 & -d^1 \\ u_3^c & 0 & u_1^c & -u^2 & -d^2 \\ u_2^c & -u_1^c & 0 & -u^3 & -d^3 \\ u^1 & u^2 & u^3 & 0 & -e^+ \\ d^1 & d^2 & d^3 & e^+ & 0 \end{bmatrix}_L \tag{23}$$

The **10** representation is written in the form of an anti-symmetric matrix with the factor $1/\sqrt{2}$ introduced for convenience of normalization (Langacker). It may be worthwhile at this stage to see cleary how the identifications of various particles in (23) arise:

$$\mathbf{5}: \psi^a = \begin{pmatrix} \phi^\alpha \\ \phi^r \end{pmatrix} \begin{matrix} \alpha = 1,\ 2,\ 3 \\ r = 4,\ 5. \end{matrix}$$

Then, when we take the tensor product 5×5 and antisymmetrize, we obtain

$$\psi^{\alpha\beta} = \frac{\phi^\alpha \phi^\beta - \phi^\beta \phi^\alpha}{\sqrt{2}}, \qquad \psi^{\alpha r} = \frac{\phi^\alpha \phi^r - \phi^r \phi^\alpha}{\sqrt{2}} \tag{24}$$

From the definition of the charge operator (16) in the fundamental representation, we see that $\psi^{\alpha\beta}$ has charge $-2/3$, and $\psi^{\alpha 4}$, $\psi^{\alpha 5}$ have charges $+2/3$ and $-1/3$, respectively. Hence, the identifications $\psi^{12} \propto u_3^c$, $\psi^{\alpha 4} \propto u_\alpha$, $\psi^{\alpha 5} \propto d_a$ etc.

2.3.2. Gauge Bosons. In a non-Abelian gauge theory, the gauge bosons belong to the *adjoint* representation which in our case is **24** contained in the direct product of 5 and 5*,

$$\mathbf{5} \times \mathbf{5^*} = \mathbf{24} + \mathbf{1} \tag{25}$$

Now we decompose **24** with respect to $SU(3) \times SU(2) \times U(1)$, we obtain

$$\mathbf{24} = (\mathbf{8}, \mathbf{1}, 0) + (\mathbf{1}, \mathbf{3}, 0) + (\mathbf{1}, \mathbf{1}, 0) + (3, 2^*, \tfrac{5}{6}) + (3^*, 2, \tfrac{5}{6}). \tag{26}$$

The $(\mathbf{8}, \mathbf{1}, 0)$ components belong to the adjoint representation of SU(3). The QCD gluons can be identified with them. Likewise the $(\mathbf{1}, \mathbf{3}, 0)$ and $(\mathbf{1}, \mathbf{1}, 0)$ components belong to the adjoint representation of $SU(2) \times U(1)$ and they will ultimately be the intermediate vector bosons (W^+, W^-, Z^o) and the photon. The remaining twelve gauge bosons $(3, 2^*, -\tfrac{5}{6})$ and $(3^*, 2, \tfrac{5}{6})$ are the new bosons which can cause transitions between color triplet and color singlet fermions, that is, between quarks and leptons. They are therefore called lepto-quarks. Note that their electric charges are fractions, (4/3, 1/3), which they have to be in order to cause a transition between a quark and a lepton.

Let the corresponding fields, the non-Abelian gauge vector potentials, be denoted by A^a in the Hermitian basis and by A_b^a in the non-Hermitian basis. (For simplicity of notation, we are suppressing the spacetime index.) The index $a(b)$ will transform like the 5 (5*) representation. Define the matrix **A** to be

$$\frac{1}{\sqrt{2}} \mathbf{A} = \sum_{a=1}^{24} T_a A^a = \frac{\lambda_a A^a}{2}, \tag{27}$$

where T_a are the Hermitian generators defined earlier. The λ's are the SU(5) generators; they correspond to Pauli matrices for SU(2) or Gell-Mann matrices for SU(3). Explicitly writing out the relations between the Hermitian and non-Hermitian forms of the generators, we can identify:

SU(3)$_C$ Generators

$$\begin{aligned}
&T^1 = \tfrac{1}{2}(T^1_2 + T^2_1), \qquad T^2 = \tfrac{i}{2}(T^1_2 - T^2_1), \qquad T^4 = \tfrac{1}{2}(T^1_3 + T^3_1),\\
&T^5 = \tfrac{i}{2}(T^1_3 - T^3_1), \qquad T^6 = \tfrac{1}{2}(T^2_3 + T^3_2), \qquad T^7 = \tfrac{i}{2}(T^2_3 - T^3_2),\\
&T^3 = \tfrac{1}{2}\,\mathrm{diag}(1, -1, 0, 0, 0) = \tfrac{1}{2}(T^1_1 - T^2_2),\\
&T^8 = \frac{1}{2\sqrt{3}}\,\mathrm{diag}(1, 1, -2, 0, 0) = \frac{1}{2\sqrt{3}}(T^1_1 + T^2_2 - 2T^3_3).
\end{aligned} \tag{28}$$

SU(2) Generators

$$\begin{aligned}
&T^9 = \tfrac{1}{2}(T^4_5 + T^5_4), \qquad T^{10} = \tfrac{i}{2}(T^4_5 - T^5_4),\\
&T^{11} = \tfrac{1}{2}\,\mathrm{diag}(0, 0, 0, 1, -1) = \tfrac{1}{2}(T^4_4 - T^5_5).
\end{aligned} \tag{29}$$

U(1) Generator

$$T_{12} = \sqrt{\tfrac{3}{5}}\,\mathrm{diag}\,(-\tfrac{1}{3}, -\tfrac{1}{3}, -\tfrac{1}{3}, \tfrac{1}{2}, \tfrac{1}{2}) = \sqrt{\tfrac{3}{5}}(-\tfrac{1}{3}T^\alpha_\alpha + \tfrac{1}{2}T^r_r). \tag{30}$$

Lepto-quark Generators

$$\begin{aligned}
&T^{13} = \tfrac{1}{2}(T^1_4 + T^4_1), \qquad T^{14} = \tfrac{i}{2}(T^1_4 - T^4_1), \qquad T^{15} = \tfrac{1}{2}(T^1_5 + T^5_1)\\
&T^{16} = \tfrac{i}{2}(T^1_5 - T^5_1), \qquad T^{17} = \tfrac{1}{2}(T^2_4 + T^4_2), \qquad T^{18} = \tfrac{i}{2}(T^2_4 - T^4_2)\\
&T^{19} = \tfrac{1}{2}(T^2_5 + T^5_3), \qquad T^{20} = \tfrac{i}{2}(T^2_5 - T^5_2), \qquad T^{21} = \tfrac{1}{2}(T^3_4 + T^4_3)\\
&T^{22} = \tfrac{i}{2}(T^3_4 - T^4_3), \qquad T^{23} = \tfrac{1}{2}(T^3_5 + T^5_3), \qquad T^{24} = \tfrac{i}{2}(T^3_5 - T^5_3).
\end{aligned} \tag{31}$$

Substituting (28)–(31) in (26), we can write

$$\frac{1}{\sqrt{2}}\mathbf{A} = A^b_a T^a_b, \tag{32}$$

where for $a \neq b$,

$$A^1_2 = \frac{A_1 - iA_2}{2}, \qquad A^2_1 = (A^1_2)^* = \frac{A_1 + iA_2}{2}, \text{ etc.} \tag{33}$$

and for $a = b$,

$$A^1_1 = \frac{1}{2}\left(A_3 + \frac{1}{\sqrt{3}}A_8 - \frac{2}{\sqrt{15}}A_{12}\right),$$

$$A^2_2 = \frac{1}{2}\left(-A_3 + \frac{1}{\sqrt{3}}A_8 - \frac{2}{\sqrt{15}}A_{12}\right),$$

$$A^3_3 = -\frac{1}{\sqrt{3}}\left(A_8 + \frac{1}{\sqrt{5}}A_{12}\right),$$

$$A_4^4 = \frac{1}{2}\left(A_{11} + \sqrt{\frac{3}{5}}A_{12}\right),$$

$$A_5^5 = \frac{1}{2}\left(-A_{11} + \sqrt{\frac{3}{5}}A_{12}\right). \tag{34}$$

Note that $A_1^1 + A_2^2 + \cdots + A_5^5 = 0$.

A_b^a thus defines the gauge field matrix which is Hermitian and traceless. To separate and identify the different interactions, let us denote them as follows:

$[A_\alpha^\beta] = [G_\alpha^\beta]$,	$\alpha, \beta = 1, 2, 3$	$SU(3)_C$ gluons
$[A_r^s] = [W, B]$,	$r, s = 4, 5$	$SU(2) \times U(1)$ electroweak vector bosons
$[A_r^\alpha, A_\alpha^r] = [L]$.		Lepto-quark.

Then, the matrix **A** will have the structure

$$\mathbf{A} = \left[\begin{array}{c:c} \mathbf{G} & \mathbf{L} \\ \hdashline \mathbf{L}^T & \mathbf{W} \end{array}\right] \tag{35}$$

where written out explicitly,

$$\mathbf{G} = \begin{bmatrix} G_1^1 - \frac{2B}{\sqrt{30}} & G_2^1 & G_3^1 \\ G_1^2 & G_2^2 - \frac{2B}{\sqrt{30}} & G_3^2 \\ G_1^3 & G_2^3 & G_3^3 - \frac{2B}{\sqrt{30}} \end{bmatrix}, \tag{36}$$

$$G_1^1 = G^3/\sqrt{2} + G^8/\sqrt{6}, \quad G_2^2 = -G^3/\sqrt{2} + G^8/\sqrt{6}, \quad G_3^3 = -2G^8/\sqrt{6}$$

$$\text{Electroweak Vector Bosons } \mathbf{W} = \begin{bmatrix} \frac{W^3}{\sqrt{2}} + \frac{3B}{\sqrt{30}} & W^+ \\ W^- & -\frac{W^3}{\sqrt{2}} + \frac{3B}{\sqrt{30}} \end{bmatrix}, \tag{37}$$

$$W^{\pm} = \frac{W^1 \mp iW^2}{\sqrt{2}}$$

$$\text{Lepto Quarks } \mathbf{L} = \begin{bmatrix} \bar{X}^1 & \bar{Y}^1 \\ \bar{X}^2 & \bar{Y}^2 \\ \bar{X}^3 & \bar{Y}^3 \end{bmatrix}. \tag{38}$$

Note that for states belonging to 5 × 5*, the charge operator Q is given by

$$Q = Q_5 + Q_{5^*},$$

where Q_5 acts on the upper index and is given by (16). Q_{5^*} acts on the lower index and is the negative of Q_5. Thus,

$$QA_a^b = (Q_5^b + Q_{5^*}^a)A_a^b$$

and it is straightforward to verify that the gluons have zero electric charge, the electroweak bosons have charges ± 1, 0, and the leptoquarks have charges ± 4/3, ±1/3.

2.3.3. The desired scalar fields, the Higgs scalars, can be constructed along similar lines. We shall defer their discussion to the next section.

2.4. Covariant Derivatives

In order to construct a Lagrangian invariant under a set of *non-Abelian local gauge* transformations, it is necessary to generalize the notion of *the ordinary* derivative ∂_μ to that of the *covariant derivative* D_μ in strict analogy with the Abelian electro-dynamics. Recall that in electrodynamics (U(1) gauge theory),

$$\partial_\mu \to \partial_\mu - ig\, A_\mu(x),$$

where $A_\mu(x)$ is the vector potential. In the non-Abelian case

$$\partial_\mu \to \partial_\mu = \partial_\mu \mathbf{1} - \frac{ig}{\sqrt{2}}\mathbf{A}_\mu, \tag{39}$$

leading to the following covariant derivatives of the fermion fields in the case of SU(5):

For the field ψ^a transforming like the fundamental representation **5**,

$$(D_\mu\psi)^a = \left(\partial_\mu \delta_b^a - \frac{ig_5}{\sqrt{2}}(\mathbf{A}_\mu)_b^a\right)\psi^b. \tag{40}$$

For χ_a transforming like **5***,

$$(D_\mu\chi)_a = \left(\partial_\mu \delta_a^b + \frac{ig_5}{\sqrt{2}}(\mathbf{A}_\mu)_a^b\right)\chi_b. \tag{41}$$

For ψ^{ab} transforming like **10**,

$$(D_\mu\psi)^{ab} = \left(\partial_\mu\psi^{ab} - \frac{ig_5}{\sqrt{2}}(\mathbf{A}_\mu)_c^a\,\psi^{cb} - \frac{ig_5}{\sqrt{2}}(\mathbf{A}_\mu)_c^b\,\psi^{ac}\right). \tag{42}$$

Finally, the required field tensor $\mathbf{F}_{\mu\nu}$ to represent the KE of the gauge fields is

$$\mathbf{F}_{\mu\nu} = \partial_\mu \mathbf{A}_\nu - \partial_\nu \mathbf{A}_\mu - \frac{ig_5}{\sqrt{2}}[\mathbf{A}_\mu, \mathbf{A}_\nu], \tag{43}$$

which can be easily shown to be gauge covariant.

2.5. *SU(5) Gauge Invariant Lagrangian; Interactions*

We can write the desired Lagrangian in the form

$$\mathscr{L} = \mathscr{L}_G + \mathscr{L}_F + \mathscr{L}_H + \mathscr{L}_{FH}, \tag{44}$$

where

$$\mathscr{L}_G = \text{KE term of the gauge fields} = -\frac{1}{4}\,\text{Tr}\,\mathbf{F}_{\mu\nu}\mathbf{F}^{\mu\nu},$$

with $\mathbf{F}^{\mu\nu} = g^{\mu\rho} g^{\nu\lambda}\mathbf{F}_{\rho\lambda}$.

$$\mathscr{L}_F = i(\chi_L)^\dagger \gamma^o (\not{D} \chi_L) + i(\psi_L)^\dagger \gamma^o (\not{D} \psi_L), \tag{45}$$

with $\not{D} = \gamma^\mu \mathbf{D}_\mu$.

Note that these 'kinetic' energy terms include gauge-interactions of the fermions along with the usual kinetic energy term $\bar{\chi}\not{\partial}\chi$. The $\mathscr{L}_H$ and $\mathscr{L}_{FH}$ terms in (44) involve the Higgs fields. We shall consider them in the next section.

From (45), using (41) and (42), we can obtain the gauge interaction Lagrangian $\mathscr{L}_I$,

$$\begin{aligned}\mathscr{L}_I = &-\frac{g_5}{\sqrt{2}}(\bar{\chi}_L)^a (\not{A})^b_a \chi_{Lb} + \\ &+ \frac{2g_5}{\sqrt{2}}(\bar{\psi}_L)_{ab} (\not{A})^a_c \psi^{cb}.\end{aligned} \tag{46}$$

It is straightforward to expand (46) and regroup the various interactions (Langacker (1981), Eq. (3.39), p. 263). Note that a single gauge coupling governs the various interactions. This is in the limit of exact SU(5).

2.6. *Appendix*

The SU(5)-gauge invariant Lagrangian (44) does not contain any mass terms. Both the fermions and the gauge bosons acquire masses through the spontaneous symmetry breaking mechanism. In such situations, it is convenient to introduce two-component Weyl spinors to represent the fermions instead of the conventional four-component spinors which obey the Dirac equation,

$$(i\gamma^\mu \partial_\mu - m)\psi = 0 \tag{A1}$$

where

$$\gamma^0 = \begin{pmatrix} I & \\ & -I \end{pmatrix}, \qquad \{\gamma^i\} = \boldsymbol{\gamma} = \begin{pmatrix} 0 & \boldsymbol{\sigma} \\ -\boldsymbol{\sigma} & 0 \end{pmatrix},$$

$$\gamma^5 = \gamma_5 = i\gamma^0\gamma^1\gamma^2\gamma^3 = \begin{pmatrix} 0 & I \\ I & 0 \end{pmatrix}. \tag{A2}$$

If we denote

$$\psi = \begin{pmatrix} \psi_1 \\ \psi_2 \end{pmatrix}, \quad \text{and} \quad \psi_\pm = \frac{1}{2}(\psi_1 \pm \psi_2),$$

Equation (A1) reduces to

$$i(\gamma^0\partial_0 + \boldsymbol{\sigma}\cdot\boldsymbol{\nabla})\psi_+ = \mathrm{m}\psi_-, \qquad \mathrm{i}(\gamma^0\partial_0 - \boldsymbol{\sigma}\cdot\boldsymbol{\nabla})\psi_- = \mathrm{m}\psi_+, \tag{A3}$$

so that if $m = 0$, the two equations are decoupled. While ψ has a definite parity, $\psi_\pm$ do not. Note that if we define the projection operators

$$P_\mathrm{L} = \tfrac{1}{2}(1 - \gamma_5), \qquad P_\mathrm{R} = \tfrac{1}{2}(1 + \gamma_5)$$

and

$$\psi_{\mathrm{L,R}} = P_{L,\mathrm{R}}\,\psi, \qquad (\bar{\psi})_{\mathrm{L,R}} = \bar{\psi}\,P_{\mathrm{L,R}}$$

then $\psi = \psi_\mathrm{L} + \psi_\mathrm{R}$.

$$\psi_\mathrm{L} = \begin{pmatrix} \psi_- \\ -\psi_- \end{pmatrix}; \qquad \psi_\mathrm{R} = \begin{pmatrix} \psi_+ \\ \psi_+ \end{pmatrix}.$$

This shows that it is more convenient to introduce the Weyl representation of the γ-matrices in which γ_5 is diagonal,

$$\gamma_5 = \begin{pmatrix} \mathrm{I} & 0 \\ 0 & -\mathrm{I} \end{pmatrix}, \qquad \gamma^0 = \begin{pmatrix} 0 & -\mathrm{I} \\ -\mathrm{I} & 0 \end{pmatrix}, \qquad \boldsymbol{\gamma} = \begin{pmatrix} 0 & \boldsymbol{\sigma} \\ -\boldsymbol{\sigma} & 0 \end{pmatrix}. \tag{A.4}$$

Then,

$$\psi_\mathrm{L} = P_\mathrm{L}\psi = \tfrac{1}{2}(1 - \gamma_5)\psi = \begin{pmatrix} 0 \\ \psi_2 \end{pmatrix}, \qquad \psi_\mathrm{R} = P_\mathrm{R} = \tfrac{1}{2}(1 + \gamma_5)\psi = \begin{pmatrix} \psi_1 \\ 0 \end{pmatrix},$$

so that ψ_L and ψ_R are effectively two-component spinors. Note that

$$\overline{(\psi_\mathrm{L})} \equiv \bar{\psi}_\mathrm{L} = (\bar{\psi})_\mathrm{R}; \qquad \bar{\psi}_\mathrm{R} \equiv \overline{(\psi_\mathrm{R})} = (\bar{\psi})_\mathrm{L}, \tag{A.5}$$

and if ψ_c is the charge conjugate field,

$$\psi_c = C\bar{\psi}^T,$$

where C is the charge conjugation operator

$$C = -C^T = -C^T = -C^{-1} = i\gamma^2\gamma^0, \tag{A.6}$$

$$\psi_{\mathrm{L,R}}\ \mathbf{C}\ \psi^c_{\mathrm{L,R}} \equiv P_{\mathrm{L,R}}\psi^c = C(\bar{\psi}_{\mathrm{R,L}})^T,$$

$$\bar{\psi}_{\mathrm{L,R}}\ \mathbf{C}\ \overline{\psi^c_{\mathrm{L,R}}} \equiv -(\psi_{\mathrm{R,L}})^T C^{-1} \tag{A.7}$$

That is,

$$(\psi^c)_{L,R} = (\psi_{R,L})^c,$$

or

$$\psi_R = -i\gamma^2(\psi^c)_L; \qquad \psi_L = -i\gamma^2(\psi^c)_R \tag{A.8}$$

which enable us to express the total four-component field in terms of only left-handed two-component fields ψ_L, $(\psi^c)_L$.

3. Spontaneous Symmetry Breaking

The SU(5) symmetric Lagrangian discussed in the previous section clearly does not correspond to the real world. The strong, electroweak, and the new baryon number violating gauge interactions are all governed by the same gauge coupling parameter g_5. The 24 gauge bosons corresponding to the adjoint representation, and the **15**-plet fermions are all massless. The high degree of symmetry has to be broken so that the strong, electromagnetic, and weak interactions manifest themselves distinguished from each other as we see them in the laboratory.

From the Glashow–Salam–Weinberg theory of electroweak interactions, we know that the appropriate way of breaking the symmetry is to utilize the spontaneous symmetry breaking mechanism (SSB) à la Higgs. This is the only successful way we know at present of introducing symmetry breaking by giving masses to some of the gauge bosons, yet maintaining the renormalizability of the theory. In our case we want

$$\mathrm{SU}(5) \xrightarrow[1^{\text{st}}\ \text{stage}]{M_x} \mathrm{SU}(3)_C \times \mathrm{SU}(2) \times \mathrm{U}(1),$$
$$\xrightarrow[2^{\text{nd}}\ \text{stage}]{M_W} \mathrm{SU}(3)_C \times \mathrm{U}(1)_{EM}.$$

That is,

$E > M_x$,	All gauge bosons and all fermions are massless. The theory is SU(5) symmetric.
$M_W < E < M_x$,	X, Y gauge bosons acquire mass. SU(5) is broken to $\mathrm{SU}(3)_C \times \mathrm{SU}(2) \times \mathrm{U}(1)$. The gluons, the electroweak bosons $W^{\pm}$, B, and the photon are massless. So are the fermions. This represents the 1st stage of SSB.
$E < M_W$,	$W^{\pm}$, B acquire mass. The symmetry is broken further to $\mathrm{SU}(3)_C \times \mathrm{U}(1)_{EM}$, with only the gluons and the photon remaining massless. The fermions acquire mass as well. This represents the 2nd stage of SSB.

In this section, we shall study how this is accomplished. The procedure involves, firstly, the choice of a suitable irreducible representation (in general, a combination of irreducible representations) and secondly a pattern of vacuum expectation values (VEV's) which minimizes a chosen Higgs potential.

3.1. *Spontaneous Symmetry Breaking*

3.1.1. 1st Stage. As we have stated before, there is no rule or principle that determines the choice of the Higgs representation, except that the desired symmetry breaking pattern is possible in that representation. We shall begin by examining the adjoint representation. Let a Higgs scalar multiplet $\mathbf{\Phi}$ belong to the adjoint representation. In analogy with Equation (26) of Section 2, let

$$\frac{1}{\sqrt{2}}\mathbf{\Phi} = \sum_{i=1}^{24} T_a \mathbf{\Phi}^a = \sum_{i=1}^{24} \frac{\lambda_a \mathbf{\Phi}^a}{2}. \tag{1}$$

Like the gauge boson matrix, we can write $\mathbf{\Phi}$ as a 5×5, traceless, Hermitian matrix. Further, by using SU(5) transformations, we can bring the matrix into a diagonal form to consider what kind of symmetry breaking pattern it can give rise to. Recall the transformation property and the definition of the covariant derivative of a field belonging to the adjoint representation,

$$D_\mu \mathbf{\Phi} = \partial_\mu \mathbf{\Phi} - \frac{ig_5}{\sqrt{2}}[\mathbf{A}_\mu, \mathbf{\Phi}]. \tag{2}$$

Written out explicitly,

$$(D_\mu \mathbf{\Phi})_a^b = \partial_\mu \mathbf{\Phi}_a^b - \frac{ig_5}{\sqrt{2}}(\mathbf{A}_\mu)_a^c \mathbf{\Phi}_a^b + \frac{ig_5}{\sqrt{2}}(\mathbf{A}_\mu)_d^b \mathbf{\Phi}_a^d. \tag{3}$$

Now consider the Lagrangian

$$\mathscr{L}_\Phi = \mathrm{Tr}(D_\mu \mathbf{\Phi})^\dagger (D^\mu \mathbf{\Phi}) + V(\mathbf{\Phi}), \tag{4}$$

where $V(\mathbf{\Phi})$, the Higgs potential, is a SU(5) invariant polynomial in $\mathbf{\Phi}$, restricted to be at most of the fourth degree. Higher than fourth degree polynomials ruin the renormalizability of the theory.

For generating gauge boson masses, we need terms that are quadratic in the gauge fields. From (3), the terms of interest are given by

$$\frac{g_5^2}{2}\mathrm{Tr}[\mathbf{A}_\mu, \mathbf{\Phi}]^\dagger [\mathbf{A}_\mu, \mathbf{\Phi}]. \tag{5}$$

Assuming $\mathbf{\Phi}$ to be in the diagonal form

$$\mathbf{\Phi} = \mathrm{diag}(\phi_1, \phi_2, \phi_3, \phi_4, \phi_5), \tag{6}$$

we ask what pattern of VEV breaks $SU(5) \to SU(3) \times SU(2) \times U(1)$. It should be clearly of the form

$$\begin{aligned}\mathbf{\Phi} &= V\ \mathrm{diag}(1,\ 1,\ 1,\ -3/2,\ -3/2)\\ &= -\sqrt{15}\,V\,T_{\mathrm{Y}}^{\mathrm{W}},\end{aligned} \tag{7}$$

where V is the VEV, $T_{\mathrm{Y}}^{\mathrm{W}}$ is the generator corrresponding to U(1) (Equation (15), Section 2). $T_{\mathrm{Y}}^{\mathrm{W}}$, being a multiple of identity in G and W spaces, commutes with the SU(3) and SU(2) generators. Noncommuting parts involve only the X and Y bosons which will then acquire mass. Thus

$$[\mathbf{A}_\mu, \mathbf{\Phi}] = V \begin{bmatrix} 0 & 0 & 0 & -\frac{5}{2}\bar{\mathrm{X}}^1 & -\frac{5}{2}\bar{\mathrm{Y}}^1 \\ 0 & 0 & 0 & -\frac{5}{2}\bar{\mathrm{X}}^2 & -\frac{5}{2}\bar{\mathrm{Y}}^2 \\ 0 & 0 & 0 & -\frac{5}{2}\bar{\mathrm{X}}^3 & -\frac{5}{2}\bar{\mathrm{Y}}^3 \\ \frac{5}{2}\mathrm{X}^1 & \frac{5}{2}\mathrm{X}^2 & \frac{5}{2}\mathrm{X}^3 & 0 & 0 \\ \frac{5}{2}\mathrm{Y}^1 & \frac{5}{2}\mathrm{Y}^2 & \frac{5}{2}\mathrm{Y}^3 & 0 & 0 \end{bmatrix},$$

and

$$\mathrm{Tr}\,[\mathbf{A}_\mu, \mathbf{\Phi}]^\dagger\,[\mathbf{A}_\mu, \mathbf{\Phi}] = \frac{25}{4}V^2(\mathrm{X}^i\mathrm{X}_i + \mathrm{Y}^i\mathrm{Y}_i), \tag{8}$$

implying that the X and Y bosons are no longer massless. Their masses are given by

$$M_{\mathrm{X}}^2 = M_{\mathrm{Y}}^2 = \frac{25}{8}g_5^2\,V^2. \tag{9}$$

M_{X}, M_{Y} set the scale of symmetry breaking at this stage. Appropriately choosing V, we can make the scale as high as we please. In this process of symmetry breaking 12 Higgs scalar would disappear in giving masses to the 12 gauge bosons. Twelve would survive as physical, scalar particles. It should also be noted that the fermions belonging to **5*** and **10*** remain massless along with the gluons and the electroweak bosons, since **24** does not occur in **5*** × **10** nor **10** × **10** and, hence, we cannot form Yukawa-like couplings of the Higgs scalars with the fermions. Recall that only such terms, if they exist, can give rise to the fermion masses when the Higgs scalars take on a nonvanishing VEV.

3.1.2. 2nd Stage. To break the symmetry further, that is, $\mathrm{SU(3)_C} \times \mathrm{SU(2)} \times \mathrm{U(1)} \to \mathrm{SU(3)_C} \times \mathrm{U(1)_{EM}}$, consider a Higgs multiplet H belonging to the fundamental representation **5** of SU(5), which consists of $(\mathbf{3}, \mathbf{1}, -\frac{1}{3})$ and $(\mathbf{1}, \mathbf{2}, \frac{1}{2})$, and can be denoted by

$$H = \begin{bmatrix} \chi^1 \\ \chi^2 \\ \chi^3 \\ \phi^+ \\ \phi^0 \end{bmatrix} \qquad \begin{aligned}\chi &= (3,\ 1,\ -\tfrac{1}{3}),\\ \phi &= (1,\ 2,\ \tfrac{1}{2}).\end{aligned} \tag{10}$$

The full Lagrangian would have the added terms $(D_\mu H)^\dagger (D^\mu H)$, $V(H)$, which are the KE term and the Higgs potential term, respectively. The KE term contains

$$\frac{g_5^2}{2}(A_\mu H)_a^\dagger (A^\mu H)^a$$

and, hence, it has the potentiality to give masses to the gauge bosons. Suppose we choose

$$\langle H \rangle = \frac{v}{\sqrt{2}} \begin{bmatrix} 0 \\ 0 \\ 0 \\ 0 \\ 1 \end{bmatrix}. \tag{11}$$

Then

$$\frac{g_5^2}{2}(A_\mu H)_a^\dagger (A^\mu H)^a \to \frac{g_5^2}{2}\frac{v^2}{2}\{\bar{Y}Y + \tfrac{1}{2}(W_1^2 + W_2^2) + \tfrac{8}{5}\tfrac{1}{2}Z^2\}, \tag{12}$$

where

$$Z = \sqrt{\frac{3}{8}}B - \sqrt{\frac{5}{8}}W_3, \tag{13}$$

which shows that the Weinberg angle θ_W is given by

$$\sin\theta_W = \sqrt{\frac{3}{8}} \tag{14}$$

and

$$M_W^2 = M_Z^2 \cos^2\theta_W = \frac{g_5^2 v^2}{4}. \tag{15}$$

Thus, three of the four electroweak gauge bosons acquire mass leaving only the $SU(3)_C$ gluons and the photon massless. The desired symmetry breaking is, therefore, attainable in principle. The question is: is it realizable by a Higgs potential as *its stable absolute minimum*? We shall consider this problem next, first by examining only the adjoint representation and then the fundamental and the adjoint together.

3.2. *Higgs Potential*

3.2.1. Potential with the adjoint representation $\mathbf{\Phi}$. The most general quartic potential involving $\mathbf{\Phi}$ belonging to the adjoint representation is of the form

$$V(\mathbf{\Phi}) = -\frac{\mu_a^2}{2}\mathrm{Tr}(\mathbf{\Phi}^2) + \frac{\lambda_a}{4}(\mathrm{Tr}(\mathbf{\Phi}^2))^2 + \frac{\alpha}{2}(\mathbf{\Phi}^4) + \frac{\beta}{3}\mathrm{Tr}(\mathbf{\Phi}^3), \tag{16}$$

where μ_a^2, λ_a, α, β are constant, arbitrary parameters. By an appropriate SU(5) transformation, we can bring $\mathbf{\Phi}$ to the diagonal form

$$\mathbf{\Phi} = \mathrm{diag}(\phi_1, \phi_2, \phi_3, \phi_4, \phi_5), \sum_n \phi_n = 0 \tag{17}$$

and, hence,

$$V(\mathbf{\Phi}) = -\frac{\mu_a^2}{2}\sum_n \phi_n^2 + \frac{\lambda_a}{4}\left(\sum_n \phi_n^2\right)^2 + \frac{\alpha}{2}\sum \phi_n^4 + \frac{\beta}{3}\sum \phi_n^3. \tag{18}$$

For the extrema, we require

$$\frac{\partial V}{\partial \phi_i} = 0. \tag{19}$$

There is a general theorem due to Ruegg (H. Ruegg, *Phys. Rev.* **D22**, 2040 (1980)) which states that (19) can be satisfied if and only if there are two distinct eigenvalues of Φ. That is, the diagonal form (17) is essentially

either (i) $\mathrm{diag}(\phi, \phi, \phi, \phi', \phi')$, $3\phi + 2\phi' = 0$
or (ii) $\mathrm{diag}(\phi, \phi, \phi, \phi, \phi')$, $4\phi + \phi' = 0$.

The alternative (ii) leads to SU(5) $\rightarrow$ SU(4) $\times$ U(1). Confining our attention to (i), we have the desired pattern (7) if $\phi = V$, $\phi' = -\frac{3}{2}V$. Then the extremum Equation (19) leads to

$$\frac{\partial V(\Phi)}{\partial \Phi} = 0 = \phi\left(-\frac{15}{2}\mu_a^2 + \left(\frac{15}{2}\right)^2 \lambda_a \phi^2 + \frac{105}{4}\alpha\phi^2 + \frac{39}{4}\beta\phi\right). \tag{20}$$

For simplicity, let us assume $\beta = 0$. Then (20) yields

$$\phi^2 = V^2 = \frac{2\mu_a^2}{15\lambda_a + 7\alpha} \tag{21}$$

and

$$V_{\min} = -\frac{15\mu_a^4}{4(15\lambda_a + 7\alpha)}. \tag{22}$$

This shows that the desired symmetry breaking pattern SU(5) $\rightarrow$ SU(3) $\times$ SU(2) $\times$ U(1) can be realized by a general Higgs potential constructed from the adjoint representation Φ.

3.2.2. Potential with the fundamental representation H. The most general potential involving H has the form

$$V(H) = -\frac{\mu_f^2}{2}H^\dagger H + \frac{\lambda_f}{4}(H^\dagger H)^2. \tag{23}$$

It is straightforward to add this term to $V(\Phi)$ in (16) and carry out the general analysis. If $SU(3)_C$ survives, breaking $SU(2) \times U(1) \to U(1)_{EM}$,

$$\langle H^i \rangle = \delta^{i5} \frac{v}{\sqrt{2}}, \tag{24}$$

where

$$v^2 = \frac{2\mu_f^2}{\lambda_f} \quad \text{and} \quad M_w^2 = \frac{g^2 v^2}{4}. \tag{25}$$

However, $V(\Phi) + V(H)$ contains two neutral color triplets, one from Φ and the other from H. In the process of symmetry breaking, one combination is gauged away contributing to the mass of Y, but the other remains massless. It is unacceptable as it can mediate proton decay resulting in too short a lifetime for the proton.

3.2.3. *Potential involving both Φ and H.* Finally let us consider the most general potential constructed from Φ and H including cross-terms between Φ and H given by $V(\Phi, H)$ where

$$V(\Phi, H) = \gamma_1 H^\dagger H \operatorname{Tr} \Phi^2 + \gamma_2 H^\dagger \Phi^2 H + \gamma_3 H^\dagger \Phi H. \tag{26}$$

If we omit the last term in (26) (invoking for instance additional symmetry $\Phi \to -\Phi$, which also leads to $\beta = 0$), the analysis of the minima of

$$V = V(\Phi) + V(H) + V(\Phi, H)$$

leads to the conclusion that an absolute minimum of V can be obtained provided

$$\langle \Phi \rangle = V \operatorname{diag}\left(1, 1, 1, -\frac{3}{2} - \frac{\varepsilon}{2}, -\frac{3}{2} + \frac{\varepsilon}{2}\right),$$

$$\langle H^a \rangle = \frac{v}{\sqrt{2}} \delta^{a5},$$

with ε, V, and v determined from the following equations,

$$\varepsilon = \frac{3}{20} \frac{\gamma_2}{\alpha} \left(\frac{v}{V}\right)^2 + O\left(\frac{v^4}{V^4}\right), \tag{27}$$

$$\mu_a^2 = \frac{15}{2} \lambda_a V^2 + \frac{7}{2} \alpha V^2 + \gamma_1 v^2 + \frac{9}{30} \gamma_2, \tag{28}$$

$$\mu_f^2 = \frac{\lambda_f}{2} v^2 + 15\gamma_1 V^2 + \left(\tfrac{9}{2} - 3\varepsilon\right)\gamma_2 V^2 \ldots \tag{29}$$

(For details see, M. Magg and Q. Shafi, *Z. Physik C., Particles and Fields* **4**, 63 (1980).)

It turns out that the mass of the surviving neutral colored triplet that can mediate in proton decay is given by

$$\mu_h^2 = -\tfrac{5}{2} \gamma_2 V^2 + O(v^2),$$

which requires that $\gamma_2 < 0$ and not have too small a value, since it is necessary that $\mu_h^2 \lessapprox M_x^2$ in order that the proton decay rate is not too fast. We also should note that the mass hierarchy between M_X and M_W requires that $v^2 <<< V^2$ ($v/V \sim 10^{-13}$). The question arises whether such a hierarchy can be achieved with a reasonable choice of parameters in the potential.

A close look at this problem reveals one of the fundamental difficulties of grand unified theories which require, in general, widely different scales of symmetry breaking. To achieve the desired hierarchy in such scales, the parameters (in Equation (20), for instance) require a delicate adjustment or fine tuning to a fantastic degree of accuracy. For most reasonable values of the parameters, one obtains $M_W \simeq M_X$. Further, even if the parameters are adjusted at the tree level, radiative corrections introduce large corrections to the scalar meson mass terms requiring fine tuning again. As a result the whole procedure of SSB has become suspect. A partial solution to this hierarchy and fine tuning problem is to be found in SUSY models in which one can arrange cancellation between symmetric boson and fermion contributions, thereby avoiding quadratically divergent contributions to the scalar meson mass terms. But SUSY models give rise to new problems.

3.3. Fermion Masses

If the original Lagrangian is to be SU(5) gauge invariant, there cannot be bare fermion mass terms in the Lagrangian. Masses for the fermions arise after SSB due to fermion-Higgs, Yukawa-type couplings. It is straightforward to verify that, in the minimal SU(5) model with only **24** and **5** Higgs representations, it is only the **5** representation that couples to the fermions in the model. The relevant part of the Lagrangian $\mathscr{L}_{FH}$ is given by

$$\mathscr{L}_{FH} = [\Gamma_{ij}(\overline{\chi_{iR}^c})_a \psi_j^{ab} H_b^\dagger + \text{h.c.}] +$$

$$+ [\Gamma'_{ij} \tfrac{1}{4} \varepsilon^{abcde} (\overline{\psi_{2R}^c})_{ab} (\psi_j)_{cd} H_e + \text{h.c.}], \tag{30}$$

where, for the sake of completeness, we have introduced the generation structure of fermions in the model with the index $i,j = 1,2,3$. The first term in (30) corresponds to ($5^* \times 10$) fermion couplings and the second corresponds to those of 10×10 fermions. When the symmetry is broken with

$$\langle H^a \rangle = \delta^{a5} v/\sqrt{2},$$

we have mass matrices

$$M_{ij}^{(d)} = \frac{v}{\sqrt{2}} \Gamma_{ij} \quad \text{and} \quad M_{ij}^{(u)} = \frac{v}{\sqrt{2}} \Gamma'_{ij}$$

in the down- and up-charge sectors, respectively. When these mass matrices are diagonalized, we can define physical, mass eigenstates for the quarks. In minimal SU(5), the bare masses thus obtained satisfy the relations,

$$m_{\mathrm{d}} = m_{\mathrm{e}}; \qquad m_{\mathrm{s}} = m_{\mu}; \qquad m_{\mathrm{b}} = m_{\tau}. \tag{31}$$

These relations between quark and lepton masses will be modified by radiative corrections. We shall study this aspect in the next section.

4. Predictions of Minimal SU(5)

In this section, we shall discuss three, what are considered to be classical predictions of SU(5). These include, (i) the value of $\sin^2\theta_{\mathrm{W}}$ in the measured low energy region, (ii) the mass of the b-quark, (iii) the proton decay and its life-time. The first two predictions combine exact SU(5) limit with renormalization group ideas to calculate the energy dependence of the coupling constants and the effective mass of the fermions. The last prediction is the most dramatic one. That these predictions follow more or less unambiguously made minimal SU(5), the most popular GUT model. Other GUT models contain additional parameters and therefore do not lead to clear cut predictions.

4.1. The Weinberg Angle; Coupling Constant Renormalization

In the standard theory of electro/weak interactions.

$$\sin^2\theta_{\mathrm{W}} = \frac{e^2}{g^2} = \frac{g'^2}{g^2 + g'^2},$$

where g, g' are the SU(2) and U(1) gauge couplings, respectively. Thus $\sin^2\theta_{\mathrm{W}}$ is a totally free parameter. In SU(5), there is only one gauge coupling parameter g_5, and the exact SU(5) limit,

$$g_5 = g_3 = g_2 = g_1, \tag{1}$$

where g_3 is the $\mathrm{SU(3)_c}$ gauge coupling and $g_2 = g$, $g_1 = g'$. The above equality follows when the generators of all the relevant subgroups are normalized in the same way, namely

$$\mathrm{Tr}(T_i T_j) = \tfrac{1}{2}\delta_{ij}. \tag{2}$$

Recall that the electric charge generator Q,

$$Q = T_3 + \sqrt{\tfrac{5}{3}}\,T_y$$

with

$$Q = \mathrm{diag}(-\tfrac{1}{2}, -\tfrac{1}{3}, -\tfrac{1}{3}, 0, 0) \quad \text{and} \quad T_y = \sqrt{\tfrac{3}{5}}\,\mathrm{diag}(-\tfrac{1}{3}, -\tfrac{1}{3}, -\tfrac{1}{3}, \tfrac{1}{2}, \tfrac{1}{2})$$

Since $\sqrt{\frac{5}{3}}\, T_y$ is the weak hypercharge generator in the standard theory, it follows that

$$g_1 = \sqrt{\tfrac{5}{3}}\, g' \tag{3}$$

and, hence,

$$\sin^2\theta_W = \frac{3/5}{1 + 3/5} = \frac{3}{8} = 0.375. \tag{4}$$

The experimental value of 0.215 ± 0.015 indicates a clear discrepancy until we realize that the relation (4) is in the exact SU(5) limit, that is for energies, $E \gg M_X$. The measured value of $\sin^2\theta_W$ is at much lower energies $E \ll M_W$, where SU(5) is broken. The couplings have departed, have 'run' away from their asymptotic values. We must evaluate them at lower energies using Equation (1) as the asymptotic boundary condition. The renormalization group (RNG) analysis provides the necessary tool and the equations that govern the energy-dependence of the coupling parameters. In what follows, we shall note a few salient features of RNG ideas, before we apply them to our problem.

(a) Perturbative calculations in relativistic fields theories lead to infinities which need to be properly subtracted by absorbing them into the redefinition or renormalization of physical quantities such as mass and charge.

(b) Physical quantities are defined by choosing certain, convenient boundary conditions at physical (or sometimes unphysical) points. In the $\lambda\phi^4$ self-interacting scalar meson theory, for instance, one defines the renormalized four-point coupling λ by

$$\Gamma_R^{(4)}(p_1, p_2, p_3, p_4) = -i\lambda \quad \text{at } p_i = 0 \quad \text{or at} \quad p_i^2 = \mu^2.$$

(c) Different ways of defining the physical quantities clearly give rise to different renormalized physical quantities. However, they must be related in a definite way if physics is not to depend upon the arbitrary choice of defining the parameters. Consider, for instance, two different renormalization schemes, R and R',

$$\phi_R(R) = Z_\phi^{-1/2}(R)\phi_0, \qquad \phi_{R'}(R') = Z_\phi^{-1/2}(R')\phi_0, \tag{5}$$

where $Z_\phi(R)$ and $Z_\phi(R')$ are infinite, wavefunction renormalization constants in the two schemes. Hence, eliminating ϕ_0,

$$\begin{aligned}\phi_R(R) &= \frac{Z_\phi^{-1/2}(R)}{Z_\phi^{-1/2}(R')}\phi_{R'}(R') \\ &= Z_\phi(R, R')\phi_{R'}(R')\end{aligned} \tag{6}$$

Now $Z_\phi(R, R')$ must be finite since it relates *two renormalized* fields. Similar considerations should hold for the coupling and mass parameters in the theory. $R \to R'$ can be viewed as a transformation. The set of all such transformations form a group which is called the renormalization group. The translation of the above statement into an analytic form leads to Gell-Mann–Low, or Callan–

Symanzic equations which give the energy dependence of the coupling parameters. We shall not derive these equations, but use the results.

(d) Coupling parameters of a non-Abelian gauge theory are governed by the equations

$$\mu \frac{dg_n}{d\mu} = \beta_n(g_n(\mu)) \simeq b_n g_n^3(\mu) \quad \text{for } |g_n| \ll 1. \tag{7}$$

Therefore,

$$\frac{1}{g_n^2(\mu)} = \text{const} - 2b_n \ln \mu \tag{8}$$

and considering the value at $\mu = M_X$ = grand unification mass,

$$\frac{1}{g_n^2(\mu)} = \frac{1}{g_n^2(M_X)} + 2b_n \ln \frac{M_X}{\mu} \quad \text{for } \mu \ll M_X. \tag{9}$$

If at M_X, $g_3 = g_2 = g_1 = g_5$, we have the basic equations in the minimal SU(5) model,

$$\frac{1}{\alpha_3(\mu)} = \frac{1}{\alpha_5(\mu)} + 8\pi b_3 \ln \frac{M_X}{\mu},$$

$$\frac{1}{\alpha_2(\mu)} = \frac{1}{\alpha_5(\mu)} + 8\pi b_2 \ln \frac{M_X}{\mu},$$

$$\frac{1}{\alpha_1(\mu)} = \frac{1}{\alpha_5(\mu)} + 8\pi b_1 \ln \frac{M_X}{\mu}, \tag{10}$$

where $\alpha_i = g_i^2/4\pi$.

The coefficients b_n are calculated in perturbation theory using 'dimensional' regularization. If we neglect the contribution due to the exchange of Higgs scalars,

$$b_n = \frac{(2N_F - 11n)}{48\pi^2} \quad \text{for } n > 2. \tag{11}$$

For $n = 1$,

$$b_1 = \frac{2N_F}{48\pi^2}, \tag{12}$$

where N_F is the number of quark flavors ($N_F = 6$ for three generations).

Using (10), (11), and (12), we can derive the following three important relations:

$$\text{(i)} \quad \sin^2\theta_W = \frac{1}{6} + \frac{5}{9}\frac{\alpha(\mu)}{\alpha_s(\mu)}, \tag{13}$$

which relates $\sin^2\theta_W$ to the measurable, electromagnetic (fine-structure) coupling parameter $\alpha(\mu)$ and the strong interaction QCD $\alpha_3(\mu) = \alpha_s(\mu)$.

$$\text{(ii)} \quad \ln \frac{M_X}{\mu} = \frac{\pi}{11}\left(\frac{1}{\alpha(\mu)} - \frac{8}{3}\frac{1}{\alpha_s(\mu)}\right), \tag{14}$$

which is a very important relation as it enables us to estimate the GUT mass scale M_X from the knowledge of $\alpha(\mu)$ and $\alpha_s(\mu)$.

$$\text{(iii)} \quad \sin^2\theta_W = \frac{3}{8} - \frac{55}{24\pi}\alpha(\mu)\ln\frac{M_X}{\mu}. \tag{15}$$

This is not an independent, but useful relation that follows from (13) and (14), which exhibits an alternate way of estimating M_X from the knowledge of $\sin^2\theta_W$ and $\alpha(\mu)$. Incidentally, it also shows that $\sin^2\theta_W < \frac{3}{8}$.

Recall that $M_W^2 = M_Z^2\cos^2\theta_W$. An accurate determination of M_W and M_Z gives an accurate determination of $\sin^2\theta_W$, which then provides a precise determination of M_X.

These derivations, complicated as they are, involve several simplifying assumptions. They neglect the scalar meson contributions and the proper threshold behavior. When they are included, the formulae become even more complicated and, to some extent, lose their predictive power. It is interesting, however, that the qualitative behavior of $\sin^2\theta_W$ and M_X that emerge from these formulae, are in accordance with experiments. From (13) and (14), we can estimate

$$\sin^2\theta_W = 0.206\begin{Bmatrix}+0.016\\-0.004\end{Bmatrix}, \qquad M_X = 3.6\begin{Bmatrix}\pm 3.4\\-3.2\end{Bmatrix}\times 10^{14}\,\text{GeV},$$

at $\mu = 80$ GeV.

In estimating the above numbers, we need the value of $\alpha(\mu)$ and $\alpha_s(\mu)$ at $\mu = 80$ GeV. In addition, we need to assume a value for the top-quark mass. We know that $\alpha^{-1}(0) = 137$. To evaluate it at 80 GeV where strong interactions are important, we need to use dispersion theoretic treatment of e^+e^- total cross-section. To evaluate $\alpha_s(\mu)$, we need the value of the QCD parameter $\bar{\Lambda}$. The above-mentioned values for $\sin^2\theta_W$ and M_X assume

$$50\text{ MeV} \leqslant \Lambda_{\overline{MS}} \leqslant 500\text{ MeV} \quad \text{and} \quad m_t = 20\text{ GeV}\,.$$

There are currently uncertainties in the values of these parameters, but the values for $\sin^2\theta_W$ and M_X are reasonably stable. They are in good agreement with experiments, and therefore constitute one of the successful predictions of the model.

4.2. *Mass of the b-Quark; Mass Renormalization and Effective Mass Operator*

The mass relations (31) discussed in the previous section are, as remarked earlier, consequences of exact SU(5) symmetry. Like the coupling constants, they undergo renormalization. From RNG considerations, the equation for 'effective' mass or 'running' mass is

$$\frac{d \ln m(\mu)}{d \ln \mu} = b_m^{(n)} g_n^2(\mu), \tag{16}$$

where $b_m^{(n)}$ is the coefficient (like b_n in the case of coupling parameters) that arises in the fermion self-energy calculation. Using Equation (9) for $g_n^2(\mu)$, we obtain

$$\frac{d \ln m(\mu)}{d \ln \mu} = \frac{g_n^2(\mu_0) b_m^{(n)}}{1 + 2b_n g_n^2(\mu_0) \ln (\mu/\mu_0)}. \tag{17}$$

The solution of (17) gives

$$\frac{m(\mu)}{m(\mu_0)} = \left[\frac{g_n(\mu)}{g_n(\mu_0)}\right]^{-b_m^{(n)}/b_n}; \tag{18}$$

where b_n are given by Equation (11) and (12). If we consider the graph in Figure

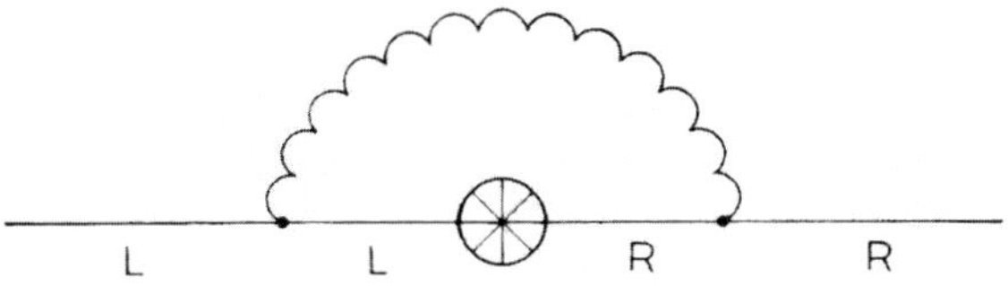

Fig. 1.

1 and note

$$\sum_a (T^a T^a)_{ij} = \frac{n^2 - 1}{2n} \delta_{ij}, \quad \text{for SU}(n) \geqslant 2, \tag{19}$$

$$(T^0)^2 = \frac{3}{5}\left(\frac{Y}{2}\right)^2 \quad \text{for U(1)}, \tag{20}$$

we get, in our case of $SU(5) \to SU(3)_c \times SU(2) \times U(1)$,

$$\left[\frac{m_{d,s,b,\dots}(\mu)}{m_{d,s,b,\dots}(M_x)}\right] = \left[\frac{g_3(\mu)}{g_3(M_x)}\right]^{\frac{8}{11-2N_f/3}} \times \left[\frac{g_2(\mu)}{g_2(M_x)}\right]^{\frac{27}{44-4N_f}} \times \left|\frac{g_1(\mu)}{g_1(M_x)}\right|^{\frac{3}{10N_f}}, \tag{21}$$

$$\left[\frac{m_{e,\mu,\tau,\dots}(\mu)}{m_{e,\mu,\tau,\dots}(M_x)}\right] = \left[\frac{g_2(\mu)}{g_2(M_x)}\right]^{\frac{27}{44-4N_f}} \times \left[\frac{g_1(\mu)}{g_1(M_x)}\right]^{-\frac{27}{10N_f}}, \tag{22}$$

which lead to

$$\frac{m_b(\mu)}{m_\tau(\mu)} = \left[\frac{g_3(\mu)}{g_5(M_x)}\right]^{\frac{8}{11-2N_f/3}} \left[\frac{g_1(\mu)}{g_5(M_x)}\right]^{\frac{3}{N_f}}, \tag{23}$$

where we have used $g_3(M_x) = g_1(M_x) = g_5(M_x)$. At $\mu = 10$ GeV, Equation (23) yields

$$\frac{m_b}{m_\tau} = 2.9 \pm 0.2,$$

to be compared with the experimental value of 2.6 – 2.9, which clearly must be regarded as a successful prediction of the minimal SU(5). However, the same considerations lead to

$$\frac{m_\mu}{m_e} = \frac{m_s}{m_d},$$

which is totally in contradiction with known considerations. m_μ/m_e is approximately 200, whereas, according to current algebra estimates, m_s/m_d is approximately 20. This discrepancy is generally blamed on the Renormalization Group Equation. It is not expected to hold for very light quarks. (For a detailed discussion of this topic, see the original paper of Buras *et al.*, *Nucl. Phys.* **B135** 66 (1978).)

4.3. *Proton Decay*

The most dramatic prediction of any grand unified theory is the instability of the proton due to baryon number-violating interactions. If the interactions are unified, if the quarks and the leptons are in the same multiplets, it is to be expected that there are B- and L-violating processes in the energy region $E > M_X$. The question is whether interactions leading to such processes can produce observable consequences in the low energy region and whether we can calculate their effects with some measure of certainty. As a grand unified theory, minimal SU(5) is well-defined; the most important feature being that there is only one GUT mass scale M_X. However, as we shall see, in calculating the life-time of the proton and the various branching ratios, it is necessary to invoke other theoretical considerations over and above those of GUTs. Predicted life-time, even in the minimal SU(5), is uncertain to the extent of two orders of magnitude.

The interaction vertices responsible for B and L violations are shown Figure 2.

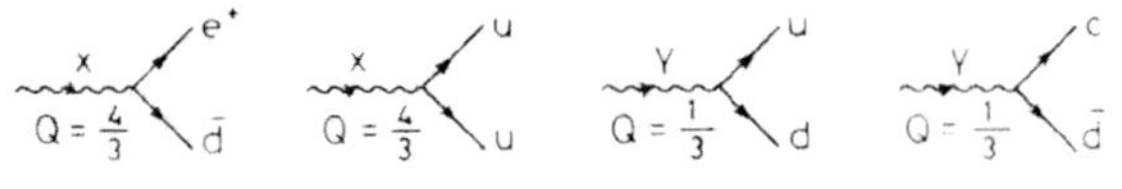

B	$-\frac{1}{3}$	$\frac{2}{3}$	$\frac{2}{3}$	$-\frac{1}{3}$
L	-1	0	0	-1
(B–L)	$\frac{2}{3}$	$\frac{2}{3}$	$\frac{2}{3}$	$\frac{2}{3}$

Fig. 2.

They involve the lepto-quark gauge bosons X and Y, leading to the interaction Lagrangian,

$$\mathscr{L}_{\substack{\Delta B \neq 0 \\ \Delta L \neq 0}} = \frac{g_5}{\sqrt{2}} \bar{X}^\alpha_\mu [\bar{d}_{R\alpha} \gamma^\mu e^+_R + \bar{d}_{L\alpha} \gamma^\mu e^+_L + \varepsilon_{\alpha\beta\gamma} \bar{u}^{c\gamma} \gamma^\mu u^\beta_L] +$$

$$+ \frac{g_5}{\sqrt{2}} \bar{Y}^\alpha_\mu [-\bar{d}_{R\alpha} \gamma^\mu \nu^c_R - \bar{u}_{L\alpha} \gamma^\mu e^+_L + \varepsilon_{\alpha\beta\gamma} \bar{u}^{c\gamma} \gamma^\mu d^\beta_L] + \text{h.c.} \tag{24}$$

The above interaction Lagrangian gives rise to tree-level graphs (Figure 3)

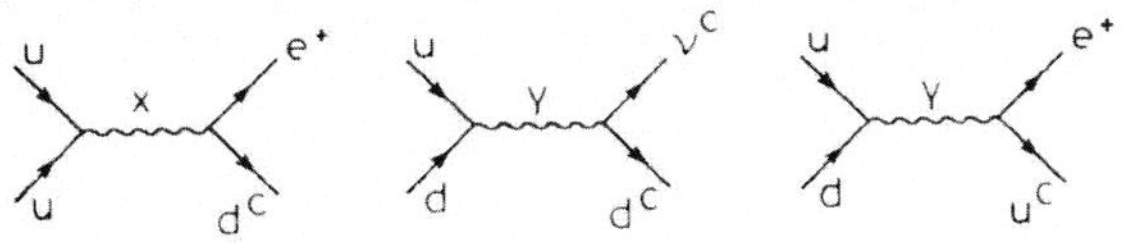

Fig. 3.

leading to decays

$$p \to e^+ X, \qquad p \to \nu^c + X \quad \text{etc.}$$

The vertices in (24) satisfy (as can be seen from the Baryon and Lepton numbers under interaction vertices) $\Delta B \neq 0$, $\Delta L \neq 0$, but $\Delta(B - L) = 0$. When we extend the model to include several families, we obtain selection rules $\Delta S = 0$, $\Delta S = -\Delta B$. These are selection rules characteristic of minimal SU(5). The tree-level graphs when extrapolated down to 1 GeV, become effectively four-fermion interactions with coupling strength $\propto g_5^2/M_x^2$. The effective interaction $\mathscr{L}_{\text{eff}}$ is given by

$$\mathscr{L}_{\text{eff}} = \frac{g_5^2}{2M_x^2} [(\varepsilon_{\alpha\beta\gamma} \overline{u^{c\gamma}_L} \gamma^\mu u^\beta_L)(2\overline{e^+_L} \gamma_\mu d^\alpha_L + \overline{e^+_R} \gamma_\mu d^\alpha_R) -$$

$$- \varepsilon_{\alpha\beta\gamma} (u^{c\gamma}_L \gamma^\mu d^\beta_L)(\nu^c_R \gamma_\mu d^\alpha_R)] + \text{h.c.} \tag{25}$$

where α, β, γ denote the color indices. The family structure is not included, but Fierz transformations and Fierz identities are used to obtain the effective interaction in the above form.

In a more general analysis one, therefore, constructs all four-fermion operators subject to SU(3) × SU(2) × U(1) invariance (S. Weinberg, *Phys. Rev. Lett.* **43**, 1566 (1979); F. Wilczek and A. Zee, *Phys. Rev. Lett.* **43**, 1571 (1979)). The lowest dimension of such an operator which violates the baryon number is six (four fermions), leading, almost on dimensional arguments to

$$\tau_{\text{proton}} = \frac{1}{\Gamma_p}, \quad \Gamma_p \propto \frac{\alpha_5^2}{M_x^4} m_p^5. \tag{26}$$

Hence,

$$\tau_p \sim \frac{M_X^4}{\alpha_5^2 m_p^5} \sim 2 \times 10^{32} \text{ yr}.$$

To perform a detailed estimate, it is necessary (1) to renormalize the effective Lagrangian written down at the mass scale M_x to bring it down to the hadron scale of 1 GeV. One has to resort to RNG techniques, as in the case of the coupling constant and mass renormalizations we discussed in Sections 1 and 2. (2) We have to find the matrix elements of the effective interaction (which is in terms of quark and lepton fields) between hadronic states. A variety of phenomenological hadron-physics techniques, SU(6), relativistic bag models, current algebra techniques based on chiral symmetry, have been used to evaluate the matrix elements of $\mathscr{L}_{\text{eff}}$ between the hadron states such as proton and pi-meson. Different methods give varying answers differing approximately by a factor of 25–30 in the proton lifetime. (For some of these details, see O. Kaymakcalan, C. H. Lo and K. C. Wali, *Phys. Rev.* **D29**, 1962, (1984).)

The lifetime depends most sensitively on M_X which, in turn, depends sensitively on the QCD parameter Λ in the RNG Equations. The current uncertainty in the value of the latter parameter leads to an uncertainty in the proton lifetime by a factor of 10^2. This situation may be improved in the future by the precise determination of the masses of M_W, M_Z and, hence, $\sin^2\theta_W$.

5. Baryon Asymmetry

As stated in the first section, astrophysics and the early Universe are the natural arena for testing grand unified theories. In this and the next and last section, we shall discuss two topics which have received a great deal of attention in recent years to illustrate how astrophysics can and does provide guidelines for GUTs.

We live in a particle-antiparticle asymmetric universe. Lucky indeed! If there was an equal amount of matter and antimatter in the Universe, we can infer from annihilation processes that the baryon number density n_B and the anti-baryon number density $n_{\bar{B}}$ being equal,

$$n_B = n_{\bar{B}} \simeq 10^{-20} n_\gamma, \tag{1}$$

where n_γ is the number density of the photons. Instead, we find $n_B \simeq 10^{-8} n_\gamma$, which indeed is much, much larger than expected, suggesting an excess of matter over antimatter in the Universe, which prevents complete annihilation.

How does this asymmetry originate? Three possible scenarios have been suggested in the literature: (i) The asymmetry is an initial condition of the Universe. The Universe was born with an asymmetry which was preserved throughout its evolution. (ii) The whole Universe is symmetric, but it is spatially nonuniform; some regions have an excess of baryons, other regions have an excess of anti-baryons. (iii) The Universe was originally symmetric; the asymmetry developed in the course of the evolution of the Universe.

The last alternative is regarded by most as the most satisfactory solution to the problem. This requires, however, as noted by Sakharov a long time ago,

(a) Baryon number nonconserving processes,

(b) CP violation,
(c) Departure from thermal equilibrium.

As discussed earlier, it is natural to expect $\Delta B \neq 0$ processes in GUTs. Since CP violation *is* observed in K-$\bar{\text{K}}$ system, it should be incorporated in a complete theory such as a GUT. The third condition (c) is provided by the expansion of the Universe. Hence, GUTs combined with the expansion and consequent cooling of the Universe as embodied in Big Bang Cosmology can, in principle, provide an answer to the long standing puzzle.

To see this in more detail, let us begin considering our Universe at temperature $kT \simeq M_P \simeq 10^{19}$ GeV ($M_P \equiv$ Planck mass). At this temperature and above, gravitational interactions dominate. They, along with other interactions, bring about thermal equilibrium for all particles containing equal numbers of quarks and antiquarks. At this stage, the baryon number of the Universe is zero.

The Universe expands; it cools and the temperature falls below that corresponding to the Planck mass. Consider the existence of some superheavy particles such as the gauge bosons X, Y, or Higgs particles that survive in the process of symmetry breaking in GUTs. Let X be the generic name for such a superheavy particle.

Now, the rates Γ_X for X decay, Γ_C for collisions, at a temperature T of the universe, are given by

$$\Gamma_X = \frac{\alpha_X M_X^2 N}{[(kT)^2 + M_X^2]^{1/2}}, \qquad \Gamma_C = \frac{\alpha_X (kT)^5 N}{[(kT)^2 + M_X^2]^2}, \tag{2}$$

where N is the number of helicity states in the theory, M_X is the mass of X, and $\alpha_X = g^2/4\pi$ if X is a gauge boson with gauge coupling g,

$$\alpha_X = \alpha_H = G_F \frac{\bar{m}^2}{4\pi}$$

if X is a Higgs boson ($\bar{m}^2$ is an average fermion mass). Typically $N \geqslant 100$, $\alpha_X \simeq 10^{-2}$, $\alpha_H \simeq 10^{-5}$, 10^{-6}.

The expansion rate of the Universe, $\dot{R}(t)/R(t) = H$, the Hubble constant, obeys the equation

$$H = \left[\frac{8\pi G}{3}\rho\right]^{1/2}, \tag{3}$$

where ρ, the proper energy density, is given by

$$\rho = N\left(\frac{m}{T}\right)\frac{\pi^2}{30}T^4. \tag{4}$$

$N(m/T)$ is the number of helicity states of particles of mass m at temperature T. (For the derivation of these equations and other details of Big Bang Cosmology, see Chapter 17.)

When the temperature T is just below the Planck temperature, Γ_X, Γ_C are much smaller than H. There is no significant decay of X. They are maintained in thermal equilibrium. The decays of X and $\bar{X}$ start becoming significant when $\Gamma_X = H$. Suppose this happens at a temperature T_D, that is, T_D given by

$$\Gamma_X = \frac{\alpha_X M_X^2 N}{[(kT_D)^2 + M_X^2]^{1/2}} = H = \frac{1.66(kT_D)^2 N^{1/2}}{m_p}, \tag{5}$$

(where we have used (3) and (4) to obtain the expression for $H(N(m/T) \sim$ constant)). Now, if $T_D \gg M_X$, the X bosons will still be in equilibrium at T_D and no asymmetry will be produced. If $T_D \ll M_X$, however, the X-particles decays upset the equilibrium distribution. For this to happen,

$$M_X > \alpha_X N^{1/2} M_p. \tag{6}$$

Gauge and Higgs bosons with masses satisfying the constraint (6) do exist in GUTs. Their baryon number violating decays can, therefore, produce the desired asymmetry.

Let X decay into two such channels. If the branching ratio of the first channel is r, by unitarity, the branching ratio of the other is $(1 - r)$. Let the baryon numbers of the two channels be B_1 and B_2, respectively. If the corresponding branching ratios of $\bar{X}$ are $\bar{r}$ and $(1 - \bar{r})$, the average baryon number produced by the decays of one pair of X and $\bar{X}$ is ΔB, where

$$\begin{aligned}\Delta B &= \tfrac{1}{2}[rB_1 + (1 - r)B_2 + \bar{r}(-B_1) + (1 - \bar{r})(-B_2)] \\ &= \frac{1}{r}(r - \bar{r})(B_1 - B_2).\end{aligned} \tag{7}$$

Note that for $\Delta B \neq 0$,

$B_1 \neq B_2 \Rightarrow$ Baryon number is not conserved,

$r \neq \bar{r} \Rightarrow$ CP is violated.

Further analysis shows that tree-level graphs are not sufficient to produce the asymmetry. CPT theorem intervenes, cancelling the contributions from X and $\bar{X}$. It is necessary to have higher-order loop graphs (Figure 4).

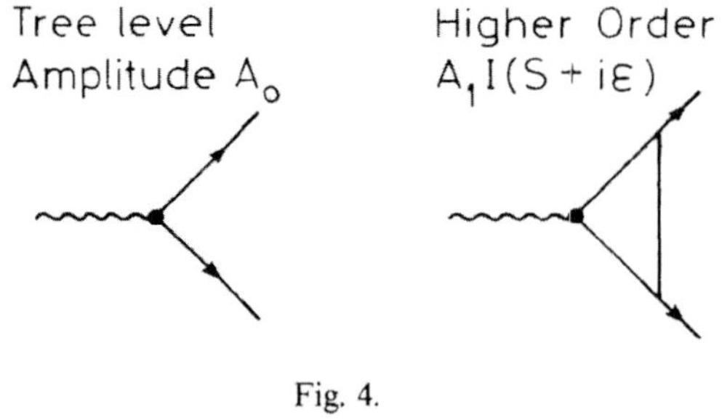

Fig. 4.

Then, we have a situation in which

X – decay amplitude $\quad A_0 + A_1 I(S + i\varepsilon)$,

$\bar{X}$ – decay amplitude $\quad A_0^* + A_1^* I(S + i\varepsilon)$,

where I is the loop integral, A_0 and A_1 are effective coupling strengths in the tree and one-loop graphs. And

$$\Delta B \propto \int d\omega [|A_0 + A_1 I(S + i\varepsilon)|^2 - |A_0^* + A_1^* I(S + i\varepsilon)|^2]$$

$$= 4 \int d\omega [\mathrm{Im}(A_0 A_1^*) \, \mathrm{Im} \, I(S + i\varepsilon)], \tag{8}$$

which shows that in addition to complex couplings (CP-violation) one needs an S-channel discontinuity to generate baryon asymmetry.

The observed asymmetry is determined by ΔB and the density of X-particles at the time of decay. From the distribution function just before the decay, the final formula is

$$n_B/n_\gamma \simeq \frac{45}{2\pi^4} \zeta(3) \frac{N_X}{N} \Delta B, \tag{9}$$

where N_X = number of X-particle states whose decays violate B and C,
N = total number of particle states,
$\zeta(3)$ = Riemann ζ function.

In minimal SU(5), the gauge couplings are real. One needs to go to higher-order loop graphs. The calculated asymmetry is too small. The color triplet of the **5**-plet Higgs is superheavy. It can decay into (quark + lepton) as well as two anti-quarks. Hence, it is a good candidate. But it turns out that interference between tree and one loop does not generate CP-violation. It is necessary to consider three loop graphs. The result is that the net baryon violation $\Delta B \sim 10^{-18}$, too small to account for the observed asymmetry. More Higgs multiplets are necessary. Thus, one needs to go beyond minimal SU(5) in order to obtain the observed value for n_B/n_γ. However, when one goes beyond minimal SU(5), the predictive power is lost. One can tune in the desired value by adjusting the parameters. The situation is the same with other GUT models such as SO(10).

6. Phase Transitions in the Early Universe

As we have seen, in GUTs the symmetries of a Lagrangian invariant under local, non-Abelian gauge transformations belonging to a group G undergo, in one or several stages, spontaneous symmetry breakdown:

$$G \to \cdots \to SU(3)_C \times SU(2) \times U(1) \to SU(3)_C \times U(1)_{EM}. \tag{1}$$

We may view this sequence of successive spontaneous symmetry breakings in the

Lagrangian describing the fundamental interactions as corresponding to successive phase transitions in the early Universe. Thus when $M_{P_l} \gg kT \gg M_X$, the Lagrangian has G$\equiv$SU(5) symmetry; we may say the Universe is in SU(5) symmetric phase. When $M_W < T < M_X$, and $G \to SU(3)_C \times SU(2) \times U(1)$, the Universe is in the $H = SU(3)_C \times SU(2) \times U(1)$ phase, and so on.

The Higgs potential described in earlier sections acquires a temperature dependence due to radiative corrections evaluated in the background of a hot Universe. The effective potential $V_{\text{eff}}(\Phi, T)$ takes the form

$$V_{\text{eff}}(\Phi, T) = V(\Phi, T=0) + V'(\Phi, T),$$

where $V(\Phi, T=0)$ is the customary Higgs potential and $V'(\Phi, T)$ contains the temperature dependence. The relevant and the essential change is that

$$-\frac{\mu^2}{2} \to -\frac{\mu_{\text{eff}}^2}{2} = -\frac{\mu^2}{2} + \sigma T^2, \tag{2}$$

where σ is a known function of all the parameters appearing in $V(\phi, T=0)$. This modification of the Higgs potential leads to the concept of a critical temperature T_c such that, if

$$T > T_c = \sqrt{\frac{\mu^2}{2\sigma}}, \qquad \langle\phi\rangle = 0 \qquad G \text{ is unbroken,}$$

$$T < T_c, \qquad \langle\phi\rangle = V \qquad G \to H,$$

with a phase transition occurring at $T = T_c$. As T falls through T_c, Φ will acquire a nonzero vacuum expectation value. The absolute minimum of V, however, is governed, not by the direction of ϕ, but only by $|\Phi|$. Consequently, the lowest energy state is not unique. There are degenerate vacua and different parts of the Universe may choose a different ground state. Fluctuations back and forth between the unique symmetric phase above T_c and ground states below T_c tend to produce spatial uniformity. However, this may be prevented by *trapped singularities* of one kind or another. The whole phenomenon is analogous to that of domain structure in ferro-magnetism and defects in solid state. It leads to topological considerations and topological classification of singularities.

God, in creating the Universe and operating it, is not only a GUT-man, but also a good condensed matter physicist; *He* is a topologist as well.

Classification of Singularities; Topology of the manifold $M = G/H$

The equivalent, degenerate vacua form a manifold $M = G/H$, where G is the unbroken symmetry group above the critical temperature T_c, and H is the surviving symmetry group below T_c. When the phase transition occurs, the universe must choose a point on the manifold M. The choice is random and may be different in different regions of space, the correlations in different parts extending up to the appropriate 'horizons'.

Consider the case when M has two or more disconnected pieces. This happens when a discrete symmetry is broken spontaneously. We shall consider two examples of such cases briefly. Both give rise to *surface singularities* or *domain walls* and put strong constraints on the models.

(A) *Model of spontaneous CP-violation.* (Zel'dovich, Kubzarev, Okun, *Zh. Eksp. Teor. Fiz.* **67**, 3–11 (1974)). One starts with a Lagrangian which is CP-invariant. However, the vacuum in the model Lagrangian is characterized by $\langle\phi\rangle = \pm\eta$. In the two domains, CP is violated; the signs of all the CP-noninvariant effects are opposite to one another in the two domains. The boundary between the two domains is what is called a domain wall.

Such domain walls are seats of energy. For reasonable values of the parameters in the model, it turns out that the walls carry massive amounts of energy ($\eta \geqslant 100$ GeV, otherwise the particle associated with the SSB becomes superlight, contradicting the known decay properties of the K-meson. The domain wall then stretched across the Universe would be 10^8 times more massive than all other known matter.) Clearly, if the model of spontaneous CP-violation is to be valid, there must be mechanisms which cause the domain walls formed to disappear at a sufficiently early stage in the evolution of the Universe. Otherwise, there will be contradictions between theory and some of the known facts about the Universe.

(B) *Peccei–Quinn symmetry and its breakdown.* Peccei–Quinn symmetry postulates a global, axial symmetry, $U(1)_{PQ}$ to solve the strong CP-violation problem that arises from the Lagrangian of quantum chromodynamics. However, when this global $U(1)_{PQ}$ breaks spontaneously, it leaves a discrete symmetry $Z(N)$ unbroken (N is the number of flavors, QCD Lagrangian has the flavor symmetry $SU_L(N) \times SU_R(N) \times U_V(1)$). The vacua are characterized by $\alpha = 2k\pi/N$, leading to domain structures separated by walls which carry massive amounts of energy. Unless appropriate constraints are put into force, contradictions between the particle theory and cosmology arise.

Other types of singularities that can arise are *line singularities* (strings) or *point singularities* (monopoles). Strings appear if M contains unshrinkable loops. And point singularities appear if M contains closed two-dimensional surfaces that cannot be shrunk to a point within M.

Topologically, the existence of singular structures in M requires that one of the homotopy groups of M be nontrivial. The elements of the nontrivial homotopy group then serve to classify the possible singularities – strings by the elements of $\Pi_1(M)$, monopoles by the elements of $\Pi_2(M)$ etc.

If the gauge group G is connected, $\Pi_0(G)$, which counts the number of disconnected pieces, is trivial, that is, $\Pi_0(G) = 1$. If it is simply connected (which

Structure	Dimension of singularity	Classified by
Domain walls	2	$\Pi_0(M)$
Strings	1	$\Pi_1(M)$
Monopoles	0	$\Pi_2(M)$

can always be done by working with the covering group), $\Pi_1(G) = 1$. If G is simple, there are isomorphisms between the relevant homotopy groups of $M = G/H$ and the unbroken symmetry subgroup H,

$$\Pi_1(M) \equiv \Pi_0(H), \qquad \Pi_2(M) \equiv \Pi_1(H)$$

and

$$\Pi_1(H) = \mathbb{Z}^k \times K,$$

where k is the number of U(1) factors in H, $\mathbb{Z}$ is the group of integers, and K is a finite group.

From the above mathematical facts, it immediately follows that if H contains at least one U(1) factor, *monopoles* must occur. Since we known that a $U(1)_{EM}$ has to exist, it follows that monopoles must make their appearance when phase transitions occur in the early Universe.

In the case $G = SU(5)$ and

$$\begin{aligned} G \to H_1 &= [SU(3) \times SU(2) \times U(1)]/Z_6 \\ \to H_2 &= [SU(3)_c \times U(1)_{EM}]/Z_3, \end{aligned}$$

There are two phase transitions and the surviving symmetries contain U(1) factors. Monopoles do occur. Estimates show that

$$\gamma_m = \left.\frac{\text{Number density of monopoles}}{\text{Number of photons}}\right|_{\text{today}} \simeq 10^{-6},$$

which is an absurdly large number. Experimental bound for γ_m is 10^{-24}. This is the famous monopole problem.

Note added in proof: The above chapter was written in the summer of 1985. Since then a lower limit of 10^{32} years on proton lifetime has been set by various experiments. This certainly rules out minimal SU(5).

Bibliography

1. *Review articles*
P. Langacker, *Phys. Rep.* **72C**, 185 (1981); R. Slansky, *Phys. Rep.* **79**, 1 (1981); P. Nair and K. C. Wali, *AIP Conference Proceedings*, No. 98; Particles and Fields Subseries No. 29, Edited by W. E. Caswell and G. A. Snow. Contains a partially annotated bibliography.

2. *Books*
Recently several good books have appeared. I have consulted the following:
Graham G. Ross, *Grand Unified Theories*, Benjamin/Cummings Publishing Co.
T. P. Cheng and L.F. Li, *Gauge Theories of Elementary Particle Physics*, Oxford University Press.

3. For brief introductions to various topics and collection of reprints of original papers, *Unity of Forces in the Universe*, Vol. I and II (A. Zee (ed.)), World Scientific.

4. For the last section, I have consulted articles by Shafi and Vilenkin in *The Very Early Universe*, Edited by G. W. Gibbons, S. W. Hawking, S. T. C. Siklos (Cambridge University Press). Also, T. W. B. Kibble, *Phys. Rev.* **67**, 183–199 (1980), *J. Phys., A: Math. Gen.* **9**, 1387 (1976).

Quark masses and mixing angles from universal seesaw mechanism

Aharon Davidson
Department of Physics, Ben-Gurion University of the Negev, Beer Sheva 84105, Israel

Stefano Ranfone
Physics Department, Syracuse University, Syracuse, New York 13244-1130
and Istituto Nazionale di Fisica Nucleare, Sezione di Pisa, Pisa, Italy

Kameshwar C. Wali
Physics Department, Syracuse University, Syracuse, New York 13244-1130
(Received 16 August 1989)

A universal seesaw mechanism is invoked to account for the observed fermion mass hierarchies. In the framework of left-right symmetry, heavy fermions (mass scale χ) which are $SU(2)_L \otimes SU(2)_R$ singlets are postulated while retaining the simplest possible Higgs system: namely, $\phi(1,2,1)_{+1} \oplus \phi(1,1,2)_{+1}$ accompanied by a left-right singlet $\sigma(1,1,1)_0$ in the standard $SU(3)_C \otimes SU(2)_L \otimes SU(2)_R \times U(1)_{B-L}$ notation. Every conventional quark and lepton is accompanied by a nonmirror singlet heavy fermion, so that the associated mass matrix is doubled and has the seesaw form usually associated only with the neutrino mass matrix. In the single-generation case, the model provides a plausible explanation for the mass hierarchy $m_{e,u,d} \sim 10^{-4} M_W$ and predicts $m_{\nu_e} m_{\nu_R} \approx m_e^2$, thus accounting for the superlightness of neutrinos. Combined with a U(1) axial symmetry, the mechanism provides a formalism in which the generations are distinguished and constraints emerge on the allowed form of mass matrices. In this paper, we consider the realistic case of three generations in a simplified version of the model in which *CP* violation does not arise from the gauge sector. Choosing the $U(1)_A$ quantum numbers so that the mass matrices are of the Fritzsch type, we calculate experimentally measured Cabibbo-Kobayashi-Maskawa matrix elements V_{us}, V_{ub}, and V_{cs} and derive their dependence on quark mass parameters. An interesting correlation between V_{us} which measures the Cabibbo angle and V_{ub} which measures the charmless decay of the b quark emerges from the model. V_{ub} is naturally suppressed if $V_{us} = \sqrt{d/s} - \sqrt{u/c}$ to a very good approximation.

I. INTRODUCTION

The situation regarding the standard theory of electroweak interactions [Glashow-Salam-Weinberg (GSW) theory] is reminiscent of the years when quantum electrodynamics (QED) was put to more and more rigorous experimental tests with the hope of finding serious discrepancies with theoretical predictions. But QED worked extremely well and its theoretical shortcomings—the infinities, the e-μ puzzle, the incalculability of the masses, and coupling constants of the theory—have remained unresolved. The standard GSW theory suffers from similar shortcomings and more. In spite of its remarkable empirical successes, the theory remains unsatisfactory for reasons so well articulated in the literature that they do not warrant repetition. The generation puzzle and fermion mass hierarchies are two of several puzzles that have attracted a great deal of attention in recent years.

The e-μ puzzle by itself was more of a theoretical and academic nature. It had few experimental consequences outside the domain of QED. In contrast, the generation puzzle presents itself in a much more extended form with consequences covering a wide range of new experimental findings associated with the physics of the new flavors. For instance, the generalized Cabibbo-type angles and phases [Kobayashi-Maskawa (KM) angles and phases] which come into play when one transforms the gauge eigenstates into mass eigenstates, govern the charm-, bottom-, and top-quark decays, weak *CP* violation, and yield quantitative estimates of the rare decay modes. The number of flavors, especially the number of light neutrinos and their masses, have important implications in astrophysics and cosmology. Thus, generally speaking, a whole range of phenomena depend on our understanding of the generation puzzle, the intragenerational interactions, the form of the mass matrices, and the symmetry-breaking mechanism. The standard theory has no answer to these questions and hence it is necessary to go beyond its framework.

Recent attempts in this direction have covered a wide range of ideas incorporating continuous and discrete symmetries,[1] radiative mechanisms for mass generation,[2] seesaw mechanisms,[3] and so on. The purpose of our paper is to explore in some detail the idea of a *universal seesaw mechanism* (USM) proposed by two of the current authors.[4,5] The mechanism is introduced within the framework of the left-right-symmetric

$SU(2)_L \otimes SU(2)_R \otimes U(1)_{B-L}$ model. The most important departure from the conventional left-right-symmetric models is the extension of the fermionic sector to include a set of new fermions[6] which are singlets under $SU(2)_L \otimes SU(2)_R$. Every conventional quark and lepton is accompanied by a new singlet partner. Also unlike in the conventional left-right-symmetric models, the Higgs system is kept minimal in that it is the straightforward generalization of the single Higgs doublet in the standard theory to left-right-symmetric combination of two Higgs doublets. The consequence of this simple idea is that the tree-level mass matrix for every fermion has the seesaw form of a $2\otimes 2$ matrix, strictly analogous to that introduced to explain the superlightness of the neutrino.[3]

In the next section we review briefly the case of a single generation. We show that the most general potential consisting of the minimal Higgs system coupled with a singlet Higgs scalar whose predominant role is to break the left-right symmetry allows, in a certain range of parameters, an absolute minimum of the potential which is consistent with a hierarchy in the vacuum expectation values (VEV's) necessary to explain the mass hierarchies in the fermionic system. In the quark and the charged leptonic sectors, the seesaw form of the mass matrices and the assumed hierarchy in the VEV's establishes the smallness of the masses of the u and d quarks and the electron. It then follows without any further assumptions a formula for the neutrino masses, namely,

$$m_{\nu_L} m_{\nu_R} \approx m_e^2 ,$$

which correlates the smallness of the neutrino mass with the smallness of the electron mass.

In Sec. III we discuss the general formalism for N generations. By imposing a global axial symmetry U(1), we distinguish the generations and show how the U(1) quantum numbers restrict the form of the mass matrices. The assumption of this extra global symmetry necessitates the doubling of the left- and right-doublet Higgs fields. We then have a seesaw form for all the fermionic mass matrices. Two cases arise: (a) the mass matrix of the extra fermions is nonsingular, in which case an effective $N\otimes N$ mass matrix can be written for the conventional light quarks; (b) the relevant mass matrix of the heavy fermions is singular. The latter instance has no simplifying feature and gives rise to a qualitatively different mass spectrum. In Sec. IV we apply these considerations to the realistic case of three generations with prescribed quantum numbers for the generations so that all the submass matrices are Fritzsch type. By assuming strict equality of the VEV's of the two left and the two right Higgs scalars separately, but the left VEV being very much smaller than the right VEV, we simplify the model. The effective light-quark mass matrix is then real and symmetric. The expressions for the mixing angles involve only the quark masses and two scale parameters. A much desired formula for the well-known Cabibbo angle emerges in terms of the quark masses; also the other two matrix elements (V_{cb} and V_{ub}) in the Cabibbo-Kobayashi-Maskawa (CKM) matrix can be easily made consistent with the known experimental information concerning the B-meson lifetime and also the known limits on its charmless decays. In fact, an extremely interesting correlation between the expression for the Cabibbo angle (the matrix element V_{us}) and V_{ub} explains naturally the suppression of the latter. Of course, by allowing the VEV's to be complex, we can introduce CP violation with a nonvanishing CP phase while retaining the good features of the results for mixing angles. We also consider in this section the singular case and discuss its implications on the mass spectrum. The final section is devoted to a summary and discussion of the results. In a follow-up paper, we propose to extend these considerations to leptons with particular interest to neutrinos, and also CP violation.

II. SINGLE-GENERATION CASE

A. General considerations

Consider the standard model of electroweak and strong interactions based on the group structure $SU(3)_C \otimes SU(2)_L \otimes SU(2)_R \otimes U(1)_{B-L}$. Within the framework of this symmetry, the familiar quarks and leptons belong to the *complex* representation

$$q_L(3,2,1)_{1/3},\quad q_R(3,1,2)_{1/3},\qquad l_L(1,2,1)_{-1},\quad l_R(1,1,2)_{-1}, \tag{2.1}$$

where $(a,b,c)_d$ label, respectively, $SU(3)_C$, $SU(2)_L$, $SU(2)_R$ representations and d the quantum number $Y=B-L$ of $U(1)_{B-L}$. In addition to the above conventional fermions, we postulate new fermions which belong to a real representation.[6] They share identical $SU(3)_C$ quantum numbers with the conventional fermions, but differ from them since they are assumed to be $SU(2)_L \otimes SU(2)_R$ singlets. We denote them by

$$U_{L,R}(3,1,1)_{4/3},\quad D_{L,R}(3,1,1)_{-2/3},\qquad N_{L,R}(1,1,1)_0,\quad E_{L,R}(1,1,1)_{-2}. \tag{2.2}$$

The $Y=B-L$ quantum numbers of the new fermions are chosen so that every ordinary fermion has a $SU(2)_L \otimes SU(2)_R$-singlet companion with identical $SU(3)_C \times U(1)_Q$ assignments $[Q=T_{3L}+T_{3R}+\frac{1}{2}(B-L)]$. The associated primary Higgs system to which the fermions couple is assumed to be

$$\phi_L(1,2,1)_{+1}+\phi_R(1,1,2)_{+1}, \tag{2.3}$$

which provides a minimal Higgs system and a natural left-right-symmetric generalization of the single Higgs doublet of the Glashow-Salam-Weinberg $SU(2)_L \times U(1)$ electroweak theory. We shall not assume at the start the conventional sources of quark-lepton masses in left-right-symmetric models: namely,

$$\phi(1,2,2)_0+\phi(1,3,1)_{-2}+\phi(1,1,3)_{-2}. \tag{2.4}$$

The above scalars can be looked upon as bilinears of our primary system (2.3), as *effective scalar fields* in the low-energy limit.[4]

However, with only (2.3) present, left-right symmetry

remains unbroken. To obtain a desired tree-level hierarchy $v_L \ll v_R$, where $\langle \phi_{L,R} \rangle = v_{L,R}$, we have to break *L-R* symmetry *explicitly*. In Ref. 4 we had assumed that such explicit *L-R*-symmetry breaking is the consequence of physics beyond $SU(3)_C \otimes SU(2)_L \otimes U(1)_{B-L}$ where it is broken only spontaneously. This was a realistic possibility that emerged from our $SU(5)_L \otimes SU(5)_R$ grand unification scheme;[7] $SU(2)_L$ and $SU(2)_R$ are not allowed to share a common evolution in that scheme if $\sin^2\theta_W$ were to conform to experimental value with three or four generations of quarks and leptons.

There is, however, an elegant way of breaking the *L-R* symmetry spontaneously that was first considered by Chang, Mohapatra, and Parida.[8] The introduction of a scalar singlet $\sigma(1,1,1)_0$ with odd parity under $L \leftrightarrow R$, even under charge conjugation, and hence odd under *CP* allows in the Higgs potential a term which can break *L-R* symmetry spontaneously. Let χ denote the vacuum expectation value of this singlet scalar: namely, $\chi = \langle \sigma(1,1,1)_0 \rangle$. We would like to postulate the hierarchy $v_L \ll v_R \ll \chi$. Assuming for the moment that we can construct a Higgs potential involving ϕ_L, ϕ_R, and σ, such that its absolute minimum allows the desired hierarchy, let us consider the quark-Higgs sector with the couplings

$$\mathcal{L}_Y = [Y_d \bar{q}_L \phi_L D_R + Y_u \bar{q}_L \tilde{\phi}_L U_R + (L \to R)] + \text{H.c.} + (Y_U \bar{U}_L U_R + Y_D \bar{D}_L D_R)\sigma + \text{H.c.} \tag{2.5}$$

In (2.5), $\tilde{\phi}_L = i\tau_2 \phi_L^*$. When σ acquires a nonvanishing vacuum expectation value χ, the singlet quarks acquire a mass of the order χ. Y_d, Y_u, Y_U, Y_D are Yukawa-type coupling constants[9] which we can assume to be real without any loss of generality.

Let

$$\langle Y_d \phi_L \rangle = Y_d v_L = V_L^d, \quad \langle Y_d \phi_R \rangle = Y_d v_R = V_R^d, \quad \langle Y_D \sigma \rangle = Y_D \chi = \chi^d . \tag{2.6}$$

Then, once the gauge symmetry is spontaneously broken down to $SU(3)_C \times U(1)_Q$, the tree-level mass matrix, in the down sector, for instance, is given by

$$M_d = \begin{bmatrix} 0 & V_L^d \\ V_R^d & \chi^d \end{bmatrix} . \tag{2.7}$$

The corresponding tree-level mass matrix in the up sector is

$$M_u = \begin{bmatrix} 0 & V_L^u \\ V_R^u & \chi^u \end{bmatrix} , \tag{2.8}$$

where

$$V_L^u = Y_u v_L^* = \frac{Y_u}{Y_d} V_L^{d*}, \quad V_R^u = Y_u v_R^* = \frac{Y_u}{Y_d} V_R^{d*} , \quad \chi^u = Y_u \chi = \frac{Y_U}{Y_D} \chi^d . \tag{2.9}$$

Thus we have a generic seesaw form for the mass matrix in both the charge sectors of the quarks.

In a similar fashion, we can write down the lepton–Higgs-boson interactions in the lepton sector with $E_{L,R}$ and $N_{L,R}$ as corresponding singlet partners of the $e_{L,R}$ and $\nu_{L,R}$. We have

$$\mathcal{L}_L = (Y_e \bar{l}_L \phi_L E_R + Y_{\nu_L} \bar{l}_L \tilde{\phi}_L N_R + L \leftrightarrow R) + \text{H.c.} + (Y'_{\nu_L} l_L^T C i\tau_2 \phi_L N_L + Y'_{\nu_R} l_R^T C i\tau_2 \phi_R N_R) + \text{H.c.} \tag{2.10}$$

Note that in (2.10), the first line involves couplings which are in exact analogy with the quark sector. The terms in the second line will give rise to Majorana mass terms, so that for the charged leptons, we have the mass matrix

$$M_e = \begin{bmatrix} 0 & V_L^e \\ V_R^e & \chi^e \end{bmatrix} , \tag{2.11}$$

whereas for the neutrino [in the space (ν_L, ν_R, N_L, N_R)], we have the mass matrix[4]

$$M_\nu = \begin{bmatrix} 0 & 0 & Y_{\nu_L} v_L & Y'_{\nu_L} v_L \\ 0 & 0 & Y_{\nu_R} v_R^* & Y'_{\nu_R} v_R^* \\ Y_{\nu_L} v_L & Y_{\nu_R} v_R^* & \chi_1 & \chi \\ Y'_{\nu_L} v_L & Y'_{\nu_R} v_R^* & \chi & \chi_2 \end{bmatrix} . \tag{2.12}$$

B. Phenomenology of the single-generation model

From (2.7), (2.8), (2.11), and (2.12) (with $\chi_1 \sim \chi_2 \sim \chi$), we identify the masses of d and u quarks, electron, and neutrinos as given by

$$m_d = \frac{V_L^d V_R^d}{\chi} = (Y_d)^2 \frac{v_L v_R}{\chi}, \quad m_u = \frac{V_L^u V_R^u}{\chi} = (Y_u)^2 \frac{v_L v_R}{\chi} , \quad m_e = \frac{V_L^e V_R^e}{\chi} = (Y_e)^2 \frac{v_L v_R}{\chi} , \tag{2.13}$$

and

$$m_{\nu_L} = \frac{Y_{\nu_L} Y'_{\nu_L} v_L^2}{\chi}, \quad m_{\nu_R} = \frac{Y_{\nu_R} Y'_{\nu_R} v_R^2}{\chi} . \tag{2.14}$$

Clearly, if all the Yukawa couplings are equal,

$$m_u = m_d = m_e \tag{2.15}$$

and

$$m_{\nu_L} m_{\nu_R} = m_e^2 . \tag{2.16}$$

However, it would be interesting to see whether we can obtain agreement with the known or expected values of quark and lepton masses *without* introducing additional hierarchies in the Y couplings. We have

$$\frac{m_u}{m_d} = \left[\frac{Y_u}{Y_d} \right]^2 \simeq \frac{1}{2} , \tag{2.17}$$

which implies $Y_u = (1/\sqrt{2}) Y_d$, and hence Y_u, Y_d are of the same order of magnitude.

Let us further note, assuming the gauge couplings $g_L = g_R = g$, that

$$M_{W_L}=\tfrac{1}{2}gv_L \quad \text{and} \quad M_{W_R}=\tfrac{1}{2}gv_R \ . \tag{2.18}$$

Let us also assume, based on the current experimental information,

$$M_{\nu_L}<18 \text{ eV} \tag{2.19}$$

and

$$10^8<\chi<10^{12} \text{ GeV} \ . \tag{2.20}$$

The above limit follows from astrophysical and cosmological considerations,[10] provided we identify[5] σ as the singlet scalar that breaks Peccei-Quinn, global, axial $U(1)_{PQ}$ symmetry as a device to avoid the strong CP violation.[11]

Then, in the case of leptons, we have, from (2.13) and (2.14),

$$\frac{m_{\nu_L}}{m_e}=\frac{Y_{\nu_L}Y'_{\nu_L}}{Y_e^2}\left[\frac{v_L}{v_R}\right], \quad \frac{m_{\nu_R}}{m_e}=\frac{Y_{\nu_R}Y'_{\nu_R}}{Y_e^2}\left[\frac{v_R}{v_L}\right]. \tag{2.21}$$

Hence, if $Y_{\nu_L}Y'_{\nu_L}/Y_e^2$ and $Y_{\nu_R}Y'_{\nu_R}/Y_e^2$ are ~ 1 as in the case of quark sector,

$$\frac{v_L}{v_R}\approx\frac{m_{\nu_L}}{m_e}\leq 3.5\times10^{-5} \tag{2.22}$$

and therefore, $v_R>v_L/(3.5\times10^{-5})\approx10^7$ GeV, where we have used $v_L=(\sqrt{2}G_F)^{-1/2}=246$ GeV. From (2.18), then

$$M_{W_R}>\frac{M_{W_L}}{3.5\times10^{-5}}\approx2.3\times10^6 \text{ GeV} \tag{2.23}$$

and, from (2.21),

$$m_{\nu_R}>14.5 \text{ GeV} \ . \tag{2.24}$$

As a final remark regarding these numbers, we note that if the Y-coupling ratios in (2.21) are ~ 1, $\chi\approx v_L^2/m_{\nu_L}\approx10^{12}$ GeV consistent with (2.20).

C. Spontaneous symmetry breaking

With the Higgs fields $\phi_L(1,2,1)_{+1}$, $\phi_R(1,1,2)_{+1}$, and $\sigma(1,1,1)_0$, the most general Higgs potential takes the form

$$\begin{aligned} V=&\Lambda[(\phi_L^\dagger\phi_L)^2+(\phi_R^\dagger\phi_R)^2]-2(M^2-\delta\sigma^2)(\phi_L^\dagger\phi_L+\phi_R^\dagger\phi_R)\\ &+2\Gamma(\phi_L^\dagger\phi_L)(\phi_R^\dagger\phi_R)\\ &+\lambda\sigma^4-2m^2\sigma^2+2\gamma\sigma(\phi_L^\dagger\phi_L-\phi_R^\dagger\phi_R) \ , \end{aligned} \tag{2.25}$$

where Λ, M, Γ, λ, m, δ, and γ are constants. Under the assumption $\sigma\leftrightarrow-\sigma$ when $L\leftrightarrow R$, V is L-R symmetric. Our object is to verify whether (2.25) allows, for a range of parameters $\Lambda,\ldots,\gamma$, the absolute minimum of V such that the desired hierarchy $v_L\ll v_R\ll\chi$ is attainable. First note the positivity conditions as σ, ϕ_L, and $\phi_R\to\infty$,

$$\lambda>0, \quad \Lambda>0, \quad \Lambda+\Gamma>0 \ . \tag{2.26}$$

Varying with respect to ϕ_L, ϕ_R, and σ, the condition for minimum of V gives

$$v_L(\Lambda v_L^2+\Gamma v_R^2-M^2+\gamma\chi+\delta\chi^2)=0 \ , \tag{2.27}$$

$$v_R(\Lambda v_R^2+\Gamma v_L^2-M^2-\gamma\chi+\delta\chi^2)=0 \ , \tag{2.28}$$

$$\lambda\chi^3-[m^2-\delta(v_L^2+v_R^2)]\chi+\frac{\gamma}{2}(v_L^2-v_R^2)=0 \ . \tag{2.29}$$

There are four cases to be studied: (i) $v_L=v_R=\chi=0$, (ii) $v_L=v_R=0$, $\chi\neq0$, (iii) $v_L=0, v_R\neq0$, and $v_L\neq0, v_R=0$, (iv) $v_L\neq0, v_R\neq0$. From (2.23)–(2.25), one gets

$$\text{case (i):} \quad V_{\min}^{(1)}=0 \ ; \tag{2.30}$$

$$\begin{aligned} &\text{case (ii):} \quad \chi^2=\frac{m^2}{\lambda} \ ,\\ &V_{\min}^{(2)}(v_L=v_R=0;\ \chi=\pm\sqrt{m^2/\lambda})=-\frac{m^4}{\lambda} \ , \end{aligned} \tag{2.31}$$

so that, for $m\neq0$, $V_{\min}^{(2)}<V_{\min}^{(1)}$.

$$\text{case (iii):} \quad \Lambda v_R^2=M^2+\gamma\chi-\delta\chi^2 \ , \tag{2.32}$$

where χ satisfies the cubic equation

$$(\lambda\Lambda-\delta^2)\chi^3+\tfrac{3}{2}\gamma\delta\chi^2-(m^2\Lambda+\tfrac{1}{2}\gamma^2-\delta M^2)\chi-\tfrac{1}{2}\gamma M^2=0 \ . \tag{2.33}$$

In terms of χ, $V_{\min}^{(3)}(v_L=0, v_R\neq0, \chi\neq0)=V_{\min}^{(3)}(v_L=0)$ is given by

$$\begin{aligned} V_{\min}^{(3)}(v_L=0)=&-\frac{M^4}{\Lambda}+\frac{\gamma\delta M^2}{4\Lambda(\lambda\Lambda-\delta^2)}\\ &-\left[m^2+\frac{\gamma^2}{2\Lambda}-\frac{\delta}{\Lambda}\left[M^2-\frac{3\gamma\delta}{4(\lambda\Lambda-\delta^2)}\right]\right]\chi^2\\ &+\frac{\gamma}{2}\left[\frac{\delta}{\lambda\Lambda-\delta^2}\left[m^2+\frac{\gamma^2}{2\Lambda}-\frac{\delta M^2}{\Lambda}\right]\right.\\ &\left.-\frac{3M^2}{\Lambda}\right]\chi \ . \end{aligned} \tag{2.34}$$

The substitution $\gamma\leftrightarrow-\gamma$ in Eqs. (2.32)–(2.34), gives corresponding expressions in the case $v_L\neq0, v_R=0, \chi\neq0$. Finally,

$$\begin{aligned} \text{case (iv):} \quad &v_L^2=\frac{M^2-\delta\chi^2}{\Lambda+\Gamma}-\frac{\gamma\chi}{\Lambda-\Gamma} \ ,\\ &v_R^2=\frac{M^2-\delta\chi^2}{\Lambda+\Gamma}+\frac{\gamma\chi}{\Lambda-\Gamma} \ , \end{aligned} \tag{2.35}$$

DAVIDSON, RANFONE, AND WALI

where

$$\left[\lambda-\frac{2\delta^2}{\Lambda+\Gamma}\right]\chi^2=m^2+\frac{\gamma^2}{\Lambda-\Gamma}-\frac{2\delta M^2}{\Lambda+\Gamma} \tag{2.36}$$

and

$$V^{(4)}_{\rm min}=-\frac{2M^4}{\Lambda+\Gamma}-\frac{\left[m^2+\frac{\gamma^2}{\Lambda-\Gamma}-\frac{2\delta M^2}{\Lambda+\Gamma}\right]^2}{\left[\lambda-\frac{2\delta^2}{\Lambda+\Gamma}\right]} . \tag{2.37}$$

Straightforward but tedious analysis shows that in all cases there is a range of parameters such that $V^{(4)}_{\rm min}<V^{(3)}_{\rm min}$, $V^{(2)}_{\rm min}$, and $v_L \ll v_R \ll \chi$.

III. MULTIGENERATIONS; GENERAL CONSIDERATIONS

As stated in the Introduction, to go beyond the single-generation case, we impose a global, axial, $U(1)_A$ symmetry at the classical level. The quantum numbers associated with this additional symmetry will distinguish the different generations[12] and provide constraints on the forms of the fermion mass matrices.

Let there be N generations of conventional fermions $f^i_{L,R}$ accompanied by $SU(2)_L\otimes SU(2)_R$-singlet partners $F^i_{L,R}$, $i=1,2,\ldots,N$ and where $f=u,d,\nu,e$, $F=U$, D,N,E (see Table I). We postulate that the Lagrangian involving these fermions and the Higgs scalars is invariant under

$$f^i_{L,R}\to e^{\pm i\theta x_i}f^i_{L,R},\quad F^i_{L,R}\to e^{\pm i\theta y_i}F^i_{L,R} , \tag{3.1}$$

where as yet unspecified $U(1)_A$ charges x_i, y_j are such that

$$x_i\neq x_j,\quad y_i\neq y_j \quad \text{for } i\neq j , \tag{3.2}$$

so that each generation has a distinct quantum number. We shall keep the Higgs structure to the minimum, but as is well known, the introduction of a global $U(1)_A$ symmetry necessitates a doubling of the minimum number of Higgs scalars.[11,12] A given scalar field ϕ and its charge-conjugate field $\tilde{\phi}=i\tau_2\phi^*$ are now distinguished because they have opposite $U(1)_A$ charges. When the symmetry is spontaneously broken and ϕ acquires a nonvanishing vacuum expectation value, ϕ and $\tilde{\phi}$ generate fermion masses in different charge sectors. If ϕ is chosen to be the doublet

$$\begin{bmatrix}\phi^+\\ \phi^0\end{bmatrix} ,$$

it contributes to the down-charge sector and $\tilde{\phi}$ contributes to the up-charge sector. In our model, we have four Higgs doublets

$$\phi_L^{h_1}(1,2,1)_1,\quad \phi_L^{h_2}(1,2,1)_1 ,$$
$$\phi_R^{-h_1}(1,1,2)_1,\quad \phi_R^{-h_2}(1,1,2)_1 ,$$

where h_i represent the $U(1)_A$ charges. It is easy to show that

$$h_1+h_2=0 \tag{3.3}$$

and, therefore, the axial charges can be normalized to ±1. Under $U(1)_A$, they transform according to

$$\phi_L^{1,-1}\to e^{\pm i\theta}\phi_L^{1,-1},\quad \phi_R^{1,-1}\to e^{\pm i\theta}\phi_R^{1,-1} . \tag{3.4}$$

The charge-conjugated fields $\tilde{\phi}^{\pm1}$ are defined by

$$\tilde{\phi}^{\pm1}=i\tau_2(\phi^{\mp1})^* . \tag{3.5}$$

Next, let $K_{ij}=x_i+y_j$. In order to have a Yukawa-type fermion–Higgs-boson vertex, we must have $K_{ij}=\pm1$. Table I describes the particle content in our model and

TABLE I. Particle content and the quantum numbers of the particles.

		$SU(3)_c$	$SU(2)_L$	$SU(2)_R$	$U(1)_{B-L}$	$U(1)_A$
Conventional quarks	$\begin{bmatrix}u\\ d\end{bmatrix}^i_L$	3	2	1	$\frac{1}{3}$	x_i
	$\begin{bmatrix}\bar{u}\\ \bar{d}\end{bmatrix}^i_L$	3*	1	2	$-\frac{1}{3}$	x_i
Heavy	U^i_L	3	1	1	$\frac{4}{3}$	y_i
singlet	D^i_L	3	1	1	$-\frac{2}{3}$	y_i
quarks	$\bar{U}^i_L$	3*	1	1	$-\frac{4}{3}$	y_i
	$\bar{D}^i_L$	3*	1	1	$\frac{2}{3}$	y_i
Higgs bosons	$\phi_L^{1,2}$	1	2	1	+1	±1
	$\phi_R^{1,2}$	1	1	2	+1	±1
	σ	1	1	1	0	0

TABLE II. Nonvanishing fermion–Higgs-boson couplings.

K_{ij}	Mass generating quark–Higgs-boson couplings	
+1	$\bar{q}^i_L \phi^{+1}_L D^j_R$	$\bar{D}^i_L \phi^{+1}_R q^j_R$
	$\bar{q}^i_L \tilde{\phi}^{+1}_L U^j_R$	$\bar{U}^i_L \tilde{\phi}^{+1}_R q^j_R$
−1	$\bar{q}^i_L \phi^{-1}_L D^j_R$	$\bar{D}^i_L \phi^{-1}_R q^j_R$
	$\bar{q}^i_L \tilde{\phi}^{-1}_L U^j_R$	$\bar{U}^i_L \tilde{\phi}^{-1}_R q^j_R$

Table II summarizes the nature of the nonvanishing type of such couplings. Note further that (i) there cannot be two identical entries in the same row or column since that would violate (3.2) and (ii) at least N out of N^2 K entries must equal ± 1, as otherwise the associated mass submatrix ($m_{ij}=0$ if $K_{ij}\neq \pm 1$) will have a vanishing determinant, leading to a zero eigenmass. We shall also ignore K patterns which allow a larger axial symmetry and consider patterns which transform into each other by $x_i \leftrightarrow x_j$, $y_i \leftrightarrow y_j$, or $\phi^1 \leftrightarrow \phi^{-1}$, as equivalent.

The general form of the mass matrix is then evident. First of all, since there are no couplings in the conventional quark sectors, the mass matrices have zero entries in these sectors. The entries in the $\bar{q}^i_L \phi_L D^j_R$ or $\bar{q}^i_L \tilde{\phi}_L U^j_R$ sectors will involve the vacuum expectation value (VEV) $\langle \phi^i_L \rangle = v^i_L$ and the appropriate fermion–Higgs-boson coupling constants. Likewise, the entries in $\bar{D}^i_L \phi_R q^j_R$ or $\bar{U}^i_L \tilde{\phi}_R q^j_R$ will involve the VEV $\langle \phi^i_R \rangle = v^i_R$. Finally, the $\bar{U}^i_L \sigma U^j_R$ sector, consisting of the singlet fermions, will contain the VEV $\langle \sigma \rangle = \chi$. Therefore, the mass matrix has the general form

$$M = \begin{bmatrix} 0 & M_L \\ M_R & M_X \end{bmatrix} , \tag{3.6}$$

where M_L, M_R, and M_X are in general $N\times N$, complex matrices. M_L (M_R) will contain the VEV's v^1_L, v^2_L (v^1_R, v^2_R), and M_X, the VEV χ.

Following our basic assumption concerning the hierarchy in the VEV's, we assume

$$v^1_L \approx v^2_L, \quad v^1_R \approx v^2_R ,$$

but

$$v^i_L \ll v^i_R \ll \chi . \tag{3.7}$$

With this assumption, we have to consider two separate cases.

(1) The nonsingular case: $\det M_X \neq 0$. Then, the effective mass matrix for the light quarks[12] is given by M_q, where

$$M_q = -M_L M_X^{-1} M_R . \tag{3.8}$$

(2) The singular case: $\det M_X = 0$. In this case, there is no simple expression for M_q and needs an exact treatment for the eigenvalues.

IV. PHENOMENOLOGY—THREE GENERATIONS

In this section we shall consider in some detail the case of $N=3$ using the general considerations put forth in the previous section. We shall show that, in spite of the apparently large number of parameters, meaningful constraints can be imposed on them. The only significant input information we use is the values of the quark masses.

A. U(1)$_A$ quantum numbers

The matrix K_{ij} consisting of the sums of the U(1)$_A$ quantum numbers has the form

$$\begin{array}{c|cccccc} & x_1 & x_2 & x_3 & y_1 & y_2 & y_3 \\ \hline x_1 & 2x_1 & x_1+x_2 & x_1+x_3 & x_1+y_1 & x_1+y_2 & x_1+y_3 \\ x_2 & x_2+x_1 & 2x_2 & x_2+x_3 & x_2+y_1 & x_2+y_2 & x_2+y_3 \\ x_3 & x_3+x_1 & x_3+x_2 & 2x_3 & x_3+y_1 & x_3+y_2 & x_3+y_3 \\ y_1 & y_1+x_1 & y_1+x_2 & y_1+x_3 & 2y_1 & y_1+y_2 & y_1+y_3 \\ y_2 & y_2+x_1 & y_2+x_2 & y_2+x_3 & y_2+y_1 & 2y_2 & y_2+y_3 \\ y_3 & y_3+x_1 & y_3+x_2 & y_3+x_3 & y_3+y_1 & y_3+y_2 & 2y_3 \end{array} \equiv \begin{bmatrix} X & L \\ R & Y \end{bmatrix} . \tag{4.1}$$

We shall restrict our considerations to the case when the mass submatrices M_L and M_R in a given charge sector are Fritzsch type; that is, in the d sector, for instance,

$$M_L^{(d)} = \begin{bmatrix} 0 & \alpha^{(d)}_{12} v^1_L & 0 \\ \alpha^{(d)}_{21} v^1_L & 0 & \alpha^{(d)}_{23} v^2_L \\ 0 & \alpha^{(d)}_{32} v^2_L & \alpha^{(d)}_{33} v^1_L \end{bmatrix}, \quad M_R^{(d)} = \begin{bmatrix} 0 & \alpha^{(d)}_{12} v^1_R & 0 \\ \alpha^{(d)}_{21} v^1_R & 0 & \alpha^{(d)}_{23} v^2_R \\ 0 & \alpha^{(d)}_{32} v^2_R & \alpha^{(d)}_{33} v^1_R \end{bmatrix} , \tag{4.2}$$

where α_{ij} are Yukawa-type fermion–Higgs-boson couplings and

$$v^{1,2}_L = \langle \phi^{1,2}_L \rangle, \quad v^{1,2}_R = \langle \phi^{1,2}_R \rangle . \tag{4.3}$$

This implies that the submatrices L and R of K have a prescribed form: namely,

$$L=\begin{bmatrix} 0 & h_1 & 0 \\ h_1 & 0 & h_2 \\ 0 & h_2 & h_1 \end{bmatrix}, \quad R=\begin{bmatrix} 0 & -h_2 & 0 \\ -h_2 & 0 & -h_1 \\ 0 & -h_1 & -h_2 \end{bmatrix}. \tag{4.4}$$

From (4.1) and (4.4),

$$x_1+y_2=x_2+y_1=h_1, \quad x_2+y_3=x_3+y_2=h_2, \quad x_3+y_3=h_1\ , \tag{4.5}$$

we find

$$x_1=x, \quad y_1=3h_1-2h_2-x, \quad x_2=2h_2-2h_1+x, \quad y_2=h_1-x, \quad x_3=h_2-h_1+x, \quad y_3=2h_1-h_2-x\ , \tag{4.6}$$

where x is an arbitrary integer and $h_1=-h_2=1$. The full K matrix then takes the form

$$K=\begin{bmatrix} 2x & 2x-4 & 2x-2 & 5 & 1 & 3 \\ 2x-4 & 2x-8 & 2x-6 & 1 & -3 & -1 \\ 2x-2 & 2x-6 & 2x-4 & 3 & -1 & 1 \\ 5 & 1 & 3 & 10-2x & 6-2x & 8-2x \\ 1 & -3 & -1 & 6-2x & 2-2x & 4-2x \\ 3 & -1 & 1 & 8-2x & 4-2x & 6-2x \end{bmatrix}. \tag{4.7}$$

Since in order that the fermion–Higgs-boson couplings exist, we need $K_{ij}=\pm 1$, we obtain the desired form for L and R. X cannot have ± 1 in any of its entry as to be expected from the fact that the assumed properties of the Higgs fields do not permit a $\bar{q}_L q_R \phi$ vertex.

Now, in order to have nonvanishing entries in Y, we should have zeros in Y. x has four possibilities, $x=2$, 3, 4, or 5, out of which only $x=3$ gives rise to a nonsingular form for the corresponding part of the mass matrix. We shall consider this case first.

B. The nonsingular case

With $x=3$,

$$K=\begin{bmatrix} 6 & 2 & 4 & 5 & 1 & 3 \\ 2 & -2 & 0 & 1 & -3 & -1 \\ 4 & 0 & 2 & 3 & -1 & 1 \\ 5 & 1 & 3 & 4 & 0 & 2 \\ 1 & -3 & -1 & 0 & -4 & -2 \\ 3 & -1 & 1 & 2 & -2 & 0 \end{bmatrix}, \tag{4.8}$$

from which we can read off the nonvanishing elements of the mass matrices. As already noted, the top-left part of the 3×3 matrix will be zero. The submass matrices in the two charge sectors are given by

$$M_L^d=\begin{bmatrix} 0 & \alpha_{12}^d v_L^1 & 0 \\ \alpha_{21}^d v_L^1 & 0 & \alpha_{23}^d v_L^2 \\ 0 & \alpha_{32}^d v_L^2 & \alpha_{33}^d v_L^1 \end{bmatrix}, \quad M_R^d=\begin{bmatrix} 0 & \alpha_{12}^d v_R^1 & 0 \\ \alpha_{21}^d v_R^1 & 0 & \alpha_{23}^d v_R^2 \\ 0 & \alpha_{32}^d v_R^2 & \alpha_{33}^d v_R^1 \end{bmatrix}, \tag{4.9}$$

$$M_L^u=\begin{bmatrix} 0 & \alpha_{12}^u v_L^{2*} & 0 \\ \alpha_{21}^u v_L^{2*} & {} & \alpha_{23}^u v_L^{1*} \\ 0 & \alpha_{32}^u v_L^{1*} & \alpha_{33}^u v_L^{2*} \end{bmatrix}, \quad M_R^u=\begin{bmatrix} 0 & \alpha_{12}^u v_R^{2*} & 0 \\ \alpha_{21}^u v_R^{2*} & 0 & \alpha_{23}^u v_R^{1*} \\ 0 & \alpha_{32}^u v_R^{1*} & \alpha_{33}^u v_R^{2*} \end{bmatrix}, \tag{4.10}$$

and

$$M_\sigma^d=\begin{bmatrix} 0 & \gamma_{12}^d \chi & 0 \\ \gamma_{21}^d \chi & 0 & 0 \\ 0 & 0 & \gamma_{33}^d \chi \end{bmatrix}, \quad M_\sigma^u=\begin{bmatrix} 0 & \gamma_{12}^u \chi & 0 \\ \gamma_{21}^u \chi & 0 & 0 \\ 0 & 0 & \gamma_{33}^u \chi \end{bmatrix}. \tag{4.11}$$

Equations (4.9)–(4.11) exhibit the correlations between the down- and up-charge sectors ($v_L^1 \leftrightarrow v_L^{2*}, v_R^1 \leftrightarrow v_R^{2*}$) and present the most general form within the framework of our model. We assume further that (a) the Y couplings are real and symmetric, (b) $\gamma_{12}\chi=\gamma_{33}\chi=\chi$, (c) the VEV's satisfy

$$v_L^i \ll v_R^i \ll \chi, \quad i=1,2\ , \tag{4.12}$$

and

$$|v_L^1| \approx |v_L^2| = v_L, \quad |v_R^1| \approx |v_R^2| = v_R\ . \tag{4.13}$$

For calculational simplicity, we shall take them to be equal. We shall neglect at the moment their phases and assume essentially that v_L, v_R are real. Hence, there will be no CP violation in the gauge sector of our model.

With the above assumptions and according to the discussion in the previous section, the mass matrix M_q for light quarks has the generic form

$$M_q = -M_L M_\sigma^{-1} M_R = \begin{bmatrix} 0 & x & 0 \\ x & z & y \\ 0 & y & w \end{bmatrix}, \qquad (4.14)$$

where, for the d sector,

$$x = (\alpha_{12}^d)^2 \frac{v_L v_R}{\chi}, \quad z = (\alpha_{23}^d)^2 \frac{v_L v_R}{\chi}, \quad w = (\alpha_{33}^d)^2 \frac{v_L v_R}{\chi},$$

and

$$y = \frac{v_L v_R}{\chi} (\alpha_{12}^d \alpha_{23}^d + \alpha_{23}^d \alpha_{33}^d), \qquad (4.15)$$

with corresponding expressions for x, y, z, w for the u sector in which $\alpha_{ij}^d \to \alpha_{ij}^u$.

If $\lambda_1, \lambda_2, \lambda_3$ are the eigenvalues of the real, symmetric mass matrix (4.14),

$$\begin{aligned} &\lambda_1 + \lambda_2 + \lambda_3 = z + w, \\ &\lambda_1\lambda_2 + \lambda_2\lambda_3 + \lambda_3\lambda_1 = -x^2 + zw - y^2, \\ &\lambda_1\lambda_2\lambda_3 = -x^2 w, \end{aligned} \qquad (4.16)$$

which lead to the solution

$$\begin{aligned} &z = (\lambda_1 + \lambda_2 + \lambda_3) - w, \\ &x^2 = -\frac{\lambda_1\lambda_2\lambda_3}{w}, \\ &wy^2 = (\lambda_1 - w)(\lambda_2 - w)(\lambda_3 - w), \end{aligned} \qquad (4.17)$$

in terms of the eigenvalues λ_1, λ_2, λ_3, and w. The positivity of x, z, w requires that one of the eigenvalues be negative. We shall assume λ_1 to be negative and identify

$$\lambda_1 = -m_1, \quad \lambda_2 = m_2, \quad \lambda_3 = m_3, \qquad (4.18)$$

where m_1, m_2, m_3 are the quark masses in the appropriate sectors. Hence,

$$\begin{aligned} &z = (-m_1 + m_2 + m_3) - w, \quad x^2 = \frac{m_1 m_2 m_3}{w}, \\ &wy^2 = (w + m_1)(w - m_2)(m_3 - w). \end{aligned} \qquad (4.19)$$

It is straightforward to compute the eigenvectors and the real, orthogonal matrix that diagonalizes M_q in terms of m_1, m_2, m_3, and w. Thus, if $OM_qO^T = \mathrm{diag}(-m_1, m_2, m_3)$,

$$O = \begin{bmatrix} \left[\frac{m_2 m_3 (w+m_1)}{w(m_2+m_1)(m_3+m_1)}\right]^{1/2} & \left[\frac{m_1 m_3 (w-m_2)}{w(m_1+m_2)(m_3-m_2)}\right]^{1/2} & \left[\frac{m_1 m_2 (m_3-w)}{w(m_3-m_2)(m_3+m_1)}\right]^{1/2} \\ -\left[\frac{m_1 (w+m_1)}{(m_1+m_2)(m_3+m_1)}\right]^{1/2} & \left[\frac{m_2 (w-m_2)}{(m_2+m_1)(m_3-m_2)}\right]^{1/2} & \left[\frac{m_3 (m_3-w)}{(m_3-m_2)(m_3+m_1)}\right]^{1/2} \\ \left[\frac{m_1 (w-m_2)(m_3-w)}{w(m_2+m_1)(m_3+m_1)}\right]^{1/2} & -\left[\frac{m_2 (w+m_1)(m_3-w)}{w(m_2+m_1)(m_3-m_2)}\right]^{1/2} & \left[\frac{m_3 (w+m_1)(w-m_2)}{w(m_3-m_2)(m_3+m_1)}\right]^{1/2} \end{bmatrix}. \qquad (4.20)$$

If we consider the quark mass parameters[13] currently believed to be established, we can expand the above matrix and to a very good approximation retain only the leading terms. Then O is given by

$$O = \begin{bmatrix} 1 & \left[\frac{m_1}{m_2}\right]^{1/2} & \left[\frac{m_1 m_2}{w m_3}\left[1-\frac{w}{m_3}\right]\right]^{1/2} \\ -\left[\frac{m_1 w}{m_2 m_3}\right]^{1/2} & \left[\frac{w}{m_3}\right]^{1/2} & \left[1-\frac{w}{m_3}\right]^{1/2} \\ \left[\frac{m_1}{m_2}\left[1-\frac{w}{m_3}\right]\right]^{1/2} & -\left[1-\frac{w}{m_3}\right]^{1/2} & \left[\frac{w}{m_3}\right]^{1/2} \end{bmatrix}. \qquad (4.21)$$

The Cabibbo-Kobayashi-Maskawa matrix V_{CKM} is then given by

$$V_{\mathrm{CKM}} = O^{(u)} O^{(d)^T}, \qquad (4.22)$$

where $O^{(u)}$ and $O^{(d)^T}$ are obtained by substituting the mass parameters in the u and d sectors, respectively, in (4.22). For simplicity, we shall use $d, s, b, \ldots,$ to represent the corresponding masses $m_d, m_s, m_b, \ldots$. Then,

$$\begin{aligned} (V_{\mathrm{CKM}})_{12} &= -\left[\frac{w_d}{b}\right]^{1/2} (\sqrt{d/s} - \sqrt{u/c}) \\ &\quad + \left[\frac{u}{t}\frac{c}{w_u}\left[1-\frac{w_u}{t}\right]\left[1-\frac{w_d}{b}\right]\right]^{1/2} \\ &\approx -\left[\frac{w_d}{b}\right]^{1/2} (\sqrt{d/s} - \sqrt{u/c}), \end{aligned} \qquad (4.23)$$

$$(V_{\rm CKM})_{13}=\left[1-\frac{w_d}{b}\right]^{1/2}(\sqrt{d/s}-\sqrt{u/c}\,)+\left[\frac{u}{t}\frac{c}{w_u}\left[1-\frac{w_u}{t}\right]\frac{w_d}{b}\right]^{1/2}\approx\left[1-\frac{w_d}{b}\right]^{1/2}(\sqrt{d/s}-\sqrt{u/c}\,)\,, \tag{4.24}$$

$$(V_{\rm CKM})_{23}=-\left[\left[1-\frac{w_d}{b}\right]\frac{w_u}{t}\right]^{1/2}(1+\sqrt{ud/cs}\,)+\left[\left[1-\frac{w_u}{t}\right]\left[\frac{w_d}{b}\right]\right]^{1/2}, \tag{4.25}$$

$$(V_{\rm CKM})_{31}=-\left[1-\frac{w_u}{t}\right]^{1/2}(\sqrt{d/s}-\sqrt{u/c}\,)+\left[\frac{d}{b}\frac{s}{w_d}\left[1-\frac{w_d}{b}\right]\frac{w_u}{t}\right]^{1/2}\approx-\left[1-\frac{w_u}{t}\right]^{1/2}(\sqrt{d/s}-\sqrt{u/c}\,)\,. \tag{4.26}$$

Equations (4.23)–(4.25) express the experimentally measured quantities V_{us}, V_{ub}, and V_{cb} in terms of the quark mass parameters and the two scale parameters w_d and w_u. Out of the three quantities, only V_{us}, which is related to the well-known Cabibbo angle, is well established experimentally. The other two are not well established owing to the uncertainties in the measurements as well as the theoretical model dependence in the interpretation of the data. The most recent Particle Data Group book gives the following ranges for the values of V_{us}, V_{ub}, and V_{cb}: V_{us}, 0.217–0.223; V_{ub}: 0.003–0.010; V_{cb}: 0.03–0.062. These uncertainties coupled with the uncertainties in the values of the quark mass parameters (especially the light quarks) make a precise quantitative comparison of (4.23)–(4.25) extremely difficult at the present stage. We look only for qualitative and semiquantitative agreement.

First of all, it is gratifying to note that the relation (4.23) for the Cabibbo angle is a consequence of the model. Indeed, in the case of two generations, $\sin\theta_C=(\sqrt{d/s}-\sqrt{u/c}\,)$ provides an extremely good relation consistent with experimental information. This suggests that in the case of three generations, the factor (w_d/b) in (4.23) should be close to unity, which implies from (4.24) that V_{ub} is naturally suppressed.[14] The dominant term in (4.25) is the second factor which can be fit to V_{cb} by choosing an appropriate value for (w_u/t). Then, (4.26) predicts V_{td} for which, at present, there is only indirect information.[15] But it is expected to be large, of the same order of magnitude as V_{cb}. That is indeed the case, as one can see from a comparison of (4.25) with (4.26).

With the range of values for the mass parameters, the factor $(\sqrt{d/s}-\sqrt{u/c}\,)$ varies from 0.1148 to 0.2382. This large variation is due to large uncertainties in d-, s-, and u-quark mass values. Taking this variation into account, we can fit, for instance, to

$$V_{us}=0.22,\quad V_{cb}=0.05,\quad V_{ub}=0.007\,, \tag{4.27}$$

by choosing

$$(\sqrt{d/s}-\sqrt{u/c}\,)=0.2211,\quad \frac{w_d}{b}=0.9995\,,\quad \frac{w_u}{t}=0.9948\,. \tag{4.28}$$

We then predict $V_{td}\approx0.016$ from (4.26).

Further, from (4.15),

$$\frac{x}{w}=\left[\frac{\alpha_{12}}{\alpha_{33}}\right]^2=\left[\frac{m_1m_2m_3}{w^3}\right]^{1/2} \tag{4.29}$$

and

$$\frac{z}{w}=\left[\frac{\alpha_{23}}{\alpha_{33}}\right]^2=\frac{m_3+m_2-m_1-w}{w}\,. \tag{4.30}$$

In the d sector, using the values for mass parameters and $w_d\approx m_b$ [from (4.28)], we find

$$\alpha_{12}^d/\alpha_{33}^d={}^4\sqrt{ds/b^2}\simeq0.086\,, \tag{4.31}$$

$$\alpha_{23}^d/\alpha_{33}^d\approx\sqrt{s/b}\simeq0.182\,, \tag{4.32}$$

which shows that the ratios of Y couplings are quite reasonable, in the sense that they do not exhibit the same degree of hierarchy as the masses. The latter would be the case in a straightforward extension of the standard model to include more generations.

Going a step further, if we use again the value $v_L=246$ GeV and $v_R/\chi=10^{-6}$, we are led to absolute values of the couplings

$$|\alpha_{12}|=12.6,\quad |\alpha_{23}|=26.7,\quad\text{and}\quad |\alpha_{33}|=147\,. \tag{4.33}$$

All these representative values for Y couplings are almost within 1 order of magnitude.

Thus, the simplified version of the model provides a satisfactory extension to three generations of quarks. Although the quark masses are not predicted, the matrix elements V_{us}, V_{cb}, V_{ub} and hence the CKM mixing angles, and also the relevant Y couplings are expressed in terms of the quark masses.

C. The singular case

We consider briefly one singular case by choosing $x=4$ in (4.7). $M_\sigma^{d,u}$ is then singular. There is no approximate form for the light-quark mass matrix in this case. The eigenvalue equation is given by

$$\lambda^6-\lambda^4(\chi^2+3V_LV_R)-2\lambda^3(V_LV_R\chi)+2\lambda^2(V_LV_R\chi^2+V_L^2V_R^2)+4\lambda V_L^2V_R^2\chi-V_L^3V_R^3=0\,. \tag{4.34}$$

Under the assumed hierarchy, we can show that the approximate eigenvalues fall into three categories:

$$\text{light: } \lambda_{1,2} \pm \frac{V_L V_R}{\sqrt{2}\chi} ,$$
$$\text{intermediate: } \lambda_{3,4} = \pm\sqrt{2V_L V_R} , \quad (4.35)$$
$$\text{heavy: } \lambda_{5,6} = \pm\chi .$$

$\lambda_{1,2}$, being the smallest, can be identified with the light quarks of the first two generations. One of λ_3, λ_4 can represent the heavier third-generation quark with one of the heavy singlet quarks being almost degenerate with it. $\lambda_{5,6}$ represent the remaining singlets.

The predicted spectrum has indeed features corresponding to reality. The quarks of the first two generations in the light of the heavy χ are near degenerate. The quark belonging to the third generation is much heavier (corresponding to the fact that b and t quarks are much heavier than their corresponding partners in the first two generations). The spectrum also predicts a heavy singlet in the mass range of the third-generation quark.

V. SUMMARY AND CONCLUSIONS

We have explored in this paper the consequences of a model based on the idea of a universal seesaw mechanism for explaining the mass hierarchies and the mixing angles in the quark sector. The basic idea is the generalization of the seesaw mechanism to generate a superlight neutrino by postulating a heavy right-handed Majorana neutrino and a Majorana mass term. Likewise, within the framework of left-right symmetry, if every fermion is accompanied by a heavy left-right singlet fermion, the mass matrix for every fermion assumes a seesaw form and gives rise to the possibility of light-mass quarks and leptons.

In the case of a single generation, the hierarchy assumption $v_L \ll v_R \ll \chi$ leads to light-quark masses and light mass charged lepton. At the same time, without any further assumption, one is lead to a superlight neutrino in a natural way. These results found earlier are further strengthened in this investigation by (1) the demonstration that the assumed hierarchy can be a consequence of spontaneous left-right-symmetry breaking caused by an odd-parity left-right-symmetric singlet, and (2) the fact that no additional hierarchy assumption in the Yukawa-type couplings between the fermions and the Higgs scalars needs to be invoked. These nice results prompt one to consider the realistic case of three generations. By postulating a global, axial U(1) symmetry, we distinguish the different generations and study the constraints imposed by the additional symmetry on the form of the mass matrices. While, in general, several forms of mass matrices are allowed, we confine our attention to the so-called Fritzsch form for the submatrices. We ignore in this investigation, *CP* violation by assuming that all the VEV's and the Y couplings are real, so that the Kobayashi-Maskawa phase is zero. This is only for the sake of simplicity and to test the viability of USM in providing a reasonably satisfactory answer for the known Cabibbo-Kobayashi-Maskawa matrix elements.

The results of the simplified version of the model are indeed encouraging. If the quark masses are assumed to be known, the results depend on two new parameters, one in each charge sector of the quarks. The known experimental information concerning the matrix elements V_{us}, V_{ub}, and V_{cb} place stringent conditions on these two parameters. There is an inner consistency in the correlations of the theoretical expressions for these matrix elements with their observed values. Further, as in the single generation case, the Y couplings do not exhibit the same degree of hierarchy as the mass parameters. Thus on the whole the model provides a promising approach to the mass hierarchy and the phenomenology of generation mixing interactions.

It must be said, however, that the model even within its own framework has considerable flexibility and several parameters. The present investigation has concentrated on reducing the dependence on the arbitrary parameters and seeking combination of parameters which are directly related to observed quantities.[16] It should also be noted that at present there are too many uncertainties in our experimental knowledge. As a result, a detailed numerical confrontation of our model (or any other model) makes it extremely difficult, perhaps even pointless. We have to wait for the discovery of the top quark and its decay characteristics, and more precise and independent measurements of all the CKM matrix elements for such a confrontation.

ACKNOWLEDGMENTS

This work was supported by the U.S. Department of Energy under Contract No. DE-FG02-85ER40231. S.R. would also like to thank "Fondazione Angelo della Riccia" for partial support. A.D. would like to thank the hospitality of the members of the Physics Department at Syracuse University.

[1]The list of references that consider SU(2)×U(1)×G_H where G_H is discrete or continuous is too long to reproduce here. See, for instance, H. Harari, in *To Fulfill a Vision,* the Jerusalem Einstein Centennial Symposium, Jerusalem, Israel, 1979, edited by Y. Ne'eman (Addison-Wesley, Reading, MA, 1980), p. 171; Ref. 4 in A. Davidson and K. C. Wali, Phys. Rev. D **21**, 787 (1980); Ref. 4 in K. C. Wali, *Weak Interactions as Probes of Unification,* proceedings of the Workshop, edited by G. B. Collins, L. N. Cheng, and J. R. Ficence (AIP Conf. Proc. No. 72) (AIP, New York, 1981); *Particles and Fields,* proceedings of 1974 Meeting of the APS Division of Particles and Fields, edited by C. A. Carlson (AIP Conf. Proc. No. 23) (AIP, New York, 1975).

[2]S. Weinberg, Phys. Rev. Lett. **29**, 388 (1972); H. Georgi and S. L. Glashow, Phys. Rev. D **6**, 2977 (1972); **7**, 2457 (1973); R. N. Mohapatra, *ibid.* **9**, 3461 (1974); S. M. Barr and A. Zee, *ibid.* **15**, 2652 (1977); **17**, 1854 (1978); S. M. Barr, *ibid.* **21**, 1424 (1980); R. Barbieri and D. V. Nanopoulos, Phys. Lett. **91B**, 369 (1980); **95B**, 43 (1980); R. Barbieri, D. V. Nanopoulos, and A. Masiero, *ibid.* **104B**, 194 (1981); R. Barbieri, D. V. Nanopoulos, and D. Wyler, *ibid.* **106B**, 303 (1981); S. M. Barr, Phys. Rev. D **24**, 1895 (1981); M. Bowick and P. Ra-

mond, Phys. Lett. **103B**, 338 (1981); S. M. Barr, Phys. Rev. D **31**, 2979 (1985); S. Balakrishna, Phys. Rev. Lett. **60**, 1602 (1988); S. Balakrishna, A. L. Kagan, and R. N. Mohapatra, Phys. Lett. B **205**, 345 (1988); S. Balakrishna and R. N. Mohapatra, *ibid.* **216**, 349 (1989).

[3]Since the original suggestion of the seesaw mechanism for understanding the superlightness of ν_L by M. Gell-Mann, P. Ramond, and R. Slansky, in *Supergravity*, edited by D. Freedman and P. Van Nieuwenhuizen (North-Holland, Amsterdam, 1979); T. Yanagida, in *Proceedings of the Workshop in Unified Theory and Baryon Number of the Universe*, Tsukuba, Ibraki, Japan, 1979, edited by O. Sawada and A. Sugamoto (National Laboratory for High Energy Physics, KEK, Tsukuba, Japan, 1979); Phys. Rev. Lett. **44**, 912 (1980), several papers have appeared in the literature extending the idea to quarks and leptons. For instance, D. Chang and R. N. Mohapatra, *ibid.* **58**, 1600 (1987); S. Rajpoot, Phys. Lett. B **191**, 122 (1987); Phys. Rev. Lett. **60**, 2003 (1988). See also Refs. 4 and 5.

[4]A. Davidson and K. C. Wali, Phys. Rev. Lett. **59**, 393 (1987).

[5]A. Davidson and K. C. Wali, Phys. Rev. Lett. **60**, 1813 (1988).

[6]The idea of introducing a heavy singlet to invoke the seesaw mechanism is contained in several papers listed in Ref. 3. What is different in the work of Refs. 4 and 5 is the generality of the idea that every conventional fermion is accompanied by a heavy nonmirror singlet companion.

[7]A. Davidson and K. C. Wali, Phys. Rev. Lett. **58**, 2623 (1987).

[8]D. Chang, R. N. Mohapatra, and M. K. Parida, Phys. Rev. Lett. **52**, 1072 (1984); Phys. Rev. D **30**, 1052 (1984).

[9]Henceforth, we shall designate such couplings as Y couplings.

[10]D. A. Dicus *et al.*, Phys. Rev. D **18**, 1829 (1978); **22**, 839 (1980); M. Fukugita, S. Watamura, and M. Yoshimura, *ibid.* **26**, 1840 (1982); N. Iwamoto, Phys. Rev. Lett. **53**, 1198 (1984); J. Preskill, M. B. Wise, and F. Wilczek, Phys. Lett. **120B**, 127 (1983); L. F. Abbott and P. Sikvie, *ibid.* **120B**, 133 (1983); M. Dine and W. Fischler, *ibid.* **120B**, 137 (1983).

[11]R. D. Peccei and H. R. Quinn, Phys. Rev. Lett. **38**, 1440 (1977); Phys. Rev. D **16**, 1791 (1977).

[12]A. Davidson and K. C. Wali, Phys. Rev. Lett. **48**, 11 (1982); A. Davidson, V. P. Nair, and K. C. Wali, Phys. Rev. D **29**, 1504 (1984); **29**, 1513 (1984).

[13]We use the following values for the mass parameters:

$$m_u = 5.1 \pm 1.5 \text{ MeV}, \quad m_d = 8.9 \pm 1.6 \text{ MeV} ,$$

$$m_c = 1.35 \pm 0.05 \text{ GeV}, \quad m_s = 0.175 \pm 0.055 \text{ GeV} ,$$

$$m_t \geq 100 \text{ GeV}, \quad m_b = 5.3 \pm 0.1 \text{ GeV} .$$

The above values are at the mass scale of 1 GeV. The top-quark-mass value implies that its physical mass is ≥ 60 GeV. For details see J. Gasser and H. Leutwyler, Phys. Rep. **87**, 77 (1982).

[14]In our model V_{ub} has a lower limit set by the condition $m_d < m_2 + m_3 - m_1$ so that $z > 0$.

[15]H. Harari, in the Proceedings of *CP* Violation in Particle Physics and Astrophysics, Chateau de Blois, France, 1989 (unpublished). Also see G. Altarelli, in the Proceedings of the 1989 International Symposium on Heavy Quark Physics, Ithaca, New York, 1989 (unpublished).

[16]The general idea here is to seek expressions for the mixing angles and Yukawa-type couplings in terms of quark mass ratios which are better determined than the masses of individual quarks. Our model is not unique in this respect. For other examples, see, for instance, Z. G. Berezhiani, Phys. Lett. **129B**, 99 (1983); **150B**, 177 (1985); F. del Aguila, G. L. Kane, and M. Quiros, Phys. Lett. B **196**, 531 (1987); M. Gronau, R. Johnson, and J. Schechter, Phys. Rev. D **39**, 1913 (1989). For a review and extensive lists of references, see Kyungsik Kang, Report No. Brown-HET-690 (unpublished).

Towards a unified treatment of Yang-Mills and Higgs fields

B. S. Balakrishna
Department of Physics, Syracuse University, Syracuse, New York 13244-1130

Feza Gürsey
Department of Physics, Yale University, New Haven, Connecticut 06511

Kameshwar C. Wali
Department of Physics, Syracuse University, Syracuse, New York 13244-1130

(Received 6 June 1991)

Starting from a noncommutative algebra $\mathcal{A}$ of the form $\mathcal{C}\otimes\mathcal{M}$, where $\mathcal{C}$ is the algebra of smooth functions on space-time and $\mathcal{M}$ is the algebra of $n\times n$ Hermitian matrices, we construct an exterior algebra of differential forms over $\mathcal{A}$. We use the one-forms of this algebra to describe Yang-Mills and Higgs fields on a similar footing and construct a Lagrangian from its two-forms. We show how, in the resulting geometrical description, a Higgs potential that leads to spontaneous symmetry breaking arises naturally. We discuss the application of this formalism to the bosonic sectors of the standard electroweak theory and a grand-unified model based on SU(5)⊗U(1).

I. INTRODUCTION

Of all the shortcomings of the standard model, the most serious one is perhaps the symmetry-breaking mechanism using fundamental scalar fields known as Higgs fields. Unlike the gauge sector involving the fermions and vector bosons, the form, the content, and the couplings in the Higgs sector are not determined by gauge principles alone. As a consequence, additional *ad hoc* assumptions seem to be indispensable in model building. There are, of course, in addition the well-known fine tuning and hierarchy problems coupled with the fact that the Higgs particle masses are not predictable. It is not surprising, therefore, that there are several attempts in the literature to eliminate some of these difficulties.

In the early 1980s, for instance, several authors [1] proposed Kaluza-Klein ideas to unify gauge and Higgs fields. In space-times with extended space degrees of freedom, the additional components of fields other than four were identified as the Higgs fields: $A_M(x^0,x^1,\ldots,x^{d-1})$ give gauge fields for $M=0,\ldots,3$ and Higgs fields for $M=4,\ldots,d-1$. By imposing suitable conditions on the dependence of the fields on the additional degrees of freedom, such theories could be reduced to four-dimensional Yang-Mills theories with built-in spontaneous symmetry breaking. Such models had the virtue of being able to predict, in the case of the Glashow-Salam-Weinberg model of electroweak interactions, the Weinberg angle, and the mass of the Higgs particle. However, the results found in specific models were not in accord with experiments. Hence, in spite of their aesthetic appeal, they were not pursued further.

More recently, some new ideas have made their appearance under the general category characterized by noncommutative geometry [2]. In one such approach advocated by Connes and applied to the standard model by Lott and Connes [3], there are more than one identical copies of space-times characterized by a discrete index $p=1,2,\ldots$. While in a given space-time the customary gauge fields provide the connections through the one-forms $A_\mu dx^\mu$, the connections between discrete space-times are identified as Higgs fields. The mixed differential forms are assembled together in the form of a matrix with graded multiplication properties [4]. An exterior derivative defined on such matrices enables one to calculate two-forms and construct a gauge-invariant Lagrangian that has some interesting properties as well as predictive power. However, from our point of view, the starting point of such models is rather arbitrary. If one demands, at the beginning, appropriate transformation properties under allowed arbitrary gauge transformations, one finds that a number of additional fields are necessary.

In this paper, we follow an alternate approach initiated by Dubois-Violette, Kerner, and Madore [5]. The central idea here, as in Kaluza-Klein theory, is to extend the space-time degrees of freedom; however, the extension is not by continuous space degrees of freedom, but by finite matrices, which are supposed to describe the internal structure of a particle. The starting point is then an associative but noncommutative algebra $\mathcal{A}$ of the form $\mathcal{C}\otimes\mathcal{M}$, where $\mathcal{C}$ is the algebra of smooth functions of space-time coordinates and $\mathcal{M}$ is the algebra of $n\times n$ matrices. By defining an exterior algebra of forms over $\mathcal{A}$, we construct a Lagrangian from the generalized two forms.

In this approach, the Higgs fields originate as $\mathcal{C}$-valued coefficients when one takes the exterior derivatives of the generators of $\mathcal{M}$. In the resulting geometrical description, Yang-Mills and Higgs fields appear on the same footing. The content of the Higgs fields is determined by the choice of the exterior derivative. Further, apart from the overall gauge coupling constant, the only free parameters in the Lagrangian are a limited number of scale parameters. They too depend on the manner in which the exterior derivative is defined. The resulting model is therefore quite restrictive. A Higgs potential with spontaneously broken symmetry appears naturally in the Lagrangian. It consists of a sum of terms, each of which is a perfect square, its absolute minimum given by the vanishing of each term. The model also implies certain relations among the coupling constants at the classical level. However, these relations, whose origin is not clear at the moment, appear to exist at only one set of mass scales in the Lagrangian. Quantum corrections induce departures from these relations [6].

In the next section, we describe the general mathematical framework. It is followed by a discussion of the most straightforward choice of the exterior derivative resulting in the models proposed by Dubois-Violette, Kerner, and Madore [5]. In Sec. III, we review our previously considered SU(2)⊗U(1) model [6] and discuss briefly another model based on embedding SU(2)⊗U(1) generators into SU(2|1) algebra. In Sec. IV, we discuss its generalization to SU(5)⊗U(1), grand unified, and explore some of its physical consequences. The final section is devoted to a summary and conclusions.

II. GENERAL FRAMEWORK

As stated in the preceding section, the noncommutative algebra $\mathcal{A}$ is made of two factors. One of them is the commutative algebra $\mathcal{C}$ of all smooth functions on the Minkowski space-time M^4. The second factor is taken to be the algebra $\mathcal{M}_n$ of complex $n\times n$ matrices. Thus

$$\mathcal{A}=\mathcal{C}\otimes\mathcal{M}_n \ . \tag{2.1}$$

One may choose the generalized Gell-Mann matrices λ_a, $a=1,\ldots,n^2-1$, along with the identity 1 (or $\lambda_0\equiv\sqrt{2/n}$) as a Hermitian basis for the algebra $\mathcal{M}_n$. Then a generic element of $\mathcal{A}$ can be written as

$$F=f_0(x)\lambda_0+f_a(x)\lambda_a \ . \tag{2.2}$$

Here x's are the coordinates of M^4. Summation over a repeated index (here a) is assumed. Two elements of $\mathcal{A}$ can be multiplied, as usual, knowing the products $\lambda_a\lambda_b$ in terms of λ's and the identity.

In the ordinary case of M^4, it is found convenient to extend the algebra $\mathcal{C}$ to the so-called exterior algebra $\mathcal{C}^*$. $\mathcal{C}^*$ can be written as a direct sum

$$\mathcal{C}^*=\mathcal{C}^0\oplus\mathcal{C}^1\oplus\cdots\oplus\mathcal{C}^4 \ . \tag{2.3}$$

$\mathcal{C}^0$ is simply $\mathcal{C}$. To define others, one introduces a set of anticommuting objects dx^μ's. Then a generic element of $\mathcal{C}^1$ is the form $f_\mu(x)dx^\mu$, and this space is called the space of one-forms. And, more generally, $\mathcal{C}^p$ is made of p-forms of the kind

$$f_{\mu_1\cdots\mu_p}(x)dx^{\mu_1}\cdots dx^{\mu_p} \ . \tag{2.4}$$

There exists a differential operator on $\mathcal{C}^*$, the so-called exterior derivative. It is defined to be $d=dx^\mu\partial_\mu$. This operator satisfied $d^2=0$ and the Leibniz rule

$$d(fg)=d(f)g+(-1)^pfd(g) \ , \tag{2.5}$$

where f and g are respectively p- and q-forms. Its action on a p-form f is given by

$$df=\partial_\mu f_{\mu_1\cdots\mu_p}(x)dx^\mu dx^{\mu_1}\cdots dx^{\mu_p} \ . \tag{2.6}$$

We wish to extend the above notions to $\mathcal{M}_n$ and hence to $\mathcal{A}$. To extend the concept of forms to the algebra $\mathcal{M}_n$, we introduce anticommuting objects $\theta_a, a=1,\ldots,n^2-1$. These are analogous to dx^μ's. For simplicity, we denote them by θ's instead of the more obviously consistent notation $d\theta$'s. Define $\mathcal{M}_n^*$ to be the direct sum

$$\mathcal{M}_n^*=\mathcal{M}_n^0\oplus\mathcal{M}_n^1\oplus\cdots\oplus\mathcal{M}_n^{n^2-1} \ . \tag{2.7}$$

$\mathcal{M}_n^0$ is simply $\mathcal{M}_n$. $\mathcal{M}_n^1$ is made of objects like $F_a\theta_a$, where F_a is an element of $\mathcal{M}_n, F_a=f_{a0}\lambda_0+f_{ab}\lambda_b$. In general, $\mathcal{M}_n^p$ is made of objects like

$$F_{a_1\cdots a_p}\theta_{a_1}\cdots\theta_{a_p} =(f_{a_1\cdots a_p0}\lambda_0+f_{a_1\cdots a_pa}\lambda_a)\theta_{a_1}\cdots\theta_{a_p} \ . \tag{2.8}$$

The exterior algebra associated with $\mathcal{A}$ can be taken to be the tensor product

$$\mathcal{A}^*=\mathcal{C}^*\otimes\mathcal{M}_n^*=\mathcal{A}^0\oplus\mathcal{A}^1\oplus\cdots\oplus\mathcal{A}^{n^2+3} \ . \tag{2.9}$$

In particular, $\mathcal{A}^p$, the space of generalized p-forms, is

$$\mathcal{A}^p=\sum_{k=0}^{p}\mathcal{C}^k\otimes\mathcal{M}_n^{p-k} \ . \tag{2.10}$$

It is assumed in the above that $\mathcal{C}^k$ and $\mathcal{M}_n^l$ vanish respectively for $k>4$ and $l>n^2-1$. Since forms can be added and multiplied, $\mathcal{A}^*$ represents an algebra, called the exterior algebra associated with $\mathcal{A}$. $\mathcal{A}^p$ consists of matrices each element of which is like a p-form in dx^μ's and θ's. In this way we seek a generalization of the usual non-Abelian gauge theories in which the differential forms are forms in $\mathcal{C}$ only. The coefficients of dx^μ's are the gauge potential matrices. By allowing space-time scalars as coefficients of θ's, in addition to space-time vectors as coefficients of dx^μ's, we can treat Higgs fields and gauge fields on a similar footing. It should be noted that, unlike in Refs. [3,4], every matrix element in $\mathcal{A}^p$ will be of a definite order as a generalized form. This way, we preserve the homogeneity of each term under multiplication. Forms of different order in diagonal and off-diagonal terms will lead to an inhomogeneous sum of terms of different orders. Gauge invariance will then require introducing additional fields or making *ad hoc* restrictions on the allowed gauge transformations.

Next we want to define an exterior derivative associated with the algebra $\mathcal{A}^*$. We have already noted that such an operator for $\mathcal{C}^*$ is $d=dx^\mu\partial_\mu$. This is constructed from the generators of the group of translations that act as automorphisms of $\mathcal{C}$. In other words,

$$e^{ia\cdot P}f(x)e^{-ia\cdot P}=f(x)+i[a\cdot P,f(x)]+\cdots =f(x)+a^\mu\partial_\mu f(x)+\cdots \ , \tag{2.11}$$

where P_μ's are the generators of translations, gives the infinitesimal change in $f(x)$ as $df(x)=a^{\mu}\partial_{\mu}f(x)$. Replacing a^{μ}'s by anticommuting objects dx^{μ}'s, we obtain the expression for the exterior derivative: $d=dx^{\mu}\partial_{\mu}$. In a similar fashion, if we can find an automorphism of the algebra $\mathcal{M}_n$, we can construct a differential operator associated with it and hence an exterior derivative for $\mathcal{M}_n^*$. This will let us generalize these notions to the full algebra $\mathcal{A}$. There is a natural choice for this automorphism, the so-called inner automorphism, given by

$$e^{i\alpha_b\lambda_b}\lambda_a e^{-i\alpha_c\lambda_c}=\lambda_a+i[\alpha_b\lambda_b,\lambda_a]+\cdots$$
$$\equiv\lambda_a+d\lambda_a+\cdots . \tag{2.12}$$

This forms the group SU(n) or SU(n)$\otimes$U(1), where the U(1) generator λ_0 drops out of the commutators. The infinitesimal change in λ_a is, thus,

$$d\lambda_a=i[\alpha_b\lambda_b,\lambda_a] . \tag{2.13}$$

This gives us a derivation E_a that acts on the elements (say, F) of the algebra $\mathcal{M}_n$ as

$$E_a(F)=\frac{m}{2}[\lambda_a,F] , \tag{2.14}$$

where m, having dimension of mass, introduces a scale into the theory (this is in analogy to ∂_{μ} that has dimension of mass in natural units). Being derived from an automorphism, this operator automatically obeys the Leibniz rule

$$E_a(FG)=E_a(F)G+FE_a(G) , \tag{2.15}$$

where F and G belong to $\mathcal{M}_n$. There exist other choices for the automorphisms as well, and hence for the derivations, that will be discussed in the next section. Unlike ∂_{μ}'s, the above operators do not commute among themselves. Instead they satisfy

$$[E_a,E_b]=imf_{abc}E_c , \tag{2.16}$$

where f_{abc}'s are the structure constants of SU(n):

$$[\lambda_a,\lambda_b]=2if_{abc}\lambda_c . \tag{2.17}$$

We thus have a set of differential operators (∂_{μ},E_a) acting on the full algebra $\mathcal{A}$. As for the exterior derivative, we follow Bowick and Gürsey [7] and construct the Becchi-Roueti-Stora-Tyutin- (BRST-) like operator Q as

$$Q=\theta_a(L_a+\tfrac{1}{2}S_a) , \tag{2.18}$$

where

$$L_a=E_a,\quad S_a=-imf_{abc}\theta_b\frac{\partial}{\partial\theta_c} . \tag{2.19}$$

L_a and S_a commute with each other, and both satisfy the commutation relations (2.16). Thus the resulting operator

$$Q=\theta_aE_a-i\frac{m}{2}f_{abc}\theta_a\theta_b\frac{\partial}{\partial\theta_c} \tag{2.20}$$

obeys $Q^2=0$. This will be the exterior derivative associated with $\mathcal{M}_n$. Now

$$D=d+Q \tag{2.21}$$

provides a natural exterior derivative for $\mathcal{A}^*$, since it follows from the automorphisms of $\mathcal{A}$, the direct product of the group of translations and that of the inner automorphisms of $\mathcal{M}_n$. θ's are taken to be anti-Hermitian to be consistent with the reality property of d. It is assumed that the objects dx^{μ}'s and θ's anticommute among themselves so that d anticommutes with Q. This and the fact that $d^2=Q^2=0$ lead to the desired property $D^2=0$.

The natural object to be investigated next is the connection one-form in $\mathcal{A}$. It has the following structure:

$$\omega=A+\Phi . \tag{2.22}$$

A is an element of $\mathcal{C}^1$ taking values in $\mathcal{M}_n$:

$$A=-igA_{\mu}dx^{\mu}$$
$$=-ig\tfrac{1}{2}[A_{\mu0}(x)\lambda_0+A_{\mu a}(x)\lambda_a]dx^{\mu} , \tag{2.23}$$

where g is the coupling constant of the theory. These will turn out to be our usual gauge fields. The Higgs field is going to be Φ. It is an element of $\mathcal{M}^1$ taking values in $\mathcal{C}$:

$$\Phi=g\Phi_a\theta_a=g\tfrac{1}{2}[\phi_{a0}(x)\lambda_0+\phi_{ab}(x)\lambda_b]\theta_a . \tag{2.24}$$

A connection one-form comes with a gauge transformation. In this case it has the infinitesimal form

$$\delta\omega=D\epsilon+\omega\epsilon-\epsilon\omega , \tag{2.25}$$

where ϵ is an element of $\mathcal{A}$ and hence has the decomposition

$$\epsilon=-ig\tfrac{1}{2}[\epsilon_0(x)\lambda_0+\epsilon_a(x)\lambda_a] . \tag{2.26}$$

In terms of A and Φ the gauge transformation becomes

$$\delta A_{\mu0}=\partial_{\mu}\epsilon_0,\quad \delta A_{\mu a}=\partial_{\mu}\epsilon_a+gf_{abc}A_{\mu b}\epsilon_c ,$$
$$\delta\phi_{a0}=0,\quad \delta\phi_{ab}=-mf_{abc}\epsilon_c+gf_{bcd}\phi_{ac}\epsilon_d . \tag{2.27}$$

We can get rid of the nonhomogeneous part in the transformation of ϕ_{ab} by shifting ϕ_{ab} to $\phi_{ab}-(m/g)\delta_{ab}$ [that is Φ to $\Phi-(m/2)\lambda_a\theta_a$]. We assume in the following that this has been done. Now it is clear from these transformations that we have SU(n)$\otimes$U(1) gauge invariance. The Higgs sector, in this approach, consists of (n^2-1) singlets and (n^2-1) adjoint representations of SU(n) all having zero U(1) charge.

The field strength or the curvature two-form is the next object to be studied. Constructing it as usual we have

$$\Omega=D\omega+\omega^2 . \tag{2.28}$$

Its components are given by

$$\Omega=\tfrac{1}{2}\Omega_{\mu\nu}dx^{\mu}dx^{\nu}+\Omega_{\mu a}dx^{\mu}\theta_a+\tfrac{1}{2}\Omega_{ab}\theta_a\theta_b , \tag{2.29}$$

where

$$\Omega_{\mu\nu}=-igF_{\mu\nu}=-ig(\partial_{\mu}A_{\nu}-\partial_{\nu}A_{\mu}-ig[A_{\mu},A_{\nu}]) ,$$
$$\Omega_{\mu a}=g\mathcal{D}_{\mu}(\Phi_a)=g(\partial_{\mu}\Phi_a-ig[A_{\mu},\Phi_a]) , \tag{2.30}$$
$$\Omega_{ab}=-ig(mf_{abc}\Phi_c+ig[\Phi_a,\Phi_b]) .$$

$F_{\mu\nu}$ gives the field strength of the SU(n)$\otimes$U(1) gauge fields. $\mathcal{D}_\mu(\phi_a)$ is the covariant derivative of the Higgs fields. As we shall see, the term Ω_{ab} is responsible for the Higgs potential.

Given the curvature two-form, one can write down a gauge-invariant Lagrangian. Consider

$$L=\frac{1}{2g^2}\,\mathrm{Tr}(\Omega_{ij}\Omega^{ij})\ , \tag{2.31}$$

where the indices i and j are assumed to be summed over the full range: the space-time and the internal (noncommuting) directions. "Tr" stands for trace in the fundamental representation of SU(n). We thus have $\mathrm{Tr}(\lambda_a\lambda_b)=2\delta_{ab}$, where a,b run from 0 to n^2-1. In terms of the components of Ω, L becomes

$$L=-\tfrac{1}{2}\mathrm{Tr}(F_{\mu\nu}F^{\mu\nu})+\mathrm{Tr}[\mathcal{D}_\mu(\Phi_a)\mathcal{D}^\mu(\Phi_a)]-V\ , \tag{2.32}$$

where the Higgs potential V is given by

$$V=\tfrac{1}{2}\mathrm{Tr}\{(mf_{abc}\Phi_c+ig[\Phi_a,\Phi_b])(mf_{abd}\Phi_d+ig[\Phi_a,\Phi_b])\}\ . \tag{2.33}$$

One of the features of the above Higgs potential is that it has a minimum that corresponds to spontaneous symmetry breaking. Being a sum of squares, V is (absolutely) minimized whenever $\Omega_{ab}=0$ or

$$mf_{abc}\Phi_c+ig[\Phi_a,\Phi_b]=0\ . \tag{2.34}$$

This equation has two solutions. One of them is the trivial one, $\Phi_a=0$, that does not lead to any spontaneous symmetry breaking. The other one

$$\Phi_a=\frac{m}{2g}\lambda_a,\ \text{ or }\ \phi_{a0}=0,\quad \phi_{ab}=\frac{m}{g}\delta_{ab}\ , \tag{2.35}$$

breaks the gauge symmetry SU(n)$\otimes$U(1) to U(1). This corresponds to SU(n) being completely broken and U(1) being untouched.

The approach outlined above is essentially that of Ref. [5]. There are some reasons why, with $n=2$, this cannot give rise to a realistic picture of the electroweak theory. First of all, we get only adjoint representations of Higgs fields. For the electroweak theory this is phenomenologically untenable, since the ρ-parameter constraint rules out the possibility of spontaneous symmetry breaking by triplet Higgs fields. Secondly, the unbroken U(1) is the same U(1) that appears explicitly in the gauge group SU(2)$\otimes$U(1). In other words, there is no mixing of this U(1) with the one from SU(2). This will lead to zero value of the Weinberg angle. In Ref. [6] an approach is presented, which, though similar to the one presented above, cures these drawbacks and leads to a more realistic picture. In the next section we first give a brief review of this and then present a more general scheme that leads to a somewhat realistic grand-unified framework.

III. APPLICATION TO ELECTROWEAK THEORY

In the previous section, we obtained a Higgs potential that leads to spontaneous symmetry breaking. In the case of $n=2$, this corresponds to SU(2)$\otimes$U(1) being broken down to U(1). As we have already noticed, the unbroken U(1) is that which appears explicitly in SU(2)$\otimes$U(1). In the basis of Pauli matrices, τ_i's and $\tau_0\equiv 1$, this means that the generator τ_0 remains untouched. We wish to modify this scenario to bring the model closer to the standard model. One may trace the origin of this problem to the definition of the exterior derivative. The minimum of the Higgs potential corresponds to a solution of $\omega=0$. This equation is invariant under the gauge transformation (2.25) whenever the gauge parameter ϵ is annihilated by D. Because the operator D happened to satisfy $D(\tau_0)=0$, a U(1) gauge symmetry generated by τ_0 survives at the end. It thus follows that those generators annihilated by D will remain unbroken in the present approach. In the following we make use of this observation and define the exterior derivative operator suitably to keep the desired generators unbroken. First we will review this approach for the case of the standard model [8], and then we will generalize it to a grand-unified framework.

We want the electric charge to survive symmetry breaking. The Gell-Mann–Nishijima formula tells us that this generator is given by $(\tau_3+Y)/2$, the hypercharge Y being essentially τ_0. Thus we assume the existence of an exterior derivative D that satisfies

$$D(\tau_0+\tau_3)=0\ . \tag{3.1}$$

We find it convenient to work with the generators of the algebra, q and $\bar{q}$, where

$$q=\tfrac{1}{2}(\tau_1+i\tau_2),\quad \bar{q}=\tfrac{1}{2}(\tau_1-i\tau_2)\ . \tag{3.2}$$

In terms of these, we may write (3.1) as $D(q\bar{q})=0$. This definition of the exterior derivative implies some identities satisfied by $D\tau$'s . For example, qDq and Dqq (and their conjugates) vanish, since

$$\begin{aligned} qDq&=qD(q\bar{q}q)=qD(q\bar{q})q+q(q\bar{q})Dq=0\ ,\\ Dqq&=Dq(q\bar{q}q)=D(qq\bar{q})q-qD(q\bar{q})q=0\ , \end{aligned} \tag{3.3}$$

using $q^2=0$ and $D(q\bar{q})=0$. Other identities are

$$\begin{aligned} &\tau_0Dq=\tau_3Dq=Dq,\quad Dq\tau_0=-Dq\tau_3\ ,\\ &D\tau_0=-D\tau_3,\quad \tau_0D\tau_3=-\tau_3D\tau_3,\quad D\tau_3q=0\ , \end{aligned} \tag{3.4}$$

and their conjugate relations. As a consequence of these relations, we will be able to work with a minimum number of Higgs fields and incorporate the Higgs doublet of the electroweak theory. The final Higgs potential will lead to a pattern of symmetry breaking that is closer to the desired one.

It is very useful to have an explicit expression for the exterior derivative. As we have noted earlier, this means looking for an automorphism of the algebra $\mathcal{M}_2$. Here we will make use of an outer automorphism that happens to be an inner automorphism of $\mathcal{M}_3$. To this end, we first embed the algebra of Pauli matrices $\mathcal{M}_2$ into that of Gell-Mann λ matrices $\mathcal{M}_3$. We identify τ_a's with the generators of I-spin λ_a's ($a=1,2,3$) and τ_0 with $Y+\frac{2}{3}$, where Y is the hypercharge $\lambda_8/\sqrt{3}$. Other elements of $\mathcal{M}_3$ are the so-called U and V spins given by

$$U_\pm=\tfrac{1}{2}(\lambda_6\pm i\lambda_7),\quad U_3=\tfrac{1}{2}[U_+,U_-]\,,$$
$$V_\pm=\tfrac{1}{2}(\lambda_4\pm i\lambda_5),\quad V_3=\tfrac{1}{2}[V_+,V_-]\,. \tag{3.5}$$

The inner automorphism we will be concerned with reads

$$e^{i\alpha_b U_b}\tau_a e^{-i\alpha_c U_c}=\tau_a+i[\alpha_b U_b,\tau_a]+\cdots$$
$$\equiv\tau_a+d\tau_a+\cdots\,. \tag{3.6}$$

This gives the infinitesimal variations $d\tau_a=i[\alpha_b U_b,\tau_a]$. This allows us to define the derivations, E's, as

$$E_\pm(F)=\frac{m}{\sqrt{2}}[U_\pm,F],\quad E_3(F)=\frac{m^2}{M}[U_3,F]\,, \tag{3.7}$$

where F is an element of $\mathcal{M}_2$, obeying the following commutation relations:

$$[E_+,E_-]=ME_3,\quad [E_3,E_\pm]=\pm\frac{m^2}{M}E_\pm\,. \tag{3.8}$$

m and M are two mass scales that need to be introduced into the theory. We have chosen to include them in Eq. (3.7) for later convenience [8]. The full exterior derivative is, as before, $D=d+Q$, where

$$Q=\theta_+E_-+\theta_-E_++\theta_3E_3$$
$$+M\theta_+\theta_-\frac{\partial}{\partial\theta_3}+\frac{m^2}{M}\theta_3\theta_+\frac{\partial}{\partial\theta_+}+\frac{m^2}{M}\theta_-\theta_3\frac{\partial}{\partial\theta_-}\,. \tag{3.9}$$

As before, we take θ_3 anti-Hermitian and $\theta_+^\dagger=-\theta_-$. One may easily check that the operator D, so defined, annihilates $\tau_0+\tau_3$ as a consequence of the fact that E's do so. Further, all the identities presented in Eqs. (3.3) and (3.4) follow, as they should.

Now we are ready to define the connection one-form or the generalized gauge potential. It has the following structure:

$$\omega=A+\frac{g}{\sqrt{2}}H\theta_-+\frac{g}{\sqrt{2}}H^\dagger\theta_++g\Delta\theta_3\,, \tag{3.10}$$

where

$$A=-igA_\mu dx^\mu=-ig\tfrac{1}{2}(B_\mu\tau_0+W_{\mu a}\tau_a)dx^\mu\,,$$
$$H=H_+V_++H_0U_+,\quad \Delta=\tfrac{1}{2}(\Delta_0\tau_0+\Delta_a\tau_a)\,, \tag{3.11}$$

and g is the coupling constant of the theory. We may note at this stage that B and W's are going to be the gauge bosons, while H's and Δ's are the scalars. This gauge potential should not be viewed as obtained from a one-form based on the algebra $\mathcal{M}_3$ by dropping certain fields. It may be worth emphasizing here that our model is based on the noncommuting algebra $\mathcal{M}_2$ and not $\mathcal{M}_3$. In writing ω, we used the basis involving U's and V's only for convenience. It is possible to rewrite the scalar sector entirely in terms of $D\tau$'s as

$$-\frac{M}{2}(\Delta_0-\Delta_3)\bar{q}Dq+\frac{M}{2}(\Delta_0+\Delta_3)qD\bar{q}$$
$$+\left[\frac{M}{2}(\Delta_0-\Delta_3)-mH_0\right]\tau_3D\tau_3$$
$$-M(\Delta_1+i\Delta_2)\tau_3D\bar{q}+mH_+qD\tau_3-\text{H.c.}\,, \tag{3.12}$$

multiplied by g/m^2. This is an expansion in a complete basis of differentials of $\mathcal{M}_2$. Other differentials ($Dq,D\bar{q},D\tau_3$, and $D\tau_0$) can be expressed in this basis:

$$Dq=-D(q\tau_3)=-Dq\tau_3-qD\tau_3\,,$$
$$D\tau_3=-D\tau_0=-D(\tau_3^2)=-D\tau_3\tau_3-\tau_3D\tau_3\,. \tag{3.13}$$

An expansion of the form (3.12) is not possible for the most general gauge potential based on $\mathcal{M}_3$.

Under a generalized gauge transformation, the gauge potential ω transforms as $\delta\omega=D\epsilon+\omega\epsilon-\epsilon\omega$, where the gauge parameter ϵ is a zero-form: $\epsilon=-ig(\epsilon_0\tau_0+\epsilon_a\tau_a)/2$. Writing this down explicitly, one finds that the theory has SU(2)⊗U(1) gauge invariance. W and B transform, respectively, as SU(2) and U(1) gauge bosons. After performing the following shifts

$$H_0\to H_0-\frac{m}{g},\quad \Delta_0\to\Delta_0+\frac{m^2}{2gM},\quad \Delta_3\to\Delta_3+\frac{m^2}{2gM}\,, \tag{3.14}$$

we find that the scalar sector transforms as the usual doublet Higgs field H with hypercharge one as well as a singlet and a triplet Higgs field, Δ_0 and Δ_i's respectively, with zero hypercharge. It is assumed in the following that these shifts have been performed.

The next step is to construct the field strength Ω. As before, we consider $\Omega=D\omega+\omega^2$, which transforms as $\delta\Omega=\Omega\epsilon-\epsilon\Omega$. Its components can be inferred from the expansion

$$\Omega=\tfrac{1}{2}\Omega_{\mu\nu}dx^\mu dx^\nu+\Omega_{\mu+}dx^\mu\theta_-+\Omega_{\mu-}dx^\mu\theta_++\Omega_{\mu3}dx^\mu\theta_3$$
$$+\Omega_{+-}\theta_-\theta_++\Omega_{+3}\theta_-\theta_3+\Omega_{3-}\theta_3\theta_+\,. \tag{3.15}$$

$\Omega_{\mu\nu}=-igF_{\mu\nu}$ gives the field strength associated with the gauge fields. The next three terms in the expansion yield covariant derivatives of the scalar fields,

$$\Omega_{\mu+}=\frac{g}{\sqrt{2}}\mathcal{D}_\mu H,\quad \Omega_{\mu-}=\Omega_{\mu+}^\dagger,\quad \Omega_{\mu3}=g\mathcal{D}_\mu\Delta\,, \tag{3.16}$$

where $\mathcal{D}_\mu=\partial_\mu-igA_\mu$ and $A_\mu\Delta$ stands for $[A_\mu,\Delta]$. The remaining three terms can be computed to be

$$\Omega_{+-}=\frac{g^2}{2}[H,H^\dagger]-gM\Delta-m^2\tau_0+\frac{m^2}{2}\,,$$
$$\Omega_{+3}=-\frac{g^2}{\sqrt{2}}\Delta H,\quad \Omega_{3-}=\Omega_{+3}^\dagger\,. \tag{3.17}$$

Just as in the preceding section, we next write down a Lagrangian that is of the form (2.31). The trace is now taken over 3×3 matrices. The indices i and j are summed over the full range that includes the θ directions (+, −, and 3). After some algebra, one obtains the usual covariant kinetic terms for the various fields,

$$-\tfrac{1}{2}\,\mathrm{Tr}(F_{\mu\nu}F^{\mu\nu})+\mathcal{D}_\mu H^\dagger \mathcal{D}^\mu H+\mathrm{Tr}(\mathcal{D}_\mu\Delta\mathcal{D}^\mu\Delta)\ , \tag{3.18}$$

and a Higgs potential $V(H,\Delta)$ given by

$$\begin{aligned}\frac{1}{g^2}V(H,\Delta)=&\frac{1}{4}\left[H^\dagger H-\frac{m^2}{g^2}\right]^2\\&+\frac{1}{2}\left[\frac{1}{2}H^\dagger H-\frac{M}{g}\Delta_0-\frac{m^2}{g^2}\right]^2\\&+\frac{1}{2}\left[\frac{1}{2}H^\dagger\tau_a H-\frac{M}{g}\Delta_a\right]^2\\&+\frac{1}{4}H^\dagger(\Delta_0+\Delta_a\tau_a)^2H\ .\end{aligned} \tag{3.19}$$

In the above expressions H is written as a two-component column vector with entries H_+ and H_0. The potential, being a sum of squares, can be minimized easily. It leads to a symmetry-broken vacuum given by

$$H_0=\frac{m}{g},\quad H_\pm=0,\quad \Delta_0=\Delta_3=-\frac{m^2}{2gM},\quad \Delta_{1,2}=0\ , \tag{3.20}$$

where only the electromagnetism survives symmetry breaking.

Coming to the mass spectrum of the model, one finds that the masses of Z and W bosons are, respectively, m and $m(1+m^2/2M^2)^{1/2}/\sqrt{2}$. Further, the ρ parameter turns out to be $1+m^2/2M^2$. Since experimentally ρ is very close to one, we require $M\gg m$. The Higgs sector gives two heavy scalars with masses M and $M(1+2m^2/M^2)^{1/2}$ and a light scalar with mass m. There are also three massless scalars, the would-be Goldstone bosons, that are eaten by the gauge bosons.

In this model we get certain relations among the coupling constants. For example, the two coupling constants g and g' that correspond to SU(2) and U(1), respectively, happen to be equal. This leads to a prediction for the Weinberg angle to be 45° from the relation $\tan\theta_W=g'/g=1$ or $\sin^2\theta_W=0.5$. We assume that such relations hold at the scale M that is, perhaps, given by the Planck scale. Because the couplings evolve, these relations get modified as we approach the electroweak scale. Assuming three generations of fermions, this brings down the value of $\sin^2\theta_W$ from 0.5 to 0.26 as we evolve from the scale M to the scale m. This is to be compared with the experimental value of 0.23. Similar evolution takes place for the Higgs self-coupling. Neglecting Yukawa couplings, this predicts the light Higgs mass to be about 80 GeV.

Before ending this section, we wish to point out that there is an alternate way of constructing the exterior derivative. Instead of embedding our algebra $\mathcal{M}_2$, loosely speaking SU(2), into SU(3) as we have done above, we now do the embedding into SU(2|1). SU(2|1) generators are essentially given by those of SU(3). However, now there is a grading involved; $U_\pm$ and $V_\pm$ are to be taken "fermionic," while all the others are to be taken "bosonic." This means replacing the commutators in Eq. (3.5) by anticommutators, thus redefining U_3 and V_3. This will change the "statistics" of $E_\pm$ as well; thus the first commutator in Eq. (3.8) will become an anticommutator. To have a fermionic BRST operator Q, we take $\theta_\pm$ to be bosonic, while θ_3 remains fermionic. We no longer need the last two terms, those proportional to $\partial/\partial\theta_\pm$, in the expression for Q to ensure $Q^2=0$:

$$Q=\theta_+E_-+\theta_-E_++\theta_3E_3-M\theta_+\theta_-\frac{\partial}{\partial\theta_3}\ . \tag{3.21}$$

With these modifications, we proceed exactly as before. The field strength Ω, corresponding to (3.15), turns out to be almost the same, Ω_{+-} being

$$\Omega_{+-}=\frac{g^2}{2}\{H,H^\dagger\}-gM\Delta-\frac{m^2}{2} \tag{3.22}$$

and the other components differing at most by a sign. In writing down the Lagrangian, it is natural to use the supertrace rather than the usual trace. This involves subtracting the third entry in the diagonal from the sum of the other two. However, the resulting Lagrangian does not involve any kinetic term for the Higgs doublet. We may thus consider taking the usual trace, though this is not natural from the point of view of SU(2|1) symmetry. This yields a Lagrangian that is identical to the one obtained before, with the same Higgs potential. All the results that follow are therefore the same, in particular, the Weinberg angle is 45° before the quantum corrections. Thus our results are markedly different from the SU(2|1) models of Ref. [9].

IV. GRAND-UNIFIED VERSION

Here we attempt to generalize the approach of the preceding section. This will lead us to a framework that is, though not completely realistic, closer to the bosonic sector of a grand-unified theory based on SU(5). As in the preceding section, we consider the noncommuting algebra $\mathcal{A}$ along with the set (∂_μ,E_a) acting as derivatives on $\mathcal{A}$. We define operators E_a by embedding the algebra of Hermitian $n\times n$ matrices ($\mathcal{M}_n$) into that of Hermitian $(n+1)\times(n+1)$ matrices ($\mathcal{M}_{n+1}$). For the sake of concreteness, we shall concentrate on $n=5$. The generalization to arbitrary n is straightforward.

In what follows, we find it convenient to utilize the non-Hermitian basis (corresponding to q and $\bar{q}$ of the preceding section) consisting of matrices T_a^b, where

$$(T_a^b)_{cd}=\delta_{ac}\delta_{bd},\quad a,b,c,d=1,\ldots,6\ . \tag{4.1}$$

These matrices form a basis for $\mathcal{M}_6$. They are normalized to satisfy $\mathrm{Tr}(T_a^bT_c^d)=\delta_{ad}\delta_{bc}$. The identity element in this basis is $\sum T_a^a$. If we restrict the indices to run from 1 to 5 only, these matrices can be regarded as forming a basis for $\mathcal{M}_5$ as well. This corresponds to an embedding of $\mathcal{M}_5$ into $\mathcal{M}_6$ that we referred to earlier. The commutation relations satisfied by these matrices are

$$[T_a^b,T_c^d]=\delta_c^bT_a^d-\delta_a^dT_c^b\ , \tag{4.2}$$

which gives the Lie algebra of U(6) [or U(5) when the indices run from 1 to 5 only].

We construct the derivatives acting on $\mathcal{M}_5$ as follows. We want SU(3) color and U(1) electromagnetism to survive spontaneous symmetry breaking, or, in other words, their generators to be annihilated by D. They are assumed to be embedded into the above algebra in the usual way: SU(3) being generated by traceless combinations of T_a^b's for $a,b=1,2,3$ and U(1) being $\mathrm{diag}(-1/3,-1/3,-1/3,1,0,0)$. We thus choose the automorphism group associated with the subalgebra SU(2)⊗U(1) that is generated by $T_5^6, T_6^5, (T_5^5-T_6^6)/2$ and S, where

$$S=\mathrm{diag}(1,1,1,-3,0,0)\ , \tag{4.3}$$

and define the derivatives, E_a's, in terms of these generators:

$$E_+(F)=\frac{m}{\sqrt{2}}[T_5^6,F],\quad E_-(F)=\frac{m}{\sqrt{2}}[T_6^5,F]\ ,$$
$$E_3(F)=\frac{m^2}{2M}[T_5^5-T_6^6,F],\quad E_4(F)=\Lambda[S,F]\ , \tag{4.4}$$

where F is an element of $\mathcal{M}_5$ and m, M, and Λ are three mass scales that need to be introduced into the theory. $E_\pm$ and E_3 satisfy the commutation relations (3.8), while E_4 commutes with them.

Next, it is straightforward to introduce anticommuting objects dx^μ's and θ's, and define the exterior algebra $\mathcal{A}^*$. The exterior derivative is, as before, $D=d+Q$, where Q is given by our earlier expression (3.9) except for an additional term, $\theta_4 E_4$. The connection one-form has the structure

$$\omega=A+\frac{g}{\sqrt{2}}H\theta_-+\frac{g}{\sqrt{2}}H^\dagger\theta_++g\Delta\theta_3+g\Sigma\theta_4\ , \tag{4.5}$$

with

$$A=-igA_\mu dx^\mu,\quad A_\mu=A_{\mu b}^{a}T_a^b,$$
$$H=H^aT_a^6,\quad \Delta=\Delta_b^aT_a^b,\quad \Sigma=\Sigma_b^aT_a^b\ , \tag{4.6}$$

where g is again the coupling constant of the theory. A's turn out to be the gauge bosons, while H's, Δ's, and Σ's are the scalars.

As before, under a generalized gauge transformation, the gauge potential ω transforms as $\delta\omega=D\epsilon+\omega\epsilon-\epsilon\omega$, where the gauge parameter ϵ has the decomposition $\epsilon=-ig\epsilon_b^aT_a^b$. There are inhomogeneous terms in the transformations of the scalar sector. Hence we shift the scalars as

$$H\to H-\frac{m}{g}T_5^6,\quad \Delta\to\Delta-\frac{m^2}{2gM}(T_5^5+T_6^6-1)\ ,$$
$$\Sigma\to\Sigma-\frac{\Lambda}{g}S\ . \tag{4.7}$$

We will assume in the following that these shifts have been performed. Writing down the transformations more explicitly, we find that the model has SU(5)⊗U(1) gauge invariance. H transforms in the fundamental representation "5" of SU(5) with a U(1) charge equal to one, while Δ and Σ (apart from singlets) transform in the adjoint representation "24" with zero U(1) charge.

The field strength, $\Omega=D\omega+\omega^2$, transforms under the gauge transformation as $\delta\Omega=\Omega\epsilon-\epsilon\Omega$. To find its components it is convenient to first decompose it as

$$\begin{aligned}\Omega=&\tfrac{1}{2}\Omega_{\mu\nu}dx^\mu dx^\nu+\Omega_{\mu+}dx^\mu\theta_-+\Omega_{\mu-}dx^\mu\theta_+\\&+\Omega_{\mu3}dx^\mu\theta_3+\Omega_{\mu4}dx^\mu\theta_4+\Omega_{+-}\theta_-\theta_+\\&+\Omega_{+3}\theta_-\theta_3+\Omega_{3-}\theta_3\theta_++\Omega_{+4}\theta_-\theta_4\\&+\Omega_{4-}\theta_4\theta_++\Omega_{34}\theta_3\theta_4\ ,\end{aligned} \tag{4.8}$$

where again $\Omega_{\mu\nu}=igF_{\mu\nu}$ gives the field strength associated with the gauge fields. The next four terms yield covariant derivatives of the scalar fields just as in Eq. (3.16). The remaining terms in the expansion turn out to be

$$\begin{aligned}&\Omega_{+-}=\frac{g^2}{2}[H,H^\dagger]-gM\Delta+m^2T_6^6-\frac{m^2}{2}\ ,\\&\Omega_{+3}=-\frac{g^2}{\sqrt{2}}\Delta H\ ,\\&\Omega_{3-}=\Omega_{+3}^\dagger\ ,\\&\Omega_{+4}=-\frac{g^2}{\sqrt{2}}\Sigma H,\quad \Omega_{4-}=\Omega_{+4}^\dagger,\quad \Omega_{34}=g^2[\Delta,\Sigma]\ .\end{aligned} \tag{4.9}$$

Continuing this approach, we next write down a Lagrangian using this field strength as in Eq. (2.31). The trace is now taken over 6×6 matrices. The indices i and j are summed over the full range: the space-time and the internal directions (+, −, 3, and 4). Along with the usual covariant kinetic terms,

$$\begin{aligned}&-\tfrac{1}{2}\mathrm{Tr}(F_{\mu\nu}F^{\mu\nu})+\mathrm{Tr}(\mathcal{D}_\mu H^\dagger\mathcal{D}^\mu H)\\&\quad+\mathrm{Tr}(\mathcal{D}_\mu\Delta\mathcal{D}^\mu\Delta)+\mathrm{Tr}(\mathcal{D}_\mu\Sigma\mathcal{D}^\mu\Sigma)\ ,\end{aligned} \tag{4.10}$$

where one obtains a Higgs potential $V(H,\Delta,\Sigma)$ given by

$$\begin{aligned}\frac{1}{g^2}V(H,\Delta,\Sigma)=&\frac{1}{4}\left[H^\dagger H-\frac{m^2}{g^2}\right]^2\\&+\mathrm{Tr}\left[\frac{1}{2}HH^\dagger-\frac{M}{g}\Delta-\frac{m^2}{2g^2}\right]^2\\&+H^\dagger\Delta^2H+H^\dagger\Sigma^2H-\mathrm{Tr}[\Delta,\Sigma]^2\ ,\end{aligned} \tag{4.11}$$

where H is written as a five-component column vector.

The Higgs potential, again being a sum of squares, can be minimized quite easily. The vacuum expectation values of H and Δ can be taken to be

$$H^\dagger=(0,0,0,0,m/g),\quad \Delta=\frac{m^2}{2gM}\mathrm{diag}(1,1,1,1,0)\ . \tag{4.12}$$

The expectation value of Σ is not determined uniquely, except for the fact that its fifth row and fifth column vanish. Radiative corrections to the Higgs potential will settle this issue, most probably making this expectation value zero. This corresponds to SU(5)⊗U(1) being broken down to SU(4)⊗U(1).

This model will be more realistic if Σ gets an expectation value of, say,

$$\Sigma=\kappa\frac{\Lambda}{g}\text{diag}(1,1,1,0,0) \tag{4.13}$$

(for some constant κ) breaking SU(5)⊗U(1) to SU(3)⊗SU(2)⊗U(1)⊗U(1) so that further breaking to SU(3)⊗U(1)⊗U(1) is achieved by the expectation value of H. To obtain the resulting gauge boson masses, we define various gauge fields by

$$A_\mu=\frac{1}{\sqrt{2}}\left[\mathcal{A}_\mu+\frac{1}{\sqrt{5}}C_\mu\right],$$
$$\mathcal{A}_\mu=\left[\begin{array}{c|cc} G-\frac{2}{\sqrt{30}}B & X & Y \\ \hline \bar{X} & \frac{1}{\sqrt{2}}W_3+\frac{3}{\sqrt{30}}B & W_+ \\ \bar{Y} & W_- & -\frac{1}{\sqrt{2}}W_3+\frac{3}{\sqrt{30}}B \end{array}\right]_\mu, \tag{4.14}$$

where X and Y stand for the leptoquark gauge bosons, while G collectively represents gluons. $W_\pm$, W_3, and B are the usual fields of the electroweak theory. C is an extra gauge boson that comes from the U(1) of our SU(5)⊗U(1) model. The mass terms for these fields can be evaluated to be

$$(\kappa\Lambda)^2\bar{X}^\mu X_\mu+\left[(\kappa\Lambda)^2+\frac{1}{2}m^2\left[1+\frac{m^2}{2M^2}\right]\right]\bar{Y}^\mu Y_\mu$$
$$+\frac{1}{2}m^2\left[1+\frac{m^2}{2M^2}\right]W_+^\mu W_{-\mu}$$
$$+\frac{1}{2}m^2\left[-\frac{1}{\sqrt{2}}W_3+\frac{3}{\sqrt{30}}B+\frac{1}{\sqrt{5}}C\right]_\mu^2 . \tag{4.15}$$

This shows that, in the neutral sector, there is only one massive state, "Z," given by the combination of W_3, B and C that appears above. It couples to diag(0,0,0,0,1). Of the two massless states orthogonal to Z, one should correspond to the photon. It is not clear at this stage, which this combination is. In the context of a flipped SU(5)⊗U(1) model [10], electric charge is embedded as diag(2/3,2/3,2/3,1,0). In this case, the photon corresponds to the combination

$$\left[\frac{3}{7}\right]^{1/2}\left[\frac{1}{\sqrt{2}}W_3-\frac{1}{\sqrt{30}}B+\frac{3}{\sqrt{5}}C\right]_\mu,$$

which is orthogonal to Z and hence massless. The second massless state,

$$\frac{2}{\sqrt{7}}\left[\frac{1}{\sqrt{2}}W_3+\frac{\sqrt{30}}{5}B-\frac{1}{2\sqrt{5}}C\right]_\mu,$$

couples to diag($-1/2,-1/2,-1/2$,1,0) and hence couples to the bosonic sector of the electroweak theory like the photon. An alternate possibility is to embed the electric charge as diag(0,0,0,1,0,) in which case the second massless state couples to diag(1,1,1,0,0) and hence decouples from the bosonic sector of the electroweak theory. Because we have not as yet incorporated fermions into the theory, it is not clear what the full implications of these results are.

V. SUMMARY AND CONCLUSIONS

The geometric formulation of the bosonic sector of nonabelian gauge theory has clearly several attractive features. Geometry dictates the origin of both the gauge fields and Higgs fields. In the extended algebra, we find the gauge fields and the Higgs fields unified and on the same footing. While the $\mathcal{C}$-valued coefficients of the anticommuting differentials dx^μ's describe the gauge fields, the $\mathcal{C}$-valued coefficients of the corresponding anticommuting θ's are the Higgs fields. These Higgs fields are determined by the subalgebra used to construct the derivation operator Q, which itself is dictated by which symmetries one wants unbroken at the end. Thus, in the case of SU(5)⊗U(1) studied in this paper, we chose a specific subalgebra SU(2)⊗U(1), where the surviving symmetry contains SU(3)⊗U(1) so that we can identify SU(3) as the SU(3) color and U(1) as the electric charge. The most interesting feature of the Higgs potential that emerges is that it consists of a sum of terms each of which is a perfect square. It defines, barring the trivial solution in which all the fields vanish, and hence none of the symmetries are broken, an absolutely minimized symmetry-broken vacuum given by the vanishing of each term in the potential.

It should be noted, however, that the Lagrangian we obtain is not the most general one allowed by gauge invariance and renormalizability. In fact, there are no quartic terms in the Higgs fields belonging to the adjoint representations. There are only the coupling terms implying the vanishing of a number of arbitrary parameters, which generally are present when one constructs a Higgs potential starting with a given set of fields. It should also be noted that the quartic coupling is equal to the only dimensionless coupling in the model, namely, the gauge coupling. These facts seem to suggest some underlying symmetries whose origin is not apparent in the present formulation. Whether these symmetries prevail and the restrictions on the couplings remain valid when one considers higher-order corrections to the potential is an extremely interesting and important question that needs investigation.

Likewise, the scale parameters we introduce are arbitrary. Making some plausible assumptions, we have derived physical results concerning the Weinberg angle and the masses of the surviving Higgs particles. These results appear to be reasonable. However, the model is far from a realistic one without fermions. Adding the fermion sector in an *ad hoc* way to the bosonic Lagrangian discussed in this paper is not the solution. A natural way, which is also geometric in origin, has to be found to include the fermions if this approach is to succeed in providing a realistic model. A possible way that suggests itself is the extension of this work to incorporate supersymmetry. Indeed, the form of the Higgs potential (a sum of squares) and the relations in the coupling constants at prescribed mass scales strongly suggests a supersymmetric framework. The nonrenormalization of certain interactions in such theories will enable us to give a meaningful interpretation of the results discussed in this paper.

ACKNOWLEDGMENTS

The work of B.S.B. and K.C.W. was supported in part by the DOE under Contract No. DE-FG02-85ER40231. The work of F.G. was supported in part by the DOE under Contract Nos. DE-FG02-85ER03074 and DE-FG02-85ER03075.

[1] D. B. Fairlie, J. Phys. G. **5**, L55 (1979); Phys. Lett. **82B**, 97 (1979). See also N. S. Manton, Nucl. Phys. **B158**, 141 (1979); and G. Chapline and N. S. Manton, *ibid.* **B184**, 391 (1981).

[2] A. Connes, in *The Interface of Mathematics and Particle Physics*, edited by D. Quillen, G. Segal, and S. Tsou (Oxford University Press, Oxford, 1990); Publ. Math. IHES, **62** (1985).

[3] A. Connes and J. Lott, Report No. IHES-P-90-23 (unpublished).

[4] R. Coquereaux, G. Esposito-Farese, and G. Vaillant, Nucl. Phys. **B353**, 689 (1991).

[5] M. Dubois-Violette, R. Kerner, and J. Madore, Class. Quantum Grav. **6**, 1709 (1989); J. Math. Phys. **31**, 323 (1990).

[6] B. S. Balakrishna, Feza Gürsey, and K. C. Wali, Phys. Lett. B **254**, 430 (1991).

[7] M. J. Bowick and Feza Gürsey, Phys. Lett. B **175**, 182 (1986); Nucl. Phys. **B283**, 331 (1987).

[8] There are some notational changes in this paper compared to Ref. [6]. The definition of U and V spins is made consistent with the literature. θ's are redefined and are now anti-Hermitian. The mass dimensions have been introduced at the beginning in the definition of the derivatives themselves. Fields have been scaled when they are defined so that they lead to properly normalized kinetic terms in the Lagrangian.

[9] Y. Ne'eman, Phys. Lett. **81B**, 190 (1979); Fairlie [1]; Y. Ne'eman and S. Sternberg, Proc. Natl. Acad. Sci. U.S.A. **77**, 3127 (1980). For recent work see Y. Ne'eman and S. Sternberg, Proc. Natl. Acad. Sci. U.S.A. **87**, 7875 (1990); R. Coquereaux, G. Esposito-Farese, and F. Scheck, Report No. IHES-P-90-100 (unpublished).

[10] See, for instance, D. V. Nanopoulos, in *Proceedings, Last Workshop on Grand Unification*, North Carolina University at Chapel Hill, edited by Paul H. Frampton (World Scientific, Singapore, 1989), p. 202.

MONOPOLES AND BLACKHOLES

Kameshwar C. Wali

Physics Department, Syracuse University
Syracuse, NY 13244–1130

The most powerful method of advance that can be suggested at present is to employ all the resources of pure mathematics in attempts to perfect and generalise the mathematical formalism that forms the existing basis of theoretical physics, and after each success in this direction, to try to interpret the new mathematical features in terms of physical entities.

Dirac, 1931

Magnetic monopoles and Blackholes are perfect examples of the above semi–philosophical/epistemological pronouncement of Dirac. Both concepts arose out of pure theoretical reasoning from the mathematical formalisms of classical electromagnetism, quantum mechanics, and the general theory of relativity, and have proved to have enormous physical significance. This essay, in honor of Louis Witten, whose friendship and affection over the past three decades has meant a great deal to me, is not intended to be an exhaustive review of the vast amount of theoretical and experimental work on this topic, but a partial account of some aspects, particularly the analogies between the theoretical underpinnings of these fascinating objects.

Dirac introduced the concept of a unit magnetic charge to explain the existence of a smallest unit of electric charge in nature in his classic paper[1] of 1931. He showed that arbitrary phases associated with quantum mechanical wave functions, when extended to include path dependent or nonintegrable phases, could be interpreted in terms of electromagnetic potentials and fields with possible point singularities. The point singularities could be identi with point magnetic charges or monopoles with a constraint on their strength g, given by $eg=(n/2)\hbar c$, where n is an integer. Seventeen years later, noting that the absence of isolated magnetic charges spoiled the symmetry that existed between electricity and magnetism in the field equations of electrodynamics, Dirac developed a more general theory of magnetic monopoles.[2] A monopole was now at the end of an infinite string which could be made unobservable by choosing suitable dynamical coordinates to describe its motion. The quantization condition was necessary in order to make the classical description of the interaction between a magnetic charge g and an electrical charge e consistent with quantum mechanics. Dirac interpreted this reciprocal relation as the quantization condition for electrical charge. The smallness of the observed unit of electrical charge implied the largeness of the unit magnetic charge and hence a strong binding between oppositely charged magnetic poles. It was natural then for them to occur in pairs.

Black holes, like magnetic monopoles, were in their origin the product of pure mathematical reasoning. Although the physical concept of a black hole goes back to the eighteenth century,[3] the reality of black holes became self-evident only when one had at one's disposal the exact solutions of Einstein's equations of general relativity. As an exact solution of Einstein–vacuum equation, an isolated black hole divides the three-dimensional space into an asymptotically flat internal and external regions separated by a convex two-dimensional null boundary surface called the event horizon. No signal originating in the interior region can escape by crossing the event horizon to the external region. Likewise, a time-like signal, once having crossed the boundary, could never reemerge in the outer region. Such an isolated black hole then could be conceived as a macroscopic soliton-like, space-time object.

Isolated static or stationary black holes have remarkably simple specifications. Static black holes are described by a single parameter family of spherically symmetric Schwarzschild solutions, the parameter characterizing the mass of the black hole. Israel's theorem assures that, given the mass of a black hole, the space-time is uniquely given by the corresponding Schwarzschild solution. Stationary black holes are necessarily axisymmetric and rotating. They are uniquely described by a two-parameter family of Kerr solutions of Einstein–vacuum equations. The two parameters characterize two physical attributes, namely, the mass and the angular momentum of the black holes. And as in the case of the static black hole, a theorem due to Robinson assures that, given the two physical parameters, the space-time is uniquely specified by the corresponding Kerr solution.

The coupled Einstein–Maxwell equations provide further simple generalizations of the Schwarzschild and Kerr black hole solutions. They are the Reissner–Nordstrom [RN] and the Kerr–Newman [KN] solutions. A two parameter family of RN solutions describes a static charged black hole with mass M and charge ($Q<M$), while a three parameter KN family of solution describes a rotating, charged black hole; the three parameters are the mass, the charge and the angular momentum per unit mass.

While such pure black hole solutions of Einstein or Einstein–Maxwell vacuum solutions are of immense importance to cosmology, particularly to the big bang origin of the universe, their practical application to astrophysics is also of great importance. Singularity theorems of Hawking and Penrose imply that a wide class of initial conditions lead unavoidably to the formation of black holes. Applied to gravitational collapse of stars, a black hole emerges as the ultimate fate of a massive star with mass greater than five or six solar masses. The existence of black holes in the astronomical universe, theoretically at least, is imperative.

One may make a similar case for the existence of pure electrodynamic magnetic monopoles suggested by Dirac from purely theoretic point of view. However, noting the lack of experimental success in finding them, Dirac, on the occasion of the conference on Magnetic Monopoles in Quantum Field Theory, wrote[4] in 1981,

I am inclined now to believe that monopoles do not exist. So many years have gone by without any encouragement from the experimental side. It will be interesting to see if your conference can produce any new angle of attack on the problem.

A new angle of attack had begun in the midseventies, with the discovery of 't Hooft–Polyakov monopoles within the framework of spontaneously broken nonabelian gauge theories. These static, finite energy, soliton-like solutions are generic to a spontaneously broken nonabelian gauge theory that ultimately has a surviving $U(1)_{EM}$. Consequently, when one considers phase transitions in the early universe brought about by spontaneous symmetry breaking of the grand unified phase of the early universe, one expects a copious production of (GUT) monopoles. The GUT monopoles are expected to be stable with mass in the range of 16 GeV. Because of such heavy mass, the GUT monopoles cannot be produced in the laboratory, but they should survive as relics of the GUT era. If monopoles exist in any substantial numbers, they would contribute to the mass density of the universe and would have significant effects in cosmology. These and related considerations have led to exciting developments in recent years. As an example, the GUT monopole production rate and their resulting density were among the motivations for invoking the idea of an inflationary universe.

From these introductory remarks, let us now turn our attention to the underlying mathematical structures. As quantum particles described by combined gauge and Higgs field configurations, nonabelian monopoles have a great deal of similarity with black holes in general relativity. In the words of 't Hooft,[5] "*A black hole can be looked upon as a particle-like solution of the equations of General Relativity.*" This is reflected in the fact that in many instances gravitational fields in general relativity and nonabelian gauge fields are intimately related. Louis Witten, for instance, demonstrated[6] that any stationary axisymmetric gravitational field yields a self-dual Yang-Mills field in an SU(2) nonabelian gauge theory. In other words, self-dual solutions of SU(2) pure gauge theory and stationary axisymmetric gravitaitonal field satisfy, with proper correspondence, the same Ernst equations. Same is true in the case of spontaneously broken SU(2) gauge theory with magnetic monopoles. Search for multiply charged monopoles leads to a set of equations with axial symmetry and these equations again, at least formally, reduce to Ernst equations. The various techniques to generate solutions to Ernst equations then are used to find the desired solutions in the case of monopoles.[7]

Let me cite another highly instructive example. Under the combined forces of Coulomb repulsion and Newtonian attraction, one can contemplate static configurations of any number of point masses at arbitrary locations. This happens when the charges (pure electric or pure magnetic), all of one sign, are equal in magnitude with the corresponding masses in appropriate units so that the Newtonian attraction is balanced by the Coulomb repulsion. It is remarkable that such a static configuration is allowed in the framework of general relativity as a solution of the Einstein–Maxwell equations. This was first shown by Weyl[8] assuming axial symmetry and later generalized by Majumdar and Papapetrou.[9] Hartle and Hawking[10] interpreted this solution as representing an assemblage of extreme Reissner–Nordstrom black holes. It is even more remarkable, as shown by Chandrasekhar,[11] that there exists a unified description of the situation in the following sense: the Majumdar–Papapetrou solution specifies a metric for a static Einstein–Maxwell space–time in terms of a scalar potential function that obeys a three–dimensional Laplace's equation. In Chandrasekhar's formulation of the problem, the scalar potential describes a metric that corresponds to a stationary axisymmetric vacuum space–time. This means that one could pass freely from the Majumdar–Papapetrou metric appropriate as a solution of the Einstein–Maxwell equations to a metric appropriate as a solution of the stationary Einstein vacuum equations, and conversely. Describing the one–to–one correspondence as a manifestation of a natural and harmonious blending of Einstein's relativity and Maxwell's electrodynamics in a single unified structure, Chandrasekhar makes an analogy with the description of a mythical hall in the Indian epic The Maha–bharata, and says,[12]

> When wandering through the great hall of general relativity, that one had believed to be Einstein's hall, is in fact a corridor leading to Maxwell's hall; and when one is certain that one is examining the gems in Einstein's hall, one has inadvertently slipped into Einstein's hall. "So matchless is the beauty" that Einstein has "imparted" to it.

Now if we consider the Yang–Mills–Higgs system and nonabelian monopoles there is an analogous static configuration with vanishing interaction between the monopoles. This happens in the case of 't Hooft–Polyakov monopoles in the Prasad–Sommerfield limit, and has a very simply physical interpretation. Vanishing of the interaction is due to the balancing of Coulomb–type repulsive forces among the magnetic charges and Yukawa–type attractive forces due to the Higgs fields.[13] A natural question that arises is what happens when such monopoles are coupled to gravity. Is there an exact solution of the Majumdar–Papapetrou–type in the case of Einstein–Yang–Mills–Higgs system?

Motivated by this equation, Balakrishna and the author[14] have considered a type of coupling of the monopoles to gravity that admits a static solution which may be viewed as a generalization of the Majumdar–Papapetrou solution to a nonabelian theory or as an extension of the Bogomol'ny equations to include gravity. Our starting point is the action

$$S=\int d^4x\,\sqrt{-g}\,\{-\frac{1}{16\pi v^2}\,R\varphi^2-\tfrac{1}{4}\,(F^a_{\mu\nu})^2+\tfrac{1}{2}\,(D_\mu\varphi^a)^2-\tfrac{\lambda}{4}\,(\varphi^2-v^2)^2\},\quad (1)$$

where R is the Ricci scalar, $F^a_{\mu\nu}$ is the field strength associated with the gauge field A^a_μ and φ^a is the Higgs field in the adjoint representation. $D_\mu\varphi^a$ is the gauge covariant derivative of φ^a and φ^2 is a short form for $\varphi^a\varphi^a$. When $\varphi^2=v^2$, the first term gives the usual Einstein Lagrangian. The rest of the terms represent the standard Yang–Mills Higgs system. The reason for the somewhat unconventional coupling of the matter field in the first term is that it leads to monopole solutions with an ansatz that resembles the Bogomol'ny equations. This happens when the dimensionless parameter $4\pi Gv^2$ becomes unity, i.e. in our units v=1. A metric analgous to that of Majumdar and Papapetrou, namely

$$ds^2=g_{\mu\nu}dx^\mu dx^\nu = V^2dt^2-\frac{1}{V^2}\,dx^idx^i,\qquad (2)$$

where V, independent of t, is a function of only the x^i's. In the background of this metric, the energy functional $\mathcal{E}$ has the form

$$\mathcal{E}=\int d^3x\,(\frac{1}{4V^2}R\varphi^2+\tfrac{1}{4}\,V^2\,F^a_{ij}F^a_{ij}+\tfrac{1}{2}\,D_i\varphi^aD_i\varphi^a),\qquad (3)$$

ignoring the λ–term. With the definition of 'electric' and magnetic fields given by

$$E_i^a = \frac{1}{V} D_i\chi^a, \quad B_i^a = \frac{V}{2} \epsilon_{ijk} F_{jk}^a, \tag{4}$$

where $x^a = V\varphi^a$, we find that the energy functional can be put in the form

$$\mathcal{E} = \frac{1}{2}\int d^3x \, [(E_i^a)^2 + (B_i^a)^2] = \frac{1}{2}\int d^3x \, (E_i^a \mp B_i^a)(E_i^a \mp B_i^a) \pm \int_a d^3x E_i^a B_i^a$$

$$\geq \pm \int d^3x \, E_i^a B_i^a, \tag{5}$$

leading to the Bogomol'ny equations $E_i^a = \pm B_i^a$ as solutions for the absolute minimum of $\mathcal{E}$ One can verify that all the equations of the coupled Einstein, Yang–Mills, and Higgs system that follow from (1) are satisfied, provided the metric function V satisfies the equation

$$D_i\left(\frac{1}{V^2} D_i\chi^a\right) = 0. \tag{6}$$

In the case of the abelian U(1) gauge group, eq. (6) reduces to $\partial^2(1/V) = 0$, and the resulting metric reduces to the Majumdar–Papapetrou solution for the Einstein–Maxwell theory. Next considering SU(2) gauge theory as in the case of 't Hooft–Polyakov monopole and specializing to the case of spherical symmetry with the ansatz,

$$\chi^a = \mp \hat{r}^a(r), \qquad A_i^a = \frac{1}{gr} \epsilon^{aij} \hat{r}^j(1 - W(r)), \tag{7}$$

where $\hat{r}$ is the unit vector along the radial direction, we find that Bogomol'ny equations lead to a system of nonlinear equations

$$r^4 F'' = e^{-2F}, \tag{8}$$

$$g/V = \frac{1}{r} + F', \tag{9}$$

where F is defined by $W = e^{-F}/r$. Eq. (8) is a special case of the so-called Lane–Emden equations that play a central role in the internal constitution of stars.[15] To our knowledge, it has not been solved analytically. However, there exists a class of solutions with the asymptotic behavior

$$F = CR+D, \quad r \to \infty \tag{10}$$

$$F = -\ln r + A\sqrt{r}\,\sin(\tfrac{\sqrt{7}}{2}\ln r + \delta), \quad r \to 0 \tag{11}$$

where $C > 0$, D, A, and δ are some constants. With $C = g$ and $M = 4\pi/g$, we get the expected asymptotic behavior for V for a monopole of mass M coinciding with its charge. The behavior of V as $r \to 0$, however is singular. V goes to zero as $\sqrt{r} \to 0$ in an oscillatory way and the area of any sphere surrounding the origin tends to zero as $r \to 0$, vanishing periodically in the process. In contrast, in the case of Majumdar–Papapetrou solution, the area tends to a constant as one approaches the event horizon of the black hole. Since $1/V$ behaves smoothly in our case, there are no event horizons to dress the singularity and hence it appears that there is a naked singularity at the origin. However, these features are specific to $4\pi Gv^2$ being unity and to get a better picture one needs to move away from this critical value. In the case of conventional Einstein, Yang–Mills, and Higgs system, some recent numerical studies do indeed reveal a rich variety of solutions.[16] If the vacuum expectation value of the Higgs field is greater than a critical value of the order of Planck mass, there exist nonsingular monopoles solutions. If it is less than the critical value, the solutions are singular at the origin, but the singularity is hidden within a horizon. Since matter fields reside outside the horizon, one may interpret such solutions as black holes within monopoles.

To conclude, the interplay between gravitational forces and gauge interactions coupled with Higgs mechanism provides a fertile ground for further insights into extended classical configurations that constitute monopoles. One may ask, for instance, whether there are, as in the case of massive stars, instabilities induced due to general relativistic effects, and whether massive monopoles become black holes by some sort of quantum radiation mechanism. If so, the monopole problem in grand unified cosmology will need a different approach. Although a clear and definitive answer needs a full–fledged quantum theory of gravity, including other interactions of nature, the study of coupled Einstein–Yang–Mills–Higgs system has provided a good theoretical laboratory for a deeper understanding of monopoles and black holes.

Acknowledgements

This work was supported by the U.S. Department of Energy under contract no. DE–FG02–85ER40231. I would like to thank Professor Carl Rosenzweig for a careful reading of the manuscript.

References and Footnotes

1. Dirac, P.A.M., *Proc. Roy. Soc.* **A133** (1931), 60.
2. Dirac, P.A.M., *Phys. Rev.* **74** (1948), 817.
3. John Michell of England is supposed to have written a paper in 1783 pointing out that a star that was sufficiently massive and compact would have such a strong gravitational field that light would not escape. Laplace independently a few years later made the same suggestion.
4. Dirac, P.A.M. in *Monopoles in Quantum Field Theory,* Proceedings of the Monopole Meeting, Trieste, Italy, December 1981 (World Scientific, p.iii).
5. 't Hooft, G., *Acta Physica Polonica,* **B19** (1988), 187.
6. Witten, L., *Phys. Rev.* **D19** (1979),718.
7. Forgács, P., Horvath, Z., Palla, L., in *Monopoles in Quantum Field Theory,* (World Scientific), p. 21.
8. Weyl, H., *Ann. Physik* **54** (1917) 117.
9. Majumdar, S.D., *Phys. Rev.* **72** (1947), 390; Papapetrous, A., *Proc. Roy. Irish Acad.*, Ser. **A51** (1947) 191.
10. Hartle, J.B. and Hawking, S.W., *Comm. Math. Phys.* **26** (1972), 87.
11. Chandrasekhar, S. "How one may explore the physical content of the general relativity," in *Proc. of the Gibbs Symposium,* Yale University, May 15–17, 1989.
12. Ibid., p. 248.
13. O'Raifeartaigh, L., Park, S.Y., and Wali, K.C., *Phys. Rev.* **D20,** (1979) 1941.
14. Balakrishna, S., and Wali, K.C., "*On Non–Abelian Monopoles Coupled to Gravity,*", SU–4240–503.

15. For a detailed discussion of this and related equations see, '*An Introduction to the Stellar Structure*,' S. Chandrasekhar, Dover Publications (1957), Chapter IV.

16. For recent work, see for instance, Lee, K, Nair, V.P., and Weinberg, E.J., *Phys. Rev.* **D45** (1992), 2751.

Towards a unified treatment of Yang-Mills and Higgs fields: A supersymmetric extension

B. S. Balakrishna
Department of Physics, Syracuse University, Syracuse, New York 13244-1130

Feza Gürsey
Department of Physics, Yale University, New Haven, Connecticut 06511

Nguyen Ai Viet* and Kameshwar C. Wali
Department of Physics, Syracuse University, Syracuse, New York 13244-11330
(Received 24 January 1992)

An approach to a unified treatment of Yang-Mills and Higgs fields was proposed previously starting from a noncommutative algebra constructed from the algebra of smooth functions on space-time and the algebra of Hermitian matrices. The formalism was further applied to the bosonic sector of the standard electroweak theory. Here, in order to incorporate fermions into this framework in a natural way, we extend the formalism to include supersymmetry. We find, however, that the scenario changes drastically in the presence of supersymmetry and the conclusions are substantially different. When applied to the standard model, this yields the supersymmetrized bosonic sector of the theory, but the inclusion of quarks and leptons requires additional superfields as in the conventional supersymmetric models.

PACS number(s): 12.10.Dm, 12.15.Cc

I. INTRODUCTION

In a previous paper [1], hereafter referred to as I, we have presented an approach to a unified treatment of Yang-Mills and Higgs fields based on a noncommutative algebra $\mathcal{A}$ of the form $\mathcal{C}\otimes\mathcal{M}$, where $\mathcal{C}$ is the algebra of smooth functions on space-time and $\mathcal{M}$ is the algebra of $n\times n$ Hermitian matrices. The one-forms of an exterior algebra of differential forms over $\mathcal{A}$ permitted us to introduce Yang-Mills gauge fields and Higgs fields on a similar footing and construct a Lagrangian from the two-forms of that algebra. In the resulting geometrical description, a Higgs potential that had spontaneous symmetry breaking arose naturally. We specialized the formalism to the bosonic sector of the standard electroweak theory and a grand unified theory based on SU(5)⊗U(1).

Several noteworthy features emerged from the formalism. The underlying geometry dictated the origin of both the gauge fields and Higgs fields. Gauge coupling and a few scale parameters were the only parameters in the theory, resulting in some relations between the generally independent couplings [2]. However, without fermions in the formalism, the models we considered were not realistic enough to draw meaningful conclusions. In the present paper, we extend the formalism to include fermions by invoking global supersymmetry. We find that this does not amount to rewriting the Lagrangian of I in terms of superfields in superspace. Instead, supersymmetry forces us to exclude some terms from the Lagrangian of I, in particular, those responsible for the Higgs potential. Further, it is found to be difficult in this approach to include any superpotential. However, as is known, supersymmetrizing the remaining terms also gives an effective potential for the Higgs that comes from the so-called D term.

We begin, as in I, with a noncommutative algebra $\mathcal{A}$ of the form $\mathcal{C}\otimes\mathcal{M}$, $\mathcal{C}$ now comprised of superfields. We construct an exterior algebra of differential forms in superspace over $\mathcal{A}$ and, as before, introduce in the one-form a vector and a scalar superfield on similar footing. From the resulting terms in the two-forms, we construct a Lagrangian which is supersymmetric. In the next section, we review briefly the basic considerations in the supersymmetric formulation that we need and proceed to consider, in Sec. III, a more realistic model by embedding SU(2) of the electroweak theory in SU(3) [3]. Section IV is devoted to a discussion of the results and prospects for a realistic model.

II. SUPERSYMMETRIC FRAMEWORK

In the standard $N=1$ global supersymmetric approach, based on differential forms in superspace [4], we have the superspace coordinates z^M, $M=(\mu,\alpha,\dot\alpha)$ given by

$$z^M=(x^\mu,\theta^\alpha,\bar\theta^{\dot\alpha}) \ , \tag{2.1}$$

and the coordinate differentials dz^M

$$dz^M=(dx^\mu,d\theta^\alpha,d\bar\theta^{\dot\alpha}) \tag{2.2}$$

with the exterior derivative d given by

$$d=dx^M\partial_M \ , \tag{2.3}$$

where $\partial_M=\partial/\partial z^M$. In terms of the supersymmetric co-

*Permanent address: Centre for Theoretical Physics, Nghiado, Tuliem, Hanoi, Vietnam.

variant derivatives, D_α and $\bar{D}_{\dot\alpha}$,

$$D_\mu=\partial_\mu\ ,\quad D_\alpha=\frac{\partial}{\partial\theta^\alpha}-i\sigma^\mu_{\alpha\dot\beta}\bar\theta^{\dot\beta}\partial_\mu\ ,$$
$$\bar D_{\dot\alpha}=\frac{\partial}{\partial\bar\theta^{\dot\alpha}}-i\theta^\beta\sigma^\mu_{\beta\dot\alpha}\partial_\mu\ , \tag{2.4}$$

and the superspace differentials, $e^M=(e^\mu,e^\alpha,\bar e^{\dot\alpha})$,

$$e^\mu=dx^\mu+i\,d\theta\,\sigma^\mu\bar\theta-i\theta\sigma^\mu d\bar\theta\ ,\quad e^\alpha=d\theta^\alpha\ ,\quad \bar e^{\dot\alpha}=d\bar\theta^{\dot\alpha}\ , \tag{2.5}$$

we have

$$d=e^M D_M=e^\mu\partial_\mu+e^\alpha D_\alpha+\bar e^{\dot\alpha}\bar D_{\dot\alpha}\ . \tag{2.6}$$

The exterior derivatives of the differential basis e^M do not vanish; instead they satisfy

$$de^\mu=-2i\,d\theta\,\sigma^\mu d\bar\theta=-2ie\sigma^\mu\bar e\ ,\quad de^\alpha=d\bar e^{\dot\alpha}=0\ . \tag{2.7}$$

Now, we extend the above superspace coordinates and their exterior derivatives to include internal symmetry generators as coordinates and their exterior derivatives given by a Becchi-Rouet-Stora-Tyutin (BRST)-type operator Q as in I. Denoting generically the generators by λ_0 and λ_a $(a=1,\ldots,n)$, we note that the superfields in the model will be functions of x, θ, $\bar\theta$, and λ. Zero-forms have the general structure $F(x,\lambda,\theta,\bar\theta)$, where

$$F(x,\lambda,\theta,\bar\theta)=f_0(x,\theta,\bar\theta)\lambda^0+f_a(x,\theta,\bar\theta)\lambda^a\ . \tag{2.8}$$

The exterior derivative has the form

$$D=d+Q\ , \tag{2.9}$$

where d is defined in Eq. (2.6) and Q is given by

$$Q=e^aE_a-i\frac{m}{2}f^c_{ab}e^ae^b\frac{\partial}{\partial e^c}\ . \tag{2.10}$$

The e^a's are a set of anticommuting "differentials" on the internal symmetry space analogous to e^M, and $E_a(F)$ are defined to be

$$E_a(F)=\frac{m}{2}[\lambda_a,F]\ . \tag{2.11}$$

The differentials e^M form a basis for one-forms in superspace. Combining them with the differentials e^a, we can write the supergauge connection one-form as

$$\omega=A+\Phi-\frac{m}{2}e^a\lambda_a\ , \tag{2.12}$$

where

$$A=e^MA_M=e^\mu A_\mu+e^\alpha A_\alpha+\bar e^{\dot\alpha}\bar A_{\dot\alpha}\ ,$$
$$\Phi=e^a\Phi_a\ , \tag{2.13}$$

and A_μ, A_α, $\bar A_{\dot\alpha}$, Φ_a are all functions of x, λ, θ, and $\bar\theta$, each of them having an expansion of the form (2.8). In the absence of supersymmetry, the one-form ω will consist of the gauge fields A_μ and the scalar fields Φ, all functions of only x's and λ's. As in I, the constant shifts introduced in (2.12) give rise to correct kinematic terms for Φ and cancel the apparently non-gauge-invariant terms in the two-forms Ω constructed from (2.12) using the definition

$$\Omega=D\omega+\omega^2\ . \tag{2.14}$$

It turns out that Ω can be written out in the form

$$\Omega=\Omega_S+\Omega_{QS}+\Omega_Q\ , \tag{2.15}$$

where

$$\Omega_S=e^MD_M(A)+A^2\ ,$$
$$\Omega_{QS}=e^M\mathcal{D}_M(\Phi)=e^Me^a\mathcal{D}_M(\Phi_a)\ , \tag{2.16}$$

with the gauge-covariant derivative $\mathcal{D}_M$ given by

$$\mathcal{D}_M=D_M+A_M\ , \tag{2.17}$$

so that

$$\mathcal{D}_M(\Phi_a)=D_M(\Phi_a)+[A_M,\Phi_a]\ . \tag{2.18}$$

Finally

$$\Omega_Q=-i\frac{m}{2}f^c_{ab}e^ae^b\Phi_c+\Phi^2\ , \tag{2.19}$$

where

$$\Phi^2=\tfrac{1}{2}e^ae^b[\Phi_a,\Phi_b]\ . \tag{2.20}$$

Thus Ω_S arises from the standard non-Abelian gauge field content in ω. Ω_{QS} originates from the cross term between A and Φ, and Ω_Q from Φ. Ω_{QS} and Ω_Q are, therefore, due to our extension of the usual non-Abelian supersymmetry to include the internal symmetry coordinates. More explicitly, the various components of Ω are given by

$$\Omega_S=-\tfrac{1}{2}e^Ne^M\Omega_{MN}$$
$$=-2ie\sigma^\mu\bar eA_\mu-e^Ne^M(D_MA_N+A_MA_N)\ ,$$
$$\Omega_{\mu\nu}=\partial_\mu A_\nu-\partial_\nu A_\mu+[A_\mu,A_\nu]\ ,$$
$$\Omega_{\alpha\beta}=D_\alpha A_\beta+D_\beta A_\alpha+\{A_\alpha,A_\beta\}\ ,$$
$$\Omega_{\mu\alpha}=\partial_\mu A_\alpha-D_\alpha A_\mu+[A_\mu,A_\alpha]\ , \tag{2.21}$$
$$\Omega_{\alpha\dot\beta}=D_\alpha\bar A_{\dot\beta}+\bar D_{\dot\beta}A_\alpha+\{A_\alpha,\bar A_{\dot\beta}\}+2i\sigma^\mu_{\alpha\dot\beta}A_\mu\ ,$$
$$\Omega_{\dot\alpha\dot\beta}=\Omega^\dagger_{\alpha\beta}\ ,\quad \Omega_{\mu\dot\alpha}=\Omega^\dagger_{\mu\alpha}\ ,\quad \Omega_{\dot\alpha\beta}=\Omega^\dagger_{\alpha\dot\beta}\ ,$$
$$\Omega_{QS}=e^Me^a\Omega_{aM}\ ,\quad \Omega_{aM}=\mathcal{D}_M(\Phi_a)\ ,$$
$$\Omega_Q=-\tfrac{1}{2}e^be^a\Omega_{ab}\ ,\quad \Omega_{ab}=-imf^c_{ab}\Phi_c+[\Phi_a,\Phi_b]\ .$$

In the conventional supersymmetric approach, it is recognized that there are too many component fields as described in (2.12). A set of self-consistent (gauge-covariant, Lorentz-invariant, and supersymmetric) constraints which do not restrict the x dependence of the component fields are thus imposed. These are

$$\Omega_{\alpha\beta}=\Omega_{\dot\alpha\dot\beta}=\Omega_{\alpha\dot\beta}=\Omega_{\dot\alpha\beta}=0\ . \tag{2.22}$$

These constraints are solved with the use of prepotentials. It is easily seen that with

$$\mathcal{D}_\alpha=e^{2V}D_\alpha e^{-2V}\ , \tag{2.23}$$

where V is an arbitrary complex superfield,

$$\Omega_{\alpha\beta}=\{\mathcal{D}_\alpha,\mathcal{D}_\beta\}=e^{2V}\{D_\alpha,D_\beta\}e^{-2V} \tag{2.24}$$

vanishes since $\{D_\alpha,D_\beta\}=0$. This solution leads to

$$\mathcal{D}_\alpha=e^{2V}D_\alpha e^{-2V}=e^{2V}e^{-2V}D_\alpha+e^{2V}D_\alpha(e^{-2V}) , \tag{2.25}$$

where the parentheses imply that D_α acts only on e^{-2V}. Because $\mathcal{D}_\alpha=D_\alpha+A_\alpha$, we thus have

$$A_\alpha=e^{2V}D_\alpha(e^{-2V}) . \tag{2.26}$$

Similarly, we can solve the constraint $\Omega_{\dot\alpha\dot\beta}=0$ and obtain $\bar A_{\dot\alpha}=e^{2U}\bar D_{\dot\alpha}(e^{-2U})$, where U is another complex superfield. In the so-called maximal approach, one allows for complex gauge transformations, that is, the gauge parameters are allowed to be complex. Then, U and V become independent superfields. We can use this freedom to fix the gauge with

$$\bar A_{\dot\alpha}=0 , \tag{2.27}$$

while A_α is nonvanishing. This gauge is known as the Wess-Zumino gauge. In this gauge $\Omega_{\dot\alpha\dot\beta}=0$ is trivially satisfied. Further, requiring the remaining constraints $\Omega_{\alpha\dot\beta}=\Omega_{\dot\alpha\beta}=0$ gives us an expression for A_μ [see Eq. (2.21)]:

$$A_\mu=\frac{i}{4}\bar\sigma_\mu^{\dot\beta\alpha}\bar D_{\dot\beta}A_\alpha=\frac{i}{4}\bar D\bar\sigma_\mu A . \tag{2.28}$$

A_μ is usually anti-Hermitian. The above result for A_μ is not so because we have allowed for complex gauge transformations. However, with V Hermitian (that will be assumed henceforth), A_μ is anti-Hermitian up to a gauge transformation by e^{2V}:

$$-A_\mu^\dagger=e^{-2V}A_\mu e^{2V}+e^{-2V}\partial_\mu(e^{2V}) . \tag{2.29}$$

Even after fixing the gauge with Eq. (2.27), we have residual gauge invariance that takes the form

$$e^{2V}\to e^{2\Lambda}e^{2V}e^{-2\bar\Lambda} , \tag{2.30}$$

where Λ is a chiral superfield (that is, $\bar D_{\dot\alpha}\Lambda=0$). It is easily seen, from Eqs. (2.26), (2.27), and (2.28), that under this

$$A_M\to e^{2\Lambda}A_M e^{-2\Lambda}+e^{2\Lambda}D_M(e^{-2\Lambda}) \tag{2.31}$$

for $M=\mu$, α, and $\dot\alpha$, as expected for a gauge transformation. This gauge symmetry can be further fixed, in a gauge that is also known as the Wess-Zumino gauge, so that the prepotential V takes the form

$$V=\theta\sigma^\mu\bar\theta V_\mu+\bar\theta^2\theta\lambda+\theta^2\bar\theta\bar\lambda+\tfrac{1}{2}\theta^2\bar\theta^2 D , \tag{2.32}$$

satisfying $V^3=0$ (λ or D should not be confused with the λ_a's or D_M's introduced earlier). This lets us rewrite the expression for A_α, Eq. (2.26), as

$$A_\alpha=e^{2V}D_\alpha(e^{-2V})=-2(D_\alpha V+[V,D_\alpha V]) . \tag{2.33}$$

Further, with the result for A_μ given by Eq. (2.28), we can express $\Omega_{\mu\dot\alpha}$ as

$$\Omega_{\mu\dot\alpha}=-\bar D_{\dot\alpha}A_\mu=-iW^\beta\sigma_{\mu\beta\dot\alpha} , \tag{2.34}$$

where we have introduced an object W_α defined by

$$W_\alpha=\tfrac{1}{8}\bar D^2 A_\alpha=-\tfrac{1}{4}\bar D^2(D_\alpha V+[V,D_\alpha V]) . \tag{2.35}$$

Using Eq. (2.32) for V, we can express W_α as

$$W_\alpha=\lambda_\alpha+\theta_\alpha D+\frac{i}{2}(\sigma^\mu\bar\sigma^\nu\theta)_\alpha V_{\nu\mu}+i(\sigma^\mu\mathcal{D}_\mu\bar\lambda)_\alpha , \tag{2.36}$$

where the fields are functions of $x-i\theta\sigma\bar\theta$. Here, unlike in the previous expressions, $\mathcal{D}_\mu=\partial_\mu-iV_\mu$. $V_{\mu\nu}$ is the field strength associated with V_μ:

$$V_{\mu\nu}=\partial_\mu V_\nu-\partial_\nu V_\mu-i[V_\mu,V_\nu] . \tag{2.37}$$

We now turn our attention to the components of Ω arising from Φ in ω, namely $\Omega_{a\alpha}$. Noting that $\mathcal{D}_\alpha\Omega_{\mu\beta}=\bar{\mathcal{D}}_{\dot\alpha}\Omega_{\mu\dot\beta}=0$ are automatically satisfied, we would like to impose

$$\mathcal{D}_\alpha\Omega_{a\beta}=\bar{\mathcal{D}}_{\dot\alpha}\Omega_{a\dot\beta}=0 \tag{2.38}$$

as additional constraints to supplement (2.22). In the Wess-Zumino gauge, $\bar A_{\dot\alpha}=0$, these new constraints imply

$$D_\alpha D_\beta(e^{-2V}\Phi_a e^{2V})=\bar D_{\dot\alpha}\bar D_{\dot\beta}(\Phi_a)=0 . \tag{2.39}$$

Further, in analogy with A_μ, we require Φ_a to be Hermitian only up to a gauge transformation by e^{2V}, that is,

$$\Phi_a^\dagger=e^{-2V}\Phi_a e^{2V} , \tag{2.40}$$

or alternately $\tilde\Phi_a=e^{-V}\Phi_a e^{V}$ is Hermitian. This, together with Eq. (2.39), leads to a solution for the superfield $\tilde\Phi$,

$$\tilde\Phi=\phi+\theta\psi+\bar\theta\bar\psi+\theta\sigma^\mu\bar\theta u_\mu+\theta^2\bar\theta\left[\frac{i}{2}\bar\sigma^\mu\mathcal{D}_\mu\psi+[\bar\lambda,\phi]\right]+\bar\theta^2\theta\left[\frac{i}{2}\sigma^\mu\mathcal{D}_\mu\bar\psi-[\lambda,\phi]\right]+\tfrac{1}{4}\theta^2\bar\theta^2(\mathcal{D}^2\phi-i[\partial^\mu V_\mu,\phi]] , \tag{2.41}$$

where we have suppressed the a subscript and the x,λ dependence of the fields. ϕ and u_μ are Hermitian, and u_μ obeys the constraint

$$\mathcal{D}^\mu u_\mu=\partial^\mu u_\mu-i[V^\mu,u_\mu]=-i[\bar\lambda,\bar\psi]-i[\lambda,\psi]+i[D,\phi] . \tag{2.42}$$

The scalar superfield thus constructed has the field content of a chiral supermultiplet. The four components of ψ (together with $\bar\psi$) are to be matched with one ϕ and three components of u_μ [one less than four due to the constraint (2.42)]. The conventional approach is to represent a chiral supermultiplet by a chiral superfield, that is, by a superfield annihilated by $\bar D_{\dot\alpha}$. Here, however, it is represented by a superfield that is not chiral.

Since $\Omega_{\mu\dot\alpha}$ and $\Omega_{a\dot\alpha}$ are annihilated by $\bar D_{\dot\beta}$, they are chiral superfields. This permits us to write down the following Lagrangian as an integral over only two θ's:

$$L=\frac{1}{8g^2}\int d^2\theta\,\mathrm{Tr}(\Omega_{\mu\dot\alpha}\Omega^{\mu\dot\alpha}+\Omega_{a\dot\alpha}\Omega^{a\dot\alpha})+\mathrm{H.c.}=L_{\mathrm{YM}}+L_{\mathrm{H}}\ , \tag{2.43}$$

where L_{YM} gives the Lagrangian for the Yang-Mills sector, while L_{H} gives that for the Higgs sector. In terms of W_α introduced earlier, we can rewrite L_{YM} as

$$\begin{aligned}L_{\mathrm{Ym}}&=\frac{1}{8g^2}\int d^2\theta\,\mathrm{Tr}(\Omega_{\mu\dot\alpha}\Omega^{\mu\dot\alpha})+\mathrm{H.c.}\\&=\frac{1}{2g^2}\int d^2\theta\,\mathrm{Tr}(W^2)+\mathrm{H.c.}\\&=2\,\mathrm{Tr}\{-\tfrac14 V_{\mu\nu}V^{\mu\nu}+i\bar\lambda\sigma^\mu\mathcal{D}_\mu\lambda+\tfrac12 D^2\}\ ,\end{aligned} \tag{2.44}$$

up to total derivatives, where in the last step we have scaled the fields by g. This gives the usual supersymmetric Yang-Mills Lagrangian. For the Higgs sector, after scaling Φ to $2g\Phi$, we get

$$\begin{aligned}L_{\mathrm{H}}&=\frac{1}{8g^2}\int d^2\theta\,\mathrm{Tr}(\Omega_{a\dot\alpha}\Omega^{a\dot\alpha})+\mathrm{H.c.}\\&=-2\int d^4\theta\,\mathrm{Tr}(\Phi^2)=-2\int d^4\theta\,\mathrm{Tr}(\bar\Phi^2)\\&=2\,\mathrm{Tr}\{\tfrac12\mathcal{D}^\mu\phi\mathcal{D}_\mu\phi+i\bar\psi\bar\sigma^\mu\mathcal{D}_\mu\psi+g[\lambda,\psi]\phi-g[\bar\lambda,\bar\psi]\phi-\tfrac12 u^\mu u_\mu\}\ ,\end{aligned} \tag{2.45}$$

where summation over a is assumed. Adding the u_μ constraint through a Lagrange multiplier ϕ', that is, adding

$$-2\,\mathrm{Tr}\{\phi'(\mathcal{D}^\mu u_\mu-ig[\bar\psi,\bar\lambda]-ig[\psi,\lambda]-ig[D,\phi])\}\ , \tag{2.46}$$

and solving for u_μ, gives us $u_\mu=\mathcal{D}_\mu\phi'$. Substituting this back into the action, we get

$$\begin{aligned}L_{\mathrm{H}}=2\,\mathrm{Tr}\{&(\mathcal{D}^\mu\phi)^\dagger\mathcal{D}_\mu\phi+i\bar\psi\bar\sigma^\mu\mathcal{D}_\mu\psi+\sqrt2 g\phi^\dagger[\lambda,\psi]\\&-\sqrt2 g\phi[\bar\lambda,\bar\psi]-g\phi^\dagger[D,\phi]\}\ ,\end{aligned} \tag{2.47}$$

where ϕ now stands for $(\phi+i\phi')/\sqrt2$. This also agrees with the usual supersymmetric Lagrangian for a chiral superfield in the adjoint representation.

The potential comes from the D term alone. Note that, unlike in the usual supersymmetric theories, we have no F terms. This has to do with the fact that the supermultiplet (associated with Φ) does not arise from a chiral superfield. Solving for D from Eq. (2.43) [sum of (2.4) and (2.47)] gives $D=g[\phi^a,\phi_a^\dagger]$, hence a quartic potential

$$\mathrm{Tr}D^2=g^2\,\mathrm{Tr}[\phi^a,\phi_a^\dagger]^2\ . \tag{2.48}$$

Thus, in the final Lagrangian, apart from the covariant kinetic terms for the various fields, one has the following interaction terms:

$$2\,\mathrm{Tr}\left\{\sqrt2 g\phi^\dagger[\lambda,\psi]-\sqrt2 g\phi[\bar\lambda,\bar\psi]-\frac{g^2}{2}\,\mathrm{Tr}[\phi^a,\phi_a^\dagger]^2\right\}\ . \tag{2.49}$$

The reader may have noticed that we have ignored the components $\Omega_{\mu\nu}$ and Ω_{ab} in (2.43) in constructing the Lagrangian. There are two reasons. First of all, they are not chiral. Hence, they cannot be simply added to our Lagrangian which has been obtained by integrating over only two θ's. One may, however, include them in an integral over all four θ's. This is not satisfactory due to the second reason, which has to do with dimension counting. With the usual assignment of dimensions (-1 for e^μ, $-\frac12$ for e^α and $\bar e^{\dot\alpha}$, -1 for e^a), we find that such an integral carries dimension 6. This will give rise to generally nonacceptable nonrenormalizable or higher derivative terms.

III. EXTENSION TO THE ELECTROWEAK THEORY

Now, as in I, we proceed to reformulate the model discussed in the preceding section to represent a more realistic situation in which the Higgs fields belong to a fundamental rather than an adjoint representation. With the standard model in mind, we consider the matrix algebra of the Pauli matrices denoted by τ's. Hence, the zero-forms are of the form

$$F(x,\tau,\theta,\bar\theta)=f_0(x,\theta,\bar\theta)\tau_0+f_i(x,\theta,\bar\theta)\tau_i \tag{3.1}$$

where $i=1,2,3$. The derivations E_a ar derived from an outer automorphism of the algebra. To obtain explicit expressions for them, we embed the τ's into an SU(3) algebra. Then the E_a's ($a=+,-,3$) are given by

$$E_a(F)=\frac{m}{2}[\lambda_a,F]\ ,\quad \lambda_\pm=\sqrt2 U_\pm\ ,\quad \lambda_3=\frac{2m}{M}U_3\ , \tag{3.2}$$

where the U's are the usual U-spin generators. With this, we obtain the "structure constants"

$$f^3_{+-}=-i\frac{M}{m}\ ,\quad f^+_{3+}=-i\frac{m}{M}\ ,\quad f^-_{3-}=i\frac{m}{M}\ . \tag{3.3}$$

The one-form is, as before, given by

$$\omega=A+\Phi-\frac{m}{2}e^a\lambda_a\ ,\quad \Phi=e^a\Phi_a\ , \tag{3.4}$$

but now a takes the values $+$, $-$, and 3, and

$$\begin{aligned}\Phi_\pm&=\frac{1}{\sqrt2}H_\pm^+V_\pm+\frac{1}{\sqrt2}H_\pm^0U_\pm\ ,\\ \Phi_3&=\tfrac12(\Delta_0\tau_0+\Delta_i\tau_i)-\frac{m^2}{2M}\ ,\end{aligned} \tag{3.5}$$

where the V's are the usual V-spin generators. The above Δ_0 differs from that of Ref. [1] by $2m^2/M$. Because we work in the maximal approach with complex gauge transformations, there are two Higgs doublets H_+ and H_- and further Δ's are complex. They are all superfields, and hence carry their superpartners. Given this Φ, we can construct the Hermitian object $\tilde{\Phi}$ as before [see Eq. (2.41)]. The relevant fields ϕ and ψ have the following decomposition:

$$\phi=e^+\phi_+ +e^-\phi_- +e^3\phi_3\ , \quad \psi=e^+\psi_+ +e^-\psi_- +e^3\psi_3\ ,$$
$$\phi_\pm=\frac{1}{\sqrt{2}}h^+_\pm V_\pm+\frac{1}{\sqrt{2}}h^0_\pm U_\pm\ ,$$
$$\psi_\pm=\frac{1}{\sqrt{2}}\chi^+_\pm V_\pm+\frac{1}{\sqrt{2}}\chi^0_\pm U_\pm\ , \tag{3.6}$$
$$\phi_3=\delta-\frac{m^2}{2M}\ , \quad \delta=\tfrac{1}{2}(\delta_0\tau_0+\delta_i\tau_i)\ ,$$
$$\psi_3=\eta=\tfrac{1}{2}(\eta_0\tau_0+\eta_i\tau_i)\ .$$

Because we have assumed ϕ' to be already absorbed into ϕ, all the component fields are complex. These are all ordinary fields, that is, they are functions of x only. With these definitions, now it is a matter of algebra to derive the final form of the Lagrangian from Eq. (2.47). Combining the fields $h_\pm$ and $\chi_\pm$ into doublets,

$$h_+=\begin{bmatrix} h^+_+ \\ h^0_+ \end{bmatrix}\ , \quad h_-=(h^0_-,h^-_-)\ ,$$
$$\chi_+=\begin{bmatrix} \chi^+_+ \\ \chi^0_+ \end{bmatrix}\ , \quad \chi_-=(\chi^0_-,\chi^-_-)\ , \tag{3.7}$$

we can write the Higgs part of the Lagrangian as

$$L_H=(\mathcal{D}^\mu h_+)^\dagger\mathcal{D}_\mu h_+ +\mathcal{D}^\mu h_-(\mathcal{D}_\mu h_-)^\dagger+2\,\mathrm{Tr}\{(\mathcal{D}^\mu\delta)^\dagger\mathcal{D}_\mu\delta)\}+i\bar{\chi}_+\bar{\sigma}^\mu\mathcal{D}_\mu\chi_+ +i\chi_-\sigma^\mu\mathcal{D}_\mu\bar{\chi}_- +2i\,\mathrm{Tr}(\bar{\eta}\bar{\sigma}^\mu\mathcal{D}_\mu\eta)$$
$$+\sqrt{2}gh^\dagger_+\lambda\chi_+ -\sqrt{2}g\chi_-\lambda h^\dagger_- +\sqrt{2}g\bar{\chi}_+\bar{\lambda}h_+ -\sqrt{2}gh_-\bar{\lambda}\bar{\chi}_- +2\sqrt{2}g\,\mathrm{Tr}\{\delta^\dagger[\lambda,\eta]\}-2\sqrt{2}g\,\mathrm{Tr}\{\delta[\bar{\lambda},\bar{\eta}]\}$$
$$-gh^\dagger_+Dh_+ +gh_-Dh^\dagger_- -2\,\mathrm{Tr}g(\delta^\dagger[D,\delta])\ . \tag{3.8}$$

The Yang-Mills part is given by our previous expression, Eq. (2.44), where now the fields are 2×2 matrices with an expansion of the form (3.1). Note that this implies, as in I, that the two gauge couplings of the electroweak theory g and g' are equal.

To find the potential, we first solve for D as before:

$$\frac{1}{g}D=\tfrac{1}{2}h_+h^\dagger_+ -\tfrac{1}{2}h^\dagger_-h_- +[\delta,\delta^\dagger]\ . \tag{3.9}$$

With this expression for D we can easily write down the potential V as

$$\frac{1}{g^2}V=\frac{1}{g^2}\mathrm{Tr}D^2=\tfrac{1}{4}(h^\dagger_+h_+)^2+\tfrac{1}{4}(h_-h^\dagger_-)^2+\mathrm{Tr}[\delta,\delta^\dagger]^2$$
$$-\tfrac{1}{2}(h^\dagger_+h^\dagger_-)(h_-h_+)+h^\dagger_+[\delta,\delta^\dagger]h_+$$
$$-h_-[\delta,\delta^\dagger]h^\dagger_-\ . \tag{3.10}$$

The minimum of this potential corresponds to $D=0$. This has a solution

$$h_+=e^{i\gamma}h^\dagger_-\ , \quad \delta_i=e^{i\kappa}\delta^\dagger_i\ , \tag{3.11}$$

where γ and κ are two arbitrary phases. This vacuum solution has zero energy, and hence does not break supersymmetry. Further, unfortunately, there are many ground states that do not say anything definite about the spontaneous breaking of the gauge symmetry. This is a typical feature of some of the supersymmetric models, and the present approach does not shed any new light on this problem.

IV. DISCUSSIONS AND CONCLUSIONS

Encouraged by the results in I, we have extended the noncommutative algebraic geometry approach to include fermions within a global supersymmetric framework. A vector and suitably defined scalar superfields provide the starting point for a one-form in superspace. The two-form that emerges has the standard supersymmetric Yang-Mills component Ω_S and two additional components Ω_{QS} and Ω_Q arising from the scalar field. The treatment of the Yang-Mills component proceeds in the standard way by imposing certain constraints that respect supersymmetry and gauge invariance on the spinor components of Ω_S. Generalization of such constraints on $\Omega_{\alpha\alpha}$ leads us to choose the original scalar field different from the conventional superfield used in supersymmetric models. By allowing complex gauge transformations and using the so-called maximal approach, we construct this field explicitly and the corresponding terms in the components of the two-forms. Dimensional considerations dictate the choice of certain terms in the two-form for the construction of a supersymmetric Lagrangian. In the component form the Lagrangian contains Yang-Mills gauge fields accompanied by spin-$\frac{1}{2}$ fermions. The kinetic and the interaction terms have the required standard transformation properties. A quartic potential in terms of the scalar component of Φ emerges, but it does not display spontaneous symmetry breakdown as in the pure bosonic treatment of the problem. Supersymmetry forbids the terms like $\Omega_{ab}\Omega^{ab}$ in the Lagrangian. Apart from dimensional considerations, such terms which would have given rise to a spontaneously broken scalar potential through the scalar components contain nonrenormalizable and higher derivative terms.

Although the approach and the models presented in this paper provide a tight framework that permits one to include fermions, gauge fields, and scalar fields on a similar footing, it is far from furnishing a realistic model that can include the known quarks, leptons, and Higgs fields. Since supersymmetry dictates that the bosonic and fermionic partners have the same quantum numbers, we cannot, for instance, identify the scalar particles and the accompanying fermions as quarks. As is well known, if we identify them with leptons, we cannot identify the scalar particles as Higgs particles which develop a vacuum expectation value since that will violate the lepton number. Further, the Higgs partners cannot have parity-violating interactions like leptons, since we are forced to include two Higgs doublets that form together a real representation. As in the so-called minimal supersymmetric standard model [5], one is thus compelled to postulate a superfield for every set of known particles—a gauge, a quark, a lepton, and a Higgs superfield resulting in a plethora of new supersymmetric partners of the known particles. In addition, there remains the problem of breaking the supersymmetry. All these problems are present in the minimal version of the supersymmetric standard model with global supersymmetry. One usually invokes supergravity to address some of these problems. It is not clear what supergravity has got to say about these problems within the framework of noncommutative geometry.

ACKNOWLEDGMENTS

The work of B.S.B. was supported in part by DOE under Contract No. DE-FG02-85ER40231. The work of F.G. was supported in part by DOE under Contract Nos. DE-FG02-85ER03074 and 03075.

[1] B. S. Balakrishna, Feza Gürsey, and Kameshwar C. Wali, Phys. Rev. D **44**, 3313 (1991), referred to as I; see also, Phys. Lett. B **254**, 430 (1991).

[2] For instance, the Higgs quartic coupling is equal to the gauge coupling.

[3] For details, see [1].

[4] J. Wess and J. Bagger, *Supersymmetry and Supergravity* (Princeton University Press, Princeton, NJ, 1983); M. F. Sohnius, Phys. Rep. **128**, Nos. 2 & 3 (1985). We differ slightly in our conventions from that of Wess and Bagger. Our space-time metric has the signature $(+,-,-,-)$. We lower and raise the spinor indices according to $\psi_\alpha=\epsilon_{\alpha\beta}\psi^\beta$ and $\psi^\alpha=-\epsilon^{\alpha\beta}\psi_\beta$, where $\epsilon^{\alpha\beta}=\epsilon_{\alpha\beta}$ and $\epsilon_{12}=-\epsilon_{21}=1$. This is also true for the dotted indices. The σ matrices are given by $\sigma^\mu=(1,\sigma^1,\sigma^2,\sigma^3)$ and $\bar\sigma^\mu=(1,-\sigma^1,-\sigma^2,-\sigma^3)$ where σ^i's are the Pauli matrices.

[5] See, for instance, R. N. Mohapatra, in *Unification and Supersymmetry* (Springer-Verlag, Berlin, 1986), p. 224.

MAGNETIC MONOPOLES IN CURVED SPACE-TIME

Kameshwar C. Wali

Department of Physics, Syracuse University,
Syracuse, NY 13244-1130

I would like to report some work in progress in collaboration with Nguyen Ai Viet. We are especially happy to present this account here in Budapest, since a great deal of pioneering work on nonabelian monopoles and their gravitational interactions has been done here by various people that includes among others, Forgacs, Horvath, Horvathy, and Palla.

Recently there has been a flurry of papers dealing with the study of Yang-Mills-Higgs system in curved space-time in an effort to include gravitational interactions in extended configurations that constitute nonabelian monopoles. We know that, in flat space-time, these quantum-like particles described by classical finite energy configurations have complex structure. Although when isolated, they have asymptotically a magnetic field of a magnetic monopole, the Coulomb interactions of two or more monopoles are altered due to the combined interactions of Yang-Mills and Higgs fields. In the well-known Prasad-Sommerfeld limit, there results a **No Interaction Theorem** [1], which has a simple physical interpretation: the repulsive Coulomb forces due to Yang-Mills fields between like magnetic charges are balanced by the attractive forces due to Higgs fields. What happens then if we include **gravity**?

At the outset, one may think that the gravitational fields will have no significant effects since they are so weak. That this is not the case has been demonstrated by Bartnik and McKinnon [2]. Their new and unexpected (numerical) solutions in the case of Einstein-Yang-Mills system do depend on the coupling between Einstein and Yang-

Mills fields independent of the strength of the gravitational fields. More recent work of several authors incorporating Higgs fields into the system has revealed solutions with new and interesting features displaying the possibility of nonabelian interactions inside the black holes and black holes inside the monopoles [3].

The work I am going to describe depends on an unconventional coupling of the Higgs field to gravity in the Einstein-Yang-Mills-Higgs system. This has the consequence of leading to solutions of the entire system of equations on the solution of a single well-known classical nonlinear equation. We think that this reduction followed by an analytic solution of the single equation (if it can be found), will provide a more precise understanding of the complex interplay between the various interactions.

Our starting point is the action

$$S = \int d^4x\sqrt{-g}\,\{\mathcal{L}_E(x) + \mathcal{L}_m(x)\}\,, \tag{1}$$

where

$$\mathcal{L}_E(x) = -\frac{R}{16\pi G}\frac{\phi^2}{v^2}, \tag{2}$$

$$\mathcal{L}_m = -\frac{1}{4}(F^a_{\mu\nu})^2 + \frac{1}{2}(D_\mu\phi^a)^2 - \frac{\lambda}{4}(\phi^2 - v^2)^2, \tag{3}$$

and R is the Ricci scalar, $g = det \mid g_{\mu\nu} \mid$, $F^a_{\mu\nu}$ are the field strengths, A^a_μ are the gauge fields, and $D_\mu = \partial_\mu + eA_\mu$ is the covariant derivative.

Note that the Lagrangian term in the Einstein-Hilbert action is modified by the factor ϕ^2/v^2, making it the unconventional coupling of the Higgs field to gravity. When $\phi^2 = v^2$, one obtains the usual Einstein-Yang-Mills Lagrangian. However, the equations of motion derived by the standard variational procedure are different from the usual ones even when one goes to the limit $\phi^2 = v^2$. Our motivation for this choice of the Lagrangian is mostly heuristic at the present, although one knows that such a type of coupling does arise in Kaluza-Klein type theories. Also in recent developments of string theories, such unconventional couplings do arise.

We seek static solutions to equations of motion that follow from (1) in the gauge $A_0^a = 0$, so that we have only magnetic fields that are nonvanishing. Recall that in flat space-time, in the Prasad-Sommerfeld limit, the energy functional $\mathcal{E}$ is

$$\begin{aligned}\mathcal{E} &= \frac{1}{2}\int d^3x\left[(E_i^a)^2 + (B_i^a)^2\right] \\ &= \frac{1}{2}\int d^3x\left[(E_i^a \mp B_i^a)(E_i^a \mp B_i^a)\right] \pm \int d^3x E_i^a B_i^a \\ &\geq \pm \int d^3x E_i^a B_i^a.\end{aligned} \tag{4}$$

Thus, the energy functional has the absolute minimum if and only if

$$E_i^a = \pm B_i^a, \tag{5}$$

which are the Bogomolnyi conditions. Explicit solutions in closed form are then obtainable as shown by Prasad and Sommerfield [4].

We seek an extension of the above type of considerations when gravity is included by going over to a curved space-time background. In a previous paper [5], motivated by the Majumdar-Papapetrou ansatz [6] in the case of Einstein-Maxwell system, Balakrishna and myself postulated a metric given by

$$ds^2 = A^2(x)dt^2 - \frac{1}{A^2(x)}dx^i dx^i. \tag{6}$$

The energy functional $\mathcal{E}$ is then found to be

$$\mathcal{E} = \int d^3x\left(\frac{1}{4A^2}R\phi^2 + \frac{1}{4}A^2 F_{ij}^a F_{ij}^a + \frac{1}{2}D_i\phi^a D_i\phi^a\right), \tag{7}$$

where we have ignored the λ-term and chosen the units so that $4\pi G v^2 = 1$. If we now define

$$E_i^a = \frac{1}{A}D_i\mathcal{X}^a, \quad B_i^a = \frac{A}{2}\epsilon_{ijk}F_{jk}^a, \quad \mathcal{X}^a = A\phi^a, \tag{8}$$

we find that $\mathcal{E}$ has exactly the same form as in the case of flat space-time. The Bogomolnyi conditions for the minimum of $\mathcal{E}$ are

$$F_{ij}^a = \pm\frac{1}{A^2}\epsilon_{ijk}D_k\mathcal{X}^a, \tag{9}$$

just as in flat space-time, except for the appearance of the metric function A^{-2} on the right hand side. One finds that the Euler-Langrange equations that follow from the variation of the action in eq. 1 are all satisfied provided

$$\phi^2 = v^2 = 1 \quad \text{or} \quad \chi^2 = A^2. \tag{10}$$

Specializing to the $SU(2)$ gauge group with 't Hooft-Polyakov ansatz,

$$\chi^a = \mp A(r)\hat{r}^a, \quad A_i^a(r) = \frac{1}{er}\epsilon_{aij}\hat{r}^j(1 - w(r)), \tag{11}$$

one is lead to a nonlinear equation

$$r^4 F''(r) = e^{-2F}, \tag{12}$$

where prime denotes differentiation with respect to r and $w(r) = \frac{1}{r}e^{-F}$. Also,

$$\frac{e}{A(r)} = \frac{1}{r} + F'(r). \tag{13}$$

The metric function $A(r)$ is given in terms of the solution of the above equation. It is interesting to note that eq. 12 is a special case of the class of equations known as Lane-Emden equations which arise in the study of polytropes or stellar masses in equilibrium under the influence of their own gravity. Equation 12 arises in the case of isothermal polytropes and is an object of extensive analysis [7]. No analytic solution is known. However, a class of solutions with the following behavior

$$F(r) \to Cr + D \quad \text{as} \quad r \to \infty,$$
$$F(r) \to -\ell nr + A\sqrt{r}\sin\left(\frac{\sqrt{7}}{2}\ell nr + \delta\right) \quad \text{as} \quad r \to 0, \tag{14}$$

where $c > 0, D, A$, and δ are constants is proved to exist and hence as $r \to \infty$, from (13) and (14), it follows that

$$A(r) \to 1 - \frac{M}{4\pi r}, \tag{15}$$

if $C = e$ and $M = 4\pi/e$. This implies, in the units we have chosen, that we have a magnetic monopole solution with the mass of the monopole being equal to its charge.

However, from (14), we observe that as $r \to 0$, $A(r)$ goes to zero as $\sqrt{r}$ in an oscillatory way. The area of any sphere surrounding the origin, $4\pi r^2/A^2$, vanishes periodically and tends to zero as r goes to zero. The event horizon has zero surface area and hence the origin is close to being a naked singularity. The Ricci scalar diverges whenever the area vanishes and therefore, to an outside observer, these divergences will appear collectively as a naked singularity. These properties may be specific to the limit $4\pi Gv^2 = 1$ or to the specific ansatz regarding the metric. Static, spherical symmetry alone permits a more general metric given by the equation

$$ds^2 = e^{2A(r)}dt^2 - e^{2B(r)}dr^2 - d\Omega, \tag{16}$$

where

$$d\Omega = r^2(d\theta^2 + \sin^2\theta d\vartheta^2).$$

A and B are time independent functions of r only. By allowing ourselves the above more general form, Nguyen Ai Viet and I seek Bogomolnyi type solutions. It is straightforward to calculate the Ricci tensor components, the Ricci scalar, the energy momentum tensor components and write the Einstein's, Yang-Mills, and Higgs equations. For my present purposes, it is sufficient to consider the energy functional $\mathcal{E}$,

$$\mathcal{E} = \int d^3x\sqrt{-g}\left[\frac{R\phi^2}{4} + \frac{1}{4}F^a_{ij}F^{aij} - \frac{1}{2}D_i\phi^a D^i\phi^a + \frac{\lambda}{4}(\phi^2 - v^2)^2\right] \tag{17}$$

and discuss how we define an "Electric Field" so that $\mathcal{E}$ assumes the Bogomolnyi form as in the previous case. We follow the procedure proposed by Comptet, Forgacs, and Horvathy [8] by postulating

$$E^a_i = (-g)^{1/4}(-g^{ii})^{1/2}(D_i + V_i)\phi^a, \tag{18}$$

where V_i is an arbitrary function of the coordinates. We define

$$B_i^a = (-g)^{1/4}(g^{jj}g^{kk})^{1/2}\frac{1}{2}\epsilon_{ijk}F_{jk}^a. \tag{19}$$

If we now require that

a) $E_i^a = \pm B_i^a$,

b) we satisfy Yang-Mills equations,

we obtain a unique V_i, which is given by

$$V_i = \partial_i A, \tag{20}$$

where as we recall $\exp(2A) = g_{00}$. In spherical coordinates, the Bogomolnyi conditions imply

$$\begin{aligned} h'(r) + A'(r)h(r) &= \frac{e^B(1 - W^2(r))}{evr^2}, \\ W'(r) &= -evh(r)W(r)e^B, \end{aligned} \tag{21}$$

where

$$\chi^a = vh(r)\hat{r}^a, \quad A_i^a(r) = \frac{1}{er}\epsilon_{aij}\hat{r}^j(1 - W(r)). \tag{22}$$

At the same time we have to satisfy the Higgs field equation, which now reduces to

$$\frac{1}{\sqrt{-\bar{g}}}\partial^\alpha\left(\sqrt{-\bar{g}}\partial_\alpha A\right) = \frac{R}{2} + \lambda(\phi^2 - v^2), \quad \bar{g} = \det \mid g_{ij} \mid \tag{23}$$

and the static Einstein field equations.

After some algebraic manipulations, the equations we have to satisfy take the form ($H = h^2$)

$$\begin{aligned} H'' + \left(\frac{3}{2}A' - B' + \frac{2}{r}\right)H' &= \frac{\lambda v^2}{4}e^{2B}(H^2 - 1) \\ H'\left(\frac{1}{2}A' - \frac{1}{r}\right) &= \frac{\lambda v^2}{2}e^{2B}(H - 1)^2 \\ A' + \frac{1}{r} &= \alpha\frac{e^B}{r}H, \quad \alpha = \text{IntegrationConstant} \\ \left(A' + \frac{1}{r}\right)^2 + 2\frac{H'}{H}\left(A' + \frac{1}{r}\right) &= \frac{e^{2B}}{r} - \frac{\lambda v^2}{2}e^{2B}\frac{(H - 1)^2}{H} \end{aligned} \tag{24}$$

The above four equations and the Bogomolnyi conditions complete the set of equations whose solutions we seek with the asymptotic conditions

$$A(r), B(r) \to 0, \quad H(r) \to 1, \quad \text{and} \quad W(r) \to 0 \quad \text{as} \quad r \to \infty. \tag{25}$$

In the Prasad Sommerfield limit, we see from eq. (24)

$$H'\left(\frac{1}{2}A' - \frac{1}{r}\right) = 0 \tag{26}$$

which implies a) $H' = 0$ or b) $A' = 2/r$.

The alternative (b) leads to divergent metric coefficients $e^{2A} = g_{00}$ and hence it is unacceptable. However, $H = 1$ which implies $f^2 = v^2$ does have consistent solutions. With $H = 1$, the set of equations (24) reduce to

$$\begin{aligned} A' + \frac{1}{r}(1 \mp e^B) &= 0, \\ A' = -\frac{1}{r}(1 \mp e^B) &= \frac{e^B}{r}\frac{1 - W^2}{ev}, \\ W' &= -eve^B W. \end{aligned} \tag{27}$$

We note that $W = 0$, provides an exact solution with

$$e^B = \frac{evr}{evr - 1}, \quad A' = -\frac{1}{r}(1 \mp e^B) \tag{28}$$

This is, however, the Abelian monopole solution first discussed by Bais and Russell, and also Cho and Freund [9]. We seek a nonabelian solution with $W \neq 0$. From (27), we find

$$\frac{W'(r)}{W(r)} = -\frac{e^2 v^2 r}{evr + W^2 - 1}. \tag{29}$$

With a change of variables $x = erv$, and $y = (x + W^2 - 1)^- 1$, we can transform (29) into the form

$$y' = 2x(1 - x)y^3 + (2x - 1)y^2, \tag{30}$$

which is known as the Abel's equation in the mathematical literature. No analytic solution is known to exist. Going back to eq. (29) with $f = W^2$ it takes the form

$$f'(x) = -\frac{2xf(x)}{x + f(x) - 1}. \tag{31}$$

We study its solutions numerically. In figure 1, we plot a typical solution for W^2, e^{2A}, e^{2B} on a logarithmic scale. Above a critical value for $W(r)$ at $r = 0$, $W(r)$ has the desired characteristic, namely, it is nonvanishing in the region surrounding the origin and rapidly (exponentially) goes to zero as $r \to \infty$. This signals *monopole solutions as in the case* of flat space-time. The metric functions e^A and e^B are determined from W^2 up to an integration constant. Different values of this constant give different metric functions. In a typical example given in Fig. 1, we see that e^A has a minimum and e^B a corresponding maximum and both approach unity as r becomes large. These are the features we expect from a black hole type solution. The minimum in g_{00} can be made arbitrarily close to zero by choosing an appropriate value for the ingegration constant. However, analytic arguments show that it cannot be exactly zero [10]. This raises the puzzling question; whether the numerical minimum of the metric coefficient g_{00} represents a real horizon. Recent numerical solutions of others need to be examined from this point of view. Clearly a great deal of further work is necessary in order to settle intriguing questions that have arisen from our study. For $W(r) <$ the critical value at $r = 0$, $W(r)$ does not have the exponential fall of and does not vanish as $r \to \infty$. Instead, it has the behavior of the dotted curve shown in figure 1. The physical interpretation of such a $W(r)$ also needs investigation.

To conclude, although the gravitational interactions are weak, they do have strong consequences on finite energy Yang-Mills-Higgs configurations. The interplay between gravitational forces and gauge interactions coupled with Higgs interactions provides a fertile ground for further insights into the structure of nonabelian monopoles. One may speculate, for instance, on the question whether there are, as in the case of massive stars, instabilities induced to general relativistic effects, and whether massive monopoles become black holes by some sort of quantum radiation mechanism. If so, it is obvious that it will have profound consequences on our idea of grand unified cosmology and the so-called monopole problem in the early universe.

Acknowledgements

This work was supported in part by the U.S. Department of Energy under contract number DE-FG02-85ER40231.

References

[1] L. O'Raifeartaigh, S.Y. Park, and K.C. Wali, Phys. Rev. **D20**, 1941 (1979).

[2] R. Bartnik and J. McKinnon, Phys. Rev. Letters **61**, 141 (1988). See also P. van Nieuwenhuizen, D. Wilkinson, and M.J. Perry, Phys. Rev. **D13**, 778 (1976).

[3] M.E. Ortiz, Phys. Rev. **D45**, 2586 (1992); K. Lee, V.P. Nair, and E.J. Weinberg, ibid. **D45**, 2751 (1992); P. Breitenlohner, P. Forgács, and D. Maison, Nucl. Phys. **B393**, 357 (1992).

[4] M.K. Prasad and C.M. Sommerfield, Phys. Rev. Lett. **35**, 760 (1975).

[5] B.S. Balakrishna and K.C. Wali, Phys. Rev. **D46**, R5228 (1992).

[6] S.D. Majumdar, Phys. Rev. **72**, 390 (1947); A. Papapetrou, Proc. R. Irish Acad. Ser. **A51**, 191 (1947).

[7] S. Chandrasekhar, An Introduction to the study of stellar structure (Dover, New York, 1957), Chap. IV.

[8] A. Comptet, P. Forgács, P.A. Horváthy, Phys. Rev. **D30**, 468 (1984).

[9] F.S. Bais and R.J. Russell, Phys. Rev. **D11**, 2692 (1975); Y.M. Cho and P.G.O. Freund, ibid., **D12**, 1588 (1975).

[10] N. Ai Viet and K.C. Wali, Magnetic monopoles in Einstein-Yang-Mills-Higgs ystems, manuscript in preparation, SU preprint SU-4240-552.

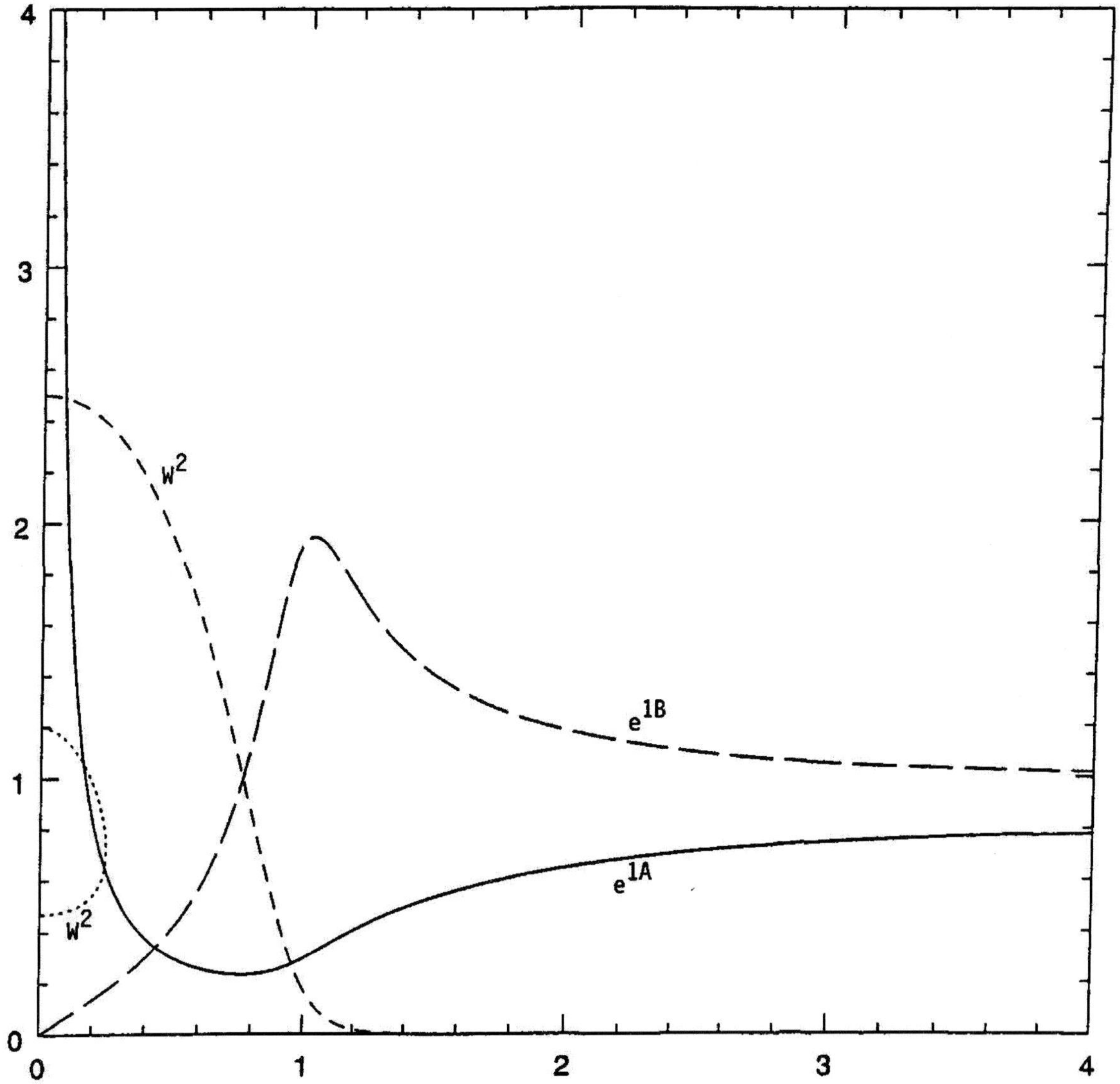

Figure 1

NON-ABELIAN MONOPOLES IN CURVED SPACE-TIME

Kameshwar C. Wali

Department of Physics, Syracuse University,
Syracuse, NY 13244-1130

In 1919, Einstein wrote a paper entitled "Do Gravitational Fields Play an Essential Role in the Structure of the Elementary Particles?", and attempted to construct a model for the electron consisting of gravitational and electromagnetic fields alone. He considered his attempt unsuccessful since he found that his model did not describe the electron uniquely. Any spherically symmetric configuration of the fields he was considering could be an equilibrium configuration. However, Einstein's question is still meaningful. For if, as we believe today, the universe in its very early phase, was governed by a grand unified theory which included all interactions, gravitational fields could be trapped inside the elementary constituents that emerged when the universe went through phase transitions. Therefore, the study of the interplay between gravitational and other interactions is important, and not surprisingly, in recent years there has been a flurry of papers dealing with this subject in the context of non-abelian monopoles in Yang-Mills-Higgs systems.

In flat space-time, non-abelian monopoles are quantum-like particles described by classical finite energy configurations that have complex structure. Although when isolated, they have asymptotically a magnetic field of a magnetic monopole, the Coulomb interactions of two or more monopoles are altered due to the combined interactions of Yang-Mills and Higgs fields. In the well-known Prasad-Sommerfeld limit, there results a **No Interaction Theorem**,[1] which has a simple physical interpretation: the repulsive Coulomb forces due to Yang-Mills fields between like magnetic charges are balanced by the attractive forces due to Higgs fields. What happens then if we include **gravity**?

At the outset, one may think that the gravitational fields will have no significant effects since they are so weak. That this is not the case has been demonstrated by Bartnik and McKinnon.[2] Their new and unexpected (numerical) solutions in the case of Einstein-Yang-Mills system do depend on the coupling between Einstein and Yang-Mills fields independent of the strength of the gravitational fields. More recent work of several authors incorporating Higgs fields into the system has revealed solutions with new and interesting features displaying the possibility of nonabelian interactions inside the black holes and black holes inside the monopoles,[3]

The work I am going to describe done in collaboration with Nguyen Ai Viet[4] depends on an unconventional coupling of the Higgs field to gravity in the Einstein-Yang-Mills-Higgs system. This has the consequence of leading to solutions of the entire system of equations on the solution of a single well-known classical nonlinear equation. We think that this reduction followed by an analytic solution of the single equation (if it can be found), will provide a more precise understanding of the complex interplay between the various interactions.

Our starting point is the action

$$S = \int d^4x \sqrt{-g}\, \{\mathcal{L}_E(x) + \mathcal{L}_m(x)\}\,, \tag{1}$$

where

$$\mathcal{L}_E(x) = -\frac{R}{16\pi G}\frac{\phi^2}{v^2}, \tag{2}$$

$$\mathcal{L}_m = -\frac{1}{4}(F^a_{\mu\nu})^2 + \frac{1}{2}(D_\mu\phi^a)^2 - \frac{\lambda}{4}(\phi^2 - v^2)^2, \tag{3}$$

and R is the Ricci scalar, $g = det \mid g_{\mu\nu} \mid$, $F^a_{\mu\nu}$ are the field strengths, A^a_μ are the gauge fields, and $D_\mu = \partial_\mu + eA_\mu$ is the covariant derivative.

Note that the Lagrangian term in the Einstein-Hilbert action is modified by the factor ϕ^2/v^2, making it the unconventional coupling of the Higgs field to gravity. When $\phi^2 = v^2$, one obtains the usual Einstein-Yang-Mills Lagrangian. However, the equations of motion derived by the standard variational procedure are different from the usual ones even when one goes to the limit $\phi^2 = v^2$. Our motivation for this choice of the Lagrangian is mostly heuristic at the present, although one knows that such a type of coupling does arise in Kaluza-Klein type theories. Also in recent developments of string theories, such unconventional couplings do arise.

We seek static solutions to equations of motion that follow from (1) in the gauge $A_0^a = 0$, so that we have only magnetic fields that are nonvanishing. Recall that in flat space-time, in the Prasad-Sommerfeld limit, the energy functional $\mathcal{E}$ is

$$\begin{aligned}\mathcal{E} &= \frac{1}{2}\int d^3x\left[(E_i^a)^2 + (B_i^a)^2\right] \\ &= \frac{1}{2}\int d^3x\,[(E_i^a \mp B_i^a)(E_i^a \mp B_i^a)] \pm \int d^3x E_i^a B_i^a \\ &\geq \pm\int d^3x E_i^a B_i^a. \end{aligned} \tag{4}$$

Thus, the energy functional has the absolute minimum if and only if

$$E_i^a = \pm B_i^a, \tag{5}$$

which are the Bogomolnyi conditions. Explicit solutions in closed form are then obtainable as shown by Prasad and Sommerfield.[5]

We seek an extension of the above type of considerations when gravity is included by going over to a curved space-time background. In a previous paper,[6] motivated by the Majumdar-Papapetrou ansatz[7] in the case of Einstein-Maxwell system, Balakrishna and myself postulated a metric given by

$$ds^2 = A^2(x)dt^2 - \frac{1}{A^2(x)}dx^i dx^i. \tag{6}$$

The energy functional $\mathcal{E}$ is then found to be

$$\mathcal{E} = \int d^3x\left(\frac{1}{4A^2}R\phi^2 + \frac{1}{4}A^2 F_{ij}^a F_{ij}^a + \frac{1}{2}D_i\phi^a D_i\phi^a\right), \tag{7}$$

where we have ignored the λ-term and chosen the units so that $4\pi G v^2 = 1$. If we now define

$$E_i^a = \frac{1}{A} D_i \mathcal{X}^a, \quad B_i^a = \frac{A}{2}\epsilon_{ijk} F_{jk}^a, \quad \mathcal{X}^a = A\phi^a, \tag{8}$$

we find that $\mathcal{E}$ has exactly the same form as in the case of flat space-time. The Bogomolnyi conditions for the minimum of $\mathcal{E}$ are

$$F_{ij}^a = \pm\frac{1}{A^2}\epsilon_{ijk} D_k \mathcal{X}^a, \tag{9}$$

just as in flat space-time, except for the appearance of the metric function A^{-2} on the right hand side. One finds that the Euler-Langrange equations that follow from the variation of the action in eq. 1 are all satisfied provided

$$\phi^2 = v^2 = 1 \text{ or } \mathcal{X}^2 = A^2. \tag{10}$$

Specializing to the $SU(2)$ gauge group with 't Hooft-Polyakov ansatz,

$$\mathcal{X}^a = \mp A(r)\hat{r}^a, \quad A_i^a(r) = \frac{1}{er}\epsilon_{aij}\hat{r}^j(1 - w(r)), \tag{11}$$

one is lead to a nonlinear equation

$$r^4 F''(r) = e^{-2F}, \tag{12}$$

where prime denotes differentiation with respect to r and $w(r) = \frac{1}{r}e^{-F}$. Also,

$$\frac{e}{A(r)} = \frac{1}{r} + F'(r). \tag{13}$$

The metric function $A(r)$ is given in terms of the solution of the above equation. It is interesting to note that eq. 12 is a special case of the class of equations known as Lane-Emden equations which arise in the study of polytropes or stellar masses in equilibrium under the influence of their own gravity. Equation 12 arises in the case of isothermal polytropes and is an object of extensive analysis.[8] No analytic solution is known. However, a class of solutions with the following behavior

$$F(r) \to Cr + D \text{ as } r \to \infty,$$

$$F(r) \to -\ell nr + A\sqrt{r}\sin\left(\frac{\sqrt{7}}{2}\ell nr + \delta\right) \quad \text{as} \quad r \to 0, \tag{14}$$

where $c > 0, D, A$, and δ are constants is proved to exist and hence as $r \to \infty$, from (13) and (14), it follows that

$$A(r) \to 1 - \frac{M}{4\pi r}, \tag{15}$$

if $C = e$ and $M = 4\pi/e$. This implies, in the units we have chosen, that we have a magnetic monopole solution with the mass of the monopole being equal to its charge.

However, from (14), we observe that as $r \to 0$, $A(r)$ goes to zero as $\sqrt{r}$ in an oscillatory way. The area of any sphere surrounding the origin, $4\pi r^2/A^2$, vanishes periodically and tends to zero as r goes to zero. The event horizon has zero surface area and hence the origin is close to being a naked singularity. The Ricci scalar diverges whenever the area vanishes and therefore, to an outside observer, these divergences will appear collectively as a naked singularity. These properties may be specific to the limit $4\pi Gv^2 = 1$ or to the specific ansatz regarding the metric. Static, spherical symmetry alone permits a more general metric in the standard coordinate system given by the equation

$$ds^2 = A^2 dt^2 - B^2(r)dr^2 - d\Omega \tag{16}$$

where

$$d\Omega = r^2(d\theta^2 + \sin^2\theta d\vartheta^2).$$

A and B are time independent functions of r only. By allowing ourselves the above more general form, Nguyen Ai Viet and I seek Bogomolnyi type solutions. It is straightforward to calculate the Ricci tensor components, the Ricci scalar, the energy momentum tensor components and write the Einstein's, Yang-Mills, and Higgs equations. For my present purposes, it is sufficient to consider the energy functional $\mathcal{E}$,

$$\mathcal{E} = \int d^3x\sqrt{-g}\left[\frac{R\phi^2}{4} + \frac{1}{4}F^a_{ij}F^{aij} - \frac{1}{2}D_i\phi^a D^i\phi^a + \frac{\lambda}{4}(\phi^2 - v^2)^2\right] \tag{17}$$

and discuss how we define an "Electric Field" so that $\mathcal{E}$ assumes the Bogomolnyi form as in the previous case. We follow the procedure proposed by Comptet, Forgacs, and Horvathy[9] by postulating

$$E_i^a = (-g)^{1/4}(-g^{ii})^{1/2}(D_i + V_i)\phi^a, \tag{18}$$

where V_i is an arbitrary function of the coordinates. We define

$$B_i^a = (-g)^{1/4}(g^{jj}g^{kk})^{1/2}\frac{1}{2}\epsilon_{ijk}F^a_{jk}. \tag{19}$$

If we now require that, we satisfy Yang-Mills equations that follow from the action given by equation (1), we obtain a unique V_i, which is given by

$$V_i = \partial_i A, \tag{20}$$

where as we recall $A^2 = g_{00}$. In spherical coordinates, the Bogomolnyi conditions imply

$$h'(r) + \frac{A'(r)}{A}h(r) = \frac{B(1 - W^2(r))}{evr^2},$$

$$W'(r) = -evh(r)W(r)B, \tag{21}$$

where

$$\chi^a = vh(r)\hat{r}^a, \quad A_i^a(r) = \frac{1}{er}\epsilon_{aij}\hat{r}^j(1 - W(r)). \tag{22}$$

At the same time we have to satisfy the Higgs field equation, which now reduces to

$$\frac{1}{\sqrt{-\bar{g}}}\partial^\alpha\left(\sqrt{-\bar{g}}\frac{\partial_\alpha A}{A}\right) = \frac{R}{2} + \lambda(\phi^2 - v^2), \quad \bar{g} = \det \mid g_{ij} \mid \tag{23}$$

and the static Einstein field equations.

After some algebraic manipulations, the equations we have to satisfy take the form

$$4\frac{h''}{h} + 4\left(\frac{h'}{h}\right)^2 + 4\frac{h'}{h}\left(\frac{2}{r} - \frac{B'}{B}\right) + 6\frac{h'}{h}\frac{A'}{A} = \frac{\lambda v^2}{2}B^2\frac{h^4 - 1}{h^2}$$

$$6\frac{h'}{h}\frac{A'}{A} - 4\frac{h'}{h}\left(\frac{C'}{C} + \frac{1}{r} + \frac{A'}{A}\right) = \lambda v^2 B^2\frac{(h^2 - 1)^2}{h^2},$$

$$\left(\frac{C'}{C}+\frac{1}{r}+\frac{A'}{A}\right)^2+4\frac{h'}{h}\left(\frac{C'}{C}+\frac{1}{r}+\frac{A'}{A}\right)=\frac{B^2}{C^2r^2}-\frac{\lambda v^2}{2}B^2\frac{(h^2-1)^2}{h^2},$$
$$\frac{C'}{C}+\frac{1}{r}+\frac{A'}{A}=\alpha\frac{B}{Cr}h^2, \tag{24}$$

where α is an arbitrary integration constant.

This set of equations together with the Bogomol'nyi equations (21) are the basic equations of our system.

The above set of equations simplify greatly if the Higgs field is frozen in the broken phase $h^2 = 1$. The first two equations in (24) are satisfied automatically. The last equation can be integrated. It coincides exactly with the previous equation, if the integration constant is chosen to be ± 1. So the set of equations to be solved reduces to

$$\frac{1}{r}+\frac{A'}{A}=\pm\frac{B}{r}, \tag{25}$$

$$\frac{A'}{A}=\frac{B}{C^2r^2}\frac{1-W^2}{vg}, \tag{26}$$

$$W'=-vgBW. \tag{27}$$

We can prove that the case with minus sign in (25) contradicts the condition on the asymptotic behavior. Introducing the new variables $x = vgr$ and $y = W^2$, after a little algebra, the case with positive sign leads us to the system of equations

$$y'=-\frac{2xy}{y-1+x}, \tag{28}$$

$$B=\frac{x}{y-1+x}, \tag{29}$$

$$\frac{A'}{A}=\frac{1-y}{x(y+x-1)}. \tag{30}$$

The eq.(28) is well-known in mathematical physics as Abel's differential equation of the second type. The Abel equation has clearly a solution $y = 0$. In this case we have an Abelian magnetic monopole which at the same time is an extreme Reissner-Nordström 'ackhole

$$B=(1-\frac{1}{x})^{-1},$$

$$A = (1 - \frac{1}{x}) \tag{31}$$

This monopole has mass $M = 4\pi v/g$. The Abel equation has no known analytic solution. We study its solutions numerically. In figure 1, we plot a typical solution for W^2, A^2, B^2 on a logarithmic scale. Above a critical value for $W(r)$ at $r = 0$, $W(r)$ has the desired characteristic, namely, it is nonvanishing in the region surrounding the origin and rapidly (exponentially) goes to zero as $r \to \infty$. This signals monopole solutions as in the case of flat space-time. The metric functions A and B are determined from W^2 up to an integration constant. Different values of this constant give different metric functions. In a typical example given in Fig. 1, we see that A has a minimum and B a corresponding maximum and both approach unity as r becomes large. These are the features we expect from a black hole type solution. The minimum in g_{00} can be made arbitrarily close to zero by choosing an appropriate value for the integration constant. However, analytic arguments show that it cannot be exactly zero.[4] This raises the puzzling question; whether the numerical minimum of the metric coefficient g_{00} represents a real horizon. Recent numerical solutions of others need to be examined from this point of view. Clearly a great deal of further work is necessary in order to settle intriguing questions that have arisen from our study. For $W(r) <$ the critical value at $r = 0$, $W(r)$ does not have the exponential fall of and does not vanish as $r \to \infty$. Instead, it has the behavior of the dotted curve shown in figure 1. The physical interpretation of such a $W(r)$ also needs investigation.

To conclude, although the gravitational interactions are weak, they do have strong consequences on finite energy Yang-Mills-Higgs configurations. The interplay between gravitational forces and gauge interactions coupled with Higgs interactions provides a fertile ground for further insights into the structure of nonabelian monopoles. One may speculate, for instance, on the question whether there are, as in the case of massive stars, instabilities induced to general relativistic effects, and whether massive monopoles become lack holes by some sort of quantum radiation mechanism. If so, it is obvious that it will have profound consequences on our idea of grand unified cosmology and the so-called monopole problem in the early universe.

Acknowledgements

This work was supported in part by the U.S. Department of Energy under contract number DE-FG02-85ER40231.

1. L. O'Raifeartaigh, S.Y. Park, and K.C. Wali, Phys. Rev. **D20**, 1941 (1979).

2. R. Bartnik and J. McKinnon, Phys. Rev. Letters **61**, 141 (1988). See also P. van Nieuwenhuizen, D. Wilkinson, and M.J. Perry, Phys. Rev. **D13**, 778 (1976).

3. M.E. Ortiz, Phys. Rev. **D45**, 2586 (1992); K. Lee, V.P. Nair, and E.J. Weinberg, ibid. **D45**, 2751 (1992); P. Breitenlohner, P. Forgács, and D. Maison, Nucl. Phys. **B393**, 357 (1992).

4. Nguyen Ai Viet and K.C. Wali, Magnetic monopoles in the Einstein-Yang-Mills System, SU preprint 4240-552.

5. M.K. Prasad and C.M. Sommerfield, Phys. Rev. Lett. **35**, 760 (1975).

6. B.S. Balakrishna and K.C. Wali, Phys. Rev. **D46**, R5228 (1992).

7. S.D. Majumdar, Phys. Rev. **72**, 390 (1947); A. Papapetrou, Proc. R. Irish Acad. Ser. **A51**, 191 (1947).

8. S. Chandrasekhar, An Introduction to the study of stellar structure (Dover, New York, 1957), Chap. IV.

9. A. Comptet, P. Forgács, P.A. Horváthy, Phys. Rev. **D30**, 468 (1984).

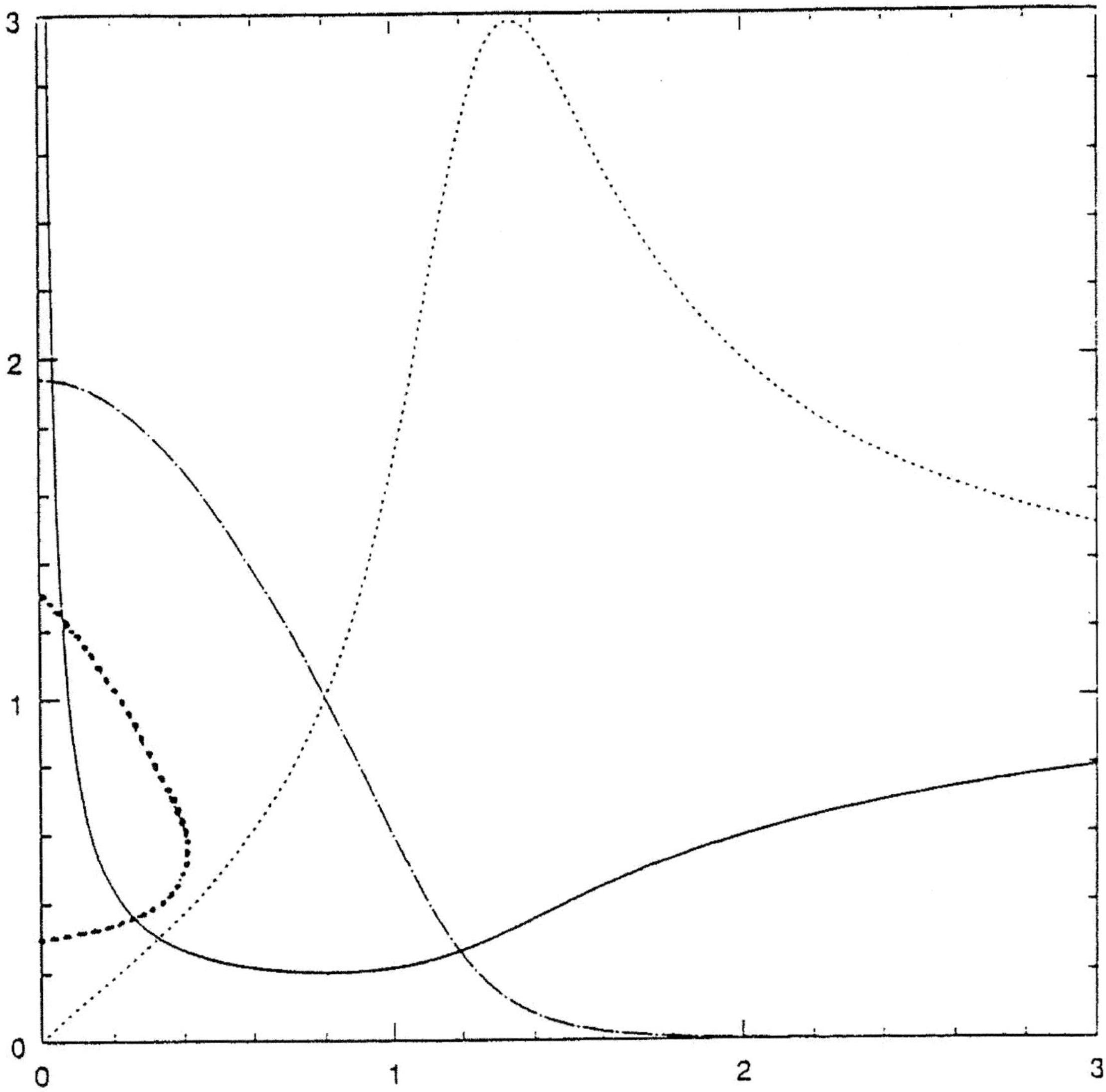

Figure 1

A monopole solution in curved space-time. The solid and dotted curves represent the metric function A and B respectively. The dashed-dotted curve is $y=w^2$. The curves are plotted as functions of $\ln(1+x)$.

Magnetic monopole solutions in a modified Einstein-Yang-Mills-Higgs system

Nguyen Ai Viet* and Kameshwar C. Wali
Physics Department, Syracuse University, Syracuse, New York 13244
(Received 1 November 1993)

We study the Yang-Mills-Higgs system within the framework of general relativity with an unconventional coupling of the scalar field to gravity. In the static situation, using a Bogomol'nyi-type analysis, we derive a positive-definite energy functional with a lower bound that is attained when the Bogomolnyi conditions are satisfied. Specializing to the gauge group SU(2) and the 't Hooft–Polyakov ansatz for the gauge and Higgs fields, we seek static, spherically symmetric solutions to the coupled system of equations together with Bogomol'nyi conditions. In both the isotropic and standard coordinate systems, in the spontaneously broken symmetry situation, we find great simplifications reducing the solutions of the coupled system to the solution of a single nonlinear differential equation, different one in each case, but well known in other contexts of physics. We find Abelian and non-Abelian monopole solutions with gravitational fields playing the role of Higgs fields in providing attraction that balances the repulsion due to the gauge fields. These solutions in general have naked singularities at the origin. But as solutions we also find extreme Reissner-Nordström black holes as well as a new non-Abelian monopole solution that has a horizon enclosing the singularity.

PACS number(s): 04.40.Nr, 04.20.Jb, 11.15.−q, 14.80.Hv

I. INTRODUCTION

Recently there has been considerable interest in the study of classical, static and spherically symmetric solutions of the coupled Einstein-Yang-Mills-Higgs [EYMH] system [1]. The main purpose of such investigations has been to understand the properties of non-Abelian monopoles such as the 't Hooft–Polyakov monopoles in the presence of gravity. The latter, which are finite energy configurations of Yang-Mills and Higgs fields, exhibit extended structure and the Coulomb interactions between two monopoles or a monopole and an antimonopole are altered by the complex interactions between the gauge fields and Higgs fields [2]. The question that arises naturally is what happens when such monopoles are thought to be in curved space-time described by a given background metric.

It is of some interest in this connection to note that, in the framework of Newtonian mechanics, a system of N mass points with like electrical (or magnetic) charges can remain in perfect static equilibrium if their masses are, in suitable units, equal to their charges. Remarkably, a generalization of this situation exists within the framework of general relativity described by the Majumdar-Papapetrou solution [3], which was shown later to represent N extreme Reissner-Nordström black holes [4]. One may, therefore, ask whether there is a corresponding generalization of Yang-Mills-Higgs monopoles. Nieuwenhuizen, Wilkinson, and Perry [5] were the first ones to investigate static solutions in the case of the EYMH system. They showed that one could construct a positive-definite energy functional for the complete system, which at its minimum, gave a regular solution for the 't Hooft–Polyakov monopole. More recently, Bartnik and McKinnon found numerical solutions [6] of the Einstein-Yang-Mills equations, which had particlelike properties and demonstrated among other things that gravitational interaction could not be dismissed as too weak to be consequential. Their solutions depended nontrivially on the coupled Yang-Mills and Einstein fields. Bais and Russell, and independently Cho and Freund found a solution [7] to the complete EYMH system, which, however, has the Abelian Dirac monopole character with a singularity at the origin. One is interested in more general solutions with non-Abelian character and in the question; whether such solutions are stable and whether they have event horizons inside or outside the core defined by the Yang-Mills-Higgs fields in the absence of gravity.

The present paper is an extension of the work by Balakrishna and Wali [8]. With a specific coupling of the Higgs field to the Einstein term in the Lagrangian describing the EYMH system, they showed that, in the static case, the energy functional can be reduced, by a Bogomol'nyi-type [9] condition, to a form that resembles the energy functional for the 't Hooft–Polyakov monopole in flat space-time. They demonstrated the existence of a non-Abelian solution, which asymptotically had all the properties of the 't Hooft–Polyakov monopole. But the solutions they found appear to have a naked singularity at the origin, and the question remained whether the singularity could be removed by an ansatz for the background metric that is more general than what was assumed or whether the singularity was because of the Higgs field being frozen at a critical value throughout space. In the present work, we address some of these issues.

In the next section, we set up our general formalism.

*Permanent address: Centre of Theoretical Physics, P.O. Box 429, Bo Ho 10000, Hanoi, Vietnam.

In Sec. III, after a brief review of the energy functional in the case of monopole in flat space-time, we consider the same problem in curved space-time. Using a Bogomol'nyi-type analysis, we find a positive-definite energy functional with a lower bound. We discuss how the total energy and mass of the monopole are related to the asymptotic behavior of the metric functions and the Higgs fields on the one hand and the topological winding number on the other hand. Two extreme cases for a gravitating monopole arise. In one case, the Higgs field is frozen to its vacuum expectation value (VEV) throughout space. The mass of the monopole is then determined by its magnetic charge and the magnitude of the VEV. In the other case, in the comoving frame, the topological charge is related to the asymptotic behavior of the Higgs field. In Sec. IV, starting from the most general static, spherically symmetric form for the metric we calculate the curvature coefficients and components of energy-momentum tensor. Using the 't Hooft–Polyakov ansatz for the gauge and Higgs fields, we add the Bogomolnyi conditions to the set of equations of motion for Einstein, Yang-Mills, and Higgs fields and come to an equivalent set of coupled differential equations. We then specialize to the case where the Higgs field is frozen everywhere to its vacuum expectation value. In isotropic and standard coordinate systems, we arrive at two well-known classical nonlinear differential equations, whose solutions determine the solutions of our system. Before solving the equations numerically we perform qualitative analysis and discuss the main features and physical implications of the solutions.

II. GENERAL FRAMEWORK

We begin by considering a modified EYMH system, defined by the action S, where

$$S = \int d^4x\sqrt{-g}\left[-\frac{1}{16\pi G v^2}R\Phi^2 - \frac{1}{4}(F^a_{\mu\nu})^2 + \frac{1}{2}(\mathcal{D}_\mu\Phi^a)^2 - \frac{\lambda}{4}(\Phi^2 - v^2)^2\right] . \quad (2.1)$$

In our conventions, $g_{\mu\nu}$ has the signature $(+ - - -)$. Indices $\mu, \nu, \ldots$ run from 0 to 3 and $i, j, \ldots$ from 1 to 3. Indices $a, b, \ldots$ run over the adjoint representation of a given gauge group. R is the Ricci scalar. $F^a_{\mu\nu}$ is the field strength associated with the gauge field A^a_μ:

$$F^a_{\mu\nu} = \partial_\mu A^a_\nu - \partial^a_\nu + g f^{abc} A^b_\mu A^c_\nu, \quad (2.2)$$

where g is the gauge coupling constant (not to be confused with the determinant of the metric) and f^{abc}'s are the structure constants of the group. The gauge-covariant derivative of the Higgs field Φ^a is given as

$$\mathcal{D}_\mu\Phi^a = \partial_\mu\Phi^a + g f^{abc} A^b_\mu \Phi^c . \quad (2.3)$$

In the broken phase of the gauge symmetry, when the Higgs field is frozen to be $\Phi^2 = v^2$, the first term gives the usual Einstein Lagrangian. The rest of the terms represent the standard Yang-Mills-Higgs system. Hereafter, we shall work in units such that

$$4\pi G v^2 = 1 , \quad (2.4)$$

and then the action is

$$S = \int d^4x\sqrt{-g}\left[-\frac{R}{4}\Phi^2 - \frac{1}{4}(F^a_{\mu\nu})^2 + \frac{1}{2}(\mathcal{D}_\mu\Phi^a)^2 - \frac{\lambda}{4}(\Phi^2 - v^2)^2\right] . \quad (2.5)$$

Varying the action (2.5) with respect to A^a_μ, Φ^a, and $g_{\mu\nu}$, respectively, we have the coupled equations for Yang-Mills and Higgs fields

$$\frac{1}{\sqrt{-g}}\mathcal{D}_\mu\left(\sqrt{-g}F^{a\mu\nu}\right) = -g f^{abc}\Phi^b\mathcal{D}^\nu\Phi^c, \quad (2.6)$$

$$\frac{1}{\sqrt{-g}}\mathcal{D}_\mu\left(\sqrt{-g}\mathcal{D}^\mu\Phi^a\right) = -\left(\frac{R}{2} + \lambda(\Phi^2 - v^2)\right)\Phi^a, \quad (2.7)$$

and the Einstein equation

$$G_{\mu\nu} = R_{\mu\nu} - \frac{1}{2}R = \frac{2}{\Phi^2}T_{\mu\nu}, \quad (2.8)$$

where the energy-momentum tensor $T_{\mu\nu}$ is given by

$$T_{\mu\nu} = g_{\mu\nu}T - F^a_{\mu\rho}F^{a\rho}_\nu + \mathcal{D}_\mu\Phi^a\mathcal{D}_\nu\Phi^a + \nabla_\mu\nabla_\nu\Phi^2 ,$$
$$T = \frac{1}{4}F^a_{\mu\nu}F^{a\mu\nu} - \frac{1}{2}\mathcal{D}_\mu\Phi^a\mathcal{D}^\mu\Phi^a + \frac{\lambda}{4}(\Phi^2 - v^2)^2 - \frac{1}{2}\Box\Phi^2 . \quad (2.9)$$

[∇_μ is the space-time covariant derivative and the D'Alembertian $\Box = \frac{1}{\sqrt{-g}}\nabla_\mu(\sqrt{-g}\nabla^\mu)$.] The last terms in $T_{\mu\nu}$ and T arise because of the coupling $R\Phi^2$ in the action [10].

III. STATIC EQUATIONS AND BOGOMOL'NYI CONDITIONS

We are interested in the static solutions of the system described in the previous section. We shall specialize to the 't Hooft–Polyakov monopole [11] with the gauge group SU(2). In the case of flat space-time, the well-known technique due to Bogomol'nyi [9] leads to first order differential equations and analytic solutions in the case of Prasad-Sommerfield limit [9], when the energy functional is saturated. The gauge group is broken to the Abelian U(1), and the magnetic charge of the configuration is related to a winding number [12]. Further, when Bogomol'nyi conditions are satisfied, one can find a connection between the asymptotic behavior of the Higgs field Φ and the topologically conserved winding number. We shall show that these features of flat space-time can be also reproduced in the case of curved space-time.

We look for static solutions of Eqs. (2.6)–(2-8), in the $A_0^a = 0$ gauge. Since

$$F_{0i}^a = 0, \quad \mathcal{D}_0\Phi^a = 0, \tag{3.1}$$

we find that Eqs. (2.6) and (2.7) reduce to

$$\frac{1}{\sqrt{-g}}\mathcal{D}_i\Big(\sqrt{-g}F^{aij}\Big) = -gf^{abc}\Phi^b\mathcal{D}^j\Phi^c, \tag{3.2}$$

$$\frac{1}{\sqrt{-g}}\mathcal{D}_i\Big(\sqrt{-g}\mathcal{D}^i\Phi^a\Big) = -\Big(\frac{R}{2} + \lambda(\Phi^2 - v^2)\Big)\Phi^a. \tag{3.3}$$

The static energy functional that follows from (2.5) in curved space-time is given by [8]

$$\mathcal{E} = \int d^3x\sqrt{-g}\Bigg[\frac{R\Phi^2}{4} + \frac{1}{4}F_{ij}^a F^{aij} - \frac{1}{2}\mathcal{D}_i\Phi^a\mathcal{D}^i\Phi^a + \frac{\lambda}{4}(\Phi^2 - v^2)^2\Bigg]. \tag{3.4}$$

Following Comtet, Forgács, and Horváthy [13], let us assume

$$\mathcal{D}_i\Phi^a + \partial_i u\Phi^a = \frac{1}{2}\sqrt{-\tilde{g}}\epsilon_{ijk}F^{ajk}, \tag{3.5}$$

where u is an arbitrary time-independent function and $\tilde{g} = \det|g_{ij}|$. Equivalently,

$$\frac{1}{2}\epsilon^{ijk}F_{jk}^a = \sqrt{-\tilde{g}}\epsilon_{ijk}F^{ajk}\ . \tag{3.6}$$

Substituting (3.6) in (3.2), and using the Bianchi identity for F^{aij}, we find that (3.2) is satisfied provided

$$\partial_i u = \frac{\partial_i\sqrt{g_{00}}}{\sqrt{g_{00}}}. \tag{3.7}$$

With (3.7), after some algebra, the Higgs field Eq. (3.3) yields

$$\frac{1}{\sqrt{-\tilde{g}}}\partial_i\Big[\sqrt{-\tilde{g}}\frac{\partial^i\sqrt{g_{00}}}{\sqrt{g_{00}}}\Big] = \frac{R}{2} + \lambda(\Phi^2 - v^2). \tag{3.8}$$

Comtet *et al.* [13] have suggested that the Bogomol'nyi Eq. (3.5) is the duality condition of a Euclidean Yang-Mills theory in curved space-time similar to the duality condition in the flat space-time. Additionally, we shall show that this condition also saturates the energy-functional. We will use Eq. (3.8) as a restriction on the degrees of freedom of our system, and then prove that the reduced energy functional under that restriction has a lower bound. Fortunately, the restriction (3.8) we choose is not imposed arbitrarily; it is an equation of motion of the system. Let us define the "electric" and "magnetic" fields as

$$\begin{aligned} E_i^a &= {}^4\!\sqrt{-g}\sqrt{-g^{ii}}\left(\mathcal{D}_i\Phi^a + \frac{\partial_i\sqrt{g_{00}}}{\sqrt{g_{00}}}\Phi^a\right) \\ &= {}^4\!\sqrt{-g}\frac{\sqrt{-g^{ii}}}{g_{00}}\mathcal{D}_i\chi^a, \\ B_i^a &= {}^4\!\sqrt{-g}\sqrt{g^{jj}g^{kk}}\frac{1}{2}\epsilon_{ijk}F_{jk}^a, \end{aligned} \tag{3.9}$$

where $\chi^a = \sqrt{g_{00}}\Phi^a$. The Bogomon'nyi conditions become

$$\mathcal{D}_i\chi^a = \sqrt{g_{00}}\frac{\epsilon^{ijk}}{\sqrt{-\tilde{g}}}F_{ij}^a. \tag{3.10}$$

Using Eqs. (3.8) and (3.9), we can express the energy-functional $\mathcal{E}$ as

$$\mathcal{E} = \frac{1}{2}\int d^3x\Bigg[(B_i^a)^2 + (E_i^a)^2 + \partial_i(\sqrt{-\tilde{g}}\partial^i\sqrt{g_{00}}\Phi^2) - \frac{\lambda}{2}\sqrt{-g}(\Phi^4 - v^4)\Bigg]. \tag{3.11}$$

Ignoring the λ term we have then

$$\mathcal{E} = \frac{1}{2}\int d^3x\Big[(B_i^a)^2 + (E_i^a)^2 + \partial_i\Big(\sqrt{-\tilde{g}}\partial^i\sqrt{g_{00}}\Phi^2\Big)\Big]. \tag{3.12}$$

The reduced energy functional (3.12) is positive definite up to a surface term. The surface term is invariant and finite, so the energy functional (3.12) is bound from below under any variations of the remaining degrees of freedom. Using the argument of Bogomolnyi, this system has solutions, which reach minimum at the bound if the Bogomolnyi conditions $E_i^a = \pm B_i^a$ are satisfied:

$$\begin{aligned} \mathcal{E} &= \frac{1}{2}\int d^3x\Big[(E_i^a \mp B_i^a)(E_i^a \mp B_i^a)\Big] + \int d^3x\Bigg[\pm E_i^a B_i^a + \frac{1}{2}\partial_i(\sqrt{-\tilde{g}}\partial^i\sqrt{g_{00}}\Phi^2)\Bigg] \\ &\geq \int d^3x\Big[\pm E_i^a B_i^a + \frac{1}{2}\partial_i\Big(\sqrt{-\tilde{g}}\partial^i\sqrt{g_{00}}\Phi^2\Big)\Big]. \end{aligned} \tag{3.13}$$

When the Bogomol'nyi conditions are satisfied, the energy-functional reaches the lower bound and we have

$$\mathcal{E} = \int d^3x\Big[\frac{-\sqrt{-\tilde{g}}}{\sqrt{g_{00}}}\mathcal{D}_i\chi^a\mathcal{D}^i\chi^a + \frac{1}{2}\partial_i(\sqrt{-\tilde{g}}\partial^i\sqrt{g_{00}}\cdot\Phi^2)\Big]. \tag{3.14}$$

Using (3.6), (3.7), and the Bianchi identity for F^{aij}, we can derive the relation

$$\mathcal{D}_i\chi^a\mathcal{D}^i\chi^a = \frac{1}{2}\frac{\sqrt{g_{00}}}{\sqrt{-\tilde{g}}}\partial_i\Big(\frac{\sqrt{-\tilde{g}}}{\sqrt{g_{00}}}\partial^i\chi^2\Big), \tag{3.15}$$

and hence

$$\mathcal{E} = \frac{1}{2}\int d^3x\partial_i\Big[\sqrt{-g}\Big(-\frac{1}{2}\frac{\partial^i g_{00}}{g_{00}}\Phi^2 - \partial^i\Phi^2\Big)\Big]$$
$$= \frac{1}{2}\int_{S_R^2} d\sigma_i\Big[-\frac{1}{2}\frac{\partial^i g_{00}}{g_{00}}\Phi^2 - \partial^i\Phi^2\Big]\sqrt{-g}. \quad (3.16)$$

Now, if we have a finite-energy configuration with finite extension and the space-time is asymptotically flat, we must have

$$\mathcal{D}\chi = 0, \quad \chi^2 = v^2, \quad (3.17)$$

and an exact analogy prevails with the flat space-time [12]. We have asymptotically an Abelian magnetic field given by

$$\mathcal{B}^i = \frac{1}{2}\frac{\epsilon^{ijk}}{\sqrt{-g}}\mathcal{F}_{jk}, \quad (3.18)$$

where

$$\mathcal{F}_{jk} = \frac{1}{2v^3 g}\epsilon^{abc}\chi^a\partial_j\chi^b\partial_k\chi^c + \partial_j A_k - \partial_k A_j, \quad (3.19)$$

with an arbitrary A_j. The magnetic charge $\mathcal{G}$ is given by

$$4\pi\mathcal{G} = \int d^3x\partial_i\mathcal{B}^i = \frac{2\pi}{v^3 g}\int_{S^2_\infty}\epsilon^{abc}\epsilon^{ijk}\chi^a\partial_j\chi^b\partial_k\chi^c d\sigma_i$$
$$= 4\pi\frac{n}{g}, \quad (3.20)$$

where n is the winding number of the mapping χ: $S^2_\infty \longrightarrow S^2$, and

$$n = \frac{1}{8\pi v^3}\int_{S^2_\infty}\epsilon^{abc}\epsilon^{ijk}\chi^a\partial_j\chi^b\partial_k\chi^c d\sigma_i. \quad (3.21)$$

We can show that, asymptotically,

$$\mathcal{E} = \int d^3x\partial_i\mathcal{B}^i = \frac{4\pi n}{g}. \quad (3.22)$$

However, if as $r \longrightarrow \infty$,

$$g_{ij} \longrightarrow 1 + \mathcal{O}\Big(\frac{1}{r}\Big),$$
$$g_{00} \longrightarrow 1 - \frac{M}{2\pi v^2 r} + \mathcal{O}\Big(\frac{1}{r^2}\Big),$$
$$\Phi^2 \longrightarrow v^2 - \frac{2v\beta}{r} + \mathcal{O}\Big(\frac{1}{r^2}\Big), \quad (3.23)$$

where M is defined as the monopole's mass, then from (3.16), we have

$$\mathcal{E} = \frac{1}{v}M + \frac{4\pi}{g}\beta. \quad (3.24)$$

From (3.22) and (3.24), we find that

$$n = \frac{1}{4\pi v}Mg + \beta. \quad (3.25)$$

We note that the mass now is not the total energy as in the case of flat space-time, but it is determined by the asymptotic behavior of g_{00}. We have two possible cases for a gravitating monopole. In the first case, the Higgs field is frozen and M is determined by the magnetic charge and the value of v

$$M = n\frac{4\pi}{g}v = 4\pi v\mathcal{G}. \quad (3.26)$$

As the topological charge is conserved, so is the mass of the monopole. It indicates that in this case the monopole is not a dissipative one. In the second case, we are in the comoving frame of reference, in which case

$$g_{00} = 1 \quad \text{and hence} \quad n = \beta, \quad (3.27)$$

and thus the topological charge is determined by the asymptotic behavior of the Higgs field.

IV. SPHERICALLY SYMMETRIC ANSATZ AND BASIC EQUATIONS

We define a spherically symmetric static metric by

$$ds^2 = A^2(r)dt^2 - B^2(r)dr^2 - C^2(r)r^2 d\Omega. \quad (4.1)$$

In fact we could always redefine the "radius" r so that the line element could be represented by two metric functions only. But we prefer the above more general form since we intend to consider both the standard and the isotropic coordinate systems. It is more convenient for our purposes to have the more general metric form (4.1) and specialize to the chosen coordinate system by suitably identifying the functions.

The Einstein tensor and the Ricci scalar curvature are then easily computed:

$$G_{00} = \frac{A^2}{B^2}\Big[\frac{1}{r^2}\Big(\frac{B^2}{C^2} - 1\Big) + \frac{2}{r}\frac{B'}{B} + 2\frac{B'}{B}\frac{C'}{C} - 2\frac{C''}{C} - \frac{6}{r}\frac{C'}{C}$$
$$-\Big(\frac{C'}{C}\Big)^2\Big],$$
$$G_{rr} = \Big(\frac{C'}{C}\Big)^2 + \frac{2}{r}\frac{C'}{C} + \frac{1}{r^2}\Big(1 - \frac{B^2}{C^2}\Big) + 2\frac{C'}{C}\frac{A'}{A} + \frac{2}{r}\frac{A'}{A},$$
$$G_{\theta\theta} = \frac{C^2r^2}{B^2}\Big(\frac{A''}{A} + \frac{1}{r}\frac{A'}{A} - \frac{A'}{A}\frac{B'}{B} + \frac{C'}{C}\frac{A'}{A} - \frac{1}{r}\frac{B'}{B}$$
$$-\frac{B'}{B}\frac{C'}{C} + \frac{C''}{C} + \frac{2}{r}\frac{C'}{C}\Big),$$
$$G_{\phi\phi} = \sin^2\theta\, G_{\theta\theta}, \quad (4.2)$$

and

$$R = \frac{2}{r^2}\Big(\frac{1}{B^2} - \frac{1}{C^2}\Big) + \frac{2}{B^2}\Big[\frac{A''}{A} + \frac{2}{r}\frac{A'}{A} - \frac{B'}{B}\frac{A'}{A} - \frac{2}{r}\frac{B'}{B}$$
$$-2\frac{B'}{B}\frac{C'}{C} + 2\frac{C'}{C}\frac{A'}{A} + 2\frac{C''}{C} + \frac{6}{r}\frac{C'}{C} + \Big(\frac{C'}{C}\Big)^2\Big]. \quad (4.3)$$

The Bogomol'nyi equations (3.5) along with (4.1) expressed in spherical polar coordinates take the form

$$\mathcal{D}_r\Phi^a + \frac{A'}{A}\Phi^a = -\frac{B}{C^2r^2\sin\theta}F^a_{\theta\phi}, \qquad (4.4)$$

$$\mathcal{D}_\theta\Phi^a = -\frac{1}{B\sin\theta}F^a_{\theta r}, \qquad (4.5)$$

$$\mathcal{D}_\phi\Phi^a = -\frac{\sin\theta}{B}F^a_{r\theta}. \qquad (4.6)$$

The 't Hooft–Polyakov spherically symmetric ansatz for Φ and A^a_μ,

$$\Phi^a = vh(r)r^a, \qquad (4.7)$$

$$A^a_i(r) = \epsilon_{iab}\frac{\hat{r}^b}{gr}[1 - W(r)], \qquad (4.8)$$

which can be transformed into spherical polar coordinates in a straightforward manner. Substituting (4.7) and (4.8) in (4.4)–(4.6), we obtain two independent equations,

$$h' + \frac{A'}{A}h = \frac{B}{C^2r^2}\frac{1-W^2}{vg},$$
$$W' = -vgBhW. \qquad (4.9)$$

Likewise, the components of the energy-momentum tensor are calculated to be

$$T_{00} = \frac{\Phi^2A^2}{B^2}\left[2\frac{C'}{C}\frac{A'}{A} + \frac{2}{r}\frac{A'}{A} - \frac{A'}{A}\frac{B'}{B} + \frac{A''}{A} - \frac{1}{2}\left(\frac{A'}{A}\right)^2 + \frac{\lambda v^2}{4}\frac{(h^2-1)^2}{h^2}\right.,$$
$$\left. +2\frac{h''}{h} + 2\left(\frac{h'}{h}\right)^2 + 2\frac{h'}{h}\left(2\frac{C'}{C} + \frac{2}{r} - \frac{B'}{B}\right) + 3\frac{A'}{A}\frac{h'}{h}\right],$$
$$T_{rr} = \Phi^2\left[-\frac{1}{2}\left(\frac{A'}{A}\right)^2 - 2\frac{h'}{h}\left(\frac{C'}{C} + \frac{A'}{A} + \frac{1}{r}\right) - \frac{\lambda v^2}{4}B^2\frac{(h^2-1)^2}{h^2}\right],$$
$$T_{\theta\theta} = \frac{C^2r^2}{B^2}\Phi^2\left[\frac{1}{2}\left(\frac{A'}{A}\right)^2 - \frac{h''}{h} - \left(\frac{h'}{h}\right)^2 - \frac{h'}{h}\left(2\frac{C'}{C} + \frac{2}{r} - \frac{B'}{B}\right) - \frac{\lambda v^2}{4}B^2\frac{(h^2-1)^2}{h^2}\right],$$
$$T_{\phi\phi} = \sin^2\theta T_{\theta\theta}\ . \qquad (4.10)$$

Then (4.2), (4.3), and (4.10) yield Einstein's equations in the static case with the ansatz (4.1). Rather than write these equations explicitly, we give below the equivalent set of four independent equations that follow from Einstein's equations and Eq. (3.8):

$$4\frac{h''}{h} + 4\left(\frac{h'}{h}\right)^2 + 4\frac{h'}{h}\left(2\frac{C'}{C} + \frac{2}{r} - \frac{B'}{B}\right) + 6\frac{h'}{h}\frac{A'}{A} = \frac{\lambda v^2}{2}B^2\frac{(h^4-1)}{h^2}, \qquad (4.11)$$

$$6\frac{h'}{h}\frac{A'}{A} - 4\frac{h'}{h}\left(\frac{C'}{C} + \frac{1}{r} + \frac{A'}{A}\right) = \lambda v^2B^2\frac{(h^2-1)^2}{h^2}, \qquad (4.12)$$

$$\left(\frac{C'}{C} + \frac{1}{r} + \frac{A'}{A}\right)^2 + 4\frac{h'}{h}\left(\frac{C'}{C} + \frac{1}{r} + \frac{A'}{A}\right) = \frac{B^2}{C^2r^2} - \frac{\lambda v^2}{2}B^2\frac{(h^2-1)^2}{h^2}, \qquad (4.13)$$

$$\frac{C'}{C} + \frac{1}{r} + \frac{A'}{A} = \alpha\frac{B}{Cr}h^2, \qquad (4.14)$$

where α is an arbitrary integration constant.

This set of equations together with the Bogomol'nyi equations (4.9) are the basic equations of our system. In the case of the Higgs vacuum, that we will discuss in the rest of this paper, the contributions of the last two terms to the energy functional in Eq. (3.11) vanish. Hence, the energy functional is saturated exactly and implies the compatibility of the Bogomol'nyi equations with the complete system. In the case of dynamical Higgs field, the question of compatibility of the Bogomol'nyi equation is difficult to establish as in the case of flat space-time. However, we note that in curved space-time, in contrast with the flat case, we still have a possibility that the third and fourth terms cancel. This ensures the compatibility of the Bogomol'nyi equations since the energy functional is again saturated. We will discuss this possibility and the general question of the compatibility of the Bogomol'nyi equation elsewhere. On the other hand, if we consider the Prasad-Sommerfield limit $\frac{\lambda}{g} \longrightarrow 0$ as in the case of flat space-time, Eq. (4.12) implies either $h = 1$ (Higgs field is frozen) or a solution for metric functions A, C that does not satisfy the asymptotic flatness condition.

A. Higgs vacuum

The above set of equations simplify greatly if the Higgs field is frozen in the broken phase $h^2 = 1$. Equations (4.11) and (4.12) are satisfied automatically. Equation (4.14) can be integrated. It coincides exactly with Eq. (4.13), if the integration constant is chosen to be ± 1. So the set of equations to be solved reduces to

$$\frac{C'}{C}+\frac{1}{r}+\frac{A'}{A}=\pm\frac{B}{Cr}\,, \tag{4.15}$$
$$\frac{A'}{A}=\frac{B}{C^2r^2}\frac{1-W^2}{vg}\,, \tag{4.16}$$
$$W'=-vgBW\,. \tag{4.17}$$

The number of unknown functions can be reduced to three since we still have the freedom to fix the scale of the radius using different cordinate systems. We note that the mass of the monopole is determined by the formula (3.26) for a given topological charge and the vacuum expectation value v of the Higgs field. The surviving Abelian magnetic field at large distances is determined completely in terms of the metric functions according to Eqs. (3.18) and (3.19)

$$\mathcal{B}=\frac{1}{2g}\frac{A^3}{BC^2r^2}\,. \tag{4.18}$$

B. Isotropic coordinate system

In the isotropic coordinate system we choose the coordinate fixing condition

$$B=C=\frac{\tilde{B}}{A}. \tag{4.19}$$

The line element in this coordinate system takes the form

$$ds^2=A^2(r)dt^2-\frac{\tilde{B}^2(r)}{A^2(r)}\left(dr^2+r^2d\Omega\right). \tag{4.20}$$

Instead of the variable r, let us introduce the new variable $x=vgr$. The system of equation (4.15)–(4.17) then reduces to

$$\frac{\tilde{B}'(x)}{\tilde{B}(x)}+\frac{1}{x}=\pm\frac{1}{x}, \tag{4.21}$$
$$\frac{A'(x)}{A^2(x)}=\frac{1-W^2(x)}{\tilde{B}(x)x^2}, \tag{4.22}$$
$$W'(x)=-\frac{\tilde{B}(x)}{A(x)}W(x), \tag{4.23}$$

where a prime now denotes differentiation with respect to x.

Equation (4.21) gives two possibilities: (i) with a positive sign on the right-hand side $\tilde{B}=$ const, which can be chosen to be unity without any loss of generality and (ii) $\tilde{B}=\frac{\text{const}}{x^2}$ with a negative sign on the right-hand side.

The first possibility has been studied by Balakrishna and Wali [8]. Their Abelian solution with $W=0$ gives

$$A=1-\frac{1}{x+1}\,, \tag{4.24}$$

with the asymptotic condition $A\longrightarrow 1$ as $x\longrightarrow\infty$. It is in reality the Majumdar-Papapetrou [3] solution representing a magnetic monopole with its mass equal to its magnetic charge or equivalently an extreme Reissner-Nordström black hole.

The non-Abelian solution corresponding to this case involves the system of nonlinear differential equations

$$r^4F''=e^{-2F},\quad \frac{g}{V}=\frac{1}{r}+F', \tag{4.25}$$

where $W=e^{-F}/r$. For the discussion of the solutions to (4.25) and their implications, the reader is referred to Ref. [8].

The second possibility leads to the equations

$$\frac{A'}{A^2}=\frac{1-W^2}{\alpha}\,,$$
$$W'=-\alpha\frac{W}{x^2A}. \tag{4.26}$$

Let us define $W=xe^{-F}$. Then, we have the nonlinear system

$$F''+\frac{2}{x}F'=e^{-2F}\,, \tag{4.27}$$
$$A=\frac{\alpha}{x(xF'-1)}. \tag{4.28}$$

Equation (4.27) is well known in mathematical physics as the classical Poisson-Boltzmann differential equation that plays a central role in the description of the thermal exploding matter [14]. The space-time is not asymptotically flat in the ordinary manner, which can be seen in the following way.

If we make the transformation $z=\frac{1}{x}$, Eq. (4.27) takes the form

$$z^4\frac{d^2F}{dz^2}=e^{-2F},\quad A=-\frac{\alpha z}{(z\frac{dF}{dz}+1)}\,, \tag{4.29}$$

which is again Eq. (4.25), which belongs to the class of Lane-Emden equations. As $z\longrightarrow 0$ (or $x\longrightarrow\infty$), we know [8]

$$F(z)\longrightarrow -\ln z+\gamma\sqrt{z}\sin\left[\frac{\sqrt{7}}{2}\ln z+\delta\right], \tag{4.30}$$

and hence,

$$A\longrightarrow\frac{-2\alpha}{\gamma}\frac{\sqrt{z}}{\sin\left[\frac{\sqrt{7}}{2}\ln z+\delta\right]+\cos\left[\frac{\sqrt{7}}{2}\ln z+\delta\right]}\longrightarrow 0 \tag{4.31}$$

instead of $A\to 1$ as $z\to 0$ or $x\to\infty$.

C. Standard coordinate system

In the standard coordinate system, the coordinate fixing condition is

$$C=1\,. \tag{4.32}$$

Again with the variable x, the equations to be solved are

$$\frac{A'}{A}+\frac{1}{x}=\pm\frac{B}{x}\,, \tag{4.33}$$

$$\frac{A'}{A} = \frac{B}{x^2}(1 - W^2) , \tag{4.34}$$

$$W' = -BW . \tag{4.35}$$

We can prove that the case with a minus sign in (4.33) contradicts the condition on asymptotic behavior. Introducing the new variable $y = W^2$, after a little algebra, the case with a positive sign leads us to the system of equations

$$y' = -\frac{2xy}{y - 1 + x}, \tag{4.36}$$

$$B = \frac{x}{y - 1 + x}, \tag{4.37}$$

$$\frac{A'}{A} = \frac{1 - y}{x(y + x - 1)}. \tag{4.38}$$

Equation (4.36) is well known in mathematical physics as Abel's differential equation of the second type. The Abel equation has clearly a solution $y = 0$. In this case we have an Abelian magnetic monopole which at the same time is an extreme Reissner-Nordström black hole

$$B = \left(1 - \frac{1}{x}\right)^{-1} , \tag{4.39}$$

$$A = \left(1 - \frac{1}{x}\right) . \tag{4.40}$$

This monopole has mass $M = 4\pi v/g$.

The Abel equation (4.41) has no known analytic solution other than $y = 0$. Before proceeding to solve it numerically, let us attempt to understand the nature of the solutions by a qualitative analysis of the equation [15]. Considering $x(t)$ and $y(t)$ as functions of a parameter t we can write the Abel equation as a pair of equations for $x(t)$ and $y(t)$ as

$$\frac{dy(t)}{dt} = -2x(t)y(t) ,$$

$$\frac{dx(t)}{dt} = y(t) + x(t) - 1 . \tag{4.41}$$

This pair has two finite singular points in the (x, y) plane at $(0, 1)$ and $(1, 0)$. The characteristic matrix at the point $(0, 1)$ has complex eigenvalues. It is described as the focus, and it can be proved that the solution curves spiral around and converge to this point. At the other point, the characteristic matrix has real eigenvalues. Such a point is described as the saddle point. The solution curves (except the curves known as separatrices) approach the saddle point and then go away from it. The separatrices in our case is the curve passing through $(0, y_{0c})$ and $(1, 0)$ and the curve $y = 0$. These features are illustrated in Fig 1.

In Fig 2, we plot the family of solutions of (4.41). Consistent with the above qualitative analysis, we find, with the initial value y_0 greater than a critical value $y_{0c} = 1.661$, the curves $y(x) = W^2(r)$ fall off exponentially for large r. With the initial value smaller than the critical value, the curves spread only to a finite value of r and return to the y axis as parts of spiral curves in the

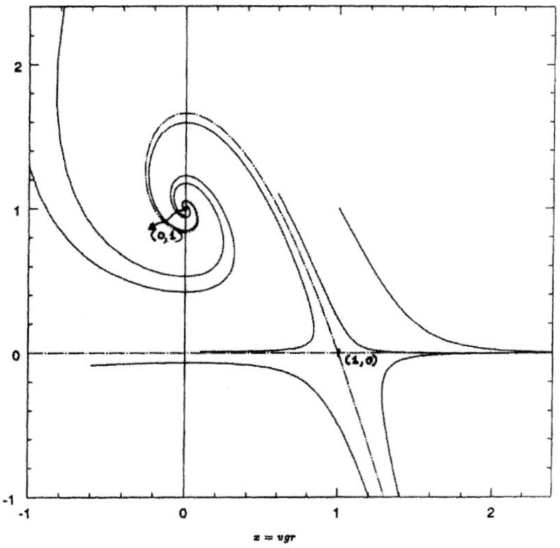

FIG. 1. Singular points in the (x, y) plane of the Abel equation's solution curves. The point $(0, 1)$ is a focus, while the one $(1, 0)$ is a saddle point with two bold dashed-dotted lines as its separatrices. Solutions spiral around a focus; solutions approach, but do not pass the saddle points.

positive quadrant of the (x, y) plane. With $y_0 = y_{0c}$, the curve passes the x axis at $x = 1$ and then continues on to negative values. Since $y = W^2$, we are not interested in negative values of y.

Since we are interested in monopole solutions, let us first concentrate our attention on the family of solutions for W^2 that fall off exponentially. In Fig. 3 we plot such

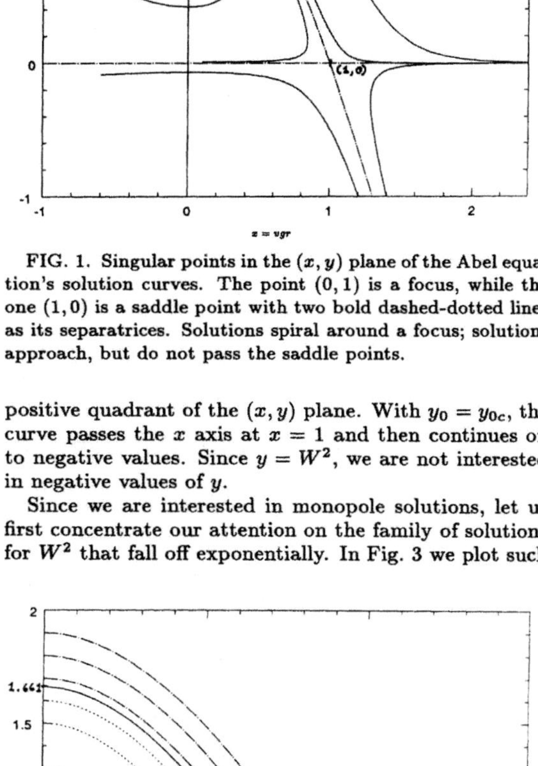

FIG. 2. Solutions of the Abel equation $y' = -\frac{2yx}{y-1+x}$. Solutions with $y_0 > y_{0c}$, fall off exponentially. Solutions with $y_0 < y_{0c}$ are shown with dotted lines. The solid line is the separatrice solution.

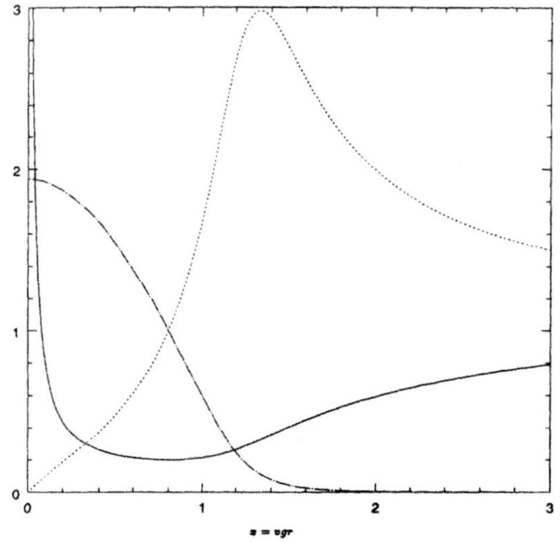

FIG. 3. A monopole solution in curved space-time. The solid and dotted curves represent the metric functions A and B, respectively. The dashed-dotted curve is $y = W^2$. The curves are plotted as function of $\ln(1 + x)$.

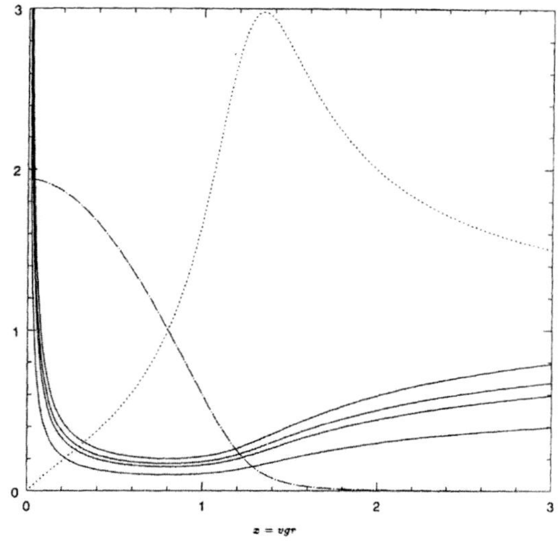

FIG. 4. Monopole solutions with the minimum of the metric function A varying. The minimum can be made arbitrarily close to zero but never zero. We can prove that if $A = 0$ at some x it has to be zero everywhere. The solid, dotted, and dashed-dotted curves represent the metric functions A, B, and the gauge field $y = W^2$, respectively. The curves are plotted as function of $\ln(1 + x)$.

a solution along with the metric functions obtained by solving (4.37) and (4.38). With the aid of (4.36) we can prove that B is regular and A has no zeros; it has a singularity at the origin. We recall from (4.10)–(4.12) that the monopole has a nontrivial, non-Abelian magnetic core when W does not vanish. Thus we have a monopole with a non-Abelian core. The singularity at the origin is not hidden by a black-hole horizon as in the case of Majumdar-Papapetrou monopole. However, it is worth noting that the metric function A has a minimum inside the core of the monopole. By changing the value of the integration constant of Eq. (4.41), this minimum can be made arbitrarily close to zero with a corresponding increase in the height of B (Fig. 4). But by analytic arguments we can show that for the exponentially falling solutions of y the metric function A which satisfies the asymptotic condition $A \longrightarrow 1$ as $x \longrightarrow \infty$ cannot strictly vanish at any finite x. This raises the following important question: can such a metric function which has a minimum arbitrarily close to zero be interpreted as a black-hole horizon? In recent numerical studies, such numerical minima are indeed interpreted as horizons.

Finally, we come to the separatrice that passes through $(0, y_{0c})$ and $(1, 0)$. Its continuation is $y = 0$ for $x \geq 1$. Taking this as a solution for y, we see that (4.38) gives

$$A(x) = 1 - \frac{1}{x} \quad \text{for} \quad x > 1, \tag{4.42}$$

where we have chosen the constant of integration to be unity so that $A(x) \longrightarrow 1$ as $x \longrightarrow \infty$ and $A(x) = 0$ at $x = 1$. It is clear from (4.37) that $B(x)$ also satisfies the asymptotic condition. It is infinite at $x = 1$ (see Fig. 5)

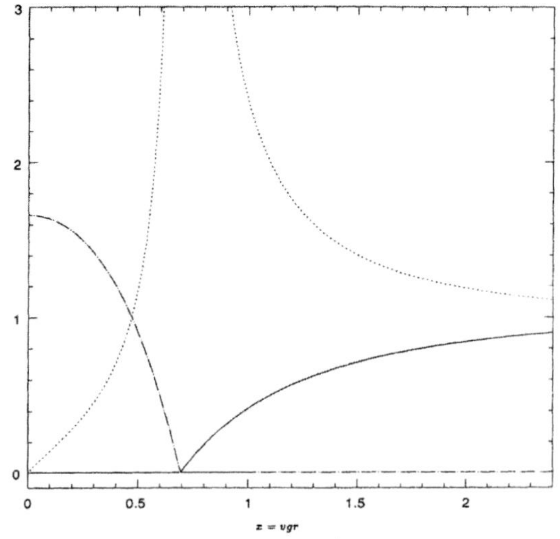

FIG. 5. A monopole-black-hole solution. The solid, dotted, and dashed-dotted curves represent the functions A, B, and $y = W^2$. The curves are plotted as function of $\ln(1 + x)$. The non-Abelian magnetic core is inside the horizon at $x = 1$. Outside of the core $W^2 = 0$, we have an Abelian field.

and is regular for $x < 1$. Now for $x \leq 1, A = 0$ is a solution. Further, we have verified that all the components of the Riemann tensor are finite and continuous at $x = 1$. The apparent singularities in these expressions all cancel as a consequence of the Abel equation (4.36). We have, therefore, a solution with a black-hole horizon at $x = 1$, surrounding the singularity at the origin. Inside the horizon, we have non-Abelian field components since $W^2 \neq 0$. Thus, the solution represents the exact analog and generalization of the Majumdar-Papapetrou solution in the case of Einstein-Maxwell system.

V. CONCLUSION

Extended objects such as non-Abelian monopoles provide a fertile ground for the study of the interplay between gravitational and gauge interactions with a spontaneously broken symmetry. Previous work by others and ours reported in this paper show conclusively that the gravitational interactions, although weak, can have significant effects on the properties of 't Hooft–Polyakov-type monopoles and therefore, cannot be ignored in quantum aspects of such extended objects.

By considering an unconventional coupling of the Higgs field to gravity, we have sought static sperically symmetric solutions to the Einstein-Yang-Mills-Higgs system. In contrast to the conventional studies of such a system with Einstein action, our unconventional coupling is better suited for the use of the Bogomol'nyi type analysis in reducing the solutions to the complex system of coupled equations to the solution of a single nonlinear equation whose solution determines the solution of the entire system. We begin with the most general spherically symmetric metric and specialize to isotropic and standard coordinate systems. In the case of the former, we extend the previous study of Balakrishna and Wali [8] to find a new type solution. The main results are in the case of the standard coordinate system. We find both Abelian and non-Abelian type solutions when the Higgs field is frozen to its vacuum expectation value.

The Abelian solution presents a generalization of the Majumdar-Papapetrou solution in the case of coupled Einstein-Maxwell system. The monopole in this case is also an extreme Reissner-Nordström black hole. In the non-Abelian case, the nonlinear equation happens to be the well-known Abel equation of the second type. We study the solutions of this equation numerically and find two types of solutions. In one family of solutions, the non-Abelian parts of the gauge fields fall off exponentially as in the case of 't Hooft–Polyakov monopoles in flat space-time. Asymptotically, we have the field of a monopole with total charge equal to its mass. However, the solution has a singularity at the origin which, strictly speaking, is not enclosed in a black-hole horizon. The metric function g_{00} has a minimum at a finite radius and can be made arbitrarily small. The corresponding metric function g_{rr} has a maximum. These features suggest the presence of a horizon. However, from analytical arguments, we can show that g_{00} cannot vanish. This leaves the physical interpretation of such a situation unclear and needs further study.

We do find a solution however, with a horizon at $vgr = x = 1$. This solution has only the Abelian component for $x \geq 1$ and behaves like a monopole for large distances with a mass $M = v4\pi/g$. Its charge is $\mathcal{G} = 1/g$. Remembering that we are working in the units $4\pi G v^2 = 1$, we can reexpress the mass as $M = \sqrt{\frac{4\pi}{G}}\frac{1}{g}$. Similarly, the magnetic charge turns out to be $\sqrt{4\pi}\mathcal{G} = \mathcal{Q} = \frac{\sqrt{4\pi}}{g}$. Thus $GM^2 = \mathcal{Q}^2$, and hence we have an extreme Reissner-Nordström black hole with magnetic charge equal to its mass. It has the unusual feature such that the metric function g_{00} vanishes for $x \leq 1$.

ACKNOWLEDGMENTS

This work was supported in part by the U.S. Department of Energy under Contract No. DE-FG02-85ER40231. Nguyen Ai Viet is grateful to Professor A. Zichichi for financial support. K.C. Wali would like to thank Professor R.F. Sawyer for the hospitality at the Physics Department, University of California, Santa Barbara where this work was partially done. He would also like to acknowledge Professor J.B. Hartle for helpful discussions.

[1] K. Lee, V.P. Nair, and E.J. Weinberg, Phys. Rev. D **45**, 2751 (1992); M.E. Ortiz, *ibid.* **45**, 2586 (1992); P. Breitenlohner, P. Forgács, and D. Maison, Nucl. Phys. **B383**, 357 (1993); P.C. Aichelburg and P. Bizon, Phys. Rev. D **48**, 607 (1993).

[2] L. O'Raifeartaigh, S.Y. Park, and K.C. Wali, Phys. Rev. D **20**, 1941 (1979). References to earlier and related work are contained in this paper.

[3] S.D. Majumdar, Phys. Rev. **72**, 390 (1947); A. Papapetrou, Proc. R. Irish Acad. Sect. A **51**, 191 (1947).

[4] J.B. Hartle and S.W. Hawking, Commun. Math. Phys. **26**, 87 (1979).

[5] P. van Nieuwenhuizen, D. Wilkinson, and M.J. Perry, Phys. Rev. D **13**, 778 (1976).

[6] R. Bartnik and J. McKinnon, Phys. Rev. Lett. **61**, 141 (1988).

[7] F.A. Bais and R.J. Russel, Phys. Rev. D **11**, 2692 (1975); Y.M. Cho and P.G.O. Freund, *ibid.* **12**, 1588 (1975).

[8] B.S. Balakrishna and K.C. Wali, Phys. Rev. D **46**, R5228 (1992).

[9] E.B. Bogomol'nyi, Yad. Fiz. **24**, 861 (1986) [Sov. J. Nucl. Phys. **24**, 449 (1976)]; M.K. Prasad and C.M. Sommerfield, Phys. Rev. Lett. **35**, 760 (1975).

[10] L. Parker, in *Recent Developments in Gravitation: Cargese 1978*, edited by M. Lévy and S. Deser, NATO Advanced Study Institute, Series B Vol. 44 (Plenum, New York, 1979), p. 219.

[11] G.'t Hooft, Nucl. Phys. **B79**, 276 (1974); A. Polyakov, Pis'ma Zh. Eksp. Teor. Fiz. **20**, 430 (1974) [JETP Lett. **20**, 194 (1974)].

[12] P. Goddard and D.I. Olive, Rep. Prog. Phys. **41**, 1357 (1978).
[13] A. Comtet, P. Forgács, and P.A. Horváthy, Phys. Rev. D **30**, 468 (1984).
[14] P.L. Chambré, J. Chem. Phys. **20**, 1795 (1952).
[15] A.A. Andronov, E.A. Leontovich, I.I. Gordon, and A.G. Maier, *Qualitative Theory of Second-order Dynamic Systems* (Wiley, New York, 1973).

Black holes with magnetic charge and quantized mass

A.Yu.Ignatiev, G.C.Joshi and Kameshwar C.Wali
Research Centre for High Energy Physics, School of Physics, University of Melbourne, Parkville 3052, Victoria, Australia
Department of Physics, Syracuse University, Syracuse, NY 13210
(Invited paper for the first editorial volume of the book series "Contemporary Fundamental Physics" by the Nova Science Publishers)

Abstract

We examine the issue of magnetic charge quantization in the presence of black holes. It is pointed out that quantization of magnetic charge can lead to the mass quantization for magnetically charged black holes. We also discuss some implications for the experimental searches of magnetically charged black holes.

The purpose of this note is to draw attention to some aspects of the interplay between gravity and the quantization of magnetic charge. The issue we are going to address is not only of fundamental theoretical importance but also can be relevant for the experimental searches of monopoles.

Let us view the problem of magnetic charge from two perspectives. First, the magnetic charge g_{RN} of the extreme Reissner-Nordstrom black hole is related to its mass according to the equation,

$$g_{RN} = \sqrt{G}M, \tag{1}$$

where M is the mass of the black hole. This relation follows from an exact solution of Einstein-Maxwell equations [1,2] , and is a purely classical result with no appeal to quantum mechanics. Similarly, an extension to coupled Einstein- Yang-Mills-Higgs system leads to non-abelian and abelian black holes with magnetic charges [3]. References to earlier and related work are contained in this paper. More recently, generalizations to slowly rotating non-Abelian black holes have been reported by M.S. Volkov and N. Straumann [4]. Second, the Dirac quantization condition [5] reads:

$$g_D = \frac{n\hbar c}{2e}. \tag{2}$$

We use the Gauss system of units where $e^2/\hbar c = \alpha$.

Now, what is the relationship between the Eqs.(1) and (2)? Should we equate the left-hand sides of the above formulas? Or are they totally independent of each other? Surprisingly, this supposedly simple question, to the best of our knowledge, does not have a straightforward answer. What we really obtain here is a dilemma: on the one hand, we can insist that the Dirac quantization condition derived in flat space-time, should be valid also in curved space-time in the presence of a black hole; on the other hand, we may conjecture that there may exist some mechanism of evading the Dirac quantization condition in the regions with strong gravitational fields. Both these alternatives have interesting consequences. Which of them is realized in Nature seems to be an open question at present. For some particular cases it has been claimed that the Dirac quantization condition remains intact in the presence of gravity [6]. However, there is no general proof that we know of, establishing the validity of the Dirac quantization condition in arbitrary curved space-time. Neither there are any models that would definitely establish a possibility of its violation. Therefore, we examine qualitatively each scenario and discuss the consequences.

If we assume that the Dirac quantization condition is valid in the presence of gravity then we can equate the left-hand sides of Eqs.(1) and (2) to obtain the equation

$$M = \frac{n\hbar c}{\sqrt{G}2e} = n\frac{M_{Pl}}{2\sqrt{\alpha}}. \tag{3}$$

This says that the masses of Reissner-Nordstrom black holes are quantized and the minimum possible mass for the extreme magnetically charged black holes is given by

$$M_{min} = 5.85M_{Pl}, \tag{4}$$

with an equi-spaced mass spectrum analogous to the energy spectrum of a quantum simple harmonic oscillator. Whether such black holes are the final stages of more massive black holes undergoing Hawking radiation and reaching a stable state is an interesting question. Anyway, theory aside, the range of magnetic monopole masses around the point $5.85M_{Pl}$ should be considered seriously in the monopole search experiments.

Now, let us consider the other alternative, namely, the possible violation of the Dirac quantization condition in the presence of gravity. Not knowing the precise nature of its violation assume that the magnetic charge of a Reissner-Nordstrom black hole obeys only one constraint, that of Eq.(1), being unrestricted by Eq.(2). Thus, we have magnetic monopoles with "dequantized" magnetic charge determined by the mass [7]. Thus there is no restriction on the range of allowed magnetic charges. However, let us confine our attention to charges between zero and one Dirac unit, $g = 68.5e$ and note that very small magnetic charges correspond to the black holes with the mass much less than the Planck mass for which the Compton wavelength $\frac{\hbar}{Mc}$ is longer than the Schwarzschild radius $\frac{2GM}{c^2}$. Therefore, the effects of quantum gravity should be quite important for such black holes. For this reason it would be safer to leave very small magnetic charges from our consideration. A particularly interesting case from the point of view of symmetry between electricity and magnetism, is $g = e$. This equality of magnetic charge of the black hole and the electric charge of the electron would occur for the black hole mass

$$M = \sqrt{\alpha} M_{Pl} = 0.08 M_{Pl} \tag{5}$$

Another special case of interest is the black hole with the Planck mass for which the magnetic charge is

$$g_{Pl} = \frac{e}{\sqrt{\alpha}} = 11.7e, \tag{6}$$

so $g_{Pl} = 1$ in dimensionless units ($\hbar = c = 1$). Black holes with the Planck mass have been called maximons in [8].

In this context, it is interesting to recall that the reported monopole observations of Ehrenhaft [9] corresponded to the values of magnetic charge *less* than the standard Dirac charge of $68.5e$, and hence they were given less credibility on the ground that they violated Dirac quantization condition.

Are there any arguments in favour of the violation of the Dirac quantization condition by the gravitational effects? Originally, the Dirac quantization condition appeared as the necessary condition for the unobservability of the Dirac string. The Dirac string is the line of a singularity in the vector potential describing the monopole. It starts at the monopole location and stretches to infinity. Analytically it can be written, for example, as

$$A_r^D = A_\theta^D = 0, \tag{7a}$$

$$A_\varphi^D = \tfrac{g}{r} \tan \tfrac{\theta}{2}. \tag{7b}$$

The magnetic field corresponding to that vector potential is also singular and has the form [10] (here, $\mathbf{n}$ is the unit vector along the string, i.e. opposite the z-axis):

$$\nabla \times \mathbf{A}^D = \frac{g}{r^3}\mathbf{r} + \mathbf{n} 4\pi g \theta(-z)\delta(x)\delta(y). \tag{8}$$

Note that to obtain the above equation rigorously one has to treat all quantities involved in the sense of distributions rather than regular functions [11]. Therefore, the string makes a singular contribution to the energy momentum tensor of the electromagnetic field. A natural question then arises: can such a string be observed by its gravitational effects? The gravitational field of a gauge string has been studied in [12] (A gauge string has a finite mass μ per unit length whereas a Dirac string has *infinite* line density). It has been found that the spacetime around such a string is locally flat, but globally it is similar to a cone. It means that the range of the polar angle ϕ (measured around the string) is not 2π but $0 \leq \phi \leq 2\pi(1-4G\mu)$. This angular deficit has been called a conical singularity. The presence of conical singularity leads to several observable effects: double images of sources located behind the string, anisotropy of the background radiation, and the creation of sheets of matter in the wake of the string. These results strongly suggest that the Dirac string *may* become observable by its gravitational effects *regardless of the value of the magnetic charge.* [1] Thus we are led to conjecture that in the presence of gravitation it may become irrelevant whether the quantization condition for the Dirac monopole is fulfilled or violated. (Of course, an alternative interpretation of the same result would be to claim that the whole concept of the Dirac monopole is inconsistent with gravity.)

Let us now turn to the situation with the t'Hooft-Polyakov type of monopoles. For these monopoles there exist two very different pictures: one uses a gauge with the "hedgehog" configuration of the scalar field; this gauge is completely free of any singularity lines (non-singular gauge). The other approach is very close to the picture of a Dirac monopole because of the presence of a string similar to the Dirac string (string gauge). These two pictures are believed to be physically equivalent because there exist a (singular) gauge transformation from one to the other. Therefore, if we choose to work in the string gauge we can again require the string to be unobservable and thus obtain the quantization condition.

Now, we would like to think of the monopole coupled to gravity. One of the fundamental questions is: will the two gauges remain physically equivalent? If they are then we can use again the same line of argument as above. Consequently, we can again conjecture that the quantization condition may become irrelevant in the presence of gravity. On the other hand, the non-singular gauge and the string gauge may become non-equivalent once the gravity is switched on, in which case we do not know the form the quantization condition takes. In other words, at the present state of our knowledge, arbitrary magnetic charges unconstrained by the Dirac quantization condition seem not to be ruled out by sound theoretical arguments. This should be kept in mind in experimental searches.

We now turn to the analysis of the experimental constraints on the flux of magnetically charged black holes ("black poles" for short). The most stringent constraints come from

---

[1]Note that there exists some evidence that the string may become visible even in flat space. In [13] the motion of a monopole in the magnetic field of an external electric current was analyzed. Roughly speaking, the physical results depend on how many times the monopole string winds up around the current line. Another configuration where the strings turn out to be a problem has been considered in [14]: a charged particle in the magnetic field of a Dirac monopole *line*. Further problems related to the Dirac quantization condition and existence of strings have been raised recently in the context of quantum field theory with monopoles, see e.g. [15–17].It is not clear whether or not the stringless formalisms of Wu-Yang and others [18–20] based on the fibre bundle theory would help in solving these problems.

the results of non- accelerator experiments devoted to the search of superheavy magnetic monopoles arising in Grand Unified Theories of strong, weak, and electromagnetic (but not including gravitational) interactions. These searches can be divided into two main classes: the first uses the effect of current induction caused by the monopole magnetic field and the second is based upon the ionizing properties of monopoles. The induction experiments give the most direct upper limits on the monopole flux because they are independent of the unknown characteristics of the monopoles such as its mass and velocity. The best upper bounds on the monopole flux F from induction experiments are:

$$F < 4.4 \times 10^{-12} cm^{-2} sr^{-1} s^{-1} \text{ [21]}, \tag{9}$$

$$F < 7.2 \times 10^{-13} cm^{-2} sr^{-1} s^{-1} \text{ [22]} \tag{10}$$

$$F < 5 \times 10^{-14} cm^{-2} sr^{-1} s^{-1} \text{ [23]} \tag{11}$$

These constraints can be taken over for the case of *quantized* black holes regardless of their mass and velocity. As for the *dequantized* black holes, however, the situation is different. The reason is that the magnetic flux through a superconducting loop is quantized and can change only in multiples of magnetic flux quantum $\phi^0 = \frac{1}{2}(4\pi \times \frac{\hbar c}{2e})$. Therefore, it is unclear to what extent the superconductive induction experiments are sensitive to the dequantized magnetic charges. A sufficient condition for a magnetic charge g to be detected in the induction experiment reads

$$4\pi g = \frac{\phi_0}{n_l}, \tag{12}$$

where n_l is a geometric factor which is specific for the loop configuration adopted in a particular experiment. For instance, for the original Cabrera's detector $n_l = 4$ since there are 4 superconducting loops connected in a series. An interesting problem is how to design an experimental setup (or use the existing setups) in such a way that it would be capable of detecting *both* quantized *and dequantized* magnetic charges (including the charges not satisfying Eq. (12)).

Let us now turn to the experiments that use the ionizing properties of monopoles. We consider first the case of quantized black holes, assuming that the black holes with $n \geq 1$ are stable. The ionization experiments are much more sensitive to the theoretical input than the induction experiments. In particular, their sensitivity crucially depends on the monopole's velocity: the monopole would be undetectable if its velocity were below the threshold, somewhere between $10^{-3}c$ and $10^{-4}c$. To find out the typical monopole velocity, we need to find the greater of the two characteristic velocities: the first is the velocity with which the monopole enters the Galaxy ($\sim 10^{-3}c$) and the second is the typical monopole velocity due to its acceleration by the galactic magnetic field, given by

$$v_n \simeq 3 \times 10^{-3} c (\frac{10^{16} GeV}{M_n})^{1/2}. \tag{13}$$

The quantized magnetically charged black holes with masses $M \gtrsim 6 \times 10^{19}$ GeV can be viewed as *ultraheavy* monopoles (as contrasted with *superheavy* GUT monopoles with the mass of the order of 10^{16} GeV). It is readily seen that the total spectrum of the quantized black poles can be divided into two parts: 1)"low-lying", low charge black poles, corresponding to $n \lesssim 600$ and 2)"very heavy ", heavily charged black poles" with $n \gtrsim 600$. For the low-lying black poles the galactic magnetic field is not strong enough to accelerate them, so their typical velocity is expected to be of the order of $10^{-3}c$. Therefore the astrophysical bounds on their flux obtained from the condition of survival of the galactic magnetic field (the Parker bound and its refinements) are *inapplicable* for them. On the contrary, for the superheavy black poles their dynamics is dominated by the galactic magnetic field, so most of them are soon ejected out of the galaxy similar to the case of the ordinary GUT monopoles. Thus, the low-lying black poles with the mass in the interval $(6 - 3600)M_{Pl}$ and, consequently, the magnetic charge in the range $(1 - 600)g_1$ seem to be more interesting experimentally. Although the Parker-type bounds do not hold for these poles, their flux can nevertheless be constrained at a similar level, $F \lesssim 10^{-15}cm^{-2}sr^{-1}s^{-1}$, from the requirement that the galactic density of the poles should not be unacceptably high. The best available limits from the ionization detectors on the ultraheavy monopole flux with velocities of the order of $10^{-3}c$ have the same order of magnitude. For instance,

$$F < 5.6 \times 10^{-15}cm^{-2}sr^{-1}s^{-1} \ [24], F < 2.7 \times 10^{-15}cm^{-2}sr^{-1}s^{-1} \ [25] \tag{14}$$

Furthermore, if one takes into account the possibility of monopole catalysis of nucleon decay (Rubakov-Callan effect) [26]) then a somewhat stronger limit can be obtained [27] (it is, however, sensitive to the assumed catalysis cross-section):

$$F < 5 \times 10^{-16}cm^{-2}sr^{-1}s^{-1}. \tag{15}$$

To summarize, we have examined the issue of magnetic charge quantization in the presence of black holes. It was pointed out that quantization of magnetic charge can lead to the mass quantization for magnetically charged black holes. We have also discussed some implications for the experimental searches of magnetically charged black holes.

This work was supported in part by a grant from National Science Foundation, Division of International Program (U.S.-Australia Cooperative Research). One of the authors (K.C.W) would also like to acknowledge the hospitality at the Institut des Hautes Etudes Scientifiques during the course of this work.

REFERENCES

[1] S.D.Majumdar, Phys.Rev. 72, 390 (1947); A.Papapetrou, Proc.R.Irish Acad. Sect.A 51, 191 (147)

[2] J.B.Hartle and S.W.Hawking, Commun.Math.Phys. 26, 87 (1979)

[3] For instance, see B.S.Balakrishna and K.C.Wali, Phys.Rev.D46, R522 (1992) N.A.Viet and K.C.Wali, Phys.Rev.D 51, 1664 (1995)

[4] M.S.Volkov and N.Straumann, ZU-TH/97 hep-th/9704026

[5] P.A.M.Dirac, Proc.Roy.Soc. (London) A133, 60 (1931)

[6] R.J.Adler, Phys.Rev. D14, 392 (1976); A.Inomata and M.Trinkala, Phys. Rev. D18, 1861 (1978); J.L.Friedman, S.Mayer and L.Parker, Phys.Rev. D17, 1957 (1978)

[7] E.B.Bogomolnyi, Yad.Phys.24, 449 (1976); M.K.Prasad and C.H.Sommerfield, Phys.Rev.Lett. 35, 760 (1975)

[8] M.A.Markov, Soviet Physics JETP 24, 584 (1967)

[9] Ehrenhaft, Z.Phys.31, 478 (1930); V.F.Mikhailov, Phys.Lett. B130, 331 (1983)

[10] G.Wentzel, Progr.Theor.Phys.Suppl. 37-38, 163 (1966)

[11] V.I.Strazhev and L.M.Tomilchik, Electrodynamics with Magnetic Charge Minsk, 1975 (in Russian)

[12] A.Vilenkin, Phys.Rev. D23, 852 (1981); R.Gregory Phys.Rev.Lett. 59, 740 (1987)

[13] H.J.Lipkin and M.Peshkin, Phys.Lett.B179, 109 (1986)

[14] J.Q.Liang and D.S.Kulshreshtha, Phys. Lett. A149, 1 (1990)

[15] H.-J. He, Z.Qiu and C.-H.Tze, Z.Phys. C 65, 175 (1994); G.I.Poulis and P.J.Mulders, Z.Phys. C67, 181 (1995)

[16] A.Yu.Ignatiev and G.C.Joshi, hep-ph/9705349, Mod.Phys.Lett. A 13, 2295 (1998) ; hep-ph/9710553, to appear in Chaos, Solitons and Fractals

[17] L.Gamberg, G.Kalbfleisch and K.A.Milton, OKHEP-98-02 hep-ph/9805365

[18] T.T.Wu and C.N.Yang, Phys. Rev. D12, 3845 (1975)

[19] A.Trautman, Int. J. Theor. Phys. 16, 561 (1977)

[20] J.L.Friedman and R.D.Sorkin, Phys.Rev. D20, 2511 (1979)

[21] Gardner, Cabrera, Huber and Taber, Phys.Rev. D44, 622 (1991)

[22] Huber, Cabrera, Taber and Gardner, Phys.Rev. D44, 636 (1991)

[23] Caplin et al. Nature 321, 402 (1986)

[24] Ahlen et al. Phys.Rev.Lett. 72, 608 (1994)

[25] Orito, Ichinose and Nakamura, Phys.Rev.Lett. 66, 1951 (1991)

[26] V.A.Rubakov, Pisma ZHETF 33, 658 (1981); Nucl.Phys.B203, 311 (1982) C.G.Callan, Phys.Rev. D25, 2141 (1982); ibid. 26, 2058 (1982)

[27] L.B.Bezrukov et al. Sov.J.Nucl.Phys. 52, 54 (1990)

Clash of symmetries on the brane

A. Davidson,[1,*] B. F. Toner,[2,†] R. R. Volkas,[2,‡] and K. C. Wali[3,§]
[1]*Physics Department, Ben-Gurion University of the Negev, Beer-Sheva 84105, Israel*
[2]*School of Physics, The University of Melbourne, Victoria 3010, Australia*
[3]*Department of Physics, Syracuse University, Syracuse, New York 13244-1130*

(Received 7 February 2002; published 5 June 2002)

If our $(3+1)$-dimensional universe is a brane or domain wall embedded in a higher dimensional space, then a phenomenon we term the "clash of symmetries" provides a new method of breaking some continuous symmetries. A global $G_{\rm cts}\otimes G_{\rm discrete}$ symmetry is spontaneously broken to $H_{\rm cts}\otimes H_{\rm discrete}$, where the continuous subgroup $H_{\rm cts}$ can be embedded in several different ways in the parent group $G_{\rm cts}$, and $H_{\rm discrete}<G_{\rm discrete}$. A certain class of topological domain wall solutions connects two vacua that are invariant under *differently embedded* $H_{\rm cts}$ subgroups. There is then enhanced symmetry breakdown to the intersection of these two subgroups on the domain wall. This is the "clash." In the brane limit, we obtain a configuration with $H_{\rm cts}$ symmetries in the bulk but the smaller intersection symmetry on the brane itself. We illustrate this idea using a permutation symmetric three-Higgs-triplet toy model exploiting the distinct $I-$, $U-$ and $V-$spin U(2) subgroups of U(3). The three disconnected portions of the vacuum manifold can be treated symmetrically through the construction of a threefold planar domain wall junction configuration, with our universe at the nexus. A possible connection with E_6 is discussed.

PACS number(s): 11.27.+d

I. INTRODUCTION

The notion of symmetry lies at the base of modern particle theory, as exemplified by the standard model. Some symmetries, such as electromagnetic gauge invariance, are manifest: the zero-temperature vacuum state and all material systems except for superconductors exhibit the symmetry in an explicit fashion. The Glashow-Weinberg-Salam $SU(2)_L\otimes U(1)_Y$ electroweak symmetry, on the other hand, is spontaneously broken: the symmetry of the Lagrangian is not shared by the vacuum state. In the standard model, the self-interactions of elementary Higgs bosons are responsible for making the vacuum asymmetric.

A common opinion is that spontaneous symmetry breaking via Higgs bosons is not completely satisfactory, because of the proliferation of parameters it brings. In the standard model these are predominantly Yukawa coupling constants, while in extended theories parameters in the Higgs potential can also abound. Hierarchies in *a priori* arbitrary parameter values can also be seen as troubling. For these reasons, and as an end in itself, a search for new ways of breaking symmetries is well motivated.

The purpose of this work is not to do away with elementary Higgs fields, but rather to show how they can induce more symmetry breakdown than allowed by conventional theory. The scope of the paper is to illustrate the basic idea through a nontrivial toy model, and then to discuss possible future developments (especially a connection with E_6).

In a conventional theory such as the standard model, the Higgs field configuration is assumed to be spatially homogeneous, with a vacuum expectation value derived by minimizing the Higgs potential. However, it is well known that completely stable solitonic configurations can also exist if the vacuum manifold has the appropriate topology [1]. Although such configurations have higher energy than the vacuum state, their topological stability allows their use as a background field. In this work, we will use domain wall configurations associated with spontaneously broken discrete symmetries. We will show how the symmetry group at the center of a domain wall can be smaller than what you get with a homogeneous vacuum configuration, through a phenomenon we term the "clash of symmetries." It arises when the symmetry group H of the vacuum manifold can be embedded in several ways within the parent group G. The enhanced symmetry breakdown is caused by the clash of the different internal orientations of H within G. We will use the $I-$, $U-$ and $V-$spin U(2) subgroups of U(3) in our toy model.

There is no observational evidence for a domain wall Higgs background of the conventional type in our $(3+1)$-dimensional universe [2]. To use the clash of symmetries for realistic model building, we will therefore ultimately have to identify our universe as a submanifold of a higher dimensional space [3,4]. If the submanifold is infinitely thin in the extra dimensions, then it is commonly called a "brane," with its complement termed the "bulk." We will identify our universe with the center of a domain wall configuration dynamically induced by Higgs fields which exist in the bulk. The symmetry group at the center of the wall is then the symmetry group of our universe. By taking the appropriate limit, the domain wall can be made infinitely thin—our universe becomes a brane. Actually, in our toy model the most theoretically appealing configuration will be a junction of three semi-infinite walls separated from each other by angles of $2\pi/3$. In this case, it seems most

*Email address: davidson@bgumail.bgu.ac.il
†Present address: Department of Physics, California Institute of Technology, Pasadena, California 91125. Email address: toner@caltech.edu
‡Email address: r.volkas@physics.unimelb.edu.au
§Email address: wali@physics.syr.edu

natural to identify our (toy) universe with the three-way intersection point, the nexus. We will call this the "three-star configuration."[1]

The proposal that we live in a domain wall was made long ago [3]. In recent times, the study of branes and/or submanifolds has become a major activity. Motivations include string theory, Regge-Teitelbaum gravity and the hierarchy problem [4]. It is interesting that our motivation to consider brane physics is the independent argument presented above. Combining the clash of symmetries idea with other brane-world ideas may be a fruitful direction for future work.

Working independently and with a completely different motivation, Pogosian and Vachaspati recently discovered a class of $SU(N)\otimes Z_2$ Higgs-adjoint kinks featuring the clash of symmetries idea [6]. They consider an $SU(N)$-adjoint Higgs field Φ, with Higgs potential

$$V(\Phi)=-m^2\mathrm{tr}(\Phi^2)+\gamma\mathrm{tr}(\Phi^3)+\lambda_1\mathrm{tr}(\Phi^4)+\lambda_2[\mathrm{tr}(\Phi^2)]^2. \quad (1)$$

In the absence of the cubic term ($\gamma=0$), there is a Z_2 phase symmetry, $\Phi\leftrightarrow-\Phi$, which is outside $SU(N)$ for odd $N\geq 5$. For example, in the region of Higgs-parameter space where SU(5) breaks to $SU(3)\otimes SU(2)\otimes U(1)$, the authors find domain wall solutions, for which the clash of symmetries results in additional symmetry breaking to $SU(2)^2\otimes U(1)^2$.[2]

In this context, quite different from the brane-world motivation, our paper will present another example of this type of kink configuration, within a $U(3)\otimes S_3$ three-Higgs-triplet model. The S_3 symmetry is a permutation symmetry acting on the three Higgs triplets. Like Pogosian and Vachaspati, we impose the discrete symmetry by hand, by restricting terms that may appear in the Higgs potential. We, however, utilize a permutation rather than phase discrete symmetry because it can be easily generalized to other groups and other Higgs representations. This is a significant difference from the Pogosian-Vachaspati scenario, and for an SU(3) model it seems to be a necessary ingredient.[3] The threefold structure of our vacuum manifold will lead us to construct the three-star wall-junction configuration mentioned above, an exercise that is also of intrinsic technical interest [5,8].

The rest of the paper is structured as follows. In the next section, we introduce the toy model and discuss its one-dimensional kink configurations. Section III is devoted to the three-star configuration. We describe a possible connection with E_6 in Sec. IV. Conclusions and future directions are aired in Sec. V. The Appendixes establish that the domain walls exhibiting the clash of symmetries will be globally stable in a region of parameter space.

II. TOY MODEL AND ITS ONE-DIMENSIONAL KINK SOLUTIONS

A. The Higgs potential and the vacuum manifold

Consider a model with three Higgs triplets $\Phi_{1,2,3}$ interacting through the potential

$$\begin{aligned}V=&-m^2(\Phi_1^\dagger\Phi_1+\Phi_2^\dagger\Phi_2+\Phi_3^\dagger\Phi_3)\\&+\kappa(\Phi_1^\dagger\Phi_1+\Phi_2^\dagger\Phi_2+\Phi_3^\dagger\Phi_3)^2+\lambda_1(\Phi_1^\dagger\Phi_1\Phi_2^\dagger\Phi_2\\&+\Phi_2^\dagger\Phi_2\Phi_3^\dagger\Phi_3+\Phi_3^\dagger\Phi_3\Phi_1^\dagger\Phi_1)+\lambda_2(\Phi_1^\dagger\Phi_2\Phi_2^\dagger\Phi_1\\&+\Phi_2^\dagger\Phi_3\Phi_3^\dagger\Phi_2+\Phi_3^\dagger\Phi_1\Phi_1^\dagger\Phi_3).\end{aligned} \quad (2)$$

The symmetry group of this potential is

$$\begin{aligned}G&=G_{\mathrm{cts}}\otimes G_{\mathrm{discrete}}\\&=SU(3)\otimes U(1)_1\otimes U(1)_2\otimes U(1)_3\otimes S_3,\end{aligned} \quad (3)$$

where the U(1)'s are individual overall phase symmetries for the Φ's. The diagonal subgroup of the U(1)'s can be merged with SU(3) to form U(3), so $G_{\mathrm{cts}}=U(3)\otimes U(1)^2$ also. The role of the discrete permutation symmetry S_3 is to provide topological stability for domain wall configurations. The λ_2 term ensures that the continuous symmetry has a common SU(3) for all three multiplets, and the sign of λ_2 will cause kinks displaying the clash of symmetries ("asymmetric kinks") to have a different energy from those that do not ("symmetric kinks"). We show below that the asymmetric kink has lower energy if $\lambda_2>0$, while the symmetric kink has lower energy if $\lambda_2<0$.

The U(1) phase symmetries are not germane to our analysis. The potential in Eq. (2) was chosen purely for simplicity. By including terms such as $\Phi_1^\dagger\Phi_2+\Phi_2^\dagger\Phi_3+\Phi_3^\dagger\Phi_1+\mathrm{H.c.}$, the symmetry group can be reduced to the more elegant $U(3)\otimes S_3$. Inclusion of such terms would change the details of our analysis but not its spirit.

To simplify the exposition, we set $\kappa=1$ in Eq. (2) by rescaling the field and spacetime coordinates and measure all mass-dimension quantities in units of $|m|$, which is equivalent to setting $m^2=1$.

A straightforward analysis shows that there exist three degenerate global minima of the form

$$\text{Vacuum I:}\quad \langle\Phi_1^\dagger\Phi_1\rangle=\frac{1}{2},\quad \langle\Phi_2^\dagger\Phi_2\rangle=\langle\Phi_3^\dagger\Phi_3\rangle=0, \quad (4)$$

$$\text{Vacuum II:}\quad \langle\Phi_2^\dagger\Phi_2\rangle=\frac{1}{2},\quad \langle\Phi_1^\dagger\Phi_1\rangle=\langle\Phi_3^\dagger\Phi_3\rangle=0, \quad (5)$$

$$\text{Vacuum III:}\quad \langle\Phi_3^\dagger\Phi_3\rangle=\frac{1}{2},\quad \langle\Phi_1^\dagger\Phi_1\rangle=\langle\Phi_2^\dagger\Phi_2\rangle=0, \quad (6)$$

---

[1] Although our toy model will not incorporate gravity, it is reassuring to note that gravity can be localized to such a brane junction in the Randall-Sundrum scenario, provided a single fine-tuning between the cosmological constant and brane tensions is satisfied [5].

[2] It is interesting to note that stable wall structures can exist even if a discrete symmetry is explicitly broken [7]. This can apply for wall configurations having an everywhere non-vanishing Higgs field, a condition satisfied by the cases of interest in this paper.

[3] It turns out that the exact SU(3) analogue of the Pogosian-Vachaspati kink does not exist because of an accidental SO(8) symmetry.

for the parameter region

$$\lambda_1>0, \quad \lambda_1+\lambda_2>0. \tag{7}$$

(Positivity of the potential requires the weaker conditions $\lambda_1>-3$ and $\lambda_1+\lambda_2>-3$.) Each global minimum of Eqs. (4)–(6) induces the spontaneous breakdowns

$$S_3 \to S_2 \equiv Z_2,$$

$$U(3)\otimes U(1)^2 \to U(2)\otimes U(1)^2. \tag{8}$$

At the level of global vacuum configurations, a $G_{\rm cts}$ transformation can always be used to bring the nonzero $\langle\Phi_i\rangle$ into the form $(1/\sqrt{2},0,0)^T$. If this is done, then the unbroken U(2) and S_2 subgroups act on the second and third entries of the triplets. The vacuum manifold consists of three disconnected pieces labeled I–III in Eqs. (4)–(6), with each piece being the set of all $G_{\rm cts}$ transforms of $(1/\sqrt{2},0,0)^T$ for the nonvanishing $\langle\Phi_i\rangle$.

B. Clash of symmetries

A kink or one-dimensional domain wall configuration interpolates between elements of I and II, or II and III, or I and III, with the vacuum states reached at spatial infinity, $z=\pm\infty$. (We will call z the coordinate perpendicular to the wall. For the purposes of the following mathematics, it does not matter how many additional spatial directions exist.) For the sake of the example, focus on I$\leftrightarrow$II kinks. Let us use our SU(3) freedom to set the vacuum I state at $z=-\infty$ to be

$$\langle\Phi_1\rangle=\left(\frac{1}{\sqrt{2}},0,0\right)^T, \quad \langle\Phi_2\rangle=(0,0,0)^T,$$

$$\langle\Phi_3\rangle=(0,0,0)^T. \tag{9}$$

The unbroken symmetry there is clearly $U(2)_{\rm I}$, where $U(2)_{\rm I,II,III}$ is defined to act on the (23,31,12) entries of the triplets.[4] *A priori*, the vacuum II state at $z=+\infty$ can be any element of piece II of the vacuum manifold. While all such kinks are in the same topological class, they are energetically distinguished by the λ_2 term in V. Only the lowest energy member of the class is guaranteed topological stability (see below).

The extreme cases are given by the symmetric kink for which

$$\Phi_1(-\infty)=\left(\frac{1}{\sqrt{2}},0,0\right)^T, \quad \Phi_2(-\infty)=(0,0,0)^T,$$

$$\Phi_1(+\infty)=(0,0,0)^T, \quad \Phi_2(+\infty)=\left(\frac{1}{\sqrt{2}},0,0\right)^T, \tag{10}$$

and by the asymmetric kink for which

$$\Phi_1(-\infty)=\left(\frac{1}{\sqrt{2}},0,0\right)^T, \quad \Phi_2(-\infty)=(0,0,0)^T,$$

$$\Phi_1(+\infty)=(0,0,0)^T, \quad \Phi_2(+\infty)=\left(0,\frac{1}{\sqrt{2}},0\right)^T, \tag{11}$$

with $\Phi_3(z)=0$ for both cases.

The basic clash of symmetries phenomenon is displayed by the asymmetric kink. The symmetric kink has $U(2)_{\rm I}$ unbroken for all z. For the asymmetric kink, the $z=-\infty$ symmetry is $U(2)_{\rm I}$ while the $z=+\infty$ symmetry is the *different* group $U(2)_{\rm II}$. At all z in between, the symmetry is reduced to

$$H_{\rm I\cap II}=U(2)_{\rm I}\cap U(2)_{\rm II}=U(1)_{\rm III}, \tag{12}$$

where $U(1)_{\rm III}$ multiplies the third entry of each Φ by the same phase. [The groups $U(1)_{\rm I,II}$ are similarly defined through cyclic permutations of the subscripts, and all three should not be confused with $U(1)_{1,2,3}$.] Because the symmetry groups at $z=\pm\infty$ have two additional U(1) factors given our simplified Higgs potential, the unbroken symmetry for $|z|<\infty$ is actually $U(1)_{\rm III}\otimes U(1)^2$. The extra generators are easily determined, and we will not display them.

C. Kink profiles

To derive the kink profiles, one solves for static z-dependent solutions to the Euler-Lagrange equations. Adopting the ansatz

$$\Phi_1=(\phi_1,0,0)^T, \quad \Phi_2=(0,\phi_2,0)^T, \quad \Phi_3=(0,0,\phi_3)^T, \tag{13}$$

with ϕ_1, ϕ_2 and ϕ_3 *real*, the equations for the asymmetric kinks are

$$\phi_1''=\phi_1[-1+2\phi_1^2+(2+\lambda_1)(\phi_2^2+\phi_3^2)]$$
$$\text{and cyclic permutations,} \tag{14}$$

where the prime denotes differentiation with respect to z. We justify the ansatz of Eq. (13) in Appendix A, where we shall see that it is both necessary and sufficient to obtain stable domain wall configurations.

Returning to our I$\leftrightarrow$II example, we set $\phi_3=0$, and rewrite the two remaining equations in terms of

$$S\equiv\phi_1+\phi_2, \quad A\equiv\phi_1-\phi_2 \tag{15}$$

to obtain

$$S''=-S+\left(1+\frac{\lambda_1}{4}\right)S^3+\left(1-\frac{\lambda_1}{4}\right)SA^2, \tag{16}$$

$$A''=-A+\left(1+\frac{\lambda_1}{4}\right)A^3+\left(1-\frac{\lambda_1}{4}\right)AS^2. \tag{17}$$

---

[4] In the old days these were called *V*-spin, *U*-spin and *I*-spin.

Notice that the value of λ_2 has no effect on the asymmetric kink profile.

The special parameter point $\lambda_1=4$ sees the equations decouple. The solutions with the correct boundary conditions are then simply

$$S(z)=\frac{1}{\sqrt{2}}, \tag{18}$$

$$A(z)=-\frac{1}{\sqrt{2}}\tanh\frac{z}{\sqrt{2}}, \tag{19}$$

or, equivalently,

$$\phi_1(z)=\frac{1}{2\sqrt{2}}\left(1-\tanh\frac{z}{\sqrt{2}}\right), \tag{20}$$

$$\phi_2(z)=\frac{1}{2\sqrt{2}}\left(1+\tanh\frac{z}{\sqrt{2}}\right). \tag{21}$$

The hyperbolic tangent function is the archetypal kink profile because it is analytically simple. If $\lambda_1\neq 4$, then kink solutions still exist but can usually be found numerically only (there is another analytic solution describing a stable kink for $\lambda_1=\infty$ which we discuss below). A feature of the hyperbolic tangent solution is that $\phi_{1,2}(z)+\phi_{1,2}(-z)=1/\sqrt{2}$.

The brane limit corresponds to the wall being infinitely thin. To access it, the mass parameter $|m|$ must be reinstated and taken to infinity (z becomes $|m|z$). In this limit

$$\phi_1(|m|z)\rightarrow\frac{\Theta(-z)}{\sqrt{2}},\quad \phi_2(|m|z)\rightarrow\frac{\Theta(+z)}{\sqrt{2}}, \tag{22}$$

where Θ is the Heaviside function. The function $(1+\tanh|m|z)/2$ is a "regularization" of $\Theta(z)$.

One can push the analysis a little further by using perturbation theory. Let $\epsilon\equiv\lambda_1-4$ be a small expansion parameter. Writing

$$S(z)=\frac{1}{\sqrt{2}}+\epsilon\,\delta S(z),$$

$$A(z)=-\frac{1}{\sqrt{2}}\tanh\frac{z}{\sqrt{2}}+\epsilon\,\delta A(z), \tag{23}$$

substituting in Eqs. (16) and (17), equating terms $O(\epsilon)$, and solving the resulting equations subject to the boundary conditions, we find that

$$\delta S(z)=\frac{1}{8\sqrt{2}}\left[\sqrt{2}z\sinh\sqrt{2}z - 2\cosh\sqrt{2}z\ln\left(2\cosh\frac{z}{\sqrt{2}}\right)+1\right], \tag{24}$$

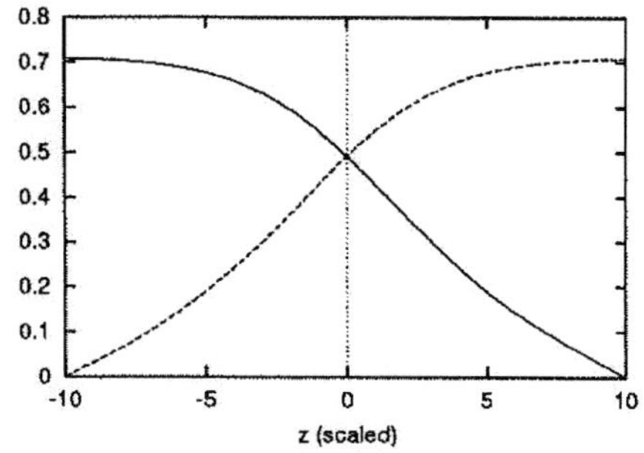

FIG. 1. Asymmetric kink profiles for $\lambda_1=0.1$. The solid line depicts ϕ_1, and the dashed line ϕ_2.

$$\delta A(z)=\frac{z}{32}\left(\tanh^2\frac{z}{\sqrt{2}}-1\right). \tag{25}$$

These results give an indication of how the profiles change when one is away from the special point but in its neighborhood. Note that $\phi_{1,2}(z)+\phi_{1,2}(-z)=1/\sqrt{2}$ no longer holds. Examples of kink solutions are exhibited in Figs. 1, 2 and 3 for $\lambda_1=0.1$, 4 and 100, respectively. We have confirmed that the perturbative results describe the curves well for sufficiently small ϵ.

We next calculate the energy per unit area σ of the wall, which is given by

$$\sigma=\int_{-\infty}^{+\infty}dz\left[(\phi_1')^2+(\phi_2')^2+V(\phi_1,\phi_2)+\frac{1}{4}\right], \tag{26}$$

where the 1/4 subtracts off the zero point energy. At the special parameter point $\lambda_1=4$, one obtains

$$\sigma=\frac{\sqrt{2}}{3}. \tag{27}$$

In Appendix B we use this result to show that the solution exhibited in Eqs. (20) and (21) is globally stable. For other parameter values, the energy can be computed numerically, as displayed in Fig. 4.

As mentioned above, there is a second analytic solution valid for all λ_1,

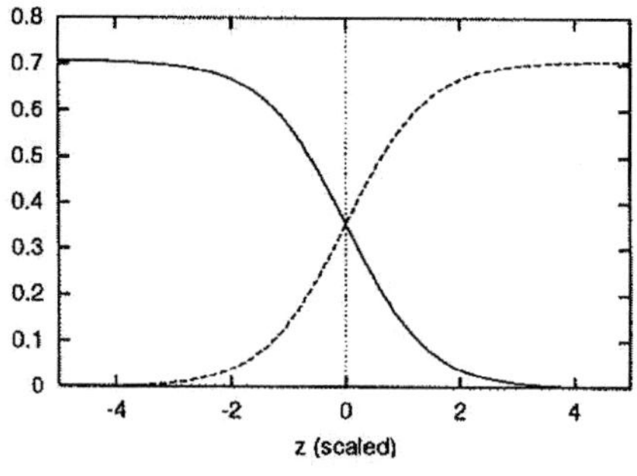

FIG. 2. As for Fig. 1 but with $\lambda_1=4$.

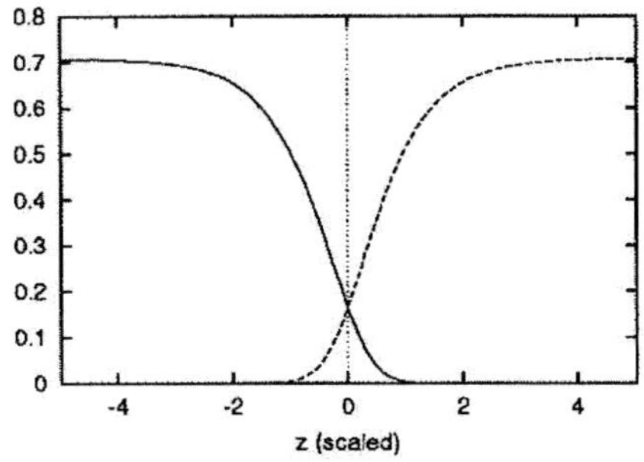

FIG. 3. As for Fig. 1 but with $\lambda_1 = 100$.

$$\phi_2(z) = \begin{cases} (1/\sqrt{2})\tanh z/\sqrt{2} & \text{for } z>0, \\ 0 & \text{for } z<0, \end{cases} \tag{28}$$

with $\phi_1(z) = \phi_2(-z)$. This solution has energy per unit area $\sigma = 2\sqrt{2}/3$ and though perturbatively unstable for all finite λ_1, it is stable in the $\lambda_1 \to \infty$ limit. This means that $2\sqrt{2}/3$ provides an analytic upper bound on σ for all values of λ_1 and, in particular, establishes the existence of finite-energy solutions.

It is apparent from Fig. 4 that σ is monotonically increasing with λ_1. We now establish this result analytically. Consider a certain value of $\lambda_1 > 0$, for which the kink solution is $\Phi_1 = (\phi_1, 0, 0)^T$, $\Phi_2 = (0, \phi_2, 0)^T$. The λ_1 term in V is then simply $\lambda_1 \phi_1^2 \phi_2^2$, which is non-negative for all z. Therefore if we reduce the value of λ_1, the same configuration (which now no longer solves the Euler-Lagrange equations) has lower energy per unit area. But the true solution by definition solves the Euler-Lagrange equations, so it necessarily has an even lower energy, energy and action minimization being equivalent for static configurations.

With these results in hand, we may compare the asymmetric kinks with their symmetric counterparts. To do that, we simply change the ansatz by moving ϕ_2 from the second to the first entry in Φ_2. The Euler-Lagrange equations are the same as Eq. (14), save for the substitution

$$\lambda_1 \to \lambda_1 + \lambda_2. \tag{29}$$

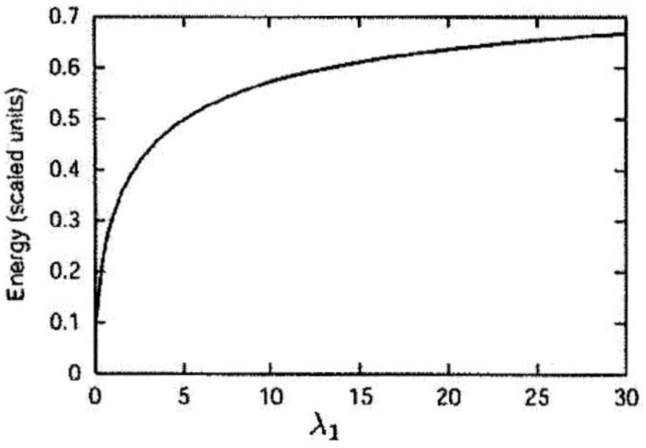

FIG. 4. Energy per unit area as a function of λ_1 for asymmetric kinks. For symmetric kinks, replace λ_1 by $\lambda_1 + \lambda_2$.

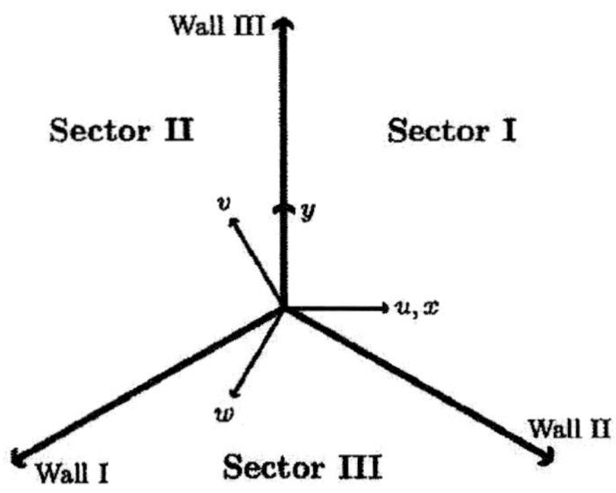

FIG. 5. The three-star domain wall junction configuration. Each wall makes a $2\pi/3$ angle with the neighboring walls. In sector I, a vacuum I state is reached asymptotically; correspondingly in sectors II and III. The two coordinate systems, (x,y) and (u,v,w), are indicated.

The solutions look very similar, except that the special "hyperbolic tangent point" is now $\lambda_1 + \lambda_2 = 4$.

So, for given values of λ_1 and λ_2 obeying Eq. (7) there are both asymmetric and symmetric kink solutions. Which one is energetically favored and therefore stable? We immediately observe that it depends simply on which of λ_1 or $\lambda_1 + \lambda_2$ is larger, i.e. on the sign of λ_2. The clash of symmetries is energetically favored if $\lambda_2 > 0$ and energetically disfavored if $\lambda_2 < 0$.

III. PLANAR WALL-JUNCTION CONFIGURATION

A. Overview and numerical solution

Our toy model was chosen to produce a vacuum manifold of three disconnected pieces I–III as per Eqs. (4)–(6). (The threefold structure is motivated by E_6, see Sec. IV below.) Each one-dimensional kink configuration, however, makes use of only two out of the three possibilities. In the context of model building, even if we are only playing with toys at this stage, it seems more natural to use all three pieces equally. Perhaps more importantly, clash-induced symmetry breaking will be enhanced through the presence of all three vacuum types.

To that end, we search for a domain wall junction configuration as depicted in Fig. 5. Three semi-infinite walls meet at a point, the origin or nexus, at angles of $2\pi/3$, dividing the two-dimensional plane into three sectors labeled I–III. Let (r, θ) be the usual plane polar coordinates. We impose boundary conditions in the obvious way: for a given θ in sector I, the configuration is required to tend to a vacuum I state as $r \to \infty$, with corresponding conditions in sectors II and III. Away from the nexus, and close to a wall, we expect the configuration to tend to a one-dimensional kink as a function of the coordinate perpendicular to the wall. Let us call this setup a "three-star." To calculate it, one must solve the equations of motion, this time using the two-dimensional Laplacian in place of d^2/dz^2 on the left-hand side of Eq. (14), static and z-independent solutions being sought.

The brane limit is most conveniently written in terms of the Mandelstam-like variables (see Fig. 5),

$$u=x, \quad v=-\frac{1}{2}x+\frac{\sqrt{3}}{2}y,$$

$$w=-u-v=-\frac{1}{2}x-\frac{\sqrt{3}}{2}y, \tag{30}$$

as

$$\phi_1 \to \Theta(u)\Theta(-w),$$

$$\phi_2 \to \Theta(v)\Theta(-u),$$

$$\phi_3 \to \Theta(w)\Theta(-v). \tag{31}$$

[It is tempting to "regularize" this configuration by replacing each Θ with a $(1+\tanh)/2$. We have checked that this suggestive form captures the spirit of the three-star we have produced numerically, but not its detail.]

There are three different types of three-stars: totally symmetric, totally asymmetric, and mixed. The symmetric star has the asymptotic vacuum states being cyclic permutations of $[\langle\Phi_1\rangle=(1/\sqrt{2},0,0)^T,\ \langle\Phi_2\rangle=(0,0,0)^T,\ \langle\Phi_3\rangle=(0,0,0)^T]$. There is no clash of symmetries anywhere for this case: the unbroken symmetry is $U(2)_I$ everywhere.

The configuration we want is the totally asymmetric star, defined by the vacuum states:

$$\text{Sector I:}\quad \langle\Phi_1\rangle=(1/\sqrt{2},0,0)^T,\quad \langle\Phi_2\rangle=(0,0,0)^T,\quad \langle\Phi_3\rangle=(0,0,0)^T,$$

$$\text{Sector II:}\quad \langle\Phi_1\rangle=(0,0,0)^T,\quad \langle\Phi_2\rangle=(0,1/\sqrt{2},0)^T,\quad \langle\Phi_3\rangle=(0,0,0)^T,$$

$$\text{Sector III:}\quad \langle\Phi_1\rangle=(0,0,0)^T,\quad \langle\Phi_2\rangle=(0,0,0)^T,\quad \langle\Phi_3\rangle=(0,0,1/\sqrt{2})^T.$$

Ignoring the superfluous U(1)'s, the clash of symmetries has the pattern:

$$H_{I\cap II}=U(2)_I\cap U(2)_{II}=U(1)_{III} \text{ along wall III},$$

$$H_{II\cap III}=U(2)_{II}\cap U(2)_{III}=U(1)_{I} \text{ along wall I},$$

$$H_{III\cap I}=U(2)_{III}\cap U(2)_{I}=U(1)_{II} \text{ along wall II}.$$

At the nexus, the symmetry is completely destroyed:

$$H_{I\cap II\cap III}=U(2)_I\cap U(2)_{II}\cap U(2)_{III}=\{1\}. \tag{32}$$

The totally asymmetric star is energetically favored over the mixed and symmetric stars for the same region of parameter space in which the asymmetric kink is favored over the symmetric one. We place our toy $(3+1)$-dimensional universe at the nexus.

Figures 6, 7 and 8 show the ϕ_1, ϕ_2 and ϕ_3 components of the numerically computed asymmetric three-star for the parameter point $\lambda_1=4$. Figure 9 displays the energy density of the three-star as a function of x and y. We expect that the three-star, defined by the $2\pi/3$ angular separation of the walls, is the lowest energy three-wall junction configuration because it minimizes the total length of the domain walls. We have checked this numerically: in our simulations, junctions with unequal angles between the walls relax to the $2\pi/3$ angular configuration we describe.

B. Partial analytical results

We now present some analytical results for two different asymptotic regimes: off-wall and near-wall, both at large r.

Note first, though, that the threefold symmetry of the star implies that if

$$\phi_1(r,\theta)=\psi(r,\theta) \tag{33}$$

then

$$\phi_2(r,\theta)=\psi\left(r,\theta-\frac{2\pi}{3}\right),\quad \phi_3(r,\theta)=\psi\left(r,\theta+\frac{2\pi}{3}\right). \tag{34}$$

Also, the function ϕ_1 must be symmetric under the reflection $v\to -v$, ϕ_2 under $w\to -w$, and ϕ_3 under $u\to -u$. This is because a sector I vacuum preserves the $\Phi_2\leftrightarrow\Phi_3$ discrete symmetry, and so on.

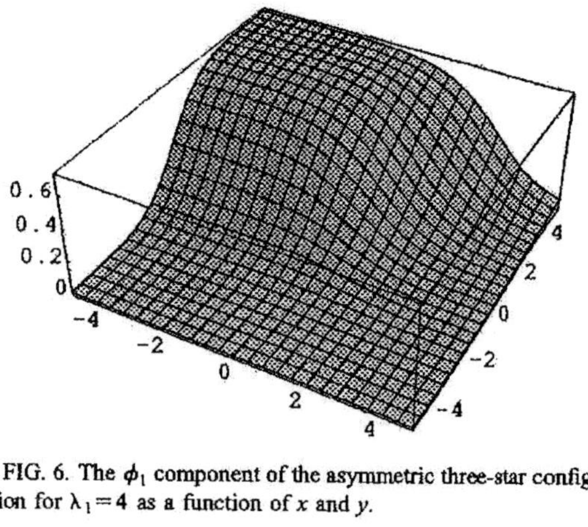

FIG. 6. The ϕ_1 component of the asymmetric three-star configuration for $\lambda_1=4$ as a function of x and y.

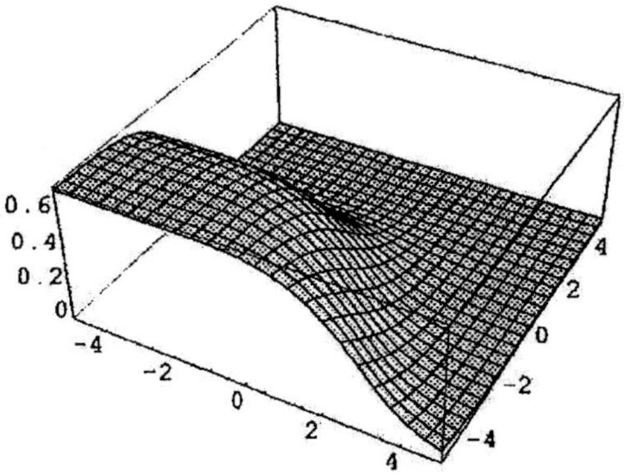

FIG. 7. The ϕ_2 component of the asymmetric three-star configuration for $\lambda_1=4$ as a function of x and y.

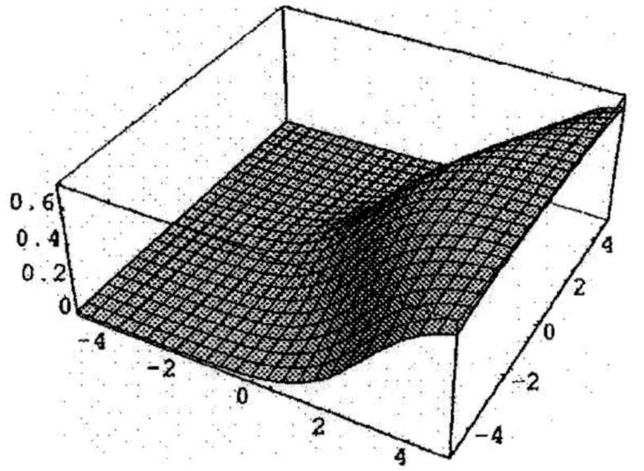

FIG. 8. The ϕ_3 component of the asymmetric three-star configuration for $\lambda_1=4$ as a function of x and y.

1. Large r, off-wall behavior

At large r, put

$$\phi_1(r,\theta)=f(\theta)+\epsilon(r,\theta), \tag{35}$$

where

$$f(\theta)=\begin{cases} 1/\sqrt{2} & \text{in sector 1,} \\ 0 & \text{elsewhere,} \end{cases} \tag{36}$$

and $\epsilon(r,\theta)\ll 1$. Equation (34) is then used to determine $\phi_{2,3}(r,\theta)$. The perturbative requirement that the function $\epsilon(r,\theta)$ be small is met off-wall and at large r.

Substituting Eq. (35) into the Euler-Lagrange equations, and equating like powers of ϵ one obtains to zeroth order

$$\frac{1}{r^2}\frac{d^2 f}{d\theta^2}(\theta) =f(\theta)\left\{-1+2f(\theta)^2+(2+\lambda_1)\left[f\left(\theta-\frac{2\pi}{3}\right)^2 +f\left(\theta+\frac{2\pi}{3}\right)^2\right]\right\}, \tag{37}$$

and to first order

$$\left[\frac{\partial^2}{\partial r^2}+\frac{1}{r}\frac{\partial}{\partial r}+\frac{1}{r^2}\frac{\partial^2}{\partial\theta^2}\right]\epsilon(r,\theta) =\left\{-1+6f(\theta)^2+(2+\lambda_1)\left[f\left(\theta-\frac{2\pi}{3}\right)^2 +f\left(\theta+\frac{2\pi}{3}\right)^2\right]\right\}\epsilon(r,\theta)+2(2+\lambda_1)f(\theta)\left[f\left(\theta-\frac{2\pi}{3}\right) \times\epsilon\left(r,\theta-\frac{2\pi}{3}\right)+f\left(\theta+\frac{2\pi}{3}\right)\epsilon\left(r,\theta+\frac{2\pi}{3}\right)\right]. \tag{38}$$

The zeroth order equation is satisfied by the f defined in Eq. (36).

The right-hand side of Eq. (38) must be treated on a sector by sector basis. It gives rise to

$$\left[\frac{\partial^2}{\partial r^2}+\frac{1}{r}\frac{\partial}{\partial r}+\frac{1}{r^2}\frac{\partial^2}{\partial\theta^2}\right]\epsilon(r,\theta) =\begin{cases} 2\epsilon(r,\theta) & \text{for } \theta \text{ in sector 1,} \\ \dfrac{\lambda_1}{2}\epsilon(r,\theta) & \text{elsewhere,} \end{cases} \tag{39}$$

remembering that we must stay off-wall.

To solve Eq. (39), we look for separated variable solutions:

$$\epsilon(r,\theta)=R(r)T(\theta). \tag{40}$$

Substituting in Eq. (39) produces

$$T''(\theta)=-n^2T(\theta), \tag{41}$$

$$R''(r)+\frac{1}{r}R'(r)=\left(k^2+\frac{n^2}{r^2}\right)R(r), \tag{42}$$

where $k^2=2$ for θ in sector 1 and $k^2=\lambda_1/2$ otherwise, and we define k to be positive. The separation constant is n^2, and we require it to be positive. Equation (41) is solved by $T(\theta)=T_n(\theta)\propto\cos(n\theta+\delta)$, where periodicity in θ requires n to be an integer, and δ is determined by requiring symmetry under $v\to -v$ (if $\theta=0$ is the $v=0$ line bisecting sector 1, then $\delta=0$).

The radial equation (42) has the general solution

$$R(r)=c_1I_n(kr)+c_2K_n(kr) \tag{43}$$

where $c_{1,2}$ are constants and I_n and K_n are modified Bessel functions. The boundary conditions require us to choose

$$R(r)\propto R_n(r)=K_n(kr), \tag{44}$$

which, for all n, has asymptotic behavior

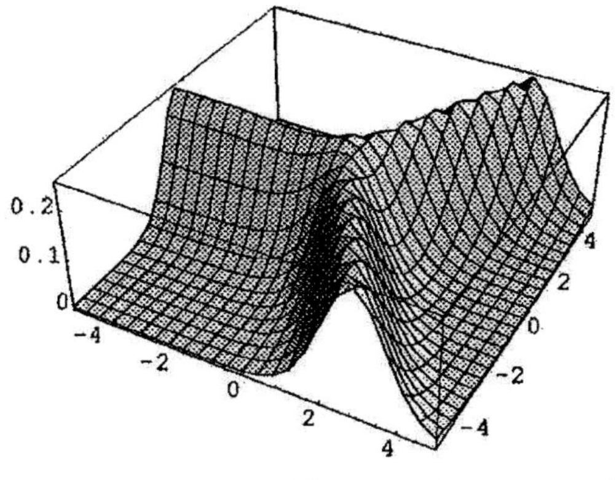

FIG. 9. The energy density of the asymmetric three-star configuration for $\lambda_1=4$ as a function of x and y.

$$R_n(r)\sim\frac{e^{-kr}}{\sqrt{kr}}. \tag{45}$$

Writing the general solution to Eq. (39) as a sum over n of $R_n(r)T_n(\theta)$ with undetermined coefficients, we find that

$$\epsilon(r,\theta)\sim\frac{e^{-kr}}{\sqrt{kr}}F(\theta), \tag{46}$$

where F is an undetermined angular function. The nature of F may well be different in sectors II and III from sector I, just as the inverse decay length k takes different values if $\lambda_1\neq 4$. Notice that $\lambda_1=4$ is the same special point that produces hyperbolic tangent kinks. We saw earlier that these kinks have the special property $\phi_{1,2}(z)+\phi_{1,2}(-z)=1/\sqrt{2}$ (taking the I$\leftrightarrow$II case). The equality of the k's on both sides of the wall is a similar special property for the star configuration.

2. *Large r, near-wall behavior*

We will now explore near-wall behavior far from the nexus, using wall II as our example. To begin with, the relevant coordinates are ρ, the radial distance directly along the wall, and w, perpendicular to the wall.

We again use a perturbative approach, writing

$$\phi_1(\rho,w)=g(w)+\delta(\rho,w), \tag{47}$$

$$\phi_2=0, \tag{48}$$

$$\phi_3(\rho,w)=g(-w)+\delta(\rho,-w), \tag{49}$$

where δ is small. The w-parity relationship between ϕ_1 and ϕ_3 is dictated by the threefold and reflection symmetries of the configuration. Although the ϕ_2 field is of order $\delta(\rho,w)$ along wall II at large ρ, it enters quadratically into the Euler-Lagrange equations for $\phi_{1,3}$ so it can be set to zero *a priori*.

Note that the regime we explore here is physically separated from the off-wall regime probed above, even though both lie far from the nexus. If we set w to some finite value and take $\rho\to\infty$, then the angular distance from the wall goes to zero: $\Delta\theta\sim w/\rho\to 0$. By contrast, the off-wall region requires finite $\Delta\theta$.

Substitution of Eqs. (47)–(49) into the Euler-Lagrange equations yields the zeroth order result

$$g''(w)=g(w)[-1+2g(w)^2+(2+\lambda_1)g(-w)^2]. \tag{50}$$

By symmetry, a similar equation with $w\to -w$ also holds. Defining

$$S_0(w)=g(w)+g(-w),\quad A_0(w)=g(w)-g(-w), \tag{51}$$

we recover Eqs. (16) and (17). This shows very clearly that the perpendicular near-wall behavior far from the origin is exactly the appropriate one-dimensional kink.

The first order analysis depends on the function g. To proceed analytically, we restrict the following to the special $\lambda_1=4$ case, so that $g(w)=[1+\tanh(w/\sqrt{2})]/2\sqrt{2}$. Setting

$$S\equiv\phi_1+\phi_3=S_0(w)+\delta S,$$

$$A\equiv\phi_1-\phi_3=A_0(w)+\delta A, \tag{52}$$

substituting in the Euler-Lagrange equations, and equating terms to first order in δS and δA we obtain:

$$\nabla^2(\delta S)=\delta S(-1+6S_0^2)=2\,\delta S, \tag{53}$$

$$\nabla^2(\delta A)=\delta A(-1+6A_0^2), \tag{54}$$

where we have also used $S_0(w)=1/\sqrt{2}$.

We now switch to polar coordinates (r,η) where $\tan\eta=w/\rho$ and $r^2=\rho^2+w^2$. Searching for separated variable solutions in these coordinates, we set

$$\delta S(r,\eta)=P(r)[W(\eta)+W(-\eta)],$$

$$\delta A(r,\eta)=P(r)[W(\eta)-W(-\eta)]. \tag{55}$$

Substitution in Eq. (53) then immediately yields

$$P''(r)+\frac{1}{r}P'(r)=\left(2+\frac{n^2}{r^2}\right)P(r), \tag{56}$$

where $n^2>0$ is the separation constant. We conclude that

$$P(r)\sim\frac{e^{-\sqrt{2}r}}{\sqrt{\sqrt{2}r}},\quad\forall n \tag{57}$$

which matches exactly the asymptotic radial behavior, Eq. (45), found earlier in the off-wall regime.

Consider now Eq. (54). Using Eq. (56) and the known function g, defining $\Delta(\eta)=W(\eta)-W(-\eta)$, we obtain

$$\frac{\Delta''(\eta)}{r^2\Delta(\eta)}=-3-\frac{n^2}{r^2}+3\tanh^2\left(\frac{r\sin\eta}{\sqrt{2}}\right). \tag{58}$$

For large r, the term depending on the separation constant n^2 is suppressed and can be omitted. For small η, we can change variables to $w=r\eta$ to get

$$\frac{\Delta''(w)}{\Delta(w)} \approx -3+3\tanh^2\frac{w}{\sqrt{2}}. \tag{59}$$

The solution with the correct antisymmetry in w is then

$$\Delta(w)\propto\left[3\tanh\frac{w}{\sqrt{2}}+\frac{w}{\sqrt{2}}\left(1-3\tanh^2\frac{w}{\sqrt{2}}\right)\right]. \tag{60}$$

The collection of results above demonstrates that one can make some progress in understanding the three-star configuration analytically, even though an exact analytic solution is at present lacking.

IV. DISCUSSION

Our toy model was chosen not only to be mathematically simple, but also because it can serve as a prototype for a more realistic theory motivated by E_6. While it is beyond the scope of this paper to explore this connection in detail, we would like to comment and speculate on possible future directions.

The most direct connection is with the maximal $SU(3)^3$ subgroup of E_6, augmented by a discrete Z_3 symmetry that rotates the SU(3) factors. The complex, anomaly-free representation

$$(3,\bar{3},1)\oplus(1,3,\bar{3})\oplus(\bar{3},1,3), \tag{61}$$

which arises from the decomposition of the 27 of E_6, naturally generalizes the three triplet Higgs boson content of the toy model. As is well known, one generation of quarks and leptons can be placed in a similar representation. It would be interesting to apply the clash of symmetries idea in this context, to see what symmetry breaking patterns can be produced.

In the future pursuit of serious brane model-building, there is no reason to restrict Higgs potentials to quartic form. If, for instance, we have a three-star configuration in mind, then the underlying spacetime is at least $5+1$ dimensional, where renormalizability requires at most cubic potentials (which are necessarily unbounded from below and thus presumably unacceptable). The question of renormalization should sensibly be deferred until such time as a connection with a proper theory of quantum gravity can be made. We recognize that many string-theoretic brane and brane junction scenarios have already been proposed [9].

If the full E_6 is considered rather than just the $SU(3)^3$ subgroup, then an important issue is domain wall stability. The Z_3 symmetry of the reduced theory, useful for wall stability, is then presumably embedded within the continuous E_6 symmetry. According to the general understanding of defect formation, the breakdown of E_6 to, first, $SU(3)^3\otimes Z_3$, and then to some smaller subgroup, will imply the existence of unstable vortex-wall hybrid structures rather than topologically stable walls. While it may be possible for the instability time scale to be very long, a bulk scalar field in the 27 of E_6 offers another natural possibility. Recognizing that singlets arise in the products $27\otimes\overline{27}$ and $27\otimes27\otimes27$, we see that a general Higgs potential will respect a Z_3 *phase* symmetry that may not be contained within E_6.

The number "3" plays a prominent role in the group theory of E_6: there are three SU(3) factors in the maximal subgroup under discussion, and there are also three ways to embed electric charge Q within the group. The latter fact has been remarked on before [10], but perhaps it has not received the attention it deserves. From the perspective of the subgroup chain $E_6\rightarrow SO(10)\otimes U(1)''\rightarrow SU(5)\otimes U(1)'\otimes U(1)''$, the three electric charge assignments correspond to the standard case where Q lies within SU(5), the flipped SU(5) case where $U(1)'$ is also involved, and the flipped SO(10) case where $U(1)''$ enters the definition of Q. Now, there is another long-standing mystery pertaining to the number three: the apparently superfluous replication of quark-lepton families. The lack of a compelling explanation despite years of thought suggests that new approaches should be seriously considered. We speculate that the three Q embeddings, the three-star configuration derived from the triply degenerate vacuum structure, and threefold quark/lepton family replication may be connected.

V. CONCLUSION

Using a model field theory comprising three U(3) Higgs triplets interacting through a permutation symmetric quartic potential, we have shown that domain wall and wall-junction solutions exist displaying the "clash of symmetries." This symmetry breaking mechanism goes beyond standard spontaneous breaking by exploiting different embeddings of isomorphic subgroups in the parent group. Our example used the *I*-, *U*- and *V*-spin U(2) subgroups of U(3). We found topologically stable domain wall solutions which asymptote to vacuum states corresponding to differently embedded unbroken U(2) subgroups on opposite sides of the wall. Non-asymptotically, the symmetry is further broken to the intersection of the asymptotically unbroken subgroups. This phenomenon has been previosuly displayed in a different model and with different motivations in Ref. [6]. We propose that such a kink-like configuration in the thin-wall or brane limit may exist in a large extra dimension, with our universe identified with the brane. In that case, some of the symmetry breaking in our universe may be due to the clash of symmetries. Increasing the number of spatial dimensions (notionally) to five, we numerically constructed a wall-junction three-star configuration that exploits the clash phenomenon to the full, with the joint or nexus identified with our (toy) universe. Future work is motivated on several fronts: a possible connection with E_6, a possible connection between the three-star and threefold family replication, and degree of freedom localization to the brane.

ACKNOWLEDGMENTS

B.F.T. was supported by the Commonwealth of Australia and the University of Melbourne. R.R.V. is supported by the

Australian Research Council and the University of Melbourne. He would very much like to thank the Ben-Gurion University for financial support during the course of some of this work. He would also like to thank his co-authors Aharon Davidson and Kamesh Wali for great hospitality at their home institutions while portions of this work were completed. K.C.W. was supported in part by a grant from NSF, Division of International Programs (U.S.-Australia Cooperative Research) and also in part by a grant from the DOE.

APPENDIX A: JUSTIFICATION OF THE KINK ANSATZ

In this appendix, we justify the asymmetric kink ansatz used in Sec. II C. We consider the case $\lambda_2>0$ and show that a globally stable kink must fit this ansatz, Eq. (13). (Analogous arguments show that the symmetric kink is globally stable for $\lambda_2<0$.)

1. The two-triplet model

We begin by considering a simpler model with just two triplets and an exchange discrete symmetry. The Higgs potential is obtained from Eq. (2) by taking $\Phi_3=0$.

Consider a general trial solution of the form

$$\Phi_i(z)=(\phi_{i,1}(z),\phi_{i,2}(z),\phi_{i,3}(z))^T, \tag{A1}$$

with $i=1,2,3$, which satisfies the asymmetric boundary conditions:

$$\Phi_1(-\infty)=\left(\frac{1}{\sqrt{2}},0,0\right)^T, \quad \Phi_2(-\infty)=(0,0,0)^T,$$

$$\Phi_1(+\infty)=(0,0,0)^T, \quad \Phi_2(+\infty)=\left(0,\frac{1}{\sqrt{2}},0\right)^T. \tag{A2}$$

Define

$$\theta_i=\sqrt{\phi_{i,1}^*\phi_{i,1}+\phi_{i,2}^*\phi_{i,2}+\phi_{i,3}^*\phi_{i,3}}, \tag{A3}$$

and consider a second trial solution of the form

$$\Theta_1(z)=(\theta_1(z),0,0)^T, \tag{A4}$$

$$\Theta_2(z)=(0,\theta_2(z),0)^T.$$

We will show that the configuration of Eq. (A5) has energy less than or equal to the energy of the initial trial solution of Eq. (A1). It is clear that this configuration satisfies the boundary conditions.

Consider first the potential energy density $V[\Phi_1,\Phi_2]$. Clearly $\Phi_1^\dagger\Phi_1=\Theta_1^\dagger\Theta_1$ and $\Phi_2^\dagger\Phi_2=\Theta_2^\dagger\Theta_2$, so we need only consider the term in the potential dependent on λ_2. But if $\lambda_2>0$, then this term is positive for $\Phi_{1,2}$ of the form of Eq. (A1) but zero for the configuration of Eq. (A4). Thus $V[\Theta_1,\Theta_2]\leqslant V[\Phi_1,\Phi_2]$.

We now turn to the kinetic energy density $T[\Theta_1,\Theta_2]$:

$$\begin{aligned}
\sum_i \Theta_i'^\dagger\Theta_i' &= \sum_i (\theta_i')^2 \\
&= \sum_i \frac{1}{4\theta_i^2}(\phi_{i,1}^*\phi_{i,1}'+\phi_{i,2}^*\phi_{i,2}'+\phi_{i,3}^*\phi_{i,3}'+\text{c.c.})^2 \\
&\leqslant \sum_i \frac{1}{\theta_i^2}(|\phi_{i,1}||\phi_{i,1}'|+|\phi_{i,2}||\phi_{i,2}'|+|\phi_{i,3}||\phi_{i,3}'|)^2 \\
&\leqslant \frac{(|\phi_{i,1}|^2+|\phi_{i,2}|^2+|\phi_{i,3}|^2)}{\theta_i^2}(|\phi_{i,1}'|^2+|\phi_{i,2}'|^2+|\phi_{i,3}'|^2) \\
&= \sum_i \Phi_i'^\dagger\Phi_i',
\end{aligned} \tag{A5}$$

by the Cauchy-Schwarz inequality. So $T[\Theta_1,\Theta_2]\leqslant T[\Phi_1,\Phi_2]$, also.

2. The three-triplet model

We now add the third triplet to the model. Staying with $\lambda_2>0$ and applying the arguments of the previous section, it suffices to consider a trial solution of the form

$$\Phi_1(z)=[\phi_1(z),0,0]^T,$$

$$\Phi_2(z)=[0,\phi_2(z),0]^T,$$

$$\Phi_3(z)=[0,0,\phi_3(z)]^T, \tag{A6}$$

with ϕ_i real. Note that the boundary conditions on ϕ_3 require it to vanish asymptotically. We now establish that this trial solution has energy greater than or equal to the alternative trial solution,

$$\Theta_1(z)=\{\sqrt{[\phi_1(z)]^2+[\phi_3(z)]^2},0,0\}^T,$$

$$\Theta_2(z)=(0,\phi_2(z),0)^T,$$

$$\Theta_3(z)=(0,0,0)^T. \tag{A7}$$

Observe that $\xi(z)\equiv\sqrt{[\phi_1(z)]^2+[\phi_3(z)]^2}$ obeys the correct boundary conditions. Now, the kinetic energy density due to ξ obeys $(\xi')^2<(\phi_1')^2+(\phi_3')^2$ by the Cauchy-Schwarz inequality. In the potential energy density function, we need only consider the λ_1 term. For trial solution Eq. (A6) this term is $\lambda_1(\phi_1^2\phi_2^2+\phi_2^2\phi_3^2+\phi_3^2\phi_1^2)$, while for trial solution Eq. (A7) it is $\lambda_1(\xi^2\phi_2^2)=\lambda_1(\phi_1^2\phi_2^2+\phi_3^2\phi_2^2)$. The latter is obviously smaller than the former.

So globally stable solutions in the two-triplet model are also globally stable in the full three-triplet model.

APPENDIX B: GLOBAL STABILITY OF THE $\lambda_1=4$ ANALYTIC SOLUTION

Consider the special parameter point $\lambda_1=4$. We prove global stability of the analytic solution, Eqs. (20) and (21), by Bogomolnyi's method [11]. The energy density $\epsilon[\phi_1,\phi_2]$ of any solution fitting the ansatz given in Eq. (13) is

$$\epsilon[\phi_1,\phi_2]=\phi_1'^2+\phi_2'^2+V(\phi_1,\phi_2)+\frac{1}{4} \tag{B1}$$

$$=\left(\phi_1'+\phi_1^2+\phi_2^2-\frac{1}{2}\right)^2+(\phi_2'+2\phi_1\phi_2)^2 + \frac{d}{dz}\left(\phi_1-\frac{2}{3}\phi_1^3-2\phi_2^2\phi_1\right). \tag{B2}$$

The first two terms of this equation are non-negative, so the total kink energy is

$$\int_{-\infty}^{\infty}\epsilon[\phi_1,\phi_2]dz\geqslant\int_{-\infty}^{\infty}dz\frac{d}{dz}\left(\phi_1-\frac{2}{3}\phi_1^3-2\phi_2^2\phi_1\right) = \frac{\sqrt{2}}{3}, \tag{B3}$$

where we have substituted for the boundary conditions. Since the analytic solution given in the main text, Eq. (27), saturates this lower bound, it is globally stable.

[1] E.P.S. Shellard and A. Vilenkin, *Cosmic Strings and Other Topological Defects* (Cambridge University Press, Cambridge, England, 1994); R. Rajaraman, *Solitons and Instantons* (North-Holland, Amsterdam, 1982), and references therein.

[2] Ya.B. Zel'dovich, I.Yu. Kobzarev, and L.B. Okun, Zh. Éksp. Teor. Fiz. **67**, 3 (1974) [Sov. Phys. JETP **40**, 1 (1974)].

[3] V.A. Rubakov and M.E. Shaposhnikov, Phys. Lett. **125B**, 136 (1983); K. Akama, *Pregeometry*, Lecture Notes in Physics Vol. 176, edited by K. Kikkawa, N. Nakanishi, and H. Nariai (Springer-Verlag, Berlin, 1983), pp. 267–271, hep-th/0001113; I. Antoniadis, Phys. Lett. B **246**, 317 (1990).

[4] N. Arkani-Hamed, S. Dimopoulos, and G.R. Dvali, Phys. Lett. B **429**, 263 (1998); L.J. Randall and R. Sundrum, Phys. Rev. Lett. **83**, 3370 (1999); **83**, 4690 (1999); T. Regge and C. Teitelboim, in Proceedings of the Marcel Grossman Conference, Trieste, 1975; A. Davidson and D. Karasik, Mod. Phys. Lett. A **13**, 2187 (1998); A. Davidson, Class. Quantum Grav. **16**, 653 (1999); I. Antoniadis, N. Arkani-Hamed, and S. Dimopoulos, Phys. Lett. B **436**, 257 (1998).

[5] C. Csaki and Y. Shirman, Phys. Rev. D **61**, 024008 (2000).

[6] L. Pogosian and T. Vachaspati, Phys. Rev. D **62**, 123506 (2000); see also T. Vachaspati, *ibid.* **63**, 105010 (2001); L. Pogosian and T. Vachaspati, *ibid.* **64**, 105023 (2001).

[7] G. Dvali, Z. Tavartkiladze, and J. Nanobashvili, Phys. Lett. B **352**, 214 (1994).

[8] T. Nihei, Phys. Rev. D **62**, 124017 (2000); S.M. Carroll, S. Hellerman, and M. Trodden, *ibid.* **61**, 065001 (2000); **62**, 044049 (2000).

[9] See, for example, L.E. Ibáñez, F. Marchesano, and R. Rabadán, J. High Energy Phys. **11**, 002 (2001).

[10] M. Bando and T. Kugo, Prog. Theor. Phys. **101**, 1313 (1999); M. Bando, T. Kugo, and K. Yoshioka, *ibid.* **104**, 211 (2000); G.W. Anderson and T. Blazek, J. Math. Phys. **41**, 4808 (2000).

[11] E.B. Bogomol'nyi, Sov. J. Nucl. Phys. **24**, No. 4 (1976); reprinted in C. Rebbi and G. Soliani, *Solitons and Particles* (World Scientific, Singapore, 1984).

Non-Abelian monopole and dyon solutions in a modified Einstein-Yang-Mills-Higgs system

A. S. Cornell* and G. C. Joshi†
Research Centre for High Energy Physics, School of Physics, University of Melbourne, Parkville, Victoria 3010, Australia

J. S. Rozowsky‡ and K. C. Wali§
Physics Department, Syracuse University, Syracuse, New York 13244
(Received 14 August 2002; published 22 May 2003)

We study a modified Yang-Mills-Higgs system coupled to Einstein gravity. The modification of the Einstein-Hilbert action involves the direct coupling of the Higgs field to the scalar curvature. In this modified system we are able to write a Bogomol'nyi type condition in curved space and demonstrate that the positive static energy functional is bounded from below. We then investigate non-Abelian spherically symmetric static solutions in a similar fashion to the 't Hooft–Polyakov monopole. After reviewing previously studied monopole solutions of this type, we extend the formalism to include electric charge and we present dyon solutions.

DOI: 10.1103/PhysRevD.67.105015 PACS number(s): 14.80.Hv, 04.70.Bw

I. INTRODUCTION

In a classic paper [1], Dirac proposed the possible existence of a magnetic monopole, the analogue of an isolated electrically charged particle. Motivated principally to restore the symmetries between electric and magnetic forces, Dirac found that the existence of a monopole provided a natural explanation for the quantization of electric charge. A description of such a monopole consistent with quantum mechanics would lead to the famous Dirac charge quantization condition

$$eg=\frac{1}{2}n\hbar c,$$

where $e^2/\hbar c$ is the fine structure constant, g is the monopole charge, and n an integer. Dirac's theory required a $U(1)$ valued gauge potential which was singular along a line (Dirac string) originating from the monopole and extending to infinity. Later Dirac's theory was reformulated by Wu and Yang [2] within the framework of fiber bundles. The singular line is avoided at the expense of introducing coordinate patches on a sphere surrounding the monopole. Then, it can be shown that there is a complete equivalence between the two descriptions [3] and the monopole emerges on a sound footing like any other particle in nature. Consequently, during recent decades, extensive effort has gone into the experimental search for monopoles, but unfortunately has yet to succeed.

In spite of the lack of experimental success, the monopole continues to thrive in the theoretical laboratory. With the pioneering work of 't Hooft and Polyakov [4], the monopole was reinvented in a new form as a finite energy, particlelike soliton in non-Abelian gauge theories with spontaneous symmetry breaking. Moreover, such objects are generic in a simply connected non-Abelian gauge theory with spontaneous symmetry breaking, which has an unbroken $U(1)$ gauge symmetry, and as such, such monopoles are expected to be produced in abundance in phase transitions of grand unified theories, which has implications for early universe cosmology. New searches for such relic monopoles are under way [5,6].

More recently, a great deal of activity has centered around monopoles in curved space-time in order to study the effects of gravity. New insights have emerged from a study of a coupled Einstein-Yang-Mills-Higgs (EYMH) system with solutions describing black holes with magnetic charge, black holes within magnetic monopoles, and magnetic monopoles within black holes [7–11]. Consequently, gravity cannot be dispensed with, arguing that the strength of its interaction is weak. The coupled set of equations for the EYMH system lead to nontrivial consequences and provided a fertile ground for the study of the interplay between gravitation and other interactions.

The present paper is an extension of the work by Nguyen and Wali [12] to include electric charge and to study dyons coupled to gravity. The starting point is a modified EYMH system with a specific coupling of the Higgs field to the Einstein term in the action. In the static case, this enables us, with the help of a Bogomol'nyi-type [9] condition, to reduce the energy functional to a form that resembles the energy functional in flat space-time and derive a lower bound on the energy and hence the mass of the dyon in terms of its electric and magnetic charges. Monopoles in this modified EYMH system have also been studied by Kao [13] in the setting of a more general Higgs potential.

In the next section, we begin with a review of the general formalism and the field equations for the coupled EYMH system. In Sec. III, we derive the field equations and the energy functional in the static case. Through the Higgs field equation, we express the Einstein term in the action in terms of metric fields, find a positive definite expression for the energy functional, and derive a lower bound on the energy. Section IV is devoted to the derivation of the basic equations in the context of a spherically symmetric static metric and the spherically symmetric 't Hooft–Polyakov ansatz for the

*Email address: a.cornell@tauon.ph.unimelb.edu.au
†Email address: joshi@physics.unimelb.edu.au
‡Email address: rozowsky@phy.syr.edu
§Email address: wali@physics.syr.edu

gauge and Higgs fields. In Sec. V, we specialize to the Higgs vacuum and find monopole and dyon solutions. The final section is devoted to some concluding remarks.

II. GENERAL FRAMEWORK: FIELD EQUATIONS

We begin by defining the action

$$S=\int d^4x\sqrt{-g}(\mathcal{L}_E+\mathcal{L}_M), \quad (2.1)$$

where

$$\mathcal{L}_E=-\frac{R-2\Lambda}{16\pi G v^2}\mathbf{\Phi}^2, \quad (2.2)$$

with R the Ricci scalar, Λ the cosmological constant, and $\mathbf{\Phi}$ the Higgs scalar field. Our metric $g_{\mu\nu}$ is chosen to have signature $(+---)$ with indicies $\mu,\nu,\ldots$ running from 0 to 3 and indicies $i,j,\ldots$ from 1 to 3, also $g=\det|g_{\mu\nu}|$. The matter content is given by

$$\mathcal{L}_M=-\frac{1}{4}g^{\mu\rho}g^{\nu\lambda}\mathbf{F}_{\mu\nu}\mathbf{F}_{\rho\lambda}+\frac{1}{2}g^{\mu\nu}(\mathcal{D}_\mu\mathbf{\Phi})(\mathcal{D}_\nu\mathbf{\Phi})$$
$$-\frac{\lambda}{4}(\mathbf{\Phi}^2-v^2)^2, \quad (2.3)$$

where

$$\mathbf{F}_{\mu\nu}=-\frac{1}{i\alpha}[\mathcal{D}_\mu,\mathcal{D}_\nu], \quad (2.4)$$

with

$$\mathcal{D}_\mu=\nabla_\mu-i\alpha[\mathbf{A}_\mu,\cdot\,]. \quad (2.5)$$

Thus $\mathbf{F}_{\mu\nu}$ is the field strength associated with the gauge field $\mathbf{A}_\mu$, α being the strength of the gauge coupling. ∇_μ is the covariant derivative with the metric compatible, torsion-free connection coefficients.

More explicitly,

$$\mathbf{F}_{\mu\nu}=\partial_\mu\mathbf{A}_\nu-\partial_\nu\mathbf{A}_\mu-i\alpha[\mathbf{A}_\mu,\mathbf{A}_\nu], \quad (2.6)$$

and in the component form,

$$F^a_{\mu\nu}=\partial_\mu A^a_\nu-\partial_\nu A^a_\mu+\alpha f^{abc}A^b_\mu A^c_\nu, \quad (2.7)$$

where f^{abc} are the structure constants of a gauge group $\mathcal{G}$, which for the most part for our purposes will be $SU(2)$. The scalar field $\mathbf{\Phi}$ belongs to the adjoint representation of $\mathcal{G}$. Its covariant derivative is given by

$$\mathcal{D}_\mu\mathbf{\Phi}=\nabla_\mu\mathbf{\Phi}-i\alpha[\mathbf{A}_\mu,\mathbf{\Phi}]$$
$$=\partial_\mu\mathbf{\Phi}-i\alpha[\mathbf{A}_\mu,\mathbf{\Phi}], \quad (2.8)$$

and in component form

$$\mathcal{D}_\mu\Phi^a=\partial_\mu\Phi^a+\alpha f^{abc}A^b_\mu\Phi^c. \quad (2.9)$$

We note that in the broken phase of the gauge symmetry, when the Higgs field $\mathbf{\Phi}$ assumes its vacuum expectation value, $\mathbf{\Phi}^2=v^2$, $\mathcal{L}_E$ in Eq. (2.2) is the conventional Einstein-Hilbert Lagrangian. $\mathcal{L}_M$ in Eq. (2.3) represents the standard Yang-Mills-Higgs Lagrangian in curved space-time.

By varying the action S with respect to $\mathbf{A}_\mu$, $\mathbf{\Phi}$, and $g_{\mu\nu}$, we obtain the coupled Yang-Mills, Higgs, and Einstein equations of motion:

$$\frac{1}{\sqrt{-g}}\mathcal{D}_\mu(\sqrt{-g}\mathbf{F}^{\mu\nu})=i\alpha[\mathbf{\Phi},\mathcal{D}^\nu\mathbf{\Phi}], \quad (2.10)$$

$$\frac{1}{\sqrt{-g}}\mathcal{D}_\mu(\sqrt{-g}\mathcal{D}^\mu\mathbf{\Phi})=\left(\frac{R-2\Lambda}{8\pi G v^2}+\lambda(\mathbf{\Phi}^2-v^2)\right)\mathbf{\Phi}, \quad (2.11)$$

and

$$G_{\mu\nu}=R_{\mu\nu}-\frac{R-2\Lambda}{2}g_{\mu\nu}=\frac{8\pi G v^2}{\mathbf{\Phi}^2}T_{\mu\nu}, \quad (2.12)$$

where the energy-momentum tensor $T_{\mu\nu}$ is given by

$$T_{\mu\nu}=-\left(\mathcal{L}_M+\frac{1}{2}\Box\mathbf{\Phi}^2\right)g_{\mu\nu}-\mathbf{F}_{\mu\rho}\cdot\mathbf{F}^\rho_\nu+\mathcal{D}_\mu\mathbf{\Phi}\cdot\mathcal{D}_\nu\mathbf{\Phi}$$
$$+\nabla_\mu\nabla_\nu\mathbf{\Phi}^2. \quad (2.13)$$

The terms involving $\mathbf{\Phi}$ in Eq. (2.13) arise because of its presence in the modified Einstein-Hilbert action [14] in Eq. (2.2). We further note that although the field equations are written in terms of the covariant derivative ∇_μ in $\mathcal{D}_\mu$, we can easily show that they can all be reduced to ordinary partial derivatives since

$$\nabla_\sigma g_{\mu\nu}=0;\quad \nabla_\mu\sqrt{-g}=\partial_\mu\sqrt{-g}-\sqrt{-g}\Gamma^\nu_{\mu\nu}=0, \quad (2.14)$$

for torsion-free, metric compatible connection coefficients $\Gamma^\lambda_{\mu\nu}=\Gamma^\lambda_{\nu\mu}$. Therefore, henceforth, in our equations,

$$\mathbf{D}_\mu=\partial_\mu\mathbf{1}-i\alpha[\mathbf{A}_\mu,\cdot\,]. \quad (2.15)$$

III. STATIC FIELD EQUATIONS; BOGOMOL'NYI BOUND

We are interested in static solutions to Eqs. (2.10)–(2.12). Setting the time derivatives of all fields equal to zero, we find Eq. (2.10) reduces to two equations:

$$\frac{1}{\sqrt{-g}}\mathbf{D}_i(\sqrt{-g}\mathbf{F}^{i0})=i\alpha[\mathbf{\Phi},\mathbf{D}^0\mathbf{\Phi}], \quad (3.1)$$

$$\mathbf{D}_0\mathbf{F}^{0j}+\frac{1}{\sqrt{-g}}\mathbf{D}_i(\sqrt{-g}\mathbf{F}^{ij})=i\alpha[\mathbf{\Phi},\mathbf{D}^j\mathbf{\Phi}]. \quad (3.2)$$

The Higgs field equation (2.11) becomes

$$\frac{1}{\sqrt{-g}}[\mathbf{D}_0(\sqrt{-g}\mathbf{D}^0\mathbf{\Phi})+\mathbf{D}_i(\sqrt{-g}\mathbf{D}^i\mathbf{\Phi})]$$
$$=-\left(\frac{R-2\Lambda}{8\pi G v^2}+\lambda(\mathbf{\Phi}^2-v^2)\right)\mathbf{\Phi}. \qquad (3.3)$$

In order to find a Bogomol'nyi-type first-order equation [15] to our problem, we make the ansatz [16],

$$\mathbf{F}_{ij}=\sqrt{-\tilde{g}}\,\epsilon_{ijk}(\mathbf{D}^k+u^k)\mathbf{\Phi}, \qquad (3.4)$$

where $u^k=\partial^k f$ is an arbitrary time-independent function, and $\tilde{g}=\det|g_{ij}|$.

With the above ansatz, we find

$$\mathbf{D}_i\mathbf{F}^{ij}=i\alpha[\mathbf{\Phi},\mathbf{D}^j\mathbf{\Phi}]+\left(\frac{\partial_i\sqrt{g_{00}}}{\sqrt{g_{00}}}-\partial_i f\right)\mathbf{F}^{ij}. \qquad (3.5)$$

Substituting Eq. (3.5) in Eq. (3.2), we have

$$[\mathbf{D}_0,\mathbf{F}^{0j}]+\left(\frac{\partial_i\sqrt{g_{00}}}{\sqrt{g_{00}}}-\partial_i f\right)\mathbf{F}^{ij}=0. \qquad (3.6)$$

For future reference, we note that the Yang-Mills equation (3.1) implies

$$\mathbf{D}_i(\sqrt{-g}\mathbf{F}^{i0})\cdot\mathbf{\Phi}=0. \qquad (3.7)$$

Using Eq. (3.4) together with Eq. (3.5), after some algebra and the use of the Bianchi identity, the Higgs field equation (3.3) reduces to

$$\mathbf{D}_0\mathbf{D}^0\mathbf{\Phi}+\left(\frac{\partial_i\sqrt{g_{00}}}{\sqrt{g_{00}}}-\partial_i f\right)\mathbf{D}^i\mathbf{\Phi}-\frac{1}{\sqrt{-\tilde{g}}}\partial_i(\sqrt{-\tilde{g}}\partial^i f)\mathbf{\Phi}$$
$$=-\left(\frac{R-2\Lambda}{8\pi G v^2}+\lambda(\mathbf{\Phi}^2-v^2)\right)\mathbf{\Phi}. \qquad (3.8)$$

In deriving the above equations, with static spherically symmetric solutions in mind, we have assumed

$$g_{00}>0, \quad g_{0i}=0, \quad \text{and} \quad -\tilde{g}>0. \qquad (3.9)$$

The static energy functional, $\mathcal{E}$, that follows from Eq. (2.1) is given by

$$\mathcal{E}=\int d^3x\frac{\sqrt{-g}}{2}\left[\frac{R-2\Lambda}{8\pi G v^2}\Phi^2-\mathbf{F}^{0i}\cdot\mathbf{F}_{0i}+\frac{1}{2}\mathbf{F}^{ij}\cdot\mathbf{F}_{ij}\right.$$
$$\left.-\mathbf{D}^i\mathbf{\Phi}\cdot\mathbf{D}_i\mathbf{\Phi}+\frac{\lambda}{2}(\mathbf{\Phi}^2-v^2)^2\right]. \qquad (3.10)$$

Next we define electric and magnetic fields, $\mathbf{E}_i$, $\mathbf{B}_i$ and corresponding $\mathbf{E}^i$, $\mathbf{B}^i$ to be given by

$$\mathbf{E}_i=\frac{\sqrt[4]{-g}}{\sqrt{g_{00}}}\mathbf{F}_{0i};\mathbf{E}^i=\sqrt[4]{-g}\sqrt{g_{00}}\mathbf{F}^{i0}, \qquad (3.11)$$

$$\mathbf{B}_i=\frac{1}{2}\sqrt[4]{-g}\sqrt{-\tilde{g}}\,\epsilon_{ijk}\mathbf{F}^{jk};\mathbf{B}^i=\frac{1}{2}\frac{\sqrt[4]{-g}}{\sqrt{-\tilde{g}}}\epsilon^{ijk}\mathbf{F}_{jk}, \qquad (3.12)$$

and

$$\boldsymbol{\chi}=\sqrt{g_{00}}\mathbf{\Phi}, \qquad (3.13)$$

where all the fields are functions of spacial coordinates only.

With these definitions and after substituting in Eq. (3.10) for $(R-2\Lambda)/8\pi G v^2$ from Eq. (3.8), we obtain

$$\mathcal{E}=\frac{1}{2}\int d^3x\left\{-\left[\mathbf{E}^i\cdot\mathbf{E}_i+\mathbf{B}^i\cdot\mathbf{B}_i+\left(\frac{\sqrt[4]{-g}}{\sqrt{g_{00}}}\mathbf{D}^i\boldsymbol{\chi}\right)\left(\frac{\sqrt[4]{-g}}{\sqrt{g_{00}}}\mathbf{D}_i\boldsymbol{\chi}\right)\right.\right.$$
$$\left.-2\frac{\sqrt{-g}}{g_{00}}\mathbf{D}_0\boldsymbol{\chi}\cdot\mathbf{D}^0\boldsymbol{\chi}\right]+\frac{1}{2}\frac{\sqrt{-g}}{g_{00}}\left(\frac{\partial^i\sqrt{g_{00}}}{\sqrt{g_{00}}}-\partial^i f\right)\partial_i\boldsymbol{\chi}^2$$
$$\left.-\frac{\lambda}{2}\sqrt{\frac{-g}{g_{00}^2}}(\boldsymbol{\chi}^4-g_{00}^2v^4)+\partial_i\left(\frac{\sqrt{-g}}{g_{00}}\boldsymbol{\chi}^2\partial^i f\right)\right\}. \qquad (3.14)$$

In the spherically symmetric case in which we are interested, with the assumed signature for the metric, we note that

$$g_{ij}=0,i\neq j \quad \text{and} \quad -g_{ii}\geqslant 0. \qquad (3.15)$$

Consequently, each of the terms in the first parentheses in Eq. (3.14) is positive. They are exact analogues of the corresponding terms in the case of flat space-time. The next two terms vanish in the Higgs vacuum, $\boldsymbol{\chi}^2=g_{00}v^2$. Finally, the last term is a total divergence, giving rise to a finite surface term. Ignoring the terms that vanish when $\boldsymbol{\chi}^2=g_{00}v^2$, we have a reduced positive-definite energy functional, $\mathcal{E}$, given by

$$\mathcal{E}=\frac{1}{2}\int d^3x\left[\mathbf{E}^i\cdot\mathbf{E}_i+\mathbf{B}^i\cdot\mathbf{B}_i+\left(\frac{\sqrt[4]{-g}}{\sqrt{g_{00}}}\mathbf{D}^i\boldsymbol{\chi}\right)\left(\frac{\sqrt[4]{-g}}{\sqrt{g_{00}}}\mathbf{D}_i\boldsymbol{\chi}\right)\right.$$
$$\left.+2\frac{\sqrt{-g}}{g_{00}}\mathbf{D}_0\boldsymbol{\chi}\cdot\mathbf{D}^0\boldsymbol{\chi}+\partial_i\left(\frac{\sqrt{-g}}{g_{00}}\boldsymbol{\chi}^2\partial^i f\right)\right]. \qquad (3.16)$$

As in the case of flat space-time [17], we can write

$$\mathcal{E}=\frac{1}{2}\int d^3x\left\{\left[\left(\mathbf{E}^i-\sin\theta\frac{\sqrt[4]{-g}}{\sqrt{g_{00}}}\mathbf{D}^i\boldsymbol{\chi}\right)\left(\mathbf{E}_i-\sin\theta\frac{\sqrt[4]{-g}}{\sqrt{g_{00}}}\mathbf{D}_i\boldsymbol{\chi}\right)\right.\right. \qquad (3.17)$$
$$\left.+\left(\mathbf{B}^i-\cos\theta\frac{\sqrt[4]{-g}}{\sqrt{g_{00}}}\mathbf{D}^i\boldsymbol{\chi}\right)\left(\mathbf{B}_i-\cos\theta\frac{\sqrt[4]{-g}}{\sqrt{g_{00}}}\mathbf{D}_i\boldsymbol{\chi}\right)\right]$$
$$+2\left[\sin\theta\frac{\sqrt[4]{-g}}{\sqrt{g_{00}}}\mathbf{E}^i\cdot\mathbf{D}_i\boldsymbol{\chi}+\cos\theta\frac{\sqrt[4]{-g}}{\sqrt{g_{00}}}\mathbf{B}^i\cdot\mathbf{D}_i\boldsymbol{\chi}\right.$$
$$\left.\left.+\frac{\sqrt{-g}}{g_{00}}\mathbf{D}_0\boldsymbol{\chi}\cdot\mathbf{D}^0\boldsymbol{\chi}+\frac{1}{2}\partial_i\left(\frac{\sqrt{-g}}{g_{00}}\boldsymbol{\chi}^2\partial^i f\right)\right]\right\}. \qquad (3.18)$$

Hence, $\mathcal{E}$, has a lower bound,

$$\mathcal{E} \geqslant \int d^3x \left[\sin\theta \frac{\sqrt[4]{-g}}{\sqrt{g_{00}}} \mathbf{E}^i \cdot \mathbf{D}_i \boldsymbol{\chi} + \cos\theta \frac{\sqrt[4]{-g}}{\sqrt{g_{00}}} \mathbf{B}^i \cdot \mathbf{D}_i \boldsymbol{\chi} \right.$$
$$\left. + \frac{\sqrt{-g}}{g_{00}} \mathbf{D}_0 \boldsymbol{\chi} \cdot \mathbf{D}^0 \boldsymbol{\chi} + \frac{1}{2} \partial_i \left(\frac{\sqrt{-g}}{g_{00}} \boldsymbol{\chi}^2 \partial^i f \right) \right], \tag{3.19}$$

reaching the lower bound when the Bogomol'nyi-type equations are satisfied, that is when

$$\mathbf{E}_i - \sin\theta \frac{\sqrt[4]{-g}}{\sqrt{g_{00}}} \mathbf{D}_i \boldsymbol{\chi} = \mathbf{B}_i - \cos\theta \frac{\sqrt[4]{-g}}{\sqrt{g_{00}}} \mathbf{D}_i \boldsymbol{\chi} = 0. \tag{3.20}$$

Now from our definition for $\mathbf{B}_i$ in Eq. (3.12),

$$\frac{\sqrt[4]{-g}}{\sqrt{g_{00}}} \mathbf{B}^i = \frac{1}{2} \epsilon^{ijk} \mathbf{F}_{jk},$$

and from the Bianchi identity, we have the conservation law,

$$\mathbf{D}_i \left(\frac{\sqrt[4]{-g}}{\sqrt{g_{00}}} \mathbf{B}^i \right) = 0. \tag{3.21}$$

Consequently,

$$\int d^3x \frac{\sqrt[4]{-g}}{\sqrt{g_{00}}} \mathbf{B}^i \cdot \mathbf{D}_i \boldsymbol{\chi} = \int d^3x \mathbf{D}_i \left(\frac{\sqrt[4]{-g}}{\sqrt{g_{00}}} \mathbf{B}^i \cdot \boldsymbol{\chi} \right). \tag{3.22}$$

Similarly,

$$\frac{\sqrt[4]{-g}}{\sqrt{g_{00}}} \mathbf{E}^i = \mathbf{F}^{i0}, \tag{3.23}$$

and

$$\mathbf{D}_i \left(\frac{\sqrt[4]{-g}}{\sqrt{g_{00}}} \mathbf{E}^i \cdot \boldsymbol{\chi} \right) = \boldsymbol{\chi} \cdot \mathbf{D}_i \mathbf{F}^{i0} + \mathbf{E}^i \cdot \frac{\sqrt[4]{-g}}{\sqrt{g_{00}}} \mathbf{D}_i \boldsymbol{\chi}$$
$$= \mathbf{E}^i \cdot \frac{\sqrt[4]{-g}}{\sqrt{g_{00}}} \mathbf{D}_i \boldsymbol{\chi}, \tag{3.24}$$

on account of Eq. (3.7).

Hence,

$$\int d^3x \left(\mathbf{E}^i \cdot \frac{\sqrt[4]{-g}}{\sqrt{g_{00}}} \mathbf{D}_i \boldsymbol{\chi} \right) = \int d^3x \mathbf{D}_i \left(\frac{\sqrt[4]{-g}}{\sqrt{g_{00}}} \mathbf{E}^i \cdot \boldsymbol{\chi} \right). \tag{3.25}$$

When the Bogomol'nyi equation (3.20) is satisfied, the lower bound on the energy functional is satisfied and we have

$$\mathcal{E} = \sin\theta \int d^3x \mathbf{D}_i \left(\frac{\sqrt[4]{-g}}{\sqrt{g_{00}}} \mathbf{E}^i \cdot \boldsymbol{\chi} \right)$$
$$+ \cos\theta \int d^3x \mathbf{D}_i \left(\frac{\sqrt[4]{-g}}{\sqrt{g_{00}}} \mathbf{B}_i \cdot \boldsymbol{\chi} \right)$$
$$+ \int d^3x \left[\frac{\sqrt{-g}}{g_{00}} \mathbf{D}_0 \boldsymbol{\chi} \cdot \mathbf{D}^0 \boldsymbol{\chi} + \frac{1}{2} \partial_i \left(\frac{\sqrt{-g}}{g_{00}} \boldsymbol{\chi}^2 \partial^i f \right) \right]. \tag{3.26}$$

Now, if we have finite-energy configurations with finite extension and asymptotically flat space-time,

$$D_\mu \boldsymbol{\Phi} = 0, \quad \boldsymbol{\Phi}^2 = v^2, \tag{3.27}$$

leading to the condition that the gauge potential $\mathbf{A}_\mu$ is given by

$$\mathbf{A}_\mu = \frac{i}{\alpha v^2} [\boldsymbol{\Phi}, \partial_\mu \boldsymbol{\Phi}] + \frac{1}{v} \boldsymbol{\Phi} W_\mu, \tag{3.28}$$

where W_μ is an arbitrary Abelian field. The field strength $\mathbf{F}_{\mu\nu}$ corresponding to the above gauge potential is

$$\mathbf{F}_{\mu\nu} = \frac{1}{v} \boldsymbol{\Phi} F_{\mu\nu},$$

where

$$F_{\mu\nu} = \frac{i}{\alpha v^3} [\partial_\mu \boldsymbol{\Phi}, \partial_\nu \boldsymbol{\Phi}] \cdot \boldsymbol{\Phi} + \partial_\mu W_\nu - \partial_\nu W_\mu. \tag{3.29}$$

In the static case, asymptotically,

$$\mathbf{A}_0 = \boldsymbol{\Phi} \frac{W_0}{v}, \quad \mathbf{A}_i = \frac{i}{\alpha v^2} [\boldsymbol{\Phi}, \partial_i \boldsymbol{\Phi}],$$

and

$$\mathbf{F}_{i0} = \boldsymbol{\Phi} \frac{1}{v} \partial_i W_0 = -\mathbf{E}_i,$$

$$\mathbf{F}_{ij} = \frac{1}{2v} \boldsymbol{\Phi} \left\{ \frac{i}{\alpha v^3} [\partial_i \boldsymbol{\Phi}, \partial_j \boldsymbol{\Phi}] \cdot \boldsymbol{\Phi} \right\}, \tag{3.30}$$

where we have assumed $W_i = 0$, $i = 1,2,3$.

The above field configurations imply that outside a finite region, the non-Abelian gauge field is purely in the direction $\boldsymbol{\Phi}$, the direction of the unbroken $U(1)$ electromagnetic field $F_{\mu\nu}$. Using Eq. (3.30), we can convert the divergence integrals of $\mathbf{E}^i$ and $\mathbf{B}^i$ into flux integrals of electric and magnetic fields over a surface at infinity and obtain

$$\mathcal{E} = \sin\theta Q_E + \cos\theta Q_M + \frac{1}{2} \int d^3x \partial_i \left(\frac{\sqrt{-g}}{g_{00}} \boldsymbol{\chi}^2 \partial^i f \right), \tag{3.31}$$

where

$$Q_E=\lim_{R\to\infty}\int_{S_R} d\sigma_R^i \frac{1}{v}(\partial_i W_0)\chi^2, \qquad (3.32)$$

and

$$Q_M=\lim_{R\to\infty}\frac{1}{2\alpha v^3}\int_{S_R} d\sigma_{Ri}\epsilon^{ijk}\boldsymbol{\chi}\cdot[\partial_j\boldsymbol{\chi},\partial_k\boldsymbol{\chi}]. \qquad (3.33)$$

Which leads to

$$Q_M=\frac{4\pi}{\alpha}n, \qquad (3.34)$$

where n, an integer, is the winding number of the mapping $\chi: S^2_\infty \to S^2$,

$$n=\frac{1}{8\pi v^3}\int_{S^2_\infty} d\sigma_i\epsilon^{ijk}\boldsymbol{\chi}\cdot[\partial_j\boldsymbol{\chi},\partial_k\boldsymbol{\chi}].$$

Q_E has no such interpretation, but its finiteness leads to a condition on W_0.

If we define

$$\sin\theta=\frac{Q_E}{\sqrt{Q_E^2+Q_M^2}}, \cos\theta=\frac{Q_M}{\sqrt{Q_E^2+Q_M^2}},$$

then

$$\mathcal{E}\geqslant\sqrt{Q_E^2+Q_M^2}+\frac{1}{2}\int d^3x\,\partial_i\left(\frac{\sqrt{-g}}{g_{00}}\chi^2\partial^i f\right), \qquad (3.35)$$

the equality holding when Eq. (3.20) is satisfied.

IV. BASIC EQUATIONS WITH SPHERICALLY SYMMETRIC ANSATZ

The considerations in the previous section are valid for any compact group $\mathcal{G}$. In this section, we shall specialize to $SU(2)$. We define a spherically symmetric static metric

$$ds^2=A^2(r)dt^2-B^2(r)dx^2-r^2(d\theta^2+\sin^2\theta d\varphi^2), \qquad (4.1)$$

and assume the spherically symmetric 't Hooft–Polyakov ansatz [4] for the gauge and Higgs fields,

$$A_0^a=\frac{\hat{x}^a}{\alpha}J(r), \qquad (4.2)$$

$$\eta_{ab}A_i^b=\epsilon_{aij}\hat{x}^j\frac{[1-W(r)])}{\alpha r}, \qquad \eta_{ab}=(-1,-1,-1), \qquad (4.3)$$

and

$$\Phi^a=\hat{x}^a vH(r). \qquad (4.4)$$

Expressed in spherical polar coordinates, our ansatz (3.4) that leads to the Bogomol'nyi-type equation takes the form

$$\mathbf{D}_r\boldsymbol{\chi}+\left(f'(r)-\frac{A'(r)}{A(r)}\right)\boldsymbol{\chi}=-\frac{A(r)B(r)}{r^2\sin\theta}\mathbf{F}_{\theta\varphi}, \qquad (4.5)$$

$$\mathbf{D}_\theta\boldsymbol{\chi}=-\frac{A(r)}{B(r)}\frac{1}{\sin\theta}\mathbf{F}_{\varphi r}, \qquad (4.6)$$

$$\mathbf{D}_\varphi\boldsymbol{\chi}=-\frac{A(r)}{B(r)}\sin\theta\mathbf{F}_{r\theta}, \qquad (4.7)$$

where the prime denotes diffentiation with respect to r.

Substituting the spherically symmetric ansatz for $\boldsymbol{\chi}$ and $\mathbf{F}$, we obtain two independent equations,

$$H'(r)+f'(r)H(r)=\frac{B(r)(1-W^2(r))}{\alpha v r^2}, \qquad (4.8)$$

$$W'(r)=-\alpha v W(r)H(r)B(r). \qquad (4.9)$$

The Yang-Mills field equations (3.1) and (3.6), and the Higgs field equation (3.8) are transformed into

$$J''(r)+\left(\frac{2}{r}-\frac{A'(r)}{A(r)}-\frac{B'(r)}{B(r)}\right)J'(r)-\frac{2B^2(r)}{r^2}J(r)W(r)^2=0, \qquad (4.10)$$

$$J^2(r)W(r)-\left(f'(r)-\frac{A'(r)}{A(r)}\right)\frac{A^2(r)}{B^2(r)}W'(r)=0, \qquad (4.11)$$

$$\left(-f'(r)+\frac{A'(r)}{A(r)}\right)\frac{H'(r)}{H(r)B^2(r)}+\frac{1}{B^2(r)}\left(\frac{B'(r)}{B(r)}-\frac{2}{r}\right)f'(r)-\frac{1}{B^2(r)}f''(r)=\frac{R-2\Lambda}{8\pi G v^2}+\lambda v^2(H^2(r)-1). \qquad (4.12)$$

We would like to remark that in the case of pure monopole in the $\mathbf{A}_0=0$ gauge then $J(r)=0$. Thus if $f'(r)=\frac{A'(r)}{A(r)}$, the Yang-Mills equations (4.10) and (4.11) are automatically satisfied and the Higgs field equation (4.12) becomes

$$\frac{1}{B^2(r)}\left[\left(\frac{B'(r)}{B(r)}-\frac{2}{r}\right)\frac{A'(r)}{A(r)}-\left(\frac{A'(r)}{A(r)}\right)'\right]$$
$$=\frac{R-2\Lambda}{8\pi G v^2}+\lambda v^2(H^2(r)-1). \tag{4.13}$$

From the metric (4.1), we calculate the components of the Einstein tensor, $G_{\mu\nu}$, and the Ricci scalar, R, and find them to be

$$G_{00}=\frac{A^2}{B^2}\left[\frac{1}{r^2}(B^2-1)+\frac{2}{r}\frac{B'}{B}\right],$$

$$G_{rr}=\frac{1}{r^2}(1-B^2)+\frac{2}{r}\frac{A'}{A},$$

$$G_{\theta\theta}=\frac{r^2}{B^2}\left(\frac{A''}{A}+\frac{1}{r}\frac{A'}{A}-\frac{A'}{A}\frac{B'}{B}-\frac{1}{r}\frac{B'}{B}\right),$$

$$G_{\varphi\varphi}=\sin^2\theta G_{\theta\theta}, \tag{4.14}$$

and

$$R=\frac{2}{r^2}\left(\frac{1}{B^2}-1\right)+\frac{2}{B^2}\left(\frac{A''}{A}+\frac{2}{r}\frac{A'}{A}-\frac{A'}{A}\frac{B'}{B}-\frac{2}{r}\frac{B'}{B}\right). \tag{4.15}$$

The rest of the components are zero. Likewise, we express the components of the energy-momentum tensor using Eqs. (4.5)–(4.7), and find

$$T_{00}=\frac{1}{\alpha^2}\left[\frac{1}{2B^2}J'^2+\frac{1}{r^2}J^2W^2+\frac{A^2}{B^2r^2}W'^2+\frac{1}{2}\frac{A^2}{r^4}(W^2-1)^2\right]$$
$$+v^2\left[\frac{1}{2}\frac{A^2}{B^2}H'^2+\frac{A^2}{r^2}W^2H^2+\frac{\lambda}{4}A^2v^2(H^2-1)^2\right], \tag{4.16}$$

$$T_{rr}=\frac{1}{\alpha^2}\left[-\frac{1}{2A^2}J'^2+\frac{B^2}{A^2}J^2W^2+\frac{1}{r^2}W'^2-\frac{B^2}{2r^4}(W^2-1)^4\right]$$
$$+v^2\left[\frac{1}{2}H'^2-\frac{B^2}{r^2}W^2H^2-\frac{\lambda}{4}B^2v^2(H^2-1)^2\right], \tag{4.17}$$

$$T_{\theta\theta}=\frac{1}{2\alpha^2}\left[\frac{1}{A^2B^2}r^2J'^2+\frac{1}{r^2}(W^2-1)^2\right]$$
$$-\frac{v^2}{2}\left[\frac{r^2}{B^2}H'^2+\frac{\lambda}{2}r^2v^2(H^2-1)^2\right], \tag{4.18}$$

$$T_{\varphi\varphi}=\sin^2\theta T_{\theta\theta}. \tag{4.19}$$

Using Eqs. (4.14) and (4.16), it is straightforward to write Einstein field equations

$$G_{\mu\nu}+\Lambda g_{\mu\nu}=\frac{8\pi G v^2}{\Phi^2}T_{\mu\nu}.$$

In what follows, we eliminate from them their dependence on $W'(r)$ and $W(r)$ by using Eqs. (4.8) and (4.9) and take linear combinations of the resulting equations. These are

$$\frac{1}{r}\left(\frac{A'}{A}+\frac{B'}{B}\right)=8\pi G\left(\frac{v^2}{2}\left(\frac{H'}{H}\right)^2+B^2\left(1-v\alpha r^2\left(\frac{H'+f'H}{B}\right)\right)\left(\frac{1}{\alpha^2r^2}\frac{J^2}{A^2H^2}+\frac{v^2}{r^2}\right)\right), \tag{4.20}$$

$$\frac{A''}{A}-\frac{A'}{A}\frac{B'}{B}+\frac{B^2-1}{r^2}=8\pi G\left[\frac{1}{\alpha^2}\frac{J'^2}{A^2H^2}+\frac{1}{2}v^2\left(\frac{H'}{H}\right)^2+\frac{v^2B^2}{r^2}+f'v^2\left(2\frac{H'}{H}+f'\right)-v^3B\alpha(H'+f'H)\right], \tag{4.21}$$

$$\frac{A''}{A}+\frac{1}{r}\left(\frac{A'}{A}-\frac{B'}{B}\right)-\frac{A'}{A}\frac{B'}{B}+\Lambda B^2=4\pi G\left[\frac{1}{\alpha^2}\frac{J'^2}{A^2H^2}+f'v^2\left(2\frac{H'}{H}+f'\right)-\frac{1}{2}B^2v^4\frac{(H^2-1)^2}{H^2}\right]. \tag{4.22}$$

V. HIGGS VACUUM: MONOPOLE AND DYON SOLUTIONS

We seek solutions to the set of coupled nonlinear equations (4.8)–(4.12) and (4.20)–(4.22) in the Higgs vacuum, that is, the Higgs field frozen to a constant, $H=1$. The equations simplify considerably. To this end, we introduce the dimensionless coordinate x and define dimensionless J,

$$x=\alpha v r, \quad \frac{J}{\alpha v}\to J. \tag{5.1}$$

We shall also work in units of the dimensionless coupling $4\pi G v^2$ set equal to unity and henceforth all the field variables are functions of x and a prime will denote derivatives with respect to x. We shall suppress their dependence on x for simplicity.

From Eqs. (4.9) and (4.11), it follows that

$$f'=\frac{A'}{A}-\frac{J^2 B}{A^2}, \tag{5.2}$$

and hence the Bogomol'nyi conditions (4.8) and (4.9) take the form

$$\frac{A'}{AB}=\frac{J^2}{A^2}+\frac{1-y}{x^2}, \tag{5.3}$$

$$B=-\frac{1}{2}\frac{y'}{y}, \tag{5.4}$$

where we have defined $y=W^2$.

Then the remaining Yang-Mills, Higgs, and Einstein field equations (4.10), (4.12), (4.20)–(4.22) are given by

$$J''+\left(\frac{2}{x}-\frac{A'}{A}-\frac{B'}{B}\right)J'-\frac{2}{x^2}B^2 J y=0, \tag{5.5}$$

$$B^2\Lambda+\frac{B^2-1}{x^2}-2\frac{A''}{A}+\frac{2}{x}\frac{B'}{B}+\frac{A'}{A}\left(\frac{A'}{A}+2\frac{B'}{B}-\frac{4}{x}\right)$$
$$=2J\frac{B}{A^2}\left[J\left(\frac{A'}{A}-\frac{1}{x}\right)-J'\right], \tag{5.6}$$

$$\frac{1}{x}\left(\frac{A'}{A}+\frac{B'}{B}\right)-\frac{2B^2}{x^2}\left(1-x^2\frac{A'}{AB}\right)$$
$$=2J^2\frac{B^2}{A^2}\left(1+\frac{1}{x^2}-\frac{A'}{AB}+\frac{J^2}{A^2}\right), \tag{5.7}$$

$$\frac{A''}{A}-\frac{A'}{A}\left(\frac{B'}{B}+2\frac{A'}{A}-2B\right)-\frac{B^2+1}{x^2}$$
$$=2\frac{J'^2}{A^2}+2J^2\frac{B^2}{A^2}\left(1-\frac{2}{B}\frac{A'}{A}+\frac{J^2}{A^2}\right), \tag{5.8}$$

$$\frac{A''}{A}+\frac{1}{x}\left(\frac{A'}{A}-\frac{B'}{B}\right)-\frac{A'}{A}\left(\frac{A'}{A}+\frac{B'}{B}\right)+\Lambda B^2$$
$$=\frac{J'^2}{A^2}-2B\frac{J^2}{A^2}\frac{A'}{A}+B^2\frac{J^4}{A^4}. \tag{5.9}$$

A. Monopole solutions

Equations for the monopole are obtained by setting $J=0$ in the above set of equations. At the outset, we note that Eq. (5.5) is automatically satisfied. By taking suitable linear combinations of Eqs. (5.6), (5.7), (5.8), and (5.9), we find

$$\Lambda(3B^2+1)=0, \tag{5.10}$$

which implies $\Lambda=0$ in order for B to be real (and have the correct assymptotic behavior). Thus, there is no monopole solution for the nonvanishing cosmological constant in our model. With $\Lambda=0$, we are left with three independent equations,

$$\frac{A'}{A}=\frac{1}{x^2}B(1-y), \quad B=-\frac{1}{2}\frac{y'}{y}, \quad B=1+x\frac{A'}{A}, \tag{5.11}$$

or equivalently,

$$y'=-\frac{2xy}{x+y-1}, \tag{5.12}$$

$$B=\frac{x}{x+y-1}, \quad \frac{A'}{A}=\frac{1-y}{x(x+y-1)}. \tag{5.13}$$

Equation (5.12) is the well known Abel's differential equation of the second type. The two equations in Eq. (5.13) are determined in terms of the solutions of Abel's equation. Abel's equation has no known analytical solution other than $y=0$, in which case, we have an Abelian magnetic monopole with metric functions given by

$$A^2=\left(1-\frac{1}{x}\right)^2, \quad B^2=\frac{1}{A^2}=\left(1-\frac{1}{x}\right)^{-2}. \tag{5.14}$$

This Abelian monopole solution is also an extreme Reissner-Nördstrom black hole with mass M and charge Q given by

$$M=\frac{4\pi G v}{\alpha}, \quad Q=\frac{M}{\alpha v}. \tag{5.15}$$

In the general case, with nonvanishing y, we can solve Eq. (5.12) numerically. Results are shown in Figs. 1–3.

We observe that we have a family of solutions that have exponentially vanishing non-Abelian components $y=W^2$. They are determined by the initial condition $y(1)=y_{\text{init}}$ (see Fig. 1). These solutions are characterized by a non-Abelian core, outside of which only an Abelian component remains. The solutions to the metric components A and B in general have a minimum and maximum, respectively, in the vicinity

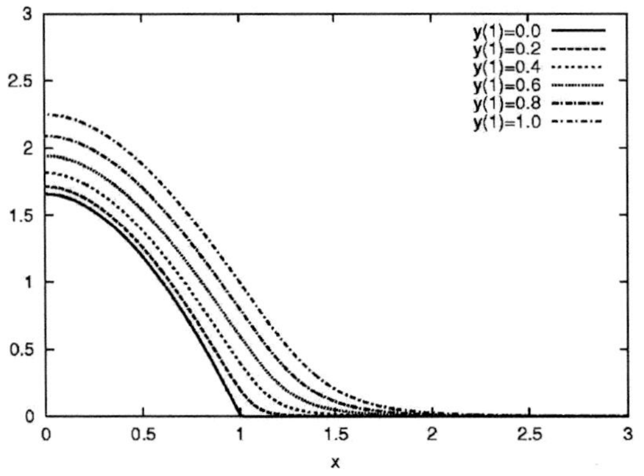

FIG. 1. Plot of family of solutions to Abel's equation, i.e., $y(x)$, for a range of initial conditions $y(1)=0$, 0.2, 0.4, 0.6, 0.8, 1.0.

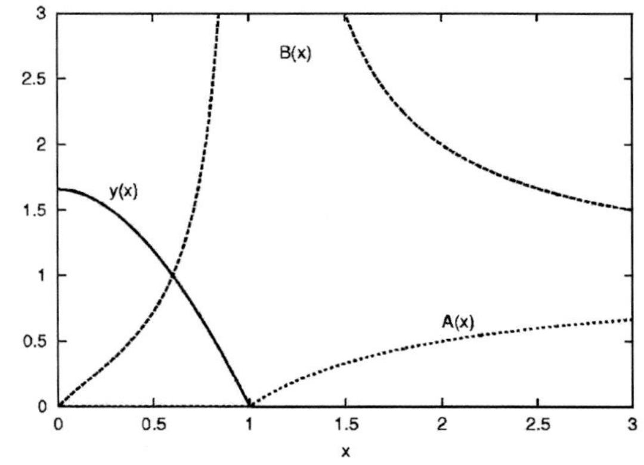

FIG. 2. Plot of $A(x)$, $B(x)$, and $y(x)$ against x for an extremal non-Abelian monopole $[y(1)=0]$. This solution is a black hole since we see the event horizon at $x=1$.

of $x=1$. From Eq. (5.13) the metric component A is dependent on an initial condition, however, once one fixes its asymptotic behavior ($A\to 1$ as $x\to\infty$) it is uniquely determined. The metric component B automatically has the correct asymptotic behavior. We see that as the initial value of y at $x=1$ approaches zero, the minimum of A also approaches zero at $x=1$ (and the maximum B at $x=1$ gets larger). The solution for $y(1)=0$ (see Fig. 2) represents the extremal case where an event horizon has formed. This extremal solution is a black hole with a non-Abelian magnetic monopole.

For the extremal solution, the non-Abelian magnetic field is confined within the black hole horizon. Thus it is only natural that the metric coefficients outside the horizon are identical to that of a Reissner-Nördstrom black hole. In Fig. 2 we see that the metric coefficient for A vanishes inside the horizon, $x\leqslant 1$. This is not a problem since A has been normalized at $x=\infty$. An observer at infinity cannot observe the interior of the black hole since for him an object takes an infinite amount of time to reach the horizon. If one choses to normalize A at the origin one would have a perfectly well defined metric inside the horizon which is infinite outside. This does not affect the determination of the non-Abelian magnetic field y which is independent of the normalization of A. This behavior is also observed in the monopole solutions of Lue and Weinberg [9].

In Fig. 3 we see a nonextremal monopole solution that is not a black hole.

B. Dyonic solutions

The set of seven coupled nonlinear equations (5.3)–(5.9) involve four functions, y, J, A, and B. We do not expect to find analytic solutions to this set of equations. Thus we are forced to look for numerical solutions.

First, however, we can analyze some features of these equations. Equation (5.7) is a quadratic equation in J^2/A^2 which has the solutions

$$2\frac{J^2}{A^2}=-\left(1+\frac{1}{x^2}-\frac{A'}{AB}\right)\pm\sqrt{\left(1-\frac{1}{x^2}+\frac{A'}{AB}\right)^2+\frac{2}{xB^2}\left(\frac{A'}{A}+\frac{B'}{B}\right)}. \tag{5.16}$$

Combining Eqs. (5.3) and (5.16) for J^2/A^2, we obtain

$$\left(1-\frac{1}{x^2}+\frac{A'}{AB}\right)+\frac{y}{x^2}=\pm\sqrt{\left(1-\frac{1}{x^2}+\frac{A'}{AB}\right)^2+\frac{2}{xB^2}\left(\frac{A'}{A}+\frac{B'}{B}\right)}. \tag{5.17}$$

By inspection, we see that if $y=0$, the case of an Abelian dyon, then the above equation implies that

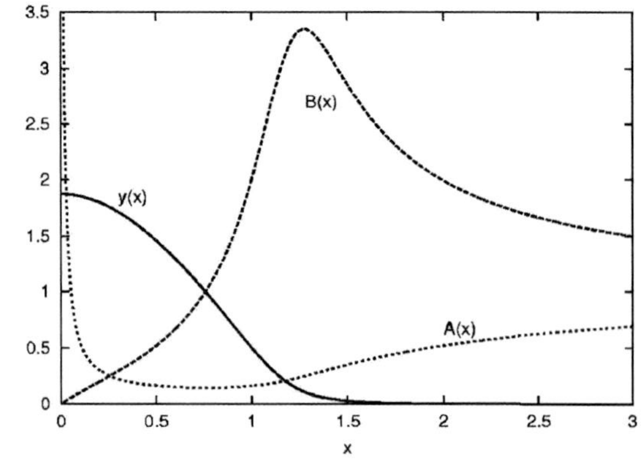

FIG. 3. Plot of $A(x)$, $B(x)$, and $y(x)$ against x for a non-Abelian monopole with the initial condition $y(1)=0.5$.

$$\frac{A'}{A}+\frac{B'}{B}=0, \tag{5.18}$$

assuming $A,B\neq 0$. This in turn implies that the product AB is constant.

Thus the Yang-Mills field equation (5.5) reduces to

$$J''+\frac{2}{x}J'=0, \tag{5.19}$$

which has the analytic solution

$$J(x)=c_1+c_2/x, \tag{5.20}$$

where c_1,c_2 are arbitrary constants.

Thus if the metric functions A and B have the correct asymptotic behavior ($A,B\to 1$ as $x\to\infty$), then inspection of Eq. (5.6) for large x shows that Λ must vanish. In addition inspection of Eq. (5.9) at large x reveals that if Λ is zero then J must vanish asymptotically, hence proving that $c_1=0$. Differentiating Eq. (5.3) and substituting into Eq. (5.9) leads to the result that

$$A^4=-c_2^2(AB)^2,$$

which is inconsistent since A^4 must be positive. Thus we have shown that there are no Abelian dyonic solutions to our equations (in contrast to the monopole case).

For the general case of non-Abelian dyon solutions with $y\neq 0$, with the mild assumption that J be asymptotically constant, examination of the large x behavior of Eqs. (5.6) and (5.9) reveals that for a consistent solution the cosmological constant must vanish (it is also a necessary condition that J vanish asymptotically). Proceeding, we find two coupled equations in y and J, whose solutions yield consistent solutions to all the equations. These are

$$y''yx^3-\frac{1}{2}y'^3(x^2-2y)-\frac{1}{2}y'^2(1-7y+2x^3)+2y'yx^2$$
$$+2yx^2(x+y')\left(y'+y\frac{J'}{J}\right)\frac{J'}{J}=0, \tag{5.21}$$

$$\left(\frac{J'}{J}\right)^4+\frac{y'}{y}\left(\frac{J'}{J}\right)^3-\left(\frac{1}{x}+\frac{y'}{2x^2}\right)\left(\frac{y'}{y}\right)^2\left(\frac{J'}{J}\right)+\frac{1}{x^2}\left(\frac{y'}{y}\right)^2$$
$$\times\left[\left(\frac{y'}{y}\right)^2\frac{(x^2-5y^2+2y-1)}{16x^2}+\frac{y'}{y}\frac{(1-5y)}{4x}-\frac{5}{4}\right]=0. \tag{5.22}$$

The solutions to y and J yield the metric functions A and B via the following equations:

$$\frac{A'}{A}=-\frac{1}{2}\frac{y'}{y}\left(1+\frac{1-3y}{x^2}\right)+2\frac{y}{y'}\left(\frac{J'}{J}+\frac{1}{2}\frac{y'}{y}\right)^2+\frac{2}{x}, \tag{5.23}$$

$$B=-\frac{1}{2}\frac{y'}{y}. \tag{5.24}$$

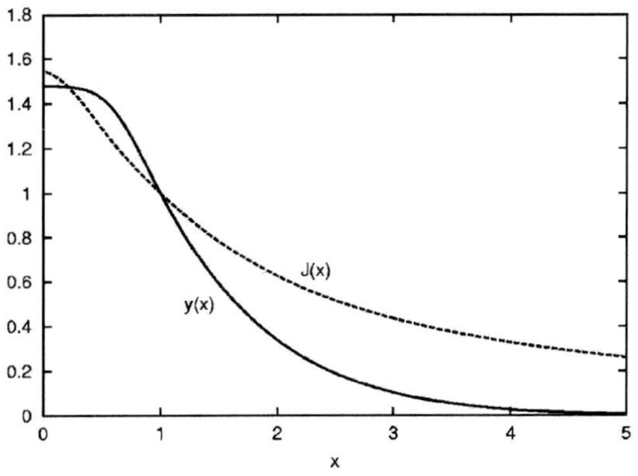

FIG. 4. Plot of $y(x)$ and $J(x)$ against x for a non-Abelian dyon with initial conditions $y(1)=1$, $y'(1)=-1$, $A(1)=2$, and $J(1)=1$.

Equation (5.22) is a quartic polynomial in J'/J in terms of y and y'. Since J is positive and the requirement that J vanish asymptotically implies that J'/J should be negative everywhere (as long as J is montonic). Thus we select the negative real root of the quartic polynomial. Equation (5.21) is a non-linear second order differential equation in y which can be readily integrated numerically using a fourth order Runge-Kutta[1] algorithm with initial conditions set at $x=1$. We attempted to understand the analytic solutions to this differential equation, however, due to its complexity we were forced to use numerical methods.

Inspection of Eqs. (5.21) and (5.22) reveals that in order for y to vanish, it can only occur at $x=1$. One can perform a similar inspection of Abel's equation in the monopole case where it is also evident that y can also only vanish at $x=1$. Thus like in the monopole case one expects that the extremal black hole solution will be reach as the initial condition for $y(x=1)\to 0$. However, unlike the monopole case there is an additional parameter, the initial value of y' at $x=1$.

Figures 4 and 5 contain plots of the solutions for y, J, A, and B for initial conditions $y(1)=1$, $y'(1)=-1$, $A(1)=2$ and $J(1)=1$. The initial value of A is set in order to obtain its correct asymptotic behavior. For these initial conditions the metric function A does not have a minimum and B does not have a maximum at finite x. However, as one adjusts the initial conditions closer to the critical value, $y(1)=0$, a minima and maxima begin to appear. This is evident in Figs. 6 and 7 where the initial conditions are $y(1)=0.5$, $y'(1)=-1.3$, $A(1)=0.4$, $J(1)=1$. From the plots it is evident

[1]We use a fourth order Runge-Kutta integration program written in $C++$ to integrate from the initial conditions at $x=1$ to the left to $x=0$ and to the right to $x=20$ in steps of $\Delta x=0.0001$. In order to check the stability of the numerics the differential equation is integrated back to $x=1$ in order to check consistency with initial conditions.

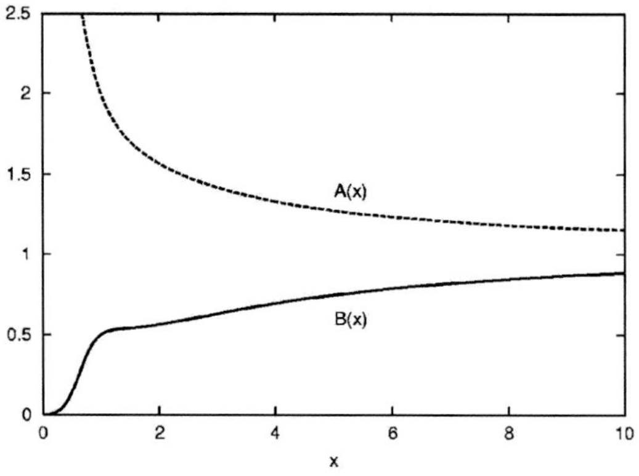

FIG. 5. Plot of $A(x)$ and $B(x)$ against x for a non-Abelian dyon with initial conditions $y(1)=1$, $y'(1)=-1$, $A(1)=2$, $J(1)=1$.

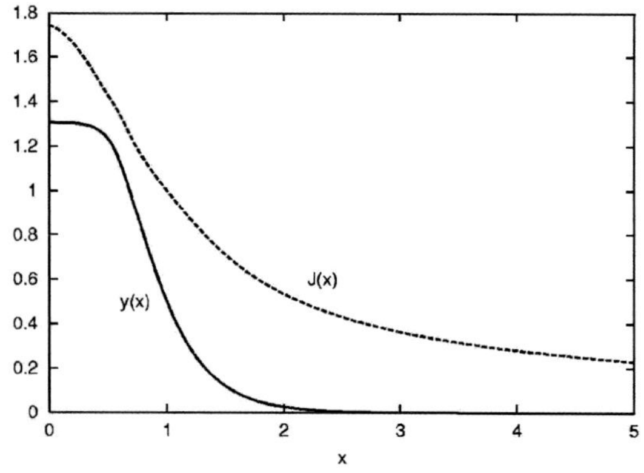

FIG. 6. Plot of $y(x)$ and $J(x)$ against x for a non-Abelian dyon with initial conditions $y(1)=0.5$, $y'(1)=-1.3$, $A(1)=0.4$, and $J(1)=1$.

that the metric functions A and B have the correct asymptotic behavior, similarly one observes that y and J vanish appropriately.

While inspecting the parameter space of initial conditions for $y'(1)$ one finds that $y'(1)\in(-2,0)$ in order for solutions to exist. For a value of $y'(1)$ in this range one finds that $y(1)$ cannot be made arbitrarily close to zero if one requires the numerical solution to be continuously defined for all x. The smallest value of $y(1)$ for a well defined solution depends on the value of $y'(1)$. Solutions for values of $y(1)$ closer to zero exhibit numerical singularities in the region $x<1$. The probable explanation for this is that unlike an extremal monopole which only has one horizon, a dyonic black hole naturally has two horizons. Thus, as the solution approaches the black hole limit the appearance of one of the horizons might occur first (meaning the metric coefficient A becomes small and B becomes large at some value of x). Also, with the existence of two horizons it is clear that our metric [Eq. (4.1)], which has positive coefficients A^2 and B^2, cannot properly describe the region between two horizons where the space-like and time-like coordinates flip. Thus, it is not unreasonable that as our dyonic solution becomes a black hole (by tuning the initial conditions closer to their critical values), the coordinates we used are not appropriate for all values of x.

VI. CONCLUSIONS

We have shown in Sec. III that, with an ansatz, Eq. (3.4), we are led to Bogomol'nyi type equations and an energy functional that is bounded from below in the case of a dyon. In flat space-time, solutions that saturate the Bogomol'nyi bound are known to exist [17]. Demonstrating this in the case of a non-Abelian dyon in curved space-time is a new result. These results follow from the form of the action in Eq. (2.1), where the Higgs scalar is directly coupled to the Ricci scalar. In the monopole case, this unconventional coupling led to a first order classic Abel's equation, the solution of which yielded solutions to all the equations in the problem [12]. In the dyon case, the problem turns out to be more complicated; nonetheless, it contains similar simplifications in contrast to the more standard approaches in the literature.

In Sec. V, we specialize in Higgs vacuum solutions to the general equations in Sec. IV. For the sake of completeness, we begin with a review of the non-Abelian monopole solutions [12]. We expand on the discussions therein. These monopole solutions depend upon initial conditions; for a specific choice the solution represents a non-Abelian extremal black hole. In addition we show that in our model there are no solutions with a nonvanishing cosmological constant in contrast to the conventional model in [10]. We also show, unlike the monopole case, that there are no Ablelian dyon solutions. We present numerical evidence for non-Abelian dyonic solutions. For a range of initial conditions the defining equations yield stable numerical results which are well-defined to the center of the dyon. However, as the initial conditions are tuned closer to critical values, the numerical results have singularities that likely correspond to the formation of horizons. We leave the full exploration of the parameter space of initial conditions and the nature of these horizons to a future publication, where in addition we would like

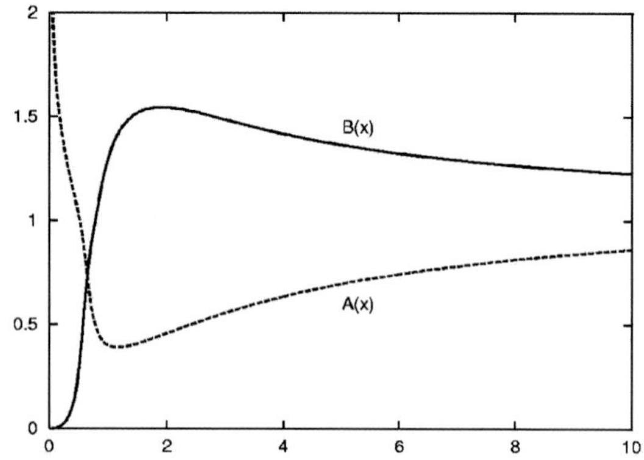

FIG. 7. Plot of $A(x)$ and $B(x)$ against x for a non-Abelian dyon with initial conditions $y(1)=0.5$, $y'(1)=-1.3$, $A(1)=0.4$, and $J(1)=1$.

to study the case where the Higgs field is not frozen to its vacuum expectation value.

Finally, to the best of our knowledge, a direct proof of Dirac charge quantization in the case of the EYMH system does not exist. However, there are compelling reasons to think that the topological considerations in the case of flat space-time can be extended to curved space-time. If so, it has some extremely interesting consequences that have been pointed out by Ignatev, Joshi, and Wali [18]. Among them are the consequences that magnetically or electrically charged black holes obey a lower bound on their mass and the consequence that is not widely appreciated, namely, the spectrum of magnetically (or electrically) charged extremal black holes are evenly spaced in mass due to charge quantization.

ACKNOWLEDGMENTS

This work was supported in part (J.S.R. and K.C.W.) by the U.S Department of Energy (DOE) under contract No. DE-FG02-85ER40237. It was also supported in part (K.C.W.) by a grant from the NSF, Division Of International Programs (U.S.-Australia Cooperative Research), No. INT-9802698. J.S.R. and K.C.W. would like to thank R. F. Sawyer for his invaluable help and suggestions in finding numerical solutions and also P. Silva for many helpful discussions.

[1] P. A. Dirac, Proc. R. Soc. London **A133**, 60 (1931).

[2] T. T. Wu and C. N. Yang, Phys. Rev. D **12**, 3845 (1975).

[3] R. A. Brandt and J. R. Primack, Phys. Rev. D **15**, 1175 (1977).

[4] G. 't Hooft, Nucl. Phys. **B79**, 276 (1974); A. M. Polyakov, Pis'ma Zh. Éksp. Teor. Fiz. **20**, 430 (1974) [JETP Lett. **20**, 194 (1974)].

[5] G. Giacomelli and L. Patrizii, hep-ex/0112009.

[6] S. D. Wick, T. W. Kephart, T. J. Weiler, and P. L. Biermann, Astropart. Phys. **18**, 663 (2003).

[7] K. M. Lee, V. P. Nair, and E. J. Weinberg, Phys. Rev. D **45**, 2751 (1992); P. Breitenlohner, P. Forgacs, and D. Maison, Nucl. Phys. **B383**, 357 (1992); P. C. Aichelburg and P. Bizon, Phys. Rev. D **48**, 607 (1993).

[8] For a review and references, see M. S. Volkov and D. V. Gal'tsov, Phys. Rep. **319**, 1 (1999).

[9] A. Lue and E. J. Weinberg, Phys. Rev. D **60**, 084025 (1999); **61**, 124003 (2000).

[10] Y. Brihaye, B. Hartmann, J. Kunz, and N. Tell, Phys. Rev. D **60**, 104016 (1999).

[11] A. R. Lugo and F. A. Schaposnik, Phys. Lett. B **467**, 43 (1999).

[12] Nguyen Ai Viet and K. C. Wali, Phys. Rev. D **51**, 1664 (1995).

[13] W. F. Kao, Phys. Rev. D **61**, 044004 (2000).

[14] L. Parker, in *Recent Developments in Gravitation*, Cargese, 1978, edited by Maurice Levy (Plenum, New York, 1979), pp. 219–273.

[15] E. B. Bogomol'nyi, Yad. Fiz. **24**, 861 (1986) [Sov. J. Nucl. Phys. **24**, 449 (1976)]; M. K. Prasad and C. M. Sommerfield, Phys. Rev. Lett. **35**, 760 (1975).

[16] A. Comtet, P. Forgacs, and P. A. Horvathy, Phys. Rev. D **30**, 468 (1984).

[17] B. Julia and A. Zee, Phys. Rev. D **11**, 2227 (1975): S. R. Coleman, S. Parke, A. Neveu, and C. M. Sommerfield, *ibid.* **15**, 544 (1977).

[18] A. Y. Ignatev, G. C. Joshi, and K. C. Wali, hep-ph/9811320.

Chiral spinors and gauge fields in noncommutative curved space-time

Nguyen Ai Viet*
Department of High-Energy Physics, CTP, Institute of Physics, National Centre of Science and Technology, Hanoi, Vietnam

Kameshwar C. Wali†
Physics Department, Syracuse University, Syracuse, New York 13244-1130, USA
(Received 12 February 2003; published 30 June 2003)

The fundamental concepts of Riemannian geometry, such as differential forms, the vielbein, metric, connection, torsion, and curvature, are generalized in the context of noncommutative geometry. This allows us to construct the Einstein-Hilbert-Cartan terms, in addition to the bosonic and fermionic ones in the Lagrangian of an action functional on noncommutative spaces. As an example, and also as a prelude to the standard model that includes gravitational interactions, we present a model of chiral spinor fields on a curved two-sheeted space-time with two distinct Abelian gauge fields. In this model, the full spectrum of the generalized metric consists of pairs of tensors, vectors, and scalar fields. They are coupled to the chiral fermions and the gauge fields leading to possible parity violation effects triggered by gravity.

DOI: 10.1103/PhysRevD.67.124029 PACS number(s): 04.50.+h, 11.10.Nx

I. INTRODUCTION

It is widely recognized that our present concepts of space and time are inadequate to provide a satisfactory basis for a unified description of all elementary particle interactions including gravity. The twin pillars of modern physics, namely, (1) general relativity, in which the dynamics of the classical continuum of space-time is coupled to the dynamics of the matter moving within it, and (2) quantum field theory, where the rules of quantization to be applied, in principle, to all degrees of freedom, including gravity, are found to be incompatible. While supersymmetric superstring theory has provided new insight, it is as yet far from being a convincing physical theory with predictable and experimentally verifiable consequences.

Both general relativity and quantum field theories assume that space-time is a continuum. A pseudo Riemannian manifold, based on such a continuum picture, provides the basis for the geometrical description of the general theory of relativity. Likewise, quantum fields and their interactions are local operators that are functions of continuous space-time coordinates. If one wants to explore the possibility of a unified quantum theory of space, time, and matter, it appears that such a continuous space-time is inadequate. Recent ideas based both on string theory and approaches to quantum gravity suggest strongly that space-time structure at the Planck scale may be discrete with noncommutative coordinates. The continuum space-time of classical physics, it is hoped, will emerge in certain limiting regimes, just as the classical behavior of quantum systems emerges in an appropriate limit.

At the present time we do not have a precisely defined noncommutative space that meets the above requirements. However, in recent years, Connes has proposed an alternate approach to the study of the structure of space-time based on noncommutative geometry (NCG) [1,2].[1] It has given rise to the description of the standard model with a geometrical interpretation of the Higgs field on the same footing as the gauge field [6,7]. Spontaneous symmetry breaking follows as a natural consequence. It enables one, in principle, to calculate some arbitrary parameters of the standard model, such as the Weinberg angle, the top quark, and Higgs boson masses [8–10].

Connes' mathematical framework is general enough to also explore the character of noncommutative spaces at the energy scales that go beyond the electroweak scale. In this framework, Connes bypasses the precise specification of the manifold as the starting point. Instead, he formulates its description in terms of an associative and involutive algebra, commutative or noncommutative. One may think of this as a generalization of the well-known theorem due to Gelfand [1,11], which states that the classical topological space based on a continuum can be completely recovered from the Abelian algebra of smooth functions. We shall discuss Connes' ideas in some detail later in this paper, but we note here that his noncommutative geometry begins with what is called a spectral triple, consisting of (i) an involutive algebra $\mathcal{A}$ of operators, commutative or noncommutative, (ii) a Hilbert space $\mathcal{H}$ as the carrier space for a representation of generalized differential forms constructed from this algebra, and (iii) a self-adjoint operator D, called the Dirac operator, acting on $\mathcal{A}$. The algebra $\mathcal{A}$ replaces and generalizes the commutative algebra of smooth functions. The Dirac operator D allows one to build a differential structure associated with any associative algebra and defines the metric.

Although the action functionals, in principle, can be constructed using the metric structure encoded in the Dirac operator, the procedure is still not unique. It allows different approaches to different problems. Thus, while the Diximier trace is used successfully to construct the action functional for the gauge sector of the standard model, other approaches

*Email address: nguyenaiviet@iop.ncst.ac.vn
†Email address: wali@physics.syr.edu

[1] See also [3–5] for a review.

have been pursued in the literature in dealing with the gravitational interactions. The Wodzicki residue, for instance, is proposed to be used as an action functional for the gravity sector in the two-sheeted space-time. It leads to the conventional Einstein-Hilbert action supplemented by a term proportional to the square of a scalar field [12,13]. The vector field that is part of the generalized metric is cancelled completely in the final action. The scalar field without a kinetic term does not propagate and causes problems when coupled to matter fields. The Diximier trace is used in other approaches with different generalizations of the Cartan formalism leading to a Brans-Dicke field coupled to gravity [14,15]. In these approaches, the vector field vanishes as a consequence of a reality condition or a consistency condition. In a more recent paper [16], Connes and Chamseddine have proposed a new universal spectral action principle governing a noncommutative space that gives rise to the Einstein-Hilbert action together with higher order terms.

Landi, Viet, and Wali [17] have adopted still another approach. It follows closely the standard Riemanian geometry and generalizes it in the case of a two-sheeted space-time that represents a discretized version of Kaluza-Klein theory in which the fifth compact circular dimension is replaced by two discrete points. By formulating the basic notions of the conventional geometry in terms of algebraic forms, they define the latter in such a way that they make sense both in the case of commutative and noncommutative situations. In their simplified version of the model in which the generalized metric is assumed to be the same on the two sheets, the resulting action functional is exactly that of the zero mode sector of the Kaluza-Klein theory. In the more general case studied by Viet and Wali [18,19], the Cartan formalism of pseudo Riemannian geometry finds its generalization to allow the most general metric that contains pairs of tensors, vectors, and scalar fields. All of these fields are present in the resulting action functional with a set of constraints on the nonvanishing torsion that generalizes the torsion free condition in the standard space-time. They find a finite spectrum of the discretized Kaluza-Klein theory without requiring the truncation of the infinite mass spectrum. In the special case [19], when the torsion free condition is imposed, the spectrum reduces to the zero mode sector coupled to dilatons with a sixth order potential and a cosmological constant [19]. Recently, Viet [20] has found a new minimal set of constraints, which together with the Cartan equations determine the connection, torsion, and curvature in terms of the pairs of tensors, vectors, and scalar fields with a simpler Einstein-Hilbert-Cartan action functional.

In this paper we extend our formalism to include a fermionic and its associated gauge boson sectors. As a prelude to the study of the standard model to include gravity, we consider a model of a left- and a right-chiral field living on a curved two-sheeted space-time and construct their action functionals. Because of the existence of different metric fields on the two sheets of space-time, we find a possible mechanism in our formalism for parity violation due to gravity.

In Sec. II, we present the basic elements of the noncommutative differential geometry after a review of the algebraic approach to the standard differential geometry. This is primarily to set the stage for the contrasting NCG approach. The formulas of the key geometric notions in the standard case are formulated in such a way that they can be used as guidelines for generalizations in the noncommutative situation. In Sec. III, continuing the same approach, we introduce metric structure and construct Riemannian geometry using the algebraic approach and generalize it to the case of noncommutative geometry à la Connes. We also discuss in this section the representation mapping π given by Connes [1] to represent operators in a given Hilbert space. This procedure will be important in the realization of a noncommutative geometric model in the rest of the paper. In Sec. IV we specialize to the algebra of differential forms and metric structure on the two-sheeted space-time. Here we follow the minimal constraints given in [20] to construct the Einstein-Hilbert-Cartan action. By following closely the formulas given in Sec. III, we establish a concrete realization of the abstract noncommutative geometry à la Connes in our formalism. In Sec. V we discuss the physical contents of the gravity sector in two-sheeted space-time. In Sec. VI we construct the action for the gauge and chiral fermion sectors. Section VII is devoted to discussions of some physical consequences of the model, and Sec. VIII is devoted to a summary and conclusions.

II. ALGEBRAIC APPROACH TO DIFFERENTIAL GEOMETRY

A. Global and algebraic approaches versus local construction

The traditional starting point of the standard differential geometry is the local construction of the vector spaces of tangent vectors and differential forms in the coordinate bases at an arbitrary local point $x \in \mathcal{M}$ [21–24]. In this construction, the differential forms are formulated as duals to the tangent vectors, which are derivatives of the functions at a point x. The differential forms form a vector space. All the geometric notions such as metric, connection, torsion, curvature, and physical fields as scalars, vectors, and tensors can be formulated conveniently in terms of differential forms. Although these notions can be defined locally, their meaning can only be understood clearly in the context of a global construction [24].

The physical fields that are defined at all points of a manifold $\mathcal{M}$, should be formulated strictly as the sections in a global framework of fiber bundles. In particular, the spinors are meaningful if and only if a global property of the manifold $\mathcal{M}$, the orientability, can be defined. The physical fields, which are defined as sections of the fiber bundles with $\mathcal{M}$ as the base manifold, have an algebraic finite projective module structure. The connection, together with other concepts such as the covariant derivative and the curvature, can also be defined in the framework of this global construction [24]. This observation is important as it paves the way to a pure algebraic approach that allows direct generalization of the ordinary geometry to NCG without reference to the concept of a point.

The concept of a point as the starting point is the main obstacle to go beyond the standard geometry and seek a unified framework of general relativity and quantum theory. In an algebraic approach, on the other hand, it is possible to start from the pure algebraic structures to formulate the basic geometric elements. Thus, the Gelfand's theorem [11], which so far has been just an alternative formulation of differential geometry, becomes the central point in such an algebraic approach to geometry without the concept of a point.

A discussion of the global approach could provide a great deal of useful insight because it can serve as a bridge between the traditional and algebraic constructions. However, such a task is beyond the scope of this paper.[2] However, from the algebraic relations that are originally derived from a local construction of the ordinary geometry, the basic geometric and physical elements can be formulated in such a way that they can be generalized directly to NCG. In the following sections, it is our goal to reformulate the basic notions of the ordinary geometry in the algebraic approach first, and then directly generalize them to the noncommutative case.

B. The ordinary differential geometry

The essential point of an algebraic approach is taking the algebra of functions as a starting point. The set of differential forms will be considered as an algebraic module, a concept more general than that of a vector space, which also makes sense when the algebra is noncommutative.

The continuous functions $f(x):\mathcal{M}\rightarrow R$ form a commutative algebra $C^\infty(\mathcal{M})$. According to Gelfand's theorem, all the information about the manifold $\mathcal{M}$ is encoded in this algebra. In other words, knowing the algebra $C^\infty(\mathcal{M})$ is equivalent to knowing the manifold $\mathcal{M}$ in the ordinary differential geometry.

The graded algebra of differential forms is then constructed with the help of a differential operator d that maps the function f as a 0-form to the 1-form df. The module of 1-forms is extended by including all the elements $df\cdot g$, where g is an arbitrary element of $C^\infty(\mathcal{M})$. The higher forms can be defined by extending the differential operator d as the mapping from n-forms onto $n+1$-forms.

The operator d is required to satisfy the following properties: (i) $d^2=0$, (ii) $d(ab)=da\cdot b+(-1)^n a\cdot db$, where a is an n-form and b is an arbitrary form (Leibnitz rule), and (iii) d is Hermitian.

In general, the differential operator can be given as any operator that satisfies the above three properties. In the particular case of ordinary differential geometry, the differential operator d is given in a finite basis of the 1-forms dx^μ ($\mu=0,1,\ldots,n$),

$$d=dx^\mu\partial_\mu. \tag{2.1}$$

The assumption about the existence of a finite basis of a module of 1-forms is essential in the noncommutative case.

The action of d on a function f as a c-number operator is given as follows:

$$df=[d,f]=dx^\mu\partial_\mu f(x). \tag{2.2}$$

An arbitrary 1-form u is also generated by the elements dx^μ:

$$u=dx^\mu u_\mu, \tag{2.3}$$

where u_μ are contained in $C^\infty(\mathcal{M})$.

The action of d on u in accordance with the properties (i)–(iii) is

$$\begin{aligned}du&=d(dx^\mu u_\mu)=d(dx^\mu)\cdot u_\mu-dx^\mu dx^\nu\partial_\nu u_\mu\\&=-dx^\mu dx^\nu\partial_\nu u_\mu.\end{aligned} \tag{2.4}$$

We can write du as a combination of an antisymmetric and a symmetric 2-forms,

$$\begin{aligned}du=&\tfrac{1}{4}(dx^\mu dx^\nu-dx^\nu dx^\mu)(\partial_\mu u_\nu-\partial_\nu u_\mu)-\tfrac{1}{4}(dx^\mu dx^\nu\\&+dx^\nu dx^\mu)(\partial_\mu u_\nu+\partial_\nu u_\mu).\end{aligned} \tag{2.5}$$

The symmetric 1-forms must be dropped since they lead to the so-called "junk forms." In a specific representation, a nonzero form can be derived from a form which is identical to zero. Junk forms are such forms. As an example, consider the form $df\cdot f-f\cdot df$ that is identical to zero; however, its differential $d(df\cdot f-f\cdot df)=-df\cdot df-df\cdot df=-2df\cdot df$ is not zero, if the symmetric form is allowed. The algebra of antisymmetric 1-forms is a closed algebra by itself.

There are various ways to avoid junk forms. The simplest way, however, is to replace an ordinary product by the wedge product

$$dx^\mu\wedge dx^\nu=\tfrac{1}{2}(dx^\mu dx^\nu-dx^\nu dx^\mu). \tag{2.6}$$

Equation (2.6) can be generalized directly for the wedge product of an arbitrary number of the elements dx^μ just by antisymmetrization.

An n-form ω is a purely antisymmetric tensor of the following form:

$$\omega=dx^{\mu_1}\wedge\ldots\wedge dx^{\mu_n}\omega_{\mu_1\ldots\mu_n}. \tag{2.7}$$

The n-forms form a vector space $\Omega_x^n(\mathcal{M})$. The wedge product is defined between an m-form ω_1 and an n-form ω_2 to form an $m+n$-form $\omega_1\wedge\omega_2$ as an antisymmetrized tensor product

$$\wedge:\Omega_x^n(\mathcal{M})\times\Omega_x^m(\mathcal{M})\rightarrow\Omega_x^{m+n}(\mathcal{M}). \tag{2.8}$$

One can define the graded algebra of the differential forms as $\Omega_x(\mathcal{M})=\oplus_n\Omega_x^n(\mathcal{M})$.

As an example of differential forms, a Lie-algebra valued gauge field belonging to a group G or gauge connection b can be given as the 1-form

$$b=dx^\mu b_\mu, \tag{2.9}$$

with its field strength as the 2-form

[2] The interested readers are referred to [24] for more details on physical applications.

$$f = db + b \wedge b = dx^\mu \wedge dx^\nu f_{\mu\nu}$$

$$= dx^\mu \wedge dx^\nu \tfrac{1}{2}\{(\partial_\mu b_\nu - \partial_\nu b_\mu) + [b_\mu, b_\nu]\}. \qquad (2.10)$$

In a "semiclassical" limit, all the forms are defined as operators on a Hilbert space $\mathcal{H}$ of spinor wave functions. The strict geometric formulation of the physical spinor fields is given as spinor bundles. On the Hilbert space $\mathcal{H}$, we can construct a γ-matrix representation by identifying the abstract element dx^μ as

$$dx^\mu \equiv \gamma^\mu, \qquad (2.11)$$

where γ^μ are the usual Dirac matrices.

In the γ-matrix representation, the differential operator d is represented by the Dirac operator

$$d = \gamma^\mu \partial_\mu . \qquad (2.12)$$

The gauge field one-form and the field strength two-form are represented, respectively, by

$$b = \gamma^\mu b_\mu \qquad (2.13)$$

and

$$f = \sigma^{\mu\nu} f_{\mu\nu}, \qquad (2.14)$$

where $\sigma^{\mu\nu} = \frac{1}{2}[\gamma^\mu, \gamma^\nu]$.

Riemannian geometry can be constructed, in accordance with the equivalence principle, by introducing a locally flat orthonormal basis $\gamma^a, a = 0,1,2,3$ and introducing a vierbein structure, which gives the transformation between the two frames

$$\gamma^\mu = e^\mu_a \gamma^a,$$

$$\gamma^a = e^a_\mu \gamma^\mu, \qquad (2.15)$$

$$e^a_\mu e^\mu_b = \delta^a_b,$$

$$e^a_\mu e^\nu_a = \delta^\mu_\nu .$$

Gravity can be introduced via the connection 1-forms

$$\omega_{ab} = \gamma^c \omega_{abc} . \qquad (2.16)$$

Two main notions that characterize the gravity structure in space-time, the torsion and curvature, can be defined by the Cartan's structure equations. The torsion 1-forms t^a are given in the first structure equation

$$t^a = d\gamma^a - \gamma^b \wedge \omega^a_b . \qquad (2.17)$$

The curvature 2-forms $r^a{}_b$, given by the second structure equation

$$r^a_b = d\omega_{ab} + \omega^a{}_c \wedge \omega^c{}_b, \qquad (2.18)$$

determines the scalar curvature completely in terms of connections.

C. Noncommutative differential geometry and the spectral triple

In the case of noncommutative geometry, we shall follow the same steps as outlined in Sec. II B. We will begin with Connes' spectral triple [1,2], which consists of (1) an involutive, unital algebra of functions $\mathcal{A}$, (2) Hilbert space $\mathcal{H}$, and (3) the Dirac operator δ that satisfies the properties (i)–(iii) in Sec. II B.

The existence of a finite basis for the module of differential forms is understood. Also, the algebra $\mathcal{A}$ of generalized functions is not necessarily commutative as in the conventional case. In this section, we shall begin with a brief review of Connes' abstract construction of universal enveloping algebra of the noncommutative differential forms.

The universal differential enveloping algebra constructed from a noncommutative algebra $\mathcal{A}$ is the differential algebra $\Omega^*(\mathcal{A})$, generated by all $a,b \in \mathcal{A}$ and a symbol δ, such that

$$\delta(1) = 0,$$

$$\delta(ab) = (\delta a)b + a(\delta b). \qquad (2.19)$$

By definition, the algebra of 0-forms $\Omega^0(\mathcal{A}) \equiv \mathcal{A}$; δa belongs to the space of universal 1-forms $\Omega^1(\mathcal{A})$, whose general element ω is a linear combination

$$\omega = \sum_i \delta a_i b_i, \quad a_i, b_i \in \mathcal{A}. \qquad (2.20)$$

The existence of a finite number of elements δa_i is guaranteed by a generalization of the vector space to the finite projective module. More details about this concept can be found in [1,3]. However, here we just remark that in the ordinary geometry there is also a finite number of elements dx^μ which are the particular forms of δa_i.

In a vector space, scalar multiplication of a vector is defined with elements of an algebraic field, usually that of the complex or real numbers. The direct generalization of a vector space of 1-forms is an $\mathcal{A}$-module, where the linear combinations with real or complex coefficients are replaced by the elements of $\mathcal{A}$ and the scalar multiplication is defined with elements of the algebra $\mathcal{A}$ from the right,

$$\omega b = \left(\sum_i \delta a_i b_i \right) b = \sum_i \delta a_i (b_i b), \quad \forall b \in \mathcal{A}. \qquad (2.21)$$

The Leibnitzian property of δ, Eq. (2.19), can be used to express left multiplication of ω by an element $b \in \mathcal{A}$,

$$b\,\omega = b\left(\sum_i \delta a_i b_i \right)$$

$$= \sum_i \delta(b a_i) b_i - \delta b\, a_i b_i, \quad \forall a_i, b_i, b \in \mathcal{A}. \qquad (2.22)$$

That is to say, the generalized 1-forms give rise to an $\mathcal{A}$-bimodule structure. Here, for the sake of definiteness, we have used the basis with δa_i on the left. However, it is always possible to use the Leibnitz rule (2.19) to rewrite a differential form with the right basis. Due to the noncommutativity of the algebra $\mathcal{A}$, the basis elements and the expansion coefficients in the right basis are not necessarily the same as in the left one as in the commutative case.

To proceed further, we note that we can multiply two 1-forms as in the commutative case. The associativity of the algebra $\mathcal{A}$ and Eq. (2.19) imply

$$(\delta b_0 a_0)(\delta b_1 a_1) = \delta b_0 (a_0 \delta b_1) a_1 = \delta b_0 \delta(a_0 b_1) a_1 - \delta b_0 \delta a_0 (b_1 a_1). \tag{2.23}$$

Continuing in this manner, we can write p-fold products of 1-forms, $\mathcal{A}$-coefficient linear combinations of which give the algebra $\Omega^p(\mathcal{A})$. A general element of $\Omega^p\mathcal{A}$ is of the form

$$\sum_i \delta a_{1i} \ldots \delta a_{pi} b_i, \quad a_{ki}, b_i \in \mathcal{A}. \tag{2.24}$$

The algebra $\Omega^p(\mathcal{A})$ is now a right $\mathcal{A}$-module being a generalization of the vector space of p-forms on a manifold. Clearly, by applying repeatedly Eq. (2.22), one may define the product of a universal p-form with a q-form

$$\Omega^p\mathcal{A} \times \Omega^q\mathcal{A} \mapsto \Omega^{p+q}\mathcal{A}. \tag{2.25}$$

Hence, the space of universal forms has the structure of a graded algebra, the universal algebra $\Omega^*(\mathcal{A}) \equiv \oplus_p \Omega^p\mathcal{A}$. The involution on $\mathcal{A}$ extends uniquely to an involution on the algebra $\Omega^*\mathcal{A}$ when we impose the condition

$$(\delta a)^* \equiv -\delta a^*. \tag{2.26}$$

To transform the graded algebra of forms $\Omega^*\mathcal{A}$ into a differential algebra, we consider the *differential* δ as a linear operator that takes

$$\Omega^p \to \Omega^{p+1} \tag{2.27}$$

by defining

$$\delta(\delta a_1 \ldots \delta a_p b) \equiv (-1)^p \delta a_1 \ldots \delta a_p \delta b. \tag{2.28}$$

Equation (2.28) implies two basic relations,

$$\delta^2 \alpha = 0, \quad \forall \alpha \in \Omega^*\mathcal{A}, \tag{2.29}$$

$$\delta(\alpha_1 \alpha_2) = (\delta\alpha_1)\alpha_2 + (-1)^{deg\alpha_1}\alpha_1 \delta\alpha_2, \quad \forall \alpha_j \in \Omega^*\mathcal{A}. \tag{2.30}$$

Thus, all the constructions of the $\mathcal{A}$-module of differential forms are in exact parallel with the standard approach. In the next section we shall present the general procedure to realize the above abstract construction in a given Hilbert space. We note that in the module of the generalized differential forms, the "junk" forms must be eliminated to guarantee that if $\delta\omega \neq 0$ the differential form ω must be a nonzero form.

D. Representation of the differential algebra on a given Hilbert space

The general constructions described in Sec. II C need to be realized by operators acting on appropriate Hilbert spaces in different applications. The operator representation of the differential forms is a direct generalization of the γ-matrix representation given in Sec. II B.

Connes [1–3] has given a general procedure to construct representations by the following graded homomorphism that preserves the involution,

$$\pi : \Omega^*\mathcal{A} \to \mathcal{L}(\mathcal{H}),$$

$$\pi_p(\delta a_1 \ldots \delta a_p b) = \prod_{i=1}^{p} [D, \pi_0(a_i)] \pi_0(b), \tag{2.31}$$

where $\mathcal{L}(\mathcal{H})$ denotes the space of bounded operators on $\mathcal{H}$ and π_0, the representation of $\mathcal{A}$ on $\mathcal{H}$. The operator δ is represented by a self-adjoint operator D, the Dirac operator, on the Hilbert space $\mathcal{H}$, with compact resolvent, such that the commutator $[D, a]$ is a bounded operator, $\forall a \in \mathcal{A}$.

In essence, differential forms, in the sense of Connes, are the images of universal forms under π. Strictly, however, a problem arises in the definition of these operator-valued forms, in that there exist universal forms α such that $\pi(\alpha) = 0$, but $\pi(\delta\alpha) \neq 0$. Once one mods out by these troublesome forms for each p, we obtain the space of operator-valued forms $\Omega^p_D(\mathcal{A})$, the analog of $\Omega^p(\mathcal{A})$. This is the problem of elimination of the "junk" forms in ordinary geometry. In two-sheeted space-time, as it will be shown in Sec. IV, this elimination can be done by a generalized wedge product in perfect parallelism with the conventional approach.

The various definitions proceed as in the preceding section. We define

$$\Omega^*_D(\mathcal{A}) \equiv \oplus \Omega^p_D(\mathcal{A}). \tag{2.32}$$

Consider Eq. (2.20) as a specific case to illustrate the representation given in Eq. (2.31). The arbitrary 1-form ω can be represented as an operator U on the Hilbert space $\mathcal{H}$ as follows:

$$U = \pi_1(\omega) = \pi_1(\delta a_i)\pi_0(b_i) = \Theta^i U_i, \tag{2.33}$$

where Θ^i are just the representations of the element δa_i in Eq. (2.20). Thus, in this general context, the basis of differential forms is given without reference to a point. To be more specific with the Dirac operator D, let F be an operator representing a generalized function w in $\Omega^0_D(\mathcal{A}) \equiv \mathcal{A}$. From Eq. (2.31) it follows that

$$DF = [D, \pi_0(w)] = \pi_1(\delta w) = \pi_1(\delta a_i w_i) = \pi_1(\delta a_i)\pi_0(w_i) = \Theta^i (DF)_i. \tag{2.34}$$

The coefficients $(DF)_i \in \mathcal{A}$ can now be regarded as derivatives but not referring to a specific point. Formally, we can define the operator D_i as

$$(DF)_i \equiv D_i F = [D_i, F]. \tag{2.35}$$

Hence, we can represent the Dirac operator in the Θ^i basis in the form

$$D = \Theta^M D_M. \tag{2.36}$$

From now on we use the latin upper case indices $M,N = 0, \ldots, n-1$ to replace the latin lower case indices i. Note that the expression of the Dirac operator D in terms of the operators D_M, implies a specific basis Θ^M. We are free to choose another basis of one-forms as a linear combination with coefficients $\in \mathcal{A}$. However, once the Dirac operator is given in a specific basis Θ^M, we can calculate derivation of forms in this basis and then transform the result to another basis whenever necessary. Having outlined the construction of the operator algebra $\Omega_D^*(\mathcal{A})$, we shall henceforth abandon explicit mention of the distinction between universal forms and their representation as operators on $\mathcal{H}$, and denote them by capital letters.

III. ALGEBRAIC FORMULATION OF RIEMANNIAN GEOMETRY WITHIN THE FRAMEWORK OF NCG

The construction of the universal algebra of differential forms based on a pseudo Riemannian manifold follows the steps discussed in Sec. II. In its application to the theory of gravitation in general relativity, we need to introduce the concept of a metric, the affine connection, the inner products of one- and two-forms, the torsion, and curvature.

The starting point of the general theory of relativity is the equivalence principle, which postulates that in an arbitrary gravitational field (curved space-time described by Riemannian geometry), it is possible to choose a locally inertial system (a flat coordinate system), such that within a sufficiently small region, the laws of special relativity prevail. The transformation between the locally flat orthonormal coordinate system and the general curved coordinate system introduces the well known vierbeins that can be regarded as gravitational fields.

The general equivalence principle allows the introduction of the metric via two sets of 1-forms Θ^A and Θ^M. The 1-forms Θ^A can be expanded in the Θ^M basis and vice versa,

$$\Theta^A = \Theta^M E_M^A, \quad \Theta^M = \Theta^A E_A^M. \tag{3.1}$$

E_M^A and E_A^M are elements of $\mathcal{A}$, satisfying

$$E_M^A E_B^M = \delta_B^A, \quad E_M^A E_A^N = \delta_N^M. \tag{3.2}$$

We define a metric tensor $\mathcal{G}^{MN}$ as the inner product of two 1-forms Θ^M and Θ^N. The inner product of 1-forms is defined as the sesquilinear functional

$$\langle , \rangle : \Omega_D^1 \times \Omega_D^1 \to \mathcal{A},$$

and

$$G^{MN} \equiv \langle \Theta^M, \Theta^N \rangle. \tag{3.3}$$

G^{MN} are functions that are elements of the algebra $\mathcal{A}$, the algebra of infinitely smooth functions on the Riemannian manifold. As the algebra $\mathcal{A}$ is unital, η^{AB} [where $\eta^{AB} = \mathrm{diag}(-1,1,\ldots,1)$] is also an element of $\mathcal{A}$. Therefore, we can choose the one-forms Θ^A so that they satisfy

$$G^{AB} \equiv \langle \Theta^A, \Theta^B \rangle = \eta^{AB}. \tag{3.4}$$

We reserve the latin uppercase letters $A,B,\ldots$ to denote the "locally flat" indices and $M,N,\ldots$ to denote the "curved" or "derivative" indices in the basis Θ^M.

From Eqs. (3.3) and (3.4) we can derive the relation

$$\langle \Theta^A, \Theta^B \rangle = \eta^{AB} = \langle \Theta^M E_M^A, \Theta^N E_N^B \rangle$$
$$= E_M^A \langle \Theta^M, \Theta^N \rangle E_N^B = E_M^A G^{MN} E_N^B, \tag{3.5}$$
$$G^{MN} = E_A^M \eta^{AB} E_B^N.$$

We note that the metric structure is determined if the curvilinear and locally orthonormal bases are given. The vierbein as a transformation matrix between the two bases determines the metric tensor. In the algebraic construction, it is more convenient to use these bases as a starting point to construct the Riemannian geometry.

If U is a general 1-form, $U = \Gamma^M U_M = \Gamma^A U_A$, the transformation between its components in the two coordinate systems is given by

$$U_M = E_M^A U_A, \quad U_A = E_A^M U_M, \tag{3.6}$$

where E_A^M is the inverse matrix of E_M^A.

In particular, in the locally orthonormal basis, the exterior derivative of arbitrary forms must be first calculated in the curvilinear basis and then transformed back to the locally flat one by using the above transformations. For instance,

$$(DF)_A = E_A^M D_M F, \tag{3.7}$$

$$(DU)_{AB} = \tfrac{1}{2} E_A^M E_B^N (D_M U_N - D_N U_M). \tag{3.8}$$

The inner product of the basis 1-forms in Eqs. (3.3) and (3.5) generates the inner product in the vector space of 1-forms between $U = \Theta^M U_M$ and $V = \Theta^M V_M$ given by

$$\langle U, V \rangle = U_M \mathcal{G}^{MN} V_N = U^N V_N. \tag{3.9}$$

The covariant derivative operator (affine connection), ∇, is given by the connection 1-forms Ω_B^A, where

$$\nabla \Theta^A = \Theta^B \otimes \Omega_B^A = \Theta^B \otimes \Theta^C \Omega_{BC}^A. \tag{3.10}$$

The Cartan structure equations define the torsion and curvature 2-forms of a given connection Ω_{AB} as follows:

$$T^A = D\Theta^A - \Theta^B \wedge \Omega_B^A, \tag{3.11}$$

$$R_B^A = D\Omega_B^A + \Omega_C^A \wedge \Omega_B^C. \tag{3.12}$$

In Eqs. (3.11) and (3.12), we need a definition of the wedge product of 1-forms. Such a definition, as explained in Sec. II, becomes specific when the representation mapping π is given. Here we only assume that the module of 2-forms is spanned by the formal wedge products $\Theta^A \wedge \Theta^B$ of 1-forms. For the ordinary space-time components, the antisymmetrization is the standard convention. However, there is no requirement that it must be the case for other generalized components. In general, there are various ways to define the wedge product, depending on the specific applications.

The Cartan structure equations (3.11) and (3.12) are not sufficient to determine connection, torsion, and curvature in terms of the vierbeins. In the conventional differential geometry, one generally imposes two additional constraints: (i) the metric compatibility condition

$$\Omega_{AB} = -\Omega_{BA} \tag{3.13}$$

and (ii) the torsion free condition

$$T^A = 0. \tag{3.14}$$

With these two constraints, the first structure equation determines the connection completely in terms of the vierbeins. The second structure equation, in turn, determines the curvature in terms of the connection.

To calculate the scalar curvature from the curvature two-form, Eq. (3.12), the inner product of 2-forms is introduced as given by

$$\langle \Theta^A \wedge \Theta^B, \Theta^C \wedge \Theta^D \rangle \equiv \eta^{AD}\eta^{BC} - \eta^{AC}\eta^{BD}. \tag{3.15}$$

This inner product can also be extended for two arbitrary 2-forms $G \equiv \Theta^A \wedge \Theta^B G_{AB}$ and $H \equiv \Theta^A \wedge \Theta^B H_{AB}$, $G_{AB}, H_{AB} \in \mathcal{A}$ as follows:

$$\langle G, H \rangle = G_{AB} \langle \Theta^A \wedge \Theta^B, \Theta^C \wedge \Theta^D \rangle H_{CD}. \tag{3.16}$$

Finally, we note that the Γ-matrix representation is given by the following homomorphism Γ:

$$\Gamma(\Theta^A) = \Gamma^A, \quad \Gamma(\Theta^M) = \Gamma^M, \tag{3.17}$$

where Γ^A are given as constant "flat" Dirac matrices. The concrete form of the "curved" Dirac matrices $\Gamma^M(x)$ is not known since it involves gravitational degrees of freedom through the vielbeins,

$$\Gamma^M(x) = E^M_A \Gamma^A. \tag{3.18}$$

Any specific assumption about the form of Γ^M independent of the above equation will be an arbitrary restriction on the metric structure.[3]

[3]In spite of this obvious fact, this kind of hidden assumption is often made in the literature without an explicit statement.

With the defining homomorphism (3.17), we can transform the various formulas in the Θ basis to the corresponding ones in the Γ-representation basis. Thus, for instance, the metric tensor is given by the inner product

$$G^{MN} = \langle \Gamma^M, \Gamma^N \rangle = \tfrac{1}{2}\,\mathrm{Tr}(\hat{\Gamma}^M \Gamma^N), \tag{3.19}$$

where the trace is taken over the spinor indices, the involutive operation "$\hat{\ }$" on the Γ matrices is to be defined, if some involutive operation is defined on the algebra $\mathcal{A}$. In Sec. IV one example of such an involutive operation in two-sheeted space-time will be shown. Other details will be given in the next section where we use explicitly the Γ-matrix representation.

IV. GEOMETRY IN TWO-SHEETED SPACE-TIME

In this section we shall apply the general formalism developed in the previous section to the case of a two-sheeted space-time with a specific choice for the spectral triple.

A. Noncommutative differential forms in two-sheeted space-time

Intuitively, two-sheeted space-time can be viewed as an extension of the physical four-dimensional space-time manifold M by a discrete, internal space $\{a,b\}$. In other words, it is a discretized version of Kaluza-Klein theory [25] where the circle is replaced by two discrete points in the fifth dimension. This discretized Kaluza-Klein theory avoids the truncation inconsistency problem of the original theory since it has a finite field content with a finite mass spectrum [18,19].

As discussed in the preceding sections, the requisite input for this model is the spectral triple, to which we apply the general techniques discussed in Secs. II C and II D. The algebra of smooth functions considered in Sec. II B becomes $C^\infty(M,\mathbf{C})$ tensored with the complex-valued functions on the set $\{a,b\}$. Clearly any such function, $f:\{a,b\} \to \mathbf{C}$ can be written $f = f_a \oplus f_b$, each summand being isomorphic to $\mathbf{C}$ itself,

$$\mathcal{A} \equiv C^\infty(\mathcal{M}) \otimes (\mathbf{C} \oplus \mathbf{C}) \cong C^\infty(\mathcal{M}, \mathbf{C}) \oplus C^\infty(\mathcal{M}, \mathbf{C}). \tag{4.1}$$

In our previous work, where we were concerned with only gravity [17,19,18], it was sufficient to have the algebra $C^\infty(R,\mathcal{M}) \oplus C^\infty(R,\mathcal{M})$. In the present paper, since we are also interested in matter fields coupled to gravity, we choose to begin with the more general algebra (4.1). However, we will find at the end, physics determines a subalgebra of Eq. (4.1) as physically relevant.

In Eq. (4.1) we have essentially doubled the algebra presented in Sec. II B, it is natural to take the representation space as the Hilbert space of square-integrable sections of a spinor bundle,

$$\mathcal{H} \equiv L^2(S,\mathcal{M}) \oplus L^2(S,\mathcal{M}). \tag{4.2}$$

The third element of the spectral triple is the generalized self-adjoint Dirac operator

$$D=\Gamma^M D_M, \quad M=0,1,2,3,5 \tag{4.3}$$

with the generalization of the gamma matrices γ^μ given by

$$\Gamma^\mu\equiv\begin{pmatrix}\gamma^\mu & 0\\ 0 & \gamma^\mu\end{pmatrix}, \quad \Gamma^5\equiv\begin{pmatrix}0 & \gamma^5\\ \gamma^5 & 0\end{pmatrix}. \tag{4.4}$$

With the basis specified, it is now possible to give an explicit form of the generalized derivatives on the functions of the Hilbert space $\mathcal{H}$. The direct generalization of the ordinary space-time derivative ∂_μ follows obviously:

$$D_\mu\equiv\begin{pmatrix}\partial_\mu & 0\\ 0 & \partial_\mu\end{pmatrix}. \tag{4.5}$$

These derivatives act along each of the space-time sheets. The derivative between the two sheets is given as

$$D_5\equiv\sigma\overline{D}_5, \tag{4.6}$$

where

$$\overline{D}_5\equiv\begin{pmatrix}0 & m\\ m & 0\end{pmatrix}, \quad \sigma\equiv\begin{pmatrix}0 & 1\\ 1 & 0\end{pmatrix}. \tag{4.7}$$

The parameter m has the dimension of mass in order to give the fifth component of space-time the same dimension as the other components. If we formally write the operator DF as

$$D(F)\equiv[D,F]=\Gamma^M(DF)_M=\Gamma^M D_M F, \tag{4.8}$$

we indeed obtain the derivative along the fifth dimension with the same dimension as other derivatives

$$D_5F=\sigma[\overline{D}_5,F]=m(F-\widetilde{F}), \tag{4.9}$$

where

$$\widetilde{F}=\begin{pmatrix}f_2 & 0\\ 0 & f_1\end{pmatrix}. \tag{4.10}$$

The operation "$\sim$" on the 0-forms is the involutive operation of our algebra of the spectral triple.

As required for the spectral triple, the operator D on $\mathcal{H}$ is self-adjoint with respect to the inner product

$$\langle\psi|D\phi\rangle=\langle D^\dagger\psi|\phi\rangle, \quad D=D^\dagger. \tag{4.11}$$

A straightforward definition of the generalized wedge product used in the previous work [17–19] is the fully antisymmetric product that truncates all the "junk" forms in a trivial way, namely,

$$\Gamma^M\wedge\Gamma^N=\frac{1}{2}(\Gamma^M\Gamma^N-\Gamma^N\Gamma^M). \tag{4.12}$$

However, in this paper, since we would like to have the quartic Higgs potential in the gauge sector, we will adopt the wedge product with the following change in Eq. (4.12), when $M,N=5$,

$$\Gamma^5\wedge\Gamma^5=1. \tag{4.13}$$

This extended definition of the wedge product has been used in noncommutative geometric constructions of the standard model. The emergence of a quartic potential, however, as we shall see later, requires the fifth dimension to be complex.

With the given spectral triple, it is straightforward to construct the algebra of forms on the two-sheeted space-time. For later use, we work out the generalized γ-matrix representations of one-forms and their derivatives. The exterior derivative of any odd form is its anticommutator with the operator D as a consequence of the Leibnitz rule for the generalized forms (2.30). Hence the exterior derivative DU of the generalized one-form $U=\Gamma^M U_M$ is

$$DU=\{D,U\}=\Gamma^M\wedge\Gamma^N(DU)_{MN}. \tag{4.14}$$

The components $(DU)_{MN}$ can be calculated with the help of the formulas (4.12) and (4.13) giving

$$(DU)_{\mu\nu}=\tfrac{1}{2}(D_\mu U_\nu-D_\nu U_\mu),$$

$$(DU)_{\mu 5}=-(DU)_{5\mu}=\tfrac{1}{2}(D_\mu U_5-D_5U_\mu), \tag{4.15}$$

$$(DU)_{55}=m(U_5+\widetilde{U}_5).$$

From Eq. (4.15), it is straightforward to show that $D^2F=0, \forall F\in\mathcal{A}$, which implies that the junk forms are, in fact, eliminated by this new wedge product [1,2].

B. Riemannian geometry on a two-sheeted space-time

The equivalence principle extended to our two-sheeted space-time requires a locally orthonormal basis. Continuing to work with the Γ representation, the locally orthonormal Θ^A given in Sec. III are now represented by Γ^A,

$$\Gamma^a\equiv\begin{pmatrix}\gamma^a & 0\\ 0 & \gamma^a\end{pmatrix}, \quad \Gamma^{\dot 5}\equiv\begin{pmatrix}0 & \gamma^5\\ \gamma^5 & 0\end{pmatrix}, \tag{4.16}$$

where γ^a and γ^5 are the usual flat Dirac matrices. By choosing γ^5 we have specialized to the two sheets of space-time of chiral spinors. We shall use a $\dot 5$ index in the orthonormal basis to distinguish it from 5 in the general case.

The inner product now can be also taken as a trace over the Clifford indices. In the representation (4.16), the following orthogonality is manifest:

$$(\Gamma^A, \Gamma^B) = \mathrm{Tr}(\Gamma^A \Gamma^B) = 2\eta^{AB} \cdot \mathbf{1}. \tag{4.17}$$

The curvilinear Θ^M will be represented by Γ^M. To obtain Γ^M and define the metric, we postulate the generalized vielbeins E_A^M, as the following diagonal matrix zero-forms:

$$E_a^\mu \equiv \begin{pmatrix} e_{1a}^\mu & 0 \\ 0 & e_{2a}^\mu \end{pmatrix}, \quad E_{\dot{5}}^\mu = 0,$$

$$E_a^5 \equiv -\begin{pmatrix} a_{1a} & 0 \\ 0 & a_{2a} \end{pmatrix} \equiv -A_a = -E_a^\mu A_\mu,$$

$$E_{\dot{5}}^5 \equiv \begin{pmatrix} \phi_1^{-1} & 0 \\ 0 & \phi_2^{-1} \end{pmatrix} \equiv \Phi^{-1}, \tag{4.18}$$

where $e_{1,2a}^\mu$ are two different vielbeins on the two sheets of space-time. Similarly, $a_{1,2}$ and $\phi_{1,2}$ are, respectively, vector and scalar fields. The dependence on the space-time variable x of the above vielbeins is understood. In what follows, we shall, for the sake of simplicity of the formulas, continue to suppress the x dependence.

The vielbeins E_A^M are invertible, giving

$$E_\mu^a \equiv \begin{pmatrix} e_{1\mu}^a & 0 \\ 0 & e_{2\mu}^a \end{pmatrix}, \quad E_5^a = 0,$$

$$E_\mu^{\dot{5}} \equiv A_\mu \Phi, \quad E_5^{\dot{5}} \equiv \begin{pmatrix} \phi_1 & 0 \\ 0 & \phi_2 \end{pmatrix} \equiv \Phi. \tag{4.19}$$

Then, the Γ^M matrices are given as $\Gamma^M = \Gamma^A E_A^M$,

$$\Gamma^\mu \equiv \begin{pmatrix} \gamma^a e_{1a}^\mu & 0 \\ 0 & \gamma^a e_{2a}^\mu \end{pmatrix}, \quad \Gamma^5 \equiv \begin{pmatrix} -a_1 & \gamma^5 \phi_2^{-1} \\ \gamma^5 \phi_1^{-1} & -a_2 \end{pmatrix}, \tag{4.20}$$

where $a_{1,2} = \gamma^a a_{1,2a}$ are two gauge connection one-forms on the two sheets.

Using the inner product defined as the trace on the Hilbert space, we can calculate the representation of the metric tensor as follows:

$$G^{MN} = \frac{1}{2}\mathrm{Tr}(\hat{\Gamma}^M \Gamma^N) = \frac{1}{2} E_A^M \mathrm{Tr}(\Gamma^A \Gamma^B) E_B^N$$

$$= E_A^M \eta^{AB} E_B^N. \tag{4.21}$$

The involutive operation, the caret on the Γ^M matrices, in this case is defined as the transpose.

In terms of the functions E_a^μ, A_μ and Φ the metric tensor can be expressed as

$$G^{\mu\nu} = G_0^{\mu\nu}, \quad G^{\mu 5} = -A^\mu,$$

$$G^{5\mu} = -A^\mu, \quad G^{55} = A^2 + \Phi^{-2}, \tag{4.22}$$

$$G_{\mu\nu} = G_{0\mu\nu} + A_\mu A_\nu \Phi^2, \quad G_{\mu 5} = A_\mu \Phi^2,$$

$$G_{5\mu} = A_\mu \Phi^2, \quad G_{55} = \Phi^2. \tag{4.23}$$

The function G_0 is a generalized function with the metrics of the two space-time sheets on the diagonal

$$G_0^{\mu\nu} \equiv \begin{pmatrix} e_{1a}^\mu \eta^{ab} e_{1b}^\nu & 0 \\ 0 & e_{2a}^\mu \eta^{ab} e_{2b}^\nu \end{pmatrix}. \tag{4.24}$$

As we see from the trace of the generalized gamma matrices, no restrictions on the vielbeins can be derived if a consistent choice of basis is selected. We can only derive the formulas of generalized metric components in terms of the given vielbein components from Eq. (4.21), which is in perfect analogy with the usual Riemannian geometry.

In terms of Γ representations, an arbitrary generalized one-form U can be represented in both Γ bases as follows:

$$U = \Gamma^A U_A = \Gamma^M U_M,$$

where U_M and U_A are diagonal matrix functions

$$U_A = \begin{pmatrix} u_{1A} & 0 \\ 0 & u_{2A} \end{pmatrix}, \quad U_M = \begin{pmatrix} u_{1M} & 0 \\ 0 & u_{2M} \end{pmatrix}. \tag{4.25}$$

The functions U_M and U_A are related to each other by the transformation laws,

$$U_M = E_M^A U_A, \quad U_A = E_A^M U_M. \tag{4.26}$$

In terms of the functions E_μ^a, A_μ, and Φ, the transformation laws can be written as

$$U_\mu = E_\mu^a U_a + A_\mu \Phi U_{\dot{5}}, \quad U_5 = \Phi U_{\dot{5}},$$

$$U_a = E_a^\mu (U_\mu - A_\mu U_5), \quad U_{\dot{5}} = \Phi^{-1} U_5. \tag{4.27}$$

In the locally orthonormal basis Γ^A, it is possible to define the involutive operation "$\sim$" on the algebra of 1-forms as follows:

$$U = \Gamma^A U_A \mapsto \tilde{U} = \Gamma^A \tilde{U}_A. \tag{4.28}$$

The "$\sim$" conjungate 1-forms will be used to construct the Lagrangian later.

The multiplication of two one-forms U and V is given by

$$UV = (\Gamma^A U_A) \cdot (\Gamma^B V_B) = (\Gamma^M U_M) \cdot (\Gamma^N U_N). \tag{4.29}$$

Here we want to address two issues concerning the multiplication of 1-forms: Firstly, as we do not know the explicit form of the matrices Γ^M to rearrange the order of terms in NCG, it is convenient to carry out the multiplication in the Γ^A basis. In fact, using the explicit form of Γ^A in Eq. (4.16), we can reorder the product in Eq. (4.29). Secondly, as the explicit forms of the basis Γ^A are known, we can define the wedge product by anti-symmetrizing the product except for the component $\Gamma^{\dot{5}} \wedge \Gamma^{\dot{5}}$, which we will keep as a formal non-vanishing basis of the "two-forms" algebra $\Omega^2(\mathcal{A})$. That is to say, we have to define the wedge product of the Γ^A matrices with the basis σ^{AB} of the "two-forms" module $\Omega^2(\mathcal{A})$, to be given by

$$\sigma^{AB} = \Gamma^A \wedge \Gamma^B = \tfrac{1}{2}[\Gamma^A, \Gamma^B], \tag{4.30}$$

when both A and B are not $\dot{5}$. And when $A=B=\dot{5}$ we have

$$\Gamma^{\dot{5}}\wedge\Gamma^{\dot{5}} \tag{4.31}$$

as a formal basis vector. This component will generate a Higgs potential for the gauge fields.

From Eq. (4.30) it is clear that

$$\Gamma^{A}\wedge\Gamma^{B}=-\Gamma^{B}\wedge\Gamma^{A}, \tag{4.32}$$

except the product $\Gamma^{\dot{5}}\wedge\Gamma^{\dot{5}}$.

Now we know how to reorder the product in Eq. (4.27) to define the wedge product of two 1-forms. The generalized functions commute with the matrices Γ^{a}. The commutation rule of a function F with the matrix $\Gamma^{\dot{5}}$ is as follows:

$$F\Gamma^{\dot{5}}=\Gamma^{\dot{5}}\tilde{F}. \tag{4.33}$$

It then follows that the product of two 1-forms U and V in Eq. (4.29) can be defined as the wedge product

$$U\wedge V=\Gamma^{A}\wedge\Gamma^{B}(U\wedge V)_{AB}, \tag{4.34}$$

and using Eqs. (4.26), we can compute the component of the wedge product $U\wedge V$ in the locally flat frame,

$$\begin{aligned}(U\wedge V)_{ab}&=\frac{1}{2}(U_aV_b-U_bV_a)\\&=\frac{1}{2}(E_a^{\mu}E_b^{\nu}-E_a^{\nu}E_b^{\mu})\\&\quad\times[U_{\mu}V_{\nu}+A_{\mu}(U_{\nu}V_5-V_{\nu}U_5)],\\(U\wedge V)_{\dot{5}a}&=-(U\wedge V)_{a\dot{5}}=\frac{1}{2}(U_{\dot{5}}V_a-\tilde{U}_aV_{\dot{5}})\\&=\frac{1}{2}\Phi^{-1}[U_5E_a^{\mu}(V_{\mu}-A_{\mu}V_5)\\&\quad-\tilde{E}_a^{\mu}V_5(\tilde{U}_{\mu}-\tilde{A}_{\mu}\tilde{U}_5)],\\(U\wedge V)_{\dot{5}\dot{5}}&=\tilde{U}_{\dot{5}}V_{\dot{5}}=(\Phi\tilde{\Phi})^{-1}\tilde{U}_5V_5.\end{aligned} \tag{4.35}$$

As a special case, we can calculate the wedge product of Γ^{M} one-forms as

$$\begin{aligned}(\Gamma^{M}\wedge\Gamma^{N})_{ab}&=\frac{1}{2}(E_a^{M}E_b^{N}-E_a^{N}E_b^{M}),\\(\Gamma^{M}\wedge\Gamma^{N})_{\dot{5}a}&=\frac{1}{2}(E_{\dot{5}}^{M}E_a^{N}-\tilde{E}_a^{M}E_{\dot{5}}^{N}),\\(\Gamma^{M}\wedge\Gamma^{N})_{\dot{5}\dot{5}}&=\tilde{E}_{\dot{5}}^{M}E_{\dot{5}}^{N}.\end{aligned} \tag{4.36}$$

The derivative DU of the 1-form U in the curved space-time basis follows from Eq. (4.15) and is given by

$$\begin{aligned}DU&=\Gamma^{M}\wedge\Gamma^{N}(DU)_{MN}=\Gamma^{A}\wedge\Gamma^{B}(DU)_{AB}\\&=\Gamma^{A}\wedge\Gamma^{B}(\Gamma^{M}\wedge\Gamma^{N})_{AB}(DU)_{MN},\end{aligned} \tag{4.37}$$

and hence

$$\begin{aligned}(DU)_{ab}&=\tfrac{1}{2}(E_a^{\mu}E_b^{\nu}-E_a^{\nu}E_b^{\mu})(D_{\mu}U_{\nu}+A_{\mu}(D_{\nu}U_5-D_5U_{\nu}),\\(DU)_{\dot{5}a}&=-(DU)_{a\dot{5}}\\&=-\frac{1}{2}\Phi^{-1}\Big[\frac{1}{2}(E_a^{\mu}+\tilde{E}_a^{\mu})(D_{\mu}U_5-D_5U_{\nu})\\&\quad+m(E_a^{\mu}A_{\mu}-\tilde{E}_a^{\mu}\tilde{A}_{\mu})(U_5+\tilde{U}_5)\Big],\\(DU)_{\dot{5}\dot{5}}&=-m(\Phi\tilde{\Phi})^{-1}(U_5+\tilde{U}_5).\end{aligned} \tag{4.38}$$

The inner product of two 2-forms

$$\begin{aligned}\langle\Gamma^{A}\wedge\Gamma^{B},\Gamma^{C}\wedge\Gamma^{D}\rangle&\equiv\tfrac{1}{4}\,\mathrm{Tr}_4(\Gamma^{A}\Gamma^{B}\Gamma^{C}\Gamma^{D})\\&=\eta^{AD}\eta^{BC}-\eta^{AC}\eta^{BD}.\end{aligned} \tag{4.39}$$

Similarly, we can give a representation of the general p-forms, multiplication rules of a p-form and a q-form, and the derivative of general p-forms. The use of the two frames is essential in such calculus. To summarize, it is convenient to carry out the multiplication in the Γ^{A} basis and the derivatives in the Γ^{M} basis and then transform to the Γ^{A} to obtain the desired components. The physical fields are given in the curved frame.

The definition of $\Gamma^{\dot{5}}\wedge\Gamma^{\dot{5}}$ is rather arbitrary but is required to retain the $\dot{5}\dot{5}$ component for a Higgs potential to survive in the gauge theory. It is more natural to introduce the discrete degrees of freedom as a pair of conjugate derivatives $D_z, D_{\bar{z}}$ as in [9,10] than just a single D_5 in the Dirac operator.

In the gravitational sector, the curvature arises from a set of connection one-forms $\{\Omega_B^A\ A,B=0,\dots,\dot{5}\}$. The connection coefficients are defined as those functions which arise in the expansion of Ω_B^A in the Γ basis, as the generalization of Eq. (3.10),

$$\Omega_B^A\equiv\Gamma^{C}\Omega_{BC}^{A}. \tag{4.40}$$

V. PHYSICAL CONTENT OF THE GRAVITY SECTOR IN TWO-SHEETED SPACE-TIME

In Sec. IV we have seen that the metric (vielbein) structure of the two-sheeted space-time contains a tensor, a vector, and a scalar, which have generalized functions in Eq. (4.18). This field content is exactly parallel to the one in the traditional Kaluza-Klein theory [25]. However, each generalized function has a pair of fields. Therefore, in principle, the gravity sector in the two-sheeted space-time contains a pair of tensors, a pair of vectors, and a pair of scalar fields.

Intuitively, as stated earlier, the two-sheeted space-time is a discretized version of the general untruncated Kaluza-Klein theory [26], where the internal space circle is replaced by two points. Therefore, intuition tells us that in general, we should expect one member of each pair to be massless and the other massive [18].

A. Minimal set of constraints and solutions of the first Cartan structure equation

The essential point is how to impose a consistent system of constraints to solve the generalized Cartan structure equations without overconstraining the metric. The second Cartan structure equation is used to calculate the curvature tensor in terms of the connections. Therefore, it does not have much to do with constraints. The first Cartan structure equation expresses a relationship between connections and torsion. Its component equations are as follows:

$$
\begin{aligned}
T_{abc} &= (D\Gamma_a)_{bc} + \tfrac{1}{2}(\Omega_{abc} - \Omega_{acb}), \\
T_{a\dot{5}b} &= (D\Gamma_a)_{\dot{5}b} + \tfrac{1}{2}(\Omega_{a\dot{5}b} - \Omega_{ab\dot{5}}), \\
T_{\dot{5}ab} &= (D\Gamma_{\dot{5}})_{ab} + \tfrac{1}{2}(\Omega_{\dot{5}ab} - \Omega_{\dot{5}ba}), \\
T_{\dot{5}\dot{5}b} &= (D\Gamma_{\dot{5}})_{\dot{5}b} + \tfrac{1}{2}(\Omega_{\dot{5}\dot{5}b} - \Omega_{\dot{5}b\dot{5}}), \qquad (5.1)\\
T_{ab\dot{5}} &= (D\Gamma_a)_{b\dot{5}} + \tfrac{1}{2}(\Omega_{ab\dot{5}} - \Omega_{a\dot{5}b}), \\
T_{\dot{5}a\dot{5}} &= (D\Gamma_{\dot{5}})_{a\dot{5}} + \tfrac{1}{2}(\Omega_{\dot{5}a\dot{5}} - \Omega_{\dot{5}\dot{5}a}), \\
T_{a\dot{5}\dot{5}} &= (D\Gamma_a)_{\dot{5}\dot{5}} + \tfrac{1}{2}(\Omega_{a\dot{5}\dot{5}} + \tilde{\Omega}_{a\dot{5}\dot{5}}), \\
T_{\dot{5}\dot{5}\dot{5}} &= (D\Gamma_{\dot{5}})_{\dot{5}\dot{5}} + \tfrac{1}{2}(\Omega_{\dot{5}\dot{5}\dot{5}} + \tilde{\Omega}_{\dot{5}\dot{5}\dot{5}}).
\end{aligned}
$$

Obviously these equations are not enough to determine both the torsion and the connection, even in the case of the ordinary geometry.

The torsion free condition $T^A = 0$ together with the metric compatibility condition turns out to be too restrictive, giving only a theory without massive excitations. However, it is possible to impose constraints in such a way as to retain all the postulated fields in the metric structure in more than one way [18–20]. In this paper we will follow the results of [20] to impose the following minimal set of constraints:

(i) The space-time torsion-free condition

$$T_{aBC} = 0, \qquad (5.2a)$$

(ii) the metric compatibility condition

$$\Omega_{AB} = -\Omega_{BA}, \qquad (5.2b)$$

(iii) an additional condition on the connections

$$\Omega_{AB\dot{5}} = 0. \qquad (5.2c)$$

With these conditions, Eq. (5.1) can be solved, giving the following solutions for the nonvanishing connection and torsion-tensor components:

$$
\begin{aligned}
\Omega_{abc} &= -(D\Gamma_a)_{bc} + (D\Gamma_b)_{ac} - (D\Gamma_c)_{ba}, \\
\Omega_{a\dot{5}b} &= -2(D\Gamma_a)_{\dot{5}b}, \\
\Omega_{\dot{5}ab} &= 2(D\Gamma_a)_{\dot{5}b}, \\
T_{\dot{5}bc} &= (D\Gamma_{\dot{5}})_{bc} + (D\Gamma_b)_{\dot{5}c} - (D\Gamma_c)_{\dot{5}b}, \qquad (5.3)\\
T_{\dot{5}\dot{5}b} &= (D\Gamma_{\dot{5}})_{\dot{5}b}, \\
T_{\dot{5}\dot{5}\dot{5}} &= (D\Gamma_{\dot{5}})_{\dot{5}\dot{5}}.
\end{aligned}
$$

The right-hand side of Eqs. containing the components of $(D\Gamma_A)$ can be expressed in terms of metric components (vielbeins) and their derivatives by the rules already stated. Thus, we have nonvanishing torsion and connection coefficients in terms of metric fields.

Here let us note that no constraints are imposed on the metric components (vielbeins). Therefore, all the six fields will contribute to the scalar curvature. If all of them have appropriate kinetic terms in the final action, they will represent themselves as physical fields of the gravity sector of our theory.

However, the fields $e^{\mu}_{1,2a}$, $a_{1,2\mu}$, and $\phi_{1,2}$ turn out not to be good variables of the theory. They do not give rise to mass eigenstates except in the case $e_1 = e_2, a_1 = a_2$, when the theory reduces exactly to the zero mode sector of the Kaluza-Klein theory with massless fields.

In the next section, we define linear combinations of these fields which lead to appropriate fields leading to mass eigenstates.

B. New physical field variables

We begin with the following linear combinations as new physical field variables:

$$
\begin{aligned}
e^{\mu}_{\pm a} &= \tfrac{1}{2}(e^{\mu}_{1a} \pm e^{\mu}_{2a}), \\
a_{\pm\mu} &= \tfrac{1}{2}(a_{1\mu} \pm a_{2\mu}), \qquad (5.4)\\
\phi_{\pm} &= \tfrac{1}{2}(\phi_1 \pm \phi_2).
\end{aligned}
$$

When $e^{\mu}_{-\nu} = a_{-\mu} = \phi_- = 0$, the theory reduces to the usual Kaluza-Klein theory with e^{μ}_{+a} as the vierbein, $a_{+\mu}$ the vector field, and ϕ_+ the Brans-Dicke scalar.

To obtain the "physical" action functional in terms of the new variables, we need to define E^{μ}_a and its inverse E^a_{μ} in terms of the new variables. This is not as straightforward as one would think.

To this end, we start with

$$E_a^\mu = \begin{pmatrix} e_{1a}^\mu & 0 \\ 0 & e_{2a}^\mu \end{pmatrix} = e_a^\mu \mathbf{1} + v_a^\mu \cdot \mathbf{r}, \tag{5.5}$$

where $v_a^\mu = e_{-a}^\mu$ and $e_a^\mu = e_{a+}^\mu$. Let $e_\mu^a = (e_a^\mu)^{-1}$ and

$$\mathbf{1} = \begin{pmatrix} 1 & 0 \\ 0 & 1 \end{pmatrix}, \quad \mathbf{r} = \begin{pmatrix} 1 & 0 \\ 0 & -1 \end{pmatrix}. \tag{5.6}$$

From the orthogonality relation of the vielbeins

$$E_a^\mu E_\mu^b = \delta_b^a \mathbf{1} \tag{5.7}$$

we obtain

$$E_\mu^a = (e_\mu^a - v_\mu^\nu s_\nu^a)\mathbf{1} + s_\mu^a \mathbf{r}, \tag{5.8}$$

where s_μ^a is a nonlinear function of v_μ^a satisfying

$$v_\mu^\nu + s_\sigma^\nu (\delta_\mu^\sigma - v_\mu^\lambda v_\lambda^\sigma) = 0, \tag{5.9}$$

and $s_\sigma^\nu = e_a^\nu s_\sigma^a$.

The tensor field $v^{\mu\nu}$ is a candidate for a massive excitation of the massless symmetric metric tensor. We can also use the metric tensors $g^{\mu\nu}$ and $g_{\mu\nu}$ to raise and lower indices and express the tensor fields v and s in a matrix notation, $v = [v_\mu^\nu]$, $s = [s_\mu^\nu]$.

The following algebraic relations are also valid:

$$vs = sv, \quad s = v(v^2 - 1)^{-1}. \tag{5.10}$$

The usual space-time component of the generalized vierbein E_a^μ and its inverse E_μ^a are determined completely in terms of the usual vierbeins e_a^μ, e_μ^a and the tensor field $v^{\mu\nu}$.

Furthermore, we can define the vector and scalar components of the metric by

$$A_\mu = a_{+\mu} \mathbf{1} + a_{-\mu} \mathbf{r},$$

$$\Phi = \phi_+ \mathbf{1} + \phi_- \mathbf{r}. \tag{5.11}$$

Hence, finally, the gravity sector of the theory contains e_a^μ, e_μ^a, and the tensor field v_ν^μ, and the vector and scalar fields $a_{\pm\mu}, \phi_\pm$. The tensor field s_ν^μ is not an independent field. It is a nonlinear combination of v_μ^ν. As we will see in the final Lagrangian, the tensor field $v^{\mu\nu}$ will be a mass eigenstate. We are, therefore, led to make the assumption, $v^{\mu\nu} = v^{\nu\mu}$, in order to have $v^{\mu\nu}$ as an independent tensor field.

As we shall also see, the signature of the kinetic terms of the fields $v^{\mu\nu}$, a_μ, and ϕ_- in the final Lagrangian built from curvature and torsion terms of our theory, requires the following field redefinitions:

$$v_\mu^a \to i v_\mu^a,$$

$$a_{-\mu} \to i a_{-\mu}, \tag{5.12}$$

$$\phi_- \to i \phi_-.$$

This field redefinition will be used hereafter in this paper.

It is convenient to introduce the following quantities in further calculations:

$$P_{a\mu\nu} = \partial_\mu (e_{a\nu} + s_{a\lambda} v_\nu^\lambda) - 2m a_{\nu-} s_{a\mu},$$

$$Q_{a\mu\nu} = \partial_\mu s_{a\nu} + 2m a_{\nu+} s_{a\mu},$$

$$X_{bc}^{[\mu,\nu]} = (e_b^\mu e_c^\nu - e_b^\nu e_c^\mu + v_b^\nu v_c^\mu - v_b^\mu v_c^\nu), \tag{5.13}$$

$$Y_{bc}^{[\mu,\nu]} = (e_b^\mu v_c^\nu - e_b^\nu v_c^\mu + v_b^\mu e_c^\nu - v_b^\nu e_c^\mu).$$

In several formulas of this theory, $P_{a\mu\nu}$ and $Q_{a\mu\nu}$ generalize the field derivatives $\partial_\mu e_{a\nu}$ and $\partial_\mu v_{a\nu}$. The $X_{ab}^{[\mu,\nu]}$ generalize the usual term $e_a^\mu e_b^\nu - e_a^\nu e_b^\mu$ which is common in general relativity. Some useful contraction formulas for $X_{ab}^{[\mu,\nu]}$ and $Y_{ab}^{[\mu,\nu]}$ can be found in Appendix A 2. These formulas allow us to write the action in a compact form.

For calculational convenience, it is useful to introduce the following projection operations for the component $f_\pm$ of a generalized function F:

$$(F)_+ = f_+ = \tfrac{1}{2}\,\mathrm{Tr}(F),$$

$$(F)_- = f_- = \tfrac{1}{2}\,\mathrm{Tr}(F \cdot \mathbf{r}). \tag{5.14}$$

The following relations are also useful in practical calculations of products of generalized functions:

$$(F \cdot G)_+ = (F)_+(G)_+ + (F)_-(G)_-,$$

$$(F \cdot G)_- = (F)_+(G)_- + (F)_-(G)_+. \tag{5.15}$$

C. Curvature, Torsion, and Lagrangian

The second structure equation is generalized in the following form:

$$R_{AB} = D\Omega_{AB} + \eta^{CD} \Omega_{AC} \wedge \Omega_{DB}. \tag{5.16}$$

With the connections known from Eq. (5.3), we can calculate all the components of the curvature in Eq. (5.16). However, we are interested in only a few components that contribute to the generalized scalar curvature,

$$R = \tfrac{1}{2}\,\mathrm{Tr}\langle \Gamma^A \wedge \Gamma^B, R_{AB} \rangle = R_{abcd+} \eta^{ad} \eta^{bc} + 2R_{a\dot{5}\dot{5}b+} \eta^{ab}$$

$$= (D\Omega_{ab})_{cd+} \eta^{ad} \eta^{bc} + (\Omega_{ae} \wedge \Omega_{bf})_{cd+} \eta^{ad} \eta^{cf} \eta^{be}$$

$$+ 2(D\Omega_{a\dot{5}})_{\dot{5}b+} \eta^{ab} + 2(\Omega_{a\dot{5}} \wedge \Omega_{\dot{5}b})_{cd+} \eta^{ad} \eta^{bc}. \tag{5.17}$$

The Lagrangian from curvature is defined as

$$\mathcal{L}_R=\frac{1}{16\pi G_N}R=\mathcal{L}_{R1}+\mathcal{L}_{R2}+\mathcal{L}_{R3}, \quad (5.18)$$

where

$$\mathcal{L}_{R1}=(16\pi G_N)^{-1}(D\Omega_{ab})_{cd+}\eta^{ad}\eta^{bc},$$

$$\mathcal{L}_{R2}=(16\pi G_N)^{-1}(\Omega_{ae}\wedge\Omega_{bf})_{cd+}\eta^{ad}\eta^{cf}\eta^{be}, \quad (5.19)$$

$$\mathcal{L}_{R3}=(8\pi G_N)^{-1}[(D\Omega_{a\dot{5}})_{\dot{5}b+}\eta^{ab} + (\Omega_{a\dot{5}}\wedge\Omega_{\dot{5}b})_{cd+}\eta^{ad}\eta^{bc}],$$

where G_N is the Newton gravitational constant. The detailed calculations of the above terms are given in Appendix A 3.

Examination of these results shows that the "curvature" Lagrangian contains the usual four-dimensional curvature (the Einstein-Hilbert term) along with a full Lagrangian for the massive tensor field $v_{\mu\nu}$. It also contains mass and interaction terms for the scalar and vector components of the metric. However, it does not contain kinetic terms for the latter. In order to obtain the complete Lagrangian, we need to take into account the contribution from the nonvanishing components of the torsion. This is given by

$$\mathcal{L}_T=-(16G_V^2)^{-1}\mathrm{Tr}\langle\tilde{T}_A,T^A\rangle=(8G_V^2)^{-1}\mathrm{Tr}(\tilde{T}_{ABC}T^{ABC})$$

$$=(4G_V^2)^{-1}(T_{ABC+}T_+^{ABC}-T_{ABC-}T_-^{ABC})$$

$$=(4G_V^2)^{-1}(T_{\dot{5}bc+}T_+^{\dot{5}bc}-T_{\dot{5}bc-}T_-^{\dot{5}bc}+2T_{\dot{5}\dot{5}b+}T_+^{\dot{5}\dot{5}b} -2T_{\dot{5}\dot{5}b-}T_-^{\dot{5}\dot{5}b}+T_{\dot{5}\dot{5}\dot{5}+}T_+^{\dot{5}\dot{5}\dot{5}})$$

$$=\mathcal{L}_{T1}+\mathcal{L}_{T2}+\mathcal{L}_{T3}, \quad (5.20)$$

where the Lagrangians $\mathcal{L}_{T1}, \mathcal{L}_{T2}$, and $\mathcal{L}_{T3}$ are given in Appendix A 4. G_V is a constant with a dimension of mass.

Finally, the Einstein-Hilbert-Cartan Lagrangian for the gravity sector in our theory is

$$\mathcal{L}_{E-H-C}=\mathcal{L}_R+\mathcal{L}_T=(16\pi G)^{-1}\mathrm{Tr}R+(16G_V^2)^{-1}\mathrm{Tr}\langle T_A,T^A\rangle$$

$$=\mathcal{L}_{R1}+\mathcal{L}_{R2}+\mathcal{L}_{R3}+\mathcal{L}_{T1}+\mathcal{L}_{T2}+\mathcal{L}_{T3}. \quad (5.21)$$

The kinetic terms of the fields v^a_μ, $a_{\mu-}$, and ϕ_- in the resulting Lagrangian require the field redefinitions defined in Eq. (5.12).

VI. MATTER FIELDS IN CURVED TWO-SHEETED SPACE-TIME

With the curved space-time described in the previous section as the background, we now proceed to construct Lagrangians for the gauge and fermionic sectors. The gauge sector will consist of two Abelian gauge fields with two Higgs scalar fields as part of a generalized one-form. The fermionic sector will consist of a left-chiral field on one sheet and a right-chiral field on the other.

A. The gauge sector

Let the one-form, U, be specialized to the gauge and Higgs fields to be associated with the chiral fermions,

$$U=B=\Gamma^M(x)B_M,$$

where

$$B_\mu=\begin{pmatrix} b_{1\mu} & 0 \\ 0 & b_{2\mu}\end{pmatrix}, \quad B_5=\begin{pmatrix} h_1 & 0 \\ 0 & h_2\end{pmatrix}, \quad (6.1)$$

or

$$B_\mu=b_{\mu+}\mathbf{1}+b_{\mu-}\mathbf{r}, \quad B_5=h_+\mathbf{1}+h_-\mathbf{r},$$

where

$$b_{\pm\mu}=\frac{1}{2}(b_{1\mu}\pm b_{2\mu}), \quad h_\pm=\frac{1}{2}(h_1\pm h_2). \quad (6.2)$$

Then, the curvature or field strengths are given by the two-form (henceforth the x dependence will be understood),

$$G=DB+B\wedge B,$$

with components, following Eqs. (4.35)–(4.38),

$$G_{ab}=\tfrac{1}{2}(E^\mu_aE^\nu_b-E^\nu_aE^\mu_b)[D_\mu B_\nu+A_\mu(D_\nu B_5-D_5B_\mu)], \quad (6.3)$$

$$G_{a\dot{5}}=\tfrac{1}{2}E^5_{\dot{5}}\bigg[\frac{1}{2}(\tilde{E}^\mu_a+E^\mu_a)(D_\mu B_5-D_5B_\mu)+m(E^\mu_aA_\mu-\tilde{E}^\mu_a\tilde{A}_\mu) \times(B_5+\tilde{B}_5)+(\tilde{E}^\mu_a\tilde{B}_\mu-E^\mu_aB_\mu+E^\mu_aA_\mu B_5 -\tilde{E}^\mu_a\tilde{A}_\mu\tilde{B}_5)B_5\bigg]=-G_{a\dot{5}}, \quad (6.4)$$

$$G_{\dot{5}\dot{5}}=\tilde{E}^5_{\dot{5}}E^5_{\dot{5}}[m(\tilde{B}_5+B_5)+\tilde{B}_5B_5]. \quad (6.5)$$

In terms of the above components, the Lagrangian is given by

$$\mathcal{L}_G=\frac{1}{2g^2}\langle\tilde{G},G\rangle=-\frac{1}{g^2}\bigg(\tilde{G}_{ab}G^{ab}+2\tilde{G}_{a\dot{5}}G^{a\dot{5}}+\frac{1}{2}\tilde{G}_{\dot{5}\dot{5}}G^{\dot{5}\dot{5}}\bigg). \quad (6.6)$$

As in the case of the torsion Lagrangian in the gravity sector, the gauge Lagrangian contains the kinetic terms for the fields $b_{-\mu}$ and h_- with a sign that requires a similar field redefinition. Therefore, we introduce the following new field variables in order to bring the gauge Lagrangian into a form, which is consistent with the standard model with two Abelian gauge fields and a Higgs scalar,

$$b_{\mu-}\leftrightarrow -ib_{\mu-},$$
$$\eta\leftrightarrow h_{+}+m+h_{-},$$
$$\overline{\eta}\leftrightarrow h_{+}+m-h_{-},$$
$$b_{\mu\nu+}=\partial_{\mu}b_{\nu+}-\partial_{\nu}b_{\mu+}, \tag{6.7}$$
$$b_{\mu\nu-}=\partial_{\mu}b_{\nu-}-\partial_{\nu}b_{\mu-},$$
$$\mathcal{D}_{\mu}\eta=\partial_{\mu}\eta-2ib_{\mu-}\eta,$$
$$\mathcal{D}_{\mu}\overline{\eta}=\partial_{\mu}\overline{\eta}+2ib_{\mu-}\overline{\eta}.$$

Substituting the even and odd projection of G_{AB} from Eqs. (6.3)–(6.5) in Eq. (6.6), we obtain the gauge sector Lagrangian $\mathcal{L}_G$. We write it in the form

$$\mathcal{L}_G=\mathcal{L}_{G1}+\mathcal{L}_{G2}+\mathcal{L}_{G3}+\mathcal{L}_{G4}, \tag{6.8}$$

where $\mathcal{L}_{G1}$ contains the kinetic terms of the vector gauge fields $b_{\mu+}$ and $b_{\mu-}$ and $\mathcal{L}_{G2}$ contains the kinetic terms of the Higgs field. $\mathcal{L}_{G3}$ contains the quartic Higgs fields. $\mathcal{L}_{G4}$ contains the remaining interaction terms of these fields with the scalar and vector fields of the gravity sector.

The Lagrangian (6.8), therefore, represents a generalization of the standard model gauge sector involving two Abelian gauge fields coupled with a complex scalar with a quartic potential in the curved two-sheeted space-time. More discussion about the physical contents of this Lagrangian will be give in Sec. VII.

B. Fermion sector

We begin with the conventional Dirac Lagrangian in curved space-time generalized to our two-sheeted space-time given by

$$\begin{aligned}\mathcal{L}_F&=i\overline{\Psi}\Gamma^{A}\left(E^{M}_{A}(D_M+iB_M)+\frac{i}{4}\Gamma^{B}\Gamma^{C}\Omega_{BCA}\right)\Psi\\&=i\overline{\Psi}\Gamma^{a}\left[\left(E^{\mu}_{a}(D_{\mu}+iB_{\mu})+\frac{i}{4}\Gamma^{b}\Gamma^{c}\Omega^{(0)}_{bca}\right.\right.\\&\quad\left.+\frac{i}{2}\Gamma^{\dot{5}}\Gamma^{c}\Omega_{\dot{5}ca}\right)+\frac{i}{4}\Gamma^{a}\Gamma^{b}\Gamma^{c}\Omega^{(1)}_{bca}+(\Gamma^{\dot{5}}\Phi^{-1}\\&\quad\left.-\Gamma^{a}E^{\mu}_{a}A_{\mu})(D_5+iB_5)\right]\Psi \qquad (6.9)\\&=\mathcal{L}_{F1}+\mathcal{L}_{F2}+\mathcal{L}_{F3}+\mathcal{L}_{F4},\end{aligned}$$

where

$$\Psi=\begin{pmatrix}\psi_L\\ \psi_R\end{pmatrix},$$

$$\psi_L=\frac{(1+\gamma_5)}{2}\psi,\qquad \psi_R=\frac{(1-\gamma_5)}{2}\psi.$$

ψ is the 4-component Dirac spinor and Ω_{BCA} are the connection coefficients. The expressions for the nonvanishing Ω_{ABC} (Ω_{abc}, $\Omega_{a\dot{5}b}$) are given in Eq. (5.3). In terms of metric component fields and their derivatives, they are

$$\Omega_{abc}=\Omega^{(0)}_{abc}+\Omega^{(1)}_{abc},$$

where

$$\Omega^{(0)}_{abc}=\frac{1}{2}(\eta_{ad}E^{\mu}_{b}E^{\nu}_{c}-\eta_{bd}E^{\mu}_{a}E^{\nu}_{c}+\eta_{cd}E^{\mu}_{a}E^{\nu}_{b})\times(D_{\mu}E^{d}_{\nu}-D_{\nu}E^{d}_{\mu}),$$

$$\Omega^{(1)}_{abc}=\frac{1}{2}(\eta_{ad}E^{\mu}_{b}E^{\nu}_{c}-\eta_{bd}E^{\mu}_{a}E^{\nu}_{c}+\eta_{cd}E^{\mu}_{a}E^{\nu}_{b})\times(A_{\nu}D_5E^{d}_{\mu}-A_{\mu}D_5E^{d}_{\nu}), \tag{6.10}$$

$$\Omega_{a\dot{5}b}=-\Omega_{\dot{5}ab}=2\eta_{ad}E^{\dot{5}}_{\dot{5}}E^{\mu}_{b}(D_5E^{d}_{\mu}). \tag{6.11}$$

The separation of Ω_{abc} into $\Omega^{(0)}_{abc}$ and $\Omega^{(1)}_{abc}$ is motivated by the consideration of the fact that $\Omega^{(0)}$ contains only the metric vierbeins and their generalizations to the two-sheeted space-time, where as $\Omega^{(1)}$ contains the additional vector and scalar fields of the gravity sector. The results of calculations of the Lagrangian $\mathcal{L}_F$ in Eq. (6.9) are given in Appendix A 6.

VII. PHYSICAL IMPLICATIONS

So far, we have concentrated on developing the mathematical formalism within the framework of Conne's NCG. The various fields we have introduced do not have the desired physical dimensions. Just as we needed to redefine the fields [Eqs. (5.12) and (6.7)] to secure the correct signs for the kinetic terms, we need to rescale them using the available dimensional parameters in our theory. To this end, we note that we have only three dimensional parameters, G_N the Newton gravitational constant, G_V the new gravitational constant from the torsion, and the parameter m with dimensions of mass. The dimensionless gauge coupling g is the only other free parameter. By fixing the standard coefficients of the kinetic terms, we determine the various coupling and masses in terms of the four parameters G_N, G_V, m, and g.

A. Kinetic terms and dimensions of the physical fields in the gravity sector

The Lagrangian of the gravity sector in Eq. (5.21) contains the vierbeins e^{μ}_{a} that define the physical metric in the form

$$L_r=\frac{1}{16\pi G_N}r, \tag{7.1}$$

where r is the scalar curvature in the conventional Riemannian geometry.

We next consider the kinetic terms of the tensor field $v^{\mu\nu}$ ($s^{\mu\nu}$) by collecting together terms quadratic in these fields and quadratic in their derivatives. They lead to the partial Lagrangian

$$L_v = -(4\pi G_N)^{-1}\nabla_\lambda v_{\mu\nu}\nabla^\lambda v^{\mu\nu} + (2\pi G_N)^{-1}\nabla_\lambda v_{\mu\nu}\nabla^\mu v^{\lambda\nu}$$
$$+(4\pi G_N)^{-1}\partial_\mu v^\nu_\nu \partial^\mu v^\rho_\rho - (2\pi G_N)^{-1}\partial_\mu v^\rho_\rho \nabla_\nu v^{\mu\nu}$$
$$-(4\phi G_N)^{-1} m^2 (v^{\mu\nu}v_{\mu\nu} - v^\mu_\mu v^\nu_\nu). \tag{7.2}$$

To obtain the correct kinetic coefficient for $v^{\mu\nu}$, we need to define

$$v^{\mu\nu} \to \sqrt{\pi G_N} v^{\mu\nu}. \tag{7.3}$$

With this redefinition, which incidently, gives the correct dimension to $v^{\mu\nu}$ (because of G_N) and yields the Fierz-Pauli Lagrangian for a spin-2 tensor field of mass m in curved spacetime.

The kinetic terms for the scalar fields are to be found from Eq. (A4.6) in Appendix A 4. The field ϕ_+ needs special treatment as in the conventional Kaluza-Klein theory, since this field tends to 1 in vacuum. The following field redefinitions will give the fields correct physical dimensions:

$$\phi_+ \to \exp(2G_V\sigma),$$
$$\phi_- \to 2G_V\phi_-. \tag{7.4}$$

The kinetic part of the scalar Lagrangian then becomes

$$\mathcal{L}_\phi = -\tfrac{1}{2}[\exp(4G_V\sigma) + 4G_V^2\phi_-^2]^{-1}$$
$$\times[\exp(2G_V\sigma)\partial_\mu\sigma\partial^\mu\sigma + \tfrac{1}{2}\partial_\mu\phi_-\partial^\mu\phi_-] - m^2(G_V)^{-1}$$
$$\times[1 + 2G_V\phi_-^2\exp(-2G_V\sigma)]^{-2}. \tag{7.5}$$

By expanding the nonlinear factors in power of σ and ϕ_-^2 and retaining the lowest order terms, we obtain

$$\mathcal{L}_\phi = -\tfrac{1}{2}\partial_\mu\sigma\partial^\mu\sigma - \tfrac{1}{2}\partial_\mu\phi_-\partial^\mu\phi_- - m^2(G_V)^{-1}\phi_-^2$$
$$-m^2(G_V)^{-1} - \ldots, \tag{7.6}$$

which represents the kinetic part of the massless Brans-Dicke scalar field σ and the massive ϕ_-. It is to be noted that it also contains a cosmological constant term m^2/G_V.

Next, we consider the kinetic and mass terms for the vector fields $a_{\mu\pm}$ to be found in Eq. (A4.1)

$$\mathcal{L}_a = -\frac{1}{4G_V^2}(a_+^{\mu\nu}a_{\mu\nu+} + a_-^{\mu\nu}a_{\mu\nu-} + 2m^2 a_-^\mu a_{\mu-}). \tag{7.7}$$

With the field redefinition

$$a_{\mu\pm} \to G_V a_{\mu\pm} \tag{7.8}$$

yields the massless vector field $a_{\mu+}$ and massive vector field $a_{\mu-}$ of mass $m/\sqrt{2}$. The constant with the dimension of mass G_V gives the scalar and vector fields correct physical dimensions in exactly the same way as the Newton gravitational G_N endows the correct dimension the massive tensor field $v^{\mu\nu}$.

B. Field redefinition in the gauge sector and gauge coupling

The vector and scalar fields in the gauge sector have the correct dimensions. However, those fields must also be redefined to have the standard kinetic terms:

$$b_{\mu\pm} \to g b_{\mu\pm},$$
$$\eta \to g\eta, \tag{7.9}$$
$$\bar{\eta} \to g\bar{\eta}.$$

The partial gauge Lagrangian that contains only gauge vector and scalar fields comes from $\mathcal{L}_{G1}$, $\mathcal{L}_{G2}$, $\mathcal{L}_{G3}$ in Appendix A 5. After redefinitions, it is given by

$$\mathcal{L}_{b\eta} = -\tfrac{1}{4}b_{\mu\nu+}b_+^{\mu\nu} - \tfrac{1}{4}b_{\mu\nu-}b_-^{\mu\nu} - (\phi_+^2 + \phi_-^2)^{-1}(\overline{\mathcal{D}^\mu\eta})$$
$$\times(\mathcal{D}_\mu\eta) - \frac{g^2}{2}(\phi_+^2 + \phi_-^2)[\bar{\eta}\eta - (m/g)^2]^2. \tag{7.10}$$

Implementing spontaneous symmetry breaking in the standard fashion, we find the mass of the surviving Higgs scalar to be $\sqrt{2}m$. The mass of $b_{\mu-}$ equals $2m$ and $b_{\mu+}$ remains massless.

From the fermionic Lagrangian (6.1) we find that the vector gauge fields $b_{\mu\pm}$ couple to the vector and axial vector currents $j_\pm^\mu$ and $j_{5\pm}^\mu$, where

$$j_+^\mu = -g\bar{\psi}\gamma^\mu\psi,$$
$$j_{5+}^\mu = -ig\sqrt{\pi G_N}\bar{\psi}\gamma^\nu\gamma^5 v_\nu^\mu\psi, \tag{7.11}$$
$$j_-^\mu = g\sqrt{\pi G_N}\bar{\psi}\gamma^\nu v_\nu^\mu\psi,$$
$$j_{5-}^\mu = -ig\bar{\psi}\gamma^\mu\gamma^5\psi.$$

When the massive gravity field vanishes, as expected, the vector gauge field $b_{\mu+}$ couples only to the vector current j_+^μ and the vector gauge field $b_{\mu-}$ couples only to the axial current j_{5-}^μ.

The massive gravity field contributes an axial vector part to the current coupled to the $b_{\mu+}$ and a vector part to the current coupled to the $b_{\mu-}$. The magnitude of these contributions is determined by the Newton gravitational constant.

C. Parity violating interactions due to extended gravity

The vector and axial vector currents of the fermionic fields coupled to the vector fields $a_{\mu\pm}$ of the gravity sector are given, respectively, by

$$J^{\mu}_{+}=-m\pi G_N G_V \overline{\psi}\gamma^a Y^{[\nu,\mu]}_{ab} s^b_\nu \psi-\frac{1}{2}\sqrt{2}gG_V\overline{\psi}\gamma^\mu(\overline{\eta}+\eta)\psi$$
$$-\frac{i}{2}\sqrt{2\pi G_N}G_V\overline{\psi}\gamma^\nu v^\mu_\nu(\overline{\eta}-\eta)\psi, \qquad (7.12)$$

and

$$J^{\mu}_{5+}=im\pi G_N G_V \overline{\psi}\gamma^a \gamma^5 X^{[\nu,\mu]}_{ab} s^b_\nu \psi-\frac{1}{2}\sqrt{2}gG_V\overline{\psi}\gamma^\mu\gamma^5$$
$$\times(\overline{\eta}+\eta)\psi+\frac{i}{2}\sqrt{2\pi G_N}G_V\overline{\psi}\gamma^\nu\gamma^5 v^\mu_\nu(\overline{\eta}-\eta)\psi. \qquad (7.13)$$

The vector current coupled to $a_{\mu-}$ is

$$J^{\mu}_{-}=-m\pi G_N G_V \overline{\psi}\gamma^a X^{[\nu,\mu]}_{ab} s^b_\nu \psi+\frac{i}{2}\sqrt{2}gG_V\overline{\psi}\gamma^\mu(\overline{\eta}-\eta)\psi$$
$$-\frac{1}{2}\sqrt{2\pi G_N}G_V\overline{\psi}\gamma^\nu v^\mu_\nu(\overline{\eta}+\eta)\psi. \qquad (7.14)$$

The axial vector current coupled to $a_{\mu-}$ is

$$J^{\mu}_{5-}=-im\pi G_N G_V \overline{\psi}\gamma^a \gamma^5 Y^{[\nu,\mu]}_{ab} s^b_\nu \psi+\frac{i}{2}\sqrt{2}gG_V\overline{\psi}\gamma^\mu$$
$$\times(\overline{\eta}-\eta)\psi+\frac{1}{2}\sqrt{2\pi G_N}G_V\overline{\psi}\gamma^\nu v^\mu_\nu(\overline{\eta}+\eta)\psi. \qquad (7.15)$$

VIII. SUMMARY AND CONCLUSIONS

The main part of this paper is the geometric formulation of a two-sheeted space-time within the framework of Conne's NCG. To start with, we have reviewed algebraic formulation of the conventional differential geometry in order to set the stage for its generalization to noncommutative case.

Our starting point for the two-sheeted space-time is the choice of the algebra $\mathcal{A}=C^\infty(C,\mathcal{M})\oplus C^\infty(C,\mathcal{M})$ as part of a spectral triple. However, our model leads us to a subalgebra where the functions on the two sheets are complex conjugates of each other. We are led to this subalgebra as the underlying mathematical structure of the two-sheeted space-time dictated by physics. Our previous pure gravity theories on a two-sheeted space-time [17–19] were constructed from a subalgebra of $\mathcal{A}$, which consisted of a pair of real functions. In the present paper, the construction of a spontaneously broken gauge theory requires that the fifth component of the gauge one-form must be a generalized complex valued function with the values on the two sheets being complex conjugates of each other. To be mathematically consistent, the same condition needs to be imposed on the differential forms of the theory. The algebra of the generalized function F of the form

$$F=\begin{pmatrix} f(x) & 0 \\ 0 & f^*(x)\end{pmatrix}, \qquad (8.1)$$

where $f(x)$ is a complex valued function, forms a closed subalgebra of $\mathcal{A}$. Therefore, the noncommutative geometry can be constructed consistently with this subalgebra to describe the physics of chiral spinors and gauge fields coupled to gravity on two-sheeted space-time. Remarkably, from the standpoint of physics, the restriction to this subalgebra is also required by the signature of the kinetic terms of the massive modes in the theory. Thus, starting from a naturally general algebra, we have ended up with a restriction to a subalgebra to have a physically meaningful theory. In the construction of the Lagrangian, the use of the involutive operation "$\sim$" plays an important role. Therefore, our results show that there is an intimate interplay between the mathematically consistent structure and physics in NCG.

The present theory also requires the definition of the wedge product to retain the Higgs quartic potential and to avoid the "junk forms" at the same time. The theory also requires a consistent involutive operator, which makes the theory consistent and the scalar products to be diagonal. The present theory also takes advantage of a minimal solution of the Cartan-Maurer structure equations, where the basic geometric quantities such as metric, connection, torsion, and curvature are completely expressed in terms of the generalized vielbeins. Physics requires that with such solutions, the contribution from torsion must be included to give kinetic terms to the physical fields. In other words, the mathematical structure of torsion is required by physics in this theory.

It is worth noting here that, although the above structure is rather unique in its simplicity and consistency, it is possible to construct the theory in an alternative way. For example, if the fully antisymmetric wedge product is desired, one may resort to a Kähler structure of the internal space consisting of a pair of complex conjugates [9,10]. In such a theory, the involutive operation can be chosen consistently to be the Hermitian conjugate. This approach, in the context of the full standard model, is currently under investigation.

From the point of view of physics, the model presented here possesses a rich and complex structure that merits further exploration in many ways. The generalized gravity sector that includes massive tensor, vector, and scalar fields, introduces simultaneous vector and axial vector couplings to matter fields. Consequently, we have the possibility of parity violating interactions due to gravity. In the context of a full standard model, we speculate that these interactions may provide the much sought *CP*-violating interactions in the early universe. This is further supported by the fact that such interactions involve G_N as the coupling. Consequently, we expect them to be small and hence of the correct order of magnitude. Incidentally, the idea of two interacting gravitons, one massless and the other massive, goes back to Isham, Salam, and Strathdee [27]. More recently, it has been the subject of study from various theoretical points of view [28,29]. It has also been pointed out by Damour and Kogan that bigravity, which is the simplest form of multigravity, arises naturally in several different physical contexts, such as brane configurations, Kaluza-Klein reductions, and noncommutative geometric models (ours is an example) [30].

In the context of cosmology, it is worth noting that the gravitational sector comes with a scalar field, an associated scalar potential, and a cosmological constant term. A scalar field and a scalar potential are invoked in the current literature to account for inflation and dark energy. Whether or not the potential in the model presented here has the desired features needs further exploration.

ACKNOWLEDGMENTS

This work was supported in part (K.C.W.) by the U.S. Department of Energy (DOE) under contract no. DE-FG02-85ER40237. A preliminary part of this work was done in collaboration with James A. Javor (James A. Javor and Kameshwar C. Wali, Construction of Action Functionals in Non-Commutative Lorenntzian Geometry, SU4240-692). One of the authors (K.C.W.) would like to thank A. Connes and J. Madore for many helpful discussions during the course of this work. He is also indebted to Joel Rozowsky for discussions and his help with the manuscript.

APPENDIX

We gather together in this appendix some formulas used in our calculations in the main text. We also present various Lagrangians needed to construct the action functionals. The field redefinitions (5.12) and (6.7) are taken into account to obtain the correct signatures for the kinetic terms in the final actions.

1. Metric components $G^{\mu\nu}$ and $G_{\mu\nu}$

Expressed in terms of the physical variables defined in Sec. V B, the even and odd components of the metric are given by

$$G^{\mu\nu}=G_{+}^{\mu\nu}\mathbf{1}+G_{-}^{\mu\nu}\mathbf{r},\quad G_{\mu\nu}=G_{\mu\nu+}\mathbf{1}+G_{\mu\nu-}\mathbf{r}, \tag{A1.1}$$

where

$$G_{+}^{\mu\nu}=g^{\mu\nu}-g^{\sigma\lambda}v_{\sigma}^{\mu}v_{\lambda}^{\nu},\quad G_{-}^{\mu\nu}=i(g^{\mu\sigma}v_{\sigma}^{\nu}+g^{\nu\sigma}v_{\sigma}^{\mu}), \tag{A1.2}$$

$$G_{\mu\nu+}=g_{\mu\nu}+(e_{\mu}^{a}\eta_{ab}s_{\lambda}^{b})v_{\nu}^{\lambda}+(e_{\nu}^{a}\eta_{ab}s_{\lambda}^{b})v_{\mu}^{\lambda}+(v_{\mu}^{\lambda}v_{\nu}^{\sigma}-\delta_{\mu}^{\lambda}\delta_{\nu}^{\sigma})s_{\lambda}^{a}\eta_{ab}s_{\sigma}^{b}, \tag{A1.3}$$

$$G_{\mu\nu-}=i(e_{\mu}^{a}\eta_{ab}s_{\nu}^{b}+e_{\nu}^{a}\eta_{ab}s_{\mu}^{b}+v_{\mu}^{\lambda}s_{\lambda}^{a}\eta_{ab}s_{\nu}^{b}+v_{\nu}^{\lambda}s_{\lambda}^{a}\eta_{ab}s_{\mu}^{b}). \tag{A1.4}$$

It is to be noted that $G^{\mu\nu}$,$G_{\mu\nu}$ are symmetric,

$$G_{\pm}^{\mu\nu}=G_{\pm}^{\nu\mu},\quad G_{\mu\nu\pm}=G_{\nu\mu\pm}. \tag{A1.5}$$

2. Components of $D\Gamma^{A}$

From

$$\Gamma^{A}=\Gamma^{M}E_{M}^{A}, \tag{A2.1}$$

Eq. (4.15) gives

$$(D\Gamma^{A})_{\mu\nu}=\tfrac{1}{2}(D_{\mu}E_{\nu}^{A}-D_{\nu}E_{\mu}^{A}),$$

$$(D\Gamma^{A})_{\mu5}=\tfrac{1}{2}(D_{\mu}E_{5}^{A}-D_{5}E_{\mu}^{A})=-(D\Gamma^{A})_{5\mu}, \tag{A2.2}$$

$$(D\Gamma^{A})_{55}=m(E_{5}^{A}+\tilde{E}_{5}^{A}).$$

Transforming Eq. (A2.2) into the locally flat frame, one obtains the components $(D\Gamma^{A})_{BC}$ as follows:

$$(D\Gamma^{a})_{bc}=\tfrac{1}{2}E_{b}^{\mu}E_{c}^{\nu}[(D_{\mu}E_{\nu}^{a}-D_{\nu}E_{\mu}^{a})+(A_{\nu}D_{5}E_{\mu}^{a}-A_{\mu}D_{5}E_{\nu}^{a})],$$

$$(D\Gamma^{a})_{b\dot{5}}=-\tfrac{1}{2}(D_{5}E_{\mu}^{a})E_{\dot{5}}^{5}(\tilde{E}_{b}^{\mu}+E_{b}^{\mu})=-(D\Gamma^{a})_{\dot{5}b},$$

$$(D\Gamma^{a})_{\dot{5}\dot{5}}=0, \tag{A2.3}$$

$$(D\Gamma^{\dot{5}})_{bc}=\tfrac{1}{2}E_{b}^{\mu}E_{c}^{\nu}\{(D_{\mu}A_{\nu}-D_{\nu}A_{\mu})\Phi+[A_{\nu}D_{5}(A_{\mu}\Phi)-A_{\mu}D_{5}(A_{\nu}\Phi)]\},$$

$$(D\Gamma^{\dot{5}})_{b\dot{5}}=\tfrac{1}{2}E_{\dot{5}}^{5}\left[\frac{1}{2}(\tilde{E}_{b}^{\mu}+E_{b}^{\mu})(D_{\mu}E_{5}^{\dot{5}}-D_{5}E_{\mu}^{\dot{5}})+m(\tilde{E}_{b}^{5}-E_{b}^{5})(E_{5}^{\dot{5}}+\tilde{E}_{5}^{\dot{5}})\right],$$

$$(D\Gamma^{\dot{5}})_{\dot{5}\dot{5}}=m\tilde{E}_{\dot{5}}^{5}E_{\dot{5}}^{5}(E_{5}^{\dot{5}}+\tilde{E}_{5}^{\dot{5}}).$$

We note that since the lower indexed Γ_{A} matrix is defined as

$$\Gamma_{A}=\eta_{AD}\Gamma^{D},$$

we use the following notation:

$$(D\Gamma_{A})_{BC}=\eta_{AD}(D\Gamma^{D})_{BC}. \tag{A2.4}$$

Using the definitions of the generalized vielbeins in Eqs. (4.18) and (4.19) and expressing them in terms of the new variables with the field redefinitions Eqs. (5.12) and (6.7), we find the even and odd components of $(D\Gamma^{A})_{BC}$. They are

$$(D\Gamma_{a})_{bc+}=\tfrac{1}{2}[X_{bc}^{[\mu,\nu]}P_{a\mu\nu}-Y_{bc}^{[\mu,\nu]}Q_{a\mu\nu}],$$

$$(D\Gamma_{a})_{bc-}=\frac{i}{2}[X_{bc}^{[\mu,\nu]}Q_{a\mu\nu}+Y_{bc}^{[\mu,\nu]}P_{a\mu\nu}], \tag{A2.5}$$

$$(D\Gamma_{a})_{b\dot{5}+}=-m\phi_{-}e_{b}^{\mu}s_{a\mu}(\phi_{+}^{2}+\phi_{-}^{2})^{-1}=-(D\Gamma_{a})_{\dot{5}b+},$$

$$(D\Gamma_{a})_{b\dot{5}-}=-im\phi_{+}e_{b}^{\mu}s_{a\mu}(\phi_{+}^{2}+\phi_{-}^{2})^{-1}=-(D\Gamma_{a})_{\dot{5}b-}, \tag{A2.6}$$

$$(D\Gamma^{a})_{\dot{5}\dot{5}\pm}=0 \tag{A2.7}$$

$$(D\Gamma_{\dot 5})_{bc+}=\tfrac{1}{2}[X_{bc}^{[\mu,\nu]}(a_{\mu\nu+}\phi_{+}-a_{\mu\nu-}\phi_{-})-Y_{bc}^{[\mu,\nu]}(a_{\mu\nu+}\phi_{-}+a_{\mu\nu-}\phi_{+})]-m(X_{bc}^{[\mu,\nu]}a_{\nu-}a_{\mu+}\phi_{-}+Y_{bc}^{[\mu,\nu]}a_{\nu+}a_{\mu-}\phi_{+}),$$

$$(D\Gamma_{\dot 5})_{bc-}=\frac{i}{2}\{[X_{bc}^{[\mu,\nu]}(a_{\mu\nu+}\phi_{-}+a_{\mu\nu-}\phi_{+})+Y_{bc}^{[\mu,\nu]}(a_{\mu\nu+}\phi_{+}-a_{\mu\nu-}\phi_{-})]+2m(X_{bc}^{[\mu,\nu]}a_{\nu+}a_{\mu-}\phi_{+}-Y_{bc}^{[\mu,\nu]}a_{\nu-}a_{\mu+}\phi_{-})\},\tag{A2.8}$$

$$(D\Gamma_{\dot 5})_{b\dot 5+}=(\phi_{+}^{2}+\phi_{-}^{2})^{-1}[\tfrac{1}{2}(\phi_{+}\partial_{\mu}\phi_{+}+\phi_{-}\partial_{\mu}\phi_{-})e_{b}^{\mu}-me_{b}^{\mu}\phi_{-}(a_{\mu-}\phi_{+}+a_{\mu+}\phi_{-})+2m\phi_{-}\phi_{+}(a_{\mu-}e_{b}^{\mu}+a_{\mu+}v_{b}^{\mu})],$$

$$(D\Gamma^{\dot 5})_{b\dot 5-}=i(\phi_{+}^{2}+\phi_{-}^{2})^{-1}[\tfrac{1}{2}(\phi_{+}\partial_{\mu}\phi_{-}-\phi_{-}\partial_{\mu}\phi_{+})e_{b}^{\mu}-me_{b}^{\mu}\phi_{+}(a_{\mu-}\phi_{+}+a_{\mu+}\phi_{-})+2m\phi_{+}^{2}(a_{\mu-}e_{b}^{\mu}+a_{\mu+}v_{b}^{\mu})],\tag{A2.9}$$

$$(D\Gamma_{\dot 5})_{\dot 5\dot 5+}=2m\phi_{+}(\phi_{+}^{2}+\phi_{-}^{2})^{-1},\tag{A2.10}$$

$$(D\Gamma^{\dot 5})_{\dot 5\dot 5-}=0,\tag{A2.11}$$

where we have defined

$$a_{\mu\nu\pm}=\tfrac{1}{2}(\partial_{\mu}a_{\nu\pm}-\partial_{\nu}a_{\mu\pm}),\tag{A2.12}$$

and $P_{a\mu\nu}$, $Q_{a\mu\nu}$, $X_{ab}^{[\mu,\nu]}$, $Y_{ab}^{[\mu,\nu]}$ are defined in Eq. (5.13).

The following formulas for $X_{bc}^{[\mu,\nu]}$ and $Y_{b}^{[\mu,\nu]}{}_{c}$ will be useful in the torsion and gauge Lagrangian calculations

$$X_{bc}^{[\mu,\nu]}X^{bc[\rho,\tau]}=2(g^{\mu\rho}g^{\nu\tau}-g^{\mu\tau}g^{\nu\rho}+2v^{\mu\tau}v^{\nu\rho}-2v^{\mu\rho}v^{\nu\tau}+\langle v^{\mu},v^{\rho}\rangle\langle v^{\nu},v^{\tau}\rangle-\langle v^{\mu},v^{\tau}\rangle\langle v^{\nu},v^{\rho}\rangle),\tag{A2.13}$$

$$X_{bc}^{[\mu,\nu]}Y^{bc[\rho,\tau]}=2(g^{\mu\rho}v^{\nu\tau}-g^{\mu\tau}v^{\nu\rho}+v^{\mu\rho}g^{\nu\tau}-v^{\mu\tau}g^{\nu\rho}+v^{\nu\rho}\langle v^{\mu},v^{\tau}\rangle-v^{\nu\tau}\langle v^{\mu},v^{\rho}\rangle+v^{\mu\tau}\langle v^{\nu},v^{\rho}\rangle-v^{\mu\rho}\langle v^{\nu},v^{\tau}\rangle),\tag{A2.14}$$

$$Y_{bc}^{[\mu,\nu]}Y^{bc[\rho,\tau]}=2(g^{\mu\rho}\langle v^{\nu},v^{\tau}\rangle-g^{\mu\tau}\langle v^{\nu},v^{\rho}\rangle+g^{\nu\tau}\langle v^{\mu},v^{\rho}\rangle-g^{\nu\rho}\langle v^{\mu},v^{\tau}\rangle+2v^{\mu\rho}v^{\nu\tau}-2v^{\mu\tau}v^{\nu\rho}).\tag{A2.15}$$

Using the formulas (A2.1)–(A2.12), it is straightforward to calculate all components of the connection, torsion, and scalar curvature.

3. Curvature Lagrangian

For the calculation of the scalar curvature in Eq. (5.17) the components Ω_{abc} and $\Omega_{a\dot 5b}$ of the connection Ω_{AB} are necessary. The explicit expressions of the projections of these components are given by

$$\Omega_{a\dot 5b+}=-2(D\Gamma_{a})_{\dot 5b+}=-2m\phi_{-}e_{b}^{\mu}s_{a\mu}(\phi_{+}^{2}+\phi_{-}^{2})^{-1},$$

$$\Omega_{a\dot 5b-}=2(D\Gamma_{a})_{\dot 5b-}=2im\phi_{+}e_{b}^{\mu}s_{a\mu}(\phi_{+}^{2}+\phi_{-}^{2})^{-1},\tag{A3.1}$$

$$\Omega_{abc+}=-[(D\Gamma_{a})_{bc+}-(a\leftrightarrow b)+(a\leftrightarrow c)]=-\tfrac{1}{2}[(X_{bc}^{[\mu,\nu]}P_{a\mu\nu}-Y_{bc}^{[\mu,\nu]}Q_{a\mu\nu})-(a\leftrightarrow b)+(a\leftrightarrow c)],\tag{A3.2}$$

$$\Omega_{abc-}=-[(D\Gamma_{a})_{bc-}-(a\leftrightarrow b)+(a\leftrightarrow c)]=-\tfrac{1}{2}[(X_{bc}^{[\mu,\nu]}Q_{a\mu\nu}+Y_{bc}^{[\mu,\nu]}P_{a\mu\nu})-(a\leftrightarrow b)+(a\leftrightarrow c)],$$

where $(a\leftrightarrow b)$ and $(a\leftrightarrow b)$ are permutations of the indices of the given tensors.

Now we can calculate the following terms:

$$(D\Omega_{ab})_{cd+}\eta^{ad}\eta^{bc}=-X^{ab[\mu,\nu]}(\tfrac{1}{2}\partial_{\mu}\Omega_{ab\nu+}+ima_{\nu-}\Omega_{ab\mu-})-iY^{ab[\mu,\nu]}(\tfrac{1}{2}\partial_{\mu}\Omega_{ab\nu-}+ima_{\nu-}\Omega_{ab\mu+}),\tag{A3.3}$$

where $\Omega_{ab\mu\pm}$ can be expressed in terms of $\Omega_{abc\pm}$ as follows:

$$\Omega_{ab\mu+}=(e_{\mu}^{c}+e_{\nu}^{c}s^{\nu\rho}v_{\rho\mu})\Omega_{abc+}+is_{\mu}^{c}\Omega_{abc-},$$

$$\Omega_{ab\mu-}=(e_{\mu}^{c}+e_{\nu}^{c}s^{\nu\rho}v_{\rho\mu})\Omega_{abc-}+is_{\mu}^{c}\Omega_{abc+}.\tag{A3.4}$$

Inserting for $\Omega_{abc\pm}$ from Eq. (A3.2) in Eqs. (A3.4), we calculate $\Omega_{ab\mu\pm}$ and then $(D\Omega_{ab})_{cd+}\eta^{ad}\eta^{bc}$, giving the following final expression:

$$\mathcal{L}_{R1}=(16\pi G_N)^{-1}(D\Omega_{ab})_{cd+}\eta^{ad}\eta^{bc},$$

$$\begin{aligned}\mathcal{L}_{R1}=&(64\pi G_N)^{-1}(e_{d\tau}+s_{d\lambda}v^{\lambda}_{\tau})[(2X^{ab[\rho,\tau]}\eta^{cd}-X^{bc[\rho,\tau]}\eta^{ad})\partial_\rho(X^{[\mu,\nu]}_{bc}P_{a\mu\nu}-Y^{[\mu,\nu]}Q_{a\mu\nu})-(2Y^{ab[\rho,\tau]}\eta^{cd}\\&-Y^{bc[\rho,\tau]}\eta^{ad})\partial_\rho(X^{[\mu,\nu]}_{bc}Q_{a\mu\nu}+Y^{[\mu,\nu]}_{bc}P_{a\mu\nu})]+(64\pi G_N)^{-1}(X^{bc[\rho,\tau]}X^{[\mu,\nu]}_{bc}\eta^{ad}+2X^{ab[\rho,\tau]}X^{d[\mu,\nu]}_{b})\\&\times[-P_{a\mu\nu}\partial_\rho(e_{d\tau}+s_{d\lambda}v^{\lambda}_{\tau})+Q_{a\mu\nu}\partial_\rho s_{d\tau}]\\&+(64\pi G_N)^{-1}(X^{bc[\rho,\tau]}Y^{[\mu,\nu]}_{bc}\eta^{ad}+2X^{ab[\rho,\tau]}Y^{d[\mu,\nu]}_{b}+Y^{bc[\rho,\tau]}X^{[\mu,\nu]}_{bc}\eta^{ad}+2Y^{ab[\rho,\tau]}X^{d[\mu,\nu]}_{b})\\&\times[P_{a\mu\nu}\partial_\rho s_{d\tau}+Q_{a\mu\nu}\partial_\rho(e_{d\tau}+s_{d\lambda}v^{\lambda}_{\tau})](64\pi G_N)^{-1}(Y^{bc[\rho,\tau]}Y^{[\mu,\nu]}_{bc}\eta^{ad}+2Y^{ab[\rho,\tau]}Y^{d[\mu,\nu]}_{b})\\&\times[P_{a\mu\nu}\partial_\rho(e_{d\tau}+s_{d\lambda}v^{\lambda}_{\tau})+Q_{a\mu\nu}\partial_\rho s_{d\tau}]\\&+m(32\pi G_N)^{-1}a_{\tau-}\{[(X^{bc[\rho,\tau]}X^{[\mu,\nu]}_{bc}-Y^{bc[\rho,\tau]}Y^{[\mu,\nu]}_{bc})\eta^{ad}-2(X^{ab[\rho,\tau]}X^{d[\mu,\nu]}_{b}-Y^{ab[\rho,\tau]}X^{d[\mu,\nu]}_{-b})]\\&\times[P_{a\mu\nu}s_{d\rho}+Q_{a\mu\nu}(e_{d\rho}+s_{d\lambda}v^{\lambda}_{\rho})]+[(Y^{bc[\rho,\tau]}X^{[\mu,\nu]}_{bc}-X^{bc[\rho,\tau]}Y^{[\mu,\nu]}_{bc})\eta^{ad}+2(Y^{ab[\rho,\tau]}X^{d[\mu,\nu]}_{b}+X^{ab[\rho,\tau]}X^{d[\mu,\nu]}_{b})]\\&\times[P_{a\mu\nu}(e_{d\rho}+s_{d\lambda}v^{\lambda}_{\rho})+Q_{a\mu\nu}s_{d\rho}]\},\end{aligned}\tag{A3.5}$$

$$\begin{aligned}\mathcal{L}_{R2}=&(16\pi G_N)^{-1}[(D\Gamma_a)_{ec+}(D\Gamma_b)_{fd+}+(D\Gamma_a)_{ec-}(D\Gamma_b)_{fd-}](\eta^{ab}\eta^{cd}\eta^{ef}+3\eta^{ad}\eta^{be}\eta^{cf}+\eta^{ac}\eta^{bd}\eta^{ef})\\=&(64\pi G_N)^{-1}(\eta^{ab}X^{cd[\mu,\nu]}X^{[\rho,\tau]}_{cd}-3X^{bc[\mu,\nu]}X^{a[\rho,\tau]}_{c}+X^{ac[\mu,\nu]}X^{b[\rho,\tau]}_{c})[(P_{a\mu\nu}P_{b\rho\tau}-Q_{a\mu\nu}Q_{b\rho\tau})]\\&+(64\pi G_N)^{-1}(\eta^{ab}Y^{cd[\mu,\nu]}Y^{[\rho,\tau]}_{cd}-3Y^{bc[\mu,\nu]}Y^{a[\rho,\tau]}_{b}+Y^{ac[\mu,\nu]}Y^{b[\rho,\tau]}_{c})[-P_{a\mu\nu}P_{b\rho\tau}+Q_{a\mu\nu}Q_{b\rho\tau}]\\&+(64\pi G_N)^{-1}[-\eta^{ab}(X^{cd[\mu,\nu]}Y^{[\rho,\tau]}_{cd}-X^{cd[\rho,\tau]}Y^{[\mu,\nu]}_{cd})+2(X^{bc[\mu,\nu]}Y^{a[\rho,\tau]}_{c}-X^{bc[\rho,\tau]}Y^{a[\mu,\nu]}_{c})]\\&\times[P_{a\mu\nu}Q_{b\rho\tau}-Q_{a\mu\nu}P_{b\rho\tau}],\end{aligned}\tag{A3.6}$$

$$\begin{aligned}\mathcal{L}_{R3}=&(8\pi G_N)^{-1}\{(D\Omega_{a\dot{5}})_{\dot{5}d+}\eta^{ad}+4[(D\Gamma_a)_{\dot{5}d+}(D\Gamma_b)_{\dot{5}c+}+(D\Gamma_a)_{\dot{5}d-}(D\Gamma_b)_{\dot{5}c-}](\eta^{ad}\eta^{bc}-\eta^{ac}\eta^{bd})\}\\=&-m^2(4\pi G_N)^{-1}(\phi_+^2+\phi_-^2)^{-2}\{(\phi_+^2-\phi_-^2)(s^{\mu\nu}s_{\mu\nu}-s^{\mu}_{\mu}s^{\nu}_{\nu})+\phi_+[\phi_+(v^{\mu\nu}s_{\mu\rho}v^{\rho\tau}s_{\tau\nu}-v^{\mu\nu}s_{\mu\nu})\\&+\phi_-s^{\mu\nu}s^{\rho}_{\mu}s_{\nu\rho}]\}.\end{aligned}\tag{A3.7}$$

4. Torsion Lagrangian

The contributions of the torsion to the total Lagrangian arise from the components $T_{\dot{5}bc}$, $T_{\dot{5}\dot{5}b}$ and $T_{\dot{5}\dot{5}\dot{5}}$. With the physical reasoning that the gravity tensor $v_{\mu\nu}$ represents a field of spin 2, we assume that the gravity tensor field $v_{\mu\nu}$ is symmetric in the indices μ,ν, i.e. $v_{\mu\nu}=v_{\nu\mu}$. This also implies $s_{\mu\nu}=s_{\nu\mu}$ and hence,

$$(D\Gamma_b)_{\dot{5}c\pm}-(D\Gamma_c)_{\dot{5}b\pm}=0.\tag{A4.1}$$

Hence, the projections of the torsion components can be expressed in terms of $D\Gamma$ as follows:

$$\begin{aligned}(T_{\dot{5}bc})_\pm&=(D\Gamma_{\dot{5}})_{bc\pm},\\(T_{\dot{5}b\dot{5}})_\pm&=(D\Gamma_{\dot{5}})_{\dot{5}b\pm},\\(T_{\dot{5}\dot{5}\dot{5}})_\pm&=(D\Gamma_{\dot{5}})_{\dot{5}\dot{5}\pm}.\end{aligned}\tag{A4.2}$$

Using the expressions of $D\Gamma$ in Appendix A 2, we obtain

$$\mathcal{L}_{T1}=(4G_V^2)^{-1}[(T_{\dot{5}bc})_+(T^{\dot{5}bc})_+-(T_{\dot{5}bc})_-(T^{\dot{5}bc})_-],\tag{A4.3}$$

$$\begin{aligned}\mathcal{L}_{T1}=&-(4G_V^2)^{-1}(X^{[\mu,\nu]}_{bc}X^{bc[\rho,\tau]}+Y^{[\mu,\nu]}_{bc}Y^{bc[\rho,\tau]})\\&\times\Big\{(\phi_+^2+\phi_-^2)(\tfrac{1}{4}a_{\mu\nu+}a_{\rho\tau_+}+\tfrac{1}{4}a_{\mu\nu-}a_{\rho\tau-}\\&+m^2a_{\mu\_}a_{\nu+}a_{\rho-}a_{\tau+})+ma_{\rho-}a_{\tau+}[2a_{\mu\nu+}\phi_+\phi_-\\&+a_{\mu\nu-}(\phi_+^2-\phi_-^2)]\Big\}+m(G_V^2)^{-1}X^{[\mu,\nu]}_{bc}Y^{bc[\rho,\tau]}\\&\times[(\phi_+^2-\phi_-^2)(a_{\mu\nu+}a_{\tau-}a_{\rho+}+a_{\nu+}a_{\mu-}a_{\rho\tau+})\\&+2\phi_+\phi_-(a_{\mu\nu-}a_{\tau+}a_{\rho-}+a_{\rho\tau-}a_{\nu-}a_{\mu+})],\end{aligned}\tag{A3}$$

$$\mathcal{L}_{T2}=(2G_V^2)^{-1}[(T_{\dot{5}\dot{5}c})_+(T^{\dot{5}\dot{5}c})_+-(T_{\dot{5}\dot{5}c})_-(T^{\dot{5}\dot{5}c})_-], \tag{A4.5}$$

$$\mathcal{L}_{T2}=-(2G_V^2)^{-1}(\phi_+^2+\phi_-^2)^{-1}\Big\{\frac{1}{4}g^{\mu\nu}(\partial_\mu\phi_+\partial_\nu\phi_+ + \partial_\mu\phi_-\partial_\nu\phi_-)+m^2[a_-^\mu a_{\mu-}(4-3\phi_+^2)+\phi_-^2 a_+^\mu a_{\mu+} -2\phi_+\phi_- a_{\mu-}a_+^\mu+4a_{\mu+}v_\rho^\mu(v^{\rho\lambda}a_{\lambda+}-a_+^\rho\phi_+\phi_- +2a_-^\rho-a_-^\rho\phi_+^2)]+m(a_-^\mu\phi_++2a_{\rho+}v^{\rho\mu}\phi_+ -a_+^\mu\phi_-)\partial_\mu\phi_-\Big\}, \tag{A4.6}$$

$$\mathcal{L}_{T3}=-(4G_V^2)^{-1}(T_{\dot{5}\dot{5}\dot{5}})_+T_+^{\dot{5}\dot{5}\dot{5}}=-\frac{m^2}{G_V^2k^2}\phi_+^2(\phi_+^2+\phi_-^2)^{-2}. \tag{A4.7}$$

5. Gauge Lagrangian

The Lagrangian terms in Eq. (6.8) can be calculated as follows:

$$\mathcal{L}_{G1}=-\frac{1}{4g^2}(g^{\mu\rho}+g^{\sigma\lambda}v_\sigma^\mu v_\lambda^\rho)(g^{\nu\tau}+g^{\sigma\lambda}v_\sigma^\nu v_\lambda^\tau)(b_{\mu\nu+}b_{\rho\tau+}+b_{\mu\nu-}b_{\rho\tau-}), \tag{A5.1}$$

$$\mathcal{L}_{G2}=-\frac{1}{g^2}(\phi_+^2+\phi_-^2)^{-1}\{g^{\mu\nu}\mathcal{D}_\mu\eta\mathcal{D}_\nu\bar{\eta}+2ig^\mu\sigma v_\sigma^\nu b_{\mu+}[(\bar{\eta}-m)\mathcal{D}_\mu\eta-(\eta-m)\mathcal{D}_\mu\bar{\eta}]+4g^{\sigma\tau}v_\sigma^\mu v_\tau^\nu b_{\mu+}b_{\nu+}(\eta-m)(\bar{\eta}-m)\}, \tag{A5.2}$$

$$\mathcal{L}_{G3}=\frac{1}{2g^2}(\phi_+^2+\phi_-)^{-1}(\bar{\eta}\eta-m^2)^2, \tag{A5.3}$$

$$\mathcal{L}_{G4}=\frac{1}{g^2}(X_{ab}^{[\mu,\nu]}X^{ab[\rho,\tau]}+Y_{ab}^{[\mu,\nu]}Y^{ab[\rho,\tau]})\Big[\{(b_{\mu\nu+}-ib_{\mu\nu-})a_\rho[\mathcal{D}_\tau\eta+2i(\eta-m)b_{\tau-}]+\text{c.c}\}+\Big(\frac{1}{2}a_\mu a_\rho^*\{\mathcal{D}_\nu\overline{\eta[\mathcal{D}_\tau\eta+2i(\eta -m)b_{\tau-}]}+4b_{\nu-}(\eta-m)(\bar{\eta}-m)\Big\}+\text{c.c}\Big)-\frac{1}{2g^2}(\phi_+^2+\phi_-^2)^{-1}g^{\mu\nu}\Big[(X_\mu+X_\mu^*)(Y_\nu+Y_\nu^*)+(X_\mu-X_\mu^*+4imv_\mu^\rho b_{\rho+})(Z_\nu -Z_\nu^*)+\frac{1}{2}(Y_\mu+Y_\mu^*)(Y_\nu+Y_\nu^*)-\frac{1}{2}(Z_\mu-Z_\mu^*)(Z_\nu-Z_\nu^*)\Big], \tag{A5.4}$$

where

$$a_\mu=a_{\mu+}+ia_{\mu-},$$
$$X_\mu=(\mathcal{D}_\mu-2iv_\mu^\nu b_{\nu+})\eta,$$
$$Y_\mu=(a_\mu+iv_\mu^\nu a_\nu)(\eta-\bar{\eta})(\eta-m), \tag{A5.5}$$
$$Z_\mu=(a_\mu-iv_\mu^\nu a_\nu)(\eta+m)(\eta+\bar{\eta}-2m),$$

and c.c denotes the complex conjugates.

6. Fermionic Lagrangian

The following identities will be useful for the calculations of the Lagrangian in Eq. (6.9):

$$\bar{\Psi}\Gamma^a A\Psi=\bar{\psi}\gamma^a(A_++\gamma^5A_-)\psi,$$
$$\bar{\Psi}\Gamma^5A\Psi=\bar{\psi}(\gamma^5A_++A_-)\psi, \tag{A6.1}$$

where

$$A=A_+\mathbf{1}+A_-\mathbf{r}.$$

To prove these identities we can take the trace over the Z_2 indices then express ψ_L,ψ_R in terms of the 4-component Dirac spinor and the parity projection operator $\frac{1}{2}(1\pm\gamma^5)$.

To calculate the terms $\mathcal{L}_{F1}$ and $\mathcal{L}_{F2}$ we use the following identity to decompose the product of three Γ matrices:

$$\Gamma^a\Gamma^b\Gamma^c=\eta^{ab}\Gamma^c-\eta^{ac}\Gamma^b+\eta^{bc}\Gamma^a+i\epsilon^{abcd}\Gamma^5\Gamma_d \tag{A6.2}$$

from where we obtain the following simplification using the symmetric properties of the tensors Ω_{BCA}:

$$\Gamma^a\Gamma^b\Gamma^c\Omega_{bca}=2\eta^{ab}\Gamma^c\Omega_{bca}+i\epsilon^{abcd}\Gamma^5\Gamma_d\Omega_{bca}. \tag{A6.3}$$

The second term in Eq. (A6.3) yields a vanishing contribution when inserted between spinors [Eq. (A6.1)]. Hence we can omit it and obtain

$$\Gamma^aA_1=\Gamma^aE_a^\mu(D_\mu+iB_\mu)+\frac{i}{4}\Gamma^a\Gamma^b\Gamma^c\Omega_{bca}^{(0)}$$

$$=\Gamma^a\left[E_a^\mu(D_\mu+iB_\mu)-\frac{i}{2}\eta^{bc}\Omega_{abc}^{(0)}\right], \tag{A6.4}$$

$$(A_1)_+=e_a^\mu(\nabla_\mu+ib_{\mu+})-iv_a^\mu b_{\mu-}-\frac{i}{2}(v_a^\mu v_b^\nu-v_a^\nu v_b^\mu)\partial_\mu e_\nu^b$$

$$+\frac{i}{2}[X_{ab}^{[\mu,\nu]}\partial_\mu(v_\nu^\lambda s_\lambda^b)+Y_{ab}^{[\mu,\nu]}\partial_\mu s_\nu^b], \tag{A6.5}$$

$$(A_1)_-=v_a^\mu(i\partial_\mu-b_{\mu+})-e_a^\mu b_{\mu-}-\tfrac{1}{2}[X_{ab}^{[\mu,\nu]}\partial_\mu s_\nu^b$$
$$+Y_{ab}^{[\mu,\nu]}\partial_\mu(e_\nu^b+v_\nu^\lambda s_\lambda^b)],$$

where the general covariant derivative $\nabla_\mu=\partial_\mu+(i/2)e_b^\nu(\partial_\mu e_\mu^b-\partial_\nu e_\mu^b)$,

$$\Gamma^a A_2=\frac{i}{4}\Gamma^a\Gamma^b\Gamma^c\Omega_{bca}^{(1)}=\frac{i}{2}\eta^{ab}\Gamma^c\Omega_{bca}^{(1)}, \tag{A6.6}$$

$$(A_2)_+=ims_\mu^b[X_{ab}^{[\mu,\nu]}a_{\nu-}+Y_{ab}^{[\mu,\nu]}a_{\nu+}],$$

$$(A_2)_-=ms_\mu^b[X_{ab}^{[\mu,\nu]}a_{\nu+}-Y_{ab}^{[\mu,\nu]}a_{\nu-}], \tag{A6.7}$$

$$\Gamma^a A_3=-\Gamma^a E_a^\mu A_\mu(D_5+iB_5), \tag{A6.8}$$

$$(A_3)_+=-\frac{i}{2}[(e_a^\mu-iv_a^\mu)\eta(a_{\mu+}+ia_{\mu-})+(e_a^\mu+iv_a^\mu)$$
$$\times\overline{\eta}(a_{\mu+}-ia_{\mu-})],$$

$$(A_3)_-=\frac{1}{2}(v_a^\mu-ie_a^\mu)\eta(a_{\mu+}+ia_{\mu-})+\frac{1}{2}(v_a^\mu+ie_a^\mu)$$
$$\times\overline{\eta}(a_{\mu+}-ia_{\mu-}), \tag{A6.9}$$

$$\Gamma^5 A_4=\tfrac{1}{2}\Gamma^5[i(\Gamma^a\Gamma^b\Omega_{b5a})+2\Phi^{-1}(D_5+iB_5)]. \tag{A6.10}$$

Since the tensor $s^{\mu\nu}$ is symmetric, the terms, which contain $\gamma^a\gamma^b$, reduce to the trace of that tensor giving

$$(A_4)_+=-i(\phi_+^2+\phi_-^2)^{-1}[\tfrac{i}{2}\eta(\phi_++i\phi_-)+\tfrac{i}{2}\overline{\eta}(\phi_+-i\phi_-)$$
$$-ms_\mu^\mu\phi_-],$$

$$(A_4)_-=(\phi_+^2+\phi_-^2)^{-1}\left[\frac{i}{2}\eta(\phi_++i\phi_-)+-\frac{i}{2}\overline{\eta}(\phi_+-i\phi_-)\right.$$
$$\left.-ms_\mu^\mu\phi_+\right]. \tag{A6.11}$$

Here, we give only the final results:

$$\mathcal{L}_{F1}=i\overline{\Psi}\left(\Gamma^a E_a^\mu(D_\mu+iB_\mu)+\frac{i}{4}\Gamma^a\Gamma^b\Gamma^c\Omega_{bca}^{(0)}\right)\Psi$$

$$=\frac{i}{2}\overline{\psi}\gamma^a\left[e_a^\mu(\nabla_\mu+ib_{\mu+})-iv_a^\mu b_{\mu-}+\frac{i}{2}(v_a^\mu v_b^\nu\right.$$
$$-v_a^\nu v_b^\mu)\partial_\mu e_\nu^b-\frac{i}{2}(X_{ab}^{[\mu,\nu]}\partial_\mu(v_\nu^\lambda s_\lambda^b)$$
$$\left.+Y_{ab}^{[\mu,\nu]}\partial_\mu s_\nu^b)\right]\psi+\frac{i}{2}\overline{\psi}\gamma^a\gamma^5\left\{v_a^\mu(i\partial_\mu-b_{\mu+})\right.$$
$$-e_a^\mu b_{\mu-}+\frac{1}{2}[X_{ab}^{[\mu,\nu]}\partial_\mu s_\nu^b$$
$$\left.+Y_{ab}^{[\mu,\nu]}\partial_\mu(e_\nu^b+v_\nu^\lambda s_\lambda^b)]\right\}\psi+\text{H.c.}, \tag{A6.12}$$

$$\mathcal{L}_{F2}=\frac{i}{4}\overline{\Psi}\Gamma^a\Gamma^b\Gamma^c\Omega_{bca}^{(1)}\Psi$$

$$=\frac{m}{2}\overline{\psi}\gamma^a(X_{ab}^{[\mu,\nu]}+i\gamma^5 Y_{ab}^{[\mu,\nu]})$$
$$\times(i\gamma^5 a_{\nu+}-a_{\nu-})s_\mu^b\psi+\text{H.c.}, \tag{A6.13}$$

$$\mathcal{L}_{F3}=i\overline{\Psi}\Gamma^a E_a^\mu A_\mu(D_5+iB_5)\Psi$$

$$=-\frac{1}{2}\overline{\psi}\gamma^a\eta(a_{\mu+}+ia_{\mu-})[(e_a^\mu-iv_a^\mu)+\gamma^5(e_a^\mu$$
$$+iv_a^\mu)]\psi+\text{H.c.}, \tag{A6.14}$$

$$\mathcal{L}_{F4}=\frac{i}{2}\overline{\Psi}\Gamma^5[(\Gamma^a\Gamma^b\Omega_{b5a})+2\Phi^{-1}(D_5+iB_5)]\Psi$$

$$=-\frac{i}{2}\overline{\psi}(\phi_+^2+\phi_-^2)^{-1}\eta(\phi_++i\phi_-)(1-\gamma^5)\psi$$
$$+\text{H.c.} \tag{A6.15}$$

[1] A. Connes, *Non-Commutative Geometry* (Academic Press, New York, 1994).

[2] A. Connes, in *The Interface of Mathematics and Particle Physics*, edited by D.G. Quillen, G.B. Segal, and S.T. Tsou (Clarendon, Oxford, 1990).

[3] R. Coquereaux, J. Geom. Phys. **6**, 425 (1989).

[4] G. Landi, *An Introduction to Noncommutative Spaces and Their Geometries* (Springer-Verlag, Berlin, 1997).

[5] J. Madore, *An Introduction to Noncommutative Differential Geometry and its Physical Applications* (Cambridge Univer-

sity Press, Cambridge, England, 1995).
[6] A. Connes and J. Lott, Nucl. Phys. B (Proc. Suppl.) **18**, 29 (1990).
[7] M. Dubois-Violette, R. Kerner, and J. Madore, J. Math. Phys. **31**, 316 (1990).
[8] J.C. Varilly and J.M. Gracia-Bondia, J. Geom. Phys. **12**, 223 (1993).
[9] Nguyen Ai Viet, in *MRST'94 What Next? Exploring the Future of High-Energy Physics*, edited by K.R. Cudel *et al.* (World Scientific, Singapore, 1994).
[10] Nguyen Ai Viet, Heavy Ion Phys. hph **1**, 263 (1995).
[11] G.F. Simmons, *Introduction to Topology and Modern Analysis* (Krieger, Melbourne, FL, 1983).
[12] W. Kalau and M. Walze, J. Geom. Phys. **16**, 327 (1995).
[13] D. Kastler, Commun. Math. Phys. **166**, 633 (1995).
[14] A.H. Chamseddine, G. Felder, and J. Fröhlich, Commun. Math. Phys. **155**, 205 (1993).
[15] A. Sitarz, Class. Quantum Grav. **11**, 2127 (1994).
[16] A.H. Chamseddine and A. Connes, Commun. Math. Phys. **186**, 731 (1997); see also V.P. Nair, Nucl. Phys. **B651**, 313 (2003); Y. Abe and V.P. Nair, hep-th/0212270, for gravitational fields on a non-commutative space.
[17] G. Landi, Nguyen Ai Viet, and K.C. Wali, Phys. Lett. B **326**, 45 (1994).
[18] N.A. Viet and K.C. Wali, Int. J. Mod. Phys. A **11**, 2403 (1996).
[19] N.A. Viet and K.C. Wali, Int. J. Mod. Phys. A **11**, 533 (1996).
[20] Nguyen Ai Viet, "Some remarks on gravity in noncommutative geometry and a new minimal set of constraints on the connection and torsion," hep-th/0303128.
[21] S. Chandrasekhar, *The Mathematical Theory of Black Holes* (Oxford University Press, New York, 1992).
[22] C.W. Misner, K.S. Thorne, and J.A. Wheeler, *Gravitation* (W.H. Freeman and Company, San Franciso, 1973).
[23] R.W. Wald, *General Gravity* (University of Chicago Press, Chicago, 1984).
[24] M. Nakahara, *Geometry, Topology and Physics* (Institute of Physics Press, Bristol, 1992).
[25] Th. Kaluza, Sitzungsber. Preuss. Akad. Wiss. Berlin, Math: Phys. **K1**, 966 (1921); O. Klein, Z. Phys. **37**, 895 (1926); Y. Thirry, Acad. Sci., Paris, C. R. **226**, 216 (1948).
[26] M.J. Duff, C.N. Pope, and K.S. Stelle, Phys. Lett. B **223**, 71 (1989).
[27] C.J. Isham, A. Salam, and J. Strathdee, Phys. Rev. D **3**, 867 (1971).
[28] N. Arkani-Hamed, H. Georgi, and M.D. Schwartz, hep-th/0210184.
[29] A.H. Chamseddine, Phys. Lett. B **557**, 247 (2003).
[30] T. Damour and I.I. Kogan, Phys. Rev. D **66**, 104024 (2002).

Domain wall solutions with Abelian gauge fields

J.S. Rozowsky [a], R.R. Volkas [b], K.C. Wali [a]

[a] *Physics Department, Syracuse University, Syracuse, NY 13244-1130, USA*
[b] *School of Physics, Research Centre for High Energy Physics, The University of Melbourne, Victoria 3010, Australia*

Received 30 May 2003; received in revised form 11 November 2003; accepted 17 November 2003

Editor: M. Cvetič

Abstract

We study kink (domain wall) solutions in a model consisting of two complex scalar fields coupled to two independent Abelian gauge fields in a Lagrangian that has $U(1) \times U(1)$ gauge plus $\mathbb{Z}_2$ discrete symmetry. We find consistent solutions such that while the $U(1)$ symmetries of the fields are preserved while in their respective vacua, they are broken on the domain wall. The gauge field solutions show that the domain wall is sandwiched between domains with constant magnetic fields.

1. Introduction

Over the last thirty years or so, the study of solitonic solutions to classical field theories has yielded many interesting results of wide relevance to particle physics, cosmology and condensed matter physics. The more recent fascination with brane-world models of particle physics and cosmology has added new motivation for these kinds of investigations. In this Letter we will study a simple model of two complex scalar or Higgs fields ϕ_1 and ϕ_2 coupling to two different $U(1)$ gauge fields $A_{1\mu}$ and $A_{2\mu}$, with the added feature of an exact discrete $\mathbb{Z}_2$ symmetry under the interchange $1 \leftrightarrow 2$. We will derive solutions to the coupled classical field equations that exhibit a kink or domain wall form for the scalar fields. The nature of the gauge field configurations self-consistently coupled to the Higgs kinks will be our primary object of study. A similar model, without the discrete exchange symmetry was studied some-time ago by Witten [1] in the context of a superconducting string solution. The model was investigated in more detail by MacKenzie [2] to show that while a symmetry is preserved in the vacuum, unexpected topological structures can arise in the interior of a domain wall. More recently, Lemperiere and Shellard [3] have reported on the behavior and stability of the superconducting currents in Witten's model.

Our own motivation for this rather abstract investigation lies with the symmetry breaking mechanism proposed in Ref. [4] in the context of brane-world models and dubbed as the "clash of symmetries". Briefly, Ref. [4] examines a toy model with Higgs fields in three triplet representations of a *global* $SU(3)$ symmetry, where a discrete permutation symmetry between the triplets is enforced. Omitting inessential complications, the vacuum states of the theory spontaneously break $SU(3)$ down to $SU(2)$, as well as spontaneously breaking the discrete symmetry. Kink solu-

E-mail addresses: rozowsky@phy.syr.edu (J.S. Rozowsky), r.volkas@physics.unimelb.edu.au (R.R. Volkas), wali@phy.syr.edu (K.C. Wali).

tions are derived that interpolate between vacua invariant under *differently embedded* $SU(2)$ subgroups.[1] For instance, one can have I-spin asymptotically preserved on one side of a domain wall, with V-spin on the other. Although the unbroken subgroups on both sides are isomorphic, the different embeddings within the parent group cause additional symmetry breakdown at all non-asymptotic points. This additional symmetry breaking is the "clash". The idea is that some of the symmetry breaking we see in our universe might be due to such a clash, if our world is indeed a brane in a higher-dimensional space.

This idea is still at the developmental stage; no realistic brane-world model building using the clash mechanism has yet been attempted, to our knowledge, though Ref. [6] reports on some recent progress. In the course of thinking about the clash of symmetries idea, however, an even simpler model field theory with $U(1)$ factors and interchange symmetries between the different sectors naturally presented itself as a useful theoretical laboratory. The model studied in this Letter arose in exactly this way, though, of course, it is also entitled to an independent existence as a simple-but-not-too-simple vehicle for the study of gauge fields coupled to domain wall Higgs configurations. From this perspective, our work is relevant to general studies of superconducting topological solitons, as in Refs. [1–3,7,8] for example. From the clash of symmetries perspective, the present exercise begins the study of the breakdown of *local* continuous symmetries.

The rest of this Letter is structured as follows: in Section 2, the model and the field equations are presented. The numerical study of kink solutions to these equations is then presented in Section 3, while Section 4 provides a physical explanation for the solutions. Section 5 contains some concluding remarks.

2. The model

Using the notation of [4] we start with the action for two complex scalar fields $\phi_{1,2}$ coupled to different $U(1)$ gauge fields $A_{1,2}$. To the overall $U(1)\times U(1)$ gauge symmetry we add a $\mathbb{Z}_2$ discrete symmetry which interchanges the scalars, $\phi_1 \leftrightarrow \phi_2$ and the gauge fields, $A_1 \leftrightarrow A_2$. The discrete symmetry makes the two gauge coupling constants equal in magnitude. The Lagrangian is

$$\mathcal{L} = -\frac{1}{4}F_1^{\mu\nu}F_{1\mu\nu} - \frac{1}{4}F_2^{\mu\nu}F_{2\mu\nu} + \left(D_1^{\mu}\phi_1\right)^*\left(D_{1\mu}\phi_1\right) + \left(D_2^{\mu}\phi_2\right)^*\left(D_{2\mu}\phi_2\right) - V(\phi_1,\phi_2), \quad (2.1)$$

where

$$V(\phi_1,\phi_2) = \lambda_1\left(\phi_1^*\phi_1 + \phi_2^*\phi_2 - v^2\right)^2 + \lambda_2\phi_1^*\phi_1\phi_2^*\phi_2. \quad (2.2)$$

The covariant derivatives in the Lagrangian are given by

$$D_{1\mu} = \partial_\mu - ieA_{1\mu}, \qquad D_{2\mu} = \partial_\mu - ieA_{2\mu}. \quad (2.3)$$

The Higgs potential admits two vacuum solutions:

$$\text{Vacuum 1:}\quad \langle\phi_1^*\phi_1\rangle = v^2, \quad \langle\phi_2^*\phi_2\rangle = 0, \quad (2.4)$$

$$\text{Vacuum 2:}\quad \langle\phi_1^*\phi_1\rangle = 0, \quad \langle\phi_2^*\phi_2\rangle = v^2. \quad (2.5)$$

These two vacua are degenerate and are the global minima of the potential for the parameter regime

$$\lambda_1 \geqslant 0 \quad\text{and}\quad \lambda_2 \geqslant 0. \quad (2.6)$$

We would like to construct domain wall solutions by requiring the scalar Higgs fields to asymptote to different respective vacua on either side of the wall. We will be interested in the behavior of the corresponding gauge fields for this kind of Higgs configuration. The boundary conditions for the scalars are

$$|\phi_1(z)| = \begin{cases} 0, & z\to-\infty, \\ v, & z\to\infty, \end{cases}$$

$$\text{and}\quad |\phi_2(z)| = \begin{cases} v, & z\to-\infty, \\ 0, & z\to\infty, \end{cases} \quad (2.7)$$

where z is the direction perpendicular to the domain wall.

It is straightforward to compute the equations of motion for the Higgs fields

$$D_{a\mu}D_a^{\mu}\phi_a = -\frac{\partial V}{\partial\phi_a^*} = -2\lambda_1\phi_a\left(\phi_a^*\phi_a + \phi_b^*\phi_b - v^2\right) - \lambda_2\phi_a\phi_b^*\phi_b, \quad (2.8)$$

[1] Qualitatively similar solutions, but to a different theory with a different motivation were discovered by Pogosian and Vachaspati in Ref. [5].

where a, b are either 1, 2 or 2, 1, respectively. The equations of motion for the gauge fields are similarly given by

$$\partial_\mu F_a^{\mu\nu} = 2e\,\mathrm{Im}\left[\phi_a^*\left(\partial^\nu - ieA_a^\nu\right)\phi_a\right]. \tag{2.9}$$

Since we are going to be looking for static domain wall solutions (i.e., static 1 + 1 solitons), we search for solutions that depend on z but are independent of all the other spatial coordinates and time t. In order to simplify our equations we make use of the temporal gauge, $A_0 = 0$. With these choices the equations of motion reduce to

$$A_{1z} = \frac{\alpha_1'}{e}, \tag{2.10}$$

$$A''_{1x,y} = 2e^2 A_{1x,y} R_1^2, \tag{2.11}$$

$$R_1'' = e^2\left(A_{1x}^2 + A_{1y}^2\right)R_1 + 2\lambda_1 R_1\left(R_1^2 + R_2^2 - \upsilon^2\right) + \lambda_2 R_1 R_2^2, \tag{2.12}$$

where prime denotes a derivative with respect to z and $\phi_a \equiv R_a(z)e^{i\alpha_a(z)}$. The corresponding equations for the fields with subscript 2 can be obtained simply by exchanging subscripts 1 and 2. We see in Eq. (2.10) that the z components of both gauge fields are pure gauge and because neither $A_z(z)$ nor $\alpha(z)$ couple to the physical degrees of the system, they can be neglected.

The coupled differential equations for this system nominally involves six degrees of freedom (one scalar and two gauge degrees of freedom for each field). However, since the x and y components of each gauge field enter quadratically into their respective Higgs field equations of motion, it is possible to rotate to a new basis $\tilde{x}$ and $\tilde{y}$ where one only needs keep track of one component of each gauge field. Note that the directions perpendicular to z in which each of the gauge fields A_1 and A_2 point are independent. We therefore have only four degrees of freedom to non-trivially solve for.

The equations we would like to solve are then

$$A_1'' = 2e^2 R_1^2 A_1, \tag{2.13}$$

$$R_1'' = e^2 A_1^2 R_1 + 2\lambda_1 R_1\left(R_1^2 + R_2^2 - \upsilon^2\right) + \lambda_2 R_1 R_2^2, \tag{2.14}$$

and $1 \leftrightarrow 2$. We have suppressed the spatial subscripts on the gauge fields, A.

For a domain wall solution the scalar fields must obey the boundary conditions in Eq. (2.7). Thus, by analyzing Eq. (2.13) we see that the gauge fields are required to have the following asymptotic behavior:

$$A_1(z \to \infty) = e^{-\sqrt{2}e\upsilon|z|} \to 0$$
$$\text{and} \quad A_2(z \to -\infty) = e^{-\sqrt{2}e\upsilon|z|} \to 0. \tag{2.15}$$

We observe that this asymptotic behavior is also consistent with Eq. (2.14). The values of $A_1(-\infty)$ and $A_2(\infty)$ are seemingly unconstrained by any of our differential equations. However, note that when $z \ll -1$ for $A_1(z)$ or when $z \gg 1$ for $A_2(z)$ the solutions become linear functions of z, the asymptotic solutions to Eq. (2.13). The linear solutions are due to the requirement that $R_1(z)$ and $R_2(z)$ vanish as $z \to -\infty, +\infty$, respectively (this is because we require them to be kink solutions). Thus, the only allowed values of $A_1(-\infty)$ and $A_2(\infty)$ are either a *constant* (corresponding to constant asymptotic behaviour) or $\pm\infty$. Consistent with this, we will also impose the boundary conditions

$$A_1'(z = -\infty) = \text{const} \neq 0$$
$$\text{and} \quad A_2'(z = +\infty) = \text{const} \neq 0. \tag{2.16}$$

The requirement that these slopes be asymptotically non-zero removes the $A_1 = A_2 = 0$ solution from our considerations. Eq. (2.16) allows the constant slopes for A_1 and A_2 to be arbitrary. If they are chosen to be unequal, it implies that the corresponding magnetic fields B_1 and B_2 are unequal, leading to a violation of the symmetry inherent in the problem and this may also cause dynamical instability of the brane as will be discussed further in Section 4. Hence, it is natural to choose the slopes to be equal. However, our numerical solutions (see Fig. 3) show that even in the asymmetrical situation, slopes of A_1 and A_2 are very nearly equal.

The coupled differential equations (2.13) and (2.14) together with the conditions of Eqs. (2.15) and (2.16) constitute our boundary value problem (BVP).

Since we shall resort to numerics to find solutions it is convenient to transform from coordinate z to u which is defined on a compact interval, $u \in [-1, 1]$, via

$$u = \tanh\left(\upsilon\sqrt{\lambda_1}\, z\right). \tag{2.17}$$

With this change of coordinates and the field rescalings

$$R_a \to v R_a, \qquad A_a \to v A_a, \tag{2.18}$$

the equations become

$$(1-u^2)^2 \frac{d^2 A_1}{du^2} - 2u(1-u^2)\frac{dA_1}{du} = 2\alpha R_1^2 A_1, \tag{2.19}$$

$$(1-u^2)^2 \frac{d^2 R_1}{du^2} - 2u(1-u^2)\frac{dR_1}{du} = \alpha A_1^2 R_1 + 2R_1(R_1^2+R_2^2-1) + \lambda R_1 R_2^2, \tag{2.20}$$

and $1 \leftrightarrow 2$. We have defined $\alpha \equiv e^2/\lambda_1$ and $\lambda \equiv \lambda_2/\lambda_1$. We see that solutions only depend on two independent coupling constants and not three. In the case of the pure Higgs model with $\alpha = 0$ (see Ref. [4]), if one takes symmetric $(R_1 + R_2)$ and anti-symmetric $(R_1 - R_2)$ linear combinations of the fields, then the differential equations decouple for the special case of $\lambda = 4$ with analytic solutions,

$$R_1 = \frac{1}{2}(1+u), \qquad R_2 = \frac{1}{2}(1-u). \tag{2.21}$$

However, this is not the case in our model for $\alpha \neq 0$.

We shall also be interested in the energy of the solutions we find, thus we need the stress energy for this system

$$T_{\mu\nu} = 2\frac{\delta \mathcal{L}}{\delta g^{\mu\nu}} - g_{\mu\nu}\mathcal{L}, \tag{2.22}$$

which for our action yields

$$\begin{aligned} T_{\mu\nu} = {} & -F_{1\mu\alpha}F_{1\nu}^{\alpha} - F_{2\mu\alpha}F_{2\nu}^{\alpha} + 2(D_{1\mu}\phi_1)^*(D_{1\nu}\phi_1) \\ & + 2(D_{2\mu}\phi_2)^*(D_{2\nu}\phi_2) \\ & + g_{\mu\nu}\bigg[\frac{1}{4}F_1^{\mu\nu}F_{1\mu\nu} + \frac{1}{4}F_2^{\mu\nu}F_{2\mu\nu} \\ & - (D_1^{\mu}\phi_1)^*(D_{1\mu}\phi_1) \\ & - (D_2^{\mu}\phi_2)^*(D_{2\mu}\phi_2) + V(\phi_1,\phi_2)\bigg]. \end{aligned} \tag{2.23}$$

The energy density is then given by the T_{00} component of the stress-energy tensor. This simplifies to

$$\begin{aligned} T_{00} = {} & \frac{1}{4}\Big[\big(A_1'(z)\big)^2 + \big(A_2'(z)\big)^2\Big] + \big(R_1'(z)\big)^2 \\ & + \big(R_2'(z)\big)^2 + e^2 A_1(z)^2 R_1(z)^2 \\ & + e^2 A_2(z)^2 R_2(z)^2 + V(R_1,R_2), \end{aligned} \tag{2.24}$$

for our static solutions and because of our gauge choice, $A_0 = 0$. Thus, in terms of the coordinate u and the rescaled fields the energy density is given by

$$\begin{aligned} \frac{T_{00}}{\lambda_1 v^4} = {} & (1-u^2)^2\bigg[\frac{(\partial_u A_1(u))^2}{4} + \frac{(\partial_u A_2(u))^2}{4} \\ & + \big(\partial_u R_1(u)\big)^2 + \big(\partial_u R_2(u)\big)^2\bigg] \\ & + \alpha A_1(u)^2 R_1(u)^2 + \alpha A_2(u)^2 R_2(u)^2 \\ & + \big(R_1(u)^2 + R_2(u)^2 - 1\big)^2 \\ & + \lambda R_1(u)^2 R_2(u)^2, \end{aligned} \tag{2.25}$$

where $\lambda_1 v^4$ sets the scale.

3. Numerical solutions

The numerical method we employ to solve these coupled differential equations is the "shooting method" using the routines from Numerical Recipes in C++ [9]. One can readily convert our system of four coupled second order differential equations to a system of eight coupled first order differential equations where the functions are: R_1, R_2, A_1, A_2, R_1', R_2', A_1' and A_2'. This is a boundary value problem with the functions R_1, R_2, A_1, A_2 specified on two boundaries but with the functions R_1', R_2', A_1', A_2' not specified on either boundary. The way the "shooting method" works is that one guesses values for the derivative functions at the left boundary ($u = -1$), then with all the functions specified on the left boundary one can numerically integrate to the right boundary. One then defines a function which measures how well the boundary conditions on the right are matched. Using this goodness of fit function one can then use a Newton–Raphson procedure to improve the guess on the left boundary for the derivatives. One can then iterate this procedure until the boundary conditions on both sides are satisfied to the desired accuracy. One potential difficulty is that if the differential equations are reasonably complicated (e.g., non-linear) then the *initial guess* might need to be reasonably good in order for the procedure to converge.

The differential equations, (2.19) and (2.20), have poles at $u = \pm 1$ when one expresses the equations as $dX/du = (1-u^2)^{-2} \times \cdots$. Since we cannot evaluate these equations at $u = \pm 1$, we set the boundaries at $u_1 = -1 + \epsilon$ and $u_2 = 1 - \epsilon$. However, because now our boundaries are not at $u = \pm 1$ ($z = \pm\infty$) we need to know the asymptotic behavior of our functions in order to set up the boundary conditions correctly.[2] For the special case of $\alpha = 0$ and $\lambda = 4$ the analytic solution, Eq. (2.21), is known from Ref. [4]. While these are not the correct solutions for general α and λ, they do exhibit the correct asymptotic behavior as $u \to \pm 1$. But as long as ϵ is sufficiently small the correct asymptotic behavior is obtained numerically. When we solve our boundary value problem numerically we shall use Eq. (2.21) to set the boundary conditions for R_1 and R_2. We also need to know the asymptotic behavior of the gauge fields near the boundaries. Substituting $A = (1 - u^2)^\beta$ into the differential equation for A (Eq. (2.19)), we can solve for β, the scaling behavior in the vicinity of the boundary. Thus

$$A_1 \sim (1-u)^{\sqrt{\alpha/2}} \sim \epsilon^{\sqrt{\alpha/2}} \quad \text{as } u \to u_2 = 1 - \epsilon, \tag{3.1}$$

$$A_2 \sim (1+u)^{\sqrt{\alpha/2}} \sim \epsilon^{\sqrt{\alpha/2}} \quad \text{as } u \to u_1 = -1 + \epsilon. \tag{3.2}$$

The values of $A_1(u_1)$ and $A_2(u_2)$ are not constrained by any of the differential equations and are therefore left as free parameters.

As mentioned before when solving a boundary value problem using the "shooting method", convergence may depend on a reasonably accurate guess of the initial conditions on the left boundary. This is the case for our set of differential equations since they have an explicit pole at $u = \pm 1$. This sensitivity gets worse as ϵ approaches zero. The method we employed to address this issue involved starting with a relatively large value of ϵ ($\epsilon = 0.5$) and incrementally reducing it to its desired value using as the initial guess for the values of the derivatives (R_1', R_2', A_1' and A_2') on the left boundary for each step the solution of the previous step.

In Figs. 1, 2 and 3 we see numerical solutions to these differential equations for a variety of couplings, α, λ and boundary conditions $A_1(u_1)$ and $A_2(u_2)$. We observe that the gauge fields A_1 and A_2 become linear functions of $\tanh^{-1}(u)$ as $u \to u_1$ and $u \to u_2$, respectively. This implies that asymptotically these gauge fields become linear functions of z, which corresponds to a constant magnetic field in the direction perpendicular to both z and $\tilde{x}$ (the direction in which the gauge field points),

$$B_{\tilde{y}} \sim \partial_z A_{\tilde{x}}(z) = \text{const.} \tag{3.3}$$

Thus the asymptotic solution (actually $\tanh(u)^{-1}$ need only be of the order of ± 2 to be in the asymptotic regime for a typical configuration) on either side of the domain wall is a constant magnetic field corresponding to the $U(1)$ fields, which point in uncorrelated directions parallel to the domain wall.

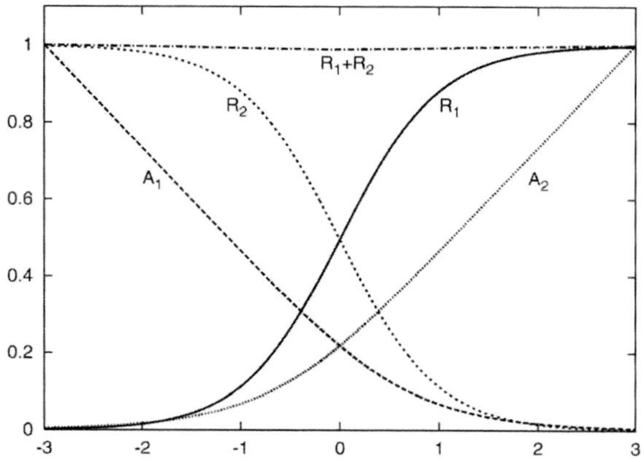

Fig. 1. Plot of R_1, A_1, R_2, A_2 and $R_1 + R_2$ against $\tanh^{-1}(u)$ for $\alpha = 1$, $\lambda = 4$. The free boundary conditions are $A_1(-1+\epsilon) = A_2(1-\epsilon) = 1$ for $\epsilon = 0.005$ which corresponds to left and right boundaries at $\tanh^{-1}(u) = \pm 3$. $R_1 + R_2$ is nearly constant for this pair of parameters.

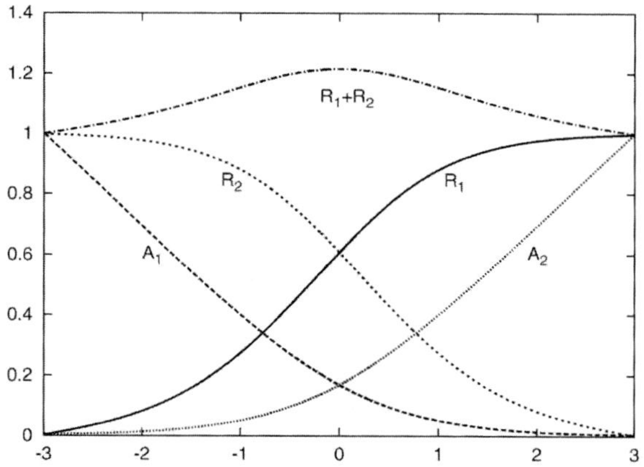

Fig. 2. Plot of R_1, A_1, R_2, A_2 and $R_1 + R_2$ against $\tanh^{-1}(u)$ for $\alpha = 1$, $\lambda = 1$. Here $A_1(-1+\epsilon) = A_2(1-\epsilon) = 1$ for $\epsilon = 0.005$.

[2] For numerical reasons we cannot just set $R_1(u_1) = 0$, $R_1(u_2) = 1, \ldots$.

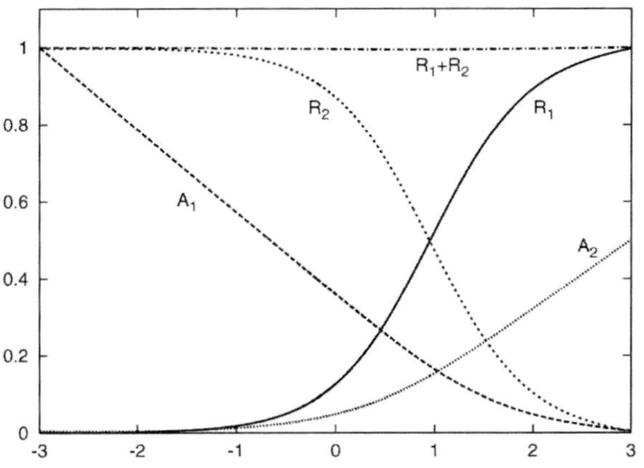

Fig. 3. Plot of R_1, A_1, R_2, A_2 and $R_1 + R_2$ against $\tanh^{-1}(u)$ for $\alpha = 1$, $\lambda = 4$. Here $A_1(-1+\epsilon) = 1$ and $A_2(1-\epsilon) = 0.5$ for $\epsilon = 0.005$. The principal effect of the asymmetrical BC's is to shift the center of the brane to the right.

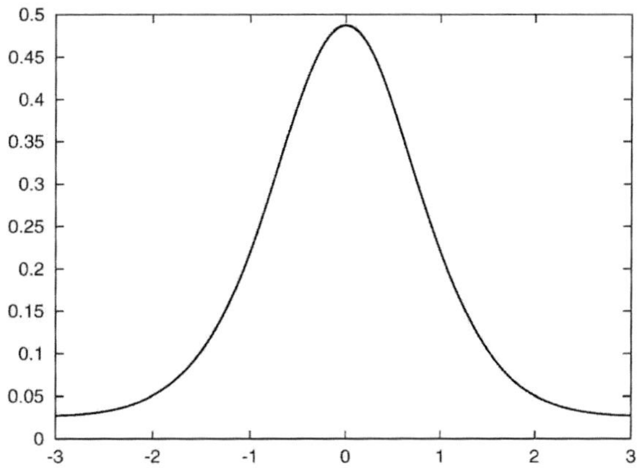

Fig. 4. Plot of the energy density against $\tanh^{-1}(u)$ for $\alpha = 1$, $\lambda = 1$. We have used the boundary conditions $A_1(-1+\epsilon) = A_2(1-\epsilon) = 1$ where $\epsilon = 0.005$.

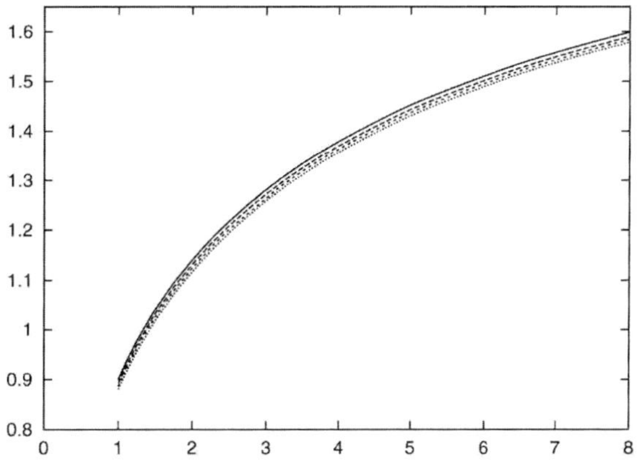

Fig. 5. Plot of the "renormalized" surface energy density against λ for $\alpha = 0.25, 0.5, 1.0, 2.0$ (from top to bottom). We have used the boundary conditions $A_1(-1+\epsilon) = A_2(1-\epsilon) = 1$ where $\epsilon = 0.001$.

These solutions have non-zero energy density away from the domain wall and thus are infinite energy configurations. The solutions where the magnetic fields are both zero corresponds to the choice of $\alpha = 0$ (i.e., no $U(1)$ gauge fields).

In Figs. 1 and 2, we have set $A_1(u_1 = -0.995) = A_2(u_2 = 0.995) = 1$. With this set of symmetric boundary conditions the domain wall is centered at $u = 0$. In Fig. 3 we see that the effect of asymmetric BC's is to shift the location of the domain wall. While not apparent in the figure the magnitudes of the uniform magnetic field far from either side of the domain wall do not exactly match. The choice of $\epsilon = 0.005$ (and $\epsilon = 0.001$ for Figs. 5 and 6) corresponds to boundaries at $\tanh(u)^{-1} = \pm 3$ (and ± 3.8). While ϵ can be made smaller at the expense of longer computing time, these values are sufficiently small for our purposes.

In Fig. 4 we see the energy density of a solution plotted as a function of the transverse direction. We see that the energy density is peaked at the center of the domain wall. If we treat the asymptotic constant magnetic field on either side of the domain wall as a background, then we can compute the energy per unit surface area of the domain wall by subtracting off the infinite energy associated with the magnetic field. In Fig. 5 the surface energy density is plotted as a function of λ for a variety of values of α (boundary conditions are $A_1(-1+\epsilon) = A_2(1-\epsilon) = 1$ at $\epsilon = 0.001$). Observe that this "renormalized" surface energy density is only weakly dependent on the value of the gauge coupling constant α. In Fig. 6 we show the subtracted energy density corresponding to the constant magnetic field as a function of λ and α. In both Figs. 5 and 6 we omit values of $\lambda < 1$ as they require a significantly smaller value for ϵ.

Our solutions are all plotted in units of $\tanh(u)^{-1}$ and not z since u is the natural variable in our system of equations, (2.19) and (2.20). The length scale $\tanh(u)^{-1}$ is dimensionless and can be converted into a physical length by dividing by $\upsilon\sqrt{\lambda_1}$. The thickness of the domain wall is typically $\sim 4/\upsilon\sqrt{\lambda_1}$

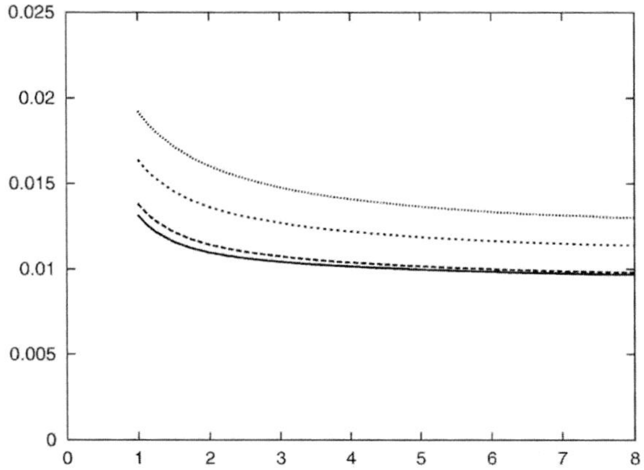

Fig. 6. Plot of the energy density of the uniform magnetic field against λ for $\alpha = 0.25, 0.5, 1.0, 2.0$ (from bottom to top). We have used the boundary conditions $A_1(-1+\epsilon) = A_2(1-\epsilon) = 1$ where $\epsilon = 0.001$.

(see Fig. 4) which can be made arbitrarily small by choosing $v\sqrt{\lambda_1} \gg 1$.

4. Discussion

The numerical solutions displayed above have a natural interpretation in terms of superconductivity. Consider, for instance, the currents associated with the $U(1)$ gauge groups,

$$J_{i\,\mu} = ie\left[\phi_i^*(\partial_\mu \phi_i) - (\partial_\mu \phi_i^*)\phi_i\right] + 2e^2 A_{i\,\mu}\phi_i^*\phi_i, \tag{4.1}$$

where $i = 1, 2$. In terms of the amplitude and phase of ϕ_i, the currents are given by

$$J_{i\,\mu} = -2eR_i^2\partial_\mu\alpha_i + 2e^2 A_{i\,\mu}R_i^2. \tag{4.2}$$

For our configurations, which depend only on z, and for which Eq. (2.10) holds, it is clear that only the x- and y-components are non-vanishing. They evaluate to

$$J_{i\,x,y}(z) = 2e^2 A_{i\,x,y}(z)R_i^2(z). \tag{4.3}$$

These steady, z-dependent current densities are uniform supercurrent densities localized to the domain wall, with the charged boson fields as the current carriers.

Eq. (4.3) shows that the currents are non-zero only when the gauge field configurations are non-zero and vice versa, so these currents are responsible for dynamically generating the magnetic fields. On the side of the wall where $R_i \neq 0$, the corresponding magnetic field is seen to decay exponentially, which is simply a Meissner effect. On the other side of the wall, where R_i is tending exponentially quickly to zero, we find the magnetic field $\vec{B}_i$ tending towards a finite, uniform configuration pointing in the plane of the wall. This is consistent with the domain wall carrying a uniform sheet of current density pointing in the $(0, A_{i\,x}, A_{i\,y}, 0)$ direction, as per Eq. (4.3). Our configurations have infinite energy because the domain wall is of infinite extent, with current densities uniformly distributed on it.

The stability or otherwise of our solutions is an important concern. While a complete stability analysis is beyond the scope of this Letter, the above considerations suggest that the geometrically symmetric solutions such as in Figs. 1 and 2 could be stable, whereas asymmetric configurations such as those of Fig. 3 are not. Let current J_1 point in the x-direction in the plane of the wall. Then Eq. (4.3) implies that A_1 also points in the same direction, so $\vec{B}_1$ is directed along the y-axis. The Lorentz force on the type 1 charge carriers lies in the negative z direction. For sector 2, similar reasoning shows that the corresponding Lorentz force on type 2 charge carriers points in the positive z direction. For symmetric boundary conditions, these forces are equal in magnitude as well as opposite in direction. This is a necessary condition for stability. For asymmetric boundary conditions, they are unequal, strongly suggesting that such configurations are unstable.

5. Conclusions

In order to further explore the idea of the "clash of symmetries" from [4], we have considered a model in which two scalar fields are coupled to their respective gauge fields in a Lagrangian which has $U(1) \times U(1)$ symmetry. We find consistent static solutions for field configurations with the vacuum conditions for the scalar fields specified by Eq. (2.7) and the implied boundary conditions for the gauge fields, Eq. (2.15). We obtain the expected kink-like solutions for the scalar fields while the two gauge fields diverge linearly on either side of the domain wall.

When we consider the idealized configuration of an infinitely thin domain wall, we have solutions such that while the $U(1)$ symmetries of the fields are preserved in their respective vacua, they are both broken on the domain wall. The gauge fields show that the domain wall is sandwiched between domains with constant magnetic fields parallel to the wall. In the case of a domain wall of finite thickness, there will be magnetic fields parallel to the wall on either side. These are associated with superconducting currents, as in the case of the superconducting string solution [1].

This model demonstrates that in addition to the breakdown of symmetries on the brane, the presence of gauge fields introduces new phenomena, such as the appearance of magnetic fields. Background magnetic fields of this kind are reminiscent of the configurations in string theory that give rise to non-commutativity of space–time coordinates. It would be very interesting to see the logical extension of this model to domain wall solutions with non-Abelian gauge fields and to study their dynamical effects in addition to symmetry breaking.

Acknowledgements

This work was supported in part (J.S.R., K.C.W.) by the US Department of Energy (DOE) under contract No. DE-FG02-85ER40237. R.R.V. was supported partially by the Australian Research Council and partially by the University of Melbourne. He thanks his coauthors for their excellent hospitality at Syracuse University where this work was begun.

References

[1] E. Witten, Nucl. Phys. B 249 (1985) 557.
[2] R. MacKenzie, Nucl. Phys. B 303 (1988) 149.
[3] Y. Lemperiere, E.P. Shellard, Nucl. Phys. B 649 (2003) 511, hep-ph/0207199.
[4] A. Davidson, B.F. Toner, R.R. Volkas, K.C. Wali, Phys. Rev. D 65 (2002) 125013, hep-th/0202042.
[5] L. Pogosian, T. Vachaspati, Phys. Rev. D 62 (2000) 123506, hep-ph/0007045;
See also: T. Vachaspati, Phys. Rev. D 63 (2001) 105010, hep-th/0102047;
L. Pogosian, T. Vachaspati, Phys. Rev. D 64 (2001) 105023, hep-th/0105128.
[6] E. Shin, R.R. Volkas, hep-ph/0309008, Phys. Rev. D, in press.
[7] J.R. Morris, Phys. Rev. D 52 (1995) 1096.
[8] G. Lazarides, Q. Shafi, Phys. Lett. B 159 (1985) 261.
[9] B.P. Flannery, W.H. Press, S.A. Teukolsky, W.T. Vetterling, Numerical Recipes in C++, second ed., Cambridge Univ. Press, New York, 2002.

Clash of Symmetries in a Brane World Picture

Kameshwar C. Wali
Physics Department, Syracuse University, Syracuse, New York

Presented at the IV International Symposium, Quantum Theory and Symmetries,
15-21 August 2005, Varna, Bulgaria

Abstract

If our (3+1) dimensional universe is a brane or domain wall embedded in a higher dimensional space, then a phenomenon that may be designated as "Clash of Symmetries" provides a new method of breaking continuous symmetries. The paper presents some non-trivial models containing the physical ideas.

1 Introductory remarks

Symmetry, as wide or as narrow as you may define its meaning, is one idea by which man through the ages has tried to comprehend and create order, beauty, and perfection

Herman Weyl in Symmetry

We may hope that the remaining two decades of our century, which began under the sign of symmetry, physicists will be able to explain how great variety of non-symmetrical forms of the real world can arise out of the beautiful,symmetrical structure of the basic equations. At the beginning of the century, Einstein taught us to understand the meaning of symmetry; we have now to learn how to break it in the most symmetrical way so that his most cherished dream, that of building a unified theory of all interactions, may one day come true.

Luigi Radicati di Brozol

These two statements express two profound and basic aspects of physics today. Weyl, in his beautiful and elegant book, described as his swan song, elaborates on the role of symmetries in Nature, in art and architecture, and then shows how they are transformed in mathematical forms to express laws of physics. Radicatti, on the other hand, stresses the equally important fact of Nature, namely, the real world does not exist in the perceived perfect symmetric form, but more naturally in its "approximate" or"broken" forms. While we can build beautiful models of elementary particles and their interactions based on exact symmetries, real world defies them and remains provocative. Whether it be the Standard Model and beyond, SUSY, String theory or any other theory, symmetry breaking poses as a fundamental problem.

In what follows, I will briefly and qualitatively describe a new approach to symmetry breaking that has been the subject of study during the past few years [1]. What I am going to describe are some non-trivial, but still in the nature of toy models. We, think, however, they do indicate a new and fruitful approach to realistic models

2 Clash of symmetries; basic idea

The conventional spontaneous symmetry breaking mechanism (SSB)in the standard model(SM) and its extensions assumes spatially homogeneous Higgs fields whose vacuum expectation values are determined by the minimization of a postulated Higgs potential. The non-vanishing constant values determine the masses of the gauge and matter fields through their couplings to the Higgs fields. The considerable arbitrariness involved in the choice of the potentials and the profusion of free parameters make this scheme unsatisfactory.

The non-commutative geometric framework of Alain Connes has provided a new and elegant scheme that elevates SSB to a new level of its understanding. This is accomplished by placing Higgs fields and gauge fields on a similar geometrical footing. This makes the models more predictive. However, this framework beyond its application to SM, requires departures from from the original rigorous approach and poses some problems that I cannot go into in this brief report.

There are, however, other types of solutions to Higgs fields that can serve as stable, static background fields: topological solitons, such as kinks, strings and monoploes. But their spatially non-homogeneous nature forbids them to be used as background fields in our 3+1 dimensional space-time universe, since they conflict with the strong evidence of large sale homogeneity. This objection, however, does not apply to brane-world models, since the non-trivial spatial dependence of the Higgs fields can be restricted to extra dimension co-ordinates only.

Consider, for instance, an extra dimension, coordinate w and topologically stable Higgs field configurations $\phi_i(w)$, some of which have kink form with respect to w. Then the pattern of symmetry breaking becomes a function of w. Suppose that the 3+1 dimensional brane world is located at $w = 0$ with a set of localized physical fields confined to the brane at $w = 0$. If the fields are strictly confined to $w = 0$, then the unbroken symmetry is the stability group of $\phi_i(w = 0)$, say a subgroup $H(w = 0)$ of some internal symmetry group G. However, in quantal(or perhaps even in classical world), one would not expect the fields absolutely confined, in which case they couple , perhaps with reduced strength, to $\phi_i(w)$ states, where w is different from $H(w = 0)$.If the stability group $\phi_i(0 < |w| < \epsilon), h(w \sim \epsilon)$ is different from $h(w = 0)$, then a rich effective symmetry breaking is possible on the brane

This happens when isomorphic groups $H(|w| = \infty)$ left unbroken at $|w| = \infty$ *can be differently embedded in the parent group G.* The break down at finite w is the intersection of the asymptotic stability groups

$$H(|w| < \infty) = H(-\infty) \cap H(+\infty) \equiv H_{clash}$$

Although $H(+\infty)$ and $H(-\infty)$ are isomorphic, their different embeddings leads to their intersection that is a smaller group.

Thus in general, if the vacuum manifold is disconnected and contains distinct vacuum states, the field can settle into different minimizing vacuum states in different spatial dimensions. Stable solitonic or kink-like domain wall configurations can exist if the vacuum manifold has the appropriate topology.

If our (3+1) dimensional universe is a brane or a domain wall, such a situation can arise, namely, a global $G_{cts} \otimes G_{discrete}$ is spontaneously broken to $H_{cts} \otimes H_{discrete}$, where H_{cts} can be embedded in several different ways in the parent $G_{cts}$and $H_{discrete} \subset G_{discrete}$.

As a consequence, a certain class of domain wall solutions connects two vacua that are invariant under differently embedded H_{cts} subgroups and there is an enhanced symmetry breakdown to the intersection of the two subgroups on the brane or domain wall. In the brane limit, H_{cts} prevails in the bulk, but smaller intersection symmetry on the brane itself.

One may call this phenomenon *CLASH OF SYMMETRIES.*

3 A model with three higgs triplets

Consider a model with three Higgs triplets $\phi_{1,2,3}$ interacting through the potential

$$V = -\sum_{i=1}^{3}(\phi_i^\dagger\phi_i) + (\sum_{i=1}^{3}\phi_i^\dagger\phi_i)^2 + V_4$$

$$V_4 = \lambda/2\sum_{i\neq j}(\phi_i^\dagger\phi_i\phi_j^\dagger\phi_j) + \sigma/2\sum_{i\neq j}(\phi_i^\dagger\phi_j\phi_j^\dagger\phi_i)$$

The symmetry group of the potential is $G = G_{cts} \otimes G_{discrete}$, where $G_{cts} = SU(3) \otimes U(1)_1 \otimes U(1)_2 \otimes U(1)_3$ and $G_{discrete} = S_3$.

Minimization of the potential yields three degenerate vacua, leading to a manifold consisting of three disconnected pieces,

$$vacuumi : (\phi_1^\dagger\phi_1) = 1/2, (\phi_2^\dagger\phi_2) = (\phi_3^\dagger\phi_3) = 0$$

with the other two obtained by permutations.

Each global minimum induces the SSB of both the parent discrete and continuous symmetries to $S_3 \rightarrow S_2$, $G_{cts} \rightarrow U(2) \otimes U(1)^2$.

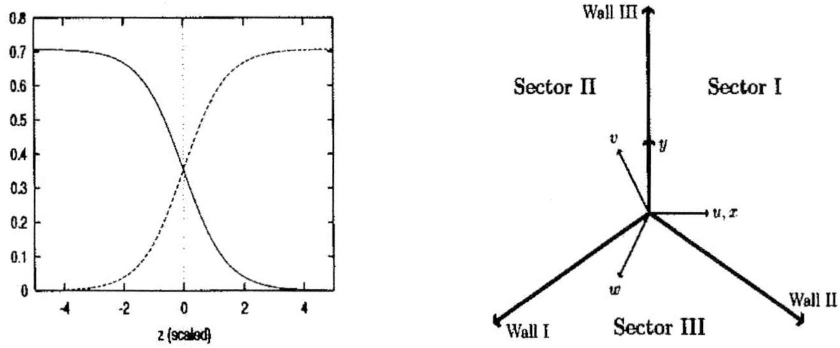

Figure 1: Left: Typical kink-type solution. Right: The three–star domain wall junction configuration.

A kink or one-dimensional domain wall configuration interpolates between the elements of disconnected vacua. A typical solution is shown in Fig. 1 (left panel).

A more interesting domain wall junction configuration is depicted Fig 1 (right panel). Three semi-infinite walls meet at a point, the origin or nexus, at angles of $2\pi/3$. Ignoring the superfluous $U(1)$'s, the clash of symmetries has the pattern:

$$H_{I\cap II} = U(2)_{II} \cap U(2)_{II} = U(1)_{III}$$

along wall III with corresponding results for $H_{II\cap III}$ and $H_{III\cap I}$

Figures of global minima are shown in Figs. 2.

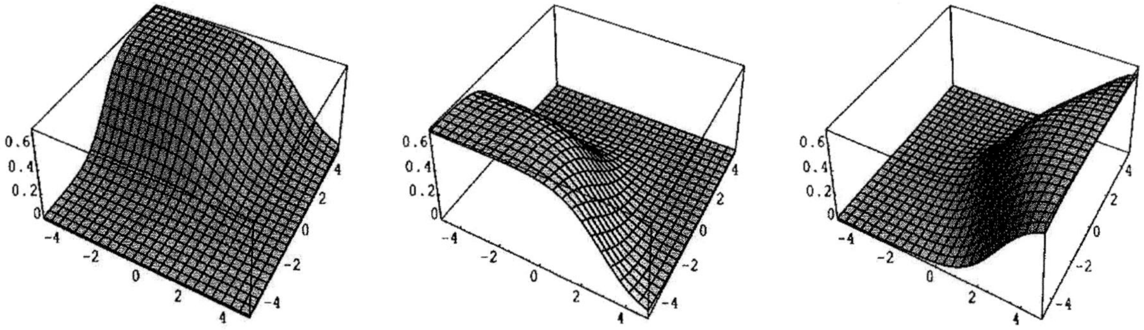

Figure 2: ϕ_1, ϕ_2, ϕ_3 components of the three–star configuration.

4 Domain wall solutions with Abelian gauge fields

The lagrangian consists of two complex scalar fields $\phi_{1.2}$ coupled to Abelian gauge fields $a_{1\mu}, a_{2\mu}$

$$\mathcal{L} = -(1/4)(\sum(F_i^{\mu\nu} F_{i\mu\nu}) + \sum(D^\mu \phi_i^\dagger) D_\mu \phi_i) - V(\phi_1, \phi_2)$$

$$V(\phi_1, \phi_2) = \lambda_1(\phi_1^\dagger \phi_1 + \phi_2^\dagger \phi_2 - v^2)^2 + \lambda_2(\phi_1^\dagger \phi_1 \phi_2^\dagger \phi_2)$$

with the overall continuous symmetries $u(1) \otimes u(1)$ and the discrete symmetry z_2.

There are two distinct vacua. With appropriate boundary conditions for the scalar fields and the implied boundary conditions for the gauge fields, we find numerical solutions for the coupled equations. We obtain expected kink solutions for the scalar fields. The gauge fields diverge linearly on either side, but fall off exponentially on opposite sides (Fig. 3). The $U(1)$symmetries are preserved in their respective vacua but broken elsewhere. The domain wall is sandwiched between domains of constant magnetic fields parallel to the wall. In the case of a domain wall with finite thickness, there will be magnetic fields parallel to the wall on either side associated with superconducting currents, as in the case of the superconducting string solutio

Thus, in addition to symmetry breaking on the brane, we find a new phenomenon such as the appearance of magnetic fields in the bulk.

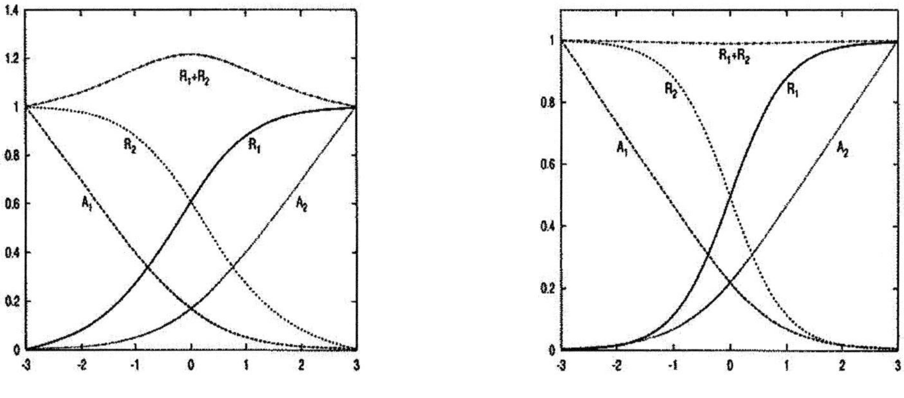

Figure 3: Plots of scalar and gauge fields for different sets of parameters

5 A global $U(1)\otimes U(1)$ model with Randall-Sundrum-like gravity

The starting point is a model with two complex scalar fields in a five dimensional space-time with the action:

$$\mathcal{S} = \int [-\kappa \mathcal{R}/2 \; - \; \mathcal{L}(\phi_1, \phi_2)]\sqrt{-g}\; d^4x \; dw,$$

where

$$\mathcal{L}(\phi_1, \phi_2) = g^{ab} \sum (D_a\phi_i)^*(D_b\phi_i) + V$$

and the metric is given by

$$ds^2 = dw^2 + e^{2f(w)}\eta^{\mu\nu} dx_\mu dx_\nu$$

In addition to the $U(1) \otimes U(1)$ symmetry, there is a discrete symmetry $\phi_1 \leftrightarrow \phi_2$. The field equations that follow are

$$2V(\phi_1, \phi_2) = -3\kappa(f'' + 4(f')^2)$$

$$\sum(\phi_i^{*}\phi_i') = -3\kappa f''$$

$$\phi_i'' + 4f'\phi_i' = (\delta V)/(\delta\phi_i),$$

where 'prime ' denotes differentiation with respect to w.

We seek static solutions of the field equations with w as the variable or alternately we attempt to find a potential that satisfies the equations. With the ansatz

$$\phi_1 = (v/\sqrt{(2)})\sqrt{(1 + \tanh \beta w)}$$

$$\phi_2 = (v/\sqrt{(2)})\sqrt{(1 - \tanh \beta w)}$$

we obtain an analytic solution

$$e^{2f(w)} = (\cosh \beta w)^{-(\beta v^2)/(6\kappa)},$$

which has the Randall-Sundrum limit

$$e^{2f(w)} \to e^{-((v^2\beta)/(6\kappa))|w|}$$

provided,

$$\beta \to \infty, v \to 0, v^2\beta \to finite\ .$$

The sextic potential that satisfies all the field equations and is both bounded from below and has the desired global minima:

$$V = -\frac{\beta^2 v^4}{24\kappa} + \frac{\beta^2}{2v^2}(1 + \frac{v^2}{3\kappa})\phi_1^2\phi_2^2 + U + W,$$

where

$$U = -\frac{\beta^2}{v^2}(\frac{3}{2} + \frac{v^2}{3\kappa})\phi_1^2\phi_2^2(\phi_1^2 + \phi_2^2 - v^2)$$

and

$$W = \zeta\frac{\beta^2}{4v^2}(\frac{3}{2} + \frac{v^2}{3\kappa})(\phi_1^2 + \phi_2^2 - v^2)^2(\eta + \frac{\phi_1^2 + \phi_2^2 - v^2}{v^2}),$$

ζ and η are parameters such that

$$\zeta \geq 1; \ \zeta\eta > 1 \ .$$

The equations are satisfied without W, but then the potential is not bounded from below. We note that there is a negative cosmological constant in the bulk

$$\lambda_5 = -\frac{\beta^2 v^4}{24\kappa}$$

In figure 4, contour plots of the potential for certain values of the parameters ζ and $|eta$ are shown to illustrate the qualitative features of the extrema of the potential.

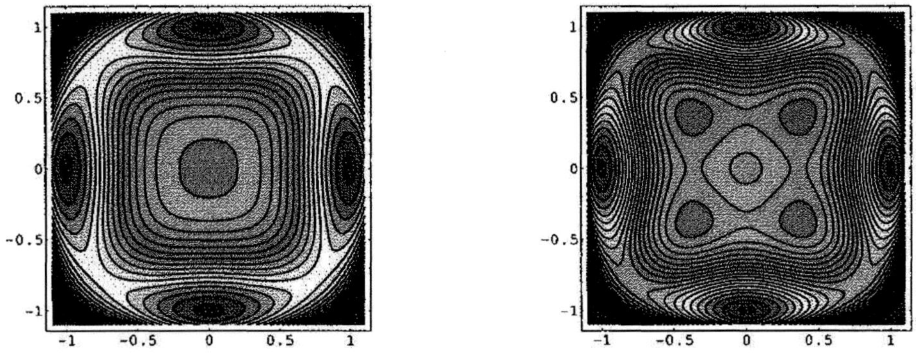

Figure 4: Contours of the qualitative features of the extrema of the potential.

6 Summary and Conclusions

The model, although in the nature of a toy model has many realistic features. It has a solution featuring clash-of-symmetric Higgs kink configurations in a $4+1$ Randall-Sundrum space-time. Gravity is localized to the dynamically generated brane,and the RS limit of the solution is well defined. For the chosen Higgs kink and metric configurations, the potential has to have a certain sextic form, whose properties we have studied in some depth.The symmetry breaking pattern varies as a function of the extra dimension coordinate w, and displays the clash of symmetries phenomenon. At all points $|w| < \infty$, both $U(1)$ are broken with alternate $U(1)'s$ restored as $w \to \pm\infty$. the spontaneous breaking of the discrete symmetry guarantees topological stability for the matter-gravity induced brane.

This work sets the stage for incorporating gravity into more complicated models displaying the clash of symmetry idea. Our eventual aim is to construct a realistic brane-world model displaying clash of of symmetries to induce spontaneous symmetry breaking.

Acknowledgements

I would like to thank Dr. Cosmin Macesanu for helpful discussions. Work supported in part by the U.S.Department of Energy (DOE) Grant No. DE-FG02-85ER40237.

References

[1] The work reported here is based on the following papers: A. Davidson, B.F. Toner, R.R.Volkas and K.C. Wali, Phys.Rev D65,125013; J.S.Rozowsky, R.R.Volkas and K.C.Wali, Physics Letters B 580 (2004)249-256; G. Dando, A. Davidson, D.P. George, R.R.Volkas and K.C. Wali, Phys.Rev D, 72, 045016 (2005) For details and references to related work, please see these papers.

MATTER FIELDS ON A TWO-SHEETED SPACE-TIME*

KAMESHWAR C. WALI

Physics Department, Syracuse University
Syracuse, NY 13244-1130 U.S.A.
email: wali@physics.syr.edu

Abstract

As a prelude to the standard model in curved space time, we present a model of a left- and a right- chiral field living on a two-sheeted space-time. The resulting action functionals include novel interactions due to extended gravity.

1 Introductory Remarks

It is widely recognized that the classical continuum picture of space-time is untenable at all scales. The twin pillars of modern physics, namely, 1). General Relativity, in which the dynamics of the classical continuum of space-time is coupled to the dynamics of the matter moving within it and 2). Quantum Field Theory, with rules of quantization to be applied, in principles, to all degrees of freedom including gravity, are found to be incompatible. Besides each pillar has its own structural fractures, such as space-time singularities in the case of general relativity and infinities and divergences in the case of quantum field theories. An ideal unified theory demands the quantization of space, time and matter all treated on the same footing. A continuum space-time at all scales seems to be inadequate for such an ideal theory. Recent ideas based both on string theory and approaches to quantum gravity suggest strongly that space-time structure at the Planck scale may be discrete with noncommutative coordinates. The continuum space-time of classical physics, it is hoped, will emerge in certain limiting regimes, just as the classical behavior of quantum systems emerges in an appropriate limit.

At the present time we do not have a precisely defined noncommutative space that meets the above requirements. However, in recent years, Connes

*This report is based on the work done in collaboration with Nguyen Ai Viet, Department of High-Energy Physics, Center for Theoretical Physics,Institute of Physics, Hanoi, Vietnam

has proposed an alternate approach to the study of the structure of space-time, based on noncommutative geometry (NCG) [1, 2]. It has given rise to the description of the Standard Model with a geometrical interpretation of the Higgs field on the same footing as the gauge field. Spontaneous symmetry breaking follows as a natural consequence. It enables one, in principle, to calculate some arbitrary parameters of the Standard Model, such as the Weinberg angle, the top quark and Higgs masses.

The main motivation for the present work is to extend Connes' formalism to include gravity. As a prelude to the full standard model, we study a simplified model consisting of a left- and a right- handed chiral spinor fields along with their respective abelian gauge fields on a two-sheeted space-time. The two-sheeted space-time can be considered as a discretized version of the Kaluza-Klein theory in which the internal space of a circle is replaced by two discrete points. In this brief report, I summarize the basics of the formalism and the summary of results. For details and references to related work, the reader is referred to the published paper [4].

2 Non-commutative differential geometry and the spectral triple on a two-sheeted space-time

In his formalism, Connes bypasses the precise specification of the manifold as the starting point. Instead, he formulates its description in terms of an associative and involutive algebra, commutative or noncommutative. One may think of this as a generalization of the well-known theorem due to Gelfand [1, 3], which states that the classical topological space based on a continuum can be completely recovered from the abelian algebra of smooth functions. The starting point in his formalism is what is called a spectral triple [Connes] : 1) an involutive, unital algebra of functions $\mathcal{A}$, 2) Hilbert space $\mathcal{H}$, and 3) a self-adjoint operator called Dirac operator $\mathcal{D}$. The existence of a finite basis for the module of differential forms is understood. Also, the the algebra $\mathcal{A}$ of generalized functions is now not necessarily commutative.

Intuitively, two sheeted space-time can be viewed as an extension of the physical four-dimensional space-time manifold M by a discrete, internal space $\{a, b\}$, a discretized version of Kaluza-Klein theory where the circle is replaced by two discrete points in the fifth dimension.

The spectral triple:

Algebra $\mathcal{A}$ is the algebra of smooth function $C^\infty(M, \mathbf{C})$ tensored with the complex-valued functions on the set $\{a, b\}$:

$$\mathcal{A} = C^\infty(M) \otimes (\mathbf{C} \oplus \mathbf{C}) \cong C^\infty(M, \mathbf{C}) \oplus C^\infty(M, \mathbf{C}). \tag{2.1}$$

Hilbert space $\mathcal{H}$ is the space of square-integrable sections of a spinor bundle,

$$\mathcal{H} \equiv L^2(M,S) \oplus L^2(M,S). \tag{2.2}$$

Dirac operator $\mathcal{D}$ is the generalized self-adjoint operator

$$D = \Gamma^M D_M, \quad M = 0,1,2,3,5 \tag{2.3}$$

with the generalization of the gamma matrices γ^μ given by

$$\Gamma^\mu \equiv \begin{pmatrix} \gamma^\mu & 0 \\ 0 & \gamma^\mu \end{pmatrix}, \quad \Gamma^5 \equiv \begin{pmatrix} 0 & \gamma^5 \\ \gamma^5 & 0 \end{pmatrix}. \tag{2.4}$$

Generalized space-time derivatives are given by

$$D_\mu \equiv \begin{pmatrix} \partial_\mu & 0 \\ 0 & \partial_\mu \end{pmatrix}. \tag{2.5}$$

These derivatives act along each of the space-time sheets. The derivative between the two sheets is given as

$$D_5 \equiv \sigma \bar{D}_5, \tag{2.6}$$

where

$$\bar{D}_5 \equiv \begin{pmatrix} 0 & m \\ m & 0 \end{pmatrix}, \quad \sigma \equiv \begin{pmatrix} 0 & 1 \\ 1 & 0 \end{pmatrix}. \tag{2.7}$$

The parameter m has the dimension of mass in order to give the fifth component of space-time the same dimension as the other components. The action of the operator DF on F is defined to be

$$D(F) \equiv [D,F] = \Gamma^M (DF)_M = \Gamma^M D_M F, \tag{2.8}$$

We define the generalized wedge product as the fully anti-symmetric product that truncates all the "junk" forms in a trivial way, namely,

$$\Gamma^M \wedge \Gamma^N = \frac{1}{2}(\Gamma^M\Gamma^N - \Gamma^N\Gamma^M). \tag{2.9}$$

except when in Eqn.(2.9), $M, N = 5$,in which case

$$\Gamma^5 \wedge \Gamma^5 = 1. \tag{2.10}$$

With the spectral triple given, it is straightforward to construct the algebra of forms in the two-sheeted space-time. The exterior derivative of any odd form is its anti-commutator with the operator D as a consequence of the Leibnitz rule. Hence the exterior derivative DU of the generalized one-form $U = \Gamma^M U_M$ is

$$DU = \{D,U\} = \Gamma^M \wedge \Gamma^N (DU)_{MN}, \tag{2.11}$$

2.1 Riemannian geometry in two-sheeted space-time

The Equivalence Principle extended to the two-sheeted space-time requires a locally orthonormal basis. Continuing to work with the Γ-representation, the locally orthonormal basis is represented by Γ^A

$$\Gamma^a \equiv \begin{pmatrix} \gamma^a & 0 \\ 0 & \gamma^a \end{pmatrix}, \qquad \Gamma^{\dot 5} \equiv \begin{pmatrix} 0 & \gamma^5 \\ \gamma^5 & 0 \end{pmatrix}, \tag{2.12}$$

where γ^a and γ^5 are the usual flat Dirac matrices. By choosing γ^5 we have specialized to the two sheets of space-time of chiral spinors. We shall use a $\dot 5$ index in the orthonormal basis to distinguish it from 5 in the general case.

The inner product is defined as the trace

$$< \Gamma^A, \Gamma^B > \; = \; Tr(\Gamma^A \Gamma^B) \; = \; 2\eta^{AB}.\mathbf{1} \tag{2.13}$$

The curvilinear basis will be represented by $\Gamma^M(x)$. To obtain $\Gamma^M(x)$ and define the metric, we postulate the generalized vielbeins $E^A_M(x)$, as the following diagonal matrix of zero-forms:

$$E^\mu_a(x) \equiv \begin{pmatrix} e^\mu_{1a}(x) & 0 \\ 0 & e^\mu_{2a}(x) \end{pmatrix}, \qquad E^\mu_{\dot 5}(x) \equiv 0 ,$$

$$E^5_a(x) \equiv - \begin{pmatrix} a_{1a} & 0 \\ 0 & a_{2a} \end{pmatrix} \equiv -A_a(x) = -E^\mu_a A_\mu ,$$

$$E^5_{\dot 5}(x) \equiv \begin{pmatrix} \phi_1^{-1}(x) & 0 \\ 0 & \phi_2^{-1}(x) \end{pmatrix} \equiv \Phi^{-1}(x) , \tag{2.14}$$

where $e^\mu_{1,2a}(x)$ are two different vielbeins on the two sheets of space-time. Similarly, $a_{1,2}(x)$ and $\phi_{1,2}(x)$ are respectively vector and scalar fields.

The vielbeins E^M_A are invertible giving,

$$E^a_\mu(x) \equiv \begin{pmatrix} e^a_{1\mu}(x) & 0 \\ 0 & e^a_{2\mu}(x) \end{pmatrix}, \qquad E^a_5(x) \equiv 0 ,$$

$$E^{\dot 5}_\mu(x) \equiv A_\mu(x)\Phi(x) , \qquad E^{\dot 5}_5(x) \equiv \begin{pmatrix} \phi_1(x) & 0 \\ 0 & \phi_2(x) \end{pmatrix} \equiv \Phi(x) . \tag{2.15}$$

Then, the $\Gamma^M(x)$ matrices are given as $\Gamma^M(x) = \Gamma^A E^M_A(x)$,

$$\Gamma^\mu(x) \equiv \begin{pmatrix} \gamma^a e^\mu_{1a}(x) & 0 \\ 0 & \gamma^a e^\mu_{2a}(x) \end{pmatrix}, \qquad \Gamma^5(x) \equiv \begin{pmatrix} -a_1(x) & \gamma^5 \phi_2^{-1}(x) \\ \gamma^5 \phi_1^{-1}(x) & -a_2(x) \end{pmatrix}, \tag{2.16}$$

where $a_{1,2}(x) = \gamma^a a_{1,2a}$ are two gauge connection one-forms on the two different sheets.

Using the inner product defined as the trace on the Hilbert space, we can calculate the representation of the metric tensor. "Metric" field components involve a pair of conventional vierbeins and, in addition, a pair of vector and a pair of scalar field.

3 Physical content of the gravity sector in two-sheeted space-time

The essential point is how to impose a consistent system of constraints to solve the generalized Cartan structure equations without over-constraining the metric. To this end, we find the following set of minimal condtions:

i) Spacetime torsion free condition

$$T_{aBC} = 0, \tag{3.1}$$

ii) Metric compatibility condition

$$\tilde{\Omega}^*_{AB} = -\Omega_{BA} \tag{3.2}$$

iii) "Reality" condition

$$\tilde{\Omega}^*_{AB} = \Omega_{AB} \tag{3.3}$$

iv) An additional condition on the connections

$$\Omega_{AB\dot{5}} = 0 \tag{3.4}$$

These conditions are sufficient to express the non-vanishing connection coefficients and torsion components in terms of metric fields. Also, more importantly, they provide Lagrangians with proper kinetic terms for all the metric component fields. .

Hence,to summarize, it is to be noted that the gravity sector of the theory contains $e^{\mu}_{a}(x), e^{a}_{\mu}(x)$, the tensor field $v^{\mu}_{\nu}(x)$, the vector and scalar fields $a_{\pm\mu}(x), \Phi_{\pm}(x)$. In the final Lagrangian, the tensor field $v^{\mu\nu}$ will be a mass eigenstate. Thus, one component of each pair of tensor, vector and scalar fields has zero mass, while the other is massive. The special case in which the components on the second sheet are identical to the ones on the first sheet, reproduces exactly the Kaluza-Klein zero mode theory.

4 Matter fields in curved two-sheeted space-time

The matter field Lagrangian consists of a gauge sector and a fermionic sector. The gauge sector contains two Abelian gauge fields and two Higgs scalar fields as part of a genaralized one-form. The fermionic sector consists of a left-chiral spinor on one sheet and a right-chiral spinor on the other.

4.1 The gauge sector

The generalized one-form:

$$U = B(x) = \Gamma^M(x)B_M(x),$$

where

$$B_\mu(x) = \begin{pmatrix} b_{1\mu}(x) & 0 \\ 0 & b_{2\mu}(x) \end{pmatrix} \ , \ B_5(x) = \begin{pmatrix} h_1(x) & 0 \\ 0 & h_2(x) \end{pmatrix}. \tag{4.1}$$

Then, the curvature or field strengths are given by the two-form (henceforth the x-dependence will be understood),

$$G = DB + B \wedge B, \tag{4.2}$$

In terms of the above components, the Lagrangian is given as follows:

$$\mathcal{L}_G = -\frac{1}{4}Tr < G, G >= -\frac{1}{4}Tr(2\tilde{G}_{ab}G^{ba} + 4\tilde{G}_{a\dot{5}}G^{\dot{5}a} + \tilde{G}_{\dot{5}\dot{5}}G^{\dot{5}\dot{5}}) \ , \tag{4.3}$$

where the trace is over the two dimensional matrices.

A straight-forward calculation results in the Lagrangian

$$\mathcal{G}_G = \mathcal{L}_{G1} + \mathrm{L}_{G2} + \mathcal{L}_{G3} + \mathcal{L}_{G4}, \tag{4.4}$$

where $\mathcal{L}_{G1}$ and $\mathcal{L}_{G2}$ contain the kinetic terms for the gauge and Higgs fields respectively; $\mathcal{L}_{G3}$ has the quartic Higgs potential of the correct form for spontaneous symmetry breaking and $\mathcal{L}_{G4}$ contains various interaction terms between the gauge sector vector and scalar fields and those from the gravity sector.

4.2 Fermion sector

With two chiral spinor fields Ψ_L on one sheet of space-time and Ψ_R on the other, the fermionic Lagrangian has the standard form,

$$\mathcal{L}_F \quad = \quad i\bar{\Psi}\Gamma^A(E_A^M(D_M + iB_M) + \tfrac{i}{4}\Gamma^B\Gamma^C\Omega_{BCA})\Psi \tag{4.5}$$

where

$$\Psi = \begin{pmatrix} \psi_L \\ \psi_R \end{pmatrix},$$

$$\psi_L = \frac{(1+\gamma_5)}{2}\psi \ , \ \psi_R = \frac{(1-\gamma_5)}{2}\psi, \tag{4.6}$$

ψ is the 4-component Dirac spinor and Ω_{BCA} are the non-vanishing connection coefficients expressed in terms of the metric fields. Consequently,this generalized Lagrangian includes gravitational interactions with the matter fields.

5 Physical implications, Summary and Conclusions

The formalism provides formal expressions for the action functionals in terms of formal fields as elements of an algebra. They need to be redefined by rescaling using the dimensional parameters, G_N ,the Newtonian constant, g_V,a new gravitational constant associated with torsion, g , the dimensionless gauge coupling and finally, m ,the parameter having the dimension of mass. Then, the requirement that the kinetic terms of each field have the standard coefficients determines the masses and coupling constants in terms of the above mentioned four parameters. On the formal side, the redefinition of the fields leads to a subalgebra of the original algebra where the fields on the two sheets are complex conjugates of each other.

To summarize the main results, one finds that the gravitational sector consists of the standard Einstein-Hilbert action, accompanied by a Fierz-Pauli Lagrangian for a massive tensor field with mass m. The mass of the vector field accompanying the zero mass vector field is $m/\sqrt{2}$. It also has a massless Brans-Dicke scalar field and massive scalar field. Among the various interaction terms, there is a scalar field potential with a cosmological constant term.

The matter field Lagrangians have the expected spontaneous symmetry breaking leading to a zero mass gauge boson accompanied by a massive one with mass $\sqrt{2}m$. More importantly, there are parity violating interactions involving matter fields and the metric component massive tensor, vector and scalar of the extended gravity sector. In the context of the full standard model, these interactions could be CP-violating and thus provide the much needed CP violation in the early universe to explain baryon-anti-baryon asymmetry.

Thus from the point of view of physics, the formalism provides a rich and complex structure with new interactions, linear and non-linear. With the extended gravity,it has cosmological implications and suggests many avenues for further research

References

[1] A. Connes, *Non-Commutative Geometry*, [Academic Press, 1994].

[2] A. Connes, "Essay on Physics and Non-Commutative Geometry", in D.G. Quillen, G.B. Segal, S.T. Tsou editors, *The Interface of Mathematics and Particle Physics*, [Clarendon Press, Oxford, 1990].

[3] G.F. Simmons, *Introduction to Topology and Modern Analysis*, [Krieger, 1983].

[4] Nguyen Ai Viet and Kameswar C. Wali, Phys. Rev. D **67**, 1240299 (2003).

Clash of symmetries in a Randall-Sundrum-like spacetime

Gareth Dando,[1,*] Aharon Davidson,[2,†] Damien P. George,[1,‡] Raymond R. Volkas,[1,§] and K.C. Wali[3,||]
[1]School of Physics, Research Centre for High Energy Physics, The University of Melbourne, Victoria 3010, Australia
[2]Physics Department, Ben-Gurion University of the Negev, Beer-Sheva 84105, Israel
[3]Physics Department, Syracuse University, Syracuse, New York 13244-1130, USA
(Received 7 July 2005; published 29 August 2005)

We present a toy model that exhibits clash-of-symmetries style Higgs field kink configurations in a Randall-Sundrum-like spacetime. The model has two complex scalar fields $\Phi_{1,2}$, with a sextic potential obeying global $U(1) \otimes U(1)$ and discrete $\Phi_1 \leftrightarrow \Phi_2$ interchange symmetries. The scalar fields are coupled to 4 + 1 dimensional gravity endowed with a bulk cosmological constant. We show that the coupled Einstein-Higgs field equations have an interesting analytic solution provided the sextic potential adopts a particular form. The 4 + 1 metric is shown to be that of a smoothed-out Randall-Sundrum type of spacetime. The thin-brane Randall-Sundrum limit, whereby the Higgs field kinks become step functions, is carefully defined in terms of the fundamental parameters in the action. The "clash-of-symmetries" feature, defined in previous papers, is manifested here through the fact that both of the U(1) symmetries are spontaneously broken at all nonasymptotic points in the extra dimension w. One of the U(1)'s is asymptotically restored as $w \rightarrow -\infty$, with the other U(1) restored as $w \rightarrow +\infty$. The spontaneously broken discrete symmetry ensures topological stability. In the gauged version of this model we find new flat-space solutions, but in the warped metric case we have been unable to find any solutions with nonzero gauge fields.

I. INTRODUCTION

The breaking of symmetries is one of the most important issues in particle physics. Nature has taught us the relevance of symmetry principles, especially through Noether's connection of global symmetries with conservation laws and the successful description of fundamental interactions via local gauge invariance. But to reconcile symmetric field equations with the real world, one needs that some of the solutions to those equations have a reduced symmetry compared to the equations themselves. This is, of course, what we mean by "spontaneous symmetry breaking (SSB)" or "hidden symmetry."

The precise mechanism by which electroweak SSB happens has yet to be experimentally determined. In the standard model (SM), the default option is to use an elementary scalar or Higgs field doublet with a quartic potential that induces a nonzero vacuum expectation value for one of the components. This will soon be tested at the CERN Large Hadron Collider through its physical-Higgs boson search programme.

The purpose of this paper is to further develop a different method of symmetry breaking called the "clash of symmetries" [1–4] (see also [5]). This mechanism is relevant for brane-world extensions to the regular 3 + 1 dimensional standard model [6,7]. The important development we report here is the rigorous incorporation of gravity into one of the toy Higgs models explored in the previous papers on this topic [3]. Since gravity is fundamental to brane-world models, this is a very welcome step forward.

The toy model we shall analyze consists of two complex scalar fields $\Phi_{1,2}$ coupled to 4 + 1 dimensional gravity endowed with a bulk cosmological constant. The action is constructed to be invariant under global $U(1) \otimes U(1)$ transformations and the discrete $\Phi_1 \leftrightarrow \Phi_2$ interchange operation. The Higgs potential we construct is sextic, and its specific form is driven by our desire to find an analytic solution to the coupled Einstein-Higgs field equations. The spacetime metric tensor we posit has the Randall-Sundrum nonfactorizable form [6], but with the warp factor a completely smooth function of the extra dimension coordinate. The precise Randall-Sundrum metric emerges by taking a certain limit. The Higgs field configurations we seek are kinklike, associated with the spontaneously broken discrete symmetry.

We also study the gauged version of this model, extending the analysis of Ref. [3]. We find additional nontrivial flat-space solutions, but despite an extensive numerical search we have been unable to find any warped metric solutions with nonzero gauge fields.

Section II reviews the clash-of-symmetries mechanism, while Sec. III defines and analyzes the ungauged version of our toy model. This is followed by an account of the gauged cases in Sec. IV. The final section is a conclusion and outlook for future work.

II. THE CLASH OF SYMMETRIES

The default SSB mechanism employed in the SM and many extensions to it utilizes a simple spatially homoge-

*Electronic address: g.dando@unimelb.edu.au
†Electronic address: davidson@bgumail.bgu.ac.il
‡Electronic address: d.george@physics.unimelb.edu.au
§Electronic address: r.volkas@physics.unimelb.edu.au
||Electronic address: wali@phy.syr.edu

neous, static and stable solution to the field equations: the Higgs fields $\phi_i(x^\mu)$ are set equal to those constant values corresponding to an absolute minimum of the Higgs potential. The locations of the nonzero constants within the Higgs multiplets then determine the stability group of the vacuum state, what we commonly call the unbroken symmetry after SSB. The nonzero Higgs fields become constant background scalar fields throughout the Universe.

There are other types of solutions to Higgs field equations that can also serve as stable, static background fields: topological solitons, such as kinks, strings, and monopoles. While there is much interest in such configurations, they are not usually considered as viable candidates for playing a major role in spontaneous symmetry breaking. This is simply because they are spatially inhomogeneous. The energy densities required for electroweak and higher symmetry breaking are so high as to be incompatible with the strong evidence for a universe that is spatially homogeneous at large scales. This objection, however, does not apply to brane-world models, because the nontrivial spatial dependence can be restricted to the extra dimension coordinates only.

For definiteness, consider the case of one extra dimension described by Cartesian coordinate w and topologically stable Higgs configurations $\phi_i(w)$, some of which have kinkform with respect to w. The pattern of spontaneous breaking then effectively becomes a function of w. The $3+1$ brane is located at, say, $w=0$. If the degrees of freedom that are confined to the brane [8] are confined absolutely, then the unbroken symmetry is the stability group of $\phi_i(w=0)$, to be denoted by $H(w=0)$ from now on (and let the internal symmetry group of the action be G). The symmetry breaking will be communicated to the brane-world denizens through appropriate interactions. But in a quantal (and perhaps even in a classical) world, one would not expect strict confinement to the brane: there should be some leakage off it into the bulk. One therefore expects the brane-localized fields to also couple, but with reduced strength, to $\phi_i(w)$ states where w is slightly different from zero. If the stability group of $\phi_i(0<|w|<\epsilon)$, call it $H(w\sim\epsilon)$, is different from $H(w=0)$, then a rather rich effective symmetry breaking outcome could ensue for the brane world. For example, a hierarchical breaking pattern $G\rightarrow H(w=0)\rightarrow H(w\sim\epsilon)$ was speculated upon in Ref. [2], where in that case $G=\mathrm{SO}(10)$, $H(w=0)=\mathrm{SU}(4)\otimes\mathrm{SU}(2)\otimes\mathrm{U}(1)$, and $H(w\sim\epsilon)=\mathrm{SU}(3)\otimes\mathrm{SU}(2)\otimes\mathrm{U}(1)^2$. The potential application to grand unified models and the gauge hierarchy puzzle is clear.

The "clash-of-symmetries" phenomenon allows the unbroken symmetry at nonasymptotic points in the extra dimension, $|w|<\infty$, to be smaller than the symmetry holding at $|w|=\infty$ [1–4]. The boundary conditions at infinity for kink configurations are set to be the Higgs vacuum expectation values (VEVs). They force the spontaneous breaking $G\rightarrow H(|w|\rightarrow\infty)$. In usual SSB, the Higgs VEV configuration is used throughout the whole of space, so the symmetry is broken to the group we are calling $H(|w|\rightarrow\infty)$ everywhere, homogeneously. In the clash-of-symmetries mechanism, this symmetry breaking pattern only holds asymptotically with respect to the extra dimension(s).

The clash of symmetries can occur if the isomorphic subgroups left unbroken at $w=-\infty$ and $w=+\infty$, $H(w\rightarrow-\infty)\equiv H(-\infty)$ and $H(w\rightarrow+\infty)\equiv H(+\infty)$, respectively, can be *differently embedded* within the parent group G. In the cases examined so far in the literature, the breakdown at finite values of w is to the *intersection* of the asymptotic stability groups,

$$H(|w|<\infty)=H(-\infty)\cap H(+\infty)\equiv H_{\mathrm{clash}}. \qquad (2.1)$$

Although $H(-\infty)$ and $H(+\infty)$ are isomorphic, their different embeddings within G lead their intersection to be a smaller group.[1] Some special values of w, most often $w=0$, may correspond to stability groups larger than H_{clash} because the spatially varying Higgs configurations may instantaneously pass through special patterns as w varies. For instance, a component of a Higgs multiplet may be an odd function of w such as $\tanh(w)$, so it would vanish at $w=0$ and thus the unbroken symmetry would be enhanced at that point. This possibility lies behind our remarks two paragraphs above about an application to the gauge hierarchy problem.

Discrete symmetries also play a vital role: their spontaneous breakdown leads the vacuum manifold to have disconnected pieces. Kinks interpolate between vacuum states from disconnected pieces. Within a given topological class of kinks, the one with lowest energy is then guaranteed to be (topologically) stable.

The easiest example of the clash of symmetries uses $G=\mathrm{SU}(3)$ spontaneously breaking to SU(2). An SU(2) subgroup can be embedded in three different ways in SU(3): I-spin, U-spin, and V-spin. Clash-of-symmetries kinks interpolate between VEVs respecting (I, U), (U, V), or (V, I) at $w=(-\infty,+\infty)$. The differently embedded asymptotic SU(2) symmetries "clash" at nonasymptotic points, leading H_{clash} to be $\mathrm{SU}(2)_I\cap\mathrm{SU}(2)_U=\{1\}$ and so on.[2]

The model-building promise of the clash of symmetries lies in the greater symmetry breaking power of spatially inhomogeneous Higgs configurations over the simple homogeneous alternative of standard SSB. We hope that this will lead to models with simpler Higgs sectors and fewer associated parameters. Some of us have already speculated that the ultimate application might be to an E_6 model,

[1] There will also typically be kink configurations that interpolate between *identically* embedded subgroups. Such kinks do not display the clash of symmetries, and one must ensure that they have higher energy than those displaying the clash.

[2] To actually implement this simple SU(3) scheme requires more work: see Ref. [1] for details.

because of the fact that there are three different E_6 generators that can serve as electric charge, corresponding to three different embeddings of electromagnetism within the full group [1]. Standard E_6 models can only use one at a time. Even more speculatively, a connection between that threefold structure and the three families of quarks and leptons might exist[3] [1].

But many preparatory studies are needed before such an ambitious vision can be realized. In this paper, we take another important step along that road: the incorporation of gravity. To facilitate the study of gravity, we begin by simplifying the group theoretic structure as much as possible without losing the essence of the clash-of-symmetries mechanism: we take $G = U(1)_1 \otimes U(1)_2$, with $H(-\infty) = U(1)_1$ and $H(+\infty) = U(1)_2$ leading to $H_{\text{clash}} = \{1\}$. The discrete symmetry we need for topological stability interchanges the two Abelian sectors.

In the next section we show that the ungauged version of this model can yield a clash-of-symmetries-style kink in a warped metric spacetime. The section after that revisits the gauged extension in both flat and curved spacetime.

III. THE UNGAUGED MODEL

A. The field equations

The model has two complex Higgs fields Φ_1 and Φ_2 coupling to 4 + 1 dimensional gravity, and it respects the $\Phi_1 \leftrightarrow \Phi_2$ discrete symmetry. The action is

$$S = -\int\Big[\frac{\kappa}{2}R + g^{MN}(\partial_M\overline{\Phi}_1\partial_N\Phi_1 + \partial_M\overline{\Phi}_2\partial_N\Phi_2) + V(\overline{\Phi}_1\Phi_1, \overline{\Phi}_2\Phi_2)\Big]\sqrt{-g}d^5x \quad (3.1)$$

where g_{MN} is the 4 + 1 dimensional metric tensor, with $M, N = 0, 1, 2, 3, 5$, R is the curvature scalar, and overbars denote complex conjugation. The matter action S_m is defined as S with the Einstein-Hilbert contribution absent. The Higgs potential V is as yet unspecified except that it must be invariant under $\Phi_1 \leftrightarrow \Phi_2$. It will have a constant piece that serves as the bulk cosmological constant. We employ the $(-, +, +, +, +)$ signature and our sign conventions are as in Weinberg [apart from our definition of $g \equiv \det(g)$] [11].

The Einstein equation found by requiring a stationary action under metric variations is

$$G^{MN} = -\frac{1}{\kappa}T^{MN} \quad (3.2)$$

where the Einstein tensor is, as usual, $G^{MN} = R^{MN} - (1/2)g^{MN}R$ with R^{MN} being the Ricci tensor. The stress-energy tensor,

$$T^{MN} \equiv \frac{2}{\sqrt{-g}}\frac{\delta S_m}{\delta g_{MN}}, \quad (3.3)$$

evaluates to

$$T^{MN} = 2g^{MP}g^{NQ}t_{PQ} - g^{MN}(g^{PQ}t_{PQ} + V), \quad (3.4)$$

where

$$t_{MN} \equiv \partial_M\overline{\Phi}_1\partial_N\Phi_1 + \partial_M\overline{\Phi}_2\partial_N\Phi_2. \quad (3.5)$$

The Higgs Euler-Lagrange equations are

$$\partial_M(\sqrt{-g}g^{MN}\partial_N\Phi_i) = \sqrt{-g}\frac{\partial V}{\partial\overline{\Phi}_i}, \quad (3.6)$$

where $i = 1, 2$, and the complex conjugate equations also hold.

We now specify our solution ansatz. We seek configurations that only depend on the extra dimension $x^5 \equiv w$. The imaginary parts of the Higgs fields are set to zero and we write

$$\Phi_i = \frac{\phi_i(w)}{\sqrt{2}}, \quad (3.7)$$

where $\phi_i(w)$ is real. The metric ansatz, of Randall-Sundrum type, is defined through the 4 + 1 dimensional line element

$$ds_5^2 = dw^2 + e^{2f(w)}ds_4^2 \quad (3.8)$$

where $f(w)$ is an as yet unknown function and $ds_4^2 = \eta_{\mu\nu}dx^\mu dx^\nu$ is the 3 + 1 dimensional Minkowski line element (our brane will be Minkowski), with $\mu, \nu = 0, 1, 2, 3$ or t, x, y, z. The field equations then reduce to

$$3f'' + 6(f')^2 = -\frac{1}{\kappa}\Big[\frac{1}{2}[(\phi_1')^2 + (\phi_2')^2] + V\Big], \quad (3.9)$$

$$6(f')^2 = \frac{1}{\kappa}\Big[\frac{1}{2}[(\phi_1')^2 + (\phi_2')^2] - V\Big], \quad (3.10)$$

$$\phi_i'' + 4f'\phi_i' = \frac{\partial V}{\partial\phi_i}, \quad (3.11)$$

where prime denotes differentiation with respect to w. Adding and subtracting Eqs. (3.9) and (3.10) gives

$$f'' + 4(f')^2 = -\frac{2}{3\kappa}V, \quad (3.12)$$

$$f'' = -\frac{1}{3\kappa}[(\phi_1')^2 + (\phi_2')^2]. \quad (3.13)$$

Note that Eq. (3.13) is independent of the potential, a feature we will use below.

We will construct a potential V that has degenerate global minima at

$$\text{vacuum 1: } \phi_1 = u, \quad \phi_2 = 0$$

$$\text{and} \quad \text{vacuum 2: } \phi_1 = 0, \quad \phi_2 = u, \quad (3.14)$$

[3]There is some similarity between the clash-of-symmetries mechanism and orbifold symmetry breaking [9]. In particular, Ref. [10] constructs a model where different subgroups of SO(10) are left unbroken on different branes, with the overall unbroken symmetry being the intersection of those individual subgroups, and with families also distributed amongst the branes. While there are similarities, it is clear that the mechanisms are not identical, because the clash of symmetries has a physical-Higgs field while the orbifold method is Higgsless.

where u would be the VEV in the standard homogeneous SSB mechanism. The vacua are related by the now spontaneously broken discrete interchange symmetry. The Higgs field boundary conditions will be

$$\phi_i(-\infty) = \text{vacuum 2} \quad \text{and} \quad \phi_i(+\infty) = \text{vacuum 1}. \tag{3.15}$$

B. Constructing the solution

In 3 + 1 dimensions without gravity, renormalizability limits us to quartic potentials. With extra dimensions, and with gravity included, this is no longer a well-motivated requirement. In fact, the only general constraint on the form of V we will impose is that it be bounded from below. A very important topic for the future of the clash-of-symmetries mechanism will indeed be the rules by which potentials should be specified.

Our more modest goal here is to provide an existence proof that kinks featuring the clash-of-symmetries form can be consistently coupled to Randall-Sundrum-like gravity. We will search for an analytically tractable model for reasons of both simplicity and elegance. To do this, it makes sense to invert the problem by first specifying the solutions $\phi_i(w)$ and $f(w)$ and then constructing a potential that gives those solutions. This, in itself, is nontrivial and interesting.

It turns out that a good choice, satisfying the boundary conditions of Eq. (3.15) and also the Higgs-potential-independent equation (3.13), is

$$\phi_1^s(w) = \frac{u}{\sqrt{2}}\sqrt{1+\tanh\beta w}, \tag{3.16}$$

$$\phi_2^s(w) = \frac{u}{\sqrt{2}}\sqrt{1-\tanh\beta w}, \tag{3.17}$$

$$f^s(w) = -\frac{u^2}{12\kappa}\ln(\cosh\beta w), \tag{3.18}$$

where β is a free parameter that specifies the brane thickness, and the superscript s denotes "solution." The Higgs configurations are the square roots of the most naïve choices, and $f^s(w)$ is essentially a regularization of $|w|$. The latter is, of course, the corresponding function in the Randall-Sundrum scenario. The smoothing out of the Randall-Sundrum cusp goes hand-in-hand with the existence of Higgs field kinks coupled to gravity, and the brane is dynamically generated rather than put in by hand. Note also that

$$\phi_1^s(w)^2 + \phi_2^s(w)^2 = u^2, \tag{3.19}$$

a feature that will turn out to be very useful.

The first inkling for what V should be comes from Eq. (3.12). It suggests that

$$V = -\frac{\beta^2 u^4}{24\kappa} + \frac{\beta^2}{2u^2}\left(1+\frac{u^2}{3\kappa}\right)\phi_1^2\phi_2^2 + U(\phi_1^2,\phi_2^2), \tag{3.20}$$

where U is a function that vanishes at the level of the solution, that is $U(\phi_1^s(w)^2, \phi_2^s(w)^2) = 0$. Notice that a negative bulk cosmological constant,

$$\Lambda_5 = -\frac{\beta^2 u^4}{24\kappa}, \tag{3.21}$$

follows from the ansatz: the bulk has anti-de Sitter qualities, as usual in Randall-Sundrum-like models, though in our case it is of course not precisely anti-de Sitter.

More information about U is supplied by the Higgs field equation (3.11). To solve it, one must specify that

$$U = -\frac{\beta^2}{u^4}\left(\frac{3}{2}+\frac{u^2}{3\kappa}\right)\phi_1^2\phi_2^2(\phi_1^2+\phi_2^2-u^2) + W(\phi_1^2+\phi_2^2-u^2), \tag{3.22}$$

where the new function W obeys

$$W(0) = 0, \tag{3.23}$$

$$\left.\frac{\partial W}{\partial \phi_{1,2}}\right|_{\phi_1^s,\phi_2^s} = 0, \tag{3.24}$$

so as not to affect the satisfaction of the field equations, Eqs. (3.12) and (3.13), respectively. As far as they are concerned, W can vanish. The resulting sextic potential, however, would not be bounded from below because the coefficient of the first term in U is negative.

We therefore construct W so that, as well as obeying Eqs. (3.23) and (3.24), it leads to a potential that is both bounded from below and has global minima at Eq. (3.14). The simplest suitable function is

$$W = \zeta\frac{\beta^2}{4u^2}\left(\frac{3}{2}+\frac{u^2}{3\kappa}\right)(\phi_1^2+\phi_2^2-u^2)^2\left(\eta+\frac{\phi_1^2+\phi_2^2-u^2}{u^2}\right) \tag{3.25}$$

where η and ζ are dimensionless parameters.

The full Higgs potential for the model is specified by Eqs. (3.20), (3.22), and (3.25). We will present a global minimization analysis of it in the next subsection. (To reinstate the complex nature of the Higgs fields, the substitution $\phi_i^2 \to 2\overline{\Phi}_i\Phi_i$ should be made, though for the forthcoming discussion we need only work with the ϕ_i.)

Before doing so, let us identify the Randall-Sundrum limit of the model. We need to identify a limit in which the warp factor, e^{2f^s}, becomes $e^{-a|w|}$ where a is some positive constant. Following Ref. [12], we note that $(\cosh\beta w)^{-(1/\beta)} \to e^{-|w|}$ as $\beta \to \infty$. Writing

$$e^{2f^s(w)} = [(\cosh\beta w)^{-1/\beta}]^{u^2\beta/6\kappa}, \tag{3.26}$$

we see that the Randall-Sundrum warp factor, $e^{-(u^2\beta/6\kappa)|w|}$, is obtained in the limit

$$\beta\to\infty, \quad u\to 0 \quad \text{such that } u^2\beta \to \text{finite constant}. \tag{3.27}$$

We also require η and ζ to be unchanged when taking the limit. Note that the Higgs field configurations, Eqs. (3.16) and (3.17), become step functions as $\beta \to \infty$, with their magnitude tending to zero because $u \to 0$. The Randall-Sundrum fine-tuning condition between the brane tension and the bulk cosmological constant is a feature of our solution. Defining the brane tension via

$$\Lambda_b \equiv \lim_{RS} \int_{-\infty}^{+\infty} \left[\frac{1}{2}[(\phi_1^{s\prime})^2 + (\phi_2^{s\prime})^2] + V(\phi_1^s, \phi_2^s)\right] \times \sqrt{-g^s} dw, \tag{3.28}$$

it is easy to show that

$$\Lambda_b^2 = -\frac{3}{2}\kappa\Lambda_5, \tag{3.29}$$

which is the Randall-Sundrum condition.

C. Minimization analysis

We now identify the parameter space region where our potential has global minima specified by Eq. (3.14).

To establish the parameter region where the potential is bounded from below, we look at the sixth-order terms in isolation. They are

$$V_6 = \lambda(\phi_1^2 + \phi_2^2)[\zeta(\phi_1^2 + \phi_2^2)^2 - 4\phi_1^2\phi_2^2], \tag{3.30}$$

where $\lambda \equiv (3\beta^2/8u^4)[1 + 2u^2/(9\kappa)] > 0$. The substitutions $\phi_1 = \chi\cos\theta$ and $\phi_2 = \chi\sin\theta$ reveal that $V_6 = \lambda\chi^6(\zeta - 4\sin^2\theta\cos^2\theta)$, from which the condition

$$\zeta > 1 \tag{3.31}$$

is trivially seen to be the only requirement to produce boundedness from below.

The minimization analysis is a four-parameter problem. For some of the local minima, both their (ϕ_1, ϕ_2) locations and the values of the potential at those points are complicated functions of the parameters. But there is a smart way to evade these unenlightening complications while yielding physically meaningful results: approach the Randall-Sundrum limit defined by Eq. (3.27). Our task then becomes a two-parameter problem in (η, ζ) that is readily visualized.

To see this, note that in Eqs. (3.20), (3.22), and (3.25), the combinations

$$1 + \frac{u^2}{3\kappa} \quad \text{and} \quad \frac{3}{2} + \frac{u^2}{3\kappa} \tag{3.32}$$

frequently appear. In the Randall-Sundrum limit, which includes $u \to 0$, these combinations simplify to 1 and 3/2, respectively. In the exact Randall-Sundrum limit, the meaning to be ascribed to the potential is obscure, so we cannot work exactly in that limit. Instead, we will approach the limit without reaching it, and write an approximation to the potential justified by the above remarks. It is

$$\begin{aligned} V \simeq V_{\text{approx}} &= -\frac{\beta^2 u^4}{24\kappa} + \frac{\beta^2}{2u^2}\phi_1^2\phi_2^2 \\ &\quad - \frac{3\beta^2}{2u^4}\phi_1^2\phi_2^2(\phi_1^2 + \phi_2^2 - u^2) \\ &\quad + \zeta\frac{3\beta^2}{8u^2}(\phi_1^2 + \phi_2^2 - u^2)^2\left(\eta + \frac{\phi_1^2 + \phi_2^2 - u^2}{u^2}\right), \\ &= \frac{3\beta^2 u^2}{8}\left[-\frac{u^2}{9\kappa} + \frac{4}{3}p_1^2p_2^2 - 4p_1^2p_2^2(p_1^2 + p_2^2 - 1)\right. \\ &\quad \left. + \zeta(p_1^2 + p_2^2 - 1)^2(\eta + p_1^2 + p_2^2 - 1)\right], \end{aligned} \tag{3.33}$$

where

$$p_i \equiv \frac{\phi_i}{u} \tag{3.34}$$

is a dimensionless Higgs field. The fact that β contributes only to an overall scale, and that κ appears solely in the constant term, reduces the analysis to a two-parameter problem in (η, ζ). While κ contributes to the potential values, it does so equally for all local minima in this approximation, which means that only η and ζ need be considered when comparing the local minima to see which are also the global minima.

The potential has three varieties, denoted (*I*, *II*, *III*), of possible local minima (or saddle points),

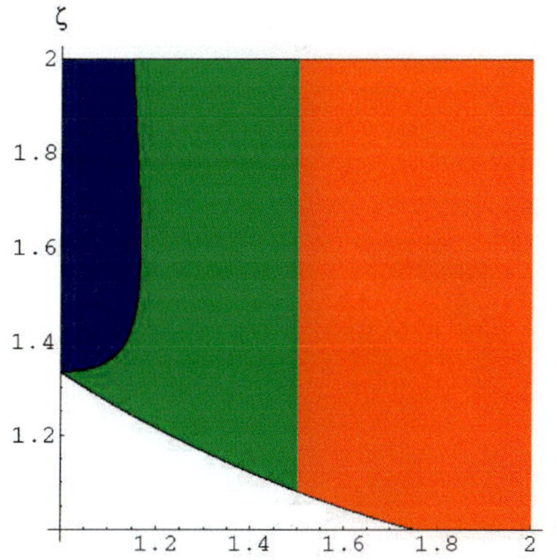

FIG. 1 (color online). The regions in the (η, ζ) plane where the approximate potential has the required discrete symmetry breaking global minimum. The potential is bounded from below for $\zeta > 1$. In the red (or medium-shaded) region, $\phi_1 = \phi_2 = 0$ is a local maximum, $\phi_1^2 = \phi_2^2 \neq 0$ is a saddle point, and $\phi_1^2 = u^2$, $\phi_2 = 0$ and $\phi_1 = 0$, $\phi_2^2 = u^2$ are degenerate global minima. The green (or light-shaded) region differs from the red (medium-shaded) only in that $\phi_1 = \phi_2 = 0$ becomes a local minimum. As η decreases further, the $\phi_1 = \phi_2 = 0$ minimum keeps falling and in the blue (or dark-shaded) region it lies below the $\phi_1^2 = \phi_2^2 \neq 0$ saddle point. For $\eta < 1$, $\phi_1 = \phi_2 = 0$ becomes the global minimum. In the unshaded region in the $\eta, \zeta > 1$ quadrant, $\phi_1 = \phi_2 \neq 0$ is the global minimum. The border between the shaded and unshaded regions is a smooth but complicated function $\zeta(\eta)$ given in the main text.

$$I \quad p_1^2 = 1, \ p_2 = 0 \ \text{degenerate with} \ p_1 = 0,$$
$$p_2^2 = 1 \ \text{(discrete symmetry breaking)}, \tag{3.35}$$

$$II \qquad p_1^2 = p_2^2 \neq 0 \quad \text{(discrete symmetry preserving)}, \tag{3.36}$$

$$III \qquad p_1 = p_2 = 0 \quad \text{(no SSB)}. \tag{3.37}$$

Which of these are actual local minima depends on (η, ζ). We want I to describe the global minima.

Of the three, it is II that has the complicated parameter dependence. In fact it is quite easy to show, even for the full potential of Eqs. (3.20), (3.22), and (3.25), that $V(I) < V(III)$ requires that

$$\eta > 1 \tag{3.38}$$

which is thus a necessary but not sufficient condition to ensure that I gives the global minima. Sufficiency requires an examination of the parameter dependence of $V(II)$ also. The results are shown in Fig. 1, where the shaded area is the Higgs parameter region where I is the global minimum.

In the red (or medium-shaded if viewing in black-and-white) region, I is a local minimum and II a saddle point with $V(I) < V(II)$, while III is a local maximum. As η decreases below 3/2, III develops into a local minimum also; the ordering is $V(I) < V(II) < V(III)$ in the green (light-shaded) region and $V(I) < V(III) < V(II)$ in the blue (dark-shaded). For $\eta < 1$, III is the global minimum so there is no SSB. The border line between the shaded and unshaded areas is given by

$$\zeta = \frac{-27 - 18\eta + 61\eta^2 + \sqrt{729 + 972\eta - 2970\eta^2 + 1900\eta^3 - 375\eta^4}}{24\eta^3}. \tag{3.39}$$

Lowering ζ has the effect of lowering $V(II)$. Below this border line, $V(II) < V(I) < V(III)$, so the symmetry preserving vacuum is the global minimum.

The border between the blue (dark-shaded) and green (light-shaded) regions is given by

$$\eta = \frac{8\zeta - 3\zeta^2 + 2\sqrt{3}\sqrt{4\zeta^2 - 7\zeta^3 + 3\zeta^4}}{3\zeta^2}. \tag{3.40}$$

Figures 2–4 show contour plots of the approximate potential for representative points in the red (medium-shaded), green (light-shaded), and blue (dark-shaded) areas of Fig. 1, respectively. The color- or shade-coding is explained in the captions. The plots illustrate the pattern of extrema described above.

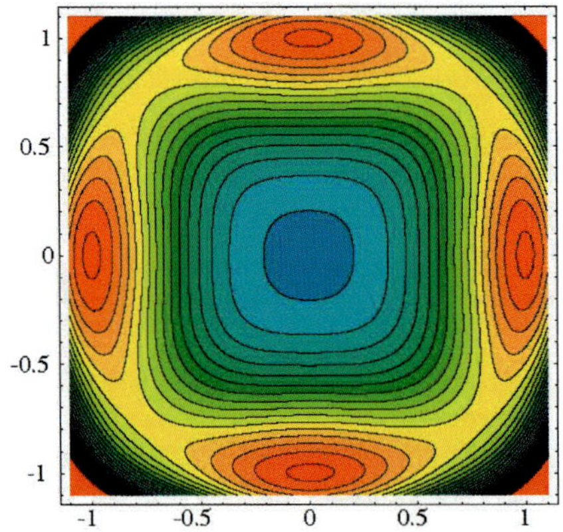

FIG. 2 (color online). Contour plot of the approximate potential for $\eta = \zeta = 1.6$ to illustrate the qualitative features of the extrema for a representative point in the red (medium-shaded) region of Fig. 1. Within this plot, red color codes for minima and blue for maxima. (These look similar in black-and-white, but the description below clarifies which is which.) The origin, $p_1 = p_2 = 0$, is a local maximum. The degenerate global minima are at $(p_1, p_2) = (0, \pm 1)$ and $(\pm 1, 0)$, while extrema II, where $p_1^2 = p_2^2$, are saddle points (light green or very-lightly shaded).

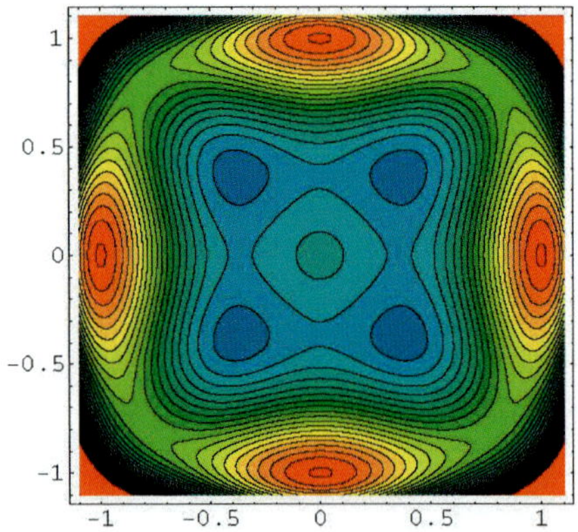

FIG. 3 (color online). Contour plot of the approximate potential for $\eta = 1.3$ and $\zeta = 1.6$ to illustrate the qualitative features of the extrema for a representative point in the green (light-shaded) region of Fig. 1. Within this plot, red color codes for minima and blue for maxima. (In black-and-white, the red region appears darker than the blue in this plot; the description below clarifies which is which.) The origin, $p_1 = p_2 = 0$, is now a local minimum, but the degenerate global minima are still at $(p_1, p_2) = (0, \pm 1)$ and $(\pm 1, 0)$. The relevant $p_1^2 = p_2^2$ extrema are again saddle points.

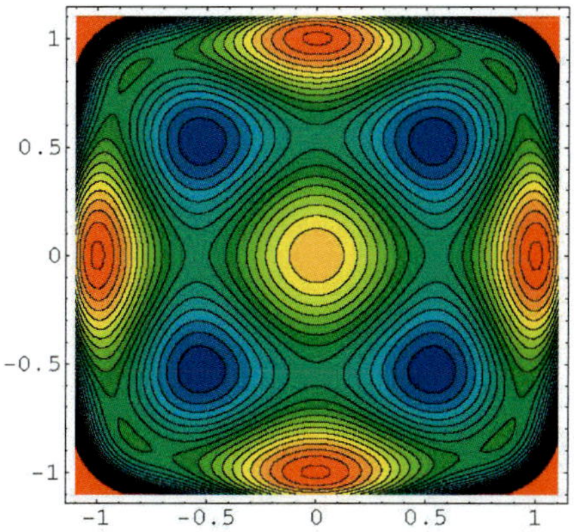

FIG. 4 (color online). Contour plot of the approximate potential for $\eta = 1.05$ and $\zeta = 1.6$ to illustrate the qualitative features of the extrema for a representative point in the blue (dark-shaded) region of Fig. 1. Within this plot, red color codes for minima and blue for maxima. (In black-and-white they appear similar, but the description below clarifies which is which.) The origin, $p_1 = p_2 = 0$, is now a deeper local minimum (yellow, or medium-shaded), having dropped below the $p_1^2 = p_2^2$ saddle points (green, or medium-shaded valleys). The local maxima are at approximately $(|p_1|, |p_2|) = (0.5, 0.5)$. The degenerate global minima are at $(p_1, p_2) = (0, \pm 1)$ and $(\pm 1, 0)$.

IV. THE GAUGED MODEL

It is natural to extend our toy model by gauging U(1) ⊗ U(1), and to study the dependence of the gauge fields on w by self-consistently solving all of the coupled Euler-Lagrange equations. In the flat-space case, Rozowsky, Volkas, and Wali [3] found some interesting solutions where the gauge field coupled to Φ_i tended towards a linear function of w on the side of the wall where $\Phi_i \to 0$, and was exponentially suppressed on the other side (Meissner effect). The linearly rising gauge field implies an asymptotically constant magnetic field, so the physical picture was of two infinite, uniform sheets (one for each sector) of supercurrent density flowing along the domain wall. In this section we will first generalize the Rozowsky *et al.* flat-space analysis and then consider the curved space problem.

A. Flat space

The Lagrangian is given by

$$\mathcal{L} = -\frac{1}{4}F_1^{MN}F_{1MN} - \frac{1}{4}F_2^{MN}F_{2MN} - (D^M\Phi_1)^*(D_M\Phi_1) - (D^M\Phi_2)^*(D_M\Phi_2) - V, \tag{4.1}$$

where

$$\begin{aligned} D_M\Phi_1 &= (\partial_M - ieA_{1M} - i\tilde{e}A_{2M})\Phi_1, \\ D_M\Phi_2 &= (\partial_M - i\tilde{e}A_{1M} - ieA_{2M})\Phi_2, \end{aligned} \tag{4.2}$$

and we take V to be quartic

$$V = \lambda_1(\overline{\Phi}_1\Phi_1 + \overline{\Phi}_2\Phi_2 - u^2)^2 + \lambda_2\overline{\Phi}_1\Phi_1\overline{\Phi}_2\Phi_2. \tag{4.3}$$

In the $\lambda_{1,2} > 0$ parameter space region, the global minima of V are

$$\begin{aligned} &\text{vacuum 1: } \overline{\Phi}_1\Phi_1 = u^2, \quad \Phi_2 = 0 \\ \text{and} \quad &\text{vacuum 2: } \Phi_1 = 0, \quad \overline{\Phi}_2\Phi_2 = u^2. \end{aligned} \tag{4.4}$$

The Lagrangian is invariant under the discrete symmetry

$$\Phi_1 \leftrightarrow \Phi_2, \qquad A_1 \leftrightarrow A_2. \tag{4.5}$$

This model generalizes that considered in Ref. [3] by having $\tilde{e} \neq 0$.

The Euler-Lagrange equations are

$$\begin{aligned} D^M D_M \Phi_i &= 2\lambda_1\Phi_i(\overline{\Phi}_i\Phi_i + \overline{\Phi}_j\Phi_j - u^2) + \lambda_2\Phi_i\overline{\Phi}_j\Phi_j, \\ \partial_M F_i^{MN} &= -2\,\mathrm{Im}(e\overline{\Phi}_i D^N\Phi_i + \tilde{e}\overline{\Phi}_j D^N\Phi_j), \end{aligned} \tag{4.6}$$

where $(i, j) = (1, 2)$ or $(2, 1)$. We now specialize to configurations that depend only on w, impose Lorentz gauge $\partial_M A_i^M = 0$, and write $\Phi_i = R_i e^{i\theta_i}$. The equations become

$$\begin{aligned} R_i'' = {}& R_i[(eA_i^\mu + \tilde{e}A_j^\mu)(eA_{i\mu} + \tilde{e}A_{j\mu})] \\ & + 2\lambda_1 R_i(R_i^2 + R_j^2 - u^2) + \lambda_2 R_i R_j^2, \end{aligned} \tag{4.7}$$

$$A_i^{\mu\prime\prime} = 2eR_i^2(eA_i^\mu + \tilde{e}A_j^\mu) + 2\tilde{e}R_j^2(eA_j^\mu + \tilde{e}A_i^\mu), \tag{4.8}$$

$$\theta_i' = -(eA_i^5 + \tilde{e}A_j^5). \tag{4.9}$$

Observe from the last of these equations that the $A = 5$ gauge field components are pure gauge, so they and the phase fields θ_i decouple from the problem.

We choose to look for solutions that do not break rotational invariance in the x, y, and z directions. This means that we can always rotate the x, y, z coordinate system so that the only nonzero component of (A_i^x, A_i^y, A_i^z) is in, say, the x direction (this is tantamount to the requirement that the ratios of these components are w independent). We also choose that the same component is nonzero for both $i = 1$ and $i = 2$. Equations (4.7) and (4.8) now simplify to

$$\begin{aligned} R_i'' = {}& R_i[-(eA_i^t + \tilde{e}A_j^t)^2 + (eA_i^x + \tilde{e}A_j^x)^2] \\ & + 2\lambda_1 R_i(R_i^2 + R_j^2 - u^2) + \lambda_2 R_i R_j^2, \end{aligned} \tag{4.10}$$

$$A_i^{t,x\prime\prime} = 2eR_i^2(eA_i^{t,x} + \tilde{e}A_j^{t,x}) + 2\tilde{e}R_j^2(eA_j^{t,x} + \tilde{e}A_i^{t,x}). \tag{4.11}$$

The final choice is whether to choose the gauge fields to be timelike $(A_i^t \neq 0, A_i^x = 0)$, lightlike $(A_i^t = A_i^x \neq 0)$, or spacelike $(A_i^t = 0, A_i^x \neq 0)$. As is obvious from Eq. (4.10), the timelike ansatz gives rise to asymptotic oscillatory behavior for the R's and is thus unacceptable. We first discuss the spacelike case.

As a boundary condition, we require that $R_{1,2}$ tend to vacua 1 and 2 on opposite sides of the wall:

$$R_1(+\infty) = u, R_2(+\infty) = 0$$
$$\text{and} \quad R_1(-\infty) = 0, R_2(-\infty) = u. \qquad (4.12)$$

The appropriate boundary conditions for the gauge fields are then determined by examining Eqs. (4.11) at the asymptotic points. As $w \longrightarrow +\infty$, we find that

$$\begin{pmatrix} A_1'' \\ A_2'' \end{pmatrix} = 2u^2 \begin{pmatrix} e^2 & e\tilde{e} \\ e\tilde{e} & \tilde{e}^2 \end{pmatrix} \begin{pmatrix} A_1 \\ A_2 \end{pmatrix}, \qquad (4.13)$$

where $A_i \equiv A_i^x$. Inputting trial asymptotic solutions of the form

$$A_{1,2} = a_{1,2} e^{-\kappa w}, \qquad (4.14)$$

we obtain an eigenvalue equation yielding

$$\kappa = \sqrt{2} u \sqrt{e^2 + \tilde{e}^2} \quad \text{and} \quad \kappa = 0. \qquad (4.15)$$

The zero eigenvalue means that one linear combination of A_1 and A_2 is always unsuppressed, and the vanishing second derivative tells us it grows linearly with w (provided its amplitude is nonzero). By the symmetry of the problem, a similar result is obtained as $w \longrightarrow -\infty$. The exponentially suppressed linear combination embodies the Meissner effect. Numerical solutions are depicted in Fig. 5.

The lightlike case is qualitatively similar to the spacelike case. But it is amusing to point out that for the parameter choice $\lambda_2 = 4\lambda_1$ an analytic solution exists. For the lightlike ansatz, the gauge fields cancel out of Eq. (4.10), and with the stated parameter choice and $e = 1$, $\tilde{e} = 0$, the resulting equations yield the solutions

$$R_{1,2}(w) = \frac{u}{2}\left[1 \pm \tanh\left(\sqrt{\lambda_1} u w\right)\right]. \qquad (4.16)$$

Inputting these results into Eq. (4.11), the analytic solutions obeying the correct boundary conditions for the gauge field functions $A_i^t = A_i^x \equiv A_i$ are

$$A_{1,2}(w) = \text{const} \times {}_2F_1\left(\frac{1+\sqrt{2}-\sqrt{3}}{2}, \frac{1+\sqrt{2}+\sqrt{3}}{2}; \times 1+\sqrt{2}; \left(1 + e^{\pm 2\sqrt{\lambda_1} u w}\right)^{-1}\right) \times \left(1 \mp \tanh\left(\sqrt{\lambda_1} u w\right)\right)^{1/\sqrt{2}}, \qquad (4.17)$$

where ${}_2F_1$ is a hypergeometric function.

B. Warped spacetime

We now generalize the above model through the inclusion of gravity [13]. The action is

$$S = -\int \left[\frac{\kappa}{2} R + g^{MN} t_{MN} + \frac{1}{4} g^{MP} g^{NQ} f_{MNPQ} + V\right] \times \sqrt{-g} d^5 x, \qquad (4.18)$$

with

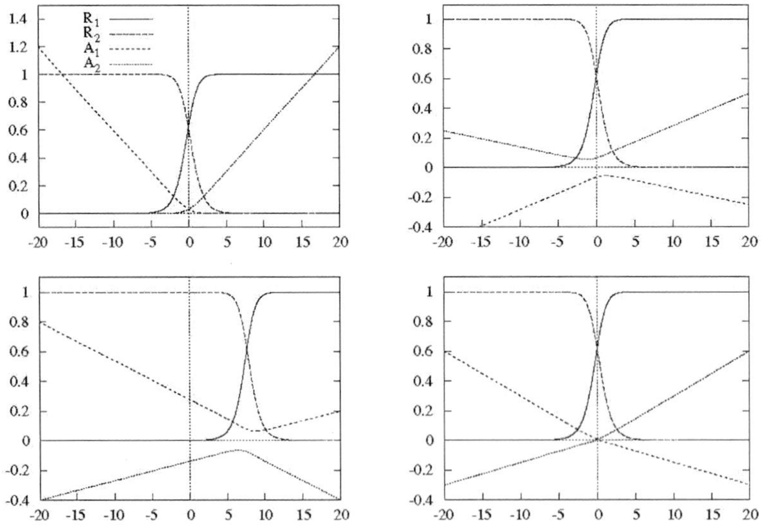

FIG. 5. Static solutions to the two scalar and two gauge fields in the U(1) ⊗ U(1) model. In all cases $\lambda_1 = \lambda_2 = e = 1$, and the fields are all plotted in units of u. In the top left plot, $\tilde{e} = 0$ as in the Rozowsky *et al.* model and the gauge fields show Meissner-like suppression under their respective scalar field. In the other three plots, $\tilde{e} = \frac{1}{2}$ and we show cases with various values for the free gauge field boundary conditions. In the two right plots, reflection symmetric boundary conditions are imposed for the pair of linear combinations $eA_i^x + \tilde{e}A_j^x$. In the bottom left plot, the gauge boundary conditions are asymmetric and their coupling to the kink shifts it to the right.

$$t_{MN} \equiv \sum_i (D_M\Phi_i)^* D_N\Phi_i, \tag{4.19}$$

$$D_M \equiv \partial_M - \sum_i \mathrm{i}Q_i A_{iM}, \tag{4.20}$$

$$f_{MNPQ} \equiv \sum_i F_{iMN}F_{iPQ}, \tag{4.21}$$

and the charges (Q_1, Q_2) of the scalar fields are $(e, \tilde{e})$ for Φ_1 and $(\tilde{e}, e)$ for Φ_2.

The Higgs and gauge field equations of motion are

$$D_M\left(\sqrt{-g}g^{MN}D_N\Phi_i\right) - \sqrt{-g}\frac{\partial V}{\partial \Phi_i^*} = 0, \tag{4.22}$$

$$\partial_M\left(\sqrt{-g}g^{MP}g^{NQ}F_{iPQ}\right) + \sqrt{-g}g^{NP}2(e\,\mathrm{Im}(\Phi_i^* D_P\Phi_i) + \tilde{e}\,\mathrm{Im}(\Phi_j^* D_P\Phi_j)) = 0, \tag{4.23}$$

while the Einstein equations, $\kappa G_{MN} = -T_{MN}$, feature

$$T_{MN} = 2t_{MN} + g^{PQ}f_{MPNQ} + g_{MN}\mathcal{L}_M, \tag{4.24}$$

$$\mathcal{L}_M = -g^{MN}t_{MN} - \tfrac{1}{4}g^{MP}g^{NQ}f_{MNPQ} - V. \tag{4.25}$$

The metric ansatz is generalized to

$$ds^2 = -e^{f(w)}dt^2 + e^{h(w)}dx^2 + e^{j(w)}(dy^2 + dz^2) + dw^2, \tag{4.26}$$

in order to have enough degrees of freedom to be able to look at sufficiently general gauge field configurations. We look for solutions of the form

$$A_{iM} = (A_i(w), B_i(w), 0, 0, Z_i(w)), \tag{4.27}$$

$$\Phi_i = R_i(w)e^{\mathrm{i}\alpha_i(w)}. \tag{4.28}$$

To simplify the algebra it is useful to define the following:

$$\mathcal{F} = \tfrac{1}{2}(f + h + 2j), \tag{4.29}$$

$$\overline{\mathcal{F}} = \tfrac{1}{2}(-f + h + 2j), \tag{4.30}$$

$$\overline{\overline{\mathcal{F}}} = \tfrac{1}{2}(f - h + 2j), \tag{4.31}$$

$$\mathcal{A}_i = eA_i + \tilde{e}A_j, \tag{4.32}$$

$$\mathcal{B}_i = eB_i + \tilde{e}B_j, \tag{4.33}$$

$$\mathcal{Z}_i = eZ_i + \tilde{e}Z_j - \alpha_i'. \tag{4.34}$$

In terms of these, the equations of motion for the gauge field components are

$$A_i'' + \overline{\mathcal{F}}'A_i' - 2eR_i^2\mathcal{A}_i - 2\tilde{e}R_j^2\mathcal{A}_j = 0, \tag{4.35}$$

$$B_i'' + \overline{\overline{\mathcal{F}}}'B_i' - 2eR_i^2\mathcal{B}_i - 2\tilde{e}R_j^2\mathcal{B}_j = 0, \tag{4.36}$$

$$(eR_i^2 + \tilde{e}R_j^2)\mathcal{Z}_i = 0. \tag{4.37}$$

Equation (4.37) implies $\mathcal{Z}_i = 0$ which we use to simplify the rest of the equations of motion. For the scalar fields we have

$$R_i'' + \mathcal{F}'R_i' + e^{-f}\mathcal{A}_i^2R_i - e^{-h}\mathcal{B}_i^2R_i - e^{-\mathrm{i}\alpha_i}\frac{\partial V}{\partial \Phi_i^*} = 0, \tag{4.38}$$

and for the three metric functions we have

$$f'' + \tfrac{1}{2}f'^2 + \tfrac{1}{3}f'h' + \tfrac{2}{3}f'j' - \tfrac{1}{3}h'j' - \tfrac{1}{3}j'^2 - \tfrac{10}{3\kappa}e^{-f}\Phi - \tfrac{2}{3\kappa}e^{-h}\Psi + \tfrac{2}{3\kappa}\Omega = 0, \tag{4.39}$$

$$h'' + \tfrac{1}{3}f'h' - \tfrac{1}{3}f'j' + \tfrac{1}{2}h'^2 + \tfrac{2}{3}h'j' - \tfrac{1}{3}j'^2 + \tfrac{2}{3\kappa}e^{-f}\Phi + \tfrac{10}{3\kappa}e^{-h}\Psi + \tfrac{2}{3\kappa}\Omega = 0, \tag{4.40}$$

$$j'' - \tfrac{1}{6}f'h' + \tfrac{1}{6}f'j' + \tfrac{1}{6}h'j' + \tfrac{5}{6}j'^2 + \tfrac{2}{3\kappa}e^{-f}\Phi - \tfrac{2}{3\kappa}e^{-h}\Psi + \tfrac{2}{3\kappa}\Omega = 0, \tag{4.41}$$

where

$$\Phi = \mathcal{A}_1^2R_1^2 + \mathcal{A}_2^2R_2^2 + \tfrac{1}{2}A_1'^2 + \tfrac{1}{2}A_2'^2, \tag{4.42}$$

$$\Psi = \mathcal{B}_1^2R_1^2 + \mathcal{B}_2^2R_2^2 + \tfrac{1}{2}B_1'^2 + \tfrac{1}{2}B_2'^2, \tag{4.43}$$

$$\Omega = R_1'^2 + R_2'^2 + V. \tag{4.44}$$

We performed an extensive numerical search for solutions to Eqs. (4.35), (4.36), (4.38), (4.39), (4.40), and (4.41), subject to the boundary conditions that the Higgs fields tend to VEVs, and that the metric functions tend to Randall-Sundrum form. For definiteness, we used the same Higgs potential as in the nongauged model, namely, as given by Eqs. (3.20), (3.22), and (3.25). The result of this investigation was the perhaps disappointing one that all solutions found required the gauge fields to vanish. Since the investigation was numerical, we cannot be certain that no nontrivial solutions exist, although we strongly suspect this to be the case.

V. CONCLUSION

We have constructed a $U(1) \otimes U(1)$ model with two complex scalar fields interchanged by a discrete symmetry coupled to 4 + 1 dimensional gravity. It has a solution featuring clash-of-symmetries-style Higgs kink configurations in a Randall-Sundrum-like spacetime. Gravity is localized to the dynamically generated (smooth) brane, and the Randall-Sundrum limit of the solution can be defined. For the chosen Higgs kink and metric configurations, the Higgs potential had to be of a certain sextic form whose properties we studied in some depth. The symmetry breaking pattern varies as a function of the extra dimension coordinate w, and displays the clash-of-symmetries phenomenon: at all points $|w| < \infty$ both U(1)'s are broken, with alternate U(1)'s restored as $w \to +\infty$ and $w \to -\infty$. The spontaneous breaking of the discrete interchange symmetry guarantees topological stability for the Higgs-gravity-induced brane. We also examined the gauged version of the theory, but we found that to have nonzero gauge fields the spacetime has to be flat.

This work sets the stage for incorporating gravity into more complicated models displaying the clash of symmetries. The eventual aim is to construct a realistic braneworld model that uses the clash of symmetries to understand spontaneous symmetry breaking in a more satisfactory fashion than obtained with the default mechanism using homogeneous Higgs vacuum expectation values.

ACKNOWLEDGMENTS

G. D. was supported in part by the Commonwealth of Australia, D. P. G. by the Puzey bequest to the University of Melbourne, R. R. V. by the Australian Research Council, and K. C. W. in part by the U.S. Department of Energy (DOE) Grant No. DE-FG02-85ER40237. R. R. V. would like to thank D. Vignaud, A. Goldwurm, and the APC group based at the Collège de France for their hospitality while this manuscript was completed.

[1] A. Davidson, B. F. Toner, R. R. Volkas, and K. C. Wali, Phys. Rev. D **65**, 125013 (2002).

[2] E. M. Shin and R. R. Volkas, Phys. Rev. D **69**, 045010 (2004).

[3] J. S. Rozowsky, R. R. Volkas, and K. C. Wali, Phys. Lett. B **580**, 249 (2004).

[4] A. Demaria and R. R. Volkas, Phys. Rev. D **71**, 105011 (2005).

[5] L. Pogosian and T. Vachaspati, Phys. Rev. D **62**, 123506 (2000); **64**, 105023 (2001); T. Vachaspati *ibid.* **64**, 105023 (2001).

[6] L. Randall and R. Sundrum, Phys. Rev. Lett. **83**, 3370 (1999); **83**, 4690 (1999).

[7] K. Akama, Lect. Notes Phys. **176**, 267 (1982); N. Arkani-Hamed, S. Dimopoulos, and G. R. Dvali, Phys. Lett. B **429**, 263 (1998); I. Antoniadis, Phys. Lett. B **246**, 377 (1990); I. Antoniadis *et al.*, *ibid.* **436**, 257 (1998).

[8] The dynamical localization of scalar, fermions, and gauge fields to branes is an important problem; see, for example, V. A. Rubakov and M. E. Shaposhnikov, Phys. Lett. B **125**, 136 (1983); C. Ringeval, P. Peter, and J. P. Uzan, Phys. Rev. D **65**, 044016 (2002); G. R. Dvali and M. A. Shifman, Phys. Lett. B **396**, 64 (1997); **407**, 452(E) (1997); S. L. Dubovsky, V. A. Rubakov, and P. G. Tinyakov, Phys. Rev. D **62**, 105011 (2000); S. Randjbar-Daemi and M. E. Shaposhnikov, Phys. Lett. B **492**, 361 (2000); E. Kh. Akhmedov, *ibid.* **521**, 79 (2001).

[9] Y. Kawamura, Prog. Theor. Phys. **103**, 613 (2000); **105**, 999 (2001); **105**, 691 (2001); A. Hebecker and J. March-Russell, Nucl. Phys. **B613**, 3 (2001); **625**, 128 (2002); L. J. Hall *et al.*, Phys. Rev. D **65**, 035008 (2002).

[10] T. Asaka, W. Buchmüller, and L. Covi, Phys. Lett. B **563**, 209 (2003); W. Buchmüller, Nucl. Phys. B Proc. Suppl. **137**, 182 (2004).

[11] S. Weinberg, *Gravitation and Cosmology: Principles and Applications of the General Theory of Relativity* (Wiley, New York, 1972).

[12] A. Davidson and P. Mannheim, hep-th/0009064; See also, O. DeWolfe, D. Z. Freedman, S. S. Gubser, and A. Karch, Phys. Rev. D **62**, 046008 (2000); C. Csaki, J. Erlich, T. J. Hollowood, and Y. Shirman, Nucl. Phys. **B581**, 309 (2000); M. Gremm, Phys. Rev. D **62**, 044017 (2000); Phys. Lett. B **478**, 434 (2000); K. Behrndt, Phys. Lett. B **487**, 30 (2000).

[13] For a six-dimensional model with gravity and gauge fields only, see Y. Brihaye, F. Clement, and B. Hartmann, Phys. Rev. D **70**, 084003 (2004).

THE HIGGS SECTOR ON A TWO-SHEETED SPACE–TIME

COSMIN MACESANU* and KAMESHWAR C. WALI†
Department of Physics, Syracuse University,
Syracuse, New York, 13244, USA
**cmacesan@physics.syr.edu*
†wali@physics.syr.edu

Received 23 January 2006

We present a general formalism based on the framework of noncommutative geometry, suitable to the study of the standard model of electroweak interactions, as well as that of more general gauge theories. Left- and right-handed chiral fields are assigned to two different sheets of space–time (a discretized version of Kaluza–Klein theory). Scalar Higgs fields find themselves treated on the same footing as the gauge fields, resulting in spontaneous symmetry breaking in a natural and predictable way. As a first test, we apply the formalism to the Standard Model, where one can predict the Higgs mass and the top Yukawa coupling. The results obtained for this case are similar with results obtained in alternate formulations. We then study the left–right symmetric model, where we show that this framework imposes interesting constraints on the type and coefficients of terms appearing in the Higgs potential.

Keywords: Higgs bosons; left–right symmetry; noncommutative geometry.

1. Introduction

The current picture of a continuum space–time at all scales underlying a smooth manifold has proved inadequate to describe all elementary particle interactions including gravity. The mathematical frameworks of both general relativity and quantum field theory of elementary particle interactions assume such a smooth manifold, and their incompatibility is one of the main signals for the inadequacy of such a continuum picture of space–time. In search of a more general framework, Connes has proposed an alternate approach based on noncommutative geometry (NCG).[1,4] The basic idea of Connes is to do away with the precise specification of an underlying manifold as the starting point. Instead, he formulates its description in terms of an associative and involutive algebra, commutative or noncommutative. One may think of this as a generalization of the well known theorem due to Gelfand, which states that the classical topological space based on a continuum can be completely recovered from the Abelian algebra of smooth functions.

Connes' ideas have been explored in several directions. Of particular relevance is their application to the standard model and beyond. In spite of the great success of the Standard Model (SM) in confronting experiments, it is recognized that it is far from a fundamental theory. One of the main drawbacks is that the mechanism of spontaneous symmetry breaking depends on the *ad hoc* addition of a Higgs sector, with a somewhat arbitrary Higgs field content and an arbitrarily chosen potential. In contrast, Connes' approach has given rise to a geometrical description of a gauge field theory in which the Higgs fields finds themselves on the same footing as the gauge fields, and where the Higgs potential has spontaneous symmetry breaking built in it.

In recent years, there has been a great deal of work beginning with Connes and Lott (Refs. 5–12 are some examples). However, most of the effort has been directed towards mathematical reformulations of Connes' original framework, which is too highly constrained to be applicable to models of current physical interest beyond the SM.[13] In our approach, which is also a reformulation, we keep close to the familiar ideas of quantum field theories and Riemannian geometry and use Connes' rigorous noncommutative geometry as a guide to construct models with interesting features from the physical point of view.

In our approach, the underlying manifold on which the theory is defined consists of the direct product between the four-dimensional (4D) Minkowski space–time and a finite number of points. Note that this is only a particular (and rather simple) realization of the more general theory of Connes. Specifically, we consider the case when the dimension of this discrete space is two; hence the name two-sheeted space–time. This framework allows a simple physical interpretation, as a model in which 4D fields live on two distinct branes embedded in a higher dimensional space; in other words, a discretized version of Kaluza–Klein theory.[14–16] As such, one can think that this theory might be derived as the 4D limit of a more fundamental theory (perhaps a string one).

We envisage the fermion left- and right-chiral fields living on two separate sheets of space–time associated with the two discrete points. The generalized gauge potential $\mathbb{A}$ (known as a connection in geometrical terms) appears as a two-by-two matrix operator on this internal space. The diagonal components couple fermion fields living on the same sheet (therefore with the same chirality) and consequently are identified with the usual 4D gauge fields. The off-diagonal components couple left- with right-handed fermion fields, and therefore can be identified as standard Higgs fields. We show that it is possible to build a suitable curvature operator (the field strength $\mathbb{F}$), and construct a consistent gauge theory by using these extended gauge operators. The Higgs potential terms appear as intrinsic components of the gauge invariant action for the Yang–Mills fields $*\mathbb{F}\mathbb{F}$.

We apply these ideas by first formulating the SM as a $\mathrm{SU}(3)_C \times \mathrm{SU}(2)_L \times \mathrm{U}(1)_Y$ gauge theory on two sheeted space–time. Due to the simplicity of the Higgs sector in this case (one Higgs field suffices), one finds that it is possible to predict the top Yukawa coupling (~ 0.8, close to but not in exact agreement with the experimentally determined value of the top mass), as well a value for Higgs mass. (These results are however tree level results, and are expected to acquire radiative corrections.)

Most of the other NCG models also have used the SM as a test case; we do not claim new results in this area, but discuss it nevertheless as a simple example, and in order to ensure that our model leads to results similar to other formulations. We think that more interesting possibilities lie in the application of this formalism to theories which may be valid at higher energy scales, such as grand unified theories based on higher gauge groups. In such theories, generally the Higgs sector can be quite complex, and in the absence of a guiding principle one must rely on additional assumptions. The approach described in this paper can provide such a guide (unlike some of the original formulations of NCG models, which were found to forbid higher symmetry groups).[13] As an example, we formulate the left–right symmetric model in this framework, and we show that one can get some interesting constraints on the type of terms one can have in the Higgs potential.

The paper is organized as follows. In Sec. 2, we briefly review Connes' abstract algebra approach and the concept of a spectral triple. In Sec. 3, we adapt this formalism to a two-sheeted space–time that may be considered as a discretized version of Kaluza–Klein theory, in which the compact circle in the fifth dimension is replaced by two discrete points. Sections 4 and 5 are devoted to the construction of SM and the left–right symmetric model in this framework. The final section is devoted to a discussion of the results and conclusions.

2. A Brief Review of Noncommutative Differential Geometry

Connes' starting point of NCG is a universal differential algebra $\Omega^*(\mathcal{A})$ constructed from an associative, commuting or noncommuting algebra $\mathcal{A}$. It is unital and involutive. This differential algebra can be thought of as being generated by the elements a, b and a symbol δ with properties $\delta 1 = 0$, $\delta(ab) = (\delta a)b + a\delta b$ for all $a,\ b \in \mathcal{A}$. The zero order forms are simply the elements of $\mathcal{A}$. Higher order forms in $\Omega^p\mathcal{A}$ are generated by "extending" the differential δ to an operator: if $a_0\delta a_1 \cdots \delta a_p$ is a p-form, then

$$\delta(a_0\delta a_1 \cdots \delta a_p) = \delta a_0\delta a_1 \cdots \delta a_p$$

is a $(p+1)$-form. This also implies $\delta^2 = 0$. For a more detailed discussion at the introductory level, see, for example, Ref. 18.

In physical applications of interest to us, the abstract algebra becomes an algebra of operators acting on a Hilbert space $\mathcal{H}$. The abstract elements $a \in \mathcal{A}$ are represented as operators through a representation $\rho(a) \equiv A$ acting on $\mathcal{H}$. The abstract symbol δ becomes the exterior derivative represented by a self-adjoint operator called the Dirac operator $\mathcal{D}$. The three elements, the algebra A, the Dirac operator $\mathcal{D}$ and the Hilbert space $\mathcal{H}$ together form a *spectral triple* according to Connes.

With the help of the Dirac operator, we can now extend the representation of $\mathcal{A}$ to its algebra of differential forms through the correspondence:

$$\rho(a_0\delta a_1 \cdots \delta a_p) = \rho(a_0)[\mathcal{D}, \rho(a_1)] \cdots [\mathcal{D}, \rho(a_p)] .$$

In this sense, the Dirac operator can be used to build a "representation" of the formal exterior derivative δ in the space of operators acting on $\mathcal{H}$; if $\phi \in \Omega(\mathcal{A})$:

$$\rho(\delta\phi) = [\mathcal{D}, \rho(\phi)] .$$

(Depending on the rank of the form ϕ and the grading properties of the algebra, one may also need to use anticommutators.)

One must be careful, however; if δ is a proper exterior derivative, one needs to have $\delta(\delta\phi) = 0$; but generally $[\mathcal{D}, [\mathcal{D}, \rho(\phi)]]$ is not zero. To get around this problem, one defines an equivalence relation on the spaces $\rho(\Omega^p(\mathcal{A}))$, so that all the operators of the form $[\mathcal{D}, [\mathcal{D}, \rho(\phi)]]$ are equivalent to the zero operator; technically, the space of operators corresponding to p-forms is $\rho(\Omega^p(\mathcal{A}))$ divided by the space of so-called junk forms $\mathcal{J} = \rho(\delta(\ker \Omega^{p-1}(\mathcal{A}))$ (this is sometimes called dividing out the junk).[a]

In our studies, $\mathcal{H}$ will consist of square integrable sections of a spinor bundle representing physical states of the fermions. Let us assume that the fermion fields form a basis for a representation of some group G. For theories defined on a smooth manifold, the gauge fields will then be associated with differential one-forms defined on the manifold. The field strength, which is used in the construction of the Lagrangian, will be a two-form. Generally, physically relevant quantum operators built out of gauge fields are elements of the differential algebra $\Omega(M)$ defined on the manifold.

For the purposes of illustration, consider the simple case of a gauge theory on the Minkowski manifold. The relevant algebra is $C^\infty(M)$. The Hilbert space $\mathcal{H}$ is the space of Dirac spinors; and the Dirac operator is the standard $\mathcal{D} = \not\partial = \gamma^\mu \partial_\mu$. Elements of $C^\infty(M)$ are in one to one correspondence with fields on the four-dimensional space–time. The space of one-forms is defined by

$$\mathbb{A} = f(x)[\not\partial, g(x)] = f(x)\partial_\mu(g)(x)\gamma^\mu \equiv A_\mu(x)\gamma^\mu ,$$

with $A_\mu(x) \in C^\infty(M)$.[b] Note that the decomposition of one-forms this way is similar to the usual writing of one-forms as $A_\mu(x)dx^\mu$, only in this case the matrices γ^μ play the role of a basis in the space of one-forms. The field strength will then be (here we use the anticommutator):

$$\begin{aligned} \mathbb{F} = d\mathbb{A} = \{\not\partial, A_\mu(x)\gamma^\mu\} &= \gamma^\nu\gamma^\mu\partial_\nu A_\mu + \gamma^\mu\gamma^\nu A_\mu\partial_\nu \\ &= \gamma^\nu\gamma^\mu\partial_\nu(A_\mu) + 2\eta^{\mu\nu}A_\mu\partial_\nu \\ &= \gamma^\nu\gamma^\mu\frac{1}{2}[\partial_\nu(A_\mu) - \partial_\mu(A_\nu)] . \end{aligned} \tag{1}$$

In going from the second to the last line, we have antisymmetrized the result with respect to indices μ, ν, and dropped the symmetric terms (the ones proportional to $\eta^{\mu\nu}$). This prescription is imposed by the requirement that if $\mathbb{A} = d\Phi$, $d\mathbb{A} = 0$.

[a] A slightly different definition of junk forms is used in the literature. However, for our purposes, the two definitions seem to give the same results (see Ref. 18).

[b] Note that the functions appearing in these relations are to be thought of as operators. Then, for example, $\partial g(x) = \partial(g)(x) + g(x)\partial$.

In other words, the equivalence relation necessary to get rid of the junk forms is: two-forms are equivalent to each other if their difference is symmetric in the μ, ν indices, and proportional to the identity matrix. Alternately, it is equivalent to using the wedge product of γ matrices as a basis in the space of higher forms; $\mathbb{F}$ can then be written as

$$\mathbb{F} = \partial_\nu(A_\mu)\gamma^\nu \wedge \gamma^\mu ,$$

with $\gamma^\nu \wedge \gamma^\mu = (\gamma^\nu\gamma^\mu - \gamma^\mu\gamma^\nu)/2$.

3. Two-Sheeted Space–Time

In the previous section we have given a brief description of the general formalism of Connes to build noncommutative field theories. There are several realizations of this formalism in practice, some of them based on Connes' original formulation,[5–8] with others based on some variation of it.[9,10,12,11] In this section (and the rest of the paper) we will discuss a model based on the approach of Refs. 14 and 15, which we call in what follows, two-sheeted space–time. The formalism is quite similar to the one in Refs. 12 and 17; however, we differ from their approach in the way we compute the curvature on the discrete space, and therefore the Higgs potential.

In our scenario, the spectral triple consists of the following.

Hilbert Space $\mathcal{H}$: The direct product of the usual Minkowski spin manifold times $\mathbb{C}^2$ ($\mathbb{C}$ being the algebra of complex numbers). Formally, it can be represented as

$$\Psi = \begin{bmatrix} \Psi_L \\ \Psi_R \end{bmatrix} ,$$

where Ψ_L, Ψ_R are each a Dirac spinor by itself; however, we can choose that each such Dirac spinor has specific helicity (either left- or right-handed) and that they live on different sheets.

Algebra $\mathcal{A}$: The most general (smooth) operator acting on such a Hilbert space is a 2×2 matrix with elements C^∞ functions. However, we choose the algebra $\mathcal{A}$ to be a subset of these matrices; specifically we require that the representations of elements of $\mathcal{A}$ be diagonal matrices:

$$\rho(a) = \begin{bmatrix} f_1(x) & 0 \\ 0 & f_2(x) \end{bmatrix} . \tag{2}$$

Dirac Operator: If we take this also to be diagonal, the resulting theory would not be very interesting; the fields living on separate sheets will not interact with each other, and in effect we will have two copies of the standard theory on a manifold. Therefore, we take the Dirac operator to have off-diagonal terms:

$$\mathcal{D} = \begin{bmatrix} i\partial\!\!\!/ & -\gamma^5 M \\ \gamma^5 M^\dagger & i\partial\!\!\!/ \end{bmatrix} , \tag{3}$$

where M is a scalar operator with dimensions of mass. With this definition, a general one-form is given by

$$\mathbb{A} = \begin{bmatrix} \gamma^\mu A_{1\mu}(x) & -\gamma^5 \Phi_1(x) \\ \gamma^5 \Phi_2(x) & \gamma^\mu A_{2\mu}(x) \end{bmatrix}, \tag{4}$$

where $A_{i\mu}$ are the gauge fields, and Φ_i are the Higgs fields of the theory.

The model described above is easily extended to incorporate non-Abelian gauge theories. In the general case, one can take the fields Ψ_L, Ψ_R as multiplets under a fundamental representation of some gauge groups G_1, G_2. Let us denote the dimension of the Ψ_L multiplet by n and the dimension of the Ψ_R multiplet by m. Then the C^∞ functions $f_1(x)$, $A_{1\mu}(x)$, and $f_2(x)$, $A_{2\mu}(x)$ will became $n \times n$ matrices, and $m \times m$, respectively. They can be written in terms of the generators of the corresponding Lie algebras T_a, T'_b:

$$\begin{aligned} f_1(x) &= f_1^a(x) T_a, \qquad & A_{1\mu}(x) &= A_{1\mu}^a(x) T_a, \\ f_2(x) &= f_2^b(x) T'_b, \qquad & A_{2\mu}(x) &= A_{2\mu}^b(x) T'_b. \end{aligned} \tag{5}$$

To support the identification of the Φ fields in (4) as Higgs fields, consider the fermionic part of the Lagrangian. With the covariant derivative defined in the usual way,

$$i\mathbb{D} = \mathcal{D} + \mathbb{A}, \tag{6}$$

one has

$$\begin{aligned} \mathcal{L}_\psi = \bar{\Psi} i\mathbb{D} \Psi &= \bar{\psi}_L(i\partial\!\!\!/ + A\!\!\!/_1)\psi_L + \bar{\psi}_R(i\partial\!\!\!/ + A\!\!\!/_2)\psi_R \\ &\quad - (\bar{\psi}_L(\Phi_1 + M)\psi_R + \bar{\psi}_R(\Phi_2 + M^\dagger)\psi_L). \end{aligned} \tag{7}$$

In order for this Lagrangian to be Hermitian, we define $\Phi_2 = \Phi_1^\dagger$.

Note that Eqs. (3), (4) are consistent with an alternative formulation,[14,c] where one writes the generalized Dirac operator and gauge one-form in terms of a decomposition over a Γ^μ, Γ^5 set of matrices:

$$\mathbb{A} = \Gamma^\mu A_\mu + \Gamma^5 \Phi, \qquad \mathcal{D} = \Gamma^\mu D_\mu + \Gamma^5 D_5,$$

with

$$A_\mu(x) = \begin{bmatrix} A_{1\mu}(x) & 0 \\ 0 & A_{2\mu}(x) \end{bmatrix}, \qquad \Phi(x) = \begin{bmatrix} \Phi_2(x) & 0 \\ 0 & \Phi_1(x) \end{bmatrix},$$

$D_\mu = \mathrm{diag}\{\partial_\mu, \partial_\mu\}$, $D_5 = \mathrm{diag}\{M^\dagger, -M\}$, and

$$\Gamma^\mu = \begin{bmatrix} \gamma^\mu & 0 \\ 0 & \gamma^\mu \end{bmatrix}, \qquad \Gamma^5 = \begin{bmatrix} 0 & -\gamma^5 \\ \gamma^5 & 0 \end{bmatrix}.$$

[c]Extensive references to related work can be found in Ref. 15.

In this approach, the Higgs fields can be viewed as components of gauge fields in a fifth (discrete) dimension. One- and two-forms can then be built by giving appropriate prescriptions for the wedge products of the Γ^μ, Γ^5 matrices. For more details, we refer the reader to Ref. 15.

In either case, the field strength is given by

$$\mathbb{F} = d\mathbb{A} + \mathbb{A} \wedge \mathbb{A}\,, \tag{8}$$

where

$$\begin{aligned} d\mathbb{A} &= \{\mathcal{D}, \mathbb{A}\} \\ &= \begin{bmatrix} i\partial\!\!\!/(A\!\!\!/_1) - M\Phi^\dagger - \Phi M^\dagger & \gamma^\mu\gamma^5(-i\partial_\mu(\Phi) + MA_{2\mu} - A_{1\mu}M) \\ \gamma^5\gamma^\mu(-i\partial_\mu(\Phi^\dagger) + M^\dagger A_{1\mu} - A_{2\mu}M^\dagger) & i\partial\!\!\!/(A\!\!\!/_2) - M^\dagger\Phi - \Phi^\dagger M \end{bmatrix}, \end{aligned} \tag{9}$$

and

$$\begin{aligned} \mathbb{A} \wedge \mathbb{A} &= \frac{1}{2}\{\mathbb{A}, \mathbb{A}\} \\ &= \begin{bmatrix} A\!\!\!/_1 A\!\!\!/_1 - \Phi\Phi^\dagger & \gamma^\mu\gamma^5(-A_{1\mu}\Phi + \Phi A_{2\mu}) \\ \gamma^5\gamma^\mu(\Phi^\dagger A_{1\mu} - A_{2\mu}\Phi^\dagger) & A\!\!\!/_2 A\!\!\!/_2 - \Phi^\dagger\Phi \end{bmatrix}. \end{aligned} \tag{10}$$

The diagonal elements of the form $\mathbb{F}$ contain $(i/2)\gamma^\mu\gamma^\nu(F_{1,2})_{\mu\nu}$, with

$$F_{1\mu\nu} = \partial_\mu A_{1\nu} - \partial_\nu A_{1\mu} - i[A_{1\mu}, A_{1\nu}]\,,$$

and the equivalent for F_2. Thus one obtains the proper covariant expressions for the field strengths of the respective gauge fields. Furthermore, it is convenient to redefine the Higgs fields as $H = \Phi + M$; then the field strength (8) can be written as

$$\mathbb{F} = \begin{bmatrix} \frac{i}{2}\gamma^\mu\gamma^\nu F_{1\mu\nu} + (MM^\dagger - HH^\dagger) & -\gamma^\mu\gamma^5(i\partial_\mu(H) - HA_{2\mu} + A_{1\mu}H) \\ -\gamma^5\gamma^\mu(i\partial_\mu(H^\dagger) - H^\dagger A_{1\mu} + A_{2\mu}H^\dagger) & \frac{i}{2}\gamma^\mu\gamma^\nu F_{2\mu\nu} + (M^\dagger M - H^\dagger H) \end{bmatrix}. \tag{11}$$

Some care has to be exercised here in treating the contributions of the Higgs field to the diagonal elements. As mentioned in the previous section, if δ is a proper exterior derivative, $\delta(\delta\alpha) = 0$ for any form α. Thus, if one takes $\mathbb{A} = d\rho$, with ρ an element of the algebra as in Eq. (2), we should have $d\mathbb{A} = dd\rho = 0$. However, from Eq. (9):

$$dd\rho = \begin{bmatrix} MM^\dagger f_1 - f_1 MM^\dagger & 0 \\ 0 & M^\dagger M f_2 - f_2 M^\dagger M \end{bmatrix} \tag{12}$$

is not zero, unless $MM^\dagger$ is proportional to the unity. Such terms are what we have characterized earlier as "junk" forms. It is easily seen that they are eliminated by taking trace of the diagonal elements in (12).

The formal justification for the trace is as follows. The elements $\mathbb{F}_{11}$, $\mathbb{F}_{22}$ of a general two-form are Hermitian matrices, which can be expressed as combinations of the traceless Lie algebra generators T and the identity matrix

$$\mathbb{F}_{11} = C_0 I + C_a T_a \,,$$

where C_a are bilinears in the Higgs field components H_{ij} (which enter in the definition of $\mathbb{F}_{11}$). Now, since the T_a generators are traceless, the function C_0 is proportional to $\mathrm{Tr}(\mathbb{F}_{11})$. Therefore, the C_0 component is zero for the part $\mathbb{F}_{11}$ in the decomposition of the form (12), while the others C_a functions typically are not (unless $MM^\dagger$ is proportional to the identity matrix).

In the space of two-forms, one can then define the components $\sim T_a$ as junk. On the quotient space we will therefore have $dd\rho = 0$, while in the expression (11) for the field strength we take the trace over the Higgs field contributions,[d] that is, replace $(MM^\dagger - HH^\dagger)$ by $\mathrm{Tr}(MM^\dagger - HH^\dagger)$ (which also ensures that the Higgs potential is gauge covariant). Note, also, that in the case when the fermion spinor space can be split into subspaces which do not mix (like lepton and quark subspaces in the SM), in order to ensure that $dd\rho = 0$, it is enough to take the trace on each subspace separately (rather than on the whole space).

The Lagrangian for the gauge sector (including the Higgs) will then be

$$\mathcal{L}_G = -\frac{1}{4}\,\mathrm{Tr}\,\mathbb{F}\mathbb{F}\,, \tag{13}$$

where the trace is first taken over the γ^μ matrices and then over the internal symmetry indices. The factor $-1/4$ in front of the trace insures that the kinetic energy for the gauge fields F_1, F_2 has the standard normalization. Then, we find

$$\mathcal{L}_G = \mathcal{L}_A + \mathcal{L}_H - V(H)\,, \tag{14}$$

where

$$\mathcal{L}_A = -\frac{1}{4}\sum_i F_{i\mu\nu}F_i^{\mu\nu} \tag{15}$$

with

$$F_{1\mu\nu} = F^a_{1\mu\nu}T_a\,, \qquad F_{2\mu\nu} = F^b_{2\mu\nu}T'_b\,,$$

with normalization $\mathrm{Tr}(T_aT_{a'}) = (1/2)\delta_{aa'}$. The kinetic energy for the Higgs fields has the form

$$\mathcal{L}_H = |\partial^\mu(H) + iHA_2^\mu - iA_1^\mu H|^2 + |\partial_\mu(H^\dagger) + iH^\dagger A_{1\mu} - iA_{2\mu}H^\dagger|^2\,, \tag{16}$$

[d]One could also drop these contributions completely, in effect setting the curvature associated to the Higgs fields to zero. However, the model obtained in this case is not very interesting, since it does not have a Higgs potential.

where we use the definition $|K|^2 = K^\dagger K$. Finally, the Higgs potential is given by

$$V(H) = (\mathrm{Tr}(MM^\dagger - HH^\dagger))^2 + (\mathrm{Tr}(M^\dagger M - H^\dagger H))^2 , \tag{17}$$

which requires that at least some Higgs fields have nonzero expectation values, thus insuring spontaneous symmetry breaking.

Note that Eq. (16) contains the proper covariant derivative for the Higgs fields H. To see this, note that the transformation properties of the Higgs fields are defined by the requirement that the fermionic Lagrangian (7) be gauge invariant. Thus, if the fermion fields Ψ_L form a multiplet supporting a fundamental representation of the gauge group G_1, so do the fields H. That is, the transformation properties of the Higgs fields under an infinitesimal gauge transformation will be the same as for the Ψ_L fermions:

$$\Psi_L(x) \to (1 + i\alpha^a T^a)\Psi_L , \qquad H(x) \to (1 + i\alpha^a T^a)H .$$

Then it is easy to verify that the derivative in Eq. (16) is the appropriate covariant derivative, and the theory is gauge invariant.

To review what we have learned so far: we have started by taking the elements of the algebra $\mathcal{A}$ to be block-diagonal matrices. However, due to the nondiagonal structure of the Dirac operator, the one-forms associated with the algebra $\mathcal{A}$ are also nondiagonal — that is, noncommutative. The diagonal elements of a one-form $\mathbb{A}$ are gauge fields, while the off-diagonal elements turn out to be Higgs fields. The resulting theory turns out to be gauge invariant with an obvious redefinition of the Higgs fields. Moreover, due to the noncommutativity of one-forms, the Higgs sector acquires a quartic potential, thus ensuring spontaneous symmetry breaking. The structure of the Higgs potential is determined by the structure of the off-diagonal terms in the Dirac operator (or, since these terms are related to the fermion masses, one can say that the Higgs potential is related to the mass spectrum of the fermion sector).

We would like to comment here on some differences between our approach and the previous ones. First, note that we do not make use of all of Connes' axioms.[4,6] These are mostly needed in order to build the scalar product on the underlying space (its definition involves the Diximier trace of the Dirac operator). By taking this space to be the Minkowski manifold times a finite number of points, we can use the standard 4D integration as a scalar product, with trace only over the internal degrees of freedom. This also allows us to work directly in Minkowski coordinates, not Euclidean ones. Also, our formalism with the conventional charge conjugation bypasses the need of the antilinear operator J to ensure the reality structure and Poincaré duality. As a consequence, our framework may allow the inclusion of larger symmetry groups than the SM (like the left–right symmetric model, which, for example, is not allowed in the Connes–Lott framework[13]). Gauge invariance of the Higgs sector in our theory follows directly from the prescription of eliminating the junk by taking the trace. Moreover, we do not link directly the entries in the matrix M to the fermion masses, and as a consequence, $\langle H \rangle = M$ is not necessarily a minimum of the Higgs potential (as it is explained in some detail in Sec. 5).

We end this section with some comments on the gauge couplings and the Higgs-fermion Yukawa couplings. In the above equations we set the gauge couplings to be one. Different values can be introduced either directly (that is, setting $i\mathbb{D} = \mathcal{D} + g\mathbb{A}$ in Eq. (6) and $\mathbb{F} = d\mathbb{A} + g\mathbb{A} \wedge \mathbb{A}$) or by generalizing the trace over $\mathbb{F}$ to include a gauge coupling matrix:

$$\mathcal{L}_G = -\frac{1}{4} \operatorname{Tr} \begin{bmatrix} 1/g_1^2 & 0 \\ 0 & 1/g_2^2 \end{bmatrix} \mathbb{F}\mathbb{F} . \tag{18}$$

Then one should use fields with proper normalization $A_1/g_1 \to A_1$, $A_2/g_2 \to A_2$, in effect setting the gauge couplings of the fields A_1, A_2 to be g_1, g_2 (one should also normalize the Higgs fields). Also, note that in the case when we have only one Higgs multiplet and one fermion multiplet, the Yukawa coupling of the Higgs with the fermions is fixed, being given by the Higgs field normalization constant. In the case when one has several fermion multiplets, one generally can (or has to) set the Yukawa coupling for each multiplet by hand; however, one still gets a sum rule relating the sum of the Yukawa couplings squared to the Higgs normalization constant. If one has several Higgs multiplets, one gets several such sum rules.

4. Standard Model on Two-Sheeted Space–Time

As the first example of the general formalism described in the previous section, let us discuss the implementation of the SM $SU(3)_C \times SU(2)_L \times U(1)_Y$ gauge theory. For simplicity, let us first restrict our analysis to a subspace of the fermion spinor space, particularly the one spanned by the u, d quark fields. Then, the components of the spinor multiplets would be

$$\Psi_L = \begin{bmatrix} u_L \\ d_L \end{bmatrix}, \qquad \Psi_R = \begin{bmatrix} u_R \\ d_R \end{bmatrix}.$$

As is well known, the Ψ_L components form a $SU(2)_L$ doublet, while the individual components of both Ψ_L and Ψ_R are $U(1)_Y$ singlets.

Then, over this subspace, the gauge fields components are

$$\begin{aligned} A_{1\mu} &= A^a_{1\mu}\tau_a + \frac{1}{2} Y_{QL} B_\mu , \\ A_{2\mu} &= \frac{1}{2} Y_{QR} B_\mu , \end{aligned} \tag{19}$$

where A^a are the $SU(2)_L$ gauge fields and B is the hypercharge field. The diagonal matrices Y_{QL}, Y_{QR} contain the hypercharges of the left- and right-type quark fields:

$$Y_{QL} = \frac{1}{3} \begin{bmatrix} 1 & 0 \\ 0 & 1 \end{bmatrix}, \qquad Y_{QR} = \begin{bmatrix} 4/3 & 0 \\ 0 & -2/3 \end{bmatrix}.$$

Note here the peculiar way we have introduced the hypercharge field. The matrix A_1 now contains fields associated with two gauge symmetries. This would generally require that A_1 be split into two parts: $A_1 \to \mathrm{diag}(A_1, B_1)$, but this is not necessary in this particular case, since the hypercharge matrix Y_{QL} is proportional to the identity matrix, and therefore commutes with the generators of the $\mathrm{SU}(2)_L$ gauge transformations. This means that the $\mathrm{SU}(2)_L$ and $\mathrm{U}(1)_Y$ gauge fields will not mix between them when computing $\mathbb{F}$ and the trace of $\mathbb{F}\mathbb{F}$. Therefore we can add them directly in the manner of (19).

The Higgs matrix for the model has the form

$$H = \begin{bmatrix} \bar{h}_0 & h_+ \\ -h_- & h_0 \end{bmatrix} = [\widetilde{\Phi}, \Phi],$$

where Φ is the $\mathrm{SU}(2)_L$ doublet and $\widetilde{\Phi}$ is its charge conjugate. Under an infinitesimal $\mathrm{U}(1)_Y$ gauge transformation $\alpha(x)$ the Higgs fields will change according to

$$H \to \left(1 + i\frac{\alpha}{2}Y_{QL}\right) H \left(1 - i\frac{\alpha}{2}Y_{QR}\right) \to H\left(1 + i\frac{\alpha}{2}(Y_{QL} - Y_{QR})\right),$$

since Y_{QL} is proportional to the identity matrix. The hypercharge matrix for the Higgs fields will then be

$$Y_H = Y_{QL} - Y_{QR} = \begin{bmatrix} -1 & 0 \\ 0 & 1 \end{bmatrix},$$

that is, the doublet Φ and antidoublet $\widetilde{\Phi}$ will have definite hypercharge $+1$ and -1 respectively. The covariant derivative for H turns out to be

$$D_\mu H = \partial_\mu H - iA^a_{1\mu}\tau_a H - iH\frac{1}{2}Y_H B,$$

or, in terms of the Φ, $\widetilde{\Phi}$ doublets:

$$D_\mu(\widetilde{\Phi}, \Phi) = \left(\left(\partial_\mu - iA^a_{1\mu}\tau_a + i\frac{1}{2}B\right)\widetilde{\Phi}, \quad \left(\partial_\mu - iA^a_{1\mu}\tau_a - i\frac{1}{2}B\right)\Phi\right). \tag{20}$$

Then,

$$\begin{aligned} \mathcal{L}_H &= 2\,\mathrm{Tr}(D^\mu H^\dagger)(D_\mu H) \\ &= 2(D^\mu \widetilde{\Phi}^\dagger)(D_\mu \widetilde{\Phi}) + 2(D^\mu \Phi^\dagger)(D_\mu \Phi) \end{aligned} \tag{21}$$

will give the standard form for the SM Higgs kinetic energy terms, although it still has to be normalized (the factor 2 in front comes from the fact that there are two equal terms in Eq. (16)).

Let us assume for a moment that this is the whole extent of the theory (that is, there are no other fermions), and work out the gauge coupling constants, Yukawa couplings and normalization factors. After multiplying with the gauge coupling matrix (as in Eq. (18)), the gauge fields have to be rescaled:

$$\begin{aligned} \frac{1}{g_1} A_1^a &\to A_1^a \,, \\ \left[\frac{1}{g_1^2} \operatorname{Tr}\left(\frac{Y_{QL}^2}{4} \right) + \frac{1}{g_2^2} \operatorname{Tr}\left(\frac{Y_{QR}^2}{4} \right) \right]^{1/2} B &\to B \,, \end{aligned} \tag{22}$$

thus setting the weak coupling constant $g = g_1$ and the hypercharge coupling constant

$$\frac{1}{g'} = \left[\frac{1}{g_1^2} \operatorname{Tr}\left(\frac{Y_{QL}^2}{4} \right) + \frac{1}{g_2^2} \operatorname{Tr}\left(\frac{Y_{QR}^2}{4} \right) \right]^{1/2} .$$

If one wishes to give different masses to the up and down type fermions, one has to introduce different Yukawa couplings in the Higgs matrix

$$H = \begin{bmatrix} \bar{h}_0 & h_+ \\ -h_- & h_0 \end{bmatrix} \begin{bmatrix} \lambda_u & 0 \\ 0 & \lambda_d \end{bmatrix} = [\lambda_u \widetilde{\Phi}, \lambda_d \Phi] \,,$$

since the vacuum expectation value (VEV) of the $\widetilde{\Phi}$ doublet gives mass to the up-type fermions, while the VEV of the Φ one gives mass to the down-type fermions. Note that this implies that the matrix M appearing in the Dirac operator has a similar structure

$$M \sim \begin{bmatrix} m\lambda_u & 0 \\ 0 & m\lambda_d \end{bmatrix} .$$

One has now to compute the rescaling factor for the Higgs field. The kinetic energy term (21) becomes

$$\mathcal{L}_H = \left(\frac{1}{g_1^2} + \frac{1}{g_2^2} \right) \lambda_u^2 (D^\mu \widetilde{\Phi}^\dagger)(D_\mu \widetilde{\Phi}) + \left(\frac{1}{g_1^2} + \frac{1}{g_2^2} \right) \lambda_d^2 (D^\mu \Phi^\dagger)(D_\mu \Phi) \,,$$

which requires the redefinition (rescaling) of H:

$$\left[\left(\frac{1}{g_1^2} + \frac{1}{g_2^2} \right) (\lambda_u^2 + \lambda_d^2) \right]^{1/2} H = \lambda H \to H \,.$$

Then, the effective couplings $\lambda_{u,d}^e$ to fermions will turn out to be

$$\lambda_u^e = \frac{\lambda_u}{\lambda} \,, \qquad \lambda_d^e = \frac{\lambda_d}{\lambda} \,,$$

and we have the sum rule mentioned in the previous section

$$(\lambda_u^e)^2 + (\lambda_d^e)^2 = \frac{g_1^2 g_2^2}{g_1^2 + g_2^2} \,. \tag{23}$$

Finally, let us consider the terms giving rise to the Higgs potential. Since the matrix M is not proportional to identity and it does not commute with a general zero-form, we should use the trace prescription in (9). Therefore,

$$\begin{aligned} V(H) &= \left(\frac{1}{g_1^2}+\frac{1}{g_2^2}\right)\left(\mathrm{Tr}[MM^\dagger - HH^\dagger]\right)^2 \\ &\to \left(\frac{1}{g_1^2}+\frac{1}{g_2^2}\right)^{-1}(h_0\bar{h}_0 + h_- h_+ - m^2)^2\,, \end{aligned} \tag{24}$$

in terms of the rescaled fields. From this, we see that the neutral Higgs component field h_0 acquires a vacuum expectation value $\langle h_0\rangle = m$ $(= v/\sqrt{2})$, and after symmetry breaking the surviving Higgs particle gets a mass

$$m_h = 2m\sqrt{(\lambda_u^e)^2 + (\lambda_d^e)^2}$$

(where the sum rule (23) has been used).

We wish now to extend the previous construction to the whole SM. One must then add leptons, and include color and flavor. For this purpose, we first extend the fermion space

$$\Psi_L = \begin{bmatrix} \begin{pmatrix} u_L \\ d_L \end{pmatrix}_\alpha \\ \begin{pmatrix} \nu_L \\ e_L \end{pmatrix} \end{bmatrix}_i\,, \qquad \Psi_R = \begin{bmatrix} \begin{pmatrix} u_R \\ d_R \end{pmatrix}_\alpha \\ \begin{pmatrix} 0 \\ e_R \end{pmatrix} \end{bmatrix}_i\,,$$

where α are the color indices, and i the flavor ones. Then the gauge one-form will split into components acting on the lepton and quark subspaces. On the lepton subspace, we have

$$\begin{aligned} A_{1\mu}^E &= \left(A_\mu^a\tau_a + \frac{1}{2}Y_{EL}B_\mu\right)\otimes \mathbf{1}_{ij}\,, \\ A_{2\mu}^E &= \left(\frac{1}{2}Y_{ER}B_\mu\right)\otimes \mathbf{1}_{ij}\,, \\ H^E &= \mathbf{\Phi}\otimes\lambda_{ij}^l\,, \end{aligned} \tag{25}$$

where Y_{EL}, Y_{ER} matrices contain the hypercharge numbers associated with the left- and right-handed charged leptons and neutrinos, and λ_{ij}^l is the Yukawa coupling matrix for leptons. On the quark subspace

$$\begin{aligned} A_{1\mu}^Q &= \left(A_\mu^a\tau_a + \frac{1}{2}Y_{QL}B_\mu\right)\otimes \mathbf{1}_{\alpha\beta}\otimes\mathbf{1}_{ij} + \mathbf{1}\otimes(G_\mu^b\Lambda_b)_{\alpha\beta}\otimes\mathbf{1}_{ij}\,, \\ A_{2\mu}^Q &= \left(\frac{1}{2}Y_{QR}B_\mu\right)\otimes \mathbf{1}_{\alpha\beta}\otimes\mathbf{1}_{ij} + \mathbf{1}\otimes(G_\mu^b\Lambda_b)_{\alpha\beta}\otimes\mathbf{1}_{ij}\,, \\ H^Q &= \widetilde{\mathbf{\Phi}}\otimes\mathbf{1}_{\alpha\beta}\otimes\lambda_{ij}^u + \mathbf{\Phi}\otimes\mathbf{1}_{\alpha\beta}\otimes\lambda_{ij}^d\,, \end{aligned} \tag{26}$$

where G^b_μ are the SU(3)$_C$ gauge bosons, and the Λ_b are the generators of the SU(3)$_C$ gauge transformation. Also, the λ^u_{ij} and λ^d_{ij} are the Yukawa coupling matrices for the up- and down-type quarks.

One can then easily verify that one obtains the SM Lagrangian. The gauge couplings for the electroweak sector are given by

$$\begin{aligned}\frac{1}{g^2} &= \frac{4N}{g_1^2}\,,\\ \frac{1}{g'^2} &= \frac{N}{g_1^2}\,\mathrm{Tr}\left(\frac{3Y_{QL}^2+Y_{EL}^2}{2}\right)+\frac{N}{g_2^2}\,\mathrm{Tr}\left(\frac{3Y_{QR}^2+Y_{ER}^2}{2}\right),\\ &= \frac{4N}{3g_1^2}+\frac{16N}{3g_2^2}\,.\end{aligned} \tag{27}$$

Here N stands for the number of generations. The Higgs kinetic energy has the form

$$\begin{aligned}\mathcal{L}_H &= \left(\frac{1}{g_1^2}+\frac{1}{g_2^2}\right) 3\,\mathrm{Tr}(\lambda^u)^2(D^\mu\widetilde{\Phi}^\dagger)(D_\mu\widetilde{\Phi})\\ &\quad+\left(\frac{1}{g_1^2}+\frac{1}{g_2^2}\right)\left(3\,\mathrm{Tr}(\lambda^d)^2+\mathrm{Tr}(\lambda^l)^2\right)(D^\mu\Phi^\dagger)(D_\mu\Phi)\,.\end{aligned} \tag{28}$$

Therefore, the constant involving in rescaling the Higgs field is

$$\lambda^2 = \left(\frac{1}{g_1^2}+\frac{1}{g_2^2}\right)\left(3\,\mathrm{Tr}(\lambda^u)^2+3\,\mathrm{Tr}(\lambda^d)^2+\mathrm{Tr}(\lambda^l)^2\right) \simeq 3\lambda_t^2\left(\frac{1}{g_1^2}+\frac{1}{g_2^2}\right),$$

if we assume that the top Yukawa coupling dominates. The resulting sum rule will predict that the effective top Yukawa coupling is

$$\lambda_t^{e2} = \frac{1}{3}\frac{g_1^2g_2^2}{g_1^2+g_2^2} = \frac{16N}{9}\frac{g^2g'^2}{g^2+g'^2}\,, \tag{29}$$

which, if one uses the values for the electroweak couplings at M_Z scale $\alpha_1(M_Z) = 1/58.97$, $\alpha_2(M_Z) = 1/29.61$ comes out to $\lambda_t^e \simeq 0.8$ (for the number of generations $N = 3$). This is to be compared with the experimentally determined value, $\lambda_t^{\rm SM} \simeq 1$.

Finally, the Higgs potential is

$$\begin{aligned}V(H) &= \left(\frac{1}{g_1^2}+\frac{1}{g_2^2}\right)(h_0\bar{h}_0+h_+h_- - \lambda^2m^2)^2\\ &\quad\times\{3[\mathrm{Tr}(\lambda^u)^2]^2+3[\mathrm{Tr}(\lambda^d)^2]^2+[\mathrm{Tr}(\lambda^l)^2]^2\}\\ &\to \frac{1}{3}\left(\frac{1}{g_1^2}+\frac{1}{g_2^2}\right)^{-1}(h_0\bar{h}_0+h_+h_- - m^2)^2\,,\end{aligned} \tag{30}$$

if the λ^u terms dominate. Then one would obtain a Higgs mass:

$$M_h \simeq 2\lambda_t^e m = 2m_t\,.$$

We note here that such relations are generic in formulations of the SM based on Connes' approach.[e] However, their exact meaning is not clear. They are not preserved under renormalization procedure; therefore one may question the consistency of these models as renormalizable field theories. However, one may also choose to interpret these relations as boundary conditions; that is, assume they are valid at some large scale M (which may be viewed as the fundamental scale at which the noncommutative approach works[f]). Then usual field theory will be a effective theory valid at lower energies; and in order to test this type of relations, one should run the value of the parameters from high to low energy scales. This type of analysis has been performed for some models inspired from Connes' ideas (see, for example, Refs. 21–23).

Our model predicts a rather large Higgs mass. (Although we would like to point out that this is an interesting value from the perspective of higher order corrections. For this value, the radiative corrections to the Higgs mass coming from fermion loops exactly cancel the corrections due to the ϕ^4 Higgs self-interaction. This might indicate interesting properties of the theory at one loop, but we will not follow this further in the present paper.) Also, the prediction for the top Yukawa coupling is not quite in agreement with the SM value for the top mass (although it is somewhat close).[g] However, this is not too troubling; as mentioned above, it is not clear at what scale we should evaluate the coupling constants appearing in Eq. (29), and loop corrections might modify this relation. Moreover, we do not expect that the SM is a fundamental theory, valid at all scales. Instead, it seems probable that the $\mathrm{SU}(3)_C \times \mathrm{SU}(2)_L \times \mathrm{U}(1)_Y$ gauge symmetry visible at M_Z scale is a remnant of the breaking of some larger symmetry, which is manifest at a higher scale. Hence, one should consider applying the framework used above to such extensions of the SM.

5. Left–Right Symmetric Model on Two-Sheeted Space–Time

In spite of the great phenomenological success of the SM as a gauge theory based on spontaneously broken symmetry $\mathrm{SU}(3)_C \times \mathrm{SU}(2)_L \times \mathrm{U}(1)_Y$, it is recognized that it has several unsatisfactory features that include, besides the proliferation of free parameters, a lack of an understanding of the origin of parity violation in the low energy region. Left–right symmetric models have always been attractive as the minimal extensions of the standard model. A natural consequence of left–right symmetric models is the existence of right-handed neutrinos, which, through

[e] The particular form these relations take depends upon the specific model under consideration, of course.

[f] One should note that by introducing a scale $M \gg M_Z$, one resurrects the hierarchy problem, that is, radiative corrections would indicate that all Higgs masses should be at scale M. Also, for such models we cannot use supersymmetry to solve this problem, since so far it is not known how to consistently include supersymmetry in the NCG approach (although some attempts exist).[20]

[g] It is possible to fix the prediction for the top mass by introducing a separate coupling for the Higgs part of the field strength (as, for example, in Ref. 14). However, this would lead to a loss of predictive power for the model.

the seesaw mechanism,[24] give rise to small mass for the left-handed neutrinos. The discovery of convincing evidence for nonzero neutrino masses gives therefore further weight to the idea that the SM may be a low energy version of such models.

One of the drawbacks of Connes' original formulation is that it does not allow the inclusion of larger gauge groups in the noncommutative algebra (see, for example Refs. 8, 19, 13). The approach of Chamseddine *et al.* permits more freedom, allowing for the construction of $SU(2)_L \times SU(2)_R$-type models, as well as SU(5) and SO(10) ones[12,25,26] (see also Ref. 27). Other realizations of the left–right symmetric model on Minkowski space–time and a finite number of discrete points in the internal space generally follow along similar paths (for example, Refs. 28 and 29). Our formulation of the $SU(2)_L \times SU(2)_R$ model has common elements with that in Refs. 12 and 28; however, we differ in our treatment of the Higgs sector, particularly in the way we deal with the junk forms. Moreover, we do not necessarily relate entries in the generalized Dirac operator to fermion masses, and therefore we do not necessarily have $\langle H \rangle = M$. This affects the derivation of the resulting Higgs potential, and can have important phenomenological consequences.

To start formulating the left–right symmetric model as a theory on two-sheeted space–time, we have to first specify the Hilbert space of spinors. One could use the same space as for the SM; however, in that case the only Higgs multiplets one can introduce are $SU(2)_L \times SU(2)_R$ bidoublets. Since one needs triplet Higgses to break the left–right symmetry (as well as to give mass to the right-handed neutrinos), we take the Hilbert space to also include the charge conjugate fields of Ψ_L, Ψ_R,

$$\Psi = \begin{bmatrix} \begin{pmatrix} \Psi_L \\ \Psi_R^c \end{pmatrix} \\ \begin{pmatrix} \Psi_R \\ \Psi_L^c \end{pmatrix} \end{bmatrix}_i ,$$

in such a way that the left-handed fields live on one sheet and the right-handed ones live on the other. The matrix structure associated with the gauge fields will then be

$$\begin{aligned} A_{1\mu} &= \begin{bmatrix} A^a_{L\mu}\tau_a + Y_L B_\mu/2 & 0 \\ 0 & A^a_{R\mu}\tau_a - Y_R B_\mu/2 \end{bmatrix}, \\ A_{2\mu} &= \begin{bmatrix} A^a_{R\mu}\tau_a + Y_R B_\mu/2 & 0 \\ 0 & A^a_{L\mu}\tau_a - Y_L B_\mu/2 \end{bmatrix}. \end{aligned} \tag{31}$$

The A^a_L, A^a_R are the gauge fields associated with the $SU(2)_L$, $SU(2)_R$ group transformations, while B is the one associated with the $U(1)_{B-L}$ group. The $Y_{L,R}$ numbers are the charges associated with $U(1)_{B-L}$ transformations for the left- and right-handed fields. They take different values on lepton and quark spaces; for leptons $Y_{EL} = Y_{ER} = -1$, and for quarks $Y_{QL} = Y_{QR} = 1/3$.

Since charge conjugate fermions are part of the spinor space, one can introduce triplet Higgses in the gauge one-form. Thus, on the lepton subspace we define

$$\mathbb{H} = \begin{bmatrix} H_1 \otimes \lambda_{ij}^{1l} & \Delta_L \otimes \lambda_{ij}^{L} \\ \Delta_R^\dagger \otimes \lambda_{ij}^{R\dagger} & H_2^c \otimes \lambda_{ij}^{2l} \end{bmatrix}, \tag{32}$$

where H_1 is a $(2, 2, 0)$ bidoublet and H_2 its conjugate,

$$H_1 = \begin{bmatrix} h_1^0 & h_2^+ \\ h_1^- & h_2^0 \end{bmatrix}, \qquad H_2 = \begin{bmatrix} h_2^{0*} & -h_1^+ \\ -h_2^- & h_1^{0*} \end{bmatrix}, \qquad H_2^c = \sigma_2 H_2^T \sigma_2,$$

and $\Delta_{L,R}$ are SU(2) triplets which couple to the left- and right-handed leptons

$$\Delta_{L,R} = \begin{bmatrix} \delta^-/\sqrt{2} & \delta^0 \\ \delta^{--} & -\delta^-/\sqrt{2} \end{bmatrix}_{L.R}.$$

The $\lambda_{ij}^{1,2l}$, $\lambda_{ij}^{R,L}$ matrices acting on the flavor space are the Yukawa couplings associated with the Dirac mass term for both the charged leptons and neutrinos (the $\lambda^{1,2l}$ matrices), and the Majorana mass terms for the right- and left-handed neutrinos (the $\lambda^{R,L}$ ones).[h] On the other hand the Higgs matrix (32) on the quark subspace will contain only the diagonal H_1, H_2^c fields (with Yukawa matrices $\lambda^{1,2q}$ associated with the quarks) since the Δ_L, Δ_R triplets do not couple to quarks. Note also that $H_2^c = H_1^\dagger$; we use this notation to remind ourselves that the mass term $\bar{\psi}_R^c H_2^c \psi_L^c$ is usually written as $\bar{\psi}_L H_2 \psi_R$.

Substituting (31) and (32) in Eq. (4) (with $\Phi_1 = \mathbb{H}$, $\Phi_2 = \mathbb{H}^\dagger$), one obtains the corresponding expressions for the field strength $\mathbb{F}$, which will give the standard Lagrangian terms for the gauge fields A_L, A_R, and B. It can be easily verified that Eq. (16) will give the right kinetic energy terms for the Higgs multiplets

$$\begin{aligned} \mathcal{L}_H &= |\partial_\mu H + iHA_{R\mu}^a \tau_a - iA_{L\mu}^a \tau_a H|^2, \\ \mathcal{L}_\Delta &= |\partial_\mu \Delta_{L,R} + i\Delta_{L,R} A_{L,R\mu}^a \tau_a - iA_{L,R\mu}^a \tau_a \Delta_{L,R} - iY_E B_\mu \Delta_{L,R}|^2, \end{aligned} \tag{33}$$

where, due to left–right symmetry, $Y_{EL} = Y_{ER} = Y_E$. The coupling constants for the SU(2) and U(1) gauge fields are given by

$$\begin{aligned} \frac{1}{g^2} &= 4N\left(\frac{1}{g_2^2} + \frac{1}{g_1^2}\right), \\ \frac{1}{g'^2} &= 2N\left(\frac{1}{g_1^2} + \frac{1}{g_2^2}\right)(3Y_Q^2 + Y_L^2). \end{aligned} \tag{34}$$

It is interesting to note that, due to left–right symmetry on each sheet, the ratio of the two coupling constants is fixed: $g^2/g'^2 = 2/3$ (of course, this is taken to be valid at the energy where this description holds). This symmetry could be broken

[h]Note that in order to have left–right symmetry in the fermion sector, one should also require that $\lambda^R = \lambda^L$, $\lambda^l = \lambda^{l\dagger}$.

by considering more complex structure for the gauge coupling matrix in (18). Also, note that the rescaling constants for the Higgs fields are

$$\begin{aligned}\lambda^2_{\Delta R} &= \mathrm{Tr}[\lambda^R \lambda^{R\dagger}]/(4Ng^2)\,,\\ \lambda^2_{\Delta L} &= \mathrm{Tr}[\lambda^L \lambda^{L\dagger}]/(4Ng^2)\,,\\ \lambda^2_H &= \mathrm{Tr}[\lambda^{1l}\lambda^{1l\dagger} + \lambda^{2l}\lambda^{2l\dagger} + 3(\lambda^{1q}\lambda^{1q\dagger} + \lambda^{2q}\lambda^{2q\dagger})]/(4Ng^2)\,.\end{aligned} \tag{35}$$

Before continuing, let us briefly review the salient features of gauge symmetry breaking in the left–right symmetric models. The theory has five gauge bosons, two charged ones (A_L and A_R) and three neutral: B, A_{0L}, A_{0R}. At some high energy, the symmetry is unbroken, and all gauge bosons are massless. However, at some large scale, the right-handed triplet Higgs acquires a vacuum expectation value $\langle \Delta_R \rangle = v_R$, and breaks left–right symmetry. One of the charged bosons (more precisely, a combination of the A_L and A_R) and one of the neutral ones acquire masses of order v_R. Since these massive bosons have not been observed at present day colliders, one infers that v_R should be above the TeV scale.[i] Furthermore, the right-handed neutrinos also acquire Majorana mass through the VEV of Δ_R. If one accepts that the smallness of left-handed neutrino masses is due to the seesaw mechanism, this would point to the scale v_R to be around 10^{14} GeV. The remaining symmetry (that of the SM $\mathrm{SU}(2)_L \times \mathrm{U}(1)_Y$) is broken at the electroweak scale by vacuum expectations for the bidoublet Higgses $\langle h_1^0 \rangle = v_1$, $\langle h_2^0 \rangle = v_2$. Note that the left-handed triplet Δ_L may also acquire a vacuum expectation value v_L; however, since this VEV gives Majorana mass to the left-handed neutrinos, v_L should be below eV scale.

A very interesting question then arises: is this pattern of symmetry breaking consistent with our formulation of the left–right symmetric model on a two-sheeted space–time? Note that, in our scenario, the Higgs potential is fixed (up to may be an overall multiplicative constant) once the matrix M appearing in the Dirac operator has been given. Therefore one can potentially hope to predict how the gauge symmetry is broken. We shall investigate this in what follows, under different sets of assumptions concerning the structure of the M matrix.

(A) We first choose a form for the matrix M suggested by the pattern of fermion masses (that is, choose a nonzero entry in the mass matrix at places where fermion masses would naturally appear). Thus, one can choose

$$M = \begin{bmatrix} \begin{pmatrix} m_1 & 0 \\ 0 & m_2 \end{pmatrix} \otimes \lambda^{1l}_{ij}\,, & \begin{pmatrix} 0 & m_L \\ 0 & 0 \end{pmatrix} \otimes \lambda^L_{ij} \\ \begin{pmatrix} 0 & 0 \\ m_R & 0 \end{pmatrix} \otimes \lambda^R_{ij}\,, & \begin{pmatrix} m_1 & 0 \\ 0 & m_2 \end{pmatrix} \otimes \lambda^{2l}_{ij} \end{bmatrix}, \tag{36}$$

[i]Constraints on the scale of v_R can also be inferred from weak interaction data; see, for example, Ref. 30 and references therein.

on the lepton subspace. On the quark subspace, we will have a similar structure (with the fermion Yukawa couplings λ^{1l}, λ^{2l} replaced by the quark ones λ^{1q}, λ^{2q}), with the difference that the off-diagonal matrices will be zero (since one generally does not introduce triplets in the quark sector, and thus one does not get a Majorana mass for the quarks).

Before computing the Higgs potential, one has to insure the $dd\rho = 0$ condition by using the trace prescription discussed in Sec. 1. Following Eq. (2), if we set $f_1 = \text{diag}\{f_{1L}, f_{1R}\}$, we obtain on the left-handed sheet

$$(dd\rho)_{11} = \begin{pmatrix} (MM^\dagger)_{11}f_{1L} - f_{1L}(MM^\dagger)_{11} & (MM^\dagger)_{12}f_{1R} - f_{1L}(MM^\dagger)_{12} \\ (MM^\dagger)_{21}f_{1L} - f_{1R}(MM^\dagger)_{22} & (MM^\dagger)_{22}f_{1R} - f_{1R}(MM^\dagger)_{22} \end{pmatrix}, \quad (37)$$

with a similar form on the right-handed sheet. In order to satisfy $dd\rho = 0$, one then should drop the off-diagonal terms and take the trace of the diagonal ones. The contribution of the Higgs terms to the diagonal elements of $\mathbb{F}$ is then

$$MM^\dagger - \mathbb{H}\mathbb{H}^\dagger \to \begin{bmatrix} \text{Tr}[H_1H_1^\dagger + \Delta_L\Delta_L^\dagger - (MM^\dagger)_{11}] & 0 \\ 0 & \text{Tr}[H_2^cH_2^{c\dagger} + \Delta_R^\dagger\Delta_R - (MM^\dagger)_{22}] \end{bmatrix}. \quad (38)$$

Now, if we assume that by suitable transformations, one can take the VEV's of the Higgs fields to be

$$\langle H_1\rangle = \begin{bmatrix} v_1 & 0 \\ 0 & v_2 \end{bmatrix}, \qquad \langle \Delta_{L,R}\rangle = \begin{bmatrix} 0 & v_{L,R} \\ 0 & 0 \end{bmatrix},$$

the Higgs potential will become

$$\begin{aligned} V(\langle\mathbb{H}\rangle) &= \left[(v_1^2 + v_2^2 - m_1^2 - m_2^2)|\lambda^{1l}|^2 + (v_L^2 - m_L^2)|\lambda^L|^2\right]^2 \\ &\quad + \left[(v_1^2 + v_2^2 - m_1^2 - m_2^2)|\lambda^{2l}|^2 + (v_R^2 - m_R^2)|\lambda^L|^2\right]^2 \\ &\quad + 3\left(v_1^2 + v_2^2 - m_1^2 - m_2^2\right)^2\left(|\lambda^{1q}|^4 + |\lambda^{2q}|^4\right), \end{aligned} \quad (39)$$

where the first two lines are contributions coming from the lepton subspace and the last line is the contribution due to the quark subspace. (Here we have used $|\lambda|^2$ as a shorthand for $\text{Tr}[\lambda\lambda^\dagger]$.) One then sees that this potential has a minimum at $v_L = m_L$, $v_R = m_R$ and $v_1^2 + v_2^2 = m_1^2 + m_2^2$. In other words, the VEV's of the triplets Δ_R, Δ_L can be fixed to m_R, m_L, but there is a degeneracy in determining the individual values of the bidoublet Higgs VEV's. If, however, one takes $m_1^2 + m_2^2$ to be at electroweak scale, both v_1 and v_2 are at this scale, and this does not create any problem with respect to the symmetry breaking pattern. So we see that we can obtain the desired symmetry breaking pattern, although one has to put in by hand a large value for m_R, very small value for m_L and electroweak scale for m_1, m_2.

Note that this result is by no means trivial. For example, if there would be no quark sector in the theory (drop the third line in Eq. (39)), one could not separate the VEV's of the bidoublets from the VEV's of the triplets. Hence there would be no way to make sure that H does not acquire a v_R scale VEV. This is due to the highly degenerate structure of the potential (there are several flat directions). In particular, note that if one adopts the pattern (36) for the mass matrix M (where nonzero entries correspond to Higgs fields that one expects to get VEV's), it would be tempting to say that $\langle \mathbb{H} \rangle = M$ will give a minimum of the potential. One could then say that the fields Φ which appear initially in the gauge one-form (4) are the Higgs fields in the broken symmetry phase of the theory.[12] However, while it is true that $V(\langle \mathbb{H} \rangle = M) = 0$, this does not make it a true vacuum, since degeneracies could exist.

(B) One might then also consider structures for the M matrix which are not linked to the fermion masses. Actually, the parameters which seem to have physical relevance are the matrices $MM^\dagger$ and $M^\dagger M$, since they appear in both, the Higgs potential and the evaluation of the $dd\rho = 0$ condition. In the most general case, the Higgs potential can be written therefore in terms of four parameters: $\mu_1^{l2} = \mathrm{Tr}(MM^\dagger)_{11}$, $\mu_2^{l2} = \mathrm{Tr}(MM^\dagger)_{22}$ on the lepton space, and $\mu_1^{q2} = \mathrm{Tr}(MM^\dagger)_{11}$, $\mu_2^{q2} = \mathrm{Tr}(MM^\dagger)_{22}$ on the quark space. One would then obtain

$$\begin{aligned} V(\langle \mathbb{H} \rangle) &= [(v_1^2 + v_2^2)|\lambda_e^{1l}|^2 + v_L^2 - \mu_1^{l2}]^2 \\ &+ [(v_1^2 + v_2^2)|\lambda_e^{2l}|^2 + v_R^2 - \mu_2^{l2}]^2 \\ &+ 3[(v_1^2 + v_2^2)|\lambda_e^{1q}|^2 - \mu_1^{q2}]^2 + 3[(v_1^2 + v_2^2)|\lambda_e^{2q}|^2 - \mu_2^{q2}]^2 \,, \end{aligned} \tag{40}$$

where we have used the VEV's rescaled by the corresponding Higgs fields factors (35), $v_{1,2} \to v_{1,2}/\lambda_H$, $v_{L,R} \to v_{L,R}/\lambda_{\Delta L,R}$, and the constants $\lambda_e^{kl,q} = \lambda^{kl,q}/\lambda_H$. Again we see that the scale of v_1, v_2 is set by the μ_1^q, μ_2^q parameters, so these have to be at electroweak scale. Interestingly, for this potential one does not have to necessarily fine tune the VEV of the Δ_L field anymore; it will be naturally driven to zero for a whole range of values for the parameter μ_1^l. Indeed, let us call v_0^2 the value of $v_1^2 + v_2^2$ for which the last line of Eq. (40) is minimized; one then sees that if $v_0^2|\lambda_e^{1l}|^2 > \mu_1^{l2}$, then the minimum of the potential requires $v_L = 0$.

(C) In the two examples shown above, the Higgs potential has a relatively simple structure (being the sum of several perfect squares). More complex potentials can be obtained in the case when the matrix M has some symmetries. For example, if M is proportional to identity in the generation space, one need not take trace in (39) over the generation indices. Then, for example, if one would compute the contribution of the $\mathbb{H}_{11}$ part on the lepton subspace, one would obtain

$$V_{11}(v_1, v_2, v_L) = \mathrm{Tr}[|(v_1^2 + v_2^2)\lambda_e^{1l}\lambda_e^{1l\dagger} + v_L^2\lambda^L\lambda^{L\dagger} - \mu_1^{l2}|^2] \,.$$

One sees that while the coefficients of the potential terms quadratic in the Higgs fields are proportional to $\mathrm{Tr}(\lambda\lambda^\dagger)$, the coefficients of the quartic terms are proportional to $\mathrm{Tr}(\lambda\lambda^\dagger\lambda\lambda^\dagger)$, and therefore somewhat independent. One can even break the symmetry between the VEV's of the ϕ_1^0 and ϕ_2^0 fields; if the M_{11} and/or M_{22} elements are taken to be diagonal in SU(2) space, then the diagonal elements of (37) will be zero, and no trace is necessary. In this case, the (partial) Higgs potential will be

$$V_{11}(H) = \mathrm{Tr}[|H_1 H_1^\dagger \otimes \lambda_e^{1l}\lambda_e^{1l\dagger} + \Delta_L\Delta_L^\dagger \otimes \lambda^L\lambda^{L\dagger} - \mu_1^{l2}\mathbf{1}\otimes\mathbf{1}|^2], \tag{41}$$

or

$$V_{11}(v_1, v_2, v_L) = \mathrm{Tr}[|v_1^2\lambda_e^{1l}\lambda_e^{1l\dagger} + v_L^2\lambda^L\lambda^{L\dagger} - \mu_1^{l2}|^2 + |v_2^2\lambda_e^{1l}\lambda_e^{1l\dagger} - \mu_1^{l2}|^2].$$

Finally, if one sets the off-diagonal elements of the M matrix to zero ($M_{12} = M_{21} = 0$), one can keep the off-diagonal elements in $\mathbb{H}\mathbb{H}^\dagger - MM^\dagger$, which will contribute to the Higgs potential a term

$$V_{12}(H) = 2\,\mathrm{Tr}[|H_1\Delta_R \otimes \lambda_e^{1l}\lambda^R + \Delta_L H_1 \otimes \lambda^L\lambda_e^{2l\dagger}|^2]. \tag{42}$$

It is instructive to compare the Higgs potential obtained in our model with the general Higgs potential discussed in Refs. 24 and 30. We see that our model can give rise to most of the terms found in Ref. 24. One might note that only H_1, $H_1^\dagger$ appears in our potential; however, this is due to the particular choice of the Higgs matrix (32); a more general choice

$$\mathbb{H} = \begin{bmatrix} H_1 \otimes \lambda^{1l} + H_2 \otimes \lambda^{3l} & \Delta_L \otimes \lambda^L \\ \Delta_R^\dagger \otimes \lambda^{R\dagger} & H_2^c \otimes \lambda^{2l} + H_1^c \otimes \lambda^{4l} \end{bmatrix} \tag{43}$$

can be made, which will introduce H_2, $H_2^\dagger$ in the potential (if desirable). The only important difference seems to be that the terms coupling the VEV's of the left- and right-triplets $\mathrm{Tr}(\Delta_L^\dagger\Delta_L)\,\mathrm{Tr}(\Delta_R^\dagger\Delta_R)$ are missing in our model. In the absence of such a term, one cannot take the Higgs potential left–right symmetric ($\mu_1^{l2} = \mu_2^{l2}$, $\lambda^L = \lambda^R$). Indeed, it turns out that for a left–right symmetric potential (as in Ref. 24) what drives v_L to zero while keeping v_R at GUT scale is just such a term $\sim v_L^2 v_R^2$, that couples the two VEV's. Lacking such a term in our model, we are forced to take μ_2^l at GUT scale, and μ_1^l at electroweak scale.

This might lead to interesting phenomenology at low energies. Indeed, if for example the Higgs potential has the form (41) with μ_1^l at electroweak scale, this will imply that the Higgs fields in the Δ_L multiplet acquire mass at the electroweak scale. Then non-SM processes (like lepton-flavor violating decays, anomalous magnetic momenta, etc.) are likely to impose significant constraints on the parameters of the model. Conversely, discovery of such particles might provide experimental evidence for this picture. We leave a more complete analysis of the phenomenological implications of this model to a future paper.

6. Conclusions

We have developed a formalism based on the framework of Connes' noncommutative geometry (NCG), with the purpose of studying the standard model of electroweak interactions and beyond. Our model is based on a two-sheeted space–time that can be thought of as a discretized version of Kaluza–Klein theory, in which the compact fifth dimension is replaced by two discrete points. The left- and right-chiral spinor fields live on the two separate sheets, while the gauge and Higgs fields are part of a generalized gauge operator represented by a 2×2 matrix acting on the internal (discrete) space.

The main virtue of this framework is that the scalar Higgs fields are an integral part of the gauge sector. Their gauge invariant kinetic parts in the Lagrangian as well as quartic forms of Higgs potentials arise naturally. Furthermore, the possibility of spontaneous symmetry breaking is built naturally into the Higgs potential.

While different NCG formulations of the SM (and other gauge theories) have been studied extensively by several authors, we have taken in this paper a less mathematical approach and focused more on the physics of the model. The formalism allows for an easy (and transparent) construction of the Higgs sector. The choice of Higgs multiplets appearing in the theory is dictated by the choice of the underlying spinor space. The Higgs potential can be written in terms of the Yukawa couplings of the fermions and the elements of the matrix M appearing in the definition of the generalized Dirac operator

Besides the predictive power in the Higgs sector, the NCG formalism leads to sum rules for the Yukawa coupling constants of the fermions. In the case of the minimal standard model, this will lead to a prediction of the top quark mass (as well as the Higgs mass). However, such results at the tree level form of the Lagrangian cannot be taken seriously unless one knows at which scale this picture holds. If NCG-inspired theories are a description of reality, one might expect them to be valid at scales close to M_{Pl}. However, at such high scales, there are good reasons to believe that the gauge group is a larger one, corresponding to a grand unified theory of which the SM is just the low energy limit.

We are led therefore to consider higher symmetries. As an example, in this paper we analyze the implications of two-sheeted space–time picture for the left–right symmetric model. Our approach predicts several specific forms possible for the Higgs potential. Interestingly, these predictions allow the desired left–right symmetry breaking pattern leading to the standard model. We discuss the various scenarios for this potential depending upon the choice and the symmetries in the M matrix of the Dirac operator. A more detailed quantitative study of these features is desirable to draw concrete conclusions. But the model has several attractive features, and fewer parameters compared with the standard left–right symmetric models.

Acknowledgments

We thank Nguyen Ai Viet and Alexandr Pinzul for many helpful discussions. This work is supported in part by the US Department of Energy grant number DE-FG02-85ER40231.

References

1. A. Connes, *Noncommutative Geometry* (Academic Press, 1994).
2. J. Madore, *An Introduction to Noncommutative Differential Geometry and Physical Applications* (Cambridge University Press, 1995).

3. G. Landi, *An Introduction to Noncommutative Spaces and their Geometry* (Springer-Verlag, Berlin, 1997).
4. A. Connes, *Commun. Math. Phys.* **182**, 155 (1996).
5. A. Connes and J. Lott, *Nucl. Phys. B* (*Proc. Suppl.*) **18**, 29 (1991).
6. A. Connes, *J. Math. Phys.* **36**, 6194 (1995).
7. J. C. Varilly and J. M. Gracia-Bondia, *J. Geom. Phys.* **12**, 223 (1993).
8. B. Iochum and T. Schucker, *Commun. Math. Phys.* **178**, 1 (1996).
9. R. Coquereaux, G. Esposito-Farese and G. Vaillant, *Nucl. Phys. B* **353**, 689 (1991).
10. R. Coquereaux, G. Esposito-Farese and F. Scheck, *Int. J. Mod. Phys. A* **7**, 6555 (1992).
11. R. Wulkenhaar, *Phys. Lett. B* **390**, 119 (1997).
12. A. H. Chamseddine, G. Felder and J. Frohlich, *Nucl. Phys. B* **395**, 672 (1993).
13. F. Lizzi, G. Mangano, G. Miele and G. Sparano, *Mod. Phys. Lett. A* **11**, 2561 (1996).
14. A. V. Nguyen, *Heavy Ion Phys.* **1**, 263 (1995).
15. N. A. Viet and K. C. Wali, *Phys. Rev. D* **67**, 124029 (2003).
16. A. H. Chamseddine, G. Felder and J. Frohlich, *Commun. Math. Phys.* **155**, 205 (1993).
17. K. Morita and Y. Okumura, *Prog. Theor. Phys.* **95**, 227 (1996).
18. T. Schucker, Geometries and forces, hep-th/9712095.
19. T. Schucker and J. M. Zylinski, *J. Geom. Phys.* **16**, 207 (1995).
20. B. S. Balakrishna, F. Gursey, A. V. Nguyen and K. C. Wali, *Phys. Rev. D* **46**, 4698 (1992).
21. E. Alvarez, J. M. Gracia-Bondia and C. P. Martin, *Phys. Lett. B* **306**, 55 (1993).
22. E. Alvarez, J. M. Gracia-Bondia and C. P. Martin, *Phys. Lett. B* **329**, 259 (1994).
23. C. P. Martin, J. M. Gracia-Bondia and J. C. Varilly, *Phys. Rep.* **294**, 363 (1998).
24. R. N. Mohapatra and G. Senjanovic, *Phys. Rev. D* **23**, 165 (1981).
25. A. H. Chamseddine, G. Felder and J. Frohlich, *Phys. Lett. B* **296**, 109 (1992).
26. A. H. Chamseddine and J. Frohlich, *Phys. Rev. D* **50**, 2893 (1994).
27. B. S. Balakrishna, F. Gursey and K. C. Wali, *Phys. Rev. D* **44**, 3313 (1991).
28. Y. Okumura and K. Morita, *Nuovo Cimento A* **109**, 311 (1996).
29. Y. Okumura, *Nuovo Cimento A* **110**, 267 (1997).
30. R. N. Mohapatra, *Unification and Supersymmetry. The Frontiers of Quark–Lepton Physics*, 3rd edn. (Springer, 2003).

$SU(5)$ grand unification on a domain-wall brane from an E_6-invariant action

Aharon Davidson,[1,*] Damien P. George,[2,+] Archil Kobakhidze,[2,‡] Raymond R. Volkas,[2,§] and Kameshwar C. Wali[3,||]

[1]*Physics Department, Ben-Gurion University, Beer-Sheva 84105, Israel*

[2]*School of Physics, Research Centre for High Energy Physics, The University of Melbourne, Victoria 3010, Australia*

[3]*Physics Department, Syracuse University, Syracuse, New York 13244-1130, USA*

(Received 22 October 2007; published 30 April 2008)

An $SU(5)$ grand unification scheme for effective 3 + 1-dimensional fields dynamically localized on a domain-wall brane is constructed. This is achieved through the confluence of the clash-of-symmetries mechanism for symmetry breaking through domain-wall formation, and the Dvali-Shifman gauge-boson localization idea. It requires an E_6 gauge-invariant action, yielding a domain-wall solution that has E_6 broken to differently embedded $SO(10) \otimes U(1)$ subgroups in the two bulk regions on opposite sides of the wall. On the wall itself, the unbroken symmetry is the intersection of the two bulk subgroups, and contains $SU(5)$. A 4 + 1-dimensional fermion family in the 27 of E_6 gives rise to localized left-handed zero modes in the $5^* \oplus 10 \oplus 1 \oplus 1$ representation of $SU(5)$. The remaining ten fermion components of the 27 are delocalized exotic states, not appearing in the effective 3 + 1-dimensional theory on the domain-wall brane. The scheme is compatible with the type-2 Randall-Sundrum mechanism for graviton localization; the single extra dimension is infinite.

I. INTRODUCTION

If our universe is a 3 + 1-dimensional brane [1–9] existing in a 4 + 1-dimensional spacetime, then the most likely field-theoretic origin for the brane is a scalar-field domain-wall (DW) or kink [1]. This generic idea is naturally compatible with the type-2 Randall-Sundrum (RS2) mechanism for producing effective 3 + 1-d gravity on the brane [9] (see, for example, Refs. [10–12] for the extension of thin-brane RS2 to a domain-wall brane). The challenge is to dynamically localize all the other ingredients necessary for a phenomenologically successful effective theory on the brane: gauge bosons, fermions, and Higgs bosons. Various localization ideas for these disparate classes of fields have been recently combined to produce an effective brane theory that is plausibly very similar to the standard model [13].

The purpose of this paper is twofold. First, we wish to point out a very elegant generic connection between the clash-of-symmetries (CoS) mechanism for symmetry breaking through domain-wall formation [14–17],[1] and the Dvali-Shifman (DS) idea for dynamical gauge-boson localization [22]. Second, we use this remarkable confluence to construct an explicit scheme that realizes an $SU(5)$ gauge-invariant effective theory on the brane. In a sense, it is a grand unified extension of the model of Ref. [13], but the way in which the Dvali-Shifman mechanism is realized is quite different, and we shall argue that it is in fact conceptually more advanced. Remarkably, this scheme immediately produces a realistic spectrum of localized fermion zero modes [23] (for a review see [24]) using the simplest possible mechanism. While it is beyond the scope of this paper to write down a complete phenomenologically acceptable domain-wall-brane localized $SU(5)$ theory, we shall conclude with brief remarks about how this could be attempted.

The clash-of-symmetries phenomenon [14–20] automatically arises when the simple Z_2 kink is extended to a theory with a continuous internal symmetry group G in addition to the discrete symmetry. Taking the scalar-field multiplet to be in a nontrivial representation of G, the domain-wall configuration spontaneously breaks G in addition to reflecting the disconnected vacuum manifold topology created by the spontaneous breaking of the discrete symmetry. Two classes of domain-wall solutions exist: those which respect the same subgroup H of G at all values of the bulk coordinate y, and those where the unbroken subgroup varies in the bulk. We shall call the first class "non-CoS domain walls," contrasted with the "CoS domain walls" of the second class. Clash-of-symmetries DWs can arise when the subgroups respected asymptotically (at $y = \pm\infty$) are isomorphic but *differently embedded* subgroups, H and H'. The symmetry group at finite y is typically the intersection $H \cap H'$, which is of course smaller than both H and H'.

The last observation provides an immediate connection with the Dvali-Shifman proposal for dynamical gauge-boson localization. The DS mechanism, as originally proposed [22], envisaged a domain-wall configuration where the full group G is restored in the bulk, but broken to H in the wall. The gauge bosons of H propagate on the

*davidson@bgu.ac.il
+d.george@physics.unimelb.edu.au
‡archilk@unimelb.edu.au
§raymondv@unimelb.edu.au
||wali@phy.syr.edu

[1]See also [18–20] for related works, and [21] for soliton-induced supersymmetry breaking.

wall either as massless Abelian gauge fields or glueballs formed from non-Abelian gauge fields. In the bulk, *all* gauge bosons have to be incorporated into massive G-glueballs.[2] Thus, the massless Abelian gauge fields on the wall have to become incorporated into massive glueballs in the bulk, and the energy cost associated with the mass gap then plausibly localizes them to the wall. This heuristic argument is bolstered by the dual-superconductivity model [26,27] for the confining bulk: the electric field lines from a source charge in the wall are repelled from the interface with the dual-superconducting bulk [28,29], just as magnetic field lines are Meissner-repelled from an ordinary superconductor. The non-Abelian gauge fields of H are also plausibly localized if the mass of the G-glueballs exceeds the mass of the H-glueballs.

The fact that the full symmetry G is asymptotically restored is clearly not a necessary condition. In the CoS situation, the brane-group $H \cap H'$ is a subgroup of both H and H', the unbroken symmetries in the two semi-infinite bulk regions. By the DS reasoning, provided H and H' contain confining non-Abelian factors, at least some of the gauge bosons of $H \cap H'$ will be localized. For a realistic theory, we need the localized gauge bosons to include those of the standard model. The model builder needs to engineer the theory to achieve this effect. While this engineering shall be the main concern in the rest of the paper, our first generic point has already been made: *the clash-of-symmetries automatically gives rise to Dvali-Shifman gauge-boson localization.* This CoS alternative realization of the DS mechanism seems conceptually neater than the original, because it can be achieved using scalars in a single irreducible representation of G. The original requires two multiplets: a G-singlet to form a kink, which in turn forces a G-multiplet to condense in the core of the wall.

We shall show that the CoS-DS confluence can naturally produce an $SU(5)$ effective theory on the brane. The basic ingredients are $G = E_6$, with the DW-producing scalar field in the adjoint or 78 representation. The groups H and H' will be the differently embedded maximal subgroups $SO(10) \otimes U(1)_E$ and $SO(10)' \otimes U(1)_{E'}$, respectively. Their intersection is $SU(5) \otimes U(1) \otimes U(1)$, with $SU(5)$ of course being a subgroup of both $SO(10)$ and $SO(10)'$. Taking both of those as confining gauge theories in the bulk, the localization of $SU(5)$ gauge bosons follows from the DS effect. The gauge fields of $U(1) \otimes U(1)$ are *not* completely localized. When 4 + 1-d fermions in the 27 of E_6 are Yukawa coupled to the scalar multiplet, we shall show that 3 + 1-d left-chiral zero modes in the phenomenologically realistic $5^* \otimes 10 \oplus 1 \oplus 1$ representation of $SU(5)$ are localized. The remaining ten fermion components remain 4 + 1-d, and are thus absent from the effective brane theory. The result that the chiralities of the zero modes come out to be phenomenologically correct is very nontrivial, as we shall explain.

We review the clash-of-symmetries idea in Sec. II. Section III describes a warm-up example featuring $SO(10)$ CoS domain walls, and explains why the extension to E_6 constructed in Sec. IV is needed. We conclude in Sec. V.

[2]The Dvali-Shifman mechanism requires a confining 4 + 1-dimensional gauge theory to live in the bulk. The issue of confinement in 4 + 1 dimensions is not completely understood, so the Dvali-Shifman mechanism in that context has the status of being a plausible conjecture. There is good lattice gauge theory evidence that pure $SU(2)$ Yang-Mills theory with an ultraviolet cutoff is confining in 4 + 1 dimensions when the gauge coupling constant exceeds a certain critical value [25]. It is thus plausible that a variety of 4 + 1-dimensional gauge theories exhibit confinement at sufficiently large coupling strength.

II. THE CLASH OF SYMMETRIES

Consider a theory (an action) whose symmetry group is the direct product of a continuous symmetry group G and a discrete symmetry Z. It is important that Z is *not* a subgroup of G. Suppose the global minima of the Higgs potential spontaneously break G to subgroup H, and simultaneously break Z to a smaller discrete group. For the sake of definiteness, we shall take the $Z = Z_2 = \{1, z \text{ s.t. } z^2 = 1\}$ example in what follows, with the Z_2 completely broken.

The vacuum manifold then consists of two disconnected copies of the coset space G/H, with the copies related by the spontaneously broken $z \in Z_2$. This is an immediate generalization of the simple Z_2 kink situation, where the vacuum manifold consists of just two disconnected points related by Z_2. Each such point is expanded into the nontrivial manifold G/H. We shall call the disconnected pieces G/H and $(G/H)_z$. The Z_2 must not be a subgroup of G for the two disconnected pieces to exist.

Let $|0\rangle$ be an element of G/H. By definition, $h|0\rangle = |0\rangle$ for all $h \in H$. Since the Higgs potential is G-invariant, if we apply a transformation $g \in G/H$ (that is, a transformation such that $g \in G$ but $g \notin H$) to $|0\rangle$, we obtain a degenerate vacuum state $|0; g\rangle \equiv g|0\rangle$. By considering all possible g's, these transformations generate the G/H piece of the vacuum manifold. Applying the nonidentity transformation $z \in Z_2$ from the discrete group, we obtain the discrete image $|0\rangle_z \equiv z|0\rangle$ of $|0\rangle$. This image is a point in the other disconnected piece $(G/H)_z$ of the full vacuum manifold. By applying all possible $g \in G/H$ to $|0\rangle_z$, the space $(G/H)_z$ is generated. Figure 1 illustrates this situation.

The degenerate vacua $|0\rangle$ and $|0; g\rangle$ respect *differently embedded* but otherwise isomorphic subgroups H and H_g, respectively. This is elementary: Let $h_1, h_2 \in H$ such that $h_1 h_2 = h_3 \in H$. Then the conjugates $g h_{1,2,3} g^{-1}$ respect the same multiplication table and hence the set gHg^{-1} is precisely H_g which is isomorphic to H but a different subset of G. If $h|0\rangle = |0\rangle$, then trivially $ghg^{-1}g|0\rangle = g|0\rangle$,

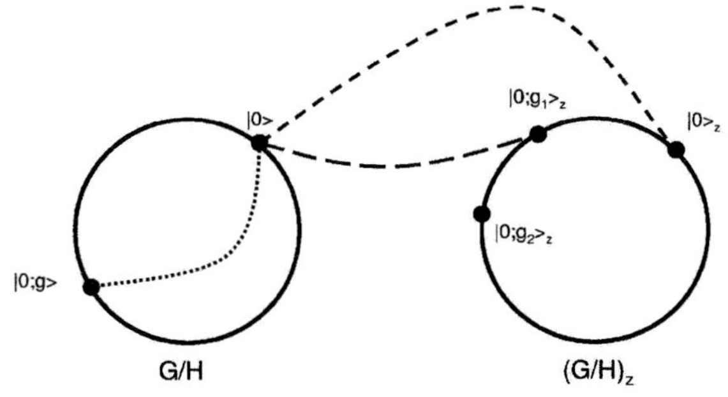

FIG. 1. The vacuum manifold of a $G \otimes Z_2 \to H$ model. The two circles schematically depict the disconnected coset spaces G/H and $(G/H)_z$. Each point along the G/H circle corresponds to a vacuum $|0;g\rangle$ for some $g \in G$, with the corresponding situation for the $(G/H)_z$ circle. The three broken lines represent possible domain-wall configurations, with the endpoints at $y = \pm\infty$ on various choices of vacua. The dotted line represents a possible nontopological domain-wall configuration. The short-dashed line represents a non-clash-of-symmetries domain-wall configuration, while the long-dashed line is a clash-of-symmetries domain wall.

which simply says that the conjugated elements preserve the other vacuum: $(ghg^{-1})|0;g\rangle = |0;g\rangle$. Similar statements are true for $(G/H)_z$.

The boundary conditions at $y = \pm\infty$ for domain-wall configurations are chosen from the vacua. If the chosen vacua are either both from G/H, or both from $(G/H)_z$, then the "domain-wall" configurations are not topologically stable: they are in the same topological class as any of the spatially homogeneous vacua $|0;g\rangle$, or respectively $|0;g\rangle_z$, and will dynamically decay to one of these vacua. They may be metastable, depending on the Higgs-potential topography,[3] so while they are of some interest we shall not consider them further in this paper.

Topologically nontrivial DW configurations have one boundary condition from G/H and the other from $(G/H)_z$. Evidently, there is an uncountable infinity of such choices, and thus potentially an uncountable infinity of DW solutions, all within the same nontrivial topological class. Figure 1 illustrates the plethora of choices. This potential richness has no analogue for the simple Z_2 kink.

Suppose that the boundary condition at $y = -\infty$ is $|0\rangle$ and at $y = +\infty$ it is $|0\rangle_z$. Then if $h|0\rangle = |0\rangle$, it also follows that

$$h|0\rangle_z \equiv hz|0\rangle = zh|0\rangle = z|0\rangle \equiv |0\rangle_z, \quad (2.1)$$

because by assumption the symmetry is $G \otimes Z$ so that $gz = zg \forall z \in Z$, $g \in G$ and hence $hz = zh$ always. Thus, the unbroken symmetry at $y = -\infty$ is precisely the same set H as at $y = +\infty$. A domain-wall configuration that interpolates between these vacua is then expected to respect the same subgroup H at all y. This is an example of a non-CoS domain wall. Clearly, taking the vacua as any pair $|0;g\rangle$ and $|0;g\rangle_z$ produces a similar outcome (the resulting configuration is nothing more than the g transform of the original one). A non-CoS domain wall is the simplest possible generalization of a Z_2 kink for a G-invariant theory.

However, there is obviously a second, more interesting possibility: if the vacuum is $|0\rangle$ at $y = -\infty$, then the vacuum at $y = +\infty$ can also be a $|0;g\rangle_z$ for $g \neq 1$. In that case, the subgroups respected asymptotically are the differently embedded but isomorphic groups H and H_g, respectively. This defines a CoS-style domain wall [14–20]. At finite y, the configuration would be expected to respect the smaller group $H \cap H_g$ due to the fact that the solution has to "reconcile" boundary conditions that have different stability groups that "clash."[4]

So, there will be an infinite family of non-CoS DWs, trivially related to each other by global transformations $g \in G/H$. They all have the same energy density, because the Hamiltonian is invariant under G. The CoS DWs have a more complicated spectrum. Consider two configurations, $\chi_1(y)$ and $\chi_2(y)$, with χ_1 interpolating between $|0\rangle$ and $|0;g_1\rangle_z$, while χ_2 interpolates between $|0\rangle$ and $|0;g_2\rangle_z$, such that $g_1 \neq g_2$. Suppose, for the moment, that G is a global but not a local symmetry. These two solutions *cannot* be transformed into each other by a *global* G-transformation, so they would be expected to have different energy densities (their configurations trace different paths through the Higgs-potential topography). As a corollary, the non-CoS solutions should have a different energy density from the CoS solutions. All these solutions are in the same topological class, so finite-energy dynamical evolution between them is allowed. Hence, the special configurations within that topological class that minimize the energy density will be topologically stable. The others should be unstable to decay to the minimum-energy configurations, which play the role of "vacua" for the "kink-sector." This general reasoning cannot tell you which configuration has the minimum energy density: you need to calculate that within a specific model. For example, in the toy model considered in Ref. [14] the sign of a Higgs-potential parameter determined whether the non-CoS or a CoS solution was energetically favored.

Suppose now that G is a gauge symmetry, and again consider the configurations discussed in the previous paragraph, together with the specification of vanishing gauge fields A_M at the solution level. The non-CoS solutions remain connected through global transformations. Two

[3] We dread to use the term "landscape".

[4] From experience, we have found that $H \cap H_g$ is the usual outcome. The specifics depend on the case considered. Sometimes there is enhanced symmetry at $y = 0$ because some of the scalar multiplet components instantaneously vanish there. This enhancement on a set of measure zero has yet to find application, although speculations exist [17,30].

CoS scalar-field configurations, χ_1 and χ_2, can be written as local G-transforms of each other. Suppose that

$$\chi_2(y) = U(y)\chi_1(y), \tag{2.2}$$

where $U(y)$ is a local-G element. Then the original first solution

$$\chi = \chi_1(y), \qquad A_M = 0 \tag{2.3}$$

is gauge equivalent to

$$\chi = \chi_2(y), \qquad A_M = -\frac{i}{e}(\partial_M U)U^\dagger, \tag{2.4}$$

where e is the gauge coupling constant, but it is *not* gauge equivalent to

$$\chi = \chi_2(y), \qquad A_M = 0, \tag{2.5}$$

which is the original second solution. Thus the two solutions $\chi = \chi_1, A_M = 0$ and $\chi = \chi_2, A_M = 0$ have different energy densities, even though the scalar-field portions are related by a local symmetry transformation. Although $A_M = -\frac{i}{e}(\partial_M U)U^\dagger$ is a pure-gauge configuration, it contributes to the energy density through the $\chi - A_M$ interaction terms.

Setting the gauge fields to zero at the solution level is basically a convenient choice of gauge, one we shall adopt from now on. Of course the solutions can be made to look very different by gauge-transforming them, but their physical consequences cannot change. This circumstance is no different from the monopole or local-string cases, where again the solutions look different in different gauges. Actually, it is no more complicated than the usual homogeneous vacuum expectation value (VEV) case. If $\langle\chi\rangle$ is a homogeneous VEV, then it can be gauge transformed to a nonhomogeneous configuration $U(x)\langle\chi\rangle$ but the scalar gradient energy is canceled by the gauge-field contribution.

The alert reader may have noticed the following: we have not proven that the minimum-energy DW configuration must have a gauge-field sector that is gauge equivalent to zero. This does indeed appear to be a loose end. We shall make the assumption that it is in fact true for the purposes of the rest of this paper. Ultimately, one could uncover its hypothetical falsity by a perturbative stability analysis for the DW, but that is well beyond the scope of the present investigation.

Finally, a technical point: The set of H_g contains an uncountable infinity of differently embedded but isomorphic subgroups. However, there is a certain useful sense in which the number of embeddings can be considered finite. Let the Cartan subalgebra $\mathcal{G}_C$ of G be a certain particular set of generators, corresponding to a particular choice of basis for the Lie algebra. If we require that the Cartan subalgebras of two subgroups H_{g_1} and H_{g_2} are both subspaces of $\mathcal{G}_C$, then the number of distinct embeddings is finite. A familiar example of this concerns the $SU(2)$ subgroups of $SU(3)$. While there is an uncountable infinity of ways of embedding $SU(2)$ in $SU(3)$, there are only three embeddings that have the $SU(2)$ Cartan subalgebras as subspaces of the given Cartan-subalgebra space of $SU(3)$. These are usually called I-spin, U-spin, and V-spin. When we say "different embeddings" below, this is what we shall mean.[5]

III. $SO(10)$ WARM-UP EXAMPLE: THE NEED FOR E_6

We now discuss a $G = SO(10)$ model that serves both as a warm-up for E_6 and explains why the extension to E_6 is necessary. We shall make use of the $O(10)$-kink analysis of Ref. [17]. While some recapitulation is necessary for completeness, we shall be as brief as possible, and the reader is referred to Ref. [17] for a detailed discussion.

Let χ be a scalar multiplet in the adjoint representation, the 45, of $SO(10)$. The most general quartic Higgs potential is

$$V = \tfrac{1}{2}\mu^2\mathrm{Tr}(\chi^2) + \tfrac{1}{4}\lambda_1\mathrm{Tr}(\chi^2)^2 + \tfrac{1}{4}\lambda_2\mathrm{Tr}(\chi^4), \tag{3.1}$$

where $\chi = f_\alpha\hat{X}^\alpha$ with the $\hat{X}$'s being matrix representations of the generators in the fundamental of $SO(10)$ while the f_α's are the components of the adjoint multiplet. The matrix χ is antisymmetric and transforms as per $\chi \to A\chi A^T$, where A is an $SO(10)$ fundamental-representation matrix. The parameter μ^2 is chosen to be positive since $\mathrm{Tr}(\chi^2)$ is negative definite. The cubic invariant $\mathrm{Tr}(\chi^3)$ identically vanishes so there is an accidental discrete Z_2 symmetry, $\chi \to -\chi$, which shall play the role of Z. It is not a subgroup of $O(10)$.

The global minimization of such a potential was performed by Li [32] (see also [33]). Using an $SO(10)$ transformation, one may always bring a VEV pattern into the standard form

$$\chi = \mathrm{diag}(f_1\epsilon, f_2\epsilon, f_3\epsilon, f_4\epsilon, f_5\epsilon), \tag{3.2}$$

where the f_i are real and

$$\epsilon \equiv i\sigma_2 = \begin{pmatrix} 0 & 1 \\ -1 & 0 \end{pmatrix}. \tag{3.3}$$

The five independent fields f_i correspond to the five generators in the $SO(10)$ Cartan subalgebra. In this basis,

[5] Note that taking linear combinations of Cartan generators to define different embeddings is in accord with Dynkin's general theory of embeddings [31]. In that formalism, the embedding of an algebra $\mathcal{H}$ into a simple or semisimple algebra $\mathcal{G}$ is fully defined by a mapping F from the Cartan subalgebra of $\mathcal{H}$ into the Cartan subalgebra of $\mathcal{G}$, as per $H_\alpha \to F(H_\alpha) = \sum_{a=1}^n F_{\alpha a}G_a$, where $H_\alpha(\alpha = 1, 2, \ldots, m)$ and $G_a(a = 1, 2, \ldots, n)$ are the Cartan generators of $\mathcal{H}$ and $\mathcal{G}$, respectively. The matrix $(F_{\alpha a})$ is the defining matrix of the embedding, and two embeddings are different if their defining matrices are different.

$$V = -\mu^2 \sum_{i=1}^{5} f_i^2 + \lambda_1 \Big(\sum_{i=1}^{5} f_i^2\Big)^2 + \frac{1}{2}\lambda_2 \sum_{i=1}^{5} f_i^4. \tag{3.4}$$

For $\lambda_2 > 0$, the global minima of V are at

$$f_i^2 = \frac{\mu^2}{10\lambda_1 + \lambda_2} \equiv f_{\rm min}^2 \forall i, \tag{3.5}$$

where we define $f_{\rm min} \equiv \sqrt{\mu^2/(10\lambda_1 + \lambda_2)}$ and the unbroken subgroup is $H = U(5)$. The values of f_i at the minima are specified up to a sign that can be chosen independently for each component

$$f_i = \pm f_{\rm min}. \tag{3.6}$$

Different choices for these signs correspond to two features: different embeddings of $U(5)$ in $SO(10)$ and also a choice of which Z_2 sector the minimum lies in.

To explore this further, let us turn to possible domain-wall configurations. Suppose that at $y = -\infty$, we choose as our boundary condition

$$\chi(-\infty) = -f_{\rm min}^{(5)} \equiv -f_{\rm min}{\rm diag}(\epsilon, \epsilon, \epsilon, \epsilon, \epsilon). \tag{3.7}$$

This defines a certain $U(5)$ unbroken at $y = -\infty$, and the VEV lies in one of the two disconnected pieces of the vacuum manifold. At $y = +\infty$, there are three choices that lie in the other piece of the vacuum manifold, disconnected from the first by the spontaneously broken Z_2:

$$\chi(+\infty) = \begin{cases} f_{\rm min}^{(5)} \equiv f_{\rm min}{\rm diag}(\epsilon, \epsilon, \epsilon, \epsilon, \epsilon) \\ f_{\rm min}^{(3,2)} \equiv f_{\rm min}{\rm diag}(\epsilon, \epsilon, \epsilon, -\epsilon, -\epsilon) \\ f_{\rm min}^{(4,1)} \equiv f_{\rm min}{\rm diag}(\epsilon, -\epsilon, -\epsilon, -\epsilon, -\epsilon). \end{cases} \tag{3.8}$$

(Permutations of the minus signs in the last two of these vacua are just a trivial rearrangement of the representation space and need not be separately considered.) Vacua with an odd number of minus signs on the right-hand side on Eq. (3.8) are continuously connected to $\chi(-\infty)$ by $SO(10)$ and shall not be considered as they would give rise to nontopological domain walls.

The three vacua in Eq. (3.8) are invariant under differently embedded subgroups of $SO(10)$: $U(5)_1$, $U(5)_2$, and $U(5)_3$. The superscripts (5), (3, 2), and (4, 1) denote the numbers of plus and minus signs in the VEVs. But they also usefully describe the unbroken symmetry of the domain wall at finite y, respectively

$$U(5), \qquad U(3) \otimes U(2) \quad \text{and} \quad U(4) \otimes U(1), \tag{3.9}$$

as we now explain.

The ansatz for domain-wall configurations that interpolate between the stated boundary conditions is $\chi(y) = h(y)\chi(-\infty) + g(y)\chi(+\infty)$, where the functions h and g obey self-evident boundary conditions. The first configuration, which interpolates between $-f_{\rm min}$ and $+f_{\rm min}$ for all components $f_i(y)$, breaks $SO(10)$ to $U(5)_1$ at all values of y, because the relative magnitudes of the components are always the same at a given y. It is a non-CoS domain wall. The second configuration has an equal-magnitude 3×3 block (of 2×2 submatrices), and an equal-magnitude 2×2 block. The unbroken symmetry is then

$$U(3) \otimes U(2) = U(5)_1 \cap U(5)_2. \tag{3.10}$$

Similarly, the third configuration's block structure leads to $U(4) \otimes U(1) = U(5)_1 \cap U(5)_3$.

The Euler-Lagrange equations

$$f_i'' = 2\Big[-\mu^2 + 2\lambda_1 \sum_{j=1}^{5} f_j^2\Big] f_i + 2\lambda_2 f_i^3, \tag{3.11}$$

with the three types of boundary conditions above may be solved numerically. However, a simple way to prove that solutions exist is to consider the $\lambda_1 = 0$ slice through parameter space. The equations can then be solved analytically to yield

$$f_i(y) = f_{\rm min} \tanh(\mu y) \ \forall i \tag{3.12}$$

for the first boundary condition choice,

$$\begin{aligned} f_i(y) &= f_{\rm min} \tanh(\mu y) \quad \text{for } i = 1, 2, 3 \quad \text{and} \\ f_i(y) &= f_{\rm min} \quad \text{for } i = 4, 5 \end{aligned} \tag{3.13}$$

for the second choice, and

$$\begin{aligned} f_i(y) &= f_{\rm min} \tanh(\mu y) \quad \text{for } i = 1 \quad \text{and} \\ f_i(y) &= f_{\rm min} \quad \text{for } i = 2, 3, 4, 5 \end{aligned} \tag{3.14}$$

for the third choice. The surface energy densities are in the ratios 5:3:1 for the first to the third solutions [17], so Eq. (3.14) gives the topologically stable configuration.

From a Dvali-Shifman point of view, this stable configuration has an unbroken $SU(4)$ on the brane that is embedded in $SU(5)_1$ on the $y < 0$ side of the wall, and $SU(5)_3$ on the $y > 0$ side. The $SU(4)$ gauge bosons are thus localized to the wall, if the Dvali-Shifman mechanism is correct, because by assumption both $SU(5)_1$ and $SU(5)_3$ are in confinement phase in their respective bulk regions. This establishes the connection between clash of symmetries and Dvali-Shifman by way of an explicit rigorously worked-out solution. It is, however, just a toy model since the symmetry breaking pattern is not what is required phenomenologically.

The second configuration, with

$$U(3) \otimes U(2) = SU(3) \otimes SU(2) \otimes U(1) \otimes U(1) \tag{3.15}$$

on the brane, is closer to what we need for a realistic model. While the analytic solution of Eq. (3.13) is unstable to dynamical evolution to Eq. (3.14), it could well be that in another region of Higgs-potential parameter space the $U(3) \otimes U(2)$ solution is the stable one. This has not been established, but let us suppose it is true. The model then still does not quite work, although it comes close.

It is certainly true that the $SU(3) \otimes SU(2)$ factor in Eq. (3.15) is Dvali-Shifman-localized, because it is a subgroup of both $SU(5)_1$ (the bulk symmetry for $y < 0$) and $SU(5)_2$ (the bulk symmetry for $y > 0$). However, there is a problem with the hypercharge gauge boson corresponding to $U(1)_Y$. To see this, we need to examine the $U(1)$ generators more closely.

The asymptotic gauge groups are

$$U(5)_1 = SU(5)_1 \otimes U(1)_{X_1} \quad \text{and} \quad U(5)_2 = SU(5)_2 \otimes U(1)_{X_2}. \tag{3.16}$$

Denote by Y_1 the hypercharge generator inside $SU(5)_1$, and Y_2 the one inside $SU(5)_2$. The two $U(1)$'s in Eq. (3.15) can be taken to be generated either by Y_1 and X_1, or by Y_2 and X_2, and each pair can be written as linear combinations of the other pair. Now, either Y_1 or Y_2 can be the physical hypercharge Y. Which one is selected will be an accident of spontaneous symmetry breaking. At some scale above the electroweak, the breaking

$$U(1)_{Y_1} \otimes U(1)_{X_1} = U(1)_{Y_2} \otimes U(1)_{X_2} \to U(1)_Y, \tag{3.17}$$

with either $Y = Y_1$ or $Y = Y_2$, will have to take place to produce an effective standard model at low energies (this will require an additional Higgs field). Suppose $Y = Y_1$ is spontaneously selected. Then the hypercharge gauge boson cannot propagate into the $y < 0$ bulk, but a component of it will propagate into the $y > 0$ bulk. The generator $Y = Y_1$ is a linear combination of Y_2 and X_2, so the hypercharge gauge field is a linear combination of the gauge fields of Y_2 and X_2. But only the Y_2 part is unable to propagate into the $y > 0$ region; the X_2 part is immune from the Dvali-Shifman effect because it is not confining. After electroweak symmetry breaking, this will imply that both the photon and Z^0 will leak into the $y > 0$ bulk, which is phenomenologically ruled out. If Y_2 happens to become the physical Y, then leakage into $y < 0$ will occur.

This structure, with localized gluons and $W^{\pm}$ bosons, but semidelocalized photons and Z^0's, almost works. But understanding its pathology also provides the cure: We need to expand the symmetry on the brane to contain a full $SU(5)$, with the physical hypercharge identified with one of its generators. Further, this brane-$SU(5)$ must be a subgroup of confining non-Abelian groups on both sides of the domain wall. These two features automatically arise when we upgrade from $SO(10)$ to E_6 as the symmetry of the action.

IV. THE E_6 DOMAIN-WALL BRANE

A. Group theory

Take a scalar-field multiplet χ in the 78-dimensional adjoint representation of E_6. In the next subsection, we shall analyze the associated Higgs potential and produce domain-wall solutions. But for now, we just need to use the fact that for a range of parameters the global minima of the potential will induce

$$E_6 \to SO(10) \otimes U(1), \tag{4.1}$$

which is a maximal subgroup. Now consider different embeddings of $SO(10) \otimes U(1)$ in E_6.[6] We shall show below that there is a pair of embeddings, which we shall call simply $SO(10) \otimes U(1)_E$ and $SO(10)' \otimes U(1)_{E'}$, that is of particular interest for model-building.[7] The domain-wall solution we shall find in the next subsection interpolates between χ VEVs that break E_6 to these different but isomorphic subgroups on opposite sides of the wall. The symmetry on the wall is then

$$[SO(10) \otimes U(1)_E] \cap [SO(10)' \otimes U(1)_{E'}] = SU(5) \otimes U(1)_E \otimes U(1)_{E'}, \tag{4.2}$$

as we shall establish. Since $SU(5) \subset SO(10) \cap SO(10)'$, the Dvali-Shifman mechanism localizes all the $SU(5)$ gauge bosons to the domain wall, including the photon and the Z^0.

Let us look at the group theory in more detail. Under

$$E_6 \to SO(10) \otimes U(1)_E \to [SU(5) \otimes U(1)_X] \otimes U(1)_E, \tag{4.3}$$

the fundamental 27-dimensional representation of E_6 branches as per

$$27 \to 1(4) + 10(-2) + 16(1) \to 1^{(0,4)} + [5^{(2,-2)} + 5^{*(-2,-2)}] + [1^{(-5,1)} + 5^{*(3,1)} + 10^{(-1,1)}]. \tag{4.4}$$

The notation for $SO(10) \otimes U(1)_E$ representations is $D(12E)$, where D is the dimension of the $SO(10)$ multiplet, and the $U(1)_E$ generator has been normalized as per

$$\mathrm{Tr}_{27}(E^2) = \tfrac{1}{2}. \tag{4.5}$$

(We use $12E$ to make the charges integers for convenience.) The $SU(5) \otimes U(1)_X \otimes U(1)_E$ notation is $D^{(4\sqrt{15}X, 12E)}$ with

$$\mathrm{Tr}_{27}(X^2) = \tfrac{1}{2}, \qquad \mathrm{Tr}_{27}(EX) = 0. \tag{4.6}$$

The second embedding is revealed by considering the linear combinations

$$X' = -\tfrac{1}{4}(X + \sqrt{15}E), \qquad E' = \tfrac{1}{4}(-\sqrt{15}X + E) \tag{4.7}$$

that correspond to

$$E_6 \to SO(10)' \otimes U(1)_{E'} \to [SU(5) \otimes U(1)_{X'}] \otimes U(1)_{E'}. \tag{4.8}$$

[6]We mean the finite number of embeddings in the sense of the final paragraph of Sec. II.

[7]The second embedding has been used in unified model building [34–37].

Rewriting the multiplets from $D^{(4\sqrt{15}X,12E)}$ notation to $D_{(4\sqrt{15}X',12E')}$ notation, we see that

$$\begin{aligned} 1^{(0,4)} &= 1_{(-5,1)} & 5^{(2,-2)} &= 5_{(2,-2)} \\ 5^{*(-2,-2)} &= 5^*_{(3,1)} & 1^{(-5,1)} &= 1_{(0,4)} \\ 5^{*(3,1)} &= 5^*_{(-2,-2)}, & 10^{(-1,1)} &= 10_{(-1,1)}, \end{aligned} \tag{4.9}$$

so the 5^*'s flip roles as do the singlets. Let us now redundantly denote the multiplets through

$$D^{(4\sqrt{15}X,12E)}_{(4\sqrt{15}X',12E')}. \tag{4.10}$$

The 10 of $SO(10)'$ is

$$5^{(2,-2)}_{(2,-2)} \oplus 5^{*(3,1)}_{(-2,-2)}, \tag{4.11}$$

whereas the 10 of the original $SO(10)$ was instead formed by

$$5^{(2,-2)}_{(2,-2)} \oplus 5^{*(-2,-2)}_{(3,1)}. \tag{4.12}$$

Similarly, the 16 of $SO(10)'$ consists of

$$1^{(0,4)}_{(-5,1)} \oplus 5^{*(-2,-2)}_{(3,1)} \oplus 10^{(-1,1)}_{(-1,1)}, \tag{4.13}$$

whereas the 16 of the original $SO(10)$ consisted of

$$1^{(-5,1)}_{(0,4)} \oplus 5^{*(3,1)}_{(-2,-2)} \oplus 10^{(-1,1)}_{(-1,1)}. \tag{4.14}$$

The $SO(10)'$ singlet is $1^{(-5,1)}_{(0,4)}$, whereas the original $SO(10)$ singlet is $1^{(0,4)}_{(-5,1)}$.

Because all higher-dimensional representations of E_6 are formed from products of 27's, the feature that some $SU(5) \otimes U(1)^2$ submultiplets flip when $(X, E) \to (X', E')$ propagates to all irreducible E_6 representations. The submultiplets can be packaged in $SO(10) \otimes U(1)_E$ multiplets, or repackaged into $SO(10)' \otimes U(1)_{E'}$ multiplets. This establishes, constructively, that the two embeddings exist, and that Eq. (4.2) is true.[8] Note that the additional $U(1)$'s are there because adjoint configurations cannot rank reduce.

Let us repeat this exercise for the adjoint of E_6:

$$78 \to 1(0) + 45(0) + 16(-3) + 16^*(3) \tag{4.15}$$

$$\begin{aligned} \to\ & 1^{(0,0)}_{(0,0)} \\ &+ \left[1^{(0,0)}_{(0,0)} + 10^{(4,0)}_{(-1,-3)} + 10^{*(-4,0)}_{(1,3)} + 24^{(0,0)}_{(0,0)}\right] \\ &+ \left[1^{(-5,-3)}_{(5,3)} + 5^{*(3,-3)}_{(3,-3)} + 10^{(-1,-3)}_{(4,0)}\right] \\ &+ \left[1^{(5,3)}_{(-5,-3)} + 5^{(-3,3)}_{(-3,3)} + 10^{*(1,3)}_{(-4,0)}\right]. \end{aligned} \tag{4.16}$$

[8] By considering additional Cartan generators beyond E and X, more embeddings of $SO(10)$ can be found. This is discussed further in the Appendix.

The flipping of roles is evidently

$$\begin{aligned} 1^{(0,0)}_{(0,0)} &\leftrightarrow 1^{(0,0)}_{(0,0)}, & 10^{*(-4,0)}_{(1,3)} &\leftrightarrow 10^{*(1,3)}_{(-4,0)}, \\ 10^{(4,0)}_{(-1,-3)} &\leftrightarrow 10^{(-1,-3)}_{(4,0)}, & 1^{(-5,-3)}_{(5,3)} &\leftrightarrow 1^{(5,3)}_{(-5,-3)}. \end{aligned} \tag{4.17}$$

The $SU(5)$ adjoint $24^{(0,0)}_{(0,0)}$ is common to both $SO(10)$ embeddings, as befits its status of being in the intersection of the two.

The two $1^{(0,0)}_{(0,0)}$ multiplets play important roles. Giving a VEV to the 1(0) in Eq. (4.15) breaks E_6 to $SO(10) \otimes U(1)_E$, while a VEV for the second singlet in Eq. (4.16) breaks E_6 to $SO(10)' \otimes U(1)_{E'}$. A clash-of-symmetries kink interpolates between these two VEVs imposed as boundary conditions. At $|y| < \infty$, both $SU(5) \otimes U(1)^2$ singlet components of the 78 have nonzero values, and this is precisely why the configuration breaks E_6 to the intersection of the two subgroups. To analyze this further, we must consider the dynamics.

B. Higgs potential and domain-wall solutions

The adjoint scalar multiplet χ shall be represented by

$$\chi = f_\alpha \hat{X}^\alpha, \qquad \alpha = 1, \ldots, 78, \tag{4.18}$$

where $\hat{X}$'s are matrix representations of the generators for the 27 of E_6, and the f's are the field components. It transforms according to

$$\chi \to U \chi U^\dagger, \tag{4.19}$$

where U is group representation matrix for the 27. We shall only be concerned with two of the 78 fields: those associated with (E, E'), equivalently (X, E) or (X', E') depending on what basis we choose for the Lie algebra.

We thus specialize to

$$\chi = f_E E + f_X X = \tilde{f}_E E + f_{E'} E' \tag{4.20}$$

with

$$\tilde{f}_E = f_E + \frac{f_X}{\sqrt{15}}, \qquad f_{E'} = -\frac{4 f_X}{\sqrt{15}}, \tag{4.21}$$

according to Eq. (4.7). The (X, E) basis is the more convenient for solving the Euler-Lagrange equations, because E and X are orthogonal as per Eq. (4.6). The (E, E') basis, however, is the simplest one for thinking about the two embeddings.

The VEVs we want for the boundary conditions are

$$(\tilde{f}_E, f_{E'}) = v(1, 0), \tag{4.22}$$

which corresponds to $E_6 \to SO(10) \otimes U(1)_E$. The other VEV is

$$(\tilde{f}_E, f_{E'}) = -v(0,1) \quad (4.23)$$

which gives $E_6 \rightarrow SO(10)' \otimes U(1)_{E'}$. The relative minus sign between Eqs. (4.22) and (4.23) comes from the breaking of a

$$\chi \rightarrow -\chi \quad (4.24)$$

discrete symmetry we shall impose on the Higgs potential, and it is *crucial* for two reasons. First, the spontaneous Z_2 breaking will ensure that our domain walls are topologically nontrivial. Second, it leads to a remarkable outcome for fermion zero-mode localization, to be explained in the next subsection.

In terms of the (X, E) basis, these same VEVs are

$$(f_X, f_E) = v(0,1) \quad \text{and} \quad v\left(\frac{\sqrt{15}}{4}, -\frac{1}{4}\right), \quad (4.25)$$

respectively.

We now need to find a Higgs potential with these two VEVs as degenerate global minima. The Higgs potential is constructed out of adjoint invariants, which according to Eqs. (4.18) and (4.19) are

$$I_n = \mathrm{Tr}(\chi^n) = \mathrm{Tr}(\hat{X}^{\alpha_1}\hat{X}^{\alpha_2}\cdots\hat{X}^{\alpha_n})f_{\alpha_1}f_{\alpha_2}\cdots f_{\alpha_n}. \quad (4.26)$$

They are simply the nth-order Casimir invariants. According to Refs. [38,39], the independent invariants are

$$I_2, I_5, I_6, I_8, I_9, I_{12}, \quad (4.27)$$

which immediately has an interesting consequence: the fact that $I_{5,9}$ are nonzero means that the discrete Z_2 of Eq. (4.24) is *not a subgroup* of E_6, because imposing it eliminates the otherwise present odd-power invariants.

It is sensible to truncate the Higgs potential at order six:

$$V = -\lambda_1 I_2 + \lambda_2 (I_2)^2 - 2304\kappa I_6 + \tfrac{4}{3}\lambda_3 (I_2)^3, \quad (4.28)$$

where some peculiar numbers and signs have been inserted for later convenience. In the extra-dimensional setting, field-theoretic models must generally be considered as effective theories valid below an ultraviolet cutoff scale Λ, because they are almost inevitably nonrenormalizable. In writing down a Higgs potential, one simply adds terms of ever higher mass dimension and truncates appropriately, given that the higher the mass dimension the more suppressed it should become. For the E_6 application, it is not helpful to truncate at fourth order, because the only fourth-order invariant is $(I_2)^2$ and I_2 is invariant under an accidental $O(78)$ symmetry. The presence of I_6 reduces the symmetry of the Higgs potential to E_6 (presumably), and eliminates a pseudo-Goldstone boson issue.[9]

Equation (4.28) is a complicated sextic in 78 fields. But to perform the global minimization analysis, one can always transform any VEV pattern to a standard form given by linear combinations of just the six generators in the Cartan subalgebra of E_6. This produces a still quite complicated sextic in six fields. To make our discussion as simple as possible, in the main body of the paper we shall further truncate to just the two Cartan-subalgebra generators of interest, and use Eq. (4.20). We extend the global minimization analysis to all six fields in the Appendix.

With just $f_{E,X} \neq 0$, the nth-order invariant simplifies to

$$I_n = \sum_{k=0}^{n} \binom{n}{n-k} \mathrm{Tr}(E^{n-k}X^k) f_E^{n-k} f_X^k. \quad (4.29)$$

The traces can be worked out by hand, because we know the matrix representations of E and X from the branching rules in Eq. (4.4). We obtain

$$I_2 = \tfrac{1}{2}(f_E^2 + f_X^2), \quad (4.30)$$

$$I_6 = \frac{1}{2304} \times \left(f_E^6 + 5f_E^4 f_X^2 + 7f_E^2 f_X^4 - \frac{48}{5\sqrt{15}} f_E f_X^5 + \frac{83}{25} f_X^6\right). \quad (4.31)$$

To understand the extrema of Eq. (4.28), it is helpful to use the polar decomposition

$$f_E = r\cos\theta, \qquad f_X = r\sin\theta. \quad (4.32)$$

The VEVs of Eq. (4.25) are then

$$(10, +)\colon \theta = 0, \quad (4.33)$$

$$(10', -)\colon \cos\theta = -\frac{1}{4}, \qquad \sin\theta = \frac{\sqrt{15}}{4}. \quad (4.34)$$

The notation $(10, +)$ means that the VEV of Eq. (4.33) induces $E_6 \rightarrow SO(10) \otimes U(1)_E$, and we have (arbitrarily) assigned it a positive Z_2 "parity" which signals that it lies in G/H rather than $(G/H)_z$. Similarly, $(10', -)$ means $E_6 \rightarrow SO(10)' \otimes U(1)_{E'}$ and it lies in $(G/H)_z$. There is another pair, with the opposite Z_2 parities:

[9] An alternative is to truncate the classical theory at fourth order, but to add a Coleman-Weinberg potential generated through quantum corrections that explicitly break the $O(78)$ [39].

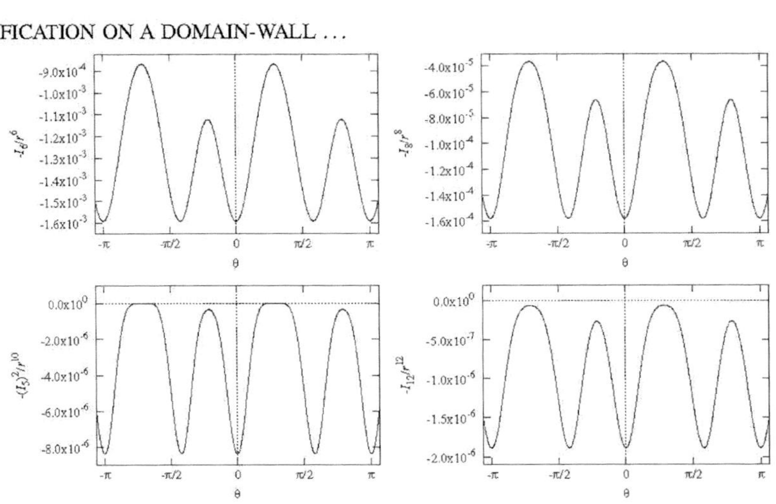

FIG. 2. The invariants $-I_6/r^6$, $-I_8/r^8$, $-(I_5)^2/r^{10}$, and $-I_{12}/r^{12}$ as functions of θ. Note the remarkable similarity. The global minima for all four, reading from left to right, are $(10, -)$, $(10', +)$, $(10, +)$, $(10', -)$, and then $(10, -)$ again (see text for explanation of this notation).

$$(10, -)\colon\ \theta = \pi, \tag{4.35}$$

$$(10', +)\colon\ \cos\theta = \frac{1}{4}, \qquad \sin\theta = -\frac{\sqrt{15}}{4}. \tag{4.36}$$

The topological CoS domain wall connects $(10, +)$ and $(10', -)$, accompanied by a CoS antidomain wall connecting $(10, -)$ and $(10', +)$. The topological non-CoS domain walls connect $(10, +)$ with $(10, -)$ [breaking E_6 to $SO(10) \otimes U(1)_E$ at all y], and $(10', +)$ with $(10', -)$ [breaking E_6 to $SO(10)' \otimes U(1)_{E'}$ at all y]. There are also nontopological configurations: (i) $(10, +)$ connected to $(10', +)$, and (ii) $(10, -)$ connected to $(10', +)$, which are both CoS-like.

Figure 2 display the invariants $-I_6/r^6$, $-I_8/r^8$, $-(I_5)^2/r^{10}$, and $-I_{12}/r^{12}$ as functions of θ. They show a remarkably similar structure. It is evident that the global minima for all four invariants are precisely the four VEVs of Eqs. (4.33), (4.34), (4.35), and (4.36). It is clear from this that choosing to truncate at the sextic level, as in Eq. (4.28), does not sacrifice much in terms of generality. We can be confident that our simplified potential leads to solutions whose qualitative characteristics would be retained were a wider class of higher-order potentials considered. In addition, the Appendix shows that there are no deeper minima than $(10, \pm)$ and $(10', \pm)$ in the whole six-dimensional Cartan domain.

The sign in front of I_6 must be negative to achieve the desired extrema as minima rather than maxima. The other terms in the Higgs potential, Eq. (4.28), are independent of θ, depending only on the radial function r. Hence, it is clear that Eq. (4.28) has the global minima we require. Figure 3 shows a contour plot of the Higgs potential for a certain parameter choice illustrating this conclusion. It is important to realize that although the minima $(10, +)$ and $(10', +)$ [similarly $(10, -)$ and $(10', -)$] look as though they are disconnected by E_6, this is just an illusion created by only plotting the two-dimensional (f_X, f_E) slice through the 78-dimensional adjoint representation space. Minima with opposite parities are definitely disconnected from each other.[10]

In the examples presented below, we further simplify the Higgs potential by setting $\lambda_2 = 0$ as this term does not play an important role. It is then easy to show that at the degenerate minima,

$$r = v \equiv \left(\frac{\lambda_1}{\lambda_3 - 22\kappa}\right)^{1/4}, \tag{4.37}$$

so we must take $\lambda_3 > 22\kappa$, and that the value of V at the minima is

$$V_{\min} = -\frac{1}{3}\sqrt{\frac{\lambda_1^3}{\lambda_3 - 22\kappa}}. \tag{4.38}$$

The latter must be subtracted from the potential

$$V \to V - V_{\min} \tag{4.39}$$

to produce finite energy densities for the domain-wall configurations. Figure 3 is a contour plot of the potential energy showing the four degenerate global minima.

[10]While it is certainly true that the Z_2 is not a subgroup of E_6, so that in general χ and $-\chi$ are disconnected from each other, one may worry that there is nevertheless an E_6 transformation that takes the specific configuration $(10, +)$ to $(10, -)$. However, this is not the case. Explicit calculation of the E_6 invariant I_5 reveals a nonzero f_E^5 term. Hence, $f_E \to -f_E$ must be outside of E_6. It does not matter that I_5 has been omitted from the Higgs potential, as it is a purely group-theoretic argument.

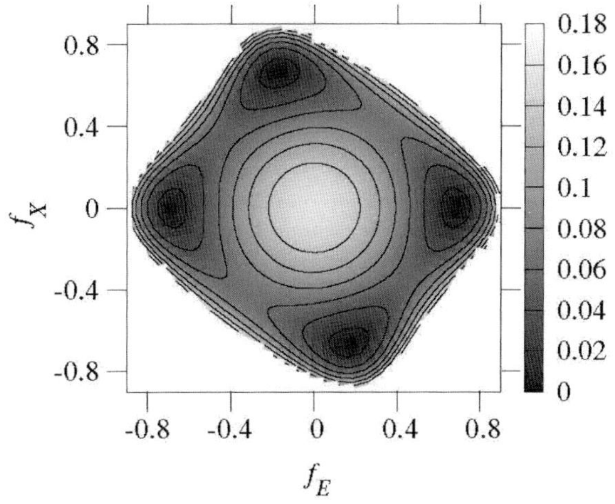

FIG. 3. Contour plot of the Higgs potential as a function of the two field components f_E and f_X. The parameters used are $\kappa = 0.8$, $\lambda_1 = 1.0$, $\lambda_2 = 0$, and $\lambda_3 = 22.0$. The darkest regions are the global minima in the order $(10, +)$, $(10', -)$, $(10, -)$, and $(10', +)$ reading anticlockwise from the rightmost minimum. The light area near the origin is a local maximum.

Having understood the global minima, we may now solve the Euler-Lagrange equations

$$f''_X(y) = \frac{\partial V}{\partial f_X}, \qquad f''_E(y) = \frac{\partial V}{\partial f_E} \tag{4.40}$$

using those VEVs as boundary conditions. Numerical solutions for CoS domain walls interpolating between $(10, +)$ at $y = -\infty$ and $(10', -)$ at $y = +\infty$ with two different parameter choices are displayed in Fig. 4.

Figure 5 depicts non-CoS domain-wall solutions for the same parameter choices. The function f_X is zero, while f_E interpolates between v and $-v$ in archetypal kink fashion. This means that the non-CoS configurations feel the large potential-energy maximum at $f_X = f_E = 0$, while the CoS configuration "skirts around" that central maximum. This immediately implies that the CoS solutions have lower energy density than the non-CoS solutions. Although they are in the same topological class, the CoS domain walls are stable while the non-CoS domain walls are unstable. Figure 6 shows a three-dimensional plot of the potential and where the two DW configurations sit with respect to the topography. There is a tall maximum at the origin, and a corrugated valley encircling it, with four low points at the VEVs. Figure 7 compares the energy densities of CoS and non-CoS domain walls.

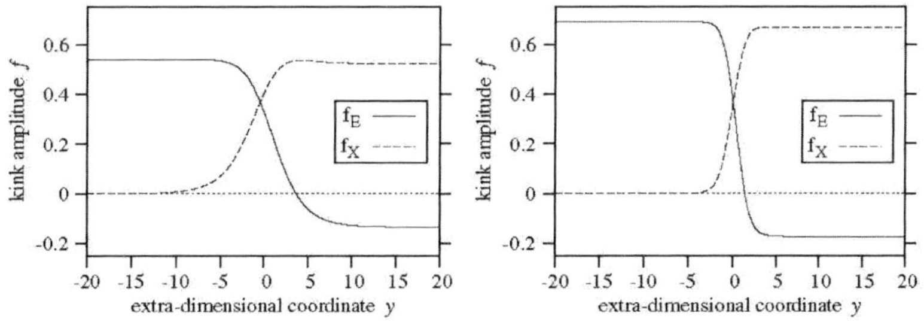

FIG. 4. Clash-of-symmetries domain-wall solutions interpolating between $(10, +)$ at $y = -\infty$ and $(10', -)$ at $y = +\infty$. The parameters used in the left plot are $\kappa = 0.2$, $\lambda_1 = 1.5$, $\lambda_2 = 0$, and $\lambda_3 = 22.0$; those in the right plot are $\kappa = 0.8$, $\lambda_1 = 1.0$, $\lambda_2 = 0$, and $\lambda_3 = 22.0$.

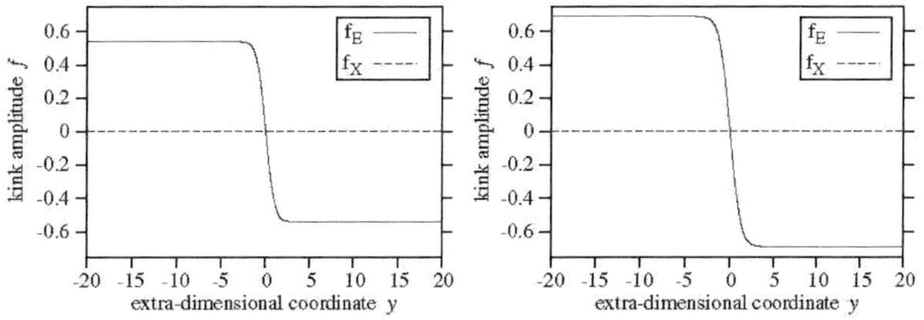

FIG. 5. Non-clash-of-symmetries domain-wall solutions interpolating between $(10, +)$ at $y = -\infty$ and $(10, -)$ at $y = +\infty$. The parameters used in the left plot are $\kappa = 0.2$, $\lambda_1 = 1.5$, $\lambda_2 = 0$, and $\lambda_3 = 22.0$; those in the right plot are $\kappa = 0.8$, $\lambda_1 = 1.0$, $\lambda_2 = 0$, and $\lambda_3 = 22.0$.

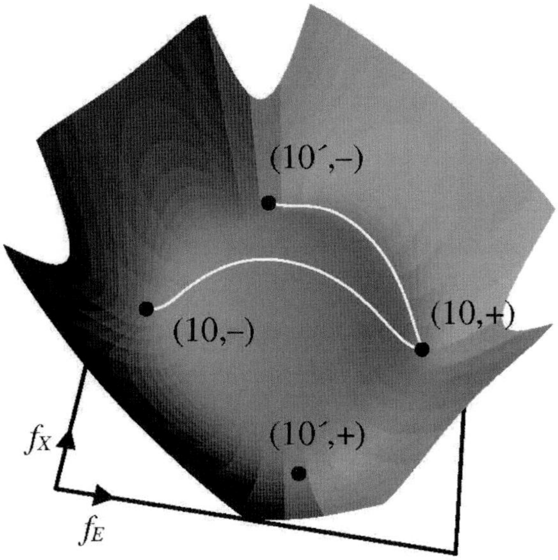

FIG. 6. Three-dimensional plot of the Higgs potential as a function of the two field components f_E and f_X. The white lines show the clash-of-symmetries domain wall (topmost) and the non-CoS domain wall (bottommost). The parameters used are $\kappa = 0.8$, $\lambda_1 = 1.0$, $\lambda_2 = 0$, and $\lambda_3 = 22.0$.

C. Fermion zero-mode localization

The CoS E_6 domain-wall solutions described above are a good starting point for the creation of domain-wall-brane models featuring $SU(5)$-invariant effective 3 + 1-d theories for localized fields. To actually create such a model, fermions, additional Higgs bosons, and gravitons have to be added. In this subsection, we demonstrate that a phenomenologically-acceptable fermion localization pattern is obtained using the simplest possible mechanism. We explain why this is a remarkable result.

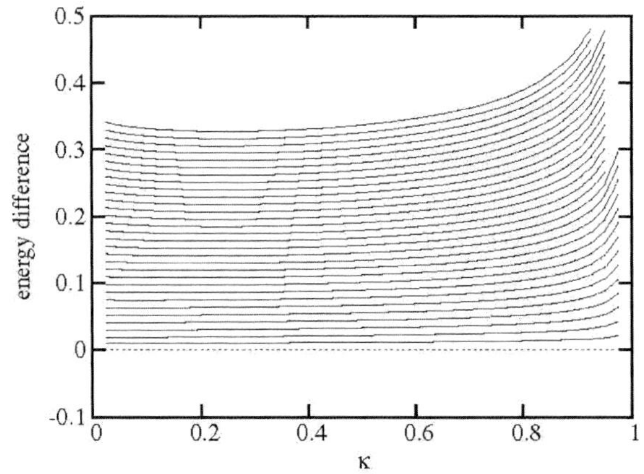

FIG. 7. The difference in energy densities between the non-CoS and CoS domain-wall solutions, $E_{\text{non-CoS}} - E_{\text{CoS}}$. We have numerically scanned through the parameter space with $0 < \kappa < 1$ along the horizontal axis, and each successive curve represents a different λ_1, beginning at $\lambda_1 = 0.05$ at the bottom and increasing in steps of 0.05 to $\lambda_1 = 1.5$ at the top. The energy difference is always positive, so the CoS domain wall has a lower energy. We set $\lambda_2 = 0$ for simplicity.

We simply Yukawa-couple a five-dimensional fermion multiplet in the 27 of E_6,

$$\Psi \sim 27, \tag{4.41}$$

to the adjoint scalar, as per

$$\mathcal{L}_Y = -h\bar{\Psi}\chi\Psi. \tag{4.42}$$

We now substitute in the background CoS DW configuration for χ and solve the resulting Dirac equations, which take the form

$$i\Gamma^M \partial_M \Psi^{(X,E)}(x^\mu, y) - h[f_X(y)X + f_E(y)E]\Psi^{(X,E)}(x^\mu, y) = 0. \tag{4.43}$$

The notation $\Psi^{(X,E)}$ signifies the component of the 27 with the specified (X, E) charges, as given in Eq. (4.4). The various components couple to different background field configurations,

$$b^{(X,E)}(y) = f_X(y)X + f_E(y)E, \tag{4.44}$$

given by the DW configuration and the charges.

The Dirac matrices are $\Gamma^M = (\gamma^\mu, -i\gamma_5)$. We search for separated-variable solutions

$$\Psi(x^\mu, y) = F(y)\psi(x^\mu), \tag{4.45}$$

demanding that ψ have definite chirality, $\gamma_5\psi = \pm\psi$, and obey the 3 + 1-d massless Dirac equation, $i\gamma^\mu\partial_\mu\psi = 0$. The solution for a profile is well known:

$$F^{(X,E)}(y) = N^{(X,E)} e^{-h\int^y b^{(X,E)}(y')dy'}, \tag{4.46}$$

where N is a normalization factor. For the profile to represent localization, it must be square integrable with respect to y. For this to happen, $b^{(X,E)}$ must pass through zero. If so and it is an increasing function of y (kinklike), then a left-(right-)handed zero mode occurs for $h > 0 (h < 0)$. If it passes through zero as a decreasing function (antikinklike), then a left-(right-)handed zero mode occurs for $h < 0 (h > 0)$.

Figure 8 show the kinklike functions $b^{(X,E)}$ for the two parameter choices we have been using as examples. Let us take h to be negative:

$$h < 0. \tag{4.47}$$

Using the notation $D^{(4\sqrt{15}X, 12E)}$ once again for the fermion multiplets, the following displays these functions and states the localization outcome, which is either "localized

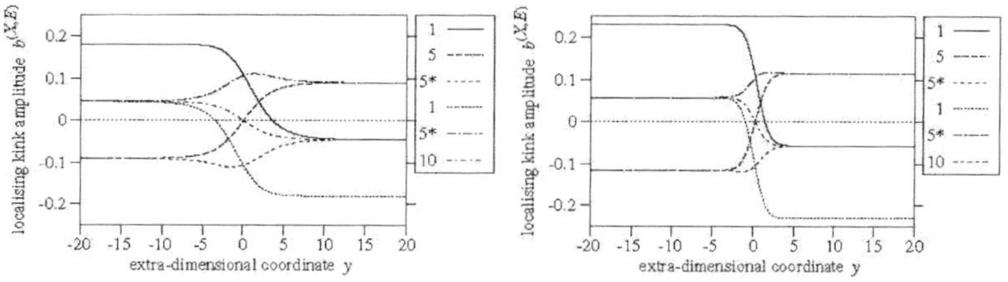

FIG. 8. Clash-of-symmetries fermion localizing profiles interpolating between $(10, +)$ at $y = -\infty$ and $(10', -)$ at $y = +\infty$. The parameters used in the left plot are $\kappa = 0.2$, $\lambda_1 = 1.5$, $\lambda_2 = 0$, and $\lambda_3 = 22.0$; those in the right plot are $\kappa = 0.8$, $\lambda_1 = 1.0$, $\lambda_2 = 0$, and $\lambda_3 = 22.0$. The top to bottom order of the $SU(5)$ fermion multiplets in the box on the right matches the order in Eq. (4.48).

as left-handed (LH) zero mode" or "localized as right-handed (RH) zero mode" or "delocalized":

$$\begin{aligned}
1^{(0,4)}&: \frac{1}{3}f_E, \quad \text{localized LH} \\
5^{(2,-2)}&: \frac{1}{2}\left(\frac{1}{\sqrt{15}}f_X - \frac{1}{3}f_E\right), \quad \text{localized RH} \\
5^{*(-2,-2)}&: -\frac{1}{2}\left(\frac{1}{\sqrt{15}}f_X + \frac{1}{3}f_E\right) \quad \text{delocalized} \\
1^{(-5,1)}&: \frac{1}{4}\left(-\frac{5}{\sqrt{15}}f_X + \frac{1}{3}f_E\right), \quad \text{localized LH} \\
5^{*(3,1)}&: \frac{1}{4}\left(\frac{3}{\sqrt{15}}f_X + \frac{1}{3}f_E\right), \quad \text{delocalized} \\
10^{(-1,1)}_{(-1,1)}&: \frac{1}{4}\left(-\frac{1}{\sqrt{15}}f_X + \frac{1}{3}f_E\right), \quad \text{localized LH.}
\end{aligned} \tag{4.48}$$

The two 5^*'s are delocalized because the associated field never goes through zero. The 5 and the 10 are localized at $y = 0$ with opposite chiralities because their background fields are kinklike and antikinklike, respectively. The two singlets are localized at nonzero y values, so the overall spectrum is "split".

This is a remarkable outcome for two reasons. First, because the 5 is localized RH, it is equivalent to a LH-localized 5^*. Thus the localized spectrum consists of LH zero-modes in the $SU(5)$ representation

$$5^* \oplus 10 \oplus 1 \oplus 1, \tag{4.49}$$

in other words one standard family plus two singlet neutrinos. Second, apart from the extra singlet, all the exotic fermions in the 27 of E_6 are delocalized and thus do not feature in the effective 3 + 1-d theory on the brane. These benign outcomes depend crucially on the boundary condition choice embodied by the CoS domain-wall solution, including the Z_2 minus sign.

Finally, there is an amusing aspect to this spectrum. It resembles a usual $SO(10)$ family plus an extra singlet. However, the LH 5^*, which is obtained from a 4 + 1-d 5, and the 10 do not come from a 16 of either $SO(10)$ or $SO(10)'$.

V. CONCLUSION

We find it extremely encouraging that, in the E_6 context, the clash-of-symmetries idea leads to good outcomes for both gauge-boson localization (assuming the Dvali-Shifman mechanism works) and fermion localization; that is the main point of this paper.

In summary, we have established a general connection between the clash-of-symmetries mechanism for simultaneous brane creation and internal-symmetry breaking with the Dvali-Shifman mechanism for gauge-boson localization. The two together provide a strong basis upon which to construct realistic domain-wall-brane models. These models should be compatible with type-2 Randall-Sundrum graviton localization (see Ref. [16] for a CoS-style toy model featuring a background warped metric).

More specifically, we have found a domain-wall solution in an E_6 adjoint-Higgs model that produces an $SU(5) \otimes U(1)^2$ symmetry on the wall itself. In one half of the bulk, the symmetry is enhanced to $SO(10) \otimes U(1)$, while in the other half of the bulk the enhancement is to $SO(10)' \otimes U(1)'$. The unprimed and primed groups are differently embedded but isomorphic subgroups of E_6. Because the brane-$SU(5)$ is contained in both $SO(10)$ and $SO(10)'$, the Dvali-Shifman localization of its gauge bosons follows. The simplest possible mechanism for fermion zero-mode localization produces a realistic spectrum, an outcome that depends on the generic features of our domain-wall configuration.

To complete a realistic model, one needs to add gravity (which is expected to be straightforward) and to arrange for the additional spontaneous symmetry breaking cascade $SU(5) \otimes U(1)^2 \to SU(3) \otimes SU(2) \otimes U(1)_Y \to SU(3) \otimes U(1)_Q$. To achieve the latter, suitable additional Higgs multiplets need to be introduced, and their background field configurations have to be nonzero inside the domain wall to trigger the additional spontaneous symmetry breaking. An example of this kind of dynamical structure is described in Ref. [13], where the dominant background domain-wall configuration breaks $SU(5)$ to $SU(3) \otimes SU(2) \otimes U(1)$, and then an additional Higgs field induces electroweak symmetry breaking inside the wall.

ACKNOWLEDGMENTS

R. R. V. and A. K. were supported by the Australian Research Council, D. P. G. by the Puzey Bequest to the University of Melbourne, and K. C. W. in part by funds provided by the U.S. Department of Energy (D.O.E.) No. DE-FG02-92ER40702. A. D. is supported in part by the Albert Einstein Chair in Theoretical Physics.

APPENDIX: FULL MINIMIZATION ANALYSIS

We do this in two steps. We first extend the Higgs-potential minimization analysis by adding a third adjoint component, associated with the Cartan subalgebra generator identified as weak hypercharge Y. This is useful because the result can be graphically visualized, and it reveals a third embedding of $SO(10)$ that is related to the two embeddings used in the main body of the text. In the second step, we report on a numerical study of the whole six-dimensional Cartan subspace.

The truncated multiplet is thus first increased to

$$\chi = f_E E + f_X X + f_Y Y, \tag{A1}$$

where

$$\begin{aligned}
E &= \frac{1}{12}\mathrm{diag}(4, -2, -2, -2, -2, -2, -2, -2, -2, -2, -2, 1, 1, 1, 1, 1, 1, 1, 1, 1, 1, 1, 1, 1, 1, 1, 1),\\
X &= \frac{1}{4\sqrt{15}}\mathrm{diag}(0, 2, 2, 2, 2, 2, -2, -2, -2, -2, -2, -5, 3, 3, 3, 3, 3, -1, -1, -1, -1, -1, -1, -1, -1, -1, -1),\\
Y &= \frac{1}{2\sqrt{10}}\mathrm{diag}\Big(0, 1, 1, -\frac{2}{3}, -\frac{2}{3}, -\frac{2}{3}, -1, -1, \frac{2}{3}, \frac{2}{3}, 0, -1, -1, \frac{2}{3}, \frac{2}{3}, \frac{2}{3}, \frac{1}{3}, \frac{1}{3}, \frac{1}{3}, \frac{1}{3}, \frac{1}{3}, \frac{1}{3}, -\frac{4}{3}, -\frac{4}{3}, -\frac{4}{3}, 2\Big),
\end{aligned} \tag{A2}$$

have been normalized as per

$$\begin{aligned}
\mathrm{Tr}(E^2) = \mathrm{Tr}(X^2) = \mathrm{Tr}(Y^2) = 1/2,\\
\mathrm{Tr}(EX) = \mathrm{Tr}(EY) = \mathrm{Tr}(XY) = 0.
\end{aligned} \tag{A3}$$

The sextic invariant is

$$I_6 = \mathrm{Tr}[(f_E E + f_X X + f_Y Y)^6]. \tag{A4}$$

To visualize its structure, we go to a spherical-polar decomposition

$$\begin{aligned}
f_E \equiv r\sin\phi\cos\theta, \qquad f_X \equiv r\sin\phi\sin\theta,\\
f_Y \equiv r\cos\phi,
\end{aligned} \tag{A5}$$

which produces

$$\begin{aligned}
I_6 = \frac{r^6}{518\,400}\Big(&710\cos^6\phi + 30\cos^4\phi\big(51 + 4\cos2\theta - 4\sqrt{15}\sin2\theta\big)\sin^2\phi + 60\sqrt{2}\cos^3\phi\sin\theta\big(2\sqrt{3} + 3\sqrt{3}\cos2\theta\\
&- \sqrt{5}\sin2\theta\big)\sin^3\phi - 45\cos^2\phi\big(-34 + 2\cos2\theta + 7\cos4\theta - 2\sqrt{15}\sin2\theta + \sqrt{15}\sin4\theta\big)\sin^4\phi\\
&+ \frac{3}{2}\times\big(440 + 15\cos2\theta + 84\cos4\theta + 11\cos6\theta - 15\sqrt{15}\sin2\theta + 12\sqrt{15}\sin4\theta - 3\sqrt{15}\sin6\theta\big)\sin^6\phi\\
&+ 144\cos^5\phi(\sqrt{10}\cos\theta\sin\phi - \sqrt{6}\sin\theta\sin\phi)\Big).
\end{aligned} \tag{A6}$$

Figure 9 plots $-I_6/r^6$ as a function of θ and ϕ. The (E, X) plane is the line $\phi = \pi/2$, along which the VEVs $(10, \pm)$ and $(10', \pm)$ can be seen. Degenerate with them are two more VEVs with $f_Y \neq 0$, located at

$$\theta = \arccos\left(-\frac{1}{2}\sqrt{\frac{5}{2}}\right), \qquad \phi = \arccos\left(-\frac{3}{\sqrt{10}}\right) \tag{A7}$$

and

$$\theta = -\arccos\left(\frac{1}{2}\sqrt{\frac{5}{2}}\right), \qquad \phi = \arccos\left(\frac{3}{\sqrt{10}}\right). \tag{A8}$$

The first VEV corresponds to a nonzero value for the adjoint component associated with the generator $-E''$, where

$$E'' = \frac{1}{4}E - \frac{1}{4}\sqrt{\frac{3}{5}}X + \frac{3}{\sqrt{10}}Y. \tag{A9}$$

As the notation suggests, this minimum breaks E_6 to yet a third differently embedded subgroup which we can call $SO(10)'' \otimes U(1)_{E''}$, with negative Z_2 parity: $(10'', -)$. The second VEV is just $(10'', +)$.

The three groups $SO(10)$, $SO(10)'$, and $SO(10)''$ share a common $SU(3) \otimes SU(2)$ subgroup, but the $SU(5)$ contained in $SO(10) \cap SO(10)'$ is *not* a subgroup of $SO(10)''$ [this is obvious, since E'' contains an admixture of Y which is a generator of that $SU(5)$]. One can imagine a domain-wall junction configuration that utilizes all three of these embeddings for boundary conditions, but such a model

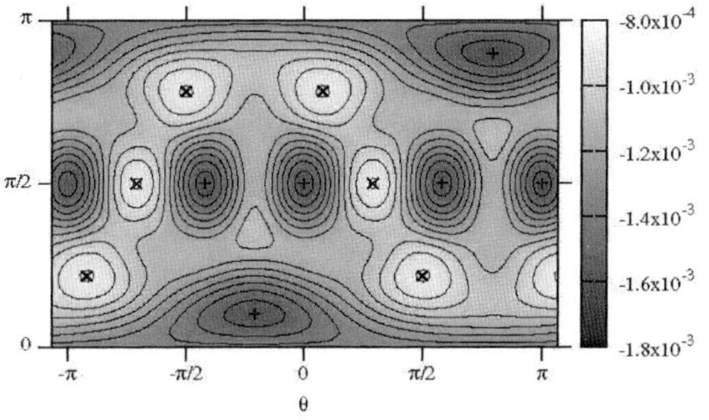

FIG. 9. Contour plot of $-I_6/r^6$ as a function of θ and ϕ. The degenerate global minima breaking E_6 to various $SO(10) \otimes U(1)$ subgroups are marked with + signs, while the local maxima are indicated with × signs. The row of minima along $\phi = \pi/2$ correspond to the global minima displayed in Fig. 2.

would have a similar photon and Z-boson leakage problem as the warm-up example of Sec. III.

The functions $-I_8/r^8$, $-(I_5)^2/r^{10}$, and $-I_{12}/r^{12}$ have exactly the same qualitative structure as $-I_6/r^6$. Thus $(10, \pm)$, $(10', \pm)$, and $(10'', \pm)$ will be the degenerate global minima for a large class of potentials. The quadratic invariant is θ, ϕ-independent,

$$I_2 = \tfrac{1}{2}(f_E^2 + f_X^2 + f_Y^2) \tag{A10}$$

so one can simply add appropriate $(I_2)^n$ terms to the potential to ensure it is bounded from below, and to generate a definite value for r at the global minima. The positions of the global minima are determined entirely from the angular structure of the nonisotropic terms.

One can extend the analysis to all six Cartan components using a six-dimensional hyperspherical polar decomposition. The six fields are represented by one modulus, r, four zenith angles $0 \le \phi_{1,2,3,4} \le \pi$, and one azimuthal angle $-\pi \le \theta < \pi$. Because the group-theoretic character of an extremum is determined entirely from the angular structure of the invariants, a numerical study can readily be performed on the finite domain $(\phi_{1,2,3,4}, \theta)$. This study confirmed that the $E_6 \to SO(10) \otimes U(1)$ VEVs are the global minima of $-I_{6,8,12}$ and $-(I_5)^2$. The additional field dimensions simply revealed more degenerate vacua, corresponding to extra embeddings of $SO(10)$ in E_6. These new embeddings must correspond, physically speaking, to choosing different $SU(3) \otimes SU(2)$ subgroups for color and isospin. The total number of $E_6 \to SO(10) \otimes U(1)$ extrema was found to be 54, consisting of 27 Z_2-related pairs. This implies that, overall, there are 27 embeddings of $SO(10) \otimes U(1)$ in E_6. Though we shall not display the results here, we have analytical expressions for the 27 linear combinations of Cartan generators that correspond to these VEVs. In the breakdown $27 \to 1 \oplus 10 \oplus 16$, these 27 linear combinations turn out to be correlated with the choice of which component to assign as the $SO(10)$ singlet in the decomposition. A deeper reason for the number 27 is perhaps the following: according to the $SU(3) \otimes SU(3) \otimes SU(3)$ maximal subgroup of E_6, there are three independent choices for the color group. The weak-isospin group can then be selected as the I-, U-, or V-spin subgroup of either of the two remaining $SU(3)$'s. This gives $3 \times 6 = 18$ choices for $SU(3) \otimes SU(2)$ embeddings. According to our previous analysis, each $SU(3) \otimes SU(2)$ is contained in the intersection of three different $SO(10)$'s, which suggests there should be $18 \times 3 = 54$ embeddings of $SO(10)$. However, recognizing that $SO(10)$ contains an $SU(2) \otimes SU(2)$ subgroup, we see that the correct number of independent embeddings is actually $54/2 = 27$.

[1] V. A. Rubakov and M. E. Shaposhnikov, Phys. Lett. **125B**, 136 (1983).
[2] K. Akama, *Proceedings of the Symposium on Gauge Theory and Gravitation, Nara, Japan, 1982*, Lecture Notes in Physics (Springer-Verlag, Berlin, 1983), Vol. 176, p. 267.
[3] M. Visser, Phys. Lett. **159B**, 22 (1985).
[4] G. W. Gibbons and D. L. Wiltshire, Nucl. Phys. **B287**, 717 (1987).
[5] N. Arkani-Hamed, S. Dimopoulos, and G. R. Dvali, Phys. Lett. B **429**, 263 (1998).
[6] I. Antoniadis, Phys. Lett. B **246**, 377 (1990).
[7] I. Antoniadis, N. Arkani-Hamed, S. Dimopoulos, and G. Dvali, Phys. Lett. B **436**, 257 (1998).
[8] L. Randall and R. Sundrum, Phys. Rev. Lett. **83**, 3370 (1999).
[9] L. Randall and R. Sundrum, Phys. Rev. Lett. **83**, 4690 (1999).
[10] M. Gremm, Phys. Lett. B **478**, 434 (2000).
[11] O. DeWolfe, D. Z. Freedman, S. S. Gubser, and A. Karch, Phys. Rev. D **62**, 046008 (2000).
[12] A. Davidson and P. D. Mannheim, arXiv:hep-th/0009064.
[13] R. Davies, D. P. George, and R. R. Volkas, arXiv:0705.1584.
[14] A. Davidson, B. F. Toner, R. R. Volkas, and K. C. Wali, Phys. Rev. D **65**, 125013 (2002).
[15] J. S. Rozowsky, R. R. Volkas, and K. C. Wali, Phys. Lett. B **580**, 249 (2004).
[16] G. Dando, A. Davidson, D. P. George, R. R. Volkas, and K. C. Wali, Phys. Rev. D **72**, 045016 (2005).
[17] E. M. Shin and R. R. Volkas, Phys. Rev. D **69**, 045010 (2004).
[18] L. Pogosian and T. Vachaspati, Phys. Rev. D **62**, 123506 (2000).
[19] T. Vachaspati, Phys. Rev. D **63**, 105010 (2001).

[20] L. Pogosian and T. Vachaspati, Phys. Rev. D **64**, 105023 (2001).

[21] G. Dvali and M. Shifman, Nucl. Phys. **B504**, 127 (1997).

[22] G.R. Dvali and M.A. Shifman, Phys. Lett. B **396**, 64 (1997).

[23] R. Jackiw and C. Rebbi, Phys. Rev. D **13**, 3398 (1976).

[24] V.A. Rubakov, Phys. Usp. **44**, 871 (2001).

[25] M. Creutz, Phys. Rev. Lett. **43**, 553 (1979).

[26] G. 't Hooft, in *Proceedings of the EPS International Conference, Palermo, 1975*, edited by A Zichichi (Editrice Compositori, Bologna, 1976).

[27] S. Mandelstam, Phys. Rep. C **23**, 245 (1976).

[28] N. Arkani-Hamed and M. Schmaltz, Phys. Lett. B **450**, 92 (1999).

[29] S.L. Dubovsky and V.A. Rubakov, Int. J. Mod. Phys. A **16**, 4331 (2001).

[30] A. Demaria and R.R. Volkas, Phys. Rev. D **71**, 105011 (2005).

[31] E. Dynkin, Mat. Sb. **30**, 349 (1952) [Amer. Math. Soc. Transl., Series 2, **6**, 111 (1957)].

[32] L.-F. Li, Phys. Rev. D **9**, 1723 (1974).

[33] O. Kaymakcalan, L. Michel, K.C. Wali, W.D. McGlinn, and L. O'Raifeartaigh, Nucl. Phys. **B267**, 203 (1986).

[34] M. Bando and T. Kugo, Prog. Theor. Phys. **101**, 1313 (1999).

[35] M. Bando, T. Kugo, and K. Yoshioka, Prog. Theor. Phys. **104**, 211 (2000).

[36] G.W. Anderson and T. Blazek, J. Math. Phys. (N.Y.) **41**, 4808 (2000).

[37] N. Maekawa and T. Yamashita, Phys. Lett. B **567**, 330 (2003).

[38] G. Racah, Lincei. Rend. Sci. Fis. Mat. Nat. **8**, 108 (1950).

[39] J.A. Harvey, Nucl. Phys. **B163**, 254 (1980).

KALUZA–KLEIN THEORY WITH TORSION CONFINED TO THE EXTRA DIMENSION

KARTHIK H. SHANKAR and KAMESHWAR C. WALI
Department of Physics, Syracuse University, Syracuse, NY 13244, USA

Received 25 January 2010
Revised 1 April 2010

Here we consider a variant of the five-dimensional Kaluza–Klein (KK) theory within the framework of Einstein–Cartan formalism that includes torsion. By imposing a set of constraints on torsion and Ricci rotation coefficients, we show that the torsion components are completely expressed in terms of the metric. Moreover, the Ricci tensor in 5D corresponds exactly to what one would obtain from torsion-free general relativity on a 4D hypersurface. The contributions of the scalar and vector fields of the standard KK theory to the Ricci tensor and the affine connections are completely nullified by the contributions from torsion. As a consequence, geodesic motions do not distinguish the torsion free 4D spacetime from a hypersurface of 5D spacetime with torsion satisfying the constraints. Since torsion is not an independent dynamical variable in this formalism, the modified Einstein equations are different from those in the general Einstein–Cartan theory. This leads to important cosmological consequences such as the emergence of cosmic acceleration.

Keywords: Modified gravity; confinement of gravity; cosmic acceleration; Kaluza–Klein theory; torsion.

In any attempt to link fundamental matter fields with intrinsic spin to gravity, it becomes necessary to extend the Riemannian spacetime to include *torsion*, defined to be the antisymmetric part of affine connection. The resulting theory of gravity, known as the Einstein–Cartan theory[1,a] treats the metric and torsion as two independent geometrical characteristics of spacetime.[2,b]

Historically, beginning with the Kaluza–Klein (KK) theory, there has been a great interest in introducing extra dimensions of spacetime to unify gravity with elementary particle interactions. In the KK theory, the scalar and vector fields, which are the extra dimensional components of the metric tensor contribute to the affine connection and the Ricci tensor, and hence modify their values from the

[a]For a historical review of concept of torsion and its role in constructing a non-Riemannian structure of spacetime.
[b]The report reviews the physical aspects of torsion including both classical and quantum aspects and the necessity of torsion to couple the spin of matter fields to spacetime.

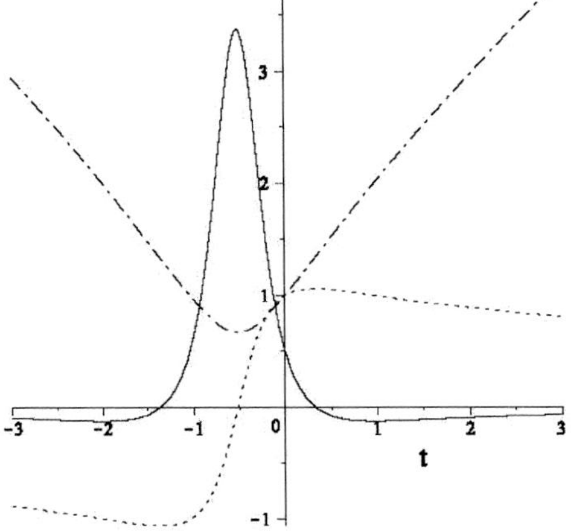

Fig. 1. The dash-dotted, dotted and solid curves denote $a(t)$, $\dot{a}(t)$, and $\ddot{a}(t)$ respectively. The time axis is in units of H_0^{-1}.

corresponding values in 4D spacetime.[3,c] The contribution of these fields to the Einstein tensor are normally interpreted as gravity induced matter.

In this work, we incorporate torsion into 5D KK theory.[4,d] The inclusion of torsion introduces free parameters in the affine connection and the Ricci tensor in addition to the contributions from extra dimensional metric components that occur in the torsion-free KK theory. In this paper we impose a minimal set of conditions so as to restrict torsion to purely extra dimensional components and determine all its components in terms of the metric. Interestingly, the imposed conditions lead to a complete cancellation between the modifications induced by the extra dimensional metric components and the contributions from the torsion. Thus the Ricci tensor in 5D spacetime in the resulting formalism is exactly the Ricci tensor in a torsion-free 4D spacetime. In the second part of the paper, we apply the action principle to derive the equations of motion of this formalism. The modified Einstein equations thus derived are finally applied to Robertson–Walker cosmology that leads to a novel expansion history for the universe (see Fig. 1).

To describe the constraints to be imposed, we start, for the sake of completeness, with a brief overview on the relationship between tetrads, torsion and affine connection.

1. Setting Up the Framework

Since the constraints we propose to impose can be more simply stated with reference to a locally flat inertial system (allowed by the equivalence principle), we proceed to define and collect together from Ref. 5, the relevant standard relations we need in

[c]For various aspects and recent developments of torsion free Kaluza–Klein theory.

[d]For an analysis of Kaluza–Klein theory in the presence of torsion.

both the coordinate and the inertial frames. Let $(i, j, k, \ldots)$ and $(A, B, \ldots)$ denote coordinate and inertial frame indices respectively, $\hat{\mathbf{e}}_i = \partial_i$ and $\hat{\boldsymbol{\theta}}^i = dx^i$ be the basis of the tangent and dual spaces at each point in spacetime. We define the corresponding inertial basis to be $\hat{\mathbf{e}}_A = e^i_{\cdot A}\hat{\mathbf{e}}_i$ and $\hat{\boldsymbol{\theta}}^A = e_i^{\cdot A}\hat{\boldsymbol{\theta}}^i$, where the vielbeins $e^i_{\cdot A}$ and $e_j^{\cdot B}$ satisfy the orthonormality conditions,

$$e_i^{\cdot A} e^j_{\cdot A} = \delta_i^j\,, \qquad e_i^{\cdot A} e^i_{\cdot B} = \delta_B^A\,. \tag{1}$$

By definition, the metric in the inertial frame is Minkowskian η_{AB}, and the metric tensor in the coordinate system is $\mathbf{g}_{ij} = e_i^{\cdot A} e_j^{\cdot B}\eta_{AB}$ and $\mathbf{g}^{ij} = e^i_{\cdot A} e^j_{\cdot B}\eta^{AB}$. The covariant derivative operator $(\tilde{\nabla})$ can be defined in terms of the coordinate basis, or equivalently in terms of the inertial frame basis,

$$\tilde{\nabla}_{\hat{\mathbf{e}}_i}\hat{\mathbf{e}}_j = \tilde{\Gamma}^k_{\cdot ij}\hat{\mathbf{e}}_k\,, \qquad \tilde{\nabla}_{\hat{\mathbf{e}}_A}\hat{\mathbf{e}}_B = \omega^C_{\cdot AB}\hat{\mathbf{e}}_C\,, \tag{2}$$

where $\tilde{\Gamma}^i_{\cdot jk}$ and $\omega^A_{\cdot BC}$ are the affine and the Ricci rotation coefficients respectively. The relationship between these two quantities follows from the transformation laws between the coordinate frame $(\hat{\mathbf{e}}_i)$ and inertial frame $(\hat{\mathbf{e}}_A)$,

$$\omega^A_{\cdot BC} = e^i_{\cdot B}(\tilde{\nabla}_{\hat{\mathbf{e}}_i} e^j_{\cdot C})e_j^{\cdot A}\,. \tag{3}$$

The affine connection by itself is not a tensor, but its antisymmetric part, the torsion, is a tensor.[e]

$$T^i_{\cdot jk} = \tilde{\Gamma}^i_{\cdot jk} - \tilde{\Gamma}^i_{\cdot kj} = 2\tilde{\Gamma}^i_{\cdot [jk]}\,. \tag{4}$$

Again, using the transformation laws between the coordinate and the inertial frames, we have

$$T^i_{\cdot jk} = e_j^{\cdot B} e_k^{\cdot C} e^i_{\cdot A} T^A_{\cdot BC}\,. \tag{5}$$

Furthermore, with the standard assumption of metric compatibility, namely $\nabla_{\hat{\mathbf{e}}_i} g_{jk} = 0$, we obtain

$$\tilde{\Gamma}^i_{\cdot jk} = \hat{\Gamma}^i_{\cdot jk} + K^i_{\cdot jk}\,, \tag{6}$$

where $\hat{\Gamma}^i_{\cdot jk}$ is the Christoffel connection,

$$\hat{\Gamma}^i_{\cdot jk} = \left\{ {}_j{}^i{}_k \right\} = \frac{1}{2}\mathbf{g}^{im}[\partial_j \mathbf{g}_{km} + \partial_k \mathbf{g}_{jm} - \partial_m \mathbf{g}_{jk}] \tag{7}$$

and $K^i_{\cdot jk}$ is the contorsion tensor.

$$K^i_{\cdot jk} = \frac{1}{2}[T^i_{\cdot jk} + T_{j\cdot k}^{\cdot i} + T_{k\cdot j}^{\cdot i}]\,. \tag{8}$$

[e]Throughout this paper, square brackets enclosing indices denote the conventional antisymmetrization, while the regular brackets enclosing the indices denotes symmetrization.

2. Constructing the 5D Spacetime

We shall now focus on applying the above formalism to a five-dimensional spacetime. Consider a foliation of the 5D spacetime in terms of a family of 4D hypersurfaces, which are parametrized by the coordinate system $\{x^\mu\}$, where $(\mu, \nu, \ldots)$ denote the coordinate indices on these hypersurfaces. Let x^5 denote the parametrization of the family, and five denote the corresponding coordinate index. The hypersurface coordinates $\{x^\mu\}$ together with x^5 will then span the entire 5D spacetime. Let the metric and its inverse on each of the hypersurfaces be $g_{\mu\nu}$ and $g^{\mu\nu}$ respectively, which can in principle depend on the x^5 coordinate. Let $\{e^{\mu}_{\cdot a}, e^{\cdot a}_{\mu}\}$ denote the tetrad system on these hypersurfaces satisfying the orthonormality relations $e^{\mu}_{\cdot a} e^{\cdot b}_{\mu} = \delta^a_b$ and $e^{\mu}_{\cdot a} e^{a}_{\cdot\nu} = \delta^\mu_\nu$. Here $(a, b, \ldots)$ denote the tetrad indices on these 4D hypersurfaces. The metric on the hypersurface is then given by $g_{\mu\nu} = e^{\cdot a}_{\mu} e^{\cdot b}_{\nu} \eta_{ab}$ and its inverse $g^{\mu\nu} = e^{\mu}_{\cdot a} e^{\nu}_{\cdot b} \eta^{ab}$.

We will now construct the vielbeins in the 5D spacetime by extending the tetrad system on the 4D hypersurfaces. We take the components of the 5D vielbeins to be $e^{i}_{\cdot A} = (e^{\mu}_{\cdot a}, e^{\mu}_{\cdot\dot{5}}, e^{5}_{\cdot a}, e^{5}_{\cdot\dot{5}})$ and $e^{\cdot A}_{i} = (e^{\cdot a}_{\mu}, e^{\cdot a}_{5}, e^{\cdot\dot{5}}_{\mu}, e^{\cdot\dot{5}}_{5})$. The index $\dot{5}$ corresponds to the fifth dimension of the inertial frame. The orthonormality relations in 5D [Eq. (1)] immediately leads to[f]

$$
\begin{aligned}
e^{\mu}_{\cdot\dot{5}} &= 0\,, \qquad e^{5}_{\cdot a} = -e^{\mu}_{\cdot a} A_\mu\,, \qquad e^{5}_{\cdot\dot{5}} = \Phi^{-1}\,, \\
e^{\cdot a}_{5} &= 0\,, \qquad e^{\cdot\dot{5}}_{\mu} = A_\mu \Phi\,, \qquad e^{\cdot\dot{5}}_{5} = \Phi\,.
\end{aligned}
\tag{9}
$$

The metric in the 5D is then given by $\mathbf{g}_{ij} = e^{\cdot A}_{i} e^{\cdot B}_{j} \eta_{AB}$ and $\mathbf{g}^{ij} = e^{i}_{\cdot A} e^{j}_{\cdot B} \eta^{AB}$.

$$
\begin{aligned}
\mathbf{g}_{\mu\nu} &= \mathrm{g}_{\mu\nu} + \epsilon A_\mu A_\nu \Phi^2\,, \qquad \mathbf{g}_{\mu 5} = \epsilon A_\mu \Phi^2\,, \qquad \mathbf{g}_{55} = \epsilon \Phi^2\,, \\
\mathbf{g}^{\mu\nu} &= \mathrm{g}^{\mu\nu}\,, \qquad \mathbf{g}^{\mu 5} = -A^\mu\,, \qquad \mathbf{g}^{55} = A_\lambda A^\lambda + \epsilon \Phi^{-2}\,.
\end{aligned}
\tag{10}
$$

The raising and lowering of indices on the vector field is done w.r.t. the 4D metric $g_{\mu\nu}$. The parameter $\epsilon = \pm 1$ denotes whether the extra dimension is space-like or time-like. Note that the induced metric on the hypersurfaces (induced by the 5D geometry), $g_{\mu\nu} + \epsilon A_\mu A_\nu \Phi^2$ is different from the 4D metric $g_{\mu\nu}$ on them, but quite evidently related by a gauge transformation.

We will now impose a set of constraints on torsion and the Ricci rotation coefficients consistent with Cartan's structure equation[5] that relates torsion and connection coefficients. With a minimal modification of the standard general relativity in mind, we chose the set of constraints.[6,g] so that the torsion components angential to the 4D hypersurface vanish (see also Ref. 7), while leaving nonvanishing tor-

[f]There exists another class of vielbeins that satisfy the orthonormality relations. But they are essentially related to Eq. (9) by gauge. Here, we choose to work with the vielbeins given by Eq. (9) to make the formalism readily comparable to the existing Kaluza–Klein literature.

[g]These minimal conditions on torsion were utilized in Viet, Wali to obtain a fully covariant action involving all the metric fields in a discretized Kaluza–Klein theory. See also N. A. Viet, arXiv:hep-th/0303128.

sion components completely determined in terms of the metric components of the 5D spacetime. In the absence of torsion, the metric compatibility condition serves to evaluate the connection coefficients (Christoffel symbols, Eq. (7)) in terms of the metric components. In the presence of torsion, the imposed constraints can be viewed as an extension to the metric compatibility condition in the spirit that they together serve to determine the connection coefficients in terms of the metric components. Furthermore, the imposed constraints are covariant conditions on the 4D hypersurface. That is, the conditions are form invariant with respect to within-hypersurface coordinate transformations (that do not mix μ, ν with the fifth coordinate).

2.1. *Condition* 1: $T^{a}_{\cdot BC} = 0$

Using Eq. (5) and noting that $e^{\mu}_{\cdot\dot{5}} = 0$, we find $T^{\mu}_{\cdot ik} = 0$. This implies that the only nonzero components of torsion are $T^{5}_{\cdot ik}$. To determine these, we impose the following condition on the Ricci rotation coefficients,

2.2. *Condition* 2: $\omega^{\dot{5}}_{\cdot BC} = 0$

This condition along with metric compatibility implies $\omega^{A}_{\cdot B\dot{5}} = 0$. We can now use Eq. (3) to write,

$$\omega^{A}_{\cdot B\dot{5}} = e^{i}_{\cdot B}(\tilde{\nabla}_{\hat{e}_i} e^{j}_{\cdot\dot{5}})e^{\cdot A}_{j} = e^{i}_{\cdot B}(\partial_i e^{j}_{\cdot\dot{5}} + \tilde{\Gamma}^{j}_{\cdot ik} e^{k}_{\cdot\dot{5}})e^{\cdot A}_{j} = 0\,. \tag{11}$$

Since $e^{\mu}_{\cdot\dot{5}} = e^{a}_{\cdot 5} = 0$, the above equation implies,

$$\tilde{\Gamma}^{\mu}_{\cdot i5} = 0\,, \qquad \tilde{\Gamma}^{5}_{\cdot i5} = -e^{\cdot\dot{5}}_{5}\partial_i e^{5}_{\cdot\dot{5}}\,. \tag{12}$$

Using the above equations along with Eq. (6), we can express the contorsion in terms of the Christoffel symbols $\hat{\Gamma}$ and the vielbeins,

$$\begin{aligned} K^{\mu}_{\cdot i5} &= -\hat{\Gamma}^{\mu}_{\cdot i5}\,, \\ K^{5}_{\cdot i5} &= -(\hat{\Gamma}^{5}_{\cdot i5} + e^{\cdot\dot{5}}_{5}\partial_i e^{5}_{\cdot\dot{5}}) = -\hat{\Gamma}^{5}_{\cdot i5} + J_i\,, \end{aligned} \tag{13}$$

where $J_i \equiv \Phi^{-1}\partial_i \Phi$.

Note that the above equations give only a subset of the components $K^{i}_{\cdot jk}$, but along with Eq. (8), they are sufficient to determine all the components of the torsion. The torsion thus obtained can in turn be substituted in Eq. (8) to determine the remaining components of contorsion. The components of torsion are

$$\begin{aligned} T^{\mu}_{\cdot ij} &= T^{5}_{\cdot 55} = 0\,, \\ T^{5}_{\cdot\mu\nu} &= 2\partial_{[\mu}A_{\nu]} + 2J_{[\mu}A_{\nu]}\,, \\ T^{5}_{\cdot\mu 5} &= J_\mu - \partial_5 A_\mu - A_\mu J_5\,. \end{aligned} \tag{14}$$

In addition to yielding the nonvanishing components of torsion in terms of the metric components, the solution to Eq. (13) also yields the following condition on the metric on the 4D hypersurfaces.

$$\partial_5 g_{\mu\nu} = 0\,. \tag{15}$$

This implies that all the hypersurfaces in the foliating family have the same 4D metric. To put things in perspective, we observe that in the standard KK theory, the assumption of cylindrical condition makes all the quantities, namely $g_{\mu\nu}$, A_μ and Φ independent of x^5. Whereas in our formulation, the constraints automatically imply $g_{\mu\nu}$ is independent of x^5, while A_μ and Φ could still depend on x^5.

With all the $K^{i}_{\cdot jk}$ determined from Eqs. (14) and (8), we now use Eq. (6) to calculate all the affine connection coefficients.

$$\begin{aligned}
&\tilde{\Gamma}^{\lambda}_{\cdot 55} = \tilde{\Gamma}^{\lambda}_{\cdot\nu 5} = \tilde{\Gamma}^{\lambda}_{\cdot 5\nu} = 0\,,\\
&\tilde{\Gamma}^{5}_{\cdot\mu\nu} = \nabla_\mu A_\nu + J_\mu A_\nu\,,\\
&\tilde{\Gamma}^{5}_{\cdot 5\mu} = \partial_5 A_\mu + J_5 A_\mu\,,\\
&\tilde{\Gamma}^{5}_{\cdot\mu 5} = J_\mu\,, \qquad \tilde{\Gamma}^{5}_{\cdot 55} = J_5\,, \qquad \tilde{\Gamma}^{\lambda}_{\cdot\mu\nu} = \Gamma^{\lambda}_{\cdot\mu\nu}\,.
\end{aligned} \tag{16}$$

Here $\Gamma^{\lambda}_{\cdot\mu\nu}$ corresponds to the Christoffel symbols (analogous to Eq. (7)) obtained from torsion-free 4D spacetime with metric $g_{\mu\nu}$. Note that the components of 5D Christoffel symbols along the hypersurface coordinates is not equal to the Christoffel symbols calculated on 4D spacetime, that is, $\hat{\Gamma}^{\lambda}_{\cdot\mu\nu} \neq \Gamma^{\lambda}_{\cdot\mu\nu}$. The symbol ∇_μ corresponds to the covariant derivative operator on the torsion-free 4D spacetime, where the Christoffel symbols are exactly the affine connection coefficients.

Substituting the above connection coefficients in the Ricci tensor defined by

$$\tilde{R}_{ik} = \partial_k \tilde{\Gamma}^{j}_{\cdot ji} - \partial_j \tilde{\Gamma}^{j}_{\cdot ki} + \tilde{\Gamma}^{j}_{\cdot km}\tilde{\Gamma}^{m}_{\cdot ji} - \tilde{\Gamma}^{j}_{\cdot jm}\tilde{\Gamma}^{m}_{\cdot ki}\,, \tag{17}$$

we find

$$\tilde{R}_{\mu\nu} = R_{\mu\nu}\,, \qquad \tilde{R}_{\mu 5} = \tilde{R}_{5\mu} = \tilde{R}_{55} = 0\,. \tag{18}$$

Here $R_{\mu\nu}$ represents the Ricci tensor on the torsion-free 4D spacetime. It also follows that the 5D Ricci scalar is exactly the same as the Ricci scalar in the torsion-free 4D space time, that is $\tilde{R} = R$. We also note, in general, the presence of torsion makes the Ricci tensor non-symmetric, but the constraints we have imposed on the torsion leaves the Ricci tensor symmetric.

It is straightforward to see that the formalism and the results obtained thus far are not specific to four and five dimensions, they can be generalized to any arbitrary D and $D+1$ dimensions. We shall now consider some implications of the formalism with respect to geodesic equations and solutions to Einstein equations.

3. Geodesic Equations

The 5D geodesic equations split into

$$\begin{gathered}
\ddot{x}^5 + \tilde{\Gamma}^{5}_{\cdot\mu\nu}\dot{x}^\mu\dot{x}^\nu + (\tilde{\Gamma}^{5}_{\cdot\mu 5} + \tilde{\Gamma}^{5}_{\cdot 5\mu})\dot{x}^\mu\dot{x}^5 + \tilde{\Gamma}^{5}_{\cdot 55}(\dot{x}^5)^2 = 0\,,\\
\ddot{x}^\lambda + \Gamma^{\lambda}_{\cdot\mu\nu}\dot{x}^\mu\dot{x}^\nu = 0\,.
\end{gathered} \tag{19}$$

We note that the components of the geodesic equations along the hypersurface are exactly the same as the geodesic equations in the torsion-free 4D spacetime. This is in contrast to the conventional KK theory where the 4D geodesic equations are modified by the presence of the fields A_μ and Φ. In this formalism, the presence of torsion completely nullifies the effect of these fields in the 4D geodesic equations. Furthermore, it is worth noting that the geodesic of a particle can be confined to a 4D hypersurface by requiring $\dot{x}^5 = 0$. From Eq. (19), we see that this requires the additional condition,

$$\tilde{\Gamma}^5_{\cdot\mu\nu} = \nabla_\mu A_\nu + J_\mu A_\nu = 0\,. \tag{20}$$

If this condition can be satisfied, then there will be no observable difference, as far as a test particle is concerned, whether we live in a torsion-free 4D spacetime or on a hypersurface within the 5D spacetime with torsion. However, it is apparent that this condition is a strong constraint requiring the vector and scalar fields satisfy the above equation for a given 4D metric. It is conceivable that for some 4D metrics, no choice of vector and scalar fields would satisfy the above constraint. In such cases, the particle would be free to move in the x^5 direction unless constrained by an external force or if the fifth coordinate is compact and small as usually assumed in most adaptations of the KK theory.

4. Einstein's Equations

To obtain the equations of motion, we need to vary the action with respect to the independent dynamical fields of the theory. In general, if one includes torsion in the theory, torsion and metric are two independent dynamical fields and consequently one needs to vary the action with respect to both of these variables.[1,4] In our approach, with torsion determined in terms of the metric, metric components are the only dynamical variables. Taking the action to be $S = \int \tilde{R}\sqrt{-\mathbf{g}}\, d^5x$, its variation yields[h]

$$\int \left[\tilde{R}\delta\sqrt{-\mathbf{g}} + \tilde{R}_{ik}\sqrt{-\mathbf{g}}\delta\mathbf{g}^{ik} + \mathbf{g}^{ik}\sqrt{-\mathbf{g}}\delta\tilde{R}_{ik}\right] d^5x\,. \tag{21}$$

One can set this variation to zero to obtain the modified Einstein equations. The first two terms give rise to the usual symmetric Einstein tensor. In the absence of torsion, the third term becomes a boundary integral that contributes nothing to the equation. While in the presence of torsion, the third term can be shown to contribute

$$\int [T^m_{\cdot jm}\mathbf{g}^{ik}\delta\tilde{\Gamma}^j_{\cdot ki} - T^m_{\cdot km}\mathbf{g}^{ik}\delta\tilde{\Gamma}^j_{\cdot ji} + T^m_{\cdot kj}\mathbf{g}^{ik}\delta\tilde{\Gamma}^j_{\cdot mi}]\sqrt{-\mathbf{g}}\, d^5x\,.$$

Treating the variation in the torsion and metric components independent, one can obtain the Einstein–Cartan equations derived by Hehl *et al.*[1] However, by

[h]We are using geometric units where $G = c = 1$. We also note that the action in general will include a matter Lagrangian term which would also contribute to the variation.

expressing $\delta\tilde{\Gamma}$ purely in terms of variation of the 5D metric components (from Eq. (16)), it turns out that the above expression can be simplified to

$$\int H_{\mu\nu}\delta g^{\mu\nu}\sqrt{-\mathbf{g}}\,d^5x\,, \tag{22}$$

where

$$\begin{aligned} H_{\mu\nu} &\equiv \nabla_{(\mu}B_{\nu)} - (\nabla\cdot B)g_{\mu\nu} + J_{(\mu}B_{\nu)} - (J\cdot B)g_{\mu\nu}\,,\\ B_\mu &\equiv T^5_{\cdot\mu 5} = J_\mu - \partial_5 A_\mu - A_\mu J_5\,, \end{aligned} \tag{23}$$

are interpreted as tensors in torsion-free 4D space. Since Eq. (22) is the only additional term that contributes to the variation of action, $H_{\mu\nu}$ is essentially the modification to the standard torsion-free Einstein tensor. The modified Einstein equations then turns out to be

$$R^\nu_\mu - \frac{1}{2}R\delta^\nu_\mu + H^\nu_\mu = \Sigma^\nu_\mu\,, \tag{24}$$

$$-A^\alpha R_{\mu\alpha} - A^\alpha H_{\mu\alpha} = \Sigma^5_\mu\,, \tag{25}$$

$$-\frac{1}{2}R = \Sigma^5_5\,. \tag{26}$$

Here Σ is the stress–tensor that one would obtain when matter fields are included in the Lagrangian prior to variation of action. Its 4D components will be the observed stress–energy tensor and can be identified with the stress–energy tensor of the usual 4D torsion free general relativity. Its conservation immediately yields

$$\nabla_\nu\Sigma^\nu_\mu = 0 \Rightarrow \nabla_\nu H^\nu_\mu = 0\,. \tag{27}$$

In general, the components Σ^5_μ and Σ^5_5 cannot be obtained from observations on the 4D hypersurface. Hence, without an explicit 5D matter Lagrangian specifying these components, it is not possible to solve Eqs. (25) and (26). Thus, we are left with just Eqs. (24) and (27) to solve for the metric components $g_{\mu\nu}$, A_μ and Φ.

5. Robertson–Walker Cosmology

To illustrate an application of the formalism, let us consider spatially flat homogeneous and isotropic universe with metric

$$ds^2 = -dt^2 + a^2(t)(dx^2 + dy^2 + dz^2)\,. \tag{28}$$

The assumption of homogeneity and isotropy of the 4D geometry requires that $A_\mu = (A_t, 0, 0, 0)$, A_t and Φ are functions of t and not the spatial coordinates. To simplify further, we shall also assume that the fields do not depend on x^5. From Eq. (14), these constraints imply that $B_\mu = J_\mu$ and the only nonvanishing component of J_μ is J_t, and of A_μ is A_t. Applying the conservation equation (Eq. (27)) yields

$$\text{(i) } \dot{\Phi} = 0, J_t = 0, \text{ or } \qquad \text{(ii) } \Phi = \dot{a}(t), J_t = \ddot{a}/\dot{a}\,. \tag{29}$$

Case (i) yields $H_{\mu\nu} = 0$, and Eq. (24) yields the usual Friedmann equation along with matter conservation.

$$\left(\frac{\dot{a}}{a}\right)^2 = \frac{8\pi}{3}\rho\,.$$

Case (ii) yields nonvanishing $H_{\mu\nu}$, which when applied to Eq. (24) gives the modified Friedmann equation

$$\left(\frac{\dot{a}}{a}\right)^2 + \left(\frac{\ddot{a}}{a}\right) = \frac{8\pi}{3}\rho\,, \tag{30}$$

along with matter conservation which implies $\rho a^3 = \rho_0$ is a constant in a matter dominated universe.

To solve the equation, we specify initial conditions at the current instant of time, $a = 1$, $\dot{a} = H_0$, the Hubble constant and $\ddot{a} = -q_0 H_0^2$, where q_0 is the current deceleration parameter. These conditions can be used to calculate ρ_0, the current matter density (including dark matter). Taking $q_0 = -0.5$, which is consistent with the current observations,[8] the solution to Eq. (30) is plotted in Fig. 1. From the dash-dotted curve (scale factor), note that the universe started to expand at $t = -0.518\,H_0^{-1}$, from a size of $a = 2/3$, prior to which it was in a contracting phase. This is in contrast to the solution of the usual Friedmann equations which yields a big bang ($a = 0$) at $t = -0.667\,H_0^{-1}$. From the solid curve, note that the acceleration is currently positive but decreasing and would become negative beyond $t = 0.319\,H_0^{-1}$. This is qualitatively consistent with the analysis of observed data in Ref. 9, but is in sharp contrast with the standard ΛCDM model which predicts that the acceleration would continue to increase forever. Clearly, more detailed studies are needed to understand the full implications of this formalism on cosmology.

In conclusion, the formalism in this paper provides a general mathematical result pertaining to affine connection and Ricci tensor in Einstein–Cartan theory in higher dimensions. It provides an alternative way to confine gravity from $(D+1)$ to D dimensions, with modifications that can be significant to cosmology.

Acknowledgments

This work was supported in part (K. C. Wali) by the U.S. Department of Energy (DOE) under contract No. DE-FG02-85ER40237.

References

1. F. W. Hehl, P. V. d. Hyde, G. D. Kerlick and J. M. Nester, *Rev. Mod. Phys.* **48**, 393 (1976).
2. I. L. Shapiro, *Phys. Rep.* **357**, 113 (2002).
3. J. M. Overduin and P. S. Wesson, *Phys. Rep.* **283**, 303 (1997).
4. M. W. Kalinowski, *Int. J. Theor. Phys.* **20**, 563 (1981).

5. M. Nakahara, *Geometry, Topology and Physics*, 2nd edn. (IOP, 2003), Secs. 7.81 and 7.84.
6. N. A. Viet and K. C. Wali, *Phys. Rev. D* **67**, 124029 (2003).
7. C. G. Bohmer and L. Fabbri, *Mod. Phys. Lett. A* **22**, 2727 (2007).
8. R. A. Daly and S. G. Djorgovski, CP937, *Supernova 1987A: 20 Years After*, eds. S. Immler, K. Weiler and R. McCray (Amer. Inst. Phys., 2007).
9. A. Shafieloo, V. Sahni and A. A. Starobinsky, arXiv:0903.5141.

Metric theory of gravity with torsion in an extra dimension

Karthik H. Shankar,[1] Anand Balaraman,[2] and Kameshwar C. Wali[3]
[1]*Center for Memory and Brain, Boston University, Boston, Massachusetts 02215, USA*
[2]*Department of Physics, Georgia Southern University, Statesboro, Georgia 30458, USA*
[3]*Department of Physics, Syracuse University, Syracuse, New York 13244, USA*
(Received 3 April 2012; published 3 July 2012)

We consider a theory of gravity with a hidden extra dimension and metric-dependent torsion. A set of physically motivated constraints are imposed on the geometry so that the torsion stays confined to the extra dimension and the extra dimension stays hidden at the level of four-dimensional geodesic motion. At the kinematic level, the theory maps onto general relativity, but the dynamical field equations that follow from the action principle deviate markedly from the standard Einstein equations. We study static spherically symmetric vacuum solutions and homogeneous-isotropic cosmological solutions that emerge from the field equations. In both cases, we find solutions of significant physical interest. Most notably, we find positive mass solutions with naked singularity that match the well-known Schwarzschild solution at large distances but lack an event horizon. In the cosmological context, we find an oscillatory scenario, in contrast to the inevitable singular big bang of the standard cosmology.

I. INTRODUCTION

Einstein viewed space-time as a pseudo-Riemannian differentiable manifold in order to generalize the special-relativistic flat space-time to include gravity. This was primarily motivated by the fact that the local flatness of the manifold structure naturally implemented his principle of equivalence. The generalization came along with the revolutionary idea that the trajectory of any freely moving test body is simply a geodesic in the curved manifold and that gravity is not a Newtonian instantaneous action force but an effect of the curvature of the space-time manifold.

The basic constituents of the manifold structure are the metric, which defines the distance between any two points of the manifold, and the connection, which defines the covariant derivative and the curvature of the manifold. Any theory of gravity should couple the dynamics of these quantities to the dynamics of the matter moving in the space-time manifold. Among the existing theories, the ensuing field equations of general relativity (GR) are perhaps the simplest. Since in GR, torsion, the antisymmetric combination of connection coefficients, is identically zero, and since GR has withstood numerous precise experimental tests [1,2], introduction of torsion has seemed superfluous except in the presence of matter with intrinsic spin as in Einstein-Cartan formulations [3–7].

However, for two major reasons, alternate theories of gravity that reduce to GR in the weak field limit are seriously pursued. The first reason is that GR leads to inevitable singularities—black holes (death of a massive star) and big bang (birth of the universe). Though it is conventionally assumed that quantization would eliminate these singularities, GR is not readily amenable to quantization. The second reason is that the standard model of cosmology based on GR requires most of the universe to be composed of unknown dark energy in order to account for various cosmological observations [8]. A common strategy to construct modified theories of gravity is to make the Lagrangian density a nontrivial function of the Ricci scalar [9,10] and use the action principle to derive the modified field equations. Another common strategy is to introduce extra dimensions while constraining the physical particles to a $(3 + 1)$-dimensional hypersurface as in the brane-world theories [11–13]. In this paper we explore a different approach by introducing metric-dependent torsion in Kaluza-Klein–type theories [14,15] with one extra dimension.

In our approach, we consider a five-dimensional (5D) manifold foliated by a family of 4D hypersurfaces, whose geometries are virtually indistinguishable from that of the 4D space-time of GR. The axis of foliation is special in the sense that there could exist nonvanishing torsion components along that dimension. We impose constraints on the connection so that any motion in the fifth dimension does not affect observations based on the geodesic motions along the 4D hypersurfaces, thus keeping the fifth dimension essentially hidden. The imposed constraints determine uniquely all the nonvanishing torsion components in terms of the 5D metric fields, making this a purely metric theory of gravity. Besides uniquely determining the torsion in the 5D geometry, the imposed constraints lead to interesting equivalence between the 5D geometry with torsion and the torsion-free 4D geometry of GR. In particular, it turns out that the components of the connection and the Ricci tensor along the 4D hypersurfaces exactly match what would arise from GR on a 4D space-time. Consequently, any test of this theory based on geodesic motions will yield the same results as GR.

Though, by construction, the extra dimension is hidden at the level of geodesic motion, its effect is clearly reflected in the field equations. The field equations are obtained by imposing the constraints on the action and varying it with respect to the metric. This leads to global solutions that are qualitatively distinct from those obtained from GR. Most notably, we find positive mass naked singularity solutions that match the Schwarzschild solution at large distances but lack an event horizon. In the cosmological context, we find oscillatory solutions, in contrast to the inevitable singular big bang of the standard cosmology.

We begin in Sec. II with a review of the general framework of the 5D geometry. Section III deals with the specification of the constraints and the determination of the torsion and connection in terms of the metric. Section IV is devoted to the derivation of modified Einstein equations from the standard action principle using the Ricci scalar as the Lagrangian density. In Sec. V, we apply the modified Einstein equations to the homogeneous and isotropic cosmology and identify numerical solutions pointing to accelerating and oscillatory solutions to the universe. In Sec. VI, we discuss static spherically symmetric vacuum solutions and demonstrate the existence of positive mass naked singularity solutions. The final section is devoted to a summary and discussion of the results.

II. GENERAL FRAMEWORK OF 5D GEOMETRY

We denote the coordinates of the 5D manifold by the Latin indices, $i, j, k, \ldots$ that take values 0, 1, 2, 3, and 5, and the coordinates along the 4D hypersurfaces by the Greek indices, $\mu, \nu, \lambda, \ldots$ that take values 0, 1, 2, and 3. Figure 1 is a schematic representation of the 5D geometry. With x^5 denoting the axis of foliation, the metric of the foliated 5D geometry has the form

$$\mathbf{g}_{ij} = \begin{bmatrix} g_{\mu\nu} + \epsilon A_\mu A_\nu \Phi^2 & \epsilon A_\mu \Phi^2 \\ \epsilon A_\nu \Phi^2 & \epsilon \Phi^2 \end{bmatrix}, \tag{1}$$

$$\mathbf{g}^{ij} = \begin{bmatrix} g^{\mu\nu} & -A^\mu \\ -A^\nu & A_\lambda A^\lambda + \epsilon \Phi^{-2} \end{bmatrix}, \tag{2}$$

$$\mathbf{g}_{\mu\nu} = g_{\mu\nu} + \epsilon A_\mu A_\nu \Phi^2, \quad \mathbf{g}_{\mu 5} = \epsilon A_\mu \Phi^2, \quad \mathbf{g}_{55} = \epsilon \Phi^2,$$
$$\mathbf{g}^{\mu\nu} = g^{\mu\nu}, \quad \mathbf{g}^{\mu 5} = -A^\mu, \quad \mathbf{g}^{55} = A_\lambda A^\lambda + \epsilon \Phi^{-2}. \tag{3}$$

Here A_μ is a 4D vector whose indices are raised and lowered with respect to the 4D metrics $g^{\mu\nu}$ and $g_{\mu\nu}$. The fifth dimension is spacelike if $\epsilon = +1$, and it is timelike if $\epsilon = -1$. Note that the 5D metric is denoted by $\mathbf{g}$, and the 4D metric is g.

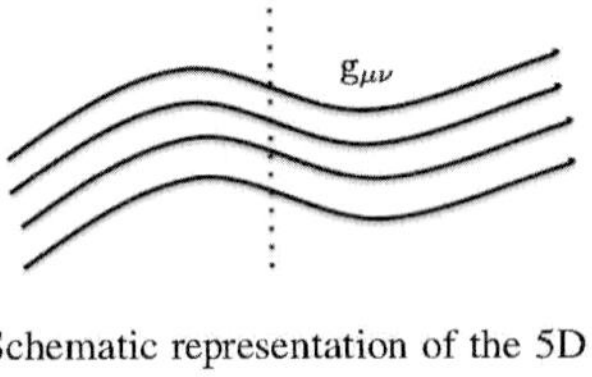

FIG. 1. Schematic representation of the 5D geometry.

Let us denote the connection in the 5D geometry by $\tilde{\Gamma}$. and its antisymmetric part, the torsion, by T..

$$T^{i}_{\cdot jk} = \tilde{\Gamma}^{i}_{\cdot jk} - \tilde{\Gamma}^{i}_{\cdot kj}. \tag{4}$$

Denoting the covariant derivative induced by the connection by $\tilde{\nabla}$, the metricity condition is expressed as $\tilde{\nabla}_k \mathbf{g}_{ij} = 0$. With the metricity condition, the connection $\tilde{\Gamma}$. can be expressed as a sum of the Levi-Civita connection $\hat{\Gamma}$. and the contorsion K.,

$$\tilde{\Gamma}^{i}_{\cdot jk} = \hat{\Gamma}^{i}_{\cdot jk} + K^{i}_{\cdot jk}, \tag{5}$$

where the Levi-Civita connection is expressed purely in terms of the metric

$$\hat{\Gamma}^{i}_{\cdot jk} = \left\{ {i \atop j\ k} \right\} = \frac{1}{2} \mathbf{g}^{im} [\partial_j \mathbf{g}_{km} + \partial_k \mathbf{g}_{jm} - \partial_m \mathbf{g}_{jk}], \tag{6}$$

and the contorsion is expressed in terms of the torsion [16]

$$K^{i}_{\cdot jk} = \tfrac{1}{2} [T^{i}_{\cdot jk} + T_{j\cdot k}^{\cdot i} + T_{k\cdot j}^{\cdot i}]. \tag{7}$$

In the absence of torsion, the connection is simply the Levi-Civita part. In order to compare the dynamics of this geometry to GR, we consider a reference space-time in four dimensions with the metric $g_{\mu\nu}$ and torsion-free 4D Levi-Civita connection $\Gamma^{\lambda}_{\cdot\mu\nu}$,

$$\Gamma^{\lambda}_{\cdot\mu\nu} = \tfrac{1}{2} g^{\lambda\alpha} [\partial_\mu g_{\alpha\nu} + \partial_\nu g_{\mu\alpha} - \partial_\alpha g_{\mu\nu}]. \tag{8}$$

This connection is different from the 4D components of the 5D Levi-Civita connection that contains additional terms $\bar{\Gamma}^{\lambda}_{\cdot\mu\nu}$ [see Eq. (A7)] that depend on the extra-dimensional metric fields A_μ and Φ. Hence the 4D components of the 5D Levi-Civita connection can be written as

$$\hat{\Gamma}^{\lambda}_{\cdot\mu\nu} = \Gamma^{\lambda}_{\cdot\mu\nu} + \bar{\Gamma}^{\lambda}_{\cdot\mu\nu}. \tag{9}$$

In the presence of torsion, with the inclusion of contorsion, the 4D components of the 5D connection take the form

$$\tilde{\Gamma}^{\lambda}_{\cdot\mu\nu} = \Gamma^{\lambda}_{\cdot\mu\nu} + \bar{\Gamma}^{\lambda}_{\cdot\mu\nu} + K^{\lambda}_{\cdot\mu\nu}. \tag{10}$$

We note that the additional terms $(\bar{\Gamma}^{\lambda}_{\cdot\mu\nu} + K^{\lambda}_{\cdot\mu\nu})$ do not generally vanish. However, in the next section we impose constraints on the connection and find that these terms do vanish.

III. CONSTRAINTS ON THE CONNECTION

With minimal modifications to standard GR in mind, we first assume that the 4D components of the connection $\tilde{\Gamma}^{\lambda}_{\cdot\mu\nu}$ are symmetric, that is, (i) $T^{\lambda}_{\cdot\mu\nu} = 0$. Next, we require that geodesic motion and its observable effects in 4D are not affected by any motion in the fifth dimension. This requirement essentially ensures that the fifth dimension stays hidden at the level of 4D geodesics. For this purpose, considering the 4D components of the geodesic equations in the 5D geometry, namely,

$$\ddot{x}^{\lambda} + \tilde{\Gamma}^{\lambda}_{\cdot\mu\nu}\dot{x}^{\mu}\dot{x}^{\nu} + (\tilde{\Gamma}^{\lambda}_{\cdot\mu 5} + \tilde{\Gamma}^{\lambda}_{\cdot 5\mu})\dot{x}^{\mu}\dot{x}^{5} + \tilde{\Gamma}^{\lambda}_{\cdot 55}(\dot{x}^{5})^{2} = 0,$$

we are led to the second constraint, (ii) $\tilde{\Gamma}^{\lambda}_{\cdot i5} = \tilde{\Gamma}^{\lambda}_{\cdot 5i} = 0$. An alternative formulation of these constraints in terms of vielbeins is worked out in [17]. These constraints are clearly not tensorial in nature because the fifth dimension is singled out. It turns out, however, that they are sufficient to determine uniquely all the nonvanishing torsion components in terms of the metric (see Appendix A for details).

$$\begin{aligned} T^{\mu}_{\cdot ij} &= 0, \\ T^{5}_{\cdot\mu\nu} &= 2\partial_{[\mu}A_{\nu]} + 2J_{[\mu}A_{\nu]}, \\ T^{5}_{\cdot\mu 5} &= J_{\mu} - \partial_{5}A_{\mu} - A_{\mu}J_{5}, \end{aligned} \tag{11}$$

where $J_i \equiv \Phi^{-1}\partial\Phi/\partial x^i$.

Using the above results for torsion and Eqs. (5)–(7), we find the connection coefficients

$$\begin{aligned} &\tilde{\Gamma}^{\lambda}_{\cdot 55} = \tilde{\Gamma}^{\lambda}_{\cdot\nu 5} = \tilde{\Gamma}^{\lambda}_{\cdot 5\nu} = 0, \qquad \tilde{\Gamma}^{5}_{\cdot\mu\nu} = \nabla_{\mu}A_{\nu} + J_{\mu}A_{\nu}, \\ &\tilde{\Gamma}^{5}_{\cdot 5\mu} = \partial_{5}A_{\mu} + J_{5}A_{\mu}, \qquad \tilde{\Gamma}^{5}_{\cdot\mu 5} = J_{\mu}, \qquad \tilde{\Gamma}^{5}_{\cdot 55} = J_{5}, \\ &\tilde{\Gamma}^{\lambda}_{\cdot\mu\nu} = \Gamma^{\lambda}_{\cdot\mu\nu}. \end{aligned} \tag{12}$$

Here ∇_{μ} is the covariant derivative operator in the torsion-free 4D geometry with metric $g_{\mu\nu}$. This connection has a very special property: its 4D components are exactly the 4D Levi-Civita connection. That is, $\bar{\Gamma}^{\lambda}_{\cdot\mu\nu}$ and $K^{\lambda}_{\cdot\mu\nu}$ in Eq. (10) exactly cancel each other.

In addition to determining the torsion and the connection in terms of the metric fields, the constraints also imply that the 4D metrics on all the hypersurfaces are identical. As a consequence, the 4D components of the connection do not depend on x^5 either,

$$\frac{\partial g_{\mu\nu}}{\partial x^{5}} = 0 \Rightarrow \frac{\partial\tilde{\Gamma}^{\lambda}_{\cdot\mu\nu}}{\partial x^{5}} = 0. \tag{13}$$

This should be contrasted with the Kaluza-Klein–type theories where it is *a priori* assumed that $g_{\mu\nu}$, A_{μ}, and Φ are independent of x^5, known as the cylindrical condition. In our framework, though $g_{\mu\nu}$ is required to be independent of x^5, A_{μ} and Φ can, in principle, depend on x^5.

Substituting the connection [Eq. (12)] in the Ricci tensor defined by

$$\tilde{R}_{ik} = \partial_{k}\tilde{\Gamma}^{j}_{\cdot ji} - \partial_{j}\tilde{\Gamma}^{j}_{\cdot ki} + \tilde{\Gamma}^{j}_{\cdot km}\tilde{\Gamma}^{m}_{\cdot ji} - \tilde{\Gamma}^{j}_{\cdot jm}\tilde{\Gamma}^{m}_{\cdot ki}, \tag{14}$$

we find

$$\tilde{R}_{\mu\nu} = R_{\mu\nu}, \qquad \tilde{R}_{\mu 5} = \tilde{R}_{5\mu} = \tilde{R}_{55} = 0. \tag{15}$$

Here $R_{\mu\nu}$ represents the Ricci tensor constructed from the torsion-free 4D Levi-Civita connection. Hence the 4D components of the Ricci tensor exactly match the Ricci tensor in GR with the metric $g_{\mu\nu}$. It also follows that the 5D Ricci scalar is exactly the same as the Ricci scalar in the torsion-free 4D space-time, that is, $\tilde{R} = R$.

An important point to emphasize is that, at the level of geometry, this framework is virtually indistinguishable from the torsion-free 4D space-time of GR. Any observable geodesic motion or geodesic deviations between particles would match what we expect based on GR. However, this is true only to the extent that the metric $g_{\mu\nu}$ is identical to the solution of the Einstein equations in GR. In Sec. VI, we will see that this is indeed true in the weak field limit for spherically symmetric vacuum solutions that are relevant for experimental observations within the solar system.

IV. ACTION PRINCIPLE AND MODIFIED EINSTEIN EQUATIONS

We start with the standard Einstein-Hilbert action with the Ricci scalar as the Lagrangian density,

$$S = \int \tilde{R}\sqrt{-\mathbf{g}}\, d^5x. \tag{16}$$

In varying the action, we note that the Ricci scalar and the connection coefficients described in the previous section are functions of the metric components alone.

$$\delta S = \int [\tilde{R}\delta\sqrt{-\mathbf{g}} + \tilde{R}_{ik}\delta\mathbf{g}^{ik}\sqrt{-\mathbf{g}}]d^5x + \int \delta\tilde{R}_{ik}\mathbf{g}^{ik}\sqrt{-\mathbf{g}}d^5x. \tag{17}$$

The first term gives rise to the usual Einstein tensor,

$$\tilde{G}_{ik} = \tilde{R}_{ik} - (1/2)\mathbf{g}_{ik}\tilde{R}.$$

In the absence of torsion, the second term becomes a boundary integral which vanishes when the variation is fixed at the boundary and hence will not contribute to the equations of motion. But in the presence of torsion, the second term gives a nonzero contribution.

From Eq. (14), we find the variations of the Ricci tensor to be

$$\begin{aligned} \delta\tilde{R}_{ik} &= \partial_{k}\delta\tilde{\Gamma}^{j}_{\cdot ji} - \partial_{j}\delta\tilde{\Gamma}^{j}_{\cdot ki} + \tilde{\Gamma}^{j}_{\cdot km}\delta\tilde{\Gamma}^{m}_{\cdot ji} + \tilde{\Gamma}^{m}_{\cdot ji}\delta\tilde{\Gamma}^{j}_{\cdot km} \\ &\quad - \tilde{\Gamma}^{j}_{\cdot jm}\delta\tilde{\Gamma}^{m}_{\cdot ki} - \tilde{\Gamma}^{m}_{\cdot ki}\delta\tilde{\Gamma}^{j}_{\cdot jm} \\ &= [\tilde{\nabla}_{k}\delta\tilde{\Gamma}^{j}_{\cdot ji} - \tilde{\nabla}_{j}\delta\tilde{\Gamma}^{j}_{\cdot ki}] + T^{m}_{\cdot kj}\delta\tilde{\Gamma}^{j}_{\cdot mi}. \end{aligned} \tag{18}$$

Then, the second term in the right-hand side (r.h.s.) of Eq. (17) takes the form

$$\begin{aligned} \int \delta\tilde{R}_{ik}\mathbf{g}^{ik}\sqrt{-\mathbf{g}}d^5x &= \int [\tilde{\nabla}_{k}(\mathbf{g}^{ik}\delta\tilde{\Gamma}^{j}_{\cdot ji}) \\ &\quad - \tilde{\nabla}_{j}(\mathbf{g}^{ik}\delta\tilde{\Gamma}^{j}_{\cdot ki})]\sqrt{-\mathbf{g}}d^5x \\ &\quad + \int \mathbf{g}^{ik}T^{m}_{\cdot kj}\delta\tilde{\Gamma}^{j}_{\cdot mi}\sqrt{-\mathbf{g}}d^5x. \end{aligned} \tag{19}$$

In deriving the above equation, we have used the metricity condition, namely, $\tilde{\nabla}_{j}\mathbf{g}^{ik} = 0$. Substituting for the covariant derivative in the first term of the r.h.s. of Eq. (19), we obtain

$$\int[\tilde{\nabla}_k(\mathbf{g}^{ik}\delta\tilde{\Gamma}^{j}_{\cdot ji}) - \tilde{\nabla}_j(\mathbf{g}^{ik}\delta\tilde{\Gamma}^{j}_{\cdot ki})]\sqrt{-\mathbf{g}}d^5x$$
$$= \int[\partial_k(\mathbf{g}^{ik}\delta\tilde{\Gamma}^{j}_{\cdot ji}\sqrt{-\mathbf{g}}) - \partial_j(\mathbf{g}^{ik}\delta\tilde{\Gamma}^{j}_{\cdot ki}\sqrt{-\mathbf{g}})]d^5x$$
$$+ \int \mathbf{g}^{ik}\delta\tilde{\Gamma}^{j}_{\cdot ji}\left[\tilde{\Gamma}^{m}_{\cdot mk} - \frac{\partial_k\sqrt{-\mathbf{g}}}{\sqrt{-\mathbf{g}}}\right]\sqrt{-\mathbf{g}}d^5x$$
$$- \int \mathbf{g}^{ik}\delta\tilde{\Gamma}^{j}_{\cdot ki}\left[\tilde{\Gamma}^{m}_{\cdot mj} - \frac{\partial_j\sqrt{-\mathbf{g}}}{\sqrt{-\mathbf{g}}}\right]\sqrt{-\mathbf{g}}d^5x. \qquad (20)$$

The first term on the r.h.s. of the above equation is a boundary term, an integral of a total divergence. This will vanish when the variation is fixed at the boundary, and hence can be ignored. The second and third terms on the r.h.s. can be simplified by noting $\tilde{\Gamma}^{m}_{\cdot mk} = \hat{\Gamma}^{m}_{\cdot mk} + T^{m}_{\cdot mk}$ and $\hat{\Gamma}^{m}_{\cdot mk} = (\partial_k\sqrt{-\mathbf{g}})/\sqrt{-\mathbf{g}}$, leading to

$$\int[\tilde{\nabla}_k(\mathbf{g}^{ik}\delta\tilde{\Gamma}^{j}_{\cdot ji}) - \tilde{\nabla}_j(\mathbf{g}^{ik}\delta\tilde{\Gamma}^{j}_{\cdot ki})]\sqrt{-\mathbf{g}}d^5x$$
$$= -\int T^{m}_{\cdot km}\mathbf{g}^{ik}\delta\tilde{\Gamma}^{j}_{\cdot ji}\sqrt{-\mathbf{g}}d^5x$$
$$+ \int T^{m}_{\cdot jm}\mathbf{g}^{ik}\delta\tilde{\Gamma}^{j}_{\cdot ki}\sqrt{-\mathbf{g}}d^5x, \qquad (21)$$

and Eq. (19) becomes

$$\int \delta\tilde{R}_{ik}\mathbf{g}^{ik}\sqrt{-\mathbf{g}}d^5x = \int[-T^{m}_{\cdot km}\mathbf{g}^{ik}\delta\tilde{\Gamma}^{j}_{\cdot ji} + T^{m}_{\cdot jm}\mathbf{g}^{ik}\delta\tilde{\Gamma}^{j}_{\cdot ki} + T^{m}_{\cdot kj}\mathbf{g}^{ik}\delta\tilde{\Gamma}^{j}_{\cdot mi}]\sqrt{-\mathbf{g}}d^5x. \qquad (22)$$

The second term on the r.h.s. of Eq. (17) is thus a function of torsion and variations in the connection given by Eq. (22). Clearly, this term vanishes if we assume that torsion is zero, and the action principle would yield the standard Einstein equations. Alternatively, if we treat the variation in the connection to be composed of independent variations in the metric and torsion, we obtain the Einstein-Cartan equations [3], which ultimately lead to zero torsion when the matter is not coupled to the connection. However, since torsion is not an independent degree of freedom in our framework but it is a function of the metric components given by Eq. (11), we first substitute its components in terms of the metric and then carry out the variation with respect to the metric. To this end, we note

$$T^{m}_{\cdot km}\mathbf{g}^{ik}\delta\tilde{\Gamma}^{j}_{\cdot ji} = T^{5}_{\cdot \nu 5}[\mathbf{g}^{\mu\nu}(\delta\tilde{\Gamma}^{\alpha}_{\cdot\alpha\mu} + \delta\tilde{\Gamma}^{5}_{\cdot 5\mu}) + \mathbf{g}^{5\nu}\delta\tilde{\Gamma}^{5}_{\cdot 55}],$$
$$T^{m}_{\cdot jm}\mathbf{g}^{ik}\delta\tilde{\Gamma}^{j}_{\cdot ki} = T^{5}_{\cdot \nu 5}\mathbf{g}^{\mu\alpha}\delta\tilde{\Gamma}^{\nu}_{\cdot\mu\alpha},$$
$$T^{m}_{\cdot kj}\mathbf{g}^{ik}\delta\tilde{\Gamma}^{j}_{\cdot mi} = T^{5}_{\cdot \nu 5}[\mathbf{g}^{\mu\nu}\delta\tilde{\Gamma}^{5}_{\cdot 5\mu} + \mathbf{g}^{5\nu}\delta\tilde{\Gamma}^{5}_{\cdot 55}]. \qquad (23)$$

Taking these terms together, Eq. (22) takes the form

$$\int \delta\tilde{R}_{ik}\mathbf{g}^{ik}\sqrt{-\mathbf{g}}d^5x = \int T^{5}_{\cdot \nu 5}[g^{\mu\alpha}\delta\Gamma^{\nu}_{\cdot\mu\alpha} - g^{\mu\nu}\delta\Gamma^{\alpha}_{\cdot\alpha\mu}]\sqrt{-\mathbf{g}}d^5x. \qquad (24)$$

Note that the variations in the connection, $\delta\Gamma^{\nu}_{\cdot\mu\alpha}$ and $\delta\Gamma^{\alpha}_{\cdot\alpha\mu}$, involve only the 4D components. Since these are the 4D Levi-Civita components that only depend on the 4D metric $g_{\mu\nu}$, the above equation takes the form

$$\int \delta\tilde{R}_{ik}\mathbf{g}^{ik}\sqrt{-\mathbf{g}}d^5x = \int H_{\mu\nu}\delta g^{\mu\nu}\sqrt{-\mathbf{g}}d^5x, \qquad (25)$$

where (see Appendix B for details)

$$H_{\mu\nu} = \nabla_{(\mu}B_{\nu)} - (\nabla\cdot B)g_{\mu\nu} + J_{(\mu}B_{\nu)} - (J\cdot B)g_{\mu\nu},$$
$$B_\mu \equiv T^{5}_{\cdot\mu 5} = J_\mu - \partial_5 A_\mu - A_\mu J_5. \qquad (26)$$

Taking together the variations in both terms in Eq. (17), we obtain the modified Einstein tensor

$$\tilde{G}_{\mu\nu} = R_{\mu\nu} - \tfrac{1}{2}(g_{\mu\nu} + A_\mu A_\nu \epsilon\Phi^2)R + H_{\mu\nu} = \Sigma_{\mu\nu},$$
$$\tilde{G}_{\mu 5} = -\tfrac{1}{2}A_\mu \epsilon\Phi^2 R = \Sigma_{\mu 5},$$
$$\tilde{G}_{55} = -\tfrac{1}{2}\epsilon\Phi^2 R = \Sigma_{55},$$

where Σ is the stress-energy tensor that arises from the variations of the assumed matter fields in the Lagrangian. Our focus being on gravity, we will not discuss the origin of Σ further. Since the physical interpretation of the stress energy is more transparent with one covariant and one contravariant index, we express the above equations in an alternate form, by noting $\tilde{G}_i{}^j = \mathbf{g}^{kj}\tilde{G}_{ik}$,

$$R_\mu{}^\nu - \tfrac{1}{2}R\delta_\mu{}^\nu + H_\mu{}^\nu = \Sigma_\mu{}^\nu, \qquad (27)$$

$$-A^\alpha R_{\mu\alpha} - A^\alpha H_{\mu\alpha} = \Sigma_\mu{}^5, \qquad (28)$$

$$0 = \Sigma_5^\mu, \qquad -\tfrac{1}{2}R = \Sigma_5^5 \qquad (29)$$

These are the modified Einstein equations in our framework. Since, by construction, the fifth dimension is hidden with respect to the observable 4D motion, the 5D components of the stress tensor, $\Sigma_\mu{}^5$ and $\Sigma_5{}^5$, are unobservable. It is not possible to solve Eqs. (28) and (29) unless these components are theoretically known from the 5D matter Lagrangian. In the present formulation, for simplicity, we shall ignore these equations as though they simply serve to evaluate the components $\Sigma_\mu{}^5$ and $\Sigma_5{}^5$, and treat only Eq. (27) with the observable 4D stress tensor to be relevant to physical solutions. In the absence of specified matter fields in the Lagrangian, an alternate way to interpret the modified Einstein equations is to regard $-H_\mu{}^\nu$ as extra-dimensionally induced matter.

When $H_\mu{}^\nu = 0$, Eq. (27) reduces to the standard Einstein equations for the 4D metric components $g_{\mu\nu}$. In this case the 4D Bianchi identity necessarily implies the conservation of matter, $\nabla_\nu\Sigma_\mu{}^\nu = 0$. But, in general, when $H_\mu{}^\nu$ is nonvanishing and dependent on the extra-dimensional metric fields A_μ and Φ, Eq. (27) by itself may not be sufficient to solve for $g_{\mu\nu}$ along with A_μ and Φ, even after fixing the gauge. However, an important physical simplification can be achieved by generalizing the cylindrical condition to assume that A_μ and Φ do not depend on x^5. With this assumption, $B_\mu = J_\mu$ and $H_\mu{}^\nu$ depends only on Φ and not on A_μ; Eq. (27) is sufficient to solve for both $g_{\mu\nu}$ and Φ. The vector A_μ can, in principle, be evaluated from Eq. (28) by setting $\Sigma_\mu{}^5$ to zero, but this would be inconsequential as A_μ is decoupled from the physically relevant equation that solves for the 4D metric $g_{\mu\nu}$. Hence, in the rest of the paper we will make the assumption of the cylindrical condition in order to explore solutions of physical interest to the modified Einstein equations.

Finally, when $H_\mu{}^\nu$ is nonvanishing, we note that $\Sigma_\mu{}^\nu$ does not necessarily have to satisfy the 4D matter conservation. However, with minimum modifications to GR and the empirical conservation laws in mind, it is reasonable to assert the conservation of $\Sigma_\mu{}^\nu$. Since the standard Einstein tensor satisfies the 4D Bianchi identity independently of $H_\mu{}^\nu$, the 4D matter conservation implies

$$\nabla_\nu \Sigma_\mu{}^\nu = 0 \Rightarrow \nabla_\nu H_\mu{}^\nu = 0. \tag{30}$$

In the reminder of the paper, we study the solutions to the modified Einstein equations [Eq. (27)] in two extremely symmetric situations, namely, the homogeneous-isotropic geometry and the static spherically symmetric geometry.

V. HOMOGENEOUS-ISOTROPIC COSMOLOGY

The 4D metric of a homogeneous and isotropic universe has the form

$$ds^2 = -dt^2 + a^2(t)\left(\frac{dr^2}{1-kr^2} + r^2 d\Omega^2\right). \tag{31}$$

The values of $k = 0, +1, -1$ correspond, respectively, to flat, closed, and hyperbolic spatial geometries. The standard Einstein tensor for this metric is given by [18]

$$\begin{aligned} G_t^t &= 3(\dot a/a)^2 + 3k/a^2, \\ G_r^r &= 2(\ddot a/a) + (\dot a/a)^2 + k/a^2, \\ G_\theta^\theta &= G_\phi^\phi = G_r^r, \end{aligned} \tag{32}$$

where an over-dot denotes a derivative with respect to time. Since the geometry is spatially homogeneous and isotropic, the metric fields including A_μ and Φ in the 5D geometry only depend on time. Hence the only nonvanishing component of J_μ is J_t. The induced matter terms given in Eq. (26) are

$$\begin{aligned} H_t{}^t &= 3J_t(\dot a/a), \\ H_r{}^r &= 2J_t(\dot a/a) + \dot J_t + J_t^2, \\ H_\theta{}^\theta &= H_\phi{}^\phi = H_r{}^r. \end{aligned} \tag{33}$$

Before writing out the modified Einstein equations, we note that the conservation equation [Eq. (30)] now reduces to

$$\begin{aligned} \nabla_\nu H_\mu{}^\nu &= 3J_t[(\ddot a/a) - (\dot a/a)J_t] = 0 \Rightarrow \\ J_t &= 0 \quad \text{or} \quad J_t = \ddot a/\dot a. \end{aligned} \tag{34}$$

These are the only two possibilities. From the definition of J_μ, this implies that either Φ is a constant, which would give rise to the usual Friedmann-Robertson-Walker (FRW) cosmology, or $\Phi = \dot a(t)$. Focusing on the latter case, $H_\mu{}^\nu$ simplifies to

$$H_t{}^t = 3\ddot a/a, \qquad H_r{}^r = 2(\ddot a/a) + (\ddot a/\dot a). \tag{35}$$

Taking the stress tensor to be that of a perfect fluid, the modified Einstein equations [Eq. (27)] take the form

$$3(\dot a/a)^2 + 3k/a^2 + 3\ddot a/a = 8\pi\rho, \tag{36}$$

$$4(\ddot a/a) + (\dot a/a)^2 + (\ddot a/\dot a) + k/a^2 = -8\pi P, \tag{37}$$

where ρ and P are the density and pressure of the 4D matter. Combining the above equations, we find

$$\dot\rho + 3(\rho + P)\dot a/a = 0, \tag{38}$$

which is just a restatement of the 4D matter conservation equation. For a matter dominated universe, $P = 0$, and consequently Eq. (38) yields

$$\rho a^3 = \text{constant} \quad \text{or} \quad \rho = \rho_o/a^3. \tag{39}$$

In effect, Eqs. (36) and (39) are sufficient to solve for $a(t)$, which needs two initial conditions along with the specification of ρ_o.

$$a^2\ddot a + a(\dot a)^2 + ka = 8\pi\rho_o/3. \tag{40}$$

Without loss of generality, we choose the current epoch to be at $t = 0$, set the current size of the universe as $a(0) = 1$, and set the unit of time such that $\dot a(0) = 1$. In this unit of time, the current value of the Hubble constant will be 1. Rather than specifying the value of ρ_o, we choose to specify the current value of $\ddot a$. The effective equation for $a(t)$ then takes the form

$$a^2\ddot a + a(\dot a)^2 + ka = 1 + k - q_o, \tag{41}$$

where $q_o = -a(0)\ddot a(0)/\dot a^2(0)$, the current value of the deceleration parameter, is the only free parameter to be specified. Figure 2 shows the behavior of $a(t)$ for various values of q_o for spatially flat and closed topologies. For the spatially flat topology, shown in the top panels of the figure, we find that the universe does not originate from a singular big bang for all $q_o < +0.5$. For the spatially closed topology shown in the bottom panels, we find oscillatory solutions for all $q_o < +1$.

Oscillatory solutions in the spatially closed topology exhibit a scale factor that oscillates between a maximum $a_{\max}$ and a minimum $a_{\min}$. The acceleration reaches a positive value in a narrow interval around $a_{\min}$, and then becomes negative for the rest of the cycle until it gets back near $a_{\min}$.

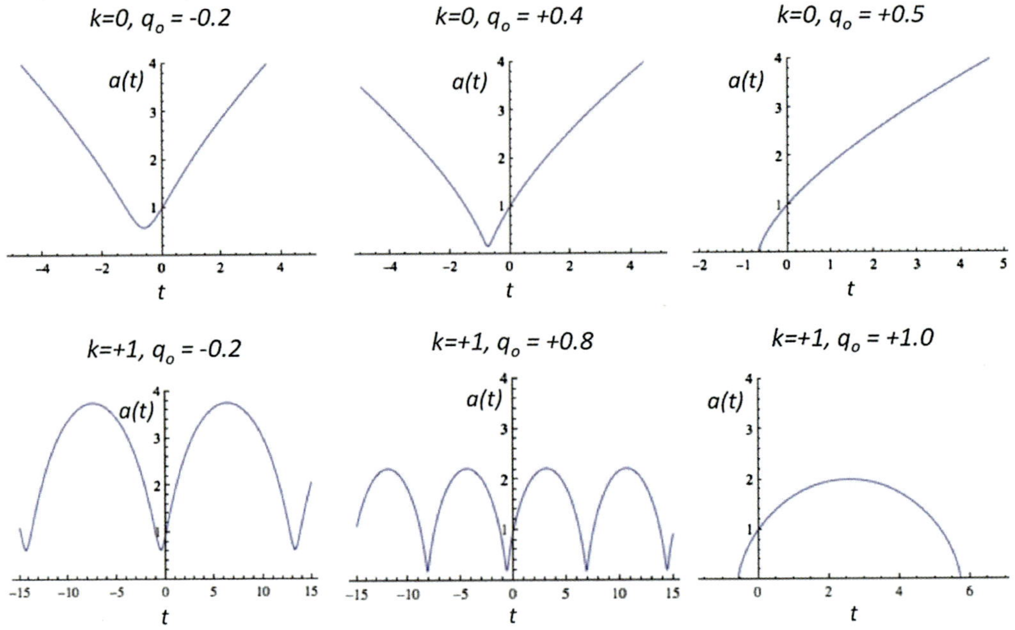

FIG. 2 (color online). Solutions for $a(t)$ for different values of the deceleration parameter for $k = 0$ and $k = +1$.

By taking the value of q_o arbitrarily close to 1, we can make $a_{\rm min}$ arbitrarily close to zero. This can be seen in the bottom-left and bottom-middle panels of Fig. 2. Thus one could construct a universe that collapses and bounces back to expand when it reaches an arbitrarily small size or, equivalently, arbitrarily high energy densities. It remains to be seen if such solutions would fit the empirical redshift data.

VI. STATIC SPHERICALLY SYMMETRIC VACUUM SOLUTIONS

The most general static spherically symmetric 4D metric has the form[1]

$$ds^2 = -A(r)dt^2 + B(r)dr^2 + r^2 d\Omega^2, \tag{42}$$

and the standard Einstein tensor for this metric is [18]

$$\begin{aligned}
G^t_t &= \frac{rB' + B^2 - B}{r^2 B^2},\\
G^r_r &= \frac{AB - rA' - A}{r^2 AB},\\
G^\theta_\theta &= \frac{2A^2B' - 2ABA' - 2rABA'' + rBA'^2 + rAA'B'}{4rA^2B^2},\\
G^\phi_\phi &= G^\theta_\theta,
\end{aligned} \tag{43}$$

[1]The scalar functions $A(r)$ and $B(r)$ defined in this section should not be confused with the vectors A_μ and B_μ defined in Eqs. (3) and (26), respectively.

where a prime in the above equations denotes a derivative with respect to r.

The additional term $H_\mu{}^\nu$ in the modified Einstein equations [Eq. (27)] depends only on Φ when the cylindrical condition is imposed on all metric components, and is given by

$$H_\mu{}^\nu = \nabla_\mu J^\nu - (\nabla \cdot J)\delta^\nu_\mu + J_\mu J^\nu - (J \cdot J)\delta^\nu_\mu. \tag{44}$$

Since $J_\mu = \Phi^{-1}\partial_\mu \Phi$, the quantity $\nabla_\mu J^\nu$ is intrinsically symmetric in μ and ν. The static spherical symmetry of the geometry implies that J_r is the only nonvanishing component, which we denote by $J(r)$. With this,

$$\begin{aligned}
H_t{}^t &= \frac{rJB' - 2B(rJ' + 2J + rJ^2)}{2rB^2},\\
H_r{}^r &= \frac{-J(rA' + 4A)}{2rAB},\\
H_\theta{}^\theta &= \frac{rJAB' - 2rABJ' - rJBA' - 2ABJ(1 + rJ)}{2rAB^2},\\
H_\phi{}^\phi &= H_\theta{}^\theta.
\end{aligned} \tag{45}$$

In order to obtain vacuum solutions, we set $\Sigma^\nu_\mu = 0$ in Eq. (27) and find the following three equations:

$$J' = -\frac{J(1+B)}{r},$$
$$A' = -\frac{2A(1-B+2rJ)}{r(2+Jr)}, \qquad (46)$$
$$B' = \frac{2B(r^2J^2 + (1+rJ)(1-B))}{r(2+Jr)}.$$

A close examination of the above equations reveals two basic properties of the function $J(r)$. First, if $J(r)$ is a constant, it has to be identically zero. Second, if $J(r)$ vanishes at some point, it has to vanish identically everywhere. The simplest solution to the coupled equations [Eq. (46)] is when $J(r)$ vanishes everywhere,

$$J(r)=0, \quad A(r)=\left(1-\frac{2M}{r}\right), \quad B(r)=\left(1-\frac{2M}{r}\right)^{-1}, \qquad (47)$$

which of course is the well-known Schwarzschild solution, as expected.

A. General solution to $J(r)$

Let $F(r) \equiv 1/rJ(r)$ when $J(r)$ is nonvanishing. Substituting for $J(r)$ in terms of $F(r)$, the coupled equations [Eq. (46)] lead to the following equation for $F(r)$,

$$r[(2F^2+F)F'' + F'^2] = F'(F+2), \qquad (48)$$

whose solution in turn determines the 4D metric functions $A(r)$ and $B(r)$. The obvious solution of Eq. (48) is $F(r) =$ constant. This leads to

$$J(r)=c/r, \quad A(r)=(r)^{-(2+4c/2+c)}, \quad B(r)=0, \qquad (49)$$

which is clearly unacceptable because $B(r)$ is identically zero.

Assuming that $F(r)$ is not a constant, we can obtain solutions to the second order differential equation [Eq. (48)]. In principle, the solution would have two integration constants that would be determined by the boundary conditions, one of which immediately follows from the form of the equation. It can be easily seen that if $F(r)$ is a solution, then $F(\lambda r)$ is also a solution for any scaling constant λ. We find a general solution in the implicit form

$$\lambda^2 r^2 = \gamma^2 |F/\gamma + \beta + 1|^{1+\beta} |F/\gamma + \beta - 1|^{1-\beta}, \qquad (50)$$

where β and γ are defined in terms of an independent arbitrary constant c.

$$\gamma = \sqrt{1+c+c^2}, \qquad \beta = \frac{1+c}{\sqrt{1+c+c^2}}. \qquad (51)$$

With λ and c as two arbitrary constants, Eq. (50) represents the general solution to the second order differential equation [Eq. (48)]. From Eq. (50), we find the derivatives of $F(r)$ to be

$$F' = \frac{1}{r}\frac{F^2 + 2(1+c)F + c}{F}, \qquad F'' = -\frac{c}{r}\frac{F'}{F^2}, \qquad (52)$$

and substituting them into Eq. (46), we obtain the metric functions $A(r)$ and $B(r)$ in terms of $F(r)$.

$$A'(r) = 2cA(r)/rF(r), \qquad (53)$$

$$B(r) = 1 + 2(1+c)/F(r) + c/F^2(r). \qquad (54)$$

In order to obtain asymptotically flat solutions, we shall impose the boundary conditions $A(r \to \infty) = 1$ and $B(r \to \infty) = 1$. To understand the behavior of the functions $A(r)$ and $B(r)$ which define the observable 4D geometry, we start with the properties of $F(r)$.

Unfortunately Eq. (50) does not yield an explicit functional form for $F(r)$ except for simple cases when c is either 0 or -1. Nevertheless, the relevant properties of $F(r)$ can be inferred from analyzing this implicit function. First note that the quantity $(1+c+c^2)$ is positive definite and β is finite and bounded for all values of c. Figure 3 plots the behavior of β to show that it asymptotically reaches $+1$ and -1 at $c = +\infty$ and $-\infty$, respectively. The following observations summarize the qualitative properties of $F(r)$.

(1) At $r = 0$, F can take one of two possible values. If $c < 0$ ($\beta < 1$), then $F(0)$ can be either $\gamma(-\beta - 1)$ or $\gamma(-\beta + 1)$, while if $c > 0$ ($\beta > 1$), then $F(0)$ can only be $\gamma(-\beta - 1)$.
(2) In the limit $r \to \infty$, F necessarily has to diverge in order to satisfy the boundary condition $B(r \to \infty) = 1$. Equation (50) then implies that for large r, $\lambda^2 r^2 = F^2$, implying that F could be either positive or negative, such that

$$F(r \to \infty) = \lambda r \Rightarrow J(r \to \infty) = \lambda^{-1}/r^2. \qquad (55)$$

The behavior of $F(r)$ at the extremities is summarized in the following table.

	$c < 0 (\beta < 1)$	$c > 0 (\beta > 1)$
$r = 0$	$F = \begin{cases} \gamma(-\beta+1) & >0 \\ \gamma(-\beta-1) & <0 \end{cases}$	$F = \gamma(-\beta - 1) < 0$
$r \to \infty$	$F = \lambda r$	$F = \lambda r$

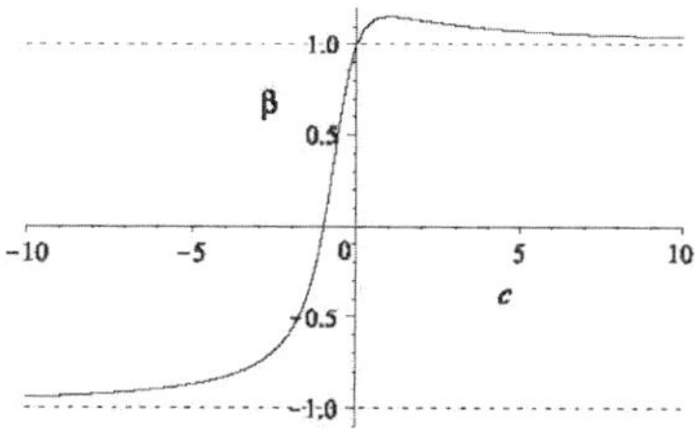

FIG. 3. β is plotted as a function of c.

(3) From Eq. (52), it can be shown that $F' = 0$ when F is either $\gamma(-\beta+1)$ or $\gamma(-\beta-1)$, which can happen only at $r = 0$. Hence F' is either positive definite or negative definite, and so $F(r)$ is a monotonic function.

(4) Though F is monotonic and finite for any finite r, it can reach zero at r_o, given by

$$\lambda^2 r_o^2 = |c| \left| \frac{\beta+1}{\beta-1} \right|^{\beta}, \tag{56}$$

and from Eq. (52), F' diverges at r_o. Considering the physical relevance of these solutions, we shall only focus on solutions that are nonvanishing everywhere. Such solutions do indeed exist for a range of parameter values. Rewriting Eq. (50) at $r = r_o$,

$$\left| \frac{F(r_o)}{\gamma(\beta+1)} + 1 \right|^{1+\beta} \left| \frac{F(r_o)}{\gamma(\beta-1)} + 1 \right|^{1-\beta} = 1, \tag{57}$$

we note that $F(r_o) = 0$ is not the only solution. Numerical plots in Fig. 4 demonstrate the existence of nonvanishing $F(r)$ solutions.

(5) From the table above, we see that (i) for $c > 0$, since $F(0)$ is negative, $F(r)$ has to be negative definite, which requires λ to be negative, and (ii) for $c < 0$, λ can be either positive or negative, making $F(r)$ either positive definite or negative definite, respectively.

The functional form of $F(r)$ described by the above five properties along with Eqs. (53) and (54) will yield the functional form of the metric functions $A(r)$ and $B(r)$.

B. Metric functions $A(r)$ and $B(r)$

With the boundary condition $A(\infty) = B(\infty) = 1$, Eqs. (53) and (54) yield

$$A(r) = \exp\left(- \int_r^{\infty} \frac{2c}{rF(r)} dr \right), \tag{58}$$

$$B(r) = 1 + 2(1+c)/F(r) + c/F^2(r). \tag{59}$$

The following observations summarize the qualitative behavior of $A(r)$ and $B(r)$.

(1) From the asymptotic behavior of $F(r) \to \lambda r$ for large r, we note that

$$A(r) = 1 - \frac{2c\lambda^{-1}}{r} + \mathcal{O}(1/r^2), \tag{60}$$

$$B(r) = 1 + \frac{2(1+c)\lambda^{-1}}{r} + \mathcal{O}(1/r^2). \tag{61}$$

Hence, when $|c| \gg 1$ and $r \to \infty$, the above solutions approximate the Schwarzschild solution with mass $M \equiv [c\lambda^{-1}]$. When both c and λ are either positive or negative, the gravity is attractive, while when one is positive and the other is negative, the gravity is repulsive.

(2) Since $F(r)$ is either positive definite or negative definite, both $A(r)$ and $B(r)$ are finite and positive for all $r > 0$. At $r = 0$, since $F(0)$ is either $\gamma(-\beta-1)$ or $\gamma(-\beta+1)$, Eq. (59) implies $B(0) = 0$.

(3) As $r \to 0$, the integral in Eq. (58) diverges as $[2c/F(0)]\ln(r)$. When $[c/F(0)]$ is positive, then $A(0) = 0$, and when $[c/F(0)]$ is negative, $A(0) = \infty$. The sign of $[c/F(0)]$ is the same as the sign of $M = [c\lambda^{-1}]$. For $M > 0$, $A(r)$ monotonically increases from $A(0) = 0$ to $A(\infty) = 1$; for $M < 0$, $A(r)$ monotonically decreases from $A(0) = +\infty$ to $A(\infty) = 1$.

(4) Irrespective of the sign of M, $B(0) = 0$ and $B(\infty) = 1$. However, $B(r)$ is not necessarily monotonic. From Eq. (59), we see that $B' = 0$ when either $F' = 0$ or when $F(r) = -c/(1+c)$. From the previous subsection, $F' \neq 0$ for all $r > 0$, but $F(r)$ could attain the value $-c/(1+c)$ for certain values of c and λ. Since $F(r)$ is a nonvanishing monotonic function taking all values from $F(0)$ to $\pm\infty$, it is straightforward to check if it would attain the value $-c/(1+c)$. When $\lambda < 0$, $F(r)$ is negative definite, and $-c/(1+c)$ needs to be a negative number less than $F(0) = \gamma(-\beta-1)$, which happens only when $c < -1$. When $\lambda > 0$, $F(r)$ is positive definite, and $-c/(1+c)$ needs to be a positive number greater than $F(0) = \gamma(-\beta+1)$, which happens only when $-1 < c < 0$.

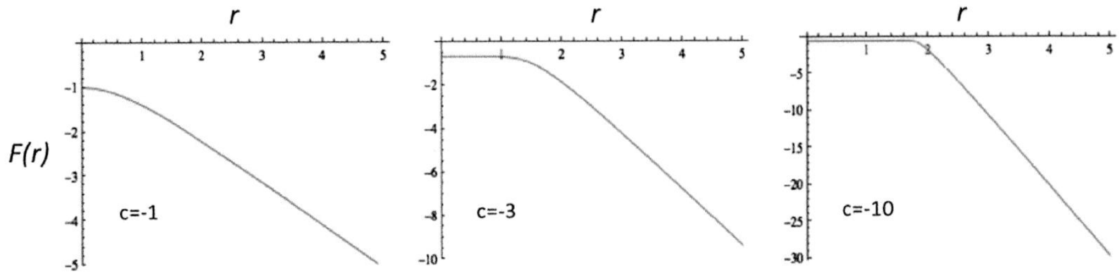

FIG. 4 (color online). Numerical solution to $F(r)$ with $\lambda = -1$.

The qualitative behavior of the functions $A(r)$ and $B(r)$ for the various allowed ranges of c and λ are shown in Fig. 5. When $M = [c\lambda^{-1}]$ is positive, $A(r)$ is a monotonically increasing function leading to attractive gravity, and it corresponds to a Schwarzschild solution at large r when $|c| \gg 1$.

C. Naked singularity at $r = 0$

An important point to note from Fig. 5 is that these solutions do not have an event horizon because both $A(r)$ and $B(r)$ are finite and positive for all $r > 0$.

Clearly, these solutions are smooth for all $r > 0$. However, the point $r = 0$ is a physical singularity. Explicit calculations show that the Ricci scalar R^{μ}_{μ} vanishes everywhere, but the quantity $R_{\mu\nu}R^{\mu\nu}$ is nonvanishing. It turns out that

$$R_{\mu\nu}R^{\mu\nu} = \frac{2(3F^2(r) + 2cF(r) + c^2)}{F^4(r)B^2(r)r^4}. \tag{62}$$

At $r = 0$, the numerator does not generally vanish, but the denominator vanishes, making $R_{\mu\nu}R^{\mu\nu}$ diverge. Hence these solutions correspond to a naked singularity at $r = 0$ with no event horizon to censor it.

VII. SUMMARY AND DISCUSSION

The metric and torsion are two independent constituents of metric compatible Riemannian geometry. Because of the immense successes of torsion-free GR, torsion has not played a significant role in theories of gravity. However, when gravity is to be included with other interactions of elementary particles with intrinsic spin, a more general theory including torsion becomes imperative [3].

In the present work, torsion is incorporated in a novel way in higher dimensional Kaluza-Klein–type theories. Here torsion is not an independent degree of freedom coupled to spin; rather, it is determined in terms of the metric through a set of physically motivated constraints, which serve (i) to confine torsion to the extra dimension, leaving the 4D space-time torsion-free, and (ii) to ensure that geodesic motions in 4D remain totally unaffected by the presence of the extra dimension. These constraints have previously been imposed in terms of veilbeins [17,19], but here it is realized that they essentially impose the requirement that the fifth dimension is hidden at the level of geodesic motion. It turns out that the nonvanishing torsion components are functions of the 5D metric components with the 4D metric $g_{\mu\nu}$ obeying the so-called cylindrical condition; namely, it is independent of x^5. In the resulting geometry, all the 4D hypersurfaces are equivalent, and the 4D components of the connection and the Ricci tensor exactly match those of the standard 4D GR. Hence, at the level of geodesics, this geometry is virtually indistinguishable from that of the standard GR.

To proceed further, we derive modified Einstein equations from the action principle with the Ricci scalar as the Lagrangian density. In this respect, an alternate approach presents itself. In the Palatini formulation of GR, the action is varied independently with respect to the connection, and in the absence of torsion, the metricity follows from the equations of motion. Recently [20,21], it has been shown that even without assuming the absence of torsion, variations of the action independently with respect to the metric and the connection lead to GR equations along with metricity, provided the matter Lagrangian is not coupled to the connection. In our case, with the constraints imposed on the connection, it is more convenient and natural to impose metricity prior to action variation. Since the entire connection is determined to be a function of the metric, we only need to vary the action with respect to the metric to obtain the modified equations, making the theory a purely metric theory of gravity. However, adopting a Palatini-style approach, one could relax the assumption of metricity and vary the action independently with respect to the metric and the connection along with the imposed constraints, which might lead to a different set of modified Einstein equations.

We apply the ensuing modified Einstein equations to study the cosmology of a homogeneous-isotropic universe. In the matter dominated phase of the universe (zero pressure), we obtain a second order differential equation for the scale factor $a(t)$, in contrast to the first order differential equation in the usual FRW cosmology. In FRW cosmology,

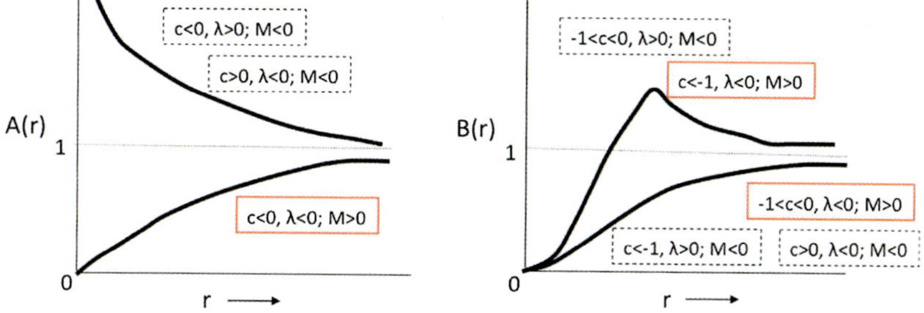

FIG. 5 (color online). Schematic behavior of $A(r)$ and $B(r)$.

the second derivative of $a(t)$ cannot be independently prescribed as an initial condition, and decelerating expansion is a necessary outcome in the absence of a cosmological constant. However, in our case, we have the choice of an initial condition for the second derivative of $a(t)$, which can be tuned to fit the observed acceleration of the universe. Figure 2 presents the behavior of $a(t)$ for various choices of the current acceleration.

In an earlier version of the present work [22], the field equations were derived differently: by varying the action with respect to the metric prior to expressing the Ricci tensor in terms of the metric. In the cosmology equations generated from those field equations, acceleration was not an independently prescribable initial condition. Chen and Jing [23] showed that those equations yield accelerating universe solutions without resorting to dark energy, and that the model not only fits the supernovae data, but also solves the cosmic age problem of old high redshift objects [24]. Whether the cosmological solutions described in the current work would fit empirical results just as well needs to be investigated.

In the case of spherically symmetric vacuum solutions to the modified Einstein equations, we find some remarkably interesting results. As is well known, in the unique Schwarzschild solution of GR, when the mass is positive, an event horizon censors the central singularity. In contrast, we find positive mass naked singularity solutions that lack an event horizon. Recently, similar positive mass solutions without a horizon have been found [25] in a simpler setting of torsion-free GR with multiple extra dimensions. It would be interesting to see how the particular vacuum solutions in the torsion induced geometry in the present case match with those solutions in the torsion-free geometry.

The existence of positive mass solutions with a naked singularity have immediate consequences on gravitational collapse, opening up the possibility of an arbitrarily large star collapsing to an arbitrarily small nonsingular state. Since trapped surfaces would not necessarily form in such collapses, finite matter pressure could be sufficient to withstand a total collapse to a singularity. This suggests a detailed analysis of such solutions by treating $-H_{\mu}{}^{\nu}$ in Eq. (27) as extra-dimensionally induced matter in standard GR. It then raises the possibility of a sufficiently strong gravitational collapse that stops short of collapsing to a singularity with a finite induced stress-energy tensor that potentially violates the weak energy condition in the region near the center. It remains to be seen if an arbitrarily small static model star with a finite stress-energy tensor can be constructed with the external geometry matching the type of solutions discussed in this paper.

In conclusion, inclusion of the torsion in the context of extra dimensions presents a novel way of obtaining modified Einstein equations that have significant physical consequences. For clarity and simplicity, we have confined the treatment to five dimensions. However, the framework can be generalized to arbitrary dimensions D, producing a torsion-free $D-1$-dimensional metric theory. Further, we could consider multiple extra dimensions and generalize the constraints so as to hide all the extra dimensions and confine the torsion in them, which would lead to a more general theory.

ACKNOWLEDGMENTS

The authors would like to thank Venky Krishnan, Ramesh Anishetty, Johannes Noller, and Stanley Deser for helpful discussions. The work was supported in part by Grant No. AFOSR-FA9550-10-1-0149 (K. H. S.) and U.S Department of Energy Grant No. DE-FG02-85ER40237 (K. C. W.).

APPENDIX A: COMPUTING THE GEOMETRIC QUANTITIES IN THE 5D MANIFOLD

In this appendix, expressions for the torsion, connection coefficients, and the Ricci tensor of the 5D geometry are computed in terms of the metric. It will be shown that the physical constraints imposed on the connection will be sufficient to determine uniquely all the components of torsion and hence the other geometric quantities in terms of the metric.

1. 5D Levi-Civita connection

The 5D Levi-Civita connection is given by

$$\hat{\Gamma}^{j}_{\cdot ik} = \tfrac{1}{2} g^{jm}(\partial_k g_{im} + \partial_i g_{km} - \partial_m g_{ik}). \tag{A1}$$

Expressing the 5D metric in terms of the 4D metric and the extra-dimensional metric fields given by Eq. (3), we find the 5D Levi-Civita connection to be

$$\begin{aligned}\hat{\Gamma}^{5}_{\cdot\mu\nu} &= \tfrac{1}{2}(\nabla_\mu A_\nu + \nabla_\nu A_\mu) + (A_\mu J_\nu + A_\nu J_\mu) \\ &\quad + \tfrac{1}{2}\epsilon\Phi^2 A^\lambda(A_\mu F_{\lambda\nu} + A_\nu F_{\lambda\mu}) + \epsilon\Phi^2 A_\mu A_\nu A^\lambda J_\lambda \\ &\quad - \tfrac{1}{2}(A^\lambda A_\lambda + \epsilon\Phi^{-2})\partial_5(\epsilon A_\mu A_\nu \Phi^2) \\ &\quad - \tfrac{1}{2}(A^\lambda A_\lambda + \epsilon\Phi^{-2})\partial_5 g_{\mu\nu},\end{aligned} \tag{A2}$$

$$\hat{\Gamma}^{\mu}_{\cdot 55} = -\epsilon\Phi^2 J^\mu + \epsilon\Phi^2\partial_5 A^\mu + \epsilon\Phi^2 A^\mu J_5 - \epsilon\Phi^2 A_\sigma \partial_5 g^{\mu\sigma}, \tag{A3}$$

$$\hat{\Gamma}^{5}_{\cdot 55} = \epsilon\Phi^2 A^\lambda(J_\lambda - \partial_5 A_\lambda) - \epsilon\Phi^2 A^\lambda A_\lambda J_5 + J_5, \tag{A4}$$

$$\begin{aligned}\hat{\Gamma}^{\mu}_{\cdot 5\nu} &= \tfrac{1}{2}\epsilon\Phi^2 F_{\nu}^{\cdot\mu} - \epsilon\Phi^2 A_\nu J^\mu + \tfrac{1}{2} g^{\mu\sigma}\partial_5(\epsilon\Phi^2 A_\sigma A_\nu) \\ &\quad + \tfrac{1}{2} g^{\mu\sigma}\partial_5 g_{\nu\sigma},\end{aligned} \tag{A5}$$

$$\begin{aligned}\hat{\Gamma}^{5}_{\cdot 5\nu} &= \tfrac{1}{2}\epsilon\Phi^2 A^\lambda F_{\lambda\nu} + \epsilon\Phi^2 A_\nu A^\lambda J_\lambda + J_\nu \\ &\quad - \tfrac{1}{2}A^\lambda\partial_5(\epsilon A_\lambda A_\nu \Phi^2) - \tfrac{1}{2}A^\lambda\partial_5 g_{\nu\lambda},\end{aligned} \tag{A6}$$

$$\hat{\Gamma}^{\lambda}_{\cdot\mu\nu} = \Gamma^{\lambda}_{\cdot\mu\nu} + \bar{\Gamma}^{\lambda}_{\cdot\mu\nu},$$
$$\bar{\Gamma}^{\lambda}_{\cdot\mu\nu} = \tfrac{1}{2}\epsilon\Phi^2(A_\nu F^{\cdot\lambda}_{\mu} + A_\mu F^{\cdot\lambda}_{\nu}) - \epsilon\Phi^2 A_\mu A_\nu J^\lambda + \tfrac{1}{2}A^\lambda\partial_5(\epsilon A_\mu A_\nu\Phi^2) + \tfrac{1}{2}A^\lambda\partial_5 g_{\mu\nu}, \tag{A7}$$

where $F_{\mu\nu} \equiv \partial_\mu A_\nu - \partial_\nu A_\mu$, $J_i \equiv \epsilon\Phi^{-1}\partial_i\epsilon\Phi$, $\Gamma^{\lambda}_{\cdot\mu\nu}$ is the 4D Levi-Civita connection obtained from the metric $g_{\mu\nu}$, and ∇_μ is the derivative operator with the 4D Levi-Civita connection. The raising and lowering of indices on $F_{\mu\nu}$, A_μ, and J_μ is performed with respect to the 4D metric.

2. Torsion components

We start with the conditions that $T^{\lambda}_{\cdot\mu\nu} = 0$ and $\tilde{\Gamma}^{\lambda}_{\cdot i5} = \tilde{\Gamma}^{\lambda}_{\cdot 5i} = 0$. Together these conditions imply $T^{\lambda}_{\cdot ij} = 0$. The remaining nonvanishing components of torsion that need to be determined are $T^{5}_{\cdot ij}$, a total of ten independent components. The conditions $\tilde{\Gamma}^{\lambda}_{\cdot 5i} = 0$, consisting of 20 equations, are sufficient to determine uniquely all the nonvanishing torsion components.

$$\tilde{\Gamma}^{\mu}_{\cdot i5} = \hat{\Gamma}^{\mu}_{\cdot i5} + K^{\mu}_{\cdot i5} = 0. \tag{A8}$$

From the above equation, the nonvanishing components of contorsion and torsion can be determined in terms of components of the 5D Levi-Civita connection $\hat{\Gamma}_{\cdot\cdot}$. First taking $i = 5$,

$$K^{\mu}_{\cdot 55} = \mathbf{g}^{\mu j}\mathbf{g}_{55}T^{5}_{\cdot j5} = \mathbf{g}^{\mu\nu}\mathbf{g}_{55}T^{5}_{\cdot\nu 5} = -\hat{\Gamma}^{\mu}_{\cdot 55}. \tag{A9}$$

From the 5D metric [Eq. (2)], we have $\mathbf{g}^{\mu\nu} = g^{\mu\nu}$. Multiplying both sides by $g_{\mu\sigma}$ and using the orthogonality relations of the metric, we obtain

$$\begin{aligned} g_{\mu\sigma}g^{\mu\nu}\mathbf{g}_{55}T^{5}_{\cdot\nu 5} &= -g_{\mu\sigma}\hat{\Gamma}^{\mu}_{\cdot 55},\\ \mathbf{g}_{55}T^{5}_{\cdot\sigma 5} &= -g_{\mu\sigma}\hat{\Gamma}^{\mu}_{\cdot 55},\\ \Rightarrow T^{5}_{\cdot\sigma 5} &= -g_{\mu\sigma}\hat{\Gamma}^{\mu}_{\cdot 55}\epsilon\Phi^{-2}. \end{aligned} \tag{A10}$$

Substituting for $\hat{\Gamma}^{\mu}_{\cdot 55}$ from Eq. (A3), we obtain

$$T^{5}_{\cdot\sigma 5} = g_{\mu\sigma}[J^\mu - \partial_5 A^\mu - A^\mu J_5 + A_\sigma\partial_5 g^{\mu\sigma}]. \tag{A11}$$

We have thus far used four equations and solved for four of the ten independent torsion components. Next, take $i = \nu$ in Eq. (A8). The contorsion components $K^{\mu}_{\cdot\nu 5}$ can be expressed in terms of the torsion components as follows:

$$\begin{aligned} K^{\mu}_{\cdot\nu 5} &= \tfrac{1}{2}\mathbf{g}^{\mu j}(\mathbf{g}_{55}T^{5}_{\cdot j\nu} + \mathbf{g}_{\nu 5}T^{5}_{\cdot j5})\\ &= \tfrac{1}{2}\mathbf{g}^{\mu 5}(\mathbf{g}_{55}T^{5}_{\cdot 5\nu}) + \tfrac{1}{2}\mathbf{g}^{\mu\sigma}(\mathbf{g}_{55}T^{5}_{\cdot\sigma\nu} + \mathbf{g}_{\nu 5}T^{5}_{\cdot\sigma 5})\\ &= \tfrac{1}{2}\epsilon A^\mu\Phi^2 T^{5}_{\cdot\nu 5} + \tfrac{1}{2}\epsilon A_\nu\Phi^2 g^{\mu\sigma}T^{5}_{\cdot\sigma 5} + \tfrac{1}{2}\epsilon\Phi^2 g^{\mu\sigma}T^{5}_{\cdot\sigma\nu}. \end{aligned} \tag{A12}$$

From $K^{\mu}_{\cdot\nu 5} = -\hat{\Gamma}^{\mu}_{\cdot\nu 5}$, and using Eq. (A10), we have

$$-2\hat{\Gamma}^{\mu}_{\cdot\nu 5} + A^\mu g_{\alpha\nu}\hat{\Gamma}^{\alpha}_{\cdot 55} + A_\nu\hat{\Gamma}^{\mu}_{\cdot 55} = \epsilon\Phi^2 g^{\mu\sigma}T^{5}_{\cdot\sigma\nu} \tag{A13}$$

and hence,

$$T^{5}_{\cdot\sigma\nu} = \epsilon\Phi^{-2}g_{\mu\sigma}[-2\hat{\Gamma}^{\mu}_{\cdot\nu 5} + A^\mu g_{\lambda\nu}\hat{\Gamma}^{\lambda}_{\cdot 55} + A_\nu\hat{\Gamma}^{\mu}_{\cdot 55}]. \tag{A14}$$

There are 16 equations. Though the torsion in the left-hand side (l.h.s.) is antisymmetric, the r.h.s. is a combination of symmetric and antisymmetric terms. Substituting for the 5D Levi-Civita connection from Sec. A 1, the above equation has the form

$$T^{5}_{\cdot\sigma\nu} = [\partial_\sigma A_\nu - \partial_\nu A_\sigma] + [J_\sigma A_\nu - J_\nu A_\sigma] + [\epsilon\Phi^{-2}\partial_5 g_{\sigma\nu} + A_\sigma A^\lambda\partial_5 g_{\lambda\nu} + A_\nu A^\lambda\partial_5 g_{\lambda\sigma}]. \tag{A15}$$

The first two terms on the r.h.s. above are antisymmetric in σ and ν, while the third term is symmetric. The antisymmetry of torsion implies that the symmetric terms on the r.h.s. must be zero, and hence

$$\partial_5 g_{\sigma\nu} = 0. \tag{A16}$$

Consequently,

$$T^{5}_{\cdot\sigma\nu} = [\partial_\sigma A_\nu - \partial_\nu A_\sigma] + [J_\sigma A_\nu - J_\nu A_\sigma]. \tag{A17}$$

Thus the 20 equations of the imposed condition [Eq. (A8)] have determined all ten independent nonvanishing components of the torsion [Eqs. (A11) and (A17)] and, in addition, have imposed a constraint on the ten independent components of the 4D metric $g_{\sigma\nu}$, making them independent of x^5.

3. The contorsion, connection, and Ricci tensor

Some components of contorsion are directly prescribed by the imposed condition, namely, $K^{\mu}_{\cdot i5} = K^{\mu}_{\cdot 5i} = -\hat{\Gamma}^{\mu}_{\cdot i5}$. The remaining components of the contorsion can be calculated from the torsion components by using Eq. (7). Since $T^{5}_{\cdot ij}$ are the only nonvanishing components of torsion, it follows that

$$\begin{aligned} K^{\mu}_{\cdot i5} &= K^{\mu}_{\cdot 5i} = -\hat{\Gamma}^{\mu}_{\cdot i5},\\ K^{5}_{\cdot i5} &= \tfrac{1}{2}T^{5}_{\cdot i5} + \tfrac{1}{2}\mathbf{g}^{5j}(\mathbf{g}_{i5}T^{5}_{\cdot j5} + \mathbf{g}_{55}T^{5}_{\cdot ji}),\\ K^{5}_{\cdot 5i} &= K^{5}_{\cdot i5} + T^{5}_{\cdot 5i},\\ K^{\mu}_{\cdot\nu\lambda} &= \tfrac{1}{2}\mathbf{g}^{\mu j}(\mathbf{g}_{\nu 5}T^{5}_{\cdot j\lambda} + \mathbf{g}_{\lambda 5}T^{5}_{\cdot j\nu}),\\ K^{5}_{\cdot\nu\lambda} &= \tfrac{1}{2}T^{5}_{\cdot\nu\lambda} + \tfrac{1}{2}\mathbf{g}^{5j}(\mathbf{g}_{\nu 5}T^{5}_{\cdot j\lambda} + \mathbf{g}_{\lambda 5}T^{5}_{\cdot j\nu}). \end{aligned} \tag{A18}$$

Substituting for the nonvanishing torsion components given by Eqs. (A11) and (A17), along with the requirement that the 4D metric is independent of the fifth dimension [Eq. (A16)], the components of contorsion are found to be

$$K^{\mu}_{\cdot i5} = K^{\mu}_{\cdot 5i} = -\hat{\Gamma}^{\mu}_{\cdot i5},$$
$$K^{5}_{\cdot i5} = -\hat{\Gamma}^{5}_{\cdot i5} + J_i, \quad K^{5}_{\cdot 5i} = K^{5}_{\cdot i5} + T^{5}_{\cdot 5i},$$
$$K^{\mu}_{\cdot \nu\lambda} = \epsilon\Phi^2 F^{\mu}_{\cdot(\lambda}A_{\nu)} + A_\lambda A_\nu J^\mu \epsilon\Phi^2 - \tfrac{1}{2}A^\mu \partial_5(\epsilon A_\nu A_\lambda \Phi^2),$$
$$K^{5}_{\cdot\nu\lambda} = -\epsilon\Phi^2 A^\sigma F_{\sigma(\lambda}A_{\nu)} - (A_\sigma J^\sigma)\epsilon A_\nu A_\lambda \Phi^2 + \tfrac{1}{2}(A_\sigma A^\sigma + \epsilon\Phi^{-2})\partial_5(\epsilon A_\nu A_\lambda \Phi^2) - A_\nu J_\lambda + \tfrac{1}{2}F_{\nu\lambda}. \tag{A19}$$

Now from Eq. (5), we obtain all the connection coefficients,

$$\begin{aligned} \tilde{\Gamma}^{5}_{\cdot\mu\nu} &= \nabla_\mu A_\nu + J_\mu A_\nu, \\ \tilde{\Gamma}^{5}_{\cdot 5\mu} &= \partial_5 A_\mu + J_5 A_\mu, \\ \tilde{\Gamma}^{5}_{\cdot\mu 5} &= J_\mu, \qquad \tilde{\Gamma}^{5}_{\cdot 55} = J_5, \\ \tilde{\Gamma}^{\lambda}_{\cdot\mu\nu} &= \Gamma^{\lambda}_{\cdot\mu\nu}, \\ \tilde{\Gamma}^{\mu}_{\cdot 55} &= \tilde{\Gamma}^{\mu}_{\cdot\nu 5} = \tilde{\Gamma}^{\mu}_{\cdot 5\nu} = 0. \end{aligned} \tag{A20}$$

Taking $i = \mu$ and $k = \nu$ in the Ricci tensor defined by Eq. (15), we have

$$\begin{aligned} \tilde{R}_{\mu\nu} = &+ \partial_\nu \tilde{\Gamma}^{\sigma}_{\cdot\sigma\mu} - \partial_\sigma \tilde{\Gamma}^{\sigma}_{\cdot\nu\mu} + \tilde{\Gamma}^{\sigma}_{\cdot\nu\lambda}\tilde{\Gamma}^{\lambda}_{\cdot\sigma\mu} - \tilde{\Gamma}^{\sigma}_{\cdot\sigma\lambda}\tilde{\Gamma}^{\lambda}_{\cdot\nu\mu} \\ &+ \partial_\nu \tilde{\Gamma}^{5}_{\cdot 5\mu} - \partial_5 \tilde{\Gamma}^{5}_{\cdot\nu\mu} + \tilde{\Gamma}^{5}_{\cdot\nu\lambda}\tilde{\Gamma}^{\lambda}_{\cdot 5\mu} - \tilde{\Gamma}^{5}_{\cdot 5\lambda}\tilde{\Gamma}^{\lambda}_{\cdot\nu\mu} \\ &+ \tilde{\Gamma}^{\sigma}_{\cdot\nu 5}\tilde{\Gamma}^{5}_{\cdot\sigma\mu} - \tilde{\Gamma}^{\sigma}_{\cdot\sigma 5}\tilde{\Gamma}^{5}_{\cdot\nu\mu} + \tilde{\Gamma}^{5}_{\cdot\nu 5}\tilde{\Gamma}^{5}_{\cdot 5\mu} - \tilde{\Gamma}^{5}_{\cdot 55}\tilde{\Gamma}^{5}_{\cdot\nu\mu}. \end{aligned} \tag{A21}$$

Since $\tilde{\Gamma}$. is the same as Γ. when all the indices are four dimensional, the first line is clearly the 4D Ricci tensor. The terms in the subsequent lines can be reexpressed in terms of the 4D covariant derivative operator as follows:

$$\tilde{R}_{\mu\nu} = R_{\mu\nu} + \nabla_\nu \tilde{\Gamma}^{5}_{\cdot 5\mu} - \partial_5 \tilde{\Gamma}^{5}_{\cdot\nu\mu} - \tilde{\Gamma}^{5}_{\cdot 55}\tilde{\Gamma}^{5}_{\cdot\nu\mu} + \tilde{\Gamma}^{5}_{\cdot\nu 5}\tilde{\Gamma}^{5}_{\cdot 5\mu}. \tag{A22}$$

Substituting for the connection $\tilde{\Gamma}$. from Eq. (A21), we find after some algebra this can be simplified as

$$\begin{aligned} \tilde{R}_{\mu\nu} &= R_{\mu\nu} + \nabla_\nu(\partial_5 A_\mu + J_5 A_\mu) - \partial_5(\nabla_\nu A_\mu + J_\nu A_\mu) \\ &\quad - J_5(\nabla_\nu A_\mu + J_\nu A_\mu) + J_\nu(\partial_5 A_\mu + J_5 A_\mu) \\ &= R_{\mu\nu} + A_\lambda \partial_5 \Gamma^{\lambda}_{\cdot\mu\nu} \\ &= R_{\mu\nu}. \end{aligned} \tag{A23}$$

Similarly, the other components of the Ricci tensor are found to be

$$\tilde{R}_{\mu 5} = \partial_5 \tilde{\Gamma}^{\sigma}_{\cdot\sigma\mu} = 0, \tag{A24}$$

$$\tilde{R}_{5\nu} = \partial_\nu \tilde{\Gamma}^{5}_{\cdot 55} - \partial_5 \tilde{\Gamma}^{5}_{\cdot\nu 5} = 0, \tag{A25}$$

$$\tilde{R}_{55} = 0. \tag{A26}$$

Note that neither the connection nor the Ricci tensor depends on the signature (ϵ) of the fifth dimension.

APPENDIX B: COMPUTING THE MODIFIED EINSTEIN TENSOR

Here we provide some intermediate steps to go from Eqs. (24) and (25) to Eq. (26) and obtain a simplified expression for $H_{\mu\nu}$. With $B_\nu \equiv T^{5}_{\cdot\nu 5}$ and $H_{\mu\nu}$ defined by

$$\int B_\nu[g^{\mu\alpha}\delta\Gamma^{\nu}_{\cdot\mu\alpha} - g^{\mu\nu}\delta\Gamma^{\alpha}_{\cdot\alpha\mu}]\sqrt{-\mathbf{g}}d^5x = \int H_{\mu\nu}\delta g^{\mu\nu}\sqrt{-\mathbf{g}}d^5x, \tag{B1}$$

we show that

$$H_{\mu\nu} = \nabla_{(\mu}B_{\nu)} - (\nabla\cdot B)g_{\mu\nu} + J_{(\mu}B_{\nu)} - (J\cdot B)g_{\mu\nu}. \tag{B2}$$

Proof.—Consider the first term in the integrand of the l.h.s. of Eq. (B1),

$$\begin{aligned} 2g^{\mu\alpha}\delta\Gamma^{\nu}_{\cdot\mu\alpha} = &\, g^{\mu\alpha}g^{\nu\lambda}(\partial_\mu\delta g_{\lambda\alpha} + \partial_\alpha\delta g_{\lambda\mu} - \partial_\lambda\delta g_{\mu\alpha}) \\ &+ g^{\mu\alpha}(\partial_\mu g_{\lambda\alpha} + \partial_\alpha g_{\lambda\mu} - \partial_\lambda g_{\mu\alpha})\delta g^{\nu\lambda}. \end{aligned} \tag{B3}$$

The variations of the covariant metric in the above equation can be reexpressed in terms of the variations of the contravariant metric using the identity $\delta g_{\alpha\lambda} = -g_{\alpha\mu}g_{\lambda\nu}\delta g^{\mu\nu}$. We note that

$$\begin{aligned} g^{\mu\alpha}g^{\nu\lambda}\partial_\mu\delta g_{\lambda\alpha} &= -g^{\nu\lambda}\partial_\mu g_{\lambda\sigma}(\delta g^{\sigma\mu}) \\ &\quad - g^{\mu\alpha}\partial_\mu g_{\beta\alpha}(\delta g^{\nu\beta}) - \partial_\mu(\delta g^{\nu\mu}), \\ g^{\mu\alpha}g^{\nu\lambda}\partial_\alpha\delta g_{\lambda\mu} &= g^{\mu\alpha}g^{\nu\lambda}\partial_\mu\delta g_{\lambda\alpha}, \\ -g^{\mu\alpha}g^{\nu\lambda}\partial_\lambda\delta g_{\mu\alpha} &= g^{\nu\lambda}\partial_\lambda g_{\mu\sigma}(\delta g^{\sigma\mu}) + g^{\nu\lambda}\partial_\lambda g_{\beta\sigma}(\delta g^{\sigma\beta}) \\ &\quad + g^{\nu\lambda}g_{\beta\sigma}\partial_\lambda(\delta g^{\sigma\beta}). \end{aligned}$$

Taken together, we obtain

$$\begin{aligned} 2g^{\mu\alpha}\delta\Gamma^{\nu}_{\cdot\mu\alpha} = &-2g^{\nu\lambda}\partial_\mu g_{\lambda\sigma}(\delta g^{\sigma\mu}) - 2\partial_\mu(\delta g^{\nu\mu}) \\ &+ 2g^{\nu\lambda}\partial_\lambda g_{\mu\sigma}(\delta g^{\sigma\mu}) + g^{\nu\lambda}g_{\beta\sigma}\partial_\lambda(\delta g^{\sigma\beta}) \\ &- g^{\mu\alpha}\partial_\lambda g_{\mu\alpha}(\delta g^{\nu\lambda}). \end{aligned} \tag{B4}$$

Next consider the second term in the integrand of the l.h.s. of Eq. (B1). Since $2\Gamma^{\alpha}_{\cdot\alpha\mu} = g^{\alpha\lambda}\partial_\mu g_{\alpha\lambda}$,

$$\begin{aligned} 2\delta\Gamma^{\alpha}_{\cdot\alpha\mu} &= (\delta g^{\alpha\lambda})\partial_\mu g_{\alpha\lambda} + g^{\alpha\lambda}\partial_\mu(\delta g_{\alpha\lambda}) \\ &= -(\delta g^{\alpha\lambda})\partial_\mu g_{\alpha\lambda} - g_{\alpha\lambda}\partial_\mu(\delta g^{\alpha\lambda}). \end{aligned} \tag{B5}$$

Using Eqs. (B4) and (B5), we find

$$\begin{aligned} &B_\nu[g^{\mu\alpha}\delta\Gamma^{\nu}_{\cdot\mu\alpha} - g^{\mu\nu}\delta\Gamma^{\alpha}_{\cdot\alpha\mu}]\sqrt{-\mathbf{g}} \\ &= B_\nu[-\partial_\mu(\delta g^{\mu\nu}) + g^{\mu\nu}g_{\alpha\beta}\partial_\mu(\delta g^{\alpha\beta})]\sqrt{-\mathbf{g}} \\ &\quad + B_\nu[-g^{\nu\lambda}\partial_\mu g_{\lambda\sigma}(\delta g^{\sigma\mu}) + \tfrac{3}{2}g^{\nu\lambda}\partial_\lambda g_{\mu\sigma}(\delta g^{\sigma\mu}) \\ &\quad - \tfrac{1}{2}g^{\mu\sigma}\partial_\lambda g_{\mu\sigma}(\delta g^{\nu\lambda})]\sqrt{-\mathbf{g}}. \end{aligned} \tag{B6}$$

The first term in the r.h.s. of the above equation contains terms with the derivatives of the variation. We note that these terms are eventually going to be integrated. By integrating them by parts and ignoring the boundary terms, the above equation takes the form

$$\begin{aligned}
&B_\nu[g^{\mu\alpha}\delta\Gamma^{\nu}_{\cdot\mu\alpha} - g^{\mu\nu}\delta\Gamma^{\alpha}_{\cdot\alpha\mu}]\sqrt{-\mathbf{g}} \\
&= [\partial_\mu(B_\nu\sqrt{-\mathbf{g}})\delta g^{\mu\nu} - \partial_\mu(g^{\mu\nu}g_{\alpha\beta}B_\nu\sqrt{-\mathbf{g}})\delta g^{\alpha\beta}] \\
&\quad + B_\nu[-g^{\nu\lambda}\partial_\mu g_{\lambda\sigma}(\delta g^{\sigma\mu}) + \tfrac{3}{2}g^{\nu\lambda}\partial_\lambda g_{\mu\sigma}(\delta g^{\sigma\mu}) \\
&\quad - \tfrac{1}{2}g^{\mu\sigma}\partial_\lambda g_{\mu\sigma}(\delta g^{\nu\lambda})]\sqrt{-\mathbf{g}}. \qquad (B7)
\end{aligned}$$

To simplify the r.h.s. of Eq. (B7) it is useful to note the following identities:

$$\partial_\mu(g^{\mu\nu}) = g^{\sigma\nu}g_{\sigma\lambda}\partial_\mu(g^{\mu\lambda}) = -g^{\sigma\nu}g^{\mu\lambda}\partial_\mu(g_{\sigma\lambda}), \quad (B8)$$

$$\partial_\mu(\sqrt{-\mathbf{g}}) = \frac{\partial_\mu \mathbf{g}}{2\mathbf{g}}\sqrt{-\mathbf{g}} = \frac{\sqrt{-\mathbf{g}}}{2}\mathbf{g}^{ij}\partial_\mu(\mathbf{g}_{ij}), \quad (B9)$$

$$\begin{aligned}
\mathbf{g}^{ij}\partial_\mu(\mathbf{g}_{ij}) &= g^{\sigma\lambda}\partial_\mu(g_{\sigma\lambda}) + g^{\sigma\lambda}\partial_\mu(\epsilon A_\sigma A_\lambda\Phi^2) \\
&\quad + 2\mathbf{g}^{\sigma 5}\partial_\mu(\epsilon A_\sigma\Phi^2) + \mathbf{g}^{55}\partial_\mu(\epsilon\Phi^2) \\
&= g^{\sigma\lambda}\partial_\mu(g_{\sigma\lambda}) + 2J_\mu. \qquad (B10)
\end{aligned}$$

Using the above identities, the r.h.s of Eq. (B7) becomes

$$\begin{aligned}
&B_\nu[g^{\mu\alpha}\delta\Gamma^{\nu}_{\cdot\mu\alpha} - g^{\mu\nu}\delta\Gamma^{\alpha}_{\cdot\alpha\mu}]\sqrt{-\mathbf{g}} \\
&= [\partial_\mu(B_\nu)\delta g^{\mu\nu} - \partial_\mu(B_\nu)g^{\mu\nu}g_{\alpha\beta}\delta g^{\alpha\beta} \\
&\quad + B_\nu g_{\alpha\beta}g^{\sigma\nu}g^{\mu\lambda}\partial_\mu(g_{\sigma\lambda})\delta g^{\alpha\beta} \\
&\quad - \tfrac{1}{2}B_\nu g^{\mu\nu}g_{\alpha\beta}g^{\sigma\lambda}\partial_\mu(g_{\sigma\lambda})\delta g^{\alpha\beta} \\
&\quad - B_\nu g^{\nu\sigma}\partial_\mu(g_{\sigma\alpha})\delta g^{\alpha\mu} + \tfrac{1}{2}B_\nu g^{\nu\sigma}\partial_\sigma(g_{\alpha\mu})\delta g^{\alpha\mu} \\
&\quad + B_\nu J_\mu\delta g^{\mu\nu} - B_\nu J_\mu g^{\mu\nu}g_{\alpha\beta}\delta g^{\alpha\beta}]\sqrt{-\mathbf{g}}. \qquad (B11)
\end{aligned}$$

Rewriting the derivatives of the metric in terms of the Levi-Civita connection, we find

$$\begin{aligned}
&B_\nu[g^{\mu\alpha}\delta\Gamma^{\nu}_{\cdot\mu\alpha} - g^{\mu\nu}\delta\Gamma^{\alpha}_{\cdot\alpha\mu}]\sqrt{-\mathbf{g}} \\
&= [\partial_\mu(B_\nu)\delta g^{\mu\nu} - \partial_\mu(B_\nu)g^{\mu\nu}g_{\alpha\beta}\delta g^{\alpha\beta} \\
&\quad - B_\nu\Gamma^{\nu}_{\cdot\mu\sigma}\delta g^{\sigma\mu} + B_\nu\Gamma^{\nu}_{\cdot\mu\sigma}g^{\sigma\mu}g_{\alpha\beta}\delta g^{\alpha\beta} \\
&\quad + B_\nu J_\mu\delta g^{\mu\nu} - B_\nu J_\mu g^{\mu\nu}g_{\alpha\beta}\delta g^{\alpha\beta}]\sqrt{-\mathbf{g}} \\
&= [\nabla_\mu B_\nu - (\nabla\cdot B)g_{\mu\nu} + J_\mu B_\nu \\
&\quad - (J\cdot B)g_{\mu\nu}]\delta g^{\mu\nu}\sqrt{-\mathbf{g}}. \qquad (B12)
\end{aligned}$$

Since the variation $\delta g^{\mu\nu}$ is symmetric in the indices μ and ν, only the symmetric part of the r.h.s. of the above equation will contribute to the equations of motion. Hence $H_{\mu\nu}$ will be given by Eq. (B2).

[1] C.M. Will, Living Rev. Relativity **9**, 3 (2006), http://relativity.livingreviews.org/Articles/lrr-2006-3.
[2] R.H. Dicke, *The Theoretical Significance of Experimental Relativity* (Gordon and Breach, New York, 1964).
[3] F.W. Hehl, P.v.d. Hyde, G.D. Kerlick, and J.M. Nester, Rev. Mod. Phys. **48**, 393 (1976).
[4] T. Watanabe and M.J. Hayashi, arXiv:gr-qc/0409029.
[5] I.L. Shapiro, Phys. Rep. **357**, 113 (2002).
[6] S. Deser and B. Zumino, Phys. Lett. **62B**, 335 (1976).
[7] N. Poplawski, Gen. Relativ. Gravit. **44**, 1007 (2012).
[8] A. Silvestri and M. Trodden, Rep. Prog. Phys. **72**, 096901 (2009).
[9] T.P. Sotiriou and V. Faraoni, Rev. Mod. Phys. **82**, 451 (2010).
[10] G.J. Olmo, Int. J. Mod. Phys. D **20**, 413 (2011).
[11] R. Maartens and K. Koyama, Living Rev. Relativity **13**, 5 (2010), http://relativity.livingreviews.org/Articles/lrr-2010-5.
[12] G.R. Dvali, G. Gabadadze, and M. Porrati, Phys. Lett. B **485**, 208 (2000).
[13] L. Randall and R. Sundrum, Phys. Rev. Lett. **83**, 3370 (1999).
[14] J.M. Overduin and P.S. Wesson, Phys. Rep. **283**, 303 (1997).
[15] M.W. Kalinowski, Int. J. Theor. Phys. **20**, 563 (1981).
[16] M. Nakahara, *Geometry, Topology and Physics* (IOP Publishing, London, 2003), 2nd ed., Secs. 7.8.1 and 7.8.4.
[17] K.H. Shankar and K.C. Wali, Mod. Phys. Lett. A **25**, 2121 (2010).
[18] R.M. Wald, *General Relativity* (University of Chicago Press, Chicago, 1984), Sec. 5.2.
[19] N.A. Viet and K.C. Wali, Phys. Rev. D **67**, 124029 (2003).
[20] T.P. Sotiriou, Classical Quantum Gravity **26**, 152001 (2009).
[21] N. Dadhich and J.M. Pons, arXiv:1010.0869v3.
[22] K.H. Shankar and K.C. Wali, arXiv:0904.4038v1 (2009).
[23] S. Chen and J. Jing, J. Cosmol. Astropart. Phys. 09 (2009) 001.
[24] A.C.S. Friaca, J.S. Alcaniz, and J.A.S. Lima, Mon. Not. R. Astron. Soc. **362**, 1295 (2005).
[25] S. Kalyana Rama, arXiv:1111.1897v3.

Einstein-Yang-Mills-Dirac systems from the discretized Kaluza-Klein theory

Nguyen Ai Viet,[1,2] Nguyen Van Dat,[1,2] Nguyen Suan Han,[2] and Kameshwar C. Wali[3]
[1]*ITI, Vietnam National University, Hanoi, Vietnam*
[2]*Physics Department, College of Natural Sciences, Vietnam National University, Hanoi, Vietnam*
[3]*Physics Department, Syracuse University, Syracuse, New York, USA*
(Dated: March 8, 2018)

A unified theory of the non-Abelian gauge interactions with gravity in the framework of a discretized Kaluaza-Kleine theory is constructed with a modified Dirac operator and wedge product. All the couplings of chiral spinors to the non-Abelian gauge fields emerge naturally as components of the couplings of the chiral spinors to the generalized gravity together with some new interactions. In particular, the currently prevalent gravity-QCD-quark and gravity-electroweak-quark-lepton models are shown to follow as special cases of the general framework.

I. INTRODUCTION

The discretized Kaluza-Klein theories(DKKT) have provided a new framework to yield a non-supersymmetric, perturbatively stable Standard Model at the TeV scale. The gauge theory of the electroweak interactions, where the symmetry is broken spontaneously with a quartic Higgs field potential, is shown to follow from a dimensional destruction by discretizing the extra internal dimensions [1, 2]. The framework and techniques can also be extended to construct an effective multi-gravity theory with finite massive gravitons up to the Planck scale [3, 4]. Recently, the ghost-free models of massive gravity and their multigravity extensions follow from considering DKKT [5]

In fact, DKKT had been proposed sometime ago [6] and developed further in [7, 8], utilizing the mathematical concepts of noncommutative geometry (NCG) *à la Connes* [9]. In this framework the compact fifth dimension of the conventional Kaluza-Klein theory [10] is replaced by two discrete points. The generalized Cartan-Einstein-Hilbert action leads to the same zero mode sector of the ordinary Kaluza-Klein theory consisting of gravity, electromagnetism and a Brans-Dicke scalar [6]. The advantage of the discretized version over the ordinary one is that it avoids the existence of the infinite towers of massive fields, which might lead to both theoretical and observational obstacles. In fact, the torsion free version [7] of this theory contains just two dilatons in addition to the zero mode sector. A more general metric compatible theory [8] leads to pairs of tensor, vector and scalar fields. In each pair, one field is massless and the other is massive.

The space-time of DKKT can also be viewed as two parallel copies of the conventional one. Connes and Lott [11] proposed a two sheeted space-time of $\mathcal{M}^4 \times Z_2$ as a realistic approach to formulate the Standard Model of electroweak and strong interactions of chiral quarks and leptons. In their approach , based on NCG, the gauge fields and the scalar Higgs fields appear on the same footing. The requires quartic Higgs potential emerges naturally to trigger the spontaneous symmetry breaking. It has many new features leading to the predictions of the parameters of the SM such as Higgs and Top quark masses [12] to name just a few. These and the inclusion of gravity with significant new results leads us to pursue the more exciting perspective of a new concept of space-time that can lay the foundation of for all the existing models and pave the way to new physics.

Along this direction, the couplings of the matter to gravity and gauge fields have been discussed in [13, 14]. Since the gauge vectors in the gravity sector are abelian, the nonabelian gauge fields have been introduced into the theory in an independent term. Since the ordinary Kaluza-Klein theory can be generalized to contain the nonabelian gauge vectors, one expect that the same situation is possible in the discretized one. Recently, Viet and Du [15] have shown that DKKT can be generalized further to have the nonabelian gauge fields as components of gravity. Interestingly, the generalization is possible only with certain limited gauge symmetries, including those of the strong and weak interactions. In other words, the space-time extended by discrete dimensions can provide explanations for the chiral symmetry of the strong interaction and for the absence of the coupling between the right handed chiral spinors and the nonabelian gauge bosons of the weak interaction. This result can be interpreted as an explanation for a model to unify all the known interactions and Higgs fields as components of gravity in the new space-time of $\mathcal{M}^4 \times Z_2 \times Z_2$ [16, 17].

Although the perspective of a complete, satisfactory unified theory of all interactions is fascinating, a unified mathematical framework for gravity and other interactions seems to be the necessary first step. This has been the main purpose of our paper. So far the approaches in DKKT theories and the construction of spontaneously broken Yang-Mills theories differ in details in formulating the necessary differential forms and constructing the necessary quartic Higgs potential, sometimes appearing ad hoc. This is avoided in the current framework, providing a consistent framework bringing together the discretized extra dimension lending to non-Abelian gauge fields and the four-dimensional continuum space-time for gravity.

The paper is organized as follows: In the next section, we set up the basic mathematical framework of NCG that generalizes the conventional differential calculus to DKKT space-time and recovers the results of Connes-Lott and other DKKT models. Section III is devoted to the gravity sector and the derivation of the Cartan-Einstein-Hilbert action including non-vanishing torsion terms subject to a set of constraints [8, 18]. In Section IV, we discuss the couplings of the chiral spinors (matter fields) to the vector and scalar fields of the generalized gravity in the extended space-time. The final section is devoted to a summary and discussion of the results.

II. GENERAL FRAMEWORK OF THE DKKT NCG SPACE-TIME

In this paper we will confine ourselves with a discretized fifth extra-dimension consisting of two points, keeping in mind that it is straight forward to generalize it to the case of arbitrary N points.

A. The spectral triple

The generalized functions now are a pair of the conventional ones defined at two different internal points. In both NCG and dimensional deconstruction approaches to DKKT [1–8], the derivation over the discrete dimension is defined as the difference of this pair of functions multiplied by a mass parameter m that carries the inverse dimension of distance between the two points.

The dimensional deconstruction approach defines the Lagrangian based on the intuitive concepts of the site and link interactions, while the NCG formalism is based on the geometric concepts.

In the general framework of NCG, the concept of space-time is freed from the notions of coordinates and derivations along the coordinates. Instead, it is defined by the spectral triple as its building blocks: a Hilbert space of the wave functions of matter fields, an algebra of fields, an algebra of functions that act as operators on the Hilbert space and an operator called as Dirac operator that encodes all the information about the space-time manifold. In LVW approach to NCG, the concept of derivative over the discrete dimension has been included in the Dirac operator to construct the geometric motions in a more transparent analogy with the conventional Riemannian geometry.

In this paper, we will adopt the Connes-Lott's two sheeted space time [11] of left and right handed chiral spinors generalized to our two arbitrary sheets [20].

i) The Hilbert space $\mathcal{H} = \mathcal{H}_L \oplus \mathcal{H}_R$. Thus, the generalized spinors in our model are direct sums of two chiral spinors of the 4-dimensions,

$$\Psi = \begin{bmatrix} \psi_L \\ \psi_R \end{bmatrix} \tag{1}$$

ii) The algebra of the function operators $\mathcal{A} = \mathcal{A}_L \oplus \mathcal{A}_R$, where $\mathcal{A}_I = \mathcal{C}^\infty(\mathcal{M}^4), I = L, R$, whose elements can be represented as the diagonal 0-form matrices $\mathcal{F}$, with $\mathcal{F}$ being elements of $\mathcal{A}$, are real functions, where

$$F(x) = \begin{bmatrix} f_L(x) & 0 \\ 0 & f_R(x) \end{bmatrix}. \tag{2}$$

iii) The Dirac operator $D = d.\mathrm{e} + \Theta$ is an extension of the usual Dirac operator $d = \gamma^\mu \partial_\mu$ in the ordinary space-time $\mathcal{M}^4$, defined by

$$D = \begin{bmatrix} \gamma^\mu \partial_\mu & -im\gamma^5 \\ im\gamma^5 & \gamma^\mu \partial_\mu \end{bmatrix}, \quad \Theta = \begin{bmatrix} 0 & -im\gamma^5 \\ im\gamma^5 & 0 \end{bmatrix}, \tag{3}$$

whose action on the generalized function F leads to

$$\begin{aligned} DF &= [D, F] \\ &= \begin{bmatrix} \gamma^\mu \partial_\mu f_L(x) & im\gamma^5(f_L(x) - f_R(x)) \\ im\gamma^5(f_L(x) - f_R(x)) & \gamma^\mu \partial_\mu f_R(x) \end{bmatrix}, \end{aligned} \tag{4}$$

The new Dirac operator leads to the desired Lagrangian of the massive and massless chiral spinors as follows

$$\mathcal{L}_\Psi = i\bar{\Psi} D \Psi = i\bar{\psi}_L \gamma^\mu \partial_\mu \psi_L + i\bar{\psi}_R \gamma^\mu \partial_\mu \psi_R + m\bar{\psi}_L \psi_R + m\bar{\psi}_R \psi_L = \bar{\psi}(i\gamma^\mu \partial_\mu + m)\psi, \tag{5}$$

B. The generalized 1-forms

Following the conventional geometry, we rewrite the exterior derivative of 0-forms in Eq.(4) in the form

$$\begin{aligned} DF &= DX^\mu [D_\mu, F(x)] + DX^5 \sigma^\dagger [D_5, F] \\ &= DX^\mu \partial_\mu F(x) + DX^5 \sigma^\dagger \partial_5 F, \end{aligned} \tag{6}$$

where

$$D_5 = \begin{bmatrix} 0 & -m \\ m & 0 \end{bmatrix}, \quad \sigma^\dagger = \begin{bmatrix} 0 & -1 \\ 1 & 0 \end{bmatrix}. \tag{7}$$

In the Dirac matrix representation, the differential elements DX^μ are represented by the following generalized

γ-matrices

$$\Gamma^\mu = \begin{bmatrix} \gamma^\mu & 0 \\ 0 & \gamma^\mu \end{bmatrix}, \Gamma^5 = \begin{bmatrix} 0 & i\gamma^5 \\ -i\gamma^5 & 0 \end{bmatrix}. \quad (8)$$

The general 1-forms U then take the form

$$U = \Gamma^\mu U_\mu + \Gamma^5 U_5 = \begin{bmatrix} \gamma^\mu u_{\mu L} & i\gamma^5 u_{5R} \\ -i\gamma^5 u_{5L} & \gamma^\mu u_{\mu R}, \end{bmatrix} \quad (9)$$

where U_M are generalized functions. So, the 1-form U contains two vectors and two scalar functions.

The exterior derivative of a function given in Eq.(4) is also a 1-form with a special condition $u_{5L} = -u_{5R}$.

For incorporating gravity, we are in particular interested in hermitian 1-forms U, with the condition $u_{5L} = u_{5R}$. Obviously, in our framework the hermitian 1-forms automatically are not derivatives of a 0-form.

The fifth derivative in Eq.(6) has been used in the "link" interaction terms of the dimensional deconstruction approach [1, 2, 5] as an ad hoc derivative over the discrete dimension.

C. Wedge products and 2-forms

To construct 2-forms from the derivative of 1-forms, one adopts generally in the continuum 4-D space, the completely anti-symmetric wedge product to avoid the so called junk-forms (for details see [9]). In our case, with the extra discretized internal space, we have a choice that leads us to the following definitions for the wedge products in Eq.(10). As we shall see, in addition to the gauge and scalar fields being on the same footings, it leads naturally to a quartic Higgs potential that is required for spontaneous symmetry breaking.

$$\begin{aligned} DX^\mu \wedge DX^\nu &= -DX^\nu \wedge DX^\mu \\ DX^\mu \wedge DX^5 &= -DX^5 \wedge DX^\mu \\ DX^5 \wedge DX^5 &\neq 0, \end{aligned} \quad (10)$$

$$\begin{aligned} (U \wedge V)_{\mu\nu} &= -(U \wedge V)_{\nu\mu} = \frac{1}{2}(U_\mu V_\nu - U_\nu V_\mu) \\ (U \wedge V)_{\mu 5} &= -(U \wedge V)_{5\mu} = \frac{1}{2}(\tilde{U}_\mu V_5 - U_5 V_\mu) \\ (U \wedge V)_{55} &= \tilde{U}_5 V_5 \end{aligned} \quad (11)$$

where the "$\sim$" operation is defined as

$$\begin{aligned} \tilde{F} &= f_+ \mathbf{e} - f_- \mathbf{r}, \; f_\pm = \frac{1}{2}(f_1 \pm f_2) \\ \mathbf{e} &= \begin{bmatrix} 1 & 0 \\ 0 & 1 \end{bmatrix}, \mathbf{r} = \begin{bmatrix} 1 & 0 \\ 0 & -1 \end{bmatrix} \end{aligned} \quad (12)$$

The exterior derivative of 1-form is given as follows

$$\begin{aligned} DU &= [D, U] = DX^M \wedge DX^N (DU)_{MN} \quad (13) \\ (DU)_{\mu\nu} &= \frac{1}{2}(\partial_\mu U_\nu - \partial_\nu U_\mu) = -(DU)_{\nu\mu} \\ (DU)_{\mu 5} &= \frac{1}{2}(\partial_\mu U_5 - m(U_\mu - \tilde{U}_\mu)) = -(DU)_{5\mu} \\ (DU)_{55} &= -m(U_{5L} + U_{5R}) \quad (14) \end{aligned}$$

It is straightforward to verify the defined wedge products lead to the de Rhams condition $D^2 = 0$ to avoid the junk forms. Also, as said before, it is motivated by the required condition that the Higgs Lagrangian contain a quartic potential.

III. SETTING UP THE CURVED SPACE-TIME

A. Vielbein and metric

Einsteins general theory of gravity begins with the assumption of a locally flat inertial system built around an orthonormal coordinate system described in terms of vierbein. Consistent with the principle of equivalence, it allows to define the metric that defines the distance between two points and affine connections and extend to curved space time. Vierbein formulation also provides the necessary spin connections to couple fermionic matter fields. We can generalize this concept to NCG by defining a locally flat frame E^A and its curvilinear one DX^M.

$$\begin{aligned} E^A = DX^M E^A_M(x) \;&,\; DX^M = E^A E^M_A(x) \\ E^A_M(x) E^N_A(x) = \delta^N_M(x) \;&,\; E^A_N(x) E^N_B(x) = \delta^A_B(x) \end{aligned} \quad (15)$$

The inner scalar products of 1-forms $\mathcal{U}$ and $\mathcal{V}$ are defined accordingly

$$\begin{aligned} &< DX^M, DX^N >= G^{MN}(x) \\ &< U, V >= U^\dagger_M G^{MN}(x) U_N \end{aligned} \quad (16)$$

The locally flat inertial coordinate system is chosen to satisfy

$$< E^A, E^B >= \eta^{AB} = diag(-1, 1, 1, 1, 1) \quad (17)$$

and using the above condition we can express the metric G^{MN} in terms of vielbeins

$$G^{MN}(x) = E^M_a(x) \eta^{ab} E^N_b(x) + E^M_{\dot{5}}(x) E^N_{\dot{5}}(x) \quad (18)$$

In the locally flat frame, with the above bases, we find the Dirac matrix representation as given by

$$\begin{aligned} DX^\mu &= \begin{bmatrix} \gamma^a e^\mu_{aL}(x) & 0 \\ 0 & \gamma^a e^\mu_{aR}(x) \end{bmatrix}, \\ DX^5 &= \begin{bmatrix} -\gamma^a e^\mu_{aL}(x) a_{\mu L}(x) & i\gamma^5 \phi^{-1}(x) \\ -i\gamma^5 \phi^{-1}(x) & -\gamma^a e^\mu_{aR}(x) a_{\mu R}(x) \end{bmatrix}, \\ E^a &= \begin{bmatrix} \gamma^a & 0 \\ 0 & \gamma^a \end{bmatrix}, \; E^{\dot{5}} = \begin{bmatrix} 0 & i\gamma^5 \\ -i\gamma^5 & 0 \end{bmatrix}. \end{aligned} \quad (19)$$

The transformation coefficients between the locally flat and curvilinear frames encode gravity are the vielbein matrices

$$\begin{aligned} E^a_\mu(x) &= \begin{bmatrix} e^a_{\mu L}(x) & 0 \\ 0 & e^a_{\mu R}(x) \end{bmatrix}, \quad E^a_{\dot 5} = 0, \quad E^{\dot 5}_{\dot 5} = \phi(x), \\ E^{\dot 5}_\mu(x) &= A_\mu(x)\phi = \begin{bmatrix} a_{\mu L}(x) & 0 \\ 0 & a_{\mu R}(x) \end{bmatrix}\phi, \end{aligned} \tag{20}$$

and their inverses

$$\begin{aligned} E^\mu_a(x) = \begin{bmatrix} e^\mu_{aL}(x) & 0 \\ 0 & e^\mu_{aR}(x) \end{bmatrix}, \quad E^\mu_{\dot 5}(x) = 0, \\ E^{\dot 5}_a(x) = -E^\mu_a(x)A_\mu(x), \; E^{\dot 5}_{\dot 5} = \phi^{-1}(x) \end{aligned} \tag{21}$$

The generalized vielbein of this theory contains two vierbeins $e^a_{I\mu}(x), I = L, R$, two possibly nonabelian vectors $a_{I\mu}(x)$ and one real Brans-Dicke scalar $\phi(x)$.

In contrast to the case of pure abelian case in Ref.[6–8], here we allow the vierbeins, vectors and scalar to have the matrix values (diagonal and non-diagonal for the possibly non abelian fields). The Z_2 structure of the fifth dimension still valid, since it represents the space-time sheets of right and left-handed chiral quark-leptons. Further dimensions leading to the matrix values can be considered as the hyperfine structure of current discretized five space-time as stated in [11, 12].

In the rest of this paper, we will use the following notations

$$\begin{aligned} h_{a\mu} &= \frac{1}{2}(E_{a\mu} - \tilde E_{a\mu})\mathbf{r} \;, \quad e^\mu_a = \frac{1}{2}(E^\mu_a + \tilde E^\mu_a) \\ g^{\mu\nu} &= e^\mu_a \eta^{ab} e^\nu_b \;, \quad \hat f_{I\mu\nu} = \partial_\mu a_{I\nu} - \partial_\nu a_{I\mu} \\ f_{I\mu\nu} &= \partial_\mu a_{I\nu} - \partial_\nu a_{I\mu} + m[a_{I\mu}, a_{I\nu}], I = \pm, L, R \\ F_{\mu\nu} &= \partial_\mu A_\nu - \partial_\nu A_\mu = \hat f_{+\mu\nu} + \hat f_{-\mu\nu}\mathbf{r} \\ a_{-\mu} &= \frac{1}{2}(A_\mu - \tilde A_\mu)\mathbf{r} \;, \quad a_{+\mu} = \frac{1}{2}(A_\mu + \tilde A_\mu) \end{aligned} \tag{22}$$

So that when $E^a_\mu = \tilde E^a_\mu$, the gravity sector reduces to the conventional one with the vielbein e^μ_a and $h_{a\mu} = 0$. In the bigravity model, one can choose $h_{a\mu}$ and e^μ_a to represent two independent fields as an alternative to $e^a_{L\mu}$ and $e^a_{R\mu}$.

B. Levi-Civita connections and Cartan's structure equations

The generalized Levi-Civita connection 1-forms Ω^A_B follow from the covariant derivative of the generalized 1-form E^A

$$\nabla E^A = E^B \otimes \Omega^A_{\;B}, \tag{23}$$

The metric compatibility and hermitian conditions require

$$\Omega^\dagger_{AB} = \Omega_{AB} = -\Omega_{BA} \tag{24}$$

The generalized Cartan's structure equations take the form

$$\begin{aligned} T^A &= DE^A - E^B \wedge \Omega^A_B & (25) \\ R^{AB} &= D\Omega^{AB} + \Omega^A_C \wedge \Omega^{CB} & (26) \end{aligned}$$

Next, we impose, as in Ref. [8] a set of minimal conditions

$$T_a = 0 \;, \; T_{\dot 5ab} = t_{\dot 5ab}\mathbf{r} \;, \; T_{\dot 5a\dot 5} = t_{\dot 5a\dot 5}\mathbf{r}, \tag{27}$$

to obtain all the Levi-Civita connections in terms of metric vielbeins

$$\begin{aligned} \Omega_{abc} &= (DE_a)_{bc} + (DE_b)_{ca} - (DE_c)_{ab} \;, \quad \Omega_{\dot 5a\dot 5} = e^\mu_a \frac{\partial_\mu \phi(x)}{\phi} \\ \Omega_{\dot 5ab} &= \frac{1}{2}E^\mu_a E^\nu_b \phi(F_{\mu\nu} - 2m(A_\nu a_{-\mu} - A_\mu a_{-\nu})\mathbf{r}) + m\phi^{-1}(3h_{a\mu}e^\mu_b - h_{b\mu}e^\mu_a)\mathbf{r} \\ &\quad -\frac{1}{4}\phi(E^\mu_a E^\nu_b F_{\mu\nu} - \tilde E^\mu_a \tilde E^\nu_b \tilde F_{\mu\nu} - 2m(E^\mu_a E^\nu_b (A_\nu a_{-\mu} - A_\mu a_{-\nu}) + \tilde E^\mu_a \tilde E^\nu_b(\tilde A_\nu a_{-\mu} - \tilde A_\mu a_{-\nu}))\mathbf{r}) \end{aligned} \tag{28}$$

Substituting the above results in Eq.(26) we obtain the components of the curvature tensors R_{AB}

$$\begin{aligned}
R_{abcd} &= (D\Omega_{ab})_{cd} + \eta^{ef}(\Omega_{ae}\wedge\Omega_{fb})_{cd} + (\Omega_{a\dot{5}}\wedge\Omega_{\dot{5}b})_{cd} = \frac{1}{2}E^{\mu}_{c}E^{\nu}_{d}(\partial_\mu(E^e_\nu\Omega_{abe}) - \partial_\nu(E^e_\mu\Omega_{abe}) \\
&+\phi(F_{\mu\nu} + m[A_\mu, A_\nu])\Omega_{ab\dot{5}} - m((A_\nu E^e_\mu - A_\mu E^e_\nu)\Omega_{abe} + (A_\mu\tilde{E}^e_\nu - A_\nu\tilde{E}^e_\mu)\tilde{\Omega}_{abe} \\
&+\phi(A_\mu\tilde{A}_\nu - A_\nu\tilde{A}_\mu)\tilde{\Omega}_{ab\dot{5}})) - \frac{1}{2}({\Omega^e}_{ac}\Omega_{ebd} - {\Omega^e}_{ad}\Omega_{ebc}) - \frac{1}{2}(\Omega_{\dot{5}ac}\Omega_{\dot{5}bd} - \Omega_{\dot{5}ad}\Omega_{\dot{5}bc}) \\
R_{\dot{5}a\dot{5}b} &= (D\Omega_{\dot{5}a})_{\dot{5}b} + \eta^{cd}(\Omega_{\dot{5}c}\wedge\Omega_{da})_{\dot{5}b} = -\frac{m}{2}\phi^{-1}e^{\mu}_{b}(E^c_\mu\Omega_{\dot{5}ac} - \tilde{E}^c_\mu\tilde{\Omega}_{\dot{5}ac}) - \frac{1}{2}\eta^{cd}\tilde{\Omega}_{\dot{5}cb}\Omega_{da\dot{5}} \\
R_{\dot{5}\dot{5}\dot{5}\dot{5}} &= \eta^{cd}(\Omega_{\dot{5}c}\wedge\Omega_{d\dot{5}})_{\dot{5}\dot{5}} = \eta^{cd}\tilde{\Omega}_{c\dot{5}\dot{5}}\Omega_{d\dot{5}\dot{5}} = g^{\mu\nu}\frac{\partial_\mu\phi}{\phi}\frac{\partial_\nu\phi}{\phi} \qquad (29)
\end{aligned}$$

C. The Hilbert-Einstein action

We begin by defining the $5D$ Ricci curvature and the covariant integration in the $5D$ space-time as follows

$$R_5 = \frac{1}{2}Tr(\eta^{AC}R_{ABCD}\eta^{BD}) \qquad (30)$$

The covariant integration in the 5D space-time is defined as follows

$$\int DX^5\sqrt{-detG} = \int dx^4\sqrt{-detg}\;\phi, \qquad (31)$$

where $detg$ is the usual 4D determinant calculated from the average metric on the two sheets $g^{\mu\nu}$ given in Eq.(22).

The generalized Hilbert-Einstein action is defined as follows

$$S_{HE}(5) = M^2_{Pl}(5)\int dx^4\sqrt{-detg}\;\phi_0\exp\left(\frac{\sigma(x)}{m_\sigma}\right)R_5, \qquad (32)$$

where m_σ is a mass parameter for giving physical dimension to the scalar field $\sigma(x)$, which is introduced by the field redefinition

$$\phi = \exp\left(\frac{\sigma(x)}{m_\sigma} + \kappa_0\right). \qquad (33)$$

Hence, the $5D$ Planck mass $M_{Pl}(5)$ is related to the $4D$ one follows.

$$M_{Pl}(5) = M_{Pl}(4)\exp\left(-\frac{\kappa_0}{2}\right) \qquad (34)$$

Hence, $M_{Pl}(5)$ can be many orders lower than the conventional Planck scale $M_{Pl}(4)$.

The generalized Ricci scalar curvature R_5 can be calculated explicitly to yield

$$\begin{aligned}
R_5 &= \frac{1}{m^2_\sigma}g^{\mu\nu}\partial_\mu\sigma(x)\partial_\nu\sigma(x) + \frac{1}{2}Tr\left(\begin{bmatrix} r_L & 0 \\ 0 & r_R\end{bmatrix}\right) - \frac{1}{2}((\Omega_{\dot{5}ac}\eta^{ac})^2 - \Omega^{\dot{5}ab}\Omega_{\dot{5}ba}) \\
&+ \frac{1}{2}E^{a\mu}E^{b\nu}(\phi(F_{\mu\nu} + m[A_\mu, A_\nu] - m(A_\mu\tilde{A}_\nu - A_\nu\tilde{A}_\mu))\Omega_{ab\dot{5}} - m((A_\nu E^e_\mu - A_\mu E^e_\nu)\Omega_{abe} \\
&+ (A_\mu\tilde{E}^e_\nu - A_\nu\tilde{E}^e_\mu)\tilde{\Omega}_{abe}))) + \frac{1}{2}Tr(-m\phi^{-1}e^{a\mu}(E^b_\mu\Omega_{\dot{5}ab} - \tilde{E}^b_\mu\tilde{\Omega}_{\dot{5}ab}) - \tilde{\Omega}^{\dot{5}ab}\Omega_{ab\dot{5}}) \qquad (35)
\end{aligned}$$

The first term in the above equation represents clearly the kinetic term for the physical field $\sigma(x)$ if one choose the mass parameter m_σ as follows

$$m_\sigma = \frac{M_{Pl}(4)}{\sqrt{2}}. \qquad (36)$$

At this point, we can conclude that the Hilbert-Einstein action defined in Eq.(32) contains bigravity, two possibly nonabelian gauge fields and one Brans-Dicke scalar. Since in this paper, we focus mainly on the Einstein-Yang-Mills-Dirac systems, we will discuss the issues related to bigravity elsewhere. The Cartan-Hilbert-Einstein's action with the contribution from the torsion is not considered in this paper, as it violates the gauge invariance.

D. Gravity and nonabelian gauge fields

Now focusing on single gravity coupled to nonabelian gauge fields and the Brans-Dicke field, we choose the vierbeins

$$e^a_{\mu L} = e^a_{\mu R} = e^a_\mu(x), \qquad (37)$$

The curvature R_5 in Eq.(35) now is reduced to the following expression

$$\begin{aligned} R_5 &= r_4 - \frac{1}{16}\phi^2 g^{\mu\rho} g^{\nu\tau}\mathcal{F}_{\mu\nu}\mathcal{F}_{\rho\tau} \\ &+ \frac{1}{2M^2_{Pl}(5)} g^{\mu\nu}\partial_\mu\sigma(x)\partial_\nu\sigma(x), \qquad (38) \\ \mathcal{F}_{\mu\nu} &= f_{L\mu\nu} + f_{R\mu\nu} + m([a_{R\nu}, a_{L\mu}] + [a_{L\nu}, a_{R\mu}]), \quad (39) \end{aligned}$$

where the gauge covariant field strengths are defined as follows

$$f_{I\mu\nu} = \partial_\mu a_{I\nu} - \partial_\nu a_{I\mu} + m[a_{I\mu}, a_{I\nu}], \; I = L, R \quad (40)$$

In general, due to the extra terms, the "field strength" $\mathcal{F}_{\mu\nu}$ is not gauge covariant. However, the gauge invariance of the curvature R_5 can kept in the following two cases:

Case 1: One of the gauge field for example $a_{R\mu}(x)$ is abelian, which is valid for the electroweak interaction. In this case $\mathcal{F}_{\mu\nu}$ reduces to the gauge covariant form.

$$\mathcal{F}_{\mu\nu} = f_{L\mu\nu} + f_{R\mu\nu} \qquad (41)$$

This case enables us to construct the Einstein-Electroweak-quark-lepton system in the next section, when we couple the generalized gravity to quark-leptons.

Case 2: If we choose the left handed gauge field to be proportional to the right one with a constant factor $a_{L\mu} = \alpha a_{R\mu} = \alpha a_\mu$, we obtain the following covariant field strength

$$\mathcal{F}_{\mu\nu} = (\alpha + 1)(\partial_\mu a_\nu - \partial_\nu a_\mu) + m(\alpha - 1)^2[a_\mu, a_\nu] \quad (42)$$

In the next Section, we will show that Case 1 enables us to unify generalized Einstein gravity and electroweak interactions of quarks and leptons. And in the Case 2, the gauge fields are the same on both sheets of space-time. The same gluons will couple to the left- and right- handed chiral quarks, thus unifying gravity and QCD. We also note that the results of Viet-Du [15] are confirmed with the new spectral triple used in the present paper.

IV. EINSTEIN-YANG-MILLS-DIRAC SYSTEMS

In this section, we specialize to the Einstein-Yang-Mills-Dirac systems emerging from the generalized gravity coupled to the chiral spinors of quarks and leptons. The gauge fields contained in the generalized gravity, will now be non-Abelian and will enable us to incorporate weak and strong interactions of the standard model with appropriate colors and flavors. The couplings of gravity, gauge fields and Brans-Dickey scalar coupled to chiral spinors emerge naturally with some new interactions in contrast to our previous papers [15–17] where these couplings were introduced as arbitrary and independent. The required total action S_{tot} in our framework is derived from the sum of the actions generalized from gravity sector S_{HE} and the gravity-chiral spinor one S_ψ,

$$S_{tot} = S_{HE}(5) + S_\Psi \qquad (43)$$

A. Generalized gravity coupled to the chiral spinors

To obtain the generalized Lagrangian that contains gravity coupled to chiral spinors, we begin with Eq.(35) and the generalized covariant Dirac operator $\mathcal{D}$, where

$$\mathcal{D} = D + \Omega = D - \frac{1}{8}\Gamma^C\Omega_{ABC}[\Gamma^A, \Gamma^B]. \qquad (44)$$

Then the generalized Dirac Lagrangian and its corresponding action are given by

$$\begin{aligned} \mathcal{L}_\Psi &= \bar{\Psi}(i\mathcal{D} + M)\Psi + h.c = \bar{\Psi}(iD + M)\Psi + \bar{\Psi}\Omega\Psi + h.c = \mathcal{L}_D + \mathcal{L}_m + \mathcal{L}_\Omega(1) + \mathcal{L}_\Omega(2), \qquad (45) \\ S_\Psi &= \int dx^4\sqrt{-detg}\exp\left(\frac{\sqrt{2}\sigma(x)}{M_{Pl}(5)} + \kappa_0\right)\mathcal{L}_\Psi. \qquad (46) \end{aligned}$$

where M is an arbitrary mass parameter to be chosen later and

$$\begin{aligned} \mathcal{L}_D &= \sum_{I=L,R}\bar{\psi}_I\gamma^a e^\mu_a(i\partial_\mu + m a_{I\mu}(x))\psi_I + h.c. \quad (47a) \\ \mathcal{L}_m &= \bar{\psi}(m\exp\left(-\frac{\sqrt{2}\sigma(x)}{M_{Pl}(4)} - \kappa_0\right) + M)\psi + h.c. \quad (47b) \\ \mathcal{L}_\Omega(1) &= -\frac{1}{8}\bar{\psi}\gamma^c\omega_{abc}[\gamma^a, \gamma^b]\psi + h.c. \qquad (47c) \end{aligned}$$

$\mathcal{L}_\Omega(2)$ ... $\sqrt{2}$ ... (47d)

If $\kappa_0 \neq 0$, the factors of the kinetic and spinor connection terms in the actions S_D and $S_\Omega(1)$ are not the usual ones.

Nevertheless, it is possible to absorb those factors by redefining the chiral spinors

$$\Psi \to e^{-\frac{\kappa_0}{2}}\Psi \qquad (48)$$

So after the redefinition of the chiral spinors in Eq.(48), retaining only the first order in $M_{Pl}^{-1}(4)$, we obtain the following results for actions from the Lagrangians in Eqs.(45) and (47a)

$$S_\Psi = S_\psi + S_\sigma + S_g + S_f \qquad (49a)$$

$$S_\psi = \int dx^4\sqrt{-detg}\ \bar{\psi}(i\gamma^a e^\mu_a\partial_\mu + me^{-\kappa_0} + M - \frac{1}{8}\gamma^c\omega_{abc}[\gamma^a,\gamma^b])\psi + h.c. \qquad (49b)$$

$$S_\sigma = \frac{\sqrt{2}}{M_{Pl}}\int dx^4\sqrt{-detg}\ \bar{\psi}[\gamma^a e^\mu_a(\partial_\mu\sigma(x) + \sigma(x)\partial_\mu) - \frac{1}{8}\gamma^c\omega_{abc}[\gamma^a,\gamma^b]\sigma(x)]\psi + h.c. \qquad (49c)$$

$$S_g = m\int dx^4\sqrt{-detg}(1 + \frac{\sqrt{2}\sigma(x)}{M_{Pl}(4)})\sum_{I=L,R}\bar{\psi}_I\gamma^a e^\mu_a a_{I\mu}(x)\psi_I + h.c. \qquad (49d)$$

$$S_f = \frac{1}{16}\int dx^4\sqrt{-detg}\ \bar{\psi}(f_{L\mu\nu} + f_{R\mu\nu})\sigma^{\mu\nu}\psi + h.c., \qquad (49e)$$

where $\sigma^{\mu\nu} = i[\gamma^\mu,\gamma^\nu]$.

Let us discuss the physical meaning of the above terms. S_ψ describes the chiral spinors in the curved 4D space-time with a mass of $me^{-\kappa_0} + M$. Since the quark-lepton masses are generated by the Higgs mechanism, we will choose $M = -me^{-\kappa_0}$ to keep the quark-lepton to be massless as in the Standard Model.

S_σ is responsible for quark-lepton pair creation of the Brans-Dicke scalar. The effect is difficult to observed experimentally, as the coupling constant is proportional to the inverse of the Plank mass.

S_g contains gauge fields and its discussion will follow in the next subsection B. The action S_f is a gauge invariant giving a significant contribution to the magnetic momentum of quark-leptons. It is not required by the first principle in the minimal coupled Yang-Mills-Dirac systems. In our framework, it is inherent. The discussion about its contributions to the magnetic momentum of quark-leptons will be discussed elsewhere.

B. The Einstein-Electroweak-Quark-Lepton system

We note that the gauge symmetry group for weak interactions in Yang-Mills approach is $SU(2)\times U(1)$ requiring a non-Abelian $SU(2)$ and an abelian $U(1)$ gauge fields. This leads us to Case 1 described in Section III in the framework the generalized gravity. Focusing on quark and lepton families including the right-handed neutrino, we have the following left- and right-handed chiral quark-lepton representation,

$$\psi_{LA} = \begin{bmatrix} q^c_L \\ l_L \end{bmatrix}_A \ , \ \psi_{RA} = \begin{bmatrix} u^c_R \\ d^c_R \\ e_R \\ \nu_R \end{bmatrix}_A \ , \qquad (50)$$

where $c = 1,2,3$ is the color index, A is the family index, the number of which we leave arbitrary.

First, we introduce the usual physical gauge fields $W_\mu(x)$ and $B_\mu(x)$ as follows

$$a_{L\mu} = \frac{1}{M_L}(W^a_\mu(x)\frac{\tau^a}{2}\otimes\mathbf{1}_4 - \frac{M_L}{M_R}\frac{Y_L}{2}B_\mu(x)\otimes\mathbf{1}_2)\otimes\mathbf{1}_{N_F},$$
$$a_{R\mu} = -\frac{1}{M_R}\frac{Y_R}{2}B_\mu(x)\otimes\mathbf{1}_{N_F}, \qquad (51)$$

where $Y_{L,R}$ are the hypercharge operators acting on the left and right-handed chiral quark-leptons. Since the gauge fields $a_{L,R\mu}$ are operators acting on the chiral quark-leptons, we must have include the corresponding unit matrices in Eqs.(51).

Now we choose the rescaling mass parameters for the gauge fields

$$M_L = \frac{m}{g} \ , \ M_R = \frac{m}{g'} \qquad (52)$$

Substituting (51) into Eq.(49d) we obtain

$$
\begin{aligned}
S_g &= \int dx^4 \sqrt{-detg}\, (g'\bar{\psi}\gamma^\mu \frac{Y}{2} B_\mu(x)\psi - g\bar{\psi}_L\gamma^\mu W^a_\mu(x)\frac{\tau^a}{2}\psi_L) \\
&+ \frac{\sqrt{2}}{M_{Pl}(4)} \int dx^4 \sqrt{-detg}\, \sigma(x)(g'\bar{\psi}\gamma^\mu \frac{Y}{2} B_\mu(x)\psi - g\bar{\psi}_L\gamma^\mu W^a_\mu(x)\frac{\tau^a}{2}\psi_L),
\end{aligned} \tag{53}
$$

where the sum is taken over the color, flavor and family indexes. We note that the first term contains the usual electroweak couplings of the gauge fields to chiral spinors in the Standard Model. The second contains interaction terms of the gauge, quarks and leptons with the Brans-Dicke scalar $\sigma(x)$.

The hyper charge operators are the same for all the families and hence we specify them to be

$$
\begin{aligned}
Y_L &= \begin{bmatrix} \frac{1}{3}\mathbf{1}_3 & 0 \\ 0 & -1 \end{bmatrix}, \\
Y_R &= \begin{bmatrix} \frac{4}{3}\mathbf{1}_3 & 0 & 0 & 0 \\ 0 & -\frac{2}{3}\mathbf{1}_3 & 0 & 0 \\ 0 & 0 & -2 & 0 \\ 0 & 0 & 0 & 0 \end{bmatrix},
\end{aligned} \tag{54}
$$

From Eqs.(51)- (38) and retaining only the first order of $\sigma(x)$ we obtain the generalized Hilbert-Einstein action for this case

$$
\begin{aligned}
S_{HE}(5) &= \int dx^4 \sqrt{-detg}((M_{Pl}^2 + \sqrt{2}M_{Pl}\sigma(x))r_4 - \frac{(M_{Pl}^2 + 3\sqrt{2}M_{Pl}\sigma(x))\phi_0^2}{128m^2}(g^2 Tr(W_{\mu\nu}W^{\mu\nu}) \\
&- \frac{g'^2}{4}\sum(2Y_L^2 + Y_R^2)B_{\mu\nu}B^{\mu\nu})),
\end{aligned} \tag{55}
$$

where the sum is taken over all the hyper charges of chiral quark-leptons and averaged on the families. Furthermore, in order to have the correct kinetic terms for the gauge fields, we find the following relations between the parameters

$$
g = \frac{4\sqrt{2}m}{M_{Pl}\phi_0}, \quad g = \frac{g'\sqrt{\sum(2Y_L^2 + Y_R^2)}}{2} = g'\sqrt{10/3}, \tag{56}
$$

where the factor 2 in front of Y_L^2 is due to the fact that the left-handed quark-leptons always appear in doublet.

These relations lead to the following simplified form of Eq.(55)

$$
\begin{aligned}
S_{HE}(5) &= \int dx^4 \sqrt{-detg}\, [M_{Pl}^2 r_4 - \frac{1}{4}(Tr(W_{\mu\nu}W^{\mu\nu}) + B_{\mu\nu}B^{\mu\nu}) \\
&+ \sqrt{2}\sigma(x)(M_{Pl}(4)r_4 - \frac{3}{4M_{Pl}(4)}(Tr(W_{\mu\nu}W^{\mu\nu}) + B_{\mu\nu}B^{\mu\nu}))]
\end{aligned} \tag{57}
$$

In the above equation, the first term is the usual kinetic term for gravity and electroweak gauge fields. The second term represents interactions of the Brans-Dicke field $\sigma(x)$ with gravity, $W_\mu(x)$ and $B_\mu(x)$ fields. It is remarkable that the gauge fields are part of the generalized gravity. So, the full action of the Einstein-Electroweak-quark-lepton system can be derived from the generalized gravity and its coupling to the chiral quark-leptons.

Let us determine the energy scale, where the above coupling constant relations become valid. From Eqs.(56) we have the prediction

$$
\sin^2\theta_W = \frac{g'^2}{g^2 + g'^2} = \frac{3}{13} = 0.23077, \tag{58}
$$

which has been given in [19] from different considerations.

Since this value is pretty close to the experimental value $\sin^2\theta_W = 0.231$, we can expect that the model is applicable for the Standard Model at currently accessible energy. It is remarkable is the gauge fields are derived from the generalized gravity. In this model, Higgs field must be added manually or generated dynamically.

C. The Einstein-QCD system

The Einstein-QCD system has the gauge group $SU(3)_{QCD}$ acting in the same way with the left and right-handed quarks. This leads us to Case 2 discussed in Section III with gauge fields as part of the generalized Hilbert-Einstein action.

The quark sector of this system is represented by the following chiral spinors

$$\psi_L = \begin{bmatrix} u_L^c \\ d_L^c \end{bmatrix} \quad , \quad \psi_R = \begin{bmatrix} u_R^c \\ d_R^c \end{bmatrix}. \tag{59}$$

Next we introduce the physical gauge fields representing color

$$a_{L\mu} = \alpha a_{R\mu} = \frac{\alpha}{m_c} C_\mu(x) \otimes \mathbf{1}_6, \tag{60}$$

where m_c is a mass parameter and $C_\mu(x) = C_\mu^a(x)\frac{\lambda^a}{2}$ is the color gauge field. The Gell Mann matrices $\lambda^a, a = 1,..8$ are used as generators of the gauge group $SU(3)_{QCD}$.

Using Eqs.(60) and Eq.(49d) we obtain the following action

$$\begin{aligned} S_g &= g_S \int dx^4 \sqrt{-detg}\, [\bar{\psi}\gamma^a e_a^\mu(x) C_\mu(x)\psi + \frac{\sqrt{2}\sigma(x)}{M_{Pl}(4)}\bar{\psi}\gamma^a e_a^\mu(x) C_\mu(x)\psi \\ &+(\alpha-1)(\bar{\psi}_L\gamma^a e_a^\mu(x) C_\mu(x)\psi_L + \frac{\sqrt{2}\sigma(x)}{M_{Pl}(4)}\bar{\psi}_L\gamma^a e_a^\mu(x) C_\mu(x)\psi_L)], \end{aligned} \tag{61}$$

where the mass parameter m_c is defined to be

$$m_c = \frac{m}{g_S}. \tag{62}$$

The first term is the coupling of between gluon gauge field and chiral quark in presence of gravity. The second term contains quark-gluon interactions with the Brans-Dicke scalar in the curved space-time. The third term violates P-conservationwhen $\alpha \neq 1$.

By retaining only the first order of $\sigma(x)$, we can explicitly calculate the generalized Hilbert-Einstein action as follows

$$\begin{aligned} S_{HE}(5) &= \int dx^4 \sqrt{-detg}\, (1 + \frac{\sqrt{2}\sigma(x)}{M_{Pl}(4)}) M_{Pl}^2(4) r_4 \\ &- \frac{M_{Pl}^2}{4}(1 + 3\frac{\sqrt{2}\sigma(x)}{M_{Pl}(x)}) g^{\mu\rho} g^{\nu\tau} (\alpha+1)^2 \frac{g_S^2\phi_0^2}{m^2} Tr(C_{\mu\nu}C_{\rho\tau}) \\ C_{\mu\nu} &= \partial_\mu C_\nu - \partial_\nu C_\mu + g_S\frac{(\alpha-1)^2}{\alpha+1}[C_\mu, C_\nu]. \end{aligned} \tag{63}$$

In order to have the correct factors of the kinetic term and to keep the gauge invariance, we must have the following relations

$$\frac{(\alpha-1)^2}{\alpha+1} = 1 \ , \ \alpha = 3 \tag{64a}$$

$$g_S = \frac{m}{4M_{Pl}(4)\phi_0} \tag{64b}$$

With the above conditions, the action and the field strength $C_{\mu\nu}$ in Eq. (63) takes the form

$$\begin{aligned} S_{HE}(5) &= \int dx^4 \sqrt{-detg}\, (M_{Pl}^2 r^4 - \frac{1}{4}Tr(C_{\mu\nu}C^{\mu\nu}) + \frac{\sqrt{2}\sigma(x)}{M_{Pl}(4)}(M_{Pl}^2(4) r_4 - \frac{3}{4}Tr C_{\mu\nu}C^{\mu\nu}), \\ C_{\mu\nu} &= \partial_\mu C_\nu - \partial_\nu C_\mu + g_S[C_\mu, C_\nu]. \end{aligned} \tag{65}$$

The first term is the usual action of gluon fields coupled to gravity. The second term represents interactions of the gluon field and the Brans-Dicke scalar in the curved space-time.

Since in Eq(64b), the strong coupling constant depends on two unknown parameters m and ϕ_0, we do not have a prediction for the coupling constant as in the Einstein-electroweak-quark-lepton model. We can use these two parameters to tune the model to be applicable at the currently accessible energy.

The new feature of the Einstein-QCD system constructed from the generalized gravity is the intrinsic P-violation. In fact, if $\alpha = 1$, the commutator in the field strength $C_{\mu\nu}$ vanishes, thus the gauge invariance can not be kept. So, the parity violation is inherent in the constructed system.

Since the currently available data suggests that the strong interaction does not violate parity at least up to the energy scale of $10^2 - 10^3 GeV$, we can choose to transfer the P-violation in Eq.(61) into the gravity sector. To do so, we must go back to our general theory with bigravity of right-handed and left-handed vierbeins in Eqs.(32) and (35). If instead of the condition (37) we assume the following

$$e^{\mu}_{L_a}(x) = \frac{1}{3} e^{\mu}_{Ra}(x) = \frac{1}{3} e^{\mu}_{a}(x), \tag{66}$$

to lead to the following parity non-violation action

$$S_g = gs \int dx^4 \sqrt{-detg}\ [\bar{\psi}\gamma^a e^{\mu}_a(x) C_\mu(x)\psi + \frac{\sqrt{2}\sigma(x)}{M_{Pl}(4)} \bar{\psi}\gamma^a e^{\mu}_a(x) C_\mu(x)\psi \tag{67}$$

Then the P-violation will appear in the action S_ψ. Although, this possibility might lead to new physical consequences, it does not contradict to the particle physics under the TeV energy range.

V. CONCLUSIONS

In this article, we have constructed the coupling of chiral spinors and generalized gravity within the framework of discretized Kaluza-Klein theory with one internal extra dimension having only two points. This is a different approach than the traditional Kaluza-Klein theory to unify gravity with the gauge fields. The discrete dimension can describe two copies of space-time of the left and right-handed chiral fermions proposed by Connes and Lott [11] or other discrete properties of quark-lepton as in Refs.[16, 17]. Starting with the Dirac operator in Eq.(3) and the not fully anti-symmetric wedge product, we have derived a general model containing a finite gravity sector with bigravity, two possibly nonabelian gauge fields and one Brans-Dicke scalar. We also recovered the same result of Viet-Du in this framework, at least one gauge field must be abelian or both gauge fields must be the same.

Thus, we are able to construct different realistic Einstein-Yang-Mills-Dirac systems out from generalized gravity coupled to chiral spinors. All the gauge couplings and kinetic terms of quark-leptons can be derived from the generalized gravity. In addition, a non-Higgs mass term together with a magnetic momentum interactions of quark-leptons is inherent in the framework. While the quark-lepton masses are zero and the value of ϕ_0 is large, the contribution of the magnetic momentum term is significant.

The Einstein-electroweak-quark-lepton system, as one special case of Viet-Du's theorem, leads to a value $\sin^2\theta_W = 0.23077$ very close to the experimental one 0.231 indicating that the model is applicable to the realistic particle physics. At the same time, with the experimental value of the coupling constant g at the electroweak energy scale, our model implies a large value of m.

The Einstein-QCD system as another case of Viet-Du's theorem, also leads to a realistic model with a new parity violation interaction, which can be transferred to the gravity sector by allowing the gravitation interacts differently with left- and right-handed chiral fermions. The parity violation coupled to gravity may be of relevance to the long standing problem of baryon-anti Baryon asymmetry.

Since the condition of Viet-Du's theorem can not hold at the same time with one single discrete extra dimension, one can consider both strong, electroweak interactions and Higgs as components of gravity in the discretized Kaluza-Klein theory with two discrete extra dimensions as proposed by one of us [16, 17]. In such an extended model, the coupling constants and Higgs mass will be related.

Thanks are due to Pham Tien Du and Do Van Thanh (Department of Physics, Hanoi University of Science, VNU) for their participation in the research group. The supports of ITI, VNU and Department of Physics, Hanoi University of Science, VNU are greatly appreciated. The work of NSH is supported by NAFOSTED under grant No 103.03-2012.02.

[1] N.Arkani Hamed, A.G.Cohen and H.Georgi, Physical Review Letters **86**(21),(2001),4757.

[2] N.Arkani Hamed, A.G.Cohen and H.Georgi, Phys. Lett. **B513**, (2001), 232. arXiv:0105239[hep-ph].

[3] N. Arkani Hamed, H. Georgi and M. D. Schwartz, Annals Phys. **305**,(2003), 96. arXiv:0210184[hep-th].

[4] C. Deffayet and J. Mourad, Int. J. Theor. Phys. **44**, (2005), 1743.

[5] C.de Rham, A.Matas and A.J.Tolley, Class.Quant.Grav. **31**,(2014), 025004. arXiv:1308.1436[hep-th]

[6] G.Landi, Nguyen Ai Viet and K.C.Wali, Phys.Letters **B326**,(1994), 45.

[7] Nguyen Ai Viet and K.C.Wali, Int.J.Mod.Phys. **11**(3), (1996), 533.

[8] Nguyen Ai Viet and K.C.Wali, Int.J.Mod.Phys. **11**(13), (1996), 2403.

[9] A.Connes, Publ.Math. IHES **62** (1986), 41; A.Connes, Noncommutative Geometry, Academic Press, San Diego, CA, (1994), 661 p.,ISBN 0-12-185860-X;A.Connes J. Math. Phys. **36** (ii),(1995), 6194.

[10] Th.Kaluza, *Sitzungsber.Preuss.Akad. Wiss.Phys.Math.Klasse* (1927) 996; O.Klein, *Z.Phs.* **37**, (1926) 895; Y.Thirry, *Comptes Rendus (Paris)* **226**, (1948) 216.

[11] A.Connes and J.Lott, Nucl.Phys.**B18**(Proc.Suppl.),(1990), 29.

[12] Nguyen Ai Viet, *Predictions of noncommutative space-time*, in *Proceedings of the MRST Workshop: What Next? Exploring The Future of High-Energy Physics*, Mc Gill University, Montréal, Canada, 1994 edited by J.R.Cuddel, K.R.Dienes and B.Margolis, World Scientific (1994); Nguyen Ai Viet, Acta Physica Hungarica New Series Heavy Ion Physics **1** (1995), 263.

[13] Nguyen Ai Viet and K.C.Wali, AIP Conference Proceedings **541** (2000), 27.

[14] Nguyen Ai Viet and K.C.Wali, Physical Review **D67**, (2003), 124029.

[15] Nguyen Ai Viet and Pham Tien Du, *Gravity and non-abelian gauge fields in noncommutative geometry*, arXiv 1510.01169 [gr-qc] (2015) (submitted to Phys.Letters B)

[16] Nguyen Ai Viet, *Electroweak, strong interactions and Higgs fields as components of gravity in noncommutative space-time*, Talk given at Rencontres du Vietnam on "Cosmology - 50 year after CMB discovery", Quy Nhon, Vietnam, August 16-22, 2015, arXiv:1510.01405 [gr-qc] (to be published)

[17] Nguyen Ai Viet, *General Relativity in noncommutative space-time as a unified framework for all interactions and the Higgs field* (2015) arXiv:1512.01643[gr-qc](to be published).

[18] Nguyen Ai Viet, Communications in Physics **24**(1), (2014), 21; Nguyen Ai Viet, *Some remarks on Gravity in noncommutative space-time and a new solution to the structure equation*, arXiv:hep-th/0303128v1(2003).

[19] J. Besprosvany, Phys.Lett. **B578**, (2004), 181

[20] Actually, in the DKKT which unifies of all interactions proposed recently, one additional discrete dimension leading to new type of spinor has been used [16, 17]

Essays on Science

What is High Energy Physics All About?

Split Face of Science: Is Science an Endangered Species?

What is High Energy Physics all about?

Kameshwar C. Wali

Institut des Hautes Etudes Scientifiques,
Bures-sur-Yvette, France

and

Physics Department, Syracuse University
Syracuse, New York

I have the uneasy feeling that after all the exciting episodes we have heard so far dealing with the history of the ZGS, my discussion will be somewhat of an anticlimax. I had no role of any significance in the conception of the ZGS or its turning on; nor in formulating its scientific programmes and policies. Hence I have no nostalgic happenings to tell you, except those associated with the "Happy Hour" every Thursday - those better wait for the happy hour to come later this evening. For the sake of the record however, I should note that the ZGS was responsible not only for the beginning of excellent high energy experimental research in the midwest, but also for the establishment of a high energy theory group whose size and stature has grown steadily. With the starting of the ZGS and the growth of the experimental program, Argonne became a central place for many theorists. Such activities as the summer visitors program, one day topical symposia, lecture series etc., were instituted and they created an exciting and stimulating atmosphere. I was indeed fortunate to have had the opportunity of spending seven crucial years of my career, here at Argonne, during that period.

I

Now the question "What is high energy physics all about?", receives generally the answer that it is the physics of the fundamental building blocks of all matter and their interactions. As such, not only is it an integral part of all physics, but also its most exciting and dominating frontier field. The subtle interplay between theory and experiment has led to many dramatic discoveries in both theory and experiment throughout its history. Antiparticles, neutrinos, pi-mesons are but a few examples of theoretical predictions, whereas neutrons, mu-mesons, strange particles, multitude of resonances are likewise a few examples of experimental discoveries that belong to the recent past. More recently we have the examples of quantum numbers like charm and beauty and particles like J/ψ and Υ (upsilon) which live a thousand or so times longer than expected from the known properties of strong interactions.

As the frontier field in search of the fundamental building

blocks and the laws that govern their binding, high energy physics has evolved through many "energy" phases. From the turn of the century to the thirties, it was the physics of the atom dealing with phenomena of a few electron volts (ev); from the thirties to the fifties, it was the physics of the nucleus in the million electron volt (Mev) region. Then onwards it has been the physics of the subnuclear particles and phenomena in billions of electron volts (Gev) range. Today we dare in high energy physics to contemplate theories and experiments in the 10^{15} Bev range! The pace at which the field progresses, the complexity and the ingenuity of the experimental techniques, the bigness (both in money and manpower) of the experimental technology is baffling and will continue to baffle not only the "outsider" but sometimes even the "insider". Think of the complexity of our present day experiments and the laboratory set ups that host them. Compared to these, the classic experiments of Thomson, Rutherford, and Millikan look like kindergarten experiments, which indeed they were, since they taught us the alphabets of the atomic and the nuclear world.

But we have come a long way from the days of the alphabets in our explorations of the atomic, nuclear and subnuclear worlds. Using higher and higher energy beams, we explore more and more of the interior of the subnuclear particles. Using different probes, we examine their different interactions. One may ask whether colliding one particle with another as violently as possible is a sensible way to know how they are made. ("To know how a swiss watch works, bang two of them and see what comes out", Feynman is supposed to have said about high energy physics). But there is no doubt we have made a great deal of progress this way. How else could we have found out, for example, that the subnuclear particles like the protons, neutrons, pi-mesons, etc., are not fundamental or elementary, but composites of simpler constituents? In what follows, I would like to describe qualitatively this simpler sub-subnuclear structure that we have currently come to believe.

II

The majority of high energy physicists believe today that we have come to the final stages in our search for the ultimate constituents. The ultimate constituents are quarks and leptons which behave as pointlike objects with no internal substructure. The quarks are the constituents of the ordinary, everyday hadrons like the pi- and K-mesons, protons and neutrons, and various mesonic and baryonic resonances. They come in three colors and at least five different flavors. A sixth flavor is expected to be discovered any day and announced in a fashion that has become customary - by calling a press conference. I should remind you, when we talk about color and flavor of these objects, we do not mean color and flavor in their every day sense of the terms. They are just convenient substitutes for a more formal mathematical description of the degrees of freedom we call internal, which we need to postulate in order to explain some of their properties. For example, the color degree of freedom obeys the exact laws of an SU(3) symmetry group; the observed hadrons are colorless combinations (SU(3) singlets) of the quarks. We do not yet know what, if any, symmetry laws govern the flavor degree of freedom. The leptons, like the quarks, come in a variety of flavors; but unlike them, exhibit no color. The presently known flavors of quarks are commonly referred to as up, down, strange, charm, and bottom (or beauty). The leptonic flavors are the familiar electrons, mu-mesons along with their distinct neutrinos and the very recently discovered heavier lepton - the τ-meson which is most probably accompanied by its own distinct neutrino like the others. There appears to be a close correspondence between the quark and the leptonic flavors:

QUARKS:	u (up)	d (down)	s (strange)	c (charm)	b (bottom or beauty)	t(?) (top or truth)
LEPTONS:	ν_e (electron neutrino)	e^- (electron)	μ^- (mu-meson)	ν_μ (mu-neutrino)	τ^- (tau-lepton)	ν_τ (tau-neutrino)

The quarks participate in all the known elementary particle interactions whereas the leptons take part only in the weak and electromagnetic interactions being immune to the strong interactions. The quanta that mediate the electromagnetic and weak interactions are the familiar photon and a set of particles ($W^\pm, Z^0$) known as the intermediate vector bosons yet to be discovered. The quanta of strong interactions are the so-called gluons. Finally to complete the list of the fundamental entities, I should also mention the necessity of the existence of at least one scalar particle called the Higgs meson.

III

Until recently one classified the elementary particle interactions into four main categories:

1) Gravitational interactions, which are all pervading, but of importance only in the domain of the large. They are responsible for the large scale organization of matter into planets, stars, galaxies, and so on. We demonstrate the irrelevance of these interactions in the microscopic domain, by usually calculating and comparing the electric and gravitational forces between, say an electron and a proton at some distance apart, and convincing our elementary physics students that the gravitational interactions are indeed negligibly small and proceed to dismiss their further consideration in elementary particle physics. We also point out to our students that this is very

ironical - the gravitational interactions have such a beautiful and elegant theory due to Einstein, and that this unique theory should stand apart and be of no relevance to our understanding of the structure of matter is such a great pity.

2) Electromagnetic interactions, which are responsible for the binding in the atoms, binding of atoms in the molecules, binding molecules and molecules to form complex inorganic and organic compounds and hence responsible for life itself as we know. They govern the domain of every day experience - "the middle ground" as Dyson likes to characterize it, which includes "atoms and electricity; light and sound: gases, liquids and solids: chairs, tables, and people". The theory that describes these interactions at the basic level is the quantum-electrodynamics. It is not as elegant as Einstein's theory of gravitation, but it is a remarkably successful theory in providing unambiguous answers with any desired degree of accuracy. Unfortunately all its successes are based on a perturbative approach which works because the fine structure constant which characterizes the strength of the electromagnetic interaction is indeed small. Also infinities occur in the calculation of physical quantities, infinities which have to be carefully identified and removed by what is called a renormalization procedure. Within the framework of such a theory, fundamental quantities like the masses and charges are to be taken as given empirical parameters with no hope of ever predicting them. So, although one would be quite happy with a theory like the quantum-electrodynamics for other interactions, one hopes eventually for a better theory.

3) Weak interactions, which were discovered in the radioactive β-decay of nuclei. They are responsible for the instability of the neutron and similar instabilities of other elementary particles. They play an important role in nuclear reactions and in the generation of energy by burning nuclear fuel in the interior of the stars. It was the weak interactions that led Pauli to postulate the neutrino - the most perplexing and the most elusive of all the elementary particles. It is in the weak interactions that we had to give up some of our cherished conservation laws, such as parity and charge conjugation combined with parity. A phenomenological model due to Fermi based on Pauli's neutrino was quite successful in explaining the low energy phenomena associated with these interactions. Although the model was based on the analogy with electromagnetic interactions, a theory like the quantum-electrodynamics could not be constructed for the weak interactions. The reason for this is that the relevant quanta (intermediate vector bosons) have to be electrically charged and massive unlike the photon which is electrically neutral and has zero mass. Any attempt to calculate higher order effects ran into grave troubles and the situation appeared hopeless till recently.

4) Strong interactions, which are responsible for the forces that

lead to stable nuclei. They have nothing to do with the electrical or weak interaction forces. It is in the domain of these interactions that one needed new concepts; the usual conservation laws such as energy-momentum and angular momentum stemming from the relativistic invariance of the interactions were not sufficient. Isotopic spin, strangeness, charm and so on had to be introduced.

One of the main characteristics of the strong interactions was the proliferation of the so-called elementary particles, namely the baryonic and mesonic resonances (which incidentally led to the expansion of Rosenfeld's wallet card to wallet book industry). This "population" explosion of the resonances made the concept of elementary particle very difficult. Which is elementary? which is composite? For a while one toyed with the idea of nuclear democracy - all are equally elementary or all are equally composite. Then higher and higher symmetries, Regge trajectories, dual models and so on. In the last two decades, the strong interaction regime has seen the birth and death (or fading away) of many theoretical breakthroughs,each claiming that a fundamental understanding was just around the corner. It did not turn out that way. Each such attempt provided only a fragmentary and only a partially satisfactory account of some observed aspects, and usually it provided only a new jargon and some new calculational tools. With so many particles or resonances around, the situation was not very different from the one around the turn of the century. The idea that the diverse observable properties of matter can be explained in terms of a few ultimate building blocks, an idea at least two thousand years old, had led to the atomic hypothesis. The invisible atoms explained successfully many laws of chemistry; they were responsible for the development of the kinetic theory of gases and were the unseen agents behind the Brownian motion. Yet the atomic model was too complicated - a closer study revealed too many different types of atoms with different chemical and physical properties, different emission and absorption spectra and so on. The idea that the atoms of different elements cannot all be different, and that they should have a common substructure became indispensable. That indeed it was so was established by Thomson by splitting the indivisible atoms and showing that they were all made from two constituents, the electrons and the nuclei.

IV

The quark model, like the atomic model, arose out of the multiplicity of the elementary particles, the resonances of the fifties and the sixties. Quarks were supposed to be the basic constituents of all the hadronic states. A fundamental triplet of quarks and an anti-triplet of anti-quarks that comes naturally with it were sufficient to explain many features of hadron spectroscopy based on Gell-Mann's eight fold way (SU(3) symmetry). The baryons and their excited states were made from three quarks, the mesons from quark-anti-quark pairs. Simple non-relativistic considerations involving these combinations in different angular momentum states went a long way in classifying a large number of states, in predicting new states and correlating their decay rates. However, such studies faced a basic difficulty. If the quarks are fermions, three identical quarks cannot exist in the lowest angular momentum state, at the same time belonging to certain SU(3) representations (e.g. 8, 10) without violating the Pauli's exclusion principle. But such states did certainly exist. To overcome this difficulty, the idea of color was introduced. Instead of one basic triplet, three basic triplets each with a different color degree of freedom (known popularly as red, white and blue) were introduced. The requirement that the observed hadrons be "colorless" combinations removed the difficulty. Then came the 1974 November revolution - the discovery of J/ψ particle followed by the discovery of "upsilon" (Υ) a few years later. These discoveries required still newer degrees of freedom. "Charm" and "bottom" (or beauty) are the currently used terms to describe these new degrees of freedom. While "charm" was anticipated to explain the absence of some decays of K-mesons, neither the "beauty" (or bottom) nor the additional lepton known as the τ-meson had compelling reasons for their existence. As I said before, the "flavor" space is the least understood, beginning from the simple question: "Is the number of such flavors of quarks finite of infinite?"

The idea that quarks were the constituents of hadrons received support from another type of experiments - deep inelastic scattering of electrons from protons. In these experiments, we probe the deep interior of the protons by electrons which interact only with the electrically charged constituents. It soon became evident that to explain the surprising experimental results, one had to imagine the proton to be made up of point-like constituents which were initially called partons. But evidence started building up which enabled the partons of the deep inelastic scattering to be identified with the quarks of the hadron spectroscopy. Today the terms partons and quarks can be used interchangeably.

V

If subnuclear particles have a substructure, our ideas about their interactions have to change. The observed interactions of the hadrons have to be derived from the interactions between quarks and quarks, quarks and leptons and so on. In the last ten years, a new theory of electromagnetic and weak interactions based on a unified description of these interactions has emerged. This new theory is mainly due to Salam and Weinberg, although many other physicists have contributed in an important way to the various ideas that have gone into synthesizing this theory. The unification provided by this theory is comparable in many respects to the synthesis of light, electricity and magnetism due to Maxwell. For example, Maxwell added the "displacement" current in empty space to the physical currents and charges in order to construct a consistent field theory that embodied Faraday's and Ampère's experimental observations. This modification introduced a conserved current which implied charge conservation. It led to the prediction of the electromagnetic waves in empty space, the velocity of propagation of which was exactly equal to the velocity of light waves. In the same way, the most important element of the new theory is the addition of a neutral current related in a specific way to the charged currents of the weak interactions. The theory is a gauge theory like the ordinary electromagnetic theory, but the combined weak and electromagnetic charges and currents are described by a non-Abelian Gauge theory based on the group $SU(2) \otimes U(1)$. The helicity states of quarks and leptons are assigned to definite representations of $SU(2)$; they, therefore, form multiplets (weak isospin multiplets) and carry a hypercharge quantum number due to $U(1)$ (weak hypercharge). The quanta of the unified field are four massless vector bosons, three of which acquire mass by a spontaneous symmetry breaking mechanism while the fourth remains massless to be identified with the photon. The spontaneous symmetry breaking known as the Higgs mechanism is accomplished by a scalar doublet which is described by four real fields. Three of the scalar fields combine with three of the transverse vector fields to make three massive vector bosons. One scalar field remains to represent a true Higgs meson. This intricate way of making the vector bosons massive preserves the renormalizability of the starting zero mass theory. The difficulties of constructing a weak interaction theory along the lines of quantum electrodynamics are resolved and we might say, we have a quantum electro/weak dynamics. Physically it means that the electromagnetic and weak interactions are delicately linked in cooperating with each other to cancel the unwanted infinities. At certain high energies, the distinction between them would disappear and we will witness new phenomena. At low energies, the theory so far has withstood all the tests and there does not appear to be any doubt concerning its validity.

Similarly the picture of strong interactions has to be modified. If quarks are the fundamental constituents, what we used to call strong interactions cannot be the primary interactions. They should be looked upon as secondary or residual interactions derivable from more basic primitive interactions between quarks, much the same way the Van der Waal's forces between neutral atoms are derived from the basic Coulomb forces. Indeed a very good candidate for such a theory has emerged recently; it is the quantum chromodynamics, which is a gauge theory like the quantum electrodynamics, but more complicated in that it is a non-Abelian Yang-Mills-type gauge theory. The quanta are eight photon-like colored "gluons" which are responsible for the strong binding of quarks into "colorless" hadrons. The non-Abelian nature of the theory predicts that at high energies, the quarks behave like free point-like constituents, consistent with the parton picture in deep inelastic scattering. At the same time, at low energies, the theory is expected to provide strong binding so that the quarks are eternally confined inside the hadrons. These two seemingly opposite pictures are referred to as "asymptotic freedom" at short distances and "infrared slavery" at large distances.

VI

Thus we have a remarkably successful gauge theory to describe electromagnetic and weak interactions in a unified way. We have a very promising candidate for strong interactions in quantum chromodynamics which is also a gauge theory. Ways of unifying electro/weak interactions and strong interactions are being seriously considered. Such a grand unified theory will be described by one gauge group characterized by a single coupling constant. Quarks and leptons will be even on a more equal footing. As a consequence, protons may not be absolutely stable! Looking even beyond, one might look for a grand, grand unified theory which includes gravitational interactions as well.

VII

This has been indeed a very brief and sketchy survey of the present day high energy physics. There are many aspects of the field that I have not touched upon. However, let me conclude by posing the question "are we close to a final theory of structure of matter?", and making a few comments on it. The answer to the question is probably "No". First of all there are too many basic fields; at least five, possibly six flavors of quarks, each flavor with three different colors; six leptons. The internal degrees of freedom like color and flavor have neither a space-time nor a dynamical origin. They have to be assumed at the outset, which suggests a possible further substructure for the quarks. Already several such substructures (Prequarks, Maons, Primons, Rishons, Quips) have been suggested in the literature. Besides, in the present theories, quark and lepton masses, various coupling constants, mixing angles and phases are arbitrary parameters. A good deal of experimental input and theoretical approximations and guess-works go into their determinations, which makes it difficult to distinguish the input assumptions from the predictions of the theory. To put it another way, more often than not, there are no precise quantitative tests by means of which one can unambiguously establish a given theory. Our present theories of structure of matter may be grandiose, abstract, complicated, and crazy but "not crazy enough", as Pauli is alleged to have said about Heisenberg's unified theory. They lack the type of basic simplicity which characterizes Einstein's Theory of Gravitation. Somehow looking at our theories, one does not have the feeling expressed by Herman Weyl about General Relativity: "It is as if a wall which separated us from the truth has collapsed. Wider expanses and greater depths are now exposed to the searching eye of knowledge, regions of which we had not even a presentiment. It has brought us much nearer to grasping the plan that underlies all physical happening".

SPLIT FACE OF SCIENCE:
IS SCIENCE AN ENDANGERED SPECIES?

Kameshwar C. Wali

I am indeed delighted to participate in this symposium in honor of Professor Louis Michel, a dear friend and mentor over the past thirty years. Like C.P. Snow, Louis is man of two cultures, the culture of pure mathematics and the culture of theoretical physics. He moves freely between the two, fraternizing with both the pure mathematicians and the physicists, bringing enlightenment to the two sides, which are as far apart as the scientists and the literary intellectuals that Snow talked about in 1956.

Let me begin with recalling a few remarks in Snow's original essay in the New Statesman (Snow 1956). Snow characterizes the traditional culture, the culture of the literary intellectuals as predominantly reactionary, behaving like "a state whose power is rapidly declining - - standing on its precarious dignity, spending far too much energy on Alexandrian intricacies, occasionally letting fly in its fits of aggressive pique quite beyond its means, too much on the defensive to show any generous imagination to the forces which inevitably must reshape it" Scientific culture, on the other hand, Snow said, "is expansive, not restrictive, confident at the roots, the more confident after its bout of Oppenheimerian self-criticism, certain that history is on its side, impatient, intolerant, creative than critical, good-natured and brash...About the whole scientific culture, there is an absence - - surprising to the outsiders - - of the feline and the oblique. Sometimes it seems that scientists relish speaking the truth, especially when it is unpleasant."

Moreover, in contrast with the attitude of the traditional culture, scientific culture, according to Snow, did not suffer from the pessimistic foreboding of human condition. The traditional culture viewed the individual human condition as tragic in that, in spite of all its triumphs and joys, the essence of it is loneliness and ultimate death. And, therefore, everyone ought to stay in their place. The scientists will not admit, because of that, the social condition must also be tragic. Because man must die, it does not follow that he must die before his time, he must suffer, he must surrender to the inevitable fate. "From that particular temptation," Snow said, "made up of defeat, self-indulgence, and moral vanity, the scientific culture is almost totally immune." Deploring the wide gulf that existed between the two cultures and the fact that neither knew the virtues of the other, Snow pleaded for a better education in the Western world, so that the scientific culture as a whole could give to humanity the moral enrichment that it was deprived of and that it most needed.

To a scientist, Snow's appraisal of scientific culture is indeed highly flattering and in resonance with most commonly held views. Perhaps it is even true to a great extent when we think of science until the early part of this century. Scientists, pure scientists that is, as distinct from engineers applied scientists, and practitioners of various technologies, were basically concerned with the workings of Nature and discovering its true laws without any desire of its mastery or manipulation, not withstanding the fact that the scientific discoveries led inevitably to applications, some for the good and some for the harm of humanity. After all the industrial revolution in the West was the product of eighteenth-century science. Enough has been said and written about its liberating as well as its harmful, tyrannizing aspects.

But, generally speaking, pure scientists could not be held responsible for the use or abuse of science in practice, since, first of all, they had no control over what they discovered in their pursuit of unraveling: the secrets of nature, and secondly, one could say justifiably that they had no control over the people at the seats of power in the society, who dictated how the discoveries should be used. Science was amoral and scientists, by and large, wanted their discoveries to be used only for the good of humanity. They were helpless, however, since society at large could not forge a moral value system that paralleled the progress made by science. If only Society cultivated the scientific values of objectivity, preponderance of human reason, rejection of dogmas, respect for others opinions, and an open mind, in contrast with the persisting ignorance, injustices and inequalities, selfishness and the exploitation by the few of the many, scientists thought that there would be no end to the real progress of humanity far beyond that contemplated by philosophers and humanists of bygone centuries.

The contention of my paper today is that the situation concerning science, its practice both in its pure and applied forms has undergone a fundamental change. It certainly continues to play a pivotal role. The triumph of the scientific method in extending the limited, controlled and simplified, and therefore predictable laboratory experience to the understanding of Nature and its control has led to its supreme domination over all spheres of human knowledge. As one report (Funtowicz and Ravetz 1992) says:

> "Common-sense experience and the skills of making and living have lost their claim to reality; they have been displaced by the theoretically constructed objects of scientific discourse, which are necessary for dealing with the invisible agents such as microbes, atoms, genes, and quasars. Just as Galileo used his telescope to transform the familiar moon into an object of applied geometry (where light and shade became peaks and valleys), the control of instruments and theories has become the power to define reality for specialists and for the public alike. Although formally democratic (since there are now no formal barriers to the training for that expertise), it is in fact a preserve of those who can engage on a prolonged and protected course of education."

To elaborate, Science today tells us not only the workings of nature beginning with the very origin of the universe and its evolution, various forces that mediate the fundamental constituents of matter, and the extreme forms matter can take, but also, in the contemporary society, it is "SCIENCE" that dictates, tells us how to live, how to think, how to solve problems, personal, social, and political. However, I would like to argue, that in the process of this dominance in human affairs, science has been corrupted. It has lost that quality of moral enrichment Snow had contemplated. Science is playing more the role of religion in the Middle Ages. As the dispensers and providers of knowledge, scientists are more like priests in the Middle Ages, and as such, they govern as well as are governed by, the power structures that support them and nourish them. Rather than exercise the cherished moral authority to lead humanity to its betterment overcoming narrow-minded national, religious, and social impediments, Science seems to serve more the very establishments that foster such impediments.

As I will try to elaborate, this dominance of science in contemporary times, has come about principally because of the unprecedented success of science in the making of the atomic bomb. I would also like to speak about a basic change that is occurring in the practice of pure science in one of its frontier fields that cuts deeply into the very heart and spirit of the traditional practice of science and makes one wonder whether true science is becoming an endangered species. The making of the atomic bomb was an unprecedented event in human history. Although wars are ancient, brutality of man-to-man and mass slaughters are as old as human history, wars in the present century have assumed a new dimension in mass killing and mass destruction. Mass killings, unlike those in the wars of the previous centuries, are not restricted to young fighting males, but extend to the whole enemy nation including men, women, and children)

Death and destruction have followed an ever-increasing scale and this has come about by the systematic enslavement of science and technology by nation states, sometimes to protect themselves from aggressors, and not rarely, to further their ambitions, territorial or otherwise. In such wars until the Second World War, scientists individually or in small groups, with rare exceptions, had faithfully responded to the call of their respective countries and met their presumed responsibilities as patriotic citizens of nation states. The First World War, for instance, consumed the brilliant British physicist Harry Moseley in the very prime of his life. And the brilliant physicist and astrophysicist Karl Schwarzschild caught an interminable disease on the war front fighting for his native Germany and died. I am sure there were countless other scientists who responded rightly or wrongly to the call of their respective countries. However, during the Second World War, the making of the bomb was a totally different scientific enterprise.

Never before had so many first-rate, talented scientists had pooled together their intellectual resources and worked with engineers and technicians to produce a weapon of death and destruction of unprecedented power. The allied scientists worked under strict secrecy and anonymity, racing against time and under the ·fear that their former colleagues and friends , now the enemy scientists, would accomplish the task before them. They took orders from the military and became exposed to a new kind of thrill and excitement, a new kind sensational association totally alien to their usual experience as scientists. And, finally, the total success of the Manhattan project, the explosion of the first atomic bomb on July 16, 1945, in the New Mexico desert, made them collectively a band of super-heroes and ushered mankind into a new era of cataclysmic fear. Before the event, to many of the participating scientists, the Bomb was a scientific curiosity.

The Los Alamos scientists had joked, and had wagered bets about its explosive power. Enrico Fermi is alleged to have offered to take wagers from his fellow scientists on whether or not the bomb would ignite the atmosphere, and if so, whether it would merely destroy New Mexico or destroy the whole world. It didn't matter to him whether the bomb went off or not. It was only an experiment to prove or disprove the feasibility of an atomic explosion. The story after the explosion was, however, different. As one of the prominent scientists of the Los Alamos group, Isidor Rabi, recalls (Rhodes 1986a), "Naturally we were jubilant over the outcome of the experiment. While this tremendous ball of flame was there before us, and we watched it, and it rolled along, it became in time diffused with the clouds.... Then it was washed out with the wind. We turned to one another and offered congratulations, for the first few minutes. Then there was a chill, which was not the morning cold; it was a chill that came to one when one thought as for instance when I thought of my wooden house in Cambridge, and my laboratory in New York, and of the millions of people living around there and this power of nature which we first understood it to be - - well there it was."

Robert Oppenheimer, the director of the Manhattan Project, looked to Bhagavadgita, the divine poem that is a part of the great Indian epic Mahabharata, to express his feelings (Rhodes 1986b),

> "We waited until the blast had passed, walked out of the shelter and then it was extremely solemn. We knew the world would not be the same. A few people laughed, a few people cried. Most people were silent. I remembered the line from the Hindu scripture, the Bhagavad-Gita: Vishnu is trying to persuade the Prince that he should do his duty and to impress him he takes on his multi-armed form and says, ' Now I am

> become Death, the destroyer of worlds.' I suppose we all thought that, one way or another."

Ironically, while the scientists were playing with their experimental bomb, the hardware of LITTLE BOY and the FAT MAN, the bombs that would destroy Hiroshima and Nagasaki within the next two weeks, were already on their way to their preassigned destinations to get ready for the job. The Los Alamos scientists had no knowledge of it. Leo Szillard was furiously running around to stop the use of the bomb on Japan. He was the genius behind the bomb, the man who was the first to think of the possibility of a chain reaction in the fall of 1933. He was the man who had persuaded Einstein to write the historic letter to President Roosevelt in 1939, and he was the one who was the master theoretician in the early stages of the making of the bomb. Once Germany was defeated, he saw no reason for the use of the bomb. He felt that the United States had a great moral responsibility in consequence of its possession of the atomic _bomb and managed to send a petition to the President with seventeen signatures of the atomic scientists that said (Rhodes 1986 c), The development of atomic power will provide the nations with new means of destruction. The atomic bombs at our disposal represent only the first step in this direction, and there is almost no limit to the destructive power which will become available in the course of their future development. Thus a nation which sets the precedent of using these newly liberated forces of nature for the purposes of destruction may have to bear the responsibility of opening the door to an era of devastation on an unimaginable scale.

We all know that the United States set that precedent. Among the signatures that were missing from the petition were those of Robert Oppenheimer, the director of the Los Alamos laboratory, and Edward Teller, the future father of the super hydrogen bomb. Teller says that he was willing to sign the petition, but Oppenheimer persuaded him not to do so, since it was not their responsibility to dictate how the bomb should be used. As he {Oppenheimer] wrote to Szillard, he felt,

> "I should do the wrong thing if I tried to say how to tie the little toe of the ghost to the bottle from which we just helped it to escape."

One wonders to what extent Oppenheimer was deeply and truly influenced by the Gita, which he had discovered when he was at Harvard and later for which, at Berkeley, he had learned Sanskrit well enough to set himself closer to the original text - - a pink copy of which apparently occupied an honored place on the bookshelf closest to his desk. In this divine poem, Prince Arjuna finds himself in the center of the battle field facing fathers and grandfathers, uncles, brothers, sons and grandsons, his former teachers and companions. those were the enemies to be killed. He was overcome with great compassion and grief and, realizing the futility of war, whether he wins or loses, becomes despondent and disarms himself by laying down his bow and arrows. Lord Krishna, the reincarnation of Vishnu, Arjuna's charioteer, then convinces him why it is right for him to fight and do his duty as a born warrior. This is not the place to digress in depth the subtle and the not.so-subtle, authoritarian arguments that Krishna uses in his discourse. At one place, for instance, appealing to a universal soul, Krishna says it is all an illusion to think that one is responsible for death and destruction, since

> "Know thou that by which all is pervaded is indestructible. Of this immutable being, no one can bring about the destruction...He who thinks that this slays and he who thinks that this is slain; both of them fail to perceive the truth; this one neither slays nor is slain...He is never born, nor does he die at any time, nor having (once) come to be will he again cease to be. He is unborn, eternal, permanent and primeval. He is not slain when the body is slain."
> (Radhakrishnan 1973)

Thus making Arjuna free of any feeling of guilt, Krishna preaches him the philosophy of action,

"'To action alone you hast thou a right and never at all to its fruits; let not the fruits of action be thy motive; neither let there be in thee any attachment to inaction."

Whether Oppenheimer believed in this philosophy of action or not, at least at the moment, he did not think it was his or the scientists1 responsibility to control or dictate the use of the bomb. Their duty was done. It is amazing to me that Oppenheimer had not realized the human destruction and misery .t h e bomb could lavish until he saw the films of Hiroshima and Nagasaki, when he was drowned in sorrow and is alleged to have said that scientist has sinned.

No doubt that this resulted in a split among the scientists based on their attitude toward the science of war, weapons of war, and international politics. Edward Teller wanted to continue the same intense effort toward the hydrogen bomb as that which went into the making of the atomic bomb. Hydrogen Bomb was his idea and he needed help. Without Oppenheimer1s help, he could neither have first rate scientists remain in the Los Alamos laboratory nor could he expect financial support from the government. Oppenheimer discouraged the idea mainly for the reasons of technical feasibility. But when rabid anticommunism and McCarthyism swept over the country, he was of accused of foot dragging because of political and ideological motives; his patriotism came under question, and he was declared a security risk and deprived of his security clearance. His former friend and collaborator testified against him. In the meantime, Szillard, Neils Bohr, and others tried desperately to bring atomic energy under international control by divesting all secrets and if possible, preventing the former Soviet Union from making its own bomb. A one-world government was the only hope from the future nuclear wars and no price was too small to pay for it. The great scientist like John von Neumann, and the great pacifist and philosopher Bertrand Russell advocated even a preventive war or threat of war against Russia to achieve that end.

We know that did not happen. Instead, the governments and the defense agencies on both sides of the Iron Curtain, influenced by the unprecedented success of the Manhattan project, continued to lure scientists pure and applied to work together to develop more and more destructive, nuclear, chemical, and biological weapons. Within four years of the Trinity test, Russia exploded its own bomb. The cold war had already begun in earnest. Scientists of each nation became soldiers aligned to carry on the interests of their respective nations. As the cold war intensified, anticommunism replaced antifascism and "Free World" scientists found a good cause to associate with. Science failed to produce the moral enrichment that Snow had contemplated. As we know, it only served to bring humanity to the brink of extinction, and survival hung only by the slender thread of the mutually assured destruction, or MAD, situation.

The end of the war meant that the scientists could go back to their civilian institutions - universities and private industries. The military was afraid of losing the talent that was assembled in Los Alamos; it was also fearful that universities in general will be unwilling to carry on secret defense related work . As a result, it began to think of new ways to keep the scientists in its fold by forming organizations such as the RAND CORPORATION and the JASON PROJECT.

Project RAND, for "Research and Development" was initiated by Army Air Force General Henry H. ("Hap") Arnold in 1945 by committing ten million dollars of leftover defense funds to Douglas Aircraft officials for research. The contract called for a "program of study and research on the broad subject of intercontinental warfare, other than surface with the object of recommending to the Air Force preferred techniques and instruments for this purpose (Poundstone 1992a). Within these loose guidelines, "The RAND's scientists were permitted to study any issue that interested them" says William Poundstone,

"with the Air Force footing the bill, regardless of whether the Air Force had any interest

> in the subject. Conversely they could refuse specific studies requested by the Air Force (occasionally doing so for "nuts-and-bolts" research that failed to whet the intellectual palates of RAND's scientists)."

"With such freedom and financial backing, it was not difficult to attract eminent social and natural scientists7 and even philosophers, to the corporation. Just to illustrate the kind of lure a scientist was offered, let me quote from the letter by John D. Williams, the head of the Project RAND to von Neumann on December 16, 1947 (Poundstone 1992b). Offering a retainer fee of $200 a month for von Neumann's services, J Williams wrote, "In practice, the members of the project with problems in your line (i.e., the wide world) could discuss with you by mail and in person. We would send you all working papers and reports of RAND which we think would interest you, expecting you to react (with frown, hint or suggestion) when you had a reaction. In this phase, the only part of your thinking time we'd like to bid for systematically is that which you spend shaving: we'd like you to pass on to us any ideas that come to you while so engaged."

In due course, as Poundstone says, RAND epitomized modem Machiavellianism. It became famous (infamous?) for its Herman Kahn and thinking about the unthinkable about the waging and consequences of, and survival in a nuclear war. Would Americans survive to continue to live as they are accustomed to with automobiles, televisions freezers, suburban houses? No one would be sure, but there was every likelihood. There were think tank projects studying problems such as what you should do with older men and women left after the nuclear attack, concluding that the survivors of nuclear attack would be best off without the old and the feeble, md U.S. policy should be to abandon them since, "the easiest way to implement a morally repugnant but socially beneficial policy is by inaction. Under stress, the managers of post-attack society would most likely resolve their problems by failing to make any special provision for the elderly, the insane and the chronically ill" (from a 1966 RAND report; Dickson 1971).

The INSTITUTE FOR DEFENSE ANALYSES set up the JASON division beginning around 1960 and enlisted a group of scientists who were described as "the cream of the scholarly community in technical fields…,.a group of America's most distinguished scientists...". Indeed, in the sixties and seventies, almost all the well-known physicists in the country were a part of JASON. A six-week summer session in some exotic resort and a few long weekends during the academic year exposed the academic scientists to the government and military heads. They were briefed on defense strategies and problems, and sought technical help. With handsome per diem and travel expenses paid, the scientists were free to work at their leisure on the problems of their choice and claim compensation for the time they had devoted to thinking or working on the defense-related problems.

As privileged members of the scientific community who had access to government secrets, they became "insiders" and inadvertently perhaps became uncritical of government policies. JASON scientists, by and large, did not belong, for instance, to the academic scientists who strongly protested the Vietnam War. The publishing of Pentagon Papers revealed for the first time the active participation of the JASON scientists in Vietnam War; the 1966 JASON summer study, for instance, gave birth to a new form of technological warfare, the automated or electronic, battlefield and the concept of what is known as the "McNamara Fence." Branded as "war physicists," some of the well-known American physicists faced protests and heckling at European summer schools. They were also confronted on the home front by an organization called the Science for People.

It also turned out that the JASON scientists had very little influence in bringing sanity to the military, which was the avowed objective on the part of some. The JASON study conclusion, for instance, of the utter ineffectiveness of strategic bombing did not have much influence. The electronic battlefield led to increased civilian deaths; it only led to the creation of new sophisticated hardware that was used by the military for their own tactical purposes. Thus the unique experience of a collective effort of scientists in making the atomic bomb during the Second World War marked a new phase in the scientific culture. The coming together of pure and applied scientists created a powerful union in the hands of governments and defense industries. Scientists were no longer silent spectators, but became scientists in action. They were as much a part of national chauvinism as other citizens; they enjoyed secrecy and privileged status in their respective countries, contrary to all the commonly cherished values in the practice of basic science. Yet, they did not gain increased control in the use or abuse of science. Decision making still remained in the hands of politicians.

The second part of this paper pertains more to my specialization, namely, High Energy Physics or Elementary Particle Physics. The developments in this frontier field are having wider implications on other fields of science and also on society.

As expounded in Professor Michel's paper in this symposium, the success of modern scientific method relies heavily on a strong interplay between experiment and theory. A vast collection of facts either through observations or experimentation is a necessary foundation. The concepts and laws that follow from the established facts allow one to make predictions of new phenomena and new observations to be tested. Prediction is a crucial part of a deep and fundamental theory since it is prediction and substantiation of the prediction that really test whether Nature works the way we think it does. A mere understanding of the facts does not constitute such a test. Neither is there short cut to our deep understanding of Nature without the help of a laborious, painstaking accumulation of facts. This latter point was beautifully expressed by Max Bom (Born 1956) in refuting Milne and Eddington who claimed that to the mind well-trained in mathematics and epistemology, the laws of Nature manifest themselves without appeal to experiment. Born said,

> "Physical laws are derived by a long inductive process, in which flashes of imagination have alternated with painful observation and interpretation of facts. I believe there is no philosophical high road in science, with epistemological signposts. No, we are in a jungle and find our way by trial and error, building our road behind as we proceed. We do not find sign posts at cross roads, but our own scouts erect them,, to help the rest My advice to those who wish to learn the art of scientific prophecy is not to rely on abstract reason, but to decipher the secret language of Nature's documents, the facts of experience."

But it so happens that at the present time, the situation in High Energy Physics is not what one envisioned. Nature(s book of documents is fast becoming inaccessible to direct experimental verification and hence the interplay between theory and experiment appears to face an impenetrable barrier. The reason for this is the fact that our theories have taken us so far in the realm of the small that for direct verification the energies needed are so huge that one cannot now, nor even in the distant future, conceive of attaining them in the laboratories. The much talked about Superconducting Super Collider is perhaps the limit. Theories then have to be judged only by the considerations of their simplicity, beauty, and aesthetic appeal, which is neither a desirable nor an unambiguous situation. Besides, the art of experimental science has already undergone a profound disturbing change.

The present high-energy experiments are mostly dedicated to experiments in the sense that they are designed to test preconceived theoretical ideas. There is little room for trial and error. Moreover, such experiments involve hundreds of scientists divided into groups and subgroups, each group or subgroup doing separate bits like assembly line workers. Such groups in size, shape, and discipline, resemble more a military organization than truth-seeking scientists. Finally, the ever increasing costs of such experiments without any immediate direct benefit to humanity pose difficult choices and problems to the society.

Concluding science in the past following the canonical scientific method of objectivity and reason, and putting to test its ideas has provided humanity not only a culture, but also immense benefits ·of technology and industry. It was always recognized that the path from scientific discoveries to benefits to the society was neither certain nor straight. And it is not clear at all whether advances in science and technology have been beneficial on the whole to the majority of human race. We still have abject poverty in the world, population explosion in developing countries, hunger, environmental degradation, depletion of natural resources, etc. There is an ever-increasing gap in the standard of living between industrialized and non-industrialized countries, the former monopolizing technological resources and exploiting the natural resources of the latter.

Scientists in the past provided the knowledge and left it to others to use it or abuse it. In recent history, they have been more a part of the society at large and have played a greater role in providing technology of war, and not much of their effort has gone into solving the real problems of the world. Although the cold war has ended, the dream of eternal peace is not there yet. Increasing alarm of discoveries in the name of being used to engage in space wars is in the air. The legacy of the Manhattan Project hangs heavily on our heads. It is imperative now that the scientific community seek to establish a new type of contact with the policymakers, based not only on demands of greater budgets for their research and vague claims of benefits of the basic scientific research but on genuine concerns of the world society at large.

Notes on writing the biography of S. Chandrasekhar

Introduction

Kameshwar C. Wali

> The simple is the seal of the true. And beauty is the splendor of truth.
>
> — *Chandra, at the conclusion of his Nobel Lecture, December 8, 1983*

Subrahmanyan Chandrasekhar, known simply as Chandra in the scientific world, is one of the foremost scientists of this century. With his prolific and wide-ranging contributions to physics, astrophysics and applied mathematics, he became and continues to be a legendary figure. However, in spite of his extraordinary accomplishments and tremendous success in contributing to our understanding of the evolutionary stages of stars and their ultimate destinies, to the vast landscape of Einstein's general relativity, and to the study of black holes, Chandra was never a household name like Feynman or Einstein. He remained a highly private individual, unknown to the world at large. The man behind the legend remained hidden. His almost ascetic, highly disciplined, organized and simplified life built around a single-minded pursuit of science made him seem unapproachable. Only a small number of people among his associates, students, admirers and close relatives were able to penetrate the seemingly impenetrable barrier. In commemorating him, I have drawn upon such people, asking them to write, not so much about Chandra's scientific triumphs, but more about Chandra as a person. In the articles that follow, you will see Chandra's rich and multi-faceted personality come alive through their encounters, anecdotes and reminiscences. For my part, I would like to offer a brief overview of Chandra's life and some of my reflections.

Chandra was born on October 19, 1910 (19–10–1910, as Chandra was fond of saying), in Lahore, Pakistan (then a part of colonial British India). His father, C.S. Ayyar, was in the government service, serving as the Deputy

Auditor-General of the Northwestern Railways. Chandra was the first son and the third child in a family of four sons and six daughters. His mother was a woman of great talent and intellectual attainments. Intensely ambitious for her children, she played a pivotal role in Chandra's pursuit of a career in pure science. Chandra's early education was at home; he was taught by his parents and private tutors. When he was twelve, his family moved to Madras. It was then that he began his regular schooling in the Hindu High School, Triplicane, during the years 1922–25, followed by his university education at the Presidency College, Madras. He graduated with a B.Sc. (Hons.) degree in 1930. Lalitha Doraiswamy, his future wife, was a fellow undergraduate studying physics. An exceptionally brilliant student throughout his student career, Chandra became nationally known by presenting an original research paper in physics as an undergraduate during the Indian Science Congress Meetings in 1929. Before he even graduated, he learned that he was awarded a Government of India Scholarship to continue his studies in England. By then he had established contact with R.H. Fowler in Cambridge which led him to decide to work under the latter's guidance for his doctoral degree. While all this was good news for a bright future, he faced at home a mother with a serious illness. The thought that, if he went to England, he might never see her again haunted him, and almost made him decide against going. However, his mother's insistence and persuasion, and her desire not to stand in the way of his future ultimately prevailed and he left for England in July of 1930. Unfortunately, his mother did not survive long enough to witness Chandra's successful career. She died in May of 1931, less than a year after Chandra's arrival in Cambridge.

The years from 1910 to 1930, the early years of Chandra in India, were turbulent ones. It was a time when India was struggling for independence from the British. At the same time British education was becoming nationally and internationally very influential in several fields, especially the sciences. While Gandhi, Nehru, Sarojini Naidu, Sardar Vallabhabhai Patel and other political leaders were becoming household names, there were men like Srinivasa Ramanujan, Meghanad Saha, Jagadish Chandra Bose, Satyendra Nath Bose, C.V. Raman and Rabindranath Tagore, who by their accomplishments in the sciences and arts had captured the imagination of the young, college-educated Indians. Their recognition by the British and the West was a source of inspiration and pride for millions of Indians who struggled in poverty but had dreams, although there were only a few who could transform them into reality. Chandra was among the fortunate few who did not have to struggle in poverty.

Gifted as he was, he developed a passion for mathematics and physics early in his life. He was hardly ten years old when he heard about the great Indian mathematician Srinivasa Ramanujan, who was to be a source of great inspiration and a role model for the rest of his life. Being the young nephew of the famous physicist Sir C.V. Raman opened the way for Chandra to get to know the senior physicists of India at the time and to experience the excitement of research. He came to know K.S. Krishnan, the principal collaborator of Raman in the discovery of the Raman effect, and through him learned about the visit of Arnold Sommerfeld to India and Madras in the fall of 1928. This led Chandra to a meeting with Sommerfeld and an introduction to the new quantum mechanics and Fermi–Dirac statistics. Chandra described his meeting with Sommerfeld as "the single most important" event in his early life which launched him on a research career (for details, see *Chandra: A Biography of S. Chandrasekhar*, University of Chicago Press, 1991, pp. 61–62).

Chandra's six years at Cambridge, England, from 1930 to 1936, first as a graduate student for three years and then as a Fellow of Trinity College, were years in which great discoveries in physics took place. They were also years of continuing British dominance in science. In 1930, on his long voyage from India to England, as a result of musing and calculations, Chandra arrived at a startling conclusion that subsequently became known as the celebrated discovery of the Chandrasekhar limit — the limit on the mass of a star that could become a white dwarf in its terminal stage. After years of hard work at developing, refining and putting the idea on a rigorous basis, Chandra presented his results at the January 1935 meeting of the Royal Astronomical Society. His findings raised challenging, fundamental questions. What happens to the more massive stars as they continue to collapse? Are there terminal stages of stars other than that of white dwarfs? Instead of getting appreciation and recognition for a fundamental discovery, Chandra unexpectedly faced what amounted to a public humiliation. Because, no sooner had he presented his paper than Sir Arthur Eddington, who had been a mentor and had followed the work closely, ridiculed the basic idea on which Chandra's conclusion was based. Eddington made it look like Chandra had gotten it all wrong. In the dramatic and unexpected encounter, young Chandra was pitted against an older, established, internationally renowned scientist whose authority, prestige and fame carried the day. More than two decades passed before the Chandrasekhar limit became an established fact and assumed its importance in astrophysical research, leading to further developments on other terminal stages of stars such as neutron stars and black holes.

In the face of this devastating experience, it was to the credit of young Chandra that he made a wise decision. Instead of a dogged fight and confrontation, he gracefully withdrew from the controversy. He stopped publishing further work in the theory of white dwarfs, and went on to research in a different area after summarizing his account in his first monograph (*An Introduction to Stellar Structure*, University of Chicago Press, Chicago, 1939; reprinted, Dover Publications, New York, 1967). This marked the beginning of Chandra's distinctive style of research throughout the rest of his life, of seeking his own perspective, and striving for a complete understanding of a particular area. When he felt he had accomplished it to his satisfaction, he presented it in a treatise or monograph; and then he abandoned the area entirely, never to return, without even maintaining a passing interest. He became inward-bound, seeking beauty and harmony in the workings of nature as exhibited in his mathematical equations.

In 1937, Chandra and Lalitha, newly married, arrived in the United States, to take up a research position at Yerkes Observatory of the University of Chicago, situated in Williams Bay, Wisconsin. Chandra immediately took on the task of developing a graduate program in astronomy and astrophysics. It wasn't too long before Chandra's reputation as a teacher, his youth and his enthusiasm for research began to attract students from all parts of the world: Paul Ledoux from Belgium; Mario Schönberg, Jorge Sahade and Carlos Cesco from Argentina; Gordon W. Wares, Ralph E. Williamson, Wasley S. Krogdahl, Margaret Kiess Krogdahl and Louis R. Henrich from the United States. These were some of Chandra's students and associates during his early years at Yerkes. Later on, after the Second World War, Guido Munch, Arthur D. Code, Donald E. Osterbrock, Esther Conwell, Jeremiah P. Ostriker and many others were to follow. In all, over fifty of them got their doctorate degree under his guidance and several of them have distinguished themselves as eminent scientists and statesmen of science. As a teacher and a lecturer, Chandra was a grand master who brought elegance and scholarship, which literally charmed the audience and kept it spellbound. The story is often told of how he drove some hundred miles between Yerkes Observatory in Williams Bay, Wisconsin, and the University of Chicago, week after week, to teach a class of two students. The administrators of the University of Chicago were not too happy with this allocation of the faculty resources; but when the whole class of Messrs. Lee and Yang got the Nobel Prize a few years later, they were quick to claim that to be an educational tradition unique to the University of Chicago. Chandra was also the sole editor of *The Astrophysical Journal* during the years 1952–1971. When he began, the journal was essentially the private property of the University of Chicago. Chandra played a decisive role in transforming it into the national journal of the American Astronomical Society and the foremost journal of its kind in the world.

The onset of the Second World War brought upon Chandra great anguish and concern for India. When Holland, Belgium, Norway and France went down to Germany in quick succession, and Greece and Turkey were trampled on, he was alarmed since he felt the security of India was threatened. However, Indian leaders felt it was the right moment to press for India's independence. Mahatma Gandhi initiated the "Quit India" movement, which led immediately to the outlawing of the Congress Party and the imprisonment of all its popular leaders, including Gandhi and Nehru. Strikes, riots and violence spread all over India. In spite of the news of these disturbing, unhappy and painful developments in India, and in spite of being a staunch supporter of India's independence, Chandra felt strong sympathy for Britain and a strong conviction that the world, including India, faced the worst danger if the Axis powers were victorious. Lalitha and Chandra participated in the civilian effort to help the British. When Japan struck at Pearl Harbor and America entered the war, Chandra joined the war effort and became a part of an outstanding group of scientists working at the Aberdeen Proving Grounds in Maryland. The group included John von Neumann, Ronald Gurney, Joseph Myer, L.H. Thomas, Martin Schwarzschild, Edwin Hubble, Robert Sachs and many others.

After the war, Chandra became more closely associated with the physics department of the University of Chicago, as he developed close collaboration with Enrico Fermi, and his researches took on new turns oriented more towards physics than astrophysics. Lalitha and Chandra subsequently moved from Williams Bay and made Chicago their permanent home. Chandra continued active research till the end of his life. After the publication of his book *The Mathematical Theory of Black Holes* in 1983, while continuing his researches on colliding gravitational waves, radial and non-radial oscillations of stars, he got interested in Newton's *Principia*. From a set of lectures, it got into the shape of a mathematical tome of some 400 pages with numerous illustrations. This was to be his last book. It came out just two months before his death. During the last few years of his life, it often appeared as though he was living for the completion of this monumental work. I also should mention that in the 1980s, his other abiding interest was getting a bust of Srinivasa Ramanujan made and presenting it to Janaki Ammal, the widow of Ramanujan. The full story of this episode is well told in Professor Askey's article.

The years I spent working with Chandra and writing his biography were the most enjoyable and creative years in my life. At the beginning, Chandra, as to many others, represented an awesome figure. I had this image of him as a glorious scientist-in-exile from India, his native country (mine as well). When I was a graduate student at the University of Wisconsin at Madison, Wisconsin, I came to know several people who knew him well. Among them was Carl Sagan, who was then a graduate student at Yerkes Observatory. My thesis advisor, Robert G. Sachs, knew Chandra during the Second World War, when he shared an office with him at the Aberdeen Proving Grounds in Maryland. Sachs told me about Chandra's tremendous mathematical power, his incredible memory, and how Chandra could repeat verbatim conversations he had or heard several years before. Carl Sagan filled me in on how students at Yerkes perceived him and how, in general, he presented a formidable personality with his scientific accomplishments. Subsequently I had occasion to meet Chandra a few times at conferences. And when he visited Syracuse to give a seminar in 1971, I experienced firsthand his overwhelming personality. During dinner conversations, he brought to life his Cambridge days with stories, anecdotes and reminiscences. Such scientists as Lord Rutherford, Sir Arthur Eddington, P.A.M. Dirac, John von Neumann — all legendary figures for our generation — came alive in Chandra's narrations. It was then that the thought of getting to know Chandra well and learn more about his life, germinated; it became persistent, and led me to approach him during the general relativity conference in Waterloo, Canada, in the summer of 1977, with the idea of writing one or two articles based on conversations with him. To my pleasant surprise, Chandra agreed to talk with me, but he warned me that I was probably wasting my time.

I began taping the first of many conversations with him on 17 December 1977. (I did not know then that soon after his return from Waterloo, Chandra had heart bypass surgery. He gave me no hint of the ordeal he had gone through just a few months before!) These conversations led me to discussions with others who had known him. Chandra also made available to me his extensive files of correspondence with other scientists, with his father and with his friends during his student days. With the surfeit of fascinating material, I could imagine how inspiring his biography would be to many. Extraordinary as it was in terms of accomplishments, it would also be an account of the hardships, struggles and sacrifices one has to make in the process. His story would be the story of students from many lands, who leave behind their native country, the comforts of their cultural surroundings and their loved ones and go in search of learning. It was then that the original idea of one or two articles gave way to writing the biography (*Chandra: A Biography of S. Chandrasekhar*, University of Chicago Press, Chicago, 1991). Chandra was always open to questions about himself and his life, and gave me complete freedom to write whatever I chose to write. It became an overwhelmingly delightful experience to be with him and be transferred most of the time to a different world, a world in which the life of the mind dominated.

Chandra combined with his extraordinary success in his scientific endeavors an equally extraordinary personality characterized by an intensity and fervor for completeness, elegance and, above everything else, gaining a personal, aesthetic perspective in his scientific work. This is best illustrated in his essay "The Series Paintings and the Landscape of General Relativity," in which he draws an analogy between the series paintings of Monet and certain equations of general relativity. In Monet's paintings, the same scene is depicted over and over again under different natural illuminations and seasonal variations. The valley, the trees and the fields, and the haystacks are the same. Superficially, they may appear boring and repetitive. However, the different paintings radiate totally different aesthetic content. In a similar fashion, the seemingly similar equations and solutions in general relativity describe vastly different physics. In concluding that essay, Chandra says he does not know if there has been any scientist who could have said what Monet said on one occasion:

> I would like to paint the way a bird sings.

But we do know of a scientist who spoke like a poet on one occasion:

> The pursuit of science has often been compared to the scaling of mountains, high and not so high. But who amongst us can hope, even in imagination, to scale the Everest and reach its summit when the sky is blue and the air is still, and in the stillness of the air survey the entire Himalayan range in the dazzling white of the snow stretching to infinity? None of us can hope for a comparable vision of nature and the universe around us, but there is nothing mean or lowly in standing in the valley below and waiting for the sun to rise over Kanchanjunga.

With such writings, often filled with parables, quotes from modern and ancient literature with his Ryerson lecture, "Shakespeare, Newton and Beethoven," and his book *Truth and Beauty*, Chandra bridged the gap between what C.P. Snow called the Two Cultures — the culture of the sciences and of the humanities.

Chandra often remembered E.A. Milne, a close friend of his earlier, Cambridge years, and quoted him as saying:

> Posterity, in time, will give us all our true measure and assign to each of us our due and humble place; and in the end it is the judgement of posterity that really matters. He really succeeds who perseveres according to his lights, unaffected by fortune, good or bad. And it is well to remember there is in general no correlation between the judgement of posterity and the judgement of contemporaries.

A year after death is not a true measure of posterity. However, a two-day symposium (December 14–15, 1996) on "Black Holes and Relativistic Stars" in Chandra's honor at the University of Chicago, and this commemorative volume, mark the beginning of that posterity's assignment to bestow on him his due place. Scores of obituaries, reviews and letters have lavished praise on Chandra — the extraordinary scientist, scholar, man of letters, humanist, rationalist, one whose life stands out for its singular dedication to science and for practicing its precepts and living up to its values to the closest possible limit in one's life. Posterity will certainly take all this into account in according him his place.

Finally, I would like to thank all the contributors for their articles illuminating different dimensions of Chandra. I have special thanks for Lalitha for her article as well as for providing this volume with pictures and photographs.

I have received invaluable help in organizing and editing this volume from Annamaria Zajec. I also would like to thank the Editor and his colleagues at World Scientific for the wonderful job they have done.

Acknowledgments

I consider myself extremely fortunate to have had the opportunity to teach and do research throughout my life. I am indebted to many colleagues and friends, in particular to Professor Robert G. Sachs, who gave me the opportunity to study at the University of Wisconsin, Madison, to complete my Doctoral degree and begin my research career. He demonstrated a strong commitment to physics as both a teacher and as a researcher. I have followed his ideas.

I have always enjoyed working closely with my collaborators and forming lasting friendships. To put it briefly in Oppenheimer's words, I could say I have experienced fully "the obvious excellences of life that physics brings to it". I owe Professor Sachs the introduction to S. Chandrasekhar that provided the opportunity to learn of Chandra's great achievements. That led me, as is well known, to my writings about Chandra and his influence on my career. The years I worked with Chandra — writing his biography were the most enjoyable and creative years of my life. During my first visit to Chandra's office, I was immediately intrigued by a photograph, which faced him from a wall opposite his desk. I asked Chandra about the photo and he launched into a story of how he obtained the artwork from the artist:

"What impressed me most about your picture was the extremely striking manner in which you visually portray one's inner feeling towards one's efforts at accomplishments: one is half-way up the ladder but the few glimmerings of structure which one sees and to which one aspires are totally inaccessible, even if one were to climb to the top of the ladder. The realization of the absolute impossibility of achieving one's goals is only enhanced by the shadow giving one an even lower feeling of one's position."

As I sat with Chandra in his office, among the books, journals, files of correspondence and sketches, conversing hours at a time, I often became transformed, and caught a glimpse of the incomparable world Res Jost so aptly described:

> "There is a secret society whose activities transcend all limits of space and time, and Chandrasekhar is one of its members. It is the ideal community of geniuses who weave and compose the fabric of our culture."

As with my other works, this one has had the support and involvement of my daughters, Alaka, Achala and Monona. It is difficult to thank my wife, Kashi, in few words. She shared my joys and patiently endured her loneliness for my work. She kept the house open to all my research colleagues from all over the world.

And finally I wish to express my thanks Heather Flaherty for preparing the manuscript (this is her fifth) and for the excellent support provided by Lakshmi Narayanan, Senior Editor at World Scientific Publishing and to the staff for their assistance in editing the book and with the publication process of this volume and all previous book.